Veterinary Care of Farm Rabbits

Springer Nature More Media App

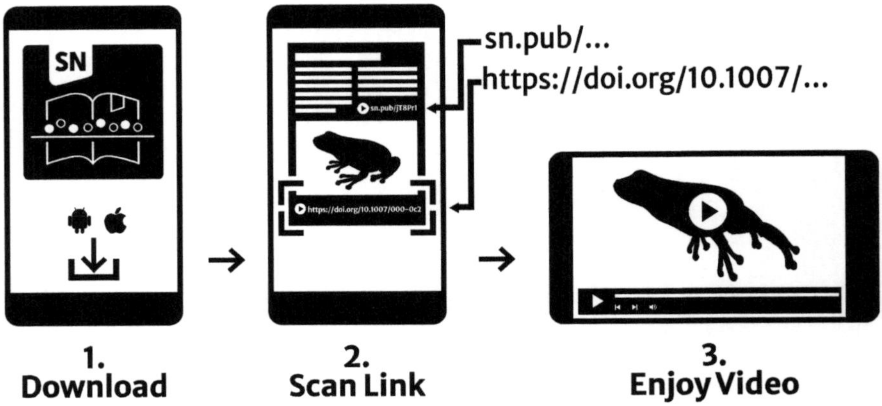

Support: customerservice@springernature.com

João Simões • José M. Monteiro
Editors

Veterinary Care of Farm Rabbits

A Complete Practice Guide to Rabbit Medicine and Production

Springer

Editors
João Simões
Departamento de Ciências Veterinárias
University of Trás-os-Montes and Alto Douro
Vila Real, Portugal

José M. Monteiro
José Azevedo Monteiro, Lda.
Vila Boa do Bispo, Portugal

This work contains media enhancements, which are displayed with a "play" icon. Material in the print book can be viewed on a mobile device by downloading the Springer Nature "More Media" app available in the major app stores. The media enhancements in the online version of the work can be accessed directly by authorized users.

ISBN 978-3-031-44541-5 ISBN 978-3-031-44542-2 (eBook)
https://doi.org/10.1007/978-3-031-44542-2

This Springer imprint is published by the registered company Springer Nature Switzerland AG
The registered company address is: Gewerbestrasse 11, 6330 Cham, Switzerland

Paper in this product is recyclable.

Caring for animals, feeding people.

Foreword

Rabbit farming has been a source of food and raw material for clothing in Mediterranean countries, leaving a deep imprint on their culture and gastronomy. Today, rabbit farming is still a major livestock activity in countries such as France, Spain, and Italy, which are the largest producers of rabbit meat next to China.

In addition to its economic value, rabbit farming also plays an important role in the rural world, providing employment and generating income in areas where other economic activities are not viable.

Over the last 50 years, rabbit farming has evolved significantly from a traditional model with little evolution to a modern and highly technological model. The use of advanced management and feeding techniques has led to increased productivity and reduced costs.

However, this modernization has had an impact on the emergence and spread of rabbit diseases. The high density of animals on modern farms, the selection of certain genetic lines and exposure to pathogens have favored the emergence and spread of diseases.

These diseases have a negative impact on animal health and welfare as well as on the profitability of rabbit farming. It is therefore essential for rabbit farmers, veterinarians, and other professionals to acquire knowledge of the different production systems, to be aware of the diseases that can affect their animals and to take appropriate preventive and control measures. Furthermore, disease prevention and control in rabbits is crucial to ensure animal welfare, food safety and public health.

In this book, the reader will find a wealth of information on the history and biology of rabbits, including their origin, breeds, and anatomical and physiological characteristics. You will also learn about the world trade in rabbit meat and fur.

A part of the book is devoted to the description of production systems, including topics such as breeding facilities, breeding and feeding management, and meat processing and safety. Important topics such as the welfare, behavior, and housing of rabbits, which are critical components in the health and care of rabbits, are also covered in detail.

Another part of the book is devoted to rabbit health management, including biosecurity, preventive medicine, and disease management. Each chapter covers a specific and fundamental aspect of commercial rabbit health, such as antimicrobial resistance, pathological processes like respiratory diseases, enteropathies, staphylococci, viral diseases, fungal infections, or parasitic processes. Hematologic and biochemical values, necropsy procedures, and sample collection are also addressed. Finally, an appendix reports abnormal findings at the slaughterhouse.

This book is of interest to rabbit farmers, veterinarians, and other professionals involved in the rabbit industry, as well as those interested in animal health and welfare. Rabbit farming remains a livestock activity of interest, and it is everyone's responsibility to work together to ensure its sustainability and contribution to rural development.

Universidad CEU Cardenal Herrera
Alfara del Patriarca, Valencia, Spain

Juan Manuel Corpa Arenas

Foreword

In March 2022, José Manuel de Azevedo Monteiro and I met in Portugal (Porto region), and he explained to me the book coordination project, in collaboration with Professor João Simões. He suggested that I review some chapters and I decided to help them gladly. I met José Manuel in November 1999, we worked together for almost 4 years and he did training/internship with me a few times.

José Manuel Monteiro and João Simões agreed with Springer to write a book on the Production and Medicine of farmed rabbits. In addition to their prolific participation in various chapters, they have coordinated 68 other authors from 10 countries in Africa, Asia, Europe, and Latin America. In the first few chapters, readers are shown the reality of the rabbit trade worldwide. In my opinion, consumer habits can be favorable if they perceive that rabbit production is optimal, in relation to nuclear aspects, such as animal welfare, including its health, the nutritional qualities of meat, and food safety. Therefore, the information that reaches rabbit producers, from books like this one, is favorable to progress.

In this book there are chapters relating to the biology of the rabbit. This is essential to interpret what is anomalous; for example, in reproduction. In addition, there are chapters related to housing, adapted to different territories and climates of the world. I also highlight the attention paid to animal welfare, both in relation to the husbandry of the rabbits and health, which is paramount. To this end, there are various chapters on diagnosis, prevention and, if necessary, treatment of diseases. The information on medical care, analysis, or necropsies also take another part of the merit of this work. Likewise, the information on slaughterhouses, meat and skin care, is complete.

This book has represented an effort for the authors and their families. Their generous effort has been worth it, for sure. Congratulations, friends!

Cunivetservice, Tarragona, Spain
April 2023

Joan Rosell, DVM PhD

Votes for this Book from the Community

The book "Veterinary Care of Farm Rabbits" is unquestionably "The Book" of Cuniculture. It is an excellent handbook produced by technicians and clinicians, in collaboration with academia, for anyone interested in learning about this unique and very addictive animal species.

It is undoubtedly a beneficial tool for veterinary clinicians and agronomists, as well as professional and amateur rabbit breeders. It addresses different diseases and their underlying causes, as well as the best management approaches and facilities to avoid them and work more efficiently.

The sector's new knowledge and techniques have experienced an astonishing exponential growth in the recent decade, significantly transforming the sector's vision and its management, demanding major continuous training of professionals specialized in this area.

This work is an example of all of this, and it gives me great pride to see the immense motivation of the text's authors, as well as the outstanding job done by the work's coordinators. I am aware of how much it costs to write a little text, and this book represents many hours of work, revisions, and minor modifications to ensure that the reader has a nice and understandable reading experience.

It gave me great pleasure to be able to read the manuscript of this book, and I hope that the readers share my excitement at seeing an initiative of this caliber and quality in our industry, and that they can enjoy a professional book that provides solutions and clear answers to their questions.

I want to thank and congratulate everyone engaged in this project for their hard work, professionalism, and, most importantly, their time.

Faculty of Veterinary Medicine
of the UAB (Autonomous
University from Barcelona), Barcelona
Spain
May 2023

Francesc Xavier Mora Igual

Contents

Part I

Rabbit Breed and Anatomy Essentials

Introductory Chapter: The Domestic Rabbit and Farm Management

João Simões and José Manuel Monteiro

Abstract

Rabbits (*Oryctolagus cuniculus*) are the only lagomorph domesticated species originating from the Mediterranean region. They are reared for multiple purposes, such as meat, fur, and hair, and serve as laboratory model animals and companion animals. The expected lifetime of rabbits can reach 10 years, but rabbits have a short reproductive life of up to an average of two years. Rabbit production is versatile and can vary from small-scale, that is, subsistence and familiar rearing, to large-scale. In the last five decades, a significant evolution of housing systems occurred using commercial rabbit lines, some of which were selected for more than 30 years to achieve high levels of production. This introductory chapter highlights relevant issues to contextualize worldwide rabbit production, focusing on high-producing animals. General zootechnical and veterinary management of rabbits as well technical-economic indexes of the rabbitries are approached.

Keywords

Rabbit breeds · Rabbit production · Genetic selection · Heard health · Technical-economic indexes

J. Simões (✉)
Animal and Veterinary Research Centre (CECAV), Associate Laboratory for Animal and Veterinary Sciences (AL4AnimalS), Department of Veterinary Sciences, School of Agricultural and Veterinary Sciences, University of Trás-os-Montes and Alto Douro (UTAD), Vila Real, Portugal
e-mail: jsimoes@utad.pt

J. M. Monteiro
José Azevedo Monteiro, Lda., Vila Boa do Bispo, Portugal

Faculty of Veterinary Medicine, Lusófona University, Lisbon, Portugal

CBIOS, Lusófona University, Lisbon, Portugal

J. Simões, J. M. Monteiro (eds.), *Veterinary Care of Farm Rabbits*,
https://doi.org/10.1007/978-3-031-44542-2_1

1 Introduction

The domestic rabbit, *Oryctolagus cuniculus* (Linnaeus, 1758), is a small mammal primarily reared for meat and fur production purposes and is highly appreciated as an animal companion. Additionally, it has been widely used for biomedical and other scientific research as a model due to its phylogenetic and genetic proximity to humans after primates (Miller et al. 2014; Sengupta and Dutta 2020).

The expected longevity of domestic rabbits (Lennox 2010), as well as wild hares in natural environments (Byrne et al. 2022), is up to 13 years. Rabbits are born with approximately 50 g of live weight, start to ingest solid feed at 14 days old, and can reach puberty early from 90 days (small breeds) with approximately 65–70% of their adult body weight. Figure 1 reports the physical and behavioral attributes of rabbits according to their lifetime.

Females show periodical (each 18 days) full receptivity to males mainly under photoperiod and nutrition influence up to approximately 3 years of age. Ovulation is induced, and after a pregnancy of 30–32 days, they can deliver 10 kits (4 to 20) on average and return immediately fertile (Varga 2014).

Even though rabbit rearing was documented centuries ago (Mediterranean region), large-scale production only started in the 1970s.

This chapter addresses several relevant aspects of rabbit management and health to contextualize the remaining 33 chapters of the present book.

2 Origin of Domestic Rabbit

Domestic rabbit (*O. cuniculus*) is a single species of the genus *Oryctolagus* ($2n = 44$; 21 autosomal chromosomes + X and Y sex chromosomes; Gustavsson 1964), family Leporidae (leporine upper lip), and order Lagomorpha. Fossil remains of lagomorphs in Europe and North America were discovered 45 million years ago (early Paleogene) (Lopez-Martinez 2008; Pelletier 2021). The family *Leporidae* is divided into 11 genera and 63 species (Ruedas et al. 2018). Hares of the genus *Sylvilagus* involve 13 species spread throughout the North American continent ($2n = 42$), and the genus *Lepus* ($2n = 48$) presents 32 species native to North America, Europe, Africa, and Asia (Robinson et al. 2002; Wilson and Reeder 2005; Somerville and Sugiyama 2021). See Catalano (1974) for colored original pictures of distinct species. Lagomorpha are considered globally widespread, except for a few zones (i.e., Antarctica, Madagascar, a few Indonesian regions, and the southern part of South America; Dalle Zotte 2014).

Domestic rabbits originate in southern Europe and North Africa (Mediterranean region). Currently, there are approximately 60 breeds originating from the (European) wild rabbit or rabbit (wild) regarding *O. cuniculus cuniculus* subspecies, which has spread throughout Europe (Branco et al. 2002) and the world. The second subspecies, *O. cuniculus algirus*, initially occupied the southwestern areas of the Iberian Peninsula and is currently spread across the Mediterranean, North Africa, and the Atlantic Islands (Branco et al. 2000; Lopez-Martinez 2008).

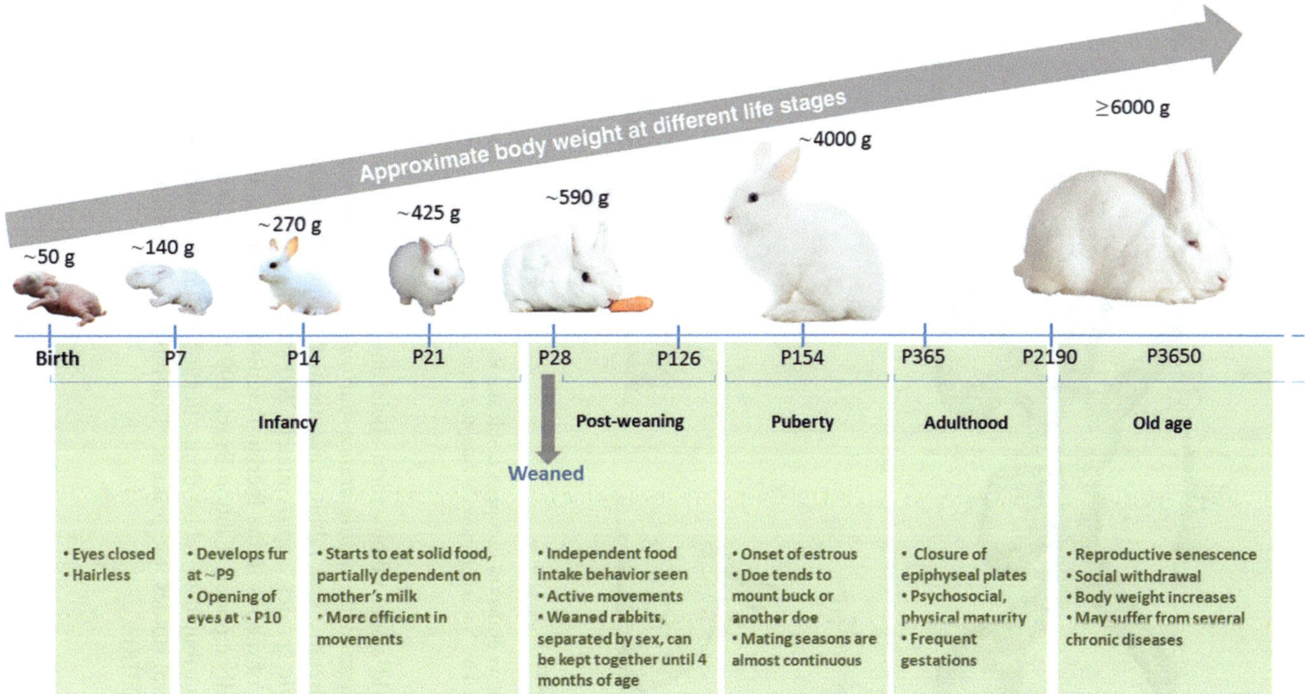

Fig. 1 Timeline of physical and behavioral attributes of rabbits (White New Zealand breed). Adapted from Dutta and Sengupta (2018) with permission from De Gruyter

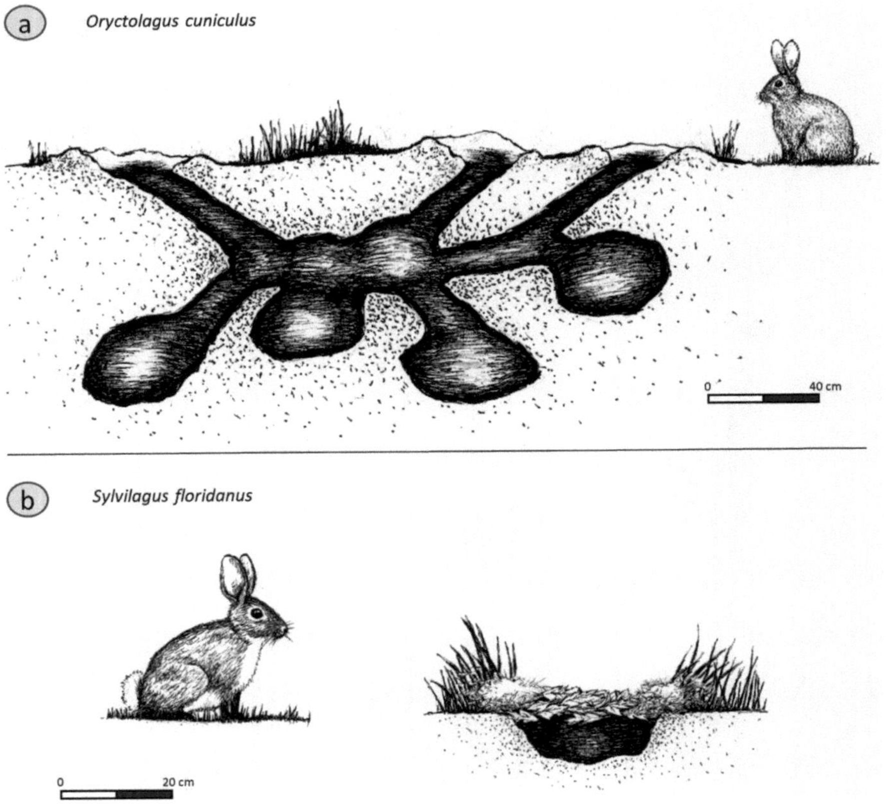

Fig. 2 Illustration of the (**a**) European rabbit and its burrow and (**b**) the eastern cottontail and its burrow. Original by Nathan Thrailkill. Adapted from Somerville and Sugiyama (2021)

At the wild stage, rabbits form social hierarchized groups living in warren, a chamber system with multi-entrance burrows (Fig. 2; Somerville and Sugiyama 2021). Apparently, true rabbit domestication occurred much more recently than other domestic species, approximately 200 years ago (Varga 2014). Nevertheless, the captivity of rabbits in the Iberian Peninsula by Romans in the first century BC was historically reported (Clutton-Brock 1999; see Lebas 2008). European convents can have contributed to rabbit domestication in the Middle Ages (Callou 2003). Despite several studies, the process of rabbits is still under discussion (see Irving-Pease et al. 2018).

3 Rabbit Breeds and Commercial Crossbreed Rabbits

The surge of "new" rabbit breeds occurred primarily by genetic selection of some races and subsequent crossbreeding (Ballan et al. 2022) for both companion and different production purposes (e.g., see Breed Standards 2021–2025—British Rabbit Council, available at https://thebritishrabbitcouncil.org and at https://arba.net/recognized-breeds/). Rabbit breeds are generally classified according to their adult live weight (Table 1), varying from approximately 1 kg to >5 kg. Lighter rabbits show puberty earliest (Vennen and Mitchell 2009; Lebas 2020).

For meat production purposes, Szendrő et al. (2012) classified rabbit breeds into five classes (Box 1).

Box 1 Categorization of Rabbit Breeds (Szendrő et al. 2012)

1. *Native (local) breeds*. Mainly reared in hot climates. Inexpensive and highly resistant to hot climate, low nutritional level, local feedstuffs). For small-scale, familiar and subsistence rabbit production.
2. *Colored (extensive) breeds*. Show an intermediate level of productivity. Used in small-scale production. For organic farming or alternative production systems, owner consumption, and local market niches.
3. *Intensive breeds*. Show good productivity. For medium- and large-scale production. New Zealand White and Californian are considered leading commercial breeds.
4. *Synthetic breeds*. A): Crossbreeding of two or more high-productivity breeds (e.g., Californian as a crossbreed of Himalayan, Chinchilla rabbit, and New Zealand White breeds). For medium- and large-scale production; B) "Hybrids," that is, commercial synthetic breeds obtained from genetic lines obtained during several generations, first to maximize productivity and currently robustness due to heterosis effect. For large-scale industrial production. For example, Zika, Hyla, Hyplus, Hycole, IRTA, and UPV lines.

Several classical breeds are reared in small-scale production (Box 1), but "hybrid" lines are currently used for industrial meat production (Szendrő et al.

Table 1 Classification of rabbit breeds according to adult live weight and their relationship with the onset of puberty

Rabbit breed	Puberty (months)	Specimens (examples)
Large (>5 kg)	>6	French Belier; Flemish Giant
Medium (3–5 kg)	5–6	New Zealand White; Californian
Small (2.5–3 kg)	3–3.5	Chinchilla, Angora

2012; ESFA 2020) using commercial crossbreeding (genetic) selection programs. Some mothers (e.g., Line V, Spain, or INRA2066, France) and paternal (e.g., Line R, Spain, or INRA1077, France) are selected for more than 30 generations (García and Argente 2020).

Box 2 Some (Synthetic) Rabbit Breeds Are Mainly Reared in Non-industrial Production Systems

- New Zealand.
- Rex.
- Flemish Giant.
- Florida White.
- Chinchilla.
- Californian.
- Champagne D´ Argent.
- Harlequin.
- Dutch rabbit.
- Satin.
- French Lop.

Commercial synthetic rabbit lines result from the crossing of two or more maternal and paternal pure breeds or lines with complementary effects, resulting in heterosis, that is, hybrid vigor (Zhang et al. 2012): the offspring show globally better traits than their parents, for example, reproductive (Ragab et al. 2016), meat (Mínguez et al. 2017), and longevity (Ragab et al. 2021) parameters.

Several different types of crossing can be used to obtain different lines of breeding females (Cheeke et al. 1987):

1. Simple or industrial [(♀A × ♂B); F1 = AB]
2. Doubles [(♀A × ♂B) × (♀C × ♂D); F2 = ABCD]
3. Three-way [♀AB × ♂C; F2 = 1/2 AB × 1/2C]
4. Absorbent [♀AB) × ♂B); F2 = 3/4B × 1/4A, . . .; the male is the absorber]
5. Synthetic [♀ABCB × ♂ABCB, . . .]

Three-way crossing is the most frequently used breeding method.

Commercial synthetic lines are selected mainly for morphological and functional traits, as well as technical-economic parameters (Box 3), and most of them can present low to moderate hereditability (h_2) and sufficient genetic variation (e.g., Zhang et al. 2012; Wang et al. 2016; Blasco et al. 2018; El Nagar et al. 2021).

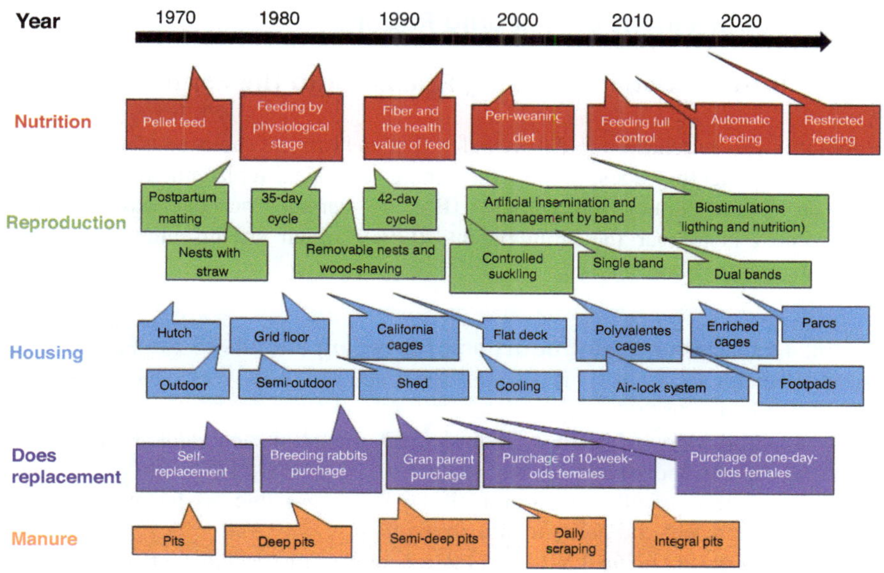

Fig. 3 Major innovations in rabbit farming from the 1970s regarding nutrition, reproduction, housing, does replacement, and manure management. Adapted from Fortun-Lamothe and Davoust (2017) with courtesy of the authors

Box 3 Morphological Traits and Measurable Functional Parameters to Consider in a Genetic Selection Program for Meat Rabbits

Morphology

Short neck; deep and wide chest; back presenting well-developed muscles; glossy hair; early and full sexual development.

Functional Parameters

Fertility; prolificity; milk yield; kindling interval; resistance to diseases (robustness); mortality; average daily gain; feed conversion; carcass yield; maternal instinct; less aggressive behavior.

The development of synthetic rabbit breeds was supported by a strong evolution of reproduction, nutrition, and feeding, as well as housing systems, during the last five decades (Fig. 3). The structure of breeding programs involves the nucleus, responsive to genetic selection and improvement of the rabbit lines (GPs) (main breeding center) and the multipliers (breeding centers) where the young crossbred does (F1) are produced.

4 Replacement of Breeding Rabbits

4.1 Introduction of Breeding Females into the Farm

Rabbitries can purchase one-day females (better farm microbiome adaptation) for adoption by suckling does, prepuberal females (classical replacement) or pregnant nulliparous does (quick replacement) (Box 4). Another popular way of female replacement is to rear the Gran parents (GPs), which are inseminated with an adequate parental line.

Box 4 Criteria for the Introduction of Breeding Rabbits to the Farm

I. Females aged 7 to 12 weeks (and males aged 12 to 15 weeks) have a better adaptation to the farm.
II. Identification (origin, age, and number) of animals and evaluation of the group and individual sanitary status.
III. Quarantine of all arrived animals, or if not possible, as far away from those present and near the air extractors.
IV. Replacement using 12–14 days pregnant does: Especially for new farms and farms previously subjected to sanitation due to contagious disease. They are 19-week-old pregnant females providing a first shortened productive cycle of the farm.
V. Replacement using up to 15-day-old females: They are placed with healthy females with high maternal behavior. Easy adaptation of kits to the microbiome of the farm reduces transport stress and is more economical. Usually, a 10% mortality rate is expected until weaning.
VI. Self-replacement of breeders from a maternal nucleus (GPs) reared on-farm.

In small-scale production systems, self-replacement according to performance records is suitable, even when synthetic breeds are reared. In this last situation, up to 30% of the breeding rabbits can be self-replaced without loss of performance.

4.2 Culling of Rabbit Does

Under intensive (9–10 kindlings per year) and semi-intensive (7–9 kindlings per year) production systems, the annual replacement of commercial rabbit does can reach 120% (Guerder 2002; Ramon and Rafel 2002; Ramon et al. 2004), usually between 80 and 120%. These values originate at approximately 10 months of reproductive life, and according to the production and sanitary status of the rabbit does. Approximately half of the cullings occur up to the third kindling (Rosell and de la Fuente 2009). Replacement costs were estimated as 3.5 €/doe and year, 0.03 €/kg live weight sold, and 1.7% of total production costs for a typical industrial rabbitry

Table 2 Target and interference levels for breeding rabbit culling and sanitary problems. Adapted from Rosell 2003 and Rosell et al. (2020)

Index variable	Target	Interference
Monthly % of adult doe culling	7	<5 >10
Monthly % of adult doe mortality	≤3	>5
Abortions (%)[a]	≤1	>2
Stillbirths (%)	≤7	>7
Pregnancies vs kindling differences (%)	<7	≥10
High fetal death risk at parturition[b]	≤7%	>9 to 10%
Small litter size, weak kits	≥ 8.5	<7.5
Kit mortality before weaning (after elimination[c]) (%)	<13 (<10)	>15
Kit mortality after weaning (from 33 to 62 days) (%)	≤7	>7

NA not applicable
[a] Percentage of abortions/services
[b] Coudert (1981) suggested a target of 2% for parturitions with all fetuses stillborn
[c] After elimination of weakest kits (5–10%)

(intensive system of 750 rabbit does housed in two bays) (Cartuche et al. 2014). A negative correlation ($r = -0.37$) was observed between monthly removal risk and the intensity of reproductive rhythm (Rosell and de la Fuente 2009). In the Iberian Peninsula, semi-intensive systems (service between 11 and 25 days, usually at 11 days after kindling) prevail, followed by extensive systems (≥32 days); intensive reproduction (<11 days) is disused (Rosell et al. 2020).

The monthly replacement of breeding rabbits is approximately 5 to 10% of females (culled to dead rabbit does females) to optimize the mean age of the group (approximately 7–8 kindlings on average) and to ensure management by band in the rabbitry. Rosell and de la Fuente (2009) calculated a 9.7% median monthly removal risk (3.4% dead and 5.7% culled females) in does. Apart from overall low productivity (Rosell and de la Fuente 2009), maternal behavior and reproductive abnormalities are major causes of does culling up to the 8th kindling (Lopez et al. 2017). Major causes of death of rabbit does are digestive (enteritis-diarrhea processes and mucoid enteropathy) and respiratory diseases, septicemia, viral hemorrhagic disease, metritis, pyometra, and uterine torsion (Rosell and de la Fuente 2016). Threshold levels for breeding rabbit culling and due to sanitary problems are reported in Table 2. From the 9th kindling, age seems to be the main cause of culling (Lopez et al. 2017). Nevertheless, the causes of culling can be categorized into zootechnical (reproduction, maternal behavior, and age), sanitary (Box 5), body condition score (cachexia and obesity), and others (e.g., trauma and other accidents) to enable their monitoring.

Box 5 Criteria for the Culling of Breeding Rabbits at the Farm

Due to Zootechnical Factors

1. Three negative pregnancy diagnoses after service.
2. Fewer than 24 kits were born alive in the first three kindlings.
3. Less than 21 weaned kits in the first three kindlings (low milk yield).
4. Two litters born dead (cannibalism; abandonment of the newborns kits).
5. Two litters killed before weaning.

Due to Sanitary Factors

1. Subcutaneous staphylococci, including pododermatitis.
2. Internal and otitis media (torticollis).
3. Marked respiratory symptoms (intense pneumonia/coryza).
4. Mastitis.
5. Metritis.
6. Abscesses.
7. Cachexia/marked weakness.

5 Main Hazards During the Production Cycle

Several alterations and diseases usually occur during the different steps of the production cycle. These occurrences have a negative impact in each production step and require hazard identification, mitigation, or prevention. According to the production cycle, breeding, kindling, lactation, litter. and weaned steps are the main issues to manage, as reported in Table 3.

6 Technical-Economic Indexes

As another profit activity, economic management considering revenues and production costs is a determinant of the sustainability of rabbitries. Both are dependent on the farm dimension and technical indexes reached for a time-lapse period, and values are usually reported for one year. See Pascual and Gómez (2021) for detailed information. Technical and economic indexes also vary according to the production system and its intensification, which usually rear the more productive synthetic rabbit breeds. Table 4 reports the threshold values of reproductive, productive, and other technical indexes for industrial semi-intensive systems (42-day production cycles).

Mondin et al. (2021) estimated fixed costs (initial investment for a 20-year depreciation time) of 448 €/m^2 for a rabbit shed from six Italian rabbitries between 2014 and 2018. This value included masonry works (318 €/m^2), climatization

Table 3 Main hazards during the different steps of rabbit production

Step of production	Hazard
Breeding (infertility)	• Errors in the dosage of hormonal products (eCG and GnRH) • Stress (general) • Failures in water administration • Extreme environmental temperatures (see Chaps. 6 and 11) • Inadequate lighting rhythm • Infectious causes • Unbalanced diet (nutritional deficiency) • Lower and high body condition scores (Rosell et al., 2000) • Pseudopregnancy (18 days)
Kindling	• Delay in delivery or retention of fetuses • Purulent uterine secretions (metritis) • Cannibalism due to: (a) Adverse environmental conditions (b) Poor sanitary status of the mother (c) Very nervous animals (d) Badly packed nest: thermal stress (e) Lack of milk production (f) Presence of rodents and other foreign animals (g) Animals in a needy or poorly fed state (h) Internal and external parasitism (i) Changes in maternal behavior (mainly primiparous does)
Lactation	• Mastitis: mainly caused by *Staphylococcus aureus*; other relevant milk pathogens involved: *Pasteurella*, *Streptococcus*, *Klebsiela*, and *Pseudomonas* • Agalactia and hypogalactia
Litter (nest)	• Cannibalism and abandonment of the newborns • Traumas (crushing) • Hygiene deficiencies in nest • Excessive cold and humidity inside the nest • Infections: Staphylococci and neonatal colibacillary diarrhea
Fattening	• Mortality due to digestive (early weaned kits), respiratory (end of fattening cycle) alterations, and non-potable water • Low live weight (<600 g) at weaning or early weaning (< 30 days) • Thermal stress • Sudden management at weaning time • Changes in drinking water • Changes in feed composition (fiber/protein imbalance causing diarrhea) • Infectious causes: enteropathies, coccidiosis, staphylococci, and pasteurellosis • Myxomatosis outbreaks

system (42 €/m^2), automatic feeding system (24 €/m^2), with silos, cleaning system (scrapers) (16 €/m^2), and electrical and water systems structures (48 €/m^2). Additionally, the estimated cost purchase per cage (10-year depreciation time) was 63 € for conventional bicellular cages (maternity cages = 45 € plus cages for rearing rabbit = 18 €), 70 € for conventional polyvalent (dual-purpose) cages, and 78 € for

Table 4 Threshold values of reproductive and performance indexes

Parameter	(Threshold) value
Breeding does and kindling	
Occupation of maternity cages	110–130%
Yearly does replacement	100–130%
Monthly mortality of females	3%
Fertility rate as kindling/AI	>80%*
No. of kindling/cage/year	8.7
Yearly mean newborns/kindling	9–10
Stillborn per kindling	7%
Alive newborns per kindling/cage/year	53–86
Suckling/weaning	
Weaned kits per kindling	8–9
Weaned kits per kindling/cage/year	44–77
Live BW at weaning (30–35 days)	650–1000 g
Mortality at weaning	10–15%
Fattening	
Mortality during growing period	2–7%
Feed conversion index	<2.5
Live weight and age at slaughter	2.2–2.7 kg at 65–75 days
No. of fattening rabbits per cage/does/year	40–75
kg (live weight) per cage/does/year	80–157
Dressing percentage of slaughter	56.5–60%
Overall feed conversion index **	<3.5
Winter	3.5
Summer	3.0
With feed restriction system	2.7–3.1

* >85% of pregnancies at 14 days after Service)
** [Total feed concentrate (kg) purchased at farm/Total sold young rabbits (kg)]
AI Artificial insemination

WRSA (World Rabbit Science Association) enriched cages. In these six farms, the total fixed costs were 0.28–0.30 € per kg of rabbit live weight.

An estimate of 6% profitability in a Czech rabbitry (2017–2019) was reported by Krupová et al. (2020). Profit results were 18 € [309 € (revenue)—291€ (cost production)] per doe per year, resulting from 0.41 € per growing rabbit and 0.15 € per kg slaughter weight of rabbits. In this rabbitry, feeding costs, including drinking water, were approximately 65%, and doe replacements were responsible for 10% of total costs (Table 5). Approximately 50% of total costs were related to weaned rabbits.

7 Conclusions

Rabbits are a well-adapted multipurpose species to different production systems. In the last five decades, synthetic rabbit lines have been developed to reach unprecedented levels of production at controlled cost. Concomitantly to the intensification of

Table 5 Estimated costs and revenues in a Czech industrial meat rabbit farm at 2017–2019 prices. Adapted from Krupová et al. (2020) with permission from Elsevier

Parameter	Value/unit	
Feed costs	*MJ DE/kg FM*	*Price in €/kg*
Lactation, pregnancy, and supplementary feed for kits	10.5	0.487
Feed in fattening		
First/second feeding phase	10.0/9.8	0.37/0.36
Diet for open does	10.0	0.37
Water requirements	(L/day per animal)	(€/1000 L)
For does	0.4	2.0
Per animal in fattening	0.25	2.0
Non-feed costs (unit)	*Does including kits to weaning*	*Animal in fattening*
Health care (€/animal[a])	1.66	0.10
Supplies and miscellaneous (€/animal per day)	0.01	0.001
Artificial insemination (€/dose)	1.00	
Disposal of dead animals (€/kg LW)	0.50	0.50
Fixed costs (€/animal per day)	0.09	0.0012
Price for purchased doe replacement (€/animal)	28.08	
Revenues	*€/kg*	
Slaughter rabbits (€/kg carcass weight)	3.26	
Culled does (€/kg live body weight)	0.97	
Dry rabbit skin (€/kg)	0.49	
Revenues from manure (€/100 kg)	0.67	
Does including kits until weaning (kg/doe and kindling interval)	28	

DE digestible energy, *FM* fresh matter, *LW* live weight
[a] Per kindling interval for does; per whole fattening period for rabbits in fattening

production, technical evolution of reproductive and nutritional traits as well as herd heath management were consistently developed.

Currently, one rabbit doe can produce more than 150 kg (live weight) of meat throughout one year. Trends of improved rabbit robustness and animal welfare maintaining this level of production, or rabbit production system resilience, are expected.

Acknowledgments The authors thank Juan Rosell for his suggestions and review of this chapter. The authors also thank Springer for the use of the following AI tools in this chapter: Dimensions Auto-Summarization 3.0 and AJE Digital Editing. This work was supported by projects UIDP/CVT/00772/2020 (CECAV) and LA/P/0059/2020 (AL4AnimalS), funded by the Portuguese Foundation for Science and Technology (FCT).

References

Ballan M, Bovo S, Schiavo G, Schiavitto M, Negrini R, Fontanesi L (2022) Genomic diversity and signatures of selection in meat and fancy rabbit breeds based on high-density marker data. Genet Sel Evol 54:3. https://doi.org/10.1186/s12711-022-00696-9

Blasco A, Nagy I, Hernández P (2018) Genetics of growth, carcass and meat quality in rabbits. Meat Sci 145:178–185. https://doi.org/10.1016/j.meatsci.2018.06.030

Branco M, Ferrand N, Monnerot M (2000) Phylogeography of the European rabbit (*Oryctolagus cuniculus*) in the Iberian Peninsula inferred from RFLP analysis of the cytochrome B gene. Heredity 85:307–317. https://doi.org/10.1046/j.1365-2540.2000.00756.x

Branco M, Monnerot M, Ferrand N, Templeton AR (2002) Postglacial dispersal of the European rabbit (Oryctolagus cuniculus) on the Iberian Peninsula reconstructed from nested clade and mismatch analyses of mitochondrial DNA genetic variation. Evolution 56:792–803. https://doi.org/10.1111/j.0014-3820.2002.tb01390.x

Byrne AW, Marnell F, Barrett D, Reid N, Hanna REB, McElroy MC, Casey M (2022) Rabbit haemorrhagic disease virus 2 (RHDV2; GI.2) in Ireland focusing on wild Irish hares (*Lepus timidus hibernicus*): an overview of the first outbreaks and contextual review. Pathogens 11:288. https://doi.org/10.3390/pathogens11030288

Callou C (2003) De la garenne au clapier. Étude archéozoologique du lapin en Europe occidentale. Publications Scientifique du Museum d'Histoire Naturelle, París, p 261. ISBN: 9782856535455

Cartuche L, Pascual M, Gòmez E, Blasco A (2014) Economic weight in rabbit meat production. World Rabbit Sci 22:165–177. https://doi.org/10.4995/wrs.2014.1747

Catalano U (1974) Mammiferi d'Italia, Spagnesi, M., De Marinis, A. M. (a cura di), 2002 - Quad. Cons. Natura, 14, Min. Ambiente - Ist. Naz. Fauna Selvatica. Available at https://www.mase.gov.it/sites/default/files/archivio/biblioteca/qcn_14.pdf. 13 Mar 2023

Cheeke PR, Patton NM, Lukefahr SD, McNitt JI (1987) The rabbit production, 6th edn. Danville, IL, Interstate Printers and Publishers

Clutton-Brock J (1999) A natural history of domesticated mammals, 2nd edn. Cambridge, Cambridge University Press. 238 pp. ISBN: 9780521634953

Coudert P (1981) Pathologie et reproduction chez le lapin. Physio-Pathologie de la Reproduction Chez le Lapin. Adeprina, INA-PG, Paris, pp 75–89

Dalle Zotte A (2014) Rabbit farming for meat purposes. Anim Front 4:62–67. https://doi.org/10.2527/af.2014-0035

Dutta S, Sengupta P (2018) Rabbits and men: relating their ages. J Basic Clin Physiol Pharmacol 29:427–435. https://doi.org/10.1515/jbcpp-2018-0002

EFSA Panel on Animal Health and Welfare (AHAW), Saxmose Nielsen S, Alvarez J, Bicout DJ, Calistri P, Depner K, Drewe JA, Garin-Bastuji B, Gonzales Rojas JL, Gortázar Schmidt C, Michel V, Miranda Chueca MÁ, Roberts HC, Sihvonen LH, Spoolder H, Stahl K, Velarde Calvo A, Viltrop A, Buijs S, Edwards S et al (2020) Health and welfare of rabbits farmed in different production systems. EFSA J 18:e05944. https://doi.org/10.2903/j.efsa.2020.5944

El Nagar AG, Baselga M, Mínguez C, Santacreu MA, Sánchez JP (2021) Functional longevity in five rabbit lines founded on different criteria: comparison at foundation and at fixed times after selection. J Anim Breed Genet 138:508–517. https://doi.org/10.1111/jbg.12543

Fortun-Lamothe L, Davoust C (2017) Innovations en élevage cunicole: Des réussites d'hier aux défis de demain. 17èmes Journées de la Recherche Cunicole, 21 et 22 novembre 2017, Le Mans, France

García ML, Argente MJ (2020) The genetic improvement in meat rabbits. In: Argente MJ, Pardo MLG, Dalton KP (eds) Lagomorpha characteristics. (p. Ch. 5). IntechOpen, London. https://doi.org/10.5772/intechopen.93896

Guerder F (2002) Conduites en bandes: de bons résultats économiques. Cuniculture 165:117–123

Gustavsson I (1964) Somatic chromosomes of the rabbit. Acta Vet Scand 5:362–369. https://doi.org/10.1186/BF03547389

Irving-Pease EK, Frantz LAF, Sykes N, Callou C, Larson G (2018) Rabbits and the specious origins of domestication. Trends Ecol Evol 33:149–152. https://doi.org/10.1016/j.tree.2017.12.009

Krupová Z, Wolfová M, Krupa E, Volek Z (2020) Economic values of rabbit traits in different production systems. Animal 14(9):1943–1951. https://doi.org/10.1017/S1751731120000683

Lebas F (2008) Méthodes et techniques d'élevage du lapin. Historique de la domestication et des méthodes d'élevage. Accessed from http://www.cuniculture.info/Docs/Elevage/Histori-01.htm. 16 Apr 2023

Lebas, F. (2020). Biology of the rabbit. Accessed from http://www.cuniculture.info/Docs/Biologie/Biology-English/Biology-Eng-01.htm. 16 Apr 2023

Lennox AM (2010) Care of the geriatric rabbit. The veterinary clinics of North America. Exotic Animal Practice 13:123–133. https://doi.org/10.1016/j.cvex.2009.09.002

Lopez S, Menard E, Favier C (2017) Analyse des causes de réforme et de mortalité des femelles reproductrices en élevage cunicole. 17èmes Journées de la Recherche Cunicole, 21 et 22 novembre 2017, Le Mans, France, pp 111–114

Lopez-Martinez N (2008) The lagomorph fossil record and the origin of the European rabbit. In: Alves PC, Ferrand N, Hackländer K (eds) Lagomorph biology. Springer, Berlin/Heidelberg. https://doi.org/10.1007/978-3-540-72446-9_3

Miller I, Rogel-Gaillard C, Spina D, Fontanesi L, de Almeida AM (2014) The rabbit as an experimental and production animal: from genomics to proteomics. Curr Protein Pept Sci 15: 134–145. https://doi.org/10.2174/1389203715666140221115135

Mínguez C, Sánchez JP, Hernández P, Ragab M, El Nagar AG, Baselga M (2017) Genetic analysis of meat quality traits in maternal lines of rabbit and their diallel cross. Meat Sci 131:1–8. https://doi.org/10.1016/j.meatsci.2017.04.005

Mondin C, Trestini S, Trocino A, Di Martino G (2021) The economics of rabbit farming: a pilot study on the impact of different housing systems. Animals 11:3040. https://doi.org/10.3390/ani11113040

Pascual M, Gómez EA (2021) Profitability in rabbit breeding. In: Argente M-J, de la Luz García Pardo M, Dalton KP (eds) Lagomorpha characteristics. IntechOpen, London. https://doi.org/10.5772/intechopen.93260

Pelletier M (2021) Morphological diversity, evolution and biogeography of early Pleistocene rabbits (genus *Oryctolagus*). Paléo 64:817–838. https://doi.org/10.1111/pala.12575

Ragab M, Sánchez JP, Mínguez C, Baselga M (2016) Crossbreeding effects on rabbit reproduction from four maternal lines of rabbits. Animal 10:1086–1092. https://doi.org/10.1017/S1751731115002918

Ragab M, Baselga M, El Nagar AG, Mínguez C, Pablo Sánchez J (2021) Genetic analysis of functional longevity in a complete diallel crossing experiment involving four maternal rabbit lines. J Anim Breed Genet 138:474–481. https://doi.org/10.1111/jbg.12520

Ramon J, Rafel O (2002) 1991–2000. Diez años de gestión global en España. II Congreso Internacional de Producción y Sanidad Animal. Expoaviga. 5–8 November, 2002. Barcelona, Spain, pp 113–117

Ramon J, Rafel O, Piles M (2004) Resultados de gestión en España. GTE 2002. Algo falla Boletín de Cunicultura 133:24–28

Robinson TJ, Yang F, Harrison WR (2002) Chromosome painting refines the history of genome evolution in hares and rabbits (order *Lagomorpha*). Cytogenet Genome Res 96:223–227. https://doi.org/10.1159/000063034

Rosell JM (2003) Technical note: health status of commercial rabbitries in the Iberian peninsula. A practitioner's study. World Rabbit Sci 11:157–159. https://doi.org/10.4995/wrs.2003.505

Rosell JM, de la Fuente LF (2009) Culling and mortality in breeding rabbits. Prev Vet Med 88:120–127. https://doi.org/10.1016/j.prevetmed.2008.08.003

Rosell JM, de la Fuente LF (2016) Causes of mortality in breeding rabbits. Prev Vet Med 127:56–63. https://doi.org/10.1016/j.prevetmed.2016.03.014

Rosell JM, Fuente LF, Carbajo MT, Fernández XM (2020) Reproductive diseases in farmed rabbit does. Animals 10:1873. https://doi.org/10.3390/ani10101873

Ruedas LA, Mora JM, Lanier HC (2018) Evolution of lagomorphs. In: Smith A, Hackländer C, Alves P, Johnston K (eds) Lagomorphs: the pikas, rabbits, and hares of the world. JHU Press, Baltimore (MD), pp 4–8

Sengupta P, Dutta S (2020) Mapping the age of laboratory rabbit strains to human. Int J Prev Med 11:194. https://doi.org/10.4103/ijpvm.IJPVM_530_18

Somerville AD, Sugiyama N (2021) Why were New World rabbits not domesticated? Anim Front 11:62–68. https://doi.org/10.1093/af/vfab026

Szendrő Z, Szendrő K, Zotte AD (2012) Management of reproduction on small, medium and large rabbit farms: a review. Asian Australas J Anim Sci 25:738–748. https://doi.org/10.5713/ajas.2012.12015

Varga M. (2014). Rabbit basic science. Textbook of Rabbit Medicine, 2nd edn Butterworth-Heinemann/Elsevier London; pp. 3–108. ISBN 978-0-7020-4979-8 https://doi.org/10.1016/B978-0-7020-4979-8.00001-7

Vennen KM, Mitchell MA (2009) Rabbits. Manual of exotic pet practice. W.B. Saunders, Missouri, pp 375–405. ISBN 9781416001195. https://doi.org/10.1016/B978-141600119-5.50017-2

Wang J, Su Y, Elzo MA, Jia X, Chen S, Lai S (2016) Comparison of carcass and meat quality traits among three rabbit breeds. Korean J Food Sci Anim Resour 36:84–89. https://doi.org/10.5851/kosfa.2016.36.1.84

Wilson DE, Reeder DM (eds) (2005) Mammal species of the world: a taxonomic and geographic reference, 3rd edn. Johns Hopkins University Press, Baltimore, MD. 2142 pp. ISBN 0-8018-8221-4

Zhang XY, Huang DP, Xie XH (2012) Hereditas 34:401–406. https://doi.org/10.3724/sp.j.1005.2012.00401

Anatomical Features of Rabbits

Carlos Venâncio, Maria da Conceição Fontes, and João Simões

Abstract

This chapter aims to describe the main anatomical features of the rabbit. A classic anatomical division of body parts and apparatus and systems, as well as comparative anatomy with other domestic mammals, is approached. Structures that have an impact on rabbits' functioning systems and behavior are described. Several anatomical particularities influencing feeding, housing, handling procedures, and clinical approaches to preserve rabbits' well-being and an adequate state of health are discussed. The terms used in this work are in accordance with the veterinary anatomical nomenclature whenever possible. The rabbit and its anatomical particularities, so far, have not been included in the published Nomina Anatomica Veterinaria, making the description in the present chapter of greater importance.

C. Venâncio (✉)
Department of Animal Science, School of Agrarian and Veterinary Sciences, University of Trás-os-Montes e Alto Douro, Vila Real, Portugal

Centre for the Research and Technology of Agro-Environmental and Biological Sciences (CITAB), Inov4Agro – Institute for Innovation, Capacity Building and Sustainability of Agri-food Production, University of Trás-os-Montes and Alto Douro, Vila Real, Portugal
e-mail: cvenanci@utad.pt

M. da Conceição Fontes · J. Simões
Department of Veterinary Sciences, School of Agricultural and Veterinary Sciences, University of Trás-os-Montes and Alto Douro (UTAD), Vila Real, Portugal

Veterinary and Animal Research Centre (CECAV), Associate Laboratory for Animal and Veterinary Sciences (AL4AnimalS), University of Trás-os-Montes and Alto Douro (UTAD), Vila Real, Portugal

J. Simões, J. M. Monteiro (eds.), *Veterinary Care of Farm Rabbits*,
https://doi.org/10.1007/978-3-031-44542-2_2

Keywords

Comparative biology · Gross anatomy · Morphology · *Oryctolagus cuniculus* · Rabbit

1 Introduction

The rabbit species comprises small mammal herbivores with over 300 breeds (Dorozynska and Maj 2021). There is weight variation according to the breed, usually between 1 and 8 kg. Hybrid breeding rabbits reared for production purposes have an average weight of 4.7 kg, ranging from 2.9 to 8.1 kg (de la Fuente and Rosell 2012). The life span of rabbits is approximately 10 years on average (Lennox 2010).

All breeds of domestic rabbits are descendants of the European wild rabbit, *Oryctolagus cuniculus*. They belong to the order Lagomorpha, which is characterized by the presence of a second small pair of upper incisors. This characteristic is particularly important to differentiate it from some species with which it shares several similarities in the order Rodentia (Brewer 2006).

Rabbits present a set of anatomical characteristics that can be found in the various organs (Box 1), which are attributed to the process of adaptation to the natural environment where their evolution occurred. This evolutionary aspect strongly affected the body surface as well as the internal apparatus and systems, with emphasis on several differences namely the enormous development of the intestine.

Box 1 Key Points of the Rabbits' Anatomical Features

- The rabbit stands plantigrade, with the whole area from hock to toe in contact with the ground, but becomes digitigrade when running.
- The hair covers almost all of its body. Only the rabbits' nose, scrotum, and inguinal areas are hairless.
- Their long, arched claws can cause scratches, particularly if their hind legs are not properly restrained.
- Long ears with short hair and highly vascularized.
- Sexual dimorphism is usually not very evident.
- Rabbits are obligate nose breathers.
- The skeleton is relatively fragile, comprising 8% of body weight (in cat, 13%).
- The muscle mass of the lumbar region and thighs are the most developed.
- The thoracic cavity is proportionally small when compared to the abdominal cavity.
- Teeth are continually growing, with a speed of approximately 2 mm per week.
- Rabbits are unable to vomit.

(continued)

Box 1 (continued)
- The cecum is sufficiently developed in length and volume.
- The intestinal wall is particularly rich in aggregates of lymphatic tissue.
- The heart is proportionally small, and the right atrioventricular valve is bicuspid and not tricuspid.
- The thymus persists in adult life.
- In males, the testicles have retraction ability from the scrotum.
- Females have two chains of mammary glands with a total of 4 to 5 pairs of nipples.
- The uterus is duplex, meaning there are two cervices leading into the vagina.
- The neonates are altricial, bald, and born blind.
- The brain has a smooth surface—the lysencephalon.
- Good olfactory capacity and keen sense of hearing.

The knowledge of these particularities is fundamental to perceive and improve the productive and reproductive performances of these animals, as well as to determine their greater susceptibility to environmental conditions and response to external agents that make them sick.

This chapter mainly aims to describe all fundamental particularities of rabbits at the anatomical level. It may still be of particular interest to those who are less familiar with veterinary anatomical nomenclature due to the differences, namely, at the level of directional terms that are different from those of humans.

2 General External Anatomy

The rabbit is a plantigrade animal and normally presents a rounded shape when resting (Fig. 1); however, when exploring the environment or running, it becomes digitigrade. In this situation, or when it rises in its pelvic limbs, its body shape is more pointed cranially. The rabbits' skin is quite thin, easily torn, and covered with soft uniform fur or hairs. Hairless areas are just the rostral extremity on the nose, the scrotum, and the inguinal areas. The normal coat is well endowed with both a short soft undercoat protected by longer guard hairs. The hair can vary in length depending on the breed, from less than 12 mm in Rex and Satin to 120 mm in Angora breed. They are usually less than 30 mm. The doe of large breeds (e.g., Flemish Giants breed) develops a prominent dewlap that is used to prepare the nest before kindling. This pendulous skin fold is particularly susceptible to wet dermatitis and dermatophytoses (Brewer 2006; O'Malley 2005).

The coloration of the coat varies greatly according to breed or varieties. The color in the wild rabbit is dusty brown with the ventral side of the abdomen and underside of the tail remaining white, which is a protective camouflaging with the

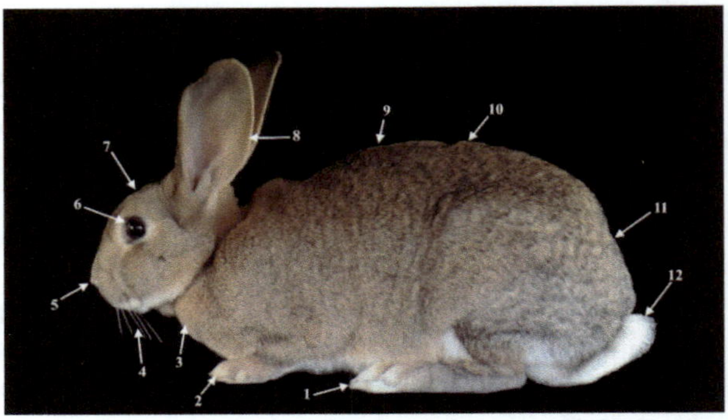

Fig. 1 The different parts of the rabbit's body. Legend: 1- Digit of hindlimb; 2- Digit of forelimb; 3- Dewlap; 4- Vibrissae; 5- Nostrils; 6- Left eye; 7- Front; 8- Ears; 9-Back; 10- Lumbar region; 11- Sacral region; 12- Tail

surroundings. The color of domestic ones is present from white, black, gray, brown, and others, also appearing coloration with meshes and various shades.

These animals do not present footpads but rather thicker skin and coarse fur on the plantar surfaces of the tarsus and palmar carpus, which contact the pavement. These areas are prone to pododermatitis if the floor is made up of thin metal meshes, as occurs in cages (Vella and Donnelly 2012).

The skin has few tubular sweat glands that are concentrated in the plantar region. There are, however, sebaceous-type glands and several other glands specialized in the production of odor secretions or scent glands. The most developed are the submandibular glands, the inguinal or perineal glands, and the anal glands, which are used for territorial and sexual marking. The submandibular glands are located on the underside of the chin and are used to rub against objects. The inguinal glands are located in the perineal region, lateral to the external genitalia; they are also called prepuce glands or clitorid glands, respectively, in males or females. These glands open into a glabrous depression in the skin, the perineal sinus, where a yellow or brown secretion is often seen. The anal glands are found lateral to the anus and, after some dissection, are a very conspicuous structure (1.5 to 2 cm). The more dominant animals of a group have larger glands since these glands are androgen-dependent and related to the level of sexual activity (O'Malley 2005; Melo and Gonzalez-Mariscal 2010; Vella and Donnelly 2012).

The head region has long funnel-like ears that represent 12% of the rabbits' body surface. The pinna is composed of a cartilaginous structure covered by short hair and is highly vascularized. On the inside of the pinnae appears quite frequently dermatitis caused by external parasites (mites) and dermatophytosis on the inside of the pinnae. In addition to hearing function, the ears are important in the thermoregulation of the animal and should not be used in its restraint due to the risk of breaking the cartilage and the pain it may cause (Brewer 2006). The rostral end of the nose, also called the rhinarium, is a hairless, Y-shaped region joining the two nasal openings (laterally) to the philtrum (ventrally), which forms a groove that divides the upper lip. The mouth opening is relatively small, with the lower lip and the characteristic split upper lip (Venâncio et al. 2007). Laterally, to these openings, rabbits present 20 to 25 long tactile hairs, the vibrissae (Brewer 2006). The eyes are

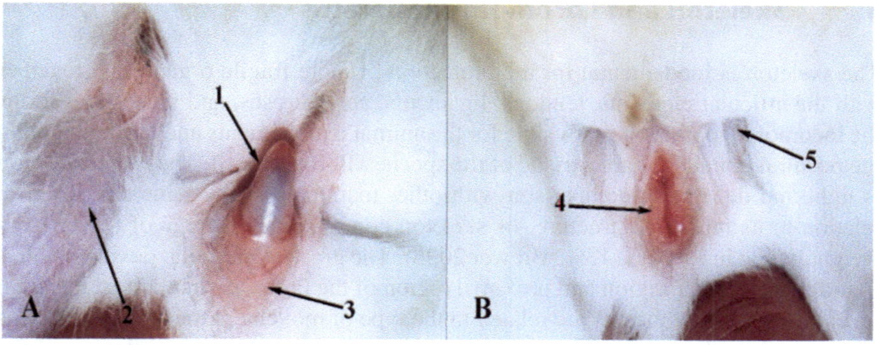

Fig. 2 External genitalia extruded: A) Male, B) Female. Legend: 1- Penis; 2- Scrotum; 3- Retracted prepuce; 4- Vulvar lips; 5- Perineal sinus

large, located laterally and on the top of the head. The eye is protected by a well-developed third eyelid in addition to the upper and lower eyelids. The third eyelid has on the inside the Harderian gland, which has a small, white upper lobe and a large, pink lower lobe. The lower lobe is also known as the deep gland of the nictitating membrane. Each eye has a lacrimal punctum with a relatively large duct (2 to 4 mm) located near the medial canthus in the lower eyelid, which easily obliterates in situations of respiratory affections. Rabbits blink spontaneously only 10 to 12 times per hour, which predisposes them to corneal ulceration and dry eye. Around the eyes, there are also long tactile hairs (O'Malley 2005).

On the trunk, they have a short neck and a well-developed abdominal region. The tail is short and recurved dorsally. Forelimbs are shorter and more mobile, with five toes, where the third toe is dominant, while the hind limbs are longer and have four toes. Long, arched claws are present on all toes, for which care must be taken in restraining the animal for the injuries they may cause (O'Malley 2005).

Sexual dimorphism is usually not very evident; however, adult males may have a wider head with well-developed muscular members and thorax, and females a more pronounced jowl. From a certain age, we can see the scrotum in the males and four, five, or six pairs of nipples (2 thoracic, 1 or 2 abdominal, and 1 or 2 inguinal) in the females. Five or six papillary ducts open in each of the nipples, with each duct corresponding to a mammary gland. Nipples are not present in bucks (Brewer 2006). The testicles descend at 10–12 weeks to scrotum, but in certain fighting situations or out of sexual activity periods, they can return to the abdominal cavity. The scrotum is found cranially to the penis. Sexual differentiation is achieved by the externalization of their external genital organs (Fig. 2). The penis protrudes as a round pink tube. The female has a slit-like orifice between vulvar lips. In young animals in which these structures are not fully developed without experience, some mistakes may occur (Brewer 2006) because there are no differences in the anogenital distance for both sexes (Lossi et al. 2016).

3 Skeleton and Locomotor Apparatus

The skeleton is fundamental for the protection of more fragile organs, and together with the articular elements, tendons, ligaments, aponeuroses, and muscles, it forms the locomotor apparatus responsible for the animal's movements and for maintaining the postural attitudes characteristic of the species (Barone 1986). The rabbit skeleton is light and delicate in comparison with other mammals of the same size, even in relation to its muscular structure. Its skeleton comprises an average of 8% of body weight, while in cats, it is 13% (Brewer 2006). The greatest muscular development is found in the lumbar region and proximal region of the limbs, particularly at the thigh level. These muscle masses are related to the type of movement the animal performs when moving, that is, through jumping, evident, also, in the natural flexion of the vertebral column and the greater length of the hind limbs. Their color is pale pink (Vella and Donnelly 2012). The rabbit has the following vertebral formula: C7, T12-13, L6-7, S4, and C15-16. The number of vertebrae varies in the thoracic-lumbar region, since the presence of only 12 thoracic vertebrae in 44% of cases stands out, contrary to the other domestic mammals that have at least 13 (Greenaway et al. 2001). Curiously, in the last thoracic vertebrae and in the first lumbar vertebrae, the ventral crest of its body is quite prominent (Barone 1986).

Restraining the rabbit should be done by holding it by the skin of its back and supporting its hind legs; otherwise, the abrupt and loose movement of these can lead to fractures in the lumbar portion of the spinal column, especially at L6-7 (Vella and Donnelly 2012).

The skeleton has particular differences related to phylogenetic evolution but is also related to the respiratory and digestive apparatus, so they will be referred to below in those apparatuses.

The similarity of its skeleton with that of the cat highlights the following morphological differences (Fig. 3):

1. Scapula more elongated and triangular in shape; its spine is situated near the cranial edge, the supraspinous fossa corresponding to approximately to 1/4 to 1/3 of the width of the scapula; the acromium presents a hamate and a suprahamate process quite developed.
2. Radius and ulna clearly fused in adult animals and significantly arched cranially.
3. Flattened ribs.
4. Lumbar vertebrae with mammillary processes equal in size to the spinous process; the transverse process is long and narrow.
5. Pelvis has a length/width relation superior to 2/1; the obturator orifice is clearly oval, and the acetabulum is situated in the middle of its length; curiously, the ilium, the ischium, and a small bone, the acetabulum bone, participate in the formation of the acetabulum, and the pubis bone is excluded.
6. The fibula is fused with the tibia and is only half as long as the tibia (Okerman 1994).

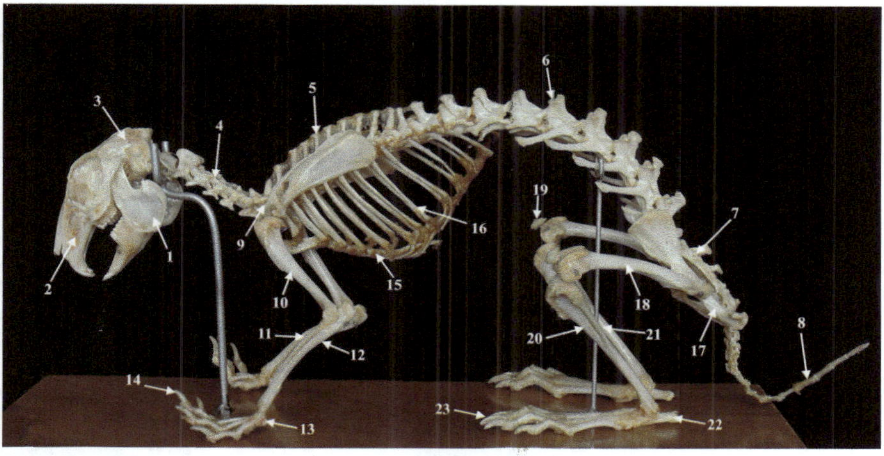

Fig. 3 Skeleton of the rabbit (left side). Legend: 1- Mandible; 2- Maxilla; 3- Cranium; 4- Cervical vertebra; 5- Thoracic vertebra; 6- Third lumbar vertebra; 7- Sacrum; 8- Coccygeal vertebrae; 9- Hamate and superhamate process of scapula; 10- Humerus; 11- Radius; 12- Ulna; 13- Carpus; 14- Digit of forelimb; 15- Sternum; 16- Rib; 17- Ischium; 18- Femur; 19- Patella; 20- Tibia; 21- Fibula; 22- Calcaneus; 23- Digit of hindlimb

4 Respiratory Apparatus

The airways can be divided into an upper part (from the nostrils to the larynx) and a lower part (consisting of the trachea and bronchial branches).

The nasal cavities are particularly well developed, occupying a large volume of the head, and they contain the vomeronasal organ and the nasal turbinates, as well as communicate with the paranasal sinuses. The sinuses occupy approximately one-third of the internal volume of the skull. These animals are obligate nasal breathers due to the rostral position of the epiglottis with respect to the soft palate. This particularity implies that any obstruction at the nasal level, such as nasal discharge in bronchopneumonia or inflammation of the nasal sinus mucosa (rhinitis), causes wheezing, especially when there is respiratory effort. In an attempt to breathe in situations of severe respiratory distress or agony, the rabbit holds its mouth open (Johnson-Delaney and Orosz 2011).

The larynx has a laryngeal ventricle on each side, with a small glottis, and its visualization is difficult because it is covered by the root of the tongue. The laryngospasm reflex is frequent in rabbits and may complicate endotracheal intubation (Barone 1997; O'Malley 2005).

The trachea consists of 45–50 rings and is almost circular in cross-section.

The lungs are bright pink in color and show lobulation with shallow fissures (Fig. 4). The right lung has four lobes (cranial, middle, caudal, and accessory), and the left lung has only two lobes (cranial lobe with cranial and caudal portions and

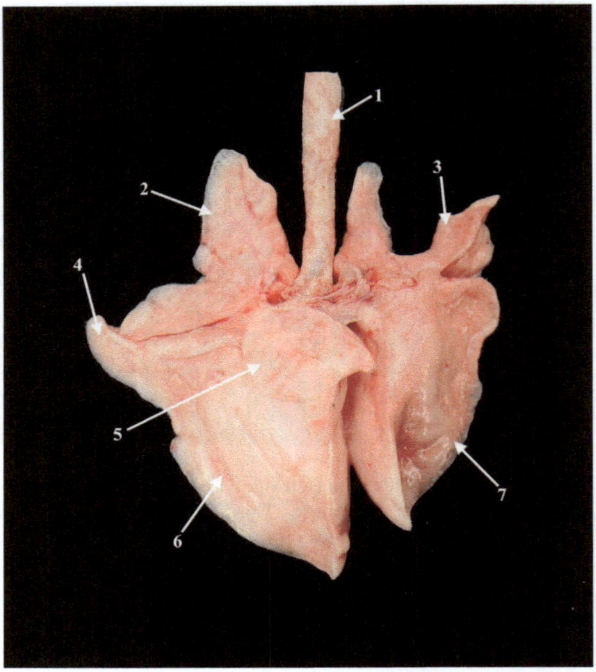

Fig. 4 Trachea and lungs of the rabbit (ventral side). Legend: 1- Trachea; 2- Cranial lobe of right lung; 3- Cranial lobe of left lung; 4- Medial lobe of right lung; 5- Accessory lobe of right lung; 6- Caudal lobe of right lung; 7- Caudal lobe of left lung

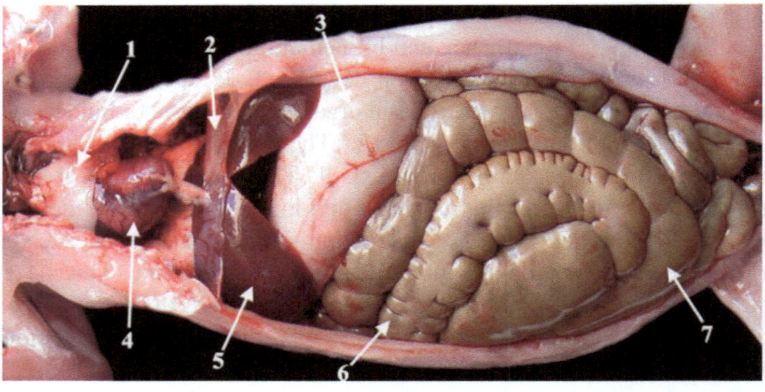

Fig. 5 Ventral view of thoracic and abdominal viscera. Legend: 1- Thymus; 2- Diaphragm; 3- Stomach; 4- Heart; 5- Liver; 6- First section of ascending colon; 7- Cecum

caudal lobe) (Barone 1997). Moreover, rabbit lungs are devoid of respiratory bronchioles and bronchial glands (Wright et al. 2008).

The thoracic cavity is small in comparison with the large abdominal cavity (Fig. 5) and breathing is largely diaphragmatic. With a normal respiratory rate at rest of 30–60 breaths per minute, the muscle contraction of the diagram is more used to promote respiratory movements than the intercostal muscles (O'Malley 2005; Varga and Harcourt-Brown 2014).

5 Digestive System

The rabbit is a monogastric herbivore, and its natural diet is based on grass, a food rich in fiber. It has a well-developed digestive system with particular emphasis on the length (4.5 to 5 m) and volume of the stomach and cecum.

The morphology of the digestive tract is completed at 9 weeks of age, with the exception of the cecal appendix, which grows up to 11 weeks. After 16 days, when solid food intake starts and the milk supply is maintained, folds appear on the wall of the cecum and colon (Fortun-Lamothe and Gidenne 2006).

5.1 Oral Cavity

The oral cavity is long and curved, narrow rostrally and wider caudally. Four pairs of major salivary glands flow into it: the parotid gland, mandibular gland, sublingual gland, and zygomatic gland. Of the minor salivary glands, the aggregation of the ventral buccal glands that form a lobe caudal to the angle of the mouth stands out. Saliva is continuously secreted, containing amylase, galactosidase, and small amounts of urea, unlike ruminants (Gargiulo et al. 1996; Davies and Davies 2003).

The tongue is long with a marked lingual torus. The dorsal surface is covered by filiform papillae, which give it a roughened surface. Among them on the rostral tongue are dispersed fungiform papillae. Near the root, there is a pair of more centrally vallate papillae, and on each side at the lateral border, there are foliate papillae with approximately 15 parallel ridges separated by grooves (Nonaka et al. 2008).

A pair of incisors in each jaw differentiates the Lagomorpha order from the Rodentia. The rostral incisor, more developed, has a small drop on the anterior surface; the other incisor, less developed, is slightly hidden caudally to the anterior incisor and has the function of protecting the palate from the potential trauma caused by the lower incisors during the cut (Crossley 1995).

The milk teeth (incisors and premolars) present at birth fall out at approximately 18 days and are replaced by permanent teeth.

The rabbit presents the following dental formula: 2/1 I, 0/0 C, 3/2 PM, and 2–3/3 M = 28 (Horowitz et al. 1973). The 28 teeth are open-rooted, continuously growing throughout life. The speed of incisor tooth growth is 2 mm per week in the maxilla and 2.4 mm in the mandible. Dental overgrowth, as well as malocclusion, which are usually observed in the first pair of incisors (Fig. 6), may result from congenital alterations in the conformation of the dental arches, in bad tooth formation, or more frequently due to the type of diet with the absence of dental abrasion (Legendre 2002). Each mandible has only one incisor. In both arches (upper and lower), there are no canines, and the premolars are separated from the incisors by a

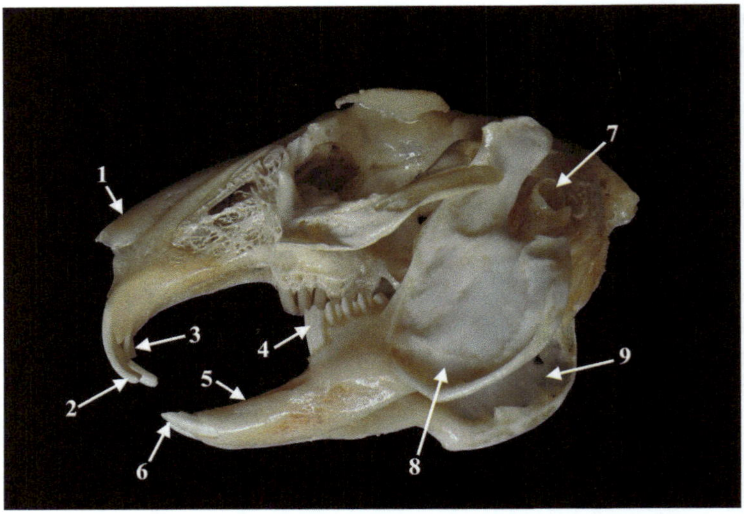

Fig. 6 Lateral view of skull and mandible (left side). Legend: 1- Nasal bone; 2- First upper incisor overgrowing (malocclusion); 3- Second upper incisor; 4- Premolar teeth; 5- Diastema; 6- Lower incisor; 7- External acoustic meatus; 8- Fossa masseterica; 9- Fossa pterygoidea

free space: the bar or diastema. Similar to equines, the maxillary dental arch is wider than the mandibular, although the teeth are arranged in parallel columns (Brewer 2006).

The various parts of the digestive tract are described in Fig. 7.

5.2 Esophagus and Stomach

The esophagus allows the movement of the bolus toward the stomach, and its three layers of striated muscles contribute to the formation of the well-developed cardiac sphincter. This does not allow vomiting or regurgitation and thus protects the esophageal mucous membrane free of mucous glands (Brewer 2006). The stomach of the rabbit has a pouch shape with a smaller and larger curvature. It is voluminous at approximately 0.1 to 0.25 liters, 15% of the volume of the gastrointestinal tract (Vella and Donnelly 2012), but not very extensible and with limited movements. It consists of a thin muscular wall (much reduced mainly in the cardiac and fundic region) and the gastric mucosa. The opening of the cardia is constituted by a well-developed cardiac sphincter that prevents vomiting. The gastric mucosa is divided internally into three distinct zones: the cardiac, fundic, and pyloric zones. The most developed and thickest is the fundic region, comprising two-thirds of the total area. The stomach is cranially in the abdominal cavity in a transversal position, with the fundus to the left and the pyloric antrum to the right. The pylorus has a developed sphincter that regulates the passage of food into the small intestine (O'Malley 2005).

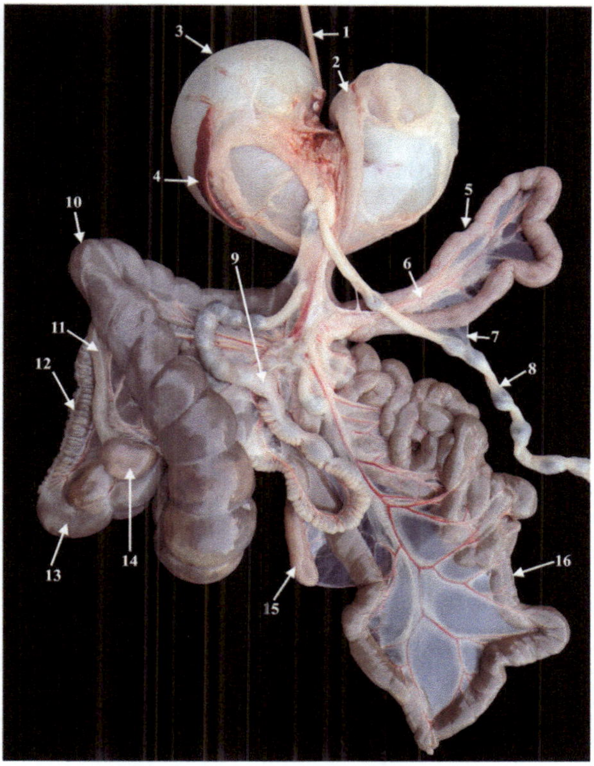

Fig. 7 Digestive tract of the rabbit. Legend: 1- Esophagus; 2- Pylorus; 3- Fundus of stomach; 4- Spleen; 5- Duodenum descendent; 6- Pancreas; 7- Duodenocolic fold; 8- Colon descendent; 9- *Fusus coli*; 10- Cecum; 11- Ileum; 12- First section of ascending colon; 13- Sacculus rotundus; 14- Ampulla coli; 15- Cecal vermiform appendix; 16- Jejunum

5.3 Small Intestine

The rabbit small intestine is approximately 3 meters long, constitutes approximately 12% of the total gastrointestinal volume, and is divided into the duodenum, jejunum, and ileum (Vella and Donnelly 2012). The duodenum, which follows the pylorus, is approximately 40 cm long and has four characteristic portions: the cranial portion, descending portion, transverse portion, and ascending portion. At its beginning, it connects the choledochal duct, and at the end of the transverse portion, it joins the pancreatic duct. The jejunum is the longest section of the small intestine, presents its typical appearance forming loops, and is slightly less thick and vascular than the duodenum (Barone 1997). The jejunum is situated on the left flank and throughout the caudal ventral part of the abdomen; Peyer's patches (aggregates of lymphoid tissue in the lamina propria) become most prominent toward the ileum. The ileum is the short terminal part to which the *plica ileocecalis* is attached. It ends into a spherical thick-walled enlargement known as the *sacculus rotundus*, which is unique to rabbits and has abundant lymphoid tissue (Fig. 8). The ileal ostium does not have an ileal papilla, and the ileal sphincter is located in the wall at the base of the cecum (Brewer 2006). This sphincter controls movement of the digesta from the ileum into

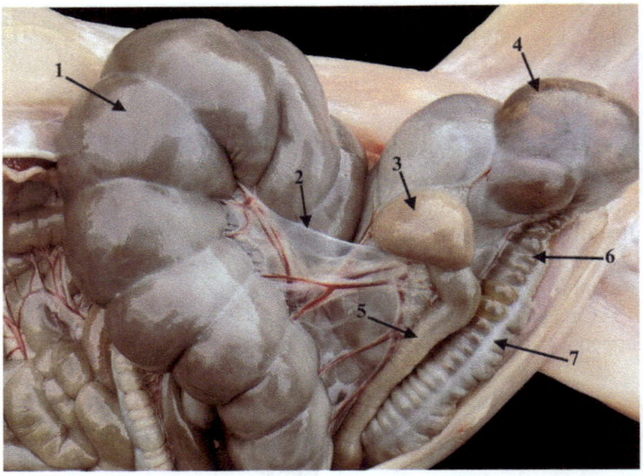

Fig. 8 Ventral view of junction between the small and large intestine of the rabbit. Legend: 1- Cecum; 2- Ileocecal fold; 3- Sacculus rotundus; 4- Ampulla coli; 5- Ileum; 6- Sacculation of the colon; 7- Band of colon

the cecum through the *sacculus rotundus* and prevents reverse flow into the ileum. It is one of the most frequent sites of intestinal obstruction by foreign bodies (Vella and Donnelly 2012).

5.4 Large Intestine

The cecum is the first part of the large intestine, and it is coiled around itself, forming a loop and a half. The adult rabbit cecum is the most voluminous portion and accounts for between 40 and 60% of the total volume of the digestive tract, with a length twice as long as the abdominal cavity. It is a blind bag that is composed of the base, the body, and the vermiform appendix of the cecum (Davies and Davies 2003). The base of the cecum is the initial portion that is connected simultaneously to the ileum and the colon. Following the base is the body of the cecum, which is approximately 27 cm long and 4–5 cm in diameter. This presents a wall structured in sacculae or saculiform rings separated by approximately 22 to 25 semilunar folds that correspond internally to a marked spiral fold. The shape of the cecum is due to this rolled-up fold consisting of a strip of smooth muscle approximately 1 m long, and this fold increases the contact surface of the mucosa with the cecal contents. The cecum is on one side united by its concave curvature to the ileum through the ileocecal fold and on the opposite convex side united by a narrow caecocolic fold to the ascending colon. In this way, the three viscera are joined together in a circle and a half. Finally, the vermiform appendix of the cecum is the terminal part of the cecum; it measures approximately 13 cm long, is narrow (1–2 cm in diameter), smooth, cylindrical, and rich in lymphatic tissue (Snipes 1978; Barone 1997).

The colon is approximately 1.5 m long and is usually divided into ascending, transverse, and descending colon. Although the characterization of these divisions is complex due to the set of loops and curvatures they present, these divisions have been maintained (Barone 1997).

The ascending colon is very long and further divided into two sections according to its external morphological characteristics. The first section is joined to the base of the cecum by a dilatation called the ampulla of the colon; it extends for approximately 10–15 cm, is joined to the cecum by the caecocolic fold, and has three series of haustrae or sacculations separated by three bands of muscular fibers, the bands of the colon (*Teniae coli*). The second section, approximately 20–30 cm long, has a single series of sacculations occupying approximately half of the colon wall, the other half being occupied by a single band of the colon (Barone 1997). The transverse colon is short, approximately 4 cm long, without sacculations, crosses from right to left and ends in a muscular thickening, typical of lagomorphs, called *fusus coli*. It has a thick mucosa and layer of richly innervated as well as vascularized muscle in this portion, which regulates the contractions of the colon and controls the production of the two types of stools. This structure divides, morphologically and functionally, the colon of the rabbit and leads to a change in nomenclature in the description of the colon, renaming it the proximal colon and distal colon instead of the ascending, transverse, and descending colon. The descending colon has a smooth, transparent, and thinner wall than the previous ones, extending for approximately 80–100 cm. The large intestine ends through the rectal ampulum at the anal canal (Davies and Davies 2003; O'Malley 2005).

5.5 Liver and Pancreas

The liver is dark purple in color and is situated between the diaphragm and the stomach. The presence of several deep fissures gives rise to a characteristic lobation formed by four main lobes: the left lobe is divided into two, the lateral lobe and the left medial lobe that has two or three tongues; the right lobe is not divided; the caudate lobe is very developed, with a papillary and a caudate process; the quadrate lobe, located ventrally between the left medial lobe and the right lobe, is small and supports the gall bladder (Fig. 9). The rabbit liver produces a large amount of bile, and the bile duct enters the duodenum separately from the pancreatic duct at 3–5 cm from the pylorus (Barone 1997; Vella and Donnelly 2012).

The pancreas is formed by a mass of relatively diffuse and isolated small lobules at the level of the mesoduodenum. Only the left pancreatic lobe is more compact and has a better-defined shape; it is found in the deep wall of the great omentum, caudal to the gastric fundus until it reaches the ventral face of the left kidney and the spleen. Only the accessory pancreatic duct is present, which opens 30–40 cm from the pylorus at the boundary between the descending and transverse parts of the duodenum (Barone 1997).

6 Urogenital Apparatus

The urogenital apparatus is made up of the urinary organs and the male or female genital organs.

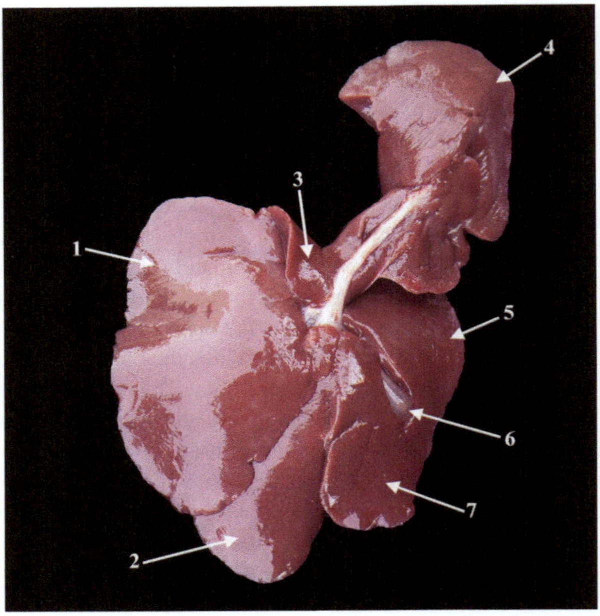

Fig. 9 Liver of the rabbit (visceral surface). Legend: 1- Left lateral lobe; 2- Left medial lobe; 3- Papillary process of caudate lobe; 4- Caudate process of caudate lobe; 5- Right lobe; 6- Gall bladder; 7- Quadrate lobe

6.1 Urinary Organs

The rabbit kidneys are smooth and unipapillary (meaning that one papilla and calyx enter the ureter) with extensive evaginations of the pelvis into the medullary tissue, as in dogs. They have the characteristic "bean" shape and a dark brown color (Fig. 10). The cranial pole of the right kidney is located at the level of the last rib, and the caudal pole reaches the third transverse process of the lumbar vertebrae. The left kidney is clearly more caudal than the right kidney and is located between the third and fifth transverse processes of the lumbar vertebrae. Both are situated in a dorsal and lateral position in the abdominal cavity. Both may be surrounded by a considerable amount of adipose tissue and can be palpated in the thoracolumbar region (Brewer 2006). The ureters flow into the urinary bladder, which is quite extensible and completely situated in the abdomen. The lateral vesical ligaments are rudimentary, but the median vesical ligament is long. The rabbit is the only known mammal in which the tubules can be separated from kidney slices with the tubular epithelium intact (O'Malley 2005; Varga and Harcourt-Brown 2014).

The rabbit depends on the kidney for calcium osmoregulation. Rabbits excrete an average of 130 ml/kg/day of alkaline urine, which is the major route for the excretion of magnesium and calcium. It is often creamy, owing to the presence of high levels of calcium carbonate, but it can also vary in color from pale yellow to dark red (often mistaken for hematuria by owners) as a consequence of the presence of endogenous and exogenous pigments, such as bile pigments and plant porphyrins (Brewer 2006; Vella and Donnelly 2012).

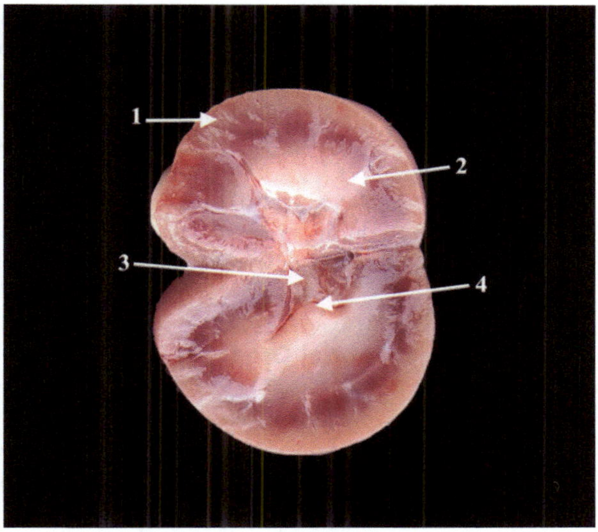

Fig. 10 Kidney of the rabbit (dorsal section). Legend: 1- Renal cortex; 2- Renal medulla; 3- Renal pelvis; 4- Renal papilla

6.2 Male Genital Organs

The genital organs are fully developed when the rabbit reaches puberty at approximately 4 to 5 months of age and sexual maturity, with the production of spermatozoids between 7–8 months earlier in smaller breeds than in larger breeds (Vella and Donnelly 2012). The sexual activity of the domestic rabbit extends throughout the year but is more intense during the spring.

The male genital organs consist of two testicles, with associated epididymis and ductus deferens, accessory genital glands, and the penis (Barone 2001).

The male has two external testicles with an elongated ovoid shape that lie on two sparsely haired scrotal sacs. In rabbits, the testicles descend into the scrotum between the 6th and 22nd postnatal days. However, they may also be endorchids due to their capacity to retract the testicles to the abdominal cavity. This is possible because the inguinal canals remain open throughout life and have a well-developed cremaster muscle (Rajfer 1982).

At the end of the ductus deferens are ampullae ductus deferens located between the vesicular gland and the urethra. Other accessory genital glands are the vesicular gland, the prostate, and the bulbourethral glands. The vesicular gland in this species is single, flattened dorso-ventrally and bilobed cranially, and its distal part is between the proprostate and ampulla ductus deferens. The prostate does not present a disseminated portion, its compact portion is well developed, and several divisions are considered. The cranial portion or proprostate (*proprostata*), the prostate (*prostata*) caudally located, and laterally the paraprostates (*paraprostata*). The proprostate and the prostate are connected by a thin septum of connective tissue and share a capsule of connective tissue, presenting two independent pairs of excretory ducts. Paraprostates are small cloverleaf- or butterfly-shaped glands in pairs or singles on either side of the pelvic urethra. Due to its separation from the other portions of the prostate, the paraprostates can also be considered independent as paraprostatic glands. The bulbourethral glands are composed of two lobes distinctly demarcated from each other, just behind the prostate on the dorsal wall of the urethra and embedded in the bulboglandular muscle (Holtz and Foote 1978; Skonieczna et al. 2019).

The penis is retroflexed (cat-like) inside the foreskin during sexual rest, and this position is reversed during erection, heading cranially. There is no os penis, and it does not have a differentiated glans (Barone 2001).

6.3 Female Genital Organs

The female genital organs are formed by the ovaries, uterine tubes, uterus, vagina, vestibule of the vagina, vulva, and clitoris. In addition to age, weight can be a good indicator to start breeding, recommending that females have reached 75–80% of live weight in adulthood (Vella and Donnelly 2012).

The ovaries are elongated and situated caudally to the kidney on each side at the level of the fifth lumbar vertebra, approximately one centimeter from the edge of the lumbar wall, approximately equidistant from the last rib and the iliac crest. The uterine tubes are relatively long (8–10 cm), with the infundibulum that is large, situated ventromedial to the ovary, which it may cover almost entirely, and the ampulla is significantly wider (3 mm) than the isthmus (1 mm) (Barone 2001).

The uterus is double, has no uterine body, and consists of two long uterine horns (10–12 cm) that communicate independently with the vagina through separate uterine coli (Fig. 11). The uterus is supported by the broad ligament of the uterus, and there is usually considerable accumulation of fatty tissue in this structure (Barone 2001).

The vagina is relatively long (4–6 cm) and large in this species, with a deep fornix (3–4 mm) around the two uterine coli. The vestibule of the vagina is long (5–6 cm), located beyond the sciatic arch and ventral of the rectum. It has two pairs of vulvar lips, the clitoris is strong, and its body is approximately 4 cm long (Barone 2001).

The rabbit can be fertilized in the days following parturition, and ovulation is induced by coitus. The type of placenta of the rabbit is discoid and hemochorial. Fetus can be felt by palpation between 12 and 14 days after mating. The gestation period is from 30 to 33 days, and the number of newborns per litter is variable, usually between 1 and 12 (O'Malley 2005).

7 Cardiovascular and Lymphatic System

The rabbit's heart has a much more cranial location than that of other domestic mammals. The area of the heart extends from the second to the fifth rib. It is located inside the pericardial cavity at the level of the middle mediastinum, as in other mammals. It has a conical shape, with its apex directed caudally and slightly to the

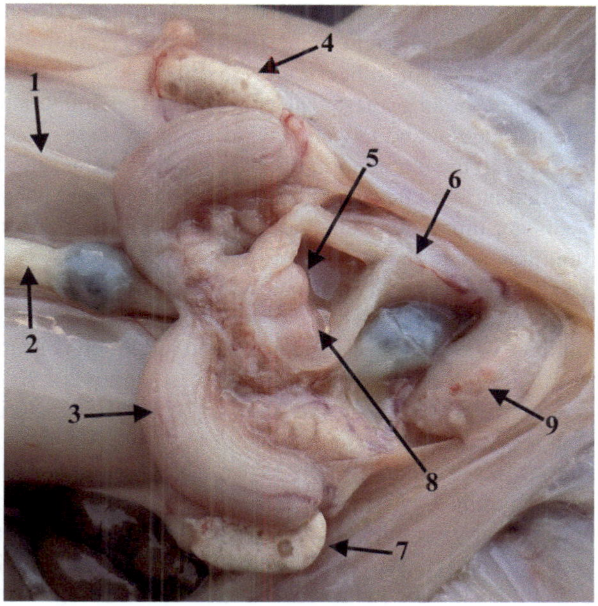

Fig. 11 Duplex uterus of the rabbit. Legend: 1- Ureter; 2- Descendent colon; 3- Right horn of the uterus; 4- Left ovary; 5- Left cervix; 6- Vagina; 7- Right ovary; 8- Right cervix; 9- Urinary bladder

left, with a clear demarcation of the atria and ventricles. The heart is relatively small, comprising only 0,3% of the total body weight, and its right atrioventricular valve has only two cusps instead of three as in other mammals (Barone 2011; Vella and Donnelly 2012). Blood flows to the heart through the left and right coronary arteries, where the caliber of the left is only slightly larger. In the left coronary artery, the circumflex branch is considerably larger than the paraconal interventricular branch and divides in the caudal region of the left ventricle wall, irrigating a greater portion of it (Flores et al. 1989; Barone 2011). Rabbit hearts have a simple conduction system, and only a small group of cells are in the sinoatrial node (Bleeker et al. 1980).

The rabbit aorta is very similar to that of domestic carnivores, with an even more marked curve in the arch, the birth of the left subclavian artery slightly farther from the brachiocephalic trunk, and the latter releasing the left common carotid artery almost at its emergence from the aorta (Barone 2011). It has a rhythmic contraction of neurogenic origin that has not been noted in other mammals (Brewer 2006). The descending aorta termination gives rise on its dorsal surface to the median sacral artery and then bifurcates into two common iliac arteries, each of which supplies, after a short course, the internal and external iliac arteries on its side (Barone 2011). The pulmonary artery is much more heavily muscled in rabbits than in other species (Brewer 2006).

A large arteriovenous anastomotic system is present in the ears for heat regulation (Brewer 2006).

At the level of the venous part, there is the caudal vena cava and two independent cranial vena cava (right and left). The right cranial vena cava has an arrangement similar to that of other domestic mammals, while the left cranial vena cava represents, among these, an exception. The left cranial vena cava receives the external jugular and left subclavian veins, bypasses the heart on the left side, passes ventrally to the termination of the pulmonary veins, and flows into the caudal part of the right atrium after receiving the greater and middle cardiac veins (Barone 2011). Veins are thin-walled and susceptible to hematoma formation, and with venipuncture, they are prone to collapse (Pariaut 2009).

The rabbit normally has a blood volume of 55–70 ml/kg, and the blood cells show some differences from those of other domestic mammals. Polychromasia and small numbers of reticulocytes and nucleated red blood cells may be observed due to their short life span and high erythrocyte turnover. Neutrophils are often called heterophils because of the staining (pink) characteristics of their cytoplasmic granules. Basophils account for up to 8–10% of the total white blood cell count (Varga and Harcourt-Brown 2014).

The lymphatic system is very similar to that of carnivores in the number of lymph nodes, their groupings (lymphocenters), and the arrangement of the main lymphatic vessels. The lymph nodes are pinkish-gray or pale pink, or even whitish, and are sometimes difficult to discern in the fat surrounding them. These are usually small in size, and the cranial mesenteric lymph nodes are the largest with approximately 2 cm (Barone 2011) (Soto-Miranda et al. 2013).

At the level of lymphatic tissue, it is important to highlight its development in several regions of the intestine, which can represent approximately 50% of the total mass of lymphoid tissue. Considering the size of the intestine in this species and the richness of its microbiota, the existence of immunological and nonimmunological intestinal defenses is considered protection against the aggression of pathogenic microorganisms (Box 2) (Schroder 1999; Kudsk 2002; Fortun-Lamothe and Gidenne 2006).

Box 2 Intestinal Defenses

Immunological (local lymphoid system):

- Peyer's patches (2 to 20 in the small intestine).
- Lymphoid tissue of the *sacculus rotundus*.
- Lymphoid tissue of the cecal appendix.

Nonimmunological:

- Peristaltic movements.
- Permanent renewal of the digestive epithelium.
- Mucus with bacteriostatic/bactericidal substances (lactoferrin, lactoperoxidase, and lysozyme).
- Stable microbiota.

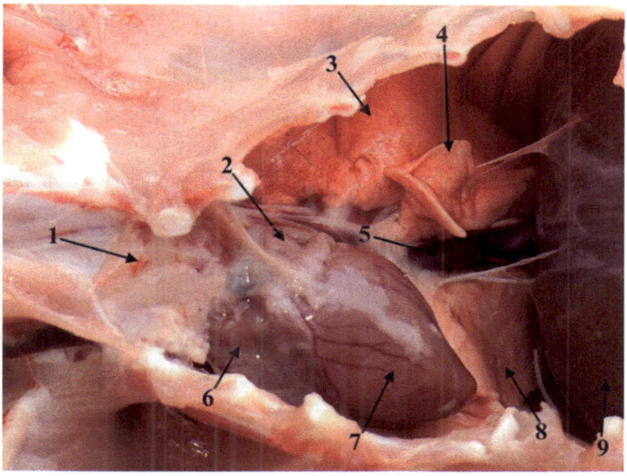

Fig. 12 Thoracic cavity (ventral view). Legend: 1- Thymus; 2- Left auricula; 3- Lobe caudal of left lung; 4- Lobe accessory of right lung; 5- Vena cava caudal; 6- Right ventricle; 7- Left ventricle; 8- Lobe caudal of right lung; 9- Diaphragm

The spleen, dark pink in color and elongated elliptical in shape, lies on the left side of the abdominal cavity joined to the stomach by the gastrosplenic ligament (Barone 2011).

The thymus reaches its maximum development at approximately three or four months, and its average weight is approximately 6 or 7 grams. The organ has the particularity of persisting large in the adult rabbit (Fig. 12); however, both cortical and medullary size have some decrease with age. Its two continuous parts, cervical and thoracic, lie cranial to the heart. The cervical part is very small only until the 6th cervical vertebra level. The thoracic part has two asymmetrical lobes separated by a clear interlobar fissure. The right lobe has a ventral and dorsal extension, but the left thoracic lobe is only in the mediastinum ventral. The left thoracic lobe is clearly larger than the right, and it has cardiac and pulmonary impressions (Barone 2011; Gaber 2017).

8 Endocrine Glands

As in other mammals, the endocrine glands of the rabbit are the pituitary, pineal gland, thyroid, parathyroids, pancreas, thymus and adrenal glands, as well as the gonads and placenta, which are also considered organs with important endocrine function but were described before (König and Liebich 2020).

The pituitary and pineal glands are both located in the brain (diencephalon). The pituitary has two major divisions, the adenohypophysis and neurohypophysis, and it is attached to the hypothalamus (König and Liebich 2020). It has a height-to-brain area ratio much higher than that in dogs or cats.

The thyroid gland consists of two lobes, right and left, connected ventrally by a thin isthmus. Each lobe is 15–20 mm in rostrocaudal length and 6–12 mm in width.

It is located between the rostral horn of the laryngeal thyroid cartilage and the 9th tracheal ring (Antakia et al. 2014).

Rabbits have four parathyroid glands, two on each side, the outer parathyroid gland and the inner parathyroid gland. These are located close to the thyroid or adjacent muscles; however, due to their relatively small size and their color, which can be confused with the thyroid or the fat in which they are often enveloped, their identification is difficult (Tan et al. 1987; Cao et al. 2021).

The thymus located in the cranial mediastinum has already been described in the lymphatic system. In the pancreas, endocrine function is only associated with the pancreatic islets.

The adrenal glands are composed of an inner medulla and outer cortex that are relatively voluminous. The rabbits' adrenal glands have an intense yellow color and are located between the cranial border of the kidneys and the caudal vena cava (Venâncio et al. 2007; Bradley et al. 2021).

9 Nervous System and Sense Organs

The central nervous system formed by the encephalon and spinal cord are protected and inside the cranial cavity and the vertebral column, respectively (König and Liebich 2020). Apart from its lining by the meninges, which extend quite caudally to the fourth sacral vertebra, the spinal cord is also considerably long to the second sacral vertebra. The brain has three major divisions: the cerebrum, the cerebellum, and the brainstem. Its most striking feature is that it is a lysencephalon, as its triangular-shaped brain has a smooth surface, without obvious grooves or circumvolutions (Fig. 13). In the midbrain, the enormous development of the rostral colliculi on its dorsal surface stands out, as well as the existence of a voluminous red nucleus internally (Barone 2004). The cerebellum is relatively voluminous and is two folds larger than long and is easily visible in a dorsocaudal view. It has a prominent paraflocculi, a small lobule of the cerebellar hemisphere, dorsal to the flocculus, and linked to the pyramis and uvula (Lossi et al. 2016). The ventral brainstem has a considerably wide trapezoidal body, as wide as the *pons*. The development of the trapezoidal body and of other structures associated with hearing,

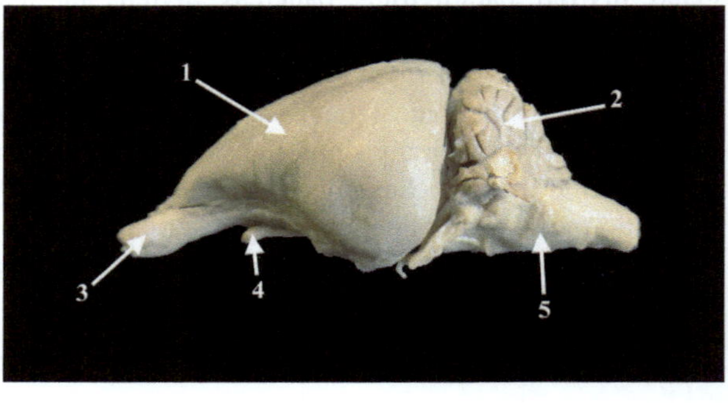

Fig. 13 Encephalon of the rabbit (left side). Legend: 1- Cerebrum; 2- Cerebellum; 3- Olfactory bulb; 4- Nerve optic; 5- Medulla oblongata

such as a relatively large tympanic bulla and long ears, are associated with their keen sense of hearing to escape from predators (Barone 2004; Lossi et al. 2016).

The olfactory bulbs are also well developed, similar to those of carnivores, extending beyond the hemispheres. This fact and the size of the nasal cavities are associated with its good olfactory capacity (Barone 2004).

The rabbit eye is compressed in the antero-posterior axis; its conjunctiva is relatively thin (10–40 μm), a cornea that occupies 30% of the eyeball is approximately 30–40 μm thick, and has a capacity of 40–43 diopters; it has a small and poorly developed ciliary body; the choroid is well developed and devoid of a tapetum; the lens has a limited capacity for accommodation (0–1.5 diopters), and in the retina, its vessels are easily visible (merangiotic) (O'Malley 2005; Labbe et al. 2006; Lossi et al. 2016). The eyes are located laterally on the head so that it has a wide angle of vision but does not visualize a small area close to the rostral extremity. This is compensated by the existence of the vibrissae (tactile hairs) lateral to the lips, thus allowing correct contact with the food. In addition, the tongue presents many lingual papillae with a taste function, which justifies the fact that they are very selective animals in the food they eat (O'Malley 2005).

10 Conclusions

From the external morphology to the different apparatuses and systems, in all of them, we can find specific features that are of interest and make this species a challenge to those who breed or care for it. Among all the differences, we must highlight the development and the particularities found in its digestive system. It is our understanding that knowledge of all these characteristics will be fundamental to carry out the most suitable management and treatment of this species, independent of the different purposes for which it is used.

Acknowledgments The authors thank Springer for the use of the AJE Digital Editing. This work was supported by projects UIDB/04033/2020 (CITAB), UIDP/CVT/00772/2020 (CECAV) and LA/P/0059/2020 (AL4AnimalS), funded by the Portuguese Foundation for Science and Technology (FCT).

References

Antakia R, Gayet P, Guillermet S, Stephenson TJ, Brown NJ, Harrison BJ, Balasubramanian SP (2014) Near infrared fluorescence imaging of rabbit thyroid and parathyroid glands. J Surg Res 192(2):480–486. https://doi.org/10.1016/j.jss.2014.05.061

Barone R (1986) In: Fréres V (ed) Anatomie comparée des mammiféres domestiques. Osteologie, vol 1, 3rd edn. Vigot Fréres

Barone R (1997) In: Fréres V (ed) Anatomie comparée des mammiféres domestiques. Splanchnologie I. Appareil digestif, appareil respiratoire, vol 3, 3rd edn. Vigot Fréres

Barone R (2001) In: Fréres V (ed) Anatomie comparée des mammiféres domestiques. Splanchnologie II. Appareil uro-génital, fœtus et ses annexes, péritoine et topographie abdominale, vol 4, 3rd edn. Vigot Fréres

Barone R (2004) In: Fréres V (ed) Anatomie comparée des mammiféres domestiques. Système nerveux central, vol 6. Vigot Fréres

Barone R (2011) In: Fréres V (ed) Anatomie comparée des mammiféres domestiques. Angiologie, vol 5, 2nd edn. Vigot Fréres

Bleeker WK, Mackaay AJ, Masson-Pevet M, Bouman LN, Becker AE (1980) Functional and morphological organization of the rabbit sinus node. Circ Res 46(1):11–22. https://doi.org/10.1161/01.res.46.1.11

Bradley AE, Wancket LM, Rinke M, Gruebbel MM, Saladino BH, Schafer K, Katsuta O, Garcia B, Chanut F, Hughes K, Nelson K, Himmel L, McInnes E, Schucker A, Uchida K (2021) International Harmonization of Nomenclature and Diagnostic Criteria (INHAND): nonproliferative and proliferative lesions of the rabbit. J Toxicol Pathol 34(3 Suppl):183S–292S. https://doi.org/10.1293/tox.34.183S

Brewer NR (2006) Biology of the rabbit. J Am Assoc Lab Anim Sci 45(1):8–24. https://www.ncbi.nlm.nih.gov/pubmed/16539330

Cao W, Su Y, Liu N, Peng Y, Diao C, Cheng R (2021) Location and vascular classification of 188 parathyroid glands in New Zealand white rabbits. Arq Bras Med Vet Zootec 73(2):377–382. https://doi.org/10.1590/1678-4162-11760

Crossley DA (1995) Clinical aspects of lagomorph dental anatomy: the rabbit (oryctolagus cuniculus). J Vet Dent 12(4):137–140. https://www.ncbi.nlm.nih.gov/pubmed/9693639

Davies RR, Davies JA (2003) Rabbit gastrointestinal physiology. Vet Clin North Am Exot Anim Pract 6(1):139–153. https://doi.org/10.1016/s1094-9194(02)00024-5

de la Fuente LF, Rosell JM (2012) Body weight and body condition of breeding rabbits in commercial units. J Anim Sci 90(9):3252–3258. https://doi.org/10.2527/jas.2011-4764

Dorozynska K, Maj D (2021) Rabbits - their domestication and molecular genetics of hair coat development and quality. Anim Genet 52(1):10–20. https://doi.org/10.1111/age.13024

Flores NA, Davies RL, Sheridan DJ (1989) Microangiographic investigation of the effects of radiographic contrast media and hyperkalaemia on coronary artery calibre in the rabbit. Q J Exp Physiol 74(2):181–195. https://doi.org/10.1113/expphysiol.1989.sp003254

Fortun-Lamothe L, Gidenne T (2006) Recent advances in digestive physiology of the growing rabbit. In: Luc Maertens P (ed) Recent advances in rabbit sciences. Institute for Agricultural and Fisheries Research

Gaber W (2017) Morphogenesis of the thymus in rabbit during prenatal and early postnatal periods. Histol Cytol Embryol 1(4):1–9

Gargiulo AM, Ceccarelli P, Pedini V (1996) The presence of granular excretory ducts in the rabbit zygomatic gland. Anat Histol Embryol J Vet Med C 25(3):175–176. https://doi.org/10.1111/j.1439-0264.1996.tb00079.x

Greenaway JB, Partlow GD, Gonsholt NL, Fisher KR (2001) Anatomy of the lumbosacral spinal cord in rabbits. J Am Anim Hosp Assoc 37(1):27–34. https://doi.org/10.5326/15473317-37-1-27

Holtz W, Foote RH (1978) The anatomy of the reproductive system in male Dutch rabbits (Oryctolagus cuniculus) with special emphasis on the accessory sex glands. J Morphol 158(1):1–20. https://doi.org/10.1002/jmor.1051580102

Horowitz SL, Weisbroth SH, Scher S (1973) Deciduous dentition in the rabbit (*Oryctolagus cuniculus*). A roentgenographic study. Arch Oral Biol 18(4):517–523. https://doi.org/10.1016/0003-9969(73)90072-1

Johnson-Delaney CA, Orosz SE (2011) Rabbit respiratory system: clinical anatomy, physiology and disease. Vet Clin North Am Exot Anim Pract 14(2):257–266., vi. https://doi.org/10.1016/j.cvex.2011.03.002

König HE, Liebich H-G (2020) Veterinary anatomy of domestic animals: textbook and colour atlas. (7th, updated and extended edition. ed.). Georg Thieme Verlag

Kudsk KA (2002) Current aspects of mucosal immunology and its influence by nutrition. Am J Surg 183(4):390–398. https://doi.org/10.1016/s0002-9610(02)00821-8

Labbe A, Liang H, Martin C, Brignole-Baudouin F, Warnet JM, Baudouin C (2006) Comparative anatomy of laboratory animal corneas with a new-generation high-resolution in vivo confocal microscope. Curr Eye Res 31(6):501–509. https://doi.org/10.1080/02713680600701513
Legendre LF (2002) Malocclusions in Guinea pigs, chinchillas and rabbits. Can Vet J 43(5): 385–390. https://www.ncbi.nlm.nih.gov/pubmed/12001507
Lennox AM (2010) Care of the geriatric rabbit. Vet Clin North Am Exot Anim Pract 13(1): 123–133. https://doi.org/10.1016/j.cvex.2009.09.002
Lossi L, D'Angelo L, De Girolamo P, Merighi A (2016) Anatomical features for an adequate choice of experimental animal model in biomedicine: II. Small laboratory rodents, rabbit, and pig. Ann Anat 204:11–28. https://doi.org/10.1016/j.aanat.2015.10.002
Melo AI, Gonzalez-Mariscal G (2010) Communication by olfactory signals in rabbits: its role in reproduction. Vitam Horm 83:351–371. https://doi.org/10.1016/S0083-6729(10)83015-8
Nonaka K, Zheng JH, Kobayashi K (2008) Comparative morphological study on the lingual papillae and their connective tissue cores in rabbits. Okajimas Folia Anat Jpn 85(2):57–66. https://doi.org/10.2535/ofaj.85.57
Okerman L (1994) Diseases of domestic rabbits. Blackwell Scientific Publications
O'Malley B (2005) Clinical anatomy and physiology of exotic species: structure and function of mammals, birds, reptiles, and amphibians. Elsevier Saunders
Pariaut R (2009) Cardiovascular physiology and diseases of the rabbit. Vet Clin North Am Exot Anim Pract 12(1):135–144., vii. https://doi.org/10.1016/j.cvex.2008.08.004
Rajfer J (1982) Endocrinological study of testicular descent in the rabbit. J Surg Res 33(2):158–163. https://doi.org/10.1016/0022-4804(82)90023-3
Schroder JM (1999) Epithelial peptide antibiotics. Biochem Pharmacol 57(2):121–134. https://doi.org/10.1016/s0006-2952(98)00226-3
Skonieczna J, Madej JP, Bedzinski R (2019) Accessory genital glands in the New Zealand white rabbit: a morphometrical and histological study. J Vet Res 63(2):251–257. https://doi.org/10.2478/jvetres-2019-0028
Snipes RL (1978) Anatomy of the rabbit cecum. Anat Embryol (Berl) 155(1):57–80. https://doi.org/10.1007/BF00315731
Soto-Miranda MA, Suami H, Chang DW (2013) Mapping superficial lymphatic territories in the rabbit. Anat Rec (Hoboken) 296(6):965–970. https://doi.org/10.1002/ar.22699
Tan SQ, Thomas D, Jao W, Bourdeau JE, Lau K (1987) Surgical thyroparathyroidectomy of the rabbit. Am J Phys 252(4 Pt 2):F761–F767. https://doi.org/10.1152/ajprenal.1987.252.4.F761
Varga M, Harcourt-Brown F (2014) Textbook of rabbit medicine: revised and edited, 2nd edn. Elsevier
Vella D, Donnelly TM (2012) Basic anatomy, physiology and husbandry of rabbits. In: Quesenberry KE, Orcutt CJ, Mans C, Carpenter JW (eds) Ferrets, rabbits, and rodents: clinical medicine and surgery, 4th edn. Elsevier, pp 131–149
Venâncio, C. F.D.; Colaço, B.; Teixeira, J.; Pinheiro, V.; Lacilla, J. (Ed.). (2007). Particularidades anatómicas do coelho. Universidade de Trás-o-Montes e Alto Douro
Wright JL, Cosio M, Churg A (2008) Animal models of chronic obstructive pulmonary disease. Am J Physiol Lung Cell Mol Physiol 295(1):L1–L15. https://doi.org/10.1152/ajplung.90200.2008

Physiological Features of Rabbits

José Júlio Martins, João Simões, Carlos Venâncio, Maria José Saavedra, and Maria da Conceição Fontes

J. J. Martins (✉)
Department of Animal Science, University of Trás-os-Montes e Alto Douro, Vila Real, Portugal

Interdisciplinary Centre of Marine and Environmental Research (CIIMAR), University of Porto, Research Group of Ecotoxicology, Stress Ecology and Environmental Health (ECOTOX), Terminal de Cruzeiros do Porto de Leixões, Matosinhos, Portugal

Animal and Veterinary Research Centre (CECAV), AL4AnimalS - Associated Laboratory for Animal and Veterinary Science , University of Trás-os-Montes and Alto Douro (UTAD), Vila Real, Portugal
e-mail: jjulio@utad.pt

J. Simões · M. da Conceição Fontes
Animal and Veterinary Research Centre (CECAV), AL4AnimalS - Associated Laboratory for Animal and Veterinary Science , University of Trás-os-Montes and Alto Douro (UTAD), Vila Real, Portugal

Department of Veterinary Sciences, School of Agricultural and Veterinary Sciences, University of Trás-os-Montes and Alto Douro (UTAD), Vila Real, Portugal

C. Venâncio
Department of Animal Science, University of Trás-os-Montes e Alto Douro, Vila Real, Portugal

Centre for the Research and Technology of Agro-Environmental and Biological Sciences (CITAB), Inov4Agro – Institute for Innovation, Capacity Building and Sustainability of Agri-food Production, University of Trás-os-Montes and Alto Douro (UTAD), Vila Real, Portugal

M. J. Saavedra
Animal and Veterinary Research Centre (CECAV), AL4AnimalS - Associated Laboratory for Animal and Veterinary Science , University of Trás-os-Montes and Alto Douro (UTAD), Vila Real, Portugal

Department of Veterinary Sciences, School of Agricultural and Veterinary Sciences, University of Trás-os-Montes and Alto Douro (UTAD), Vila Real, Portugal

Centre for the Research and Technology of Agro-Environmental and Biological Sciences (CITAB), Inov4Agro – Institute for Innovation, Capacity Building and Sustainability of Agri-food Production, University of Trás-os-Montes and Alto Douro (UTAD), Vila Real, Portugal

J. Simões, J. M. Monteiro (eds.), *Veterinary Care of Farm Rabbits*,
https://doi.org/10.1007/978-3-031-44542-2_3

Abstract

This chapter aims to review the main aspects and particularities of rabbit physiology. Several specific physiological features result from their adaptation to the environment, which are evident in both wild and domestic rabbits. Rabbit behavior and sensory system, thermoregulation and integument, cardiovascular, respiratory, urinary, reproductive, and digestive systems, and digestion physiology are discussed. Rabbits are induced ovulators but are influenced by environmental cues such as photoperiod and nutrition and are mostly sensitive to heat stress. They are highly efficient food convertors that promote the phenomenon of cecotrophy, which is a characteristic of lagomorphs. These main particularities are of relevance for understanding the needs that must be taken into consideration when breeding and caring for this species.

Keywords

Physiology · *Oryctolagus cuniculus* · Lagomorphs · Behavior · Cecotrophy · Thermoregulation · Induced ovulation

1 Introduction

The ancestors of domestic rabbits (Genus *Oryctolagus*) most likely evolved from the European rabbit, *Oryctolagus cuniculus Linnaeus, 1758,* in the middle Pleistocene period, which spread to nearby Mediterranean areas (Fox 1974; McNitt et al. 2013; Pelletier 2021). Although the first records of its association with humans go back to Romans, they were only considered truly domesticated approximately 200 years ago (Corbet 1986; Varga 2014). Despite all the years of domestication, rabbits still have a highly developed prey instinct (Sohn and Couto 2012) having evolved to be constantly vigilant, lightweight, and fast-moving, with a highly efficient digestive system (Meredith 2014).

The geographical distribution of rabbits was dramatically influenced by humans, since they were introduced in new areas, so that it could settle, reproduce, and become a food resource, for instance, on islands or shipping routes. Being highly adaptable animals, wherever they found suitable conditions, they established stable populations. Nevertheless, in some worldwide regions, such as North America, a significant diversity of rabbit species has developed instead of the relatively stable European rabbit (*O. cuniculus*) in Europe (Somerville and Sugiyama 2021). In other regions, such as Australia and New Zealand, they reproduced in such numbers that they became a pest, several decades after being introduced (Alves et al. 2022). The modern domesticated rabbit retains the main characteristics of the wild European rabbit, despite the differences in weight, color, coating, and behavior (Varga 2014).

Rabbits are classified as lagomorphs, and although they look like rodents in some morphological aspects, they differ in structural traits and serological data, showing more affinities with hoofed animals (Artiodactyla) (Nowak 1999). Modern lagomorphs are divided into two families, the family Leporidae, consisting of rabbits and hares, and the family Ochotonidae, which includes pikas (Meredith 2014). The family Leporidae includes many species that belong to eleven genera, *Brachylagus, Lepus, Bunolagus, Caprolagus, Nesolagus, Oryctolagus, Pentalagus, Poelagus, Pronolagus, Romerolagus,* and *Sylvilagus* (McNitt et al. 2013). The species *Bunolagus monticularis,* and *Sylvilagus mansuetus* are considered to be in a Critically Endangered situation (IUCN 2023), and others are considered to be endangered species.

Some basic biological data is reported in Box 1.

Box 1 Basic Physiological Data for Rabbits. Adapted from Varga (2014) with Permission from Elsevier

- Life span: 6–13 years
- Rectal temperature: 38.5–40 °C (101.3–104 °F)
- Heart rate: 130–325 bpm (beats per minute)
- Respiratory rate: 32–60 bpm
- Tidal volume: 20 ml (4–6 mL/kg)
- Urine volume: 20–250 mL/kg/24 h (median: 130 mL/kg/24 h).
- Blood volume: 55–65 mL/kg
- Gastrointestinal transit time: 4–5 h
- Optimum environmental temperature: 15–20 °C (65–70 °F)
- Water intake: 50–100 mL/kg/24 h

This chapter aims to describe the fundamental particularities of rabbit physiology. Understanding the normal physiology of this species is essential for its successful management and disease prevention.

2 Behavior and Sensory Physiology

Wild European rabbits (*O. cuniculus*) are mainly nocturnal, gregarious, and live in groups of two to eight adults, plus juveniles (McBride 1988). In the group, there is a hierarchy, where older and larger males are dominant and defend the territory (Nowak 1999). Females dig burrows for protection and nesting (Lockley 1976). Frequently, these burrows are large and complex and can be 3 m deep and 45 m long (Nowak 1999). When males get older, they may be replaced in the hierarchy by younger, fitter males, becoming solitary (Lockley 1976). Although females are less aggressive than males, they defend their nest aggressively (Lockley 1976). Sexually active males may dispute territory, and in captivity, they must be kept separate. Injury, infertility, and pseudopregnancy can occur in animals housed in groups, especially if established after sexual maturity (McNitt et al. 2013).

Their territories, and members of the group, are marked by pheromones or by urine, especially by the males (Lockley 1976). They do this by using three specialized scent glands located in the chin, groin, and anal area (Mykytowycz 1968). Urine is also used for recognition purposes, and males spray it on females and young rabbits of their group. Additionally, females may attack their own young if they have been smeared with urine from other rabbits (Mykytowycz 1968). Rabbits defend themselves through bites and scratches with claws. Its agonistic behavior is characterized by hitting the hind limbs in a way to dissuade the adversary, as well as pursuit/combat, tail wagging, urinating, and squealing.

Many behavioral and physiological processes of rabbits show a marked diurnal rhythm (Crowell-Davis 2021). One example is feeding: wild rabbits are more active in the late afternoon. At this time, they emerge from their burrows to feed, socialize, explore their territory, and mate. The next morning, they return to the burrows (McBride 1988). Another example is the lagomorph characteristic behavior of cecotrophy. This behavior, which we will discuss later, is characterized by the ingestion of cecotrophs directly from the anus. These are soft capsules of cecal material rich in nutrients that are produced a few hours after feeding and during a quiet period of time (Varga 2014).

In nature, rabbits are prey animals, and their anatomy reflects this (Jenkins 2001). The visual field of rabbits is 190° for each eye (Hillyer 1994). This wide field of vision is possible because eyes are located laterally and on the top of the head. This positioning of the eyes allows them to detect any threat but makes it impossible for the animal to see what it is eating since it cannot see what is beneath their nose. As a consequence, food selection and ingestion depend on smell and tactile information received from the sensitive vibrissae of the nose and lips (Varga 2014).

Their sight in a dark place is good, but it is generally considered that rabbit color sense is poor (Vella and Donnelly 2012; Meredith 2014). Its retina consists of a double system of rods (recognizing luminosity) and cones (recognizing colors). Its pupillary dilation is wide, increasing its sensitivity (approximately 8 times more than that of humans) to light. In this way, they are very sensitive to the blue and green colors of twilight and have good nocturnal vision (Harkness et al. 2010).

Rabbits use their vibrissae and sensitivity of their lips, as well as smell and taste, during foraging (Meredith 2014). Vibrissae are also used during orientation in nests. Smell and taste are more important in identifying members of the own breeding group than vision (Morimoto 2009).

Their ears are conspicuous, with a large surface area, and rabbits' ear sense is very good. Additionally, the ears are highly vascular and play an important role in thermoregulation (Meredith 2014).

3 Thermoregulation and Integument

Rabbits, as homeothermic animals, are able to keep their body temperature practically constant, between 38.5 and 39.5 °C, regardless of fluctuations in environmental temperature, within certain limits (Brewer and Cruise 1994). Nonetheless, they are extremely sensitive to heat (Batchelor 1999; Oladimeji et al. 2022). They do not sweat, and their salivation and panting are inefficient (O'Malley 2005). Only a few sweat glands are located inside the mouth (Yagci et al. 2006). Since they are unable to sweat or pant (Varga 2014), their thermoregulation depends largely on their ears, which have a countercurrent arteriovenous shunt: vasoconstriction or vasodilation occurs when the ambient temperature is lower or higher than the body temperature, respectively (Brewer and Cruise 1994).

When facing low ambient temperatures, the rabbit increases its food consumption and tries to reduce its body surface to decrease heat loss. In contrast, in the presence of excessive heat, the rabbit accelerates respiratory movements and evaporates water through the lung surface (O'Malley 2005).

High temperatures inhibit drinking, which can lead to dehydration. On the other hand, rabbits are sensitive to low humidity levels, but high humidity levels do not appear to be problematic (O'Malley 2005).

At birth, young rabbits (50–60 g) are hairless and have difficulty maintaining their body temperature since the mother just stays in the nest once, and only for a few minutes, to nurse them (Morimoto 2009). To maintain their body temperature and obtain a more favorable place to suckle, they huddle together (Hull and Hull 1982; McNitt et al. 2013). This behavior is complemented by heat production by metabolizing triglycerides in the brown adipose tissue (O'Malley 2005), a process used by newborn mammals to produce heat by nonshivering thermogenesis (Blumberg and Sokoloff 1998; Cannon and Nedergaard 2004; McNitt et al. 2013).

Rabbits exhibit three types of hair: undercoat (produced by the secondary hair follicles), long guard hairs (produced by the primary hair follicles), and short guard hairs, or awns (produced by the lateral primary hair follicles) (Cheeke 1987; O'Malley 2005). The differences in size, types, and color of hair produced the different rabbit breeds (O'Malley 2005). In a healthy animal, the different types of hair trap a layer of air close to the skin. This air layer rapidly warms, insulating the animal (Cheeke 1987; O'Malley 2005).

4 Respiratory Physiology

Rabbits are obligate nose breathers (Cruise and Nathan 1994; Meredith 2014). The entrance of each nostril is equipped with sensory pads, which makes rabbits very sensitive to touch on their nostrils (Nowak 1999). The nostril twitches at a very high rate, between 20 and 150/min, depending on the level of excitement (Brewer and Cruise 1994; Meredith 2014). This twitching will stop when the animal is very relaxed or anesthetized (Meredith 2014). The acute sense of smell of rabbits is explained by the vomeronasal organ and the olfactory epithelium in the turbinate bones (Brewer and Cruise 1994; Cruise and Nathan 1994; O'Malley 2005; Johnson-Delaney and Orosz 2011). As in other species, the airways harmonize the temperature and humidity of the inhaled air, filter any impurities it contains, and concomitantly intervene in the functions of smell and phonation.

At rest, rabbits mainly use the diaphragm contraction to breath (diaphragmatic breathing), not using their intercostal muscles (Brewer and Cruise 1994; Meredith 2014). The flow volume of air to the left lung is higher than that to the right due to the

Table 1 Respiratory parameters of rabbits. Adapted from Morimoto (2009) with permission from Springer Nature

Parameter	Value
Total lung capacity	111 ± 14.7 mL
Minute volume	0.6 L/min
Tidal volume	4–6 mL/kg
Arterial oxygen	12.6% ~ 15.8% volume
pO_2	85 ~ 102 mmHg
pCO_2	20 ~ 46 Torr
HCO_3	12 ~ 24 mmol/L

lower resistance of the proximal airways per unit volume (Yokoyama 1979). Additionally, lung volume increases with age in contrast to that of humans, in which it decreases (Morimoto 2009). Some respiratory parameters of rabbits are presented in Table 1.

5 Cardiovascular Physiology

Heart rate can vary from 180 to 250 beats per minute (O'Malley 2005). In rabbits, the electric potential of electrocardiogram is low, and during surgery operation, some authors report arrhythmia, down of blood and pulse pressure (Harkness and Wagner 1995; Morimoto 2009). The rabbit heart is relatively resistant to oxidative damage (Matsuki et al. 1990), making it more resistant to stunning. Circulatory parameters and hematologic values are presented in Chapter "Insights into Clinical Pathology of Rabbits".

The aorta exhibits rhythmic contractions of neurogenic origin, and the pulmonary artery is more muscular and thicker than that in cats or dogs (Meredith 2014). Additionally, the carotid artery of the rabbit is more compliant than that of the dog and has a greater ratio of elastin to collagen (Cox 1978). The external jugular vein is the main vessel for blood returning from the head (Meredith 2014).

Most likely because of its high metabolic rate, rabbit red blood cells have a short life span. Their neutrophils are called heterophils and stain pink (O'Malley 2005). Lymphocytes are the most prevalent leukocytes, reaching 60% (Harkness and Wagner 1995).

6 Urinary Physiology

Typical parameters of the urine of New Zealand White (NZW) rabbits are presented in Table 2.

Rabbits need to drink, on average, 120 mL/kg body weight (BW)/day (Cheeke 1987; Harkness and Wagner 1995; O'Malley 2005). The water intake depends on the composition (moisture) and quantity of feed and ambient temperature (Brewer and Cruise 1994; O'Malley 2005).

Table 2 Urine parameters of New Zealand White rabbits. Adapted from Morimoto (2009) with permission from Springer Nature

Parameter	Value
Volume (mL/day)	20.0–350
Gravity	1.003–1.036
pH	7.60–8.80
Total protein (g/day)	0.74–1.86
Urea nitrogen (g/day)	1.20–1.50
Uric acid (mg/day)	4.00–6.00
Creatinine (mg/day)	20.0–80.0
Ca (mg/day)	12.1–19.0
Cl (mg/day)	190–300
Mg (mg/day)	0.65–4.20
P (mg/day)	10.0–60.0
K (mg/day)	40.0–55.0
Na (mg/day)	50.0–70.0
Glomerular filtration rate (ml/min/kg)	7

Rabbit kidneys are unipapillate, and not all glomeruli are active at one time (Brewer and Cruise 1994; Cruise and Nathan 1994; O'Malley 2005). To activate dormant glomeruli, a well-hydrated rabbit does not need to increase renal plasma flow and glomerular filtration rate (Brewer and Cruise 1994; Cruise and Nathan 1994; O'Malley 2005).

One other characteristic is the lack of carbonic anhydrase, an enzyme that interconverts carbon dioxide and bicarbonate in other animals. This, together with the high production of bicarbonate by bacterial fermentation in the digestive system, makes rabbits more susceptible to metabolic alkalosis (Brewer and Cruise 1994; O'Malley 2005). Consequently, their urine is alkaline (Cheeke 1987; O'Malley 2005; Melillo 2007; Jenkins 2008).

Rabbits are also more susceptible to acid–base imbalance when bicarbonate levels are low. The reason for this is the lack of the renal ammonium buffering system: in other animals, in the case of metabolic acidosis, the ammonia levels are high, and the ammonia combines with hydrogen ions, which are then excreted as the buffer ammonium (Cheeke 1987; Brewer and Cruise 1994; O'Malley 2005).

The urine concentration ability of rabbits is lower than that of other animals, and consequently, they excrete large amounts of urine, approximately 130 mL/kg/day of urine (Mancinelli and Lord 2014). Urine volumes are affected by several factors, namely, individual and environmental factors. In some cases, traces of glucose and protein can be present (Jenkins 2008). Triple phosphate and calcium carbonate crystals are present in high numbers, but bacteria, epithelial cells and casts are absent in most cases (O'Malley 2005).

The main route for excretion of magnesium and calcium is the urine (Kennedy 1965). The high levels of ammonium phosphate and calcium carbonate monohydrate precipitates present explain its typical cloudy cream color (Flatt and Carpenter 1971; Morimoto 2009). Rabbits fed a 10% calcium carbonate diet excrete approximately 60% of the calcium carbonate in their urine (Cheeke and Amberg 1973). Urine color can vary from yellow to red (O'Malley 2005; Melillo 2007; Jenkins 2008) as a consequence of the presence of endogenous and exogenous pigments, such as bile pigments and plant porphyrins (Vella and Donnelly 2012; Meredith 2014), the latter depending on the ingestion of certain food items. For example, certain plant (beetroot, cabbage, broccoli, and dandelions) pigments can lead to the production of a bright red urine that can be confused with hemorrhage (Cheeke 1987; Melillo 2007; Jenkins 2008; Varga 2014). In young rabbits, however, urine is typically clear, although healthy young rabbits may have albuminuria (Morimoto 2009).

7 Reproductive Physiology

Rabbits reach sexual maturity by 4–4.5 months or at 4.5–5 months old in smaller and larger breeds, respectively (Batchelor 1999; Lebas 2000). Starting 48 hours after giving birth, the does exhibit a post-partum estrus, and the rabbit reproductive productivity can easily reach more than 60 kits per year (Lebas 1997).

In males, spermatogenesis begins between days 40 and 50 of life, but variation between breeds occurs (Morton et al. 1986; Lebas 1997). The testicular tubes are active from approximately day 84, and the first spermatozoa appear in the ejaculate at approximately 110 days (Lebas 1997).

The spermatogenesis duration is approximately 49 days, and spermatogenesis can produce 100–160 million spermatozoids daily (Foote and Carney 2000), with seasonal and genetic variations. Spermatozoid storage occurs in the epididymis (mainly in the tail) and *ductus deferens*, but their storage capacity is low, approximately 20 and 10% of the total, respectively (Orgebin-Crist 1965).

The volume of each ejaculate is approximately 0.3–0.6 ml. The concentration of spermatozoa in the ejaculate can vary from 150 to 500×10^6/ml, but volume and concentration can vary, depending mainly on the attempts made to copulate and the number of ejaculations per day (Lebas 1997).

In females, the first follicles appear in the ovary by the 13th day of age, and the first antrum follicles appear at approximately 65–70 days (Lebas 1997). The onset of puberty depends mainly on the breed and on body development. For most does, puberty arises when the female reaches 70–75% of its mature weight (Lebas 1997; Varga 2014).

After puberty, under the influence of FSH (follicle stimulating hormone), the ovarian follicles develop and produce estrogens, making the female receptive to the male. Follicular development takes place in waves of 5–10 follicles at the same time and at the same development stage in each ovary. If ovulation does not occur, these follicles regress, the estrogen levels diminish, and consequently, the female becomes less sexually receptive. After 4 days, the process is repeated (Varga 2014).

Rabbits are induced ovulators, and ovulation takes place 10–13 hours after coitus (Lebas 1997; O'Malley 2005; Varga 2014). The main reproductive events are reported in Table 3. In animal production, several methods can be used to induce ovulation without coitus, namely, mechanical manipulation of the vagina and

Table 3 Main reproductive events

Event	Time
Copula (last for 30 to 70 seconds)	0 hours
Hypothalamic release of GnRH	0/10 minutes
Maximum level of LH	1–2 hours
Sperm arrival to the oviduct	2–5 hours
Ovulation	8–13 hours
Fertilization	12–15 hours
Descent of the embryos to the uterus	72–96 hours
Implantation of embryos in uterine horns	5–8 days
Placentation (chorioallantoic and hemochorial)	10–17 days
Fetus up to 1 cm and 2 g weight	14 days
Fetus up to 4 cm and 6 g weight	20 days
Fetus up to 6 cm and 23 g weight	25 days
Parturition	30–31 (29 up to 36) days
Exclusive lactation	18 days *post-partum*
End of natural lactation	45 days *post-partum*

administration of luteinizing hormone (LH) or gonadotropin releasing hormone (GnRH) (Lebas 1997).

Does can mate as early as 10–12 weeks of age, but normally this does not produce ovulation (Lebas 1997). The sexual behavior of a doe is different from that of most mammal females because it can stay in heat for many days, and a pregnant female may accept mating during pregnancy (Harkness and Wagner 1995; Lebas 1997). This behavior does not affect the pregnancy or the developing embryos, and there is no superfetation (two simultaneous pregnancies), as in hares (Lebas 1997; Varga 2014).

When ovulation occurs, the positioning of the oviduct pavilion is such that the liberated oocytes will enter in it. The sperm are deposited in the upper part of the vagina, and spermatozoa will reach the distal ampulla, near the isthmus, approximately 30 minutes later, and in the process, their fertilization capability is enabled (Lebas 1997).

The eggs reach the uterus approximately 72 hours after ovulation. Implantation takes place at the blastocyst stage in the uterine horns. Progesterone levels continuously rise from the third to 15th days after mating. After this, the levels stabilize and finally drop before parturition (Lebas 1997). The placenta develops with the fetus, becoming larger until parturition. Embryo losses during pregnancy occur in the last 15 days of pregnancy and are significant, eventually reaching 30–40% (Lebas 1997; O'Malley 2005). The reasons behind this are related to the viability of the embryos and their position in the uterine horns, the season of the year, and the physiological condition of the doe, especially its age (Lebas 1997). Gestational maintenance depends on the levels of progesterone, which is secreted by the *corpora lutea* (Varga 2014). The average gestation period is 31–32 days (O'Malley 2005; Elliott and Lord 2014; Varga 2014).

There is the possibility that unfertilized ova may produce a pseudopregnancy, lasting for 15–18 days. This condition is uncommon in natural mating production systems (Lebas 1997; Varga 2014) and can be caused by infertile mating or the presence of a nearby male (O'Malley 2005). In a pseudopregnancy, the uterus and mammary glands grow under the influence of progesterone secreted by the corpus luteum (O'Malley 2005).

At the end of gestation, under the influence of the secretion of prolactin and the increased ratio of estrogen versus progesterone, most does build a nest with the available materials (Lebas 1997). Does build their nest from hay, straw, or other bedding material and pluck fur from their dewlap, abdomen, and sides (Hrapkiewicz et al. 2013; Varga 2014). Apparently, corticosteroid secretion by the adrenal glands of the fetuses, as well as $PGF_{2\alpha}$ prostaglandins, is involved in the onset of parturition (Lebas 1997; O'Malley 2005).

Parturition frequently occurs in the morning and takes between 15 and 30 minutes in most cases, depending on the litter size (Lebas 1997; O'Malley 2005; Varga 2014). After kindling, the uterus reduces its weight by more than half or less in 48 hours (Lebas 1997).

Kits are altricial, with their eyes closed and ear canals sealed, and weigh approximately 40–50 g (Harkness and Wagner 1995; Nowak 1999; Varga 2014). Other authors report a birthweight of approximately 65 g (Morimoto 2009). Kits are able to take solid food at approximately 20 days of age, and at approximately 4 weeks of age, they can leave the nest (Morimoto 2009).

During pregnancy, estrogen and progesterone inhibit the secretion of prolactin, but when kindling takes place, the levels of progesterone drop quickly. This change, together with the release of oxytocin, stimulates the secretion of prolactin, allowing for the amount of milk in the mammary glands. The stimulus of suckling leads to the release of oxytocin, the pressure in the mammary gland rises, and the kits can suckle. The amount of oxytocin released is directly related to litter size (Lebas 1997).

The doe nurses its litter once (Morimoto 2009; Varga 2014) or twice (Varga 2014) a day for 2–5 minutes each time. When being nursed, kits suck the equivalent of 20% of their bodyweight (Varga 2014). Doe's milk is more concentrated than cow's milk, except for lactose (O'Malley 2005; Varga 2014). The milk protein and fat levels increase after the third week of lactation, but the low level of lactose reaches near zero by the 30th day of lactation (Lebas 1997). Daily milk production goes from 30 to 50 grams at the beginning of lactation to 200 to 250 grams in the third week, dropping significantly afterwards (Lebas 1997).

8 The Digestive System and Digestion

The digestive system of the rabbit is ready to digest a large amount of fibrous food at a high transit rate (Varga 2014). Rabbits are extremely efficient herbivores, being capable of using fiber and protein in the diet with an efficiency 2.5–4 times that of other herbivores: p. ex., it can digest and absorb up to 80% of the protein present in the plant. If given a chance to choose, the rabbit prefers succulent and tender plants or parts of plants with a low content of fiber and high levels of protein and carbohydrates (Vella and Donnelly 2012).

8.1 Ingestion, Chewing, and Salivary Glands

The large visual field of rabbits does not include the area in front of the mouth (Brewer 2006). Food selection and ingestion are determined by smell from tactile information gathered from the sensitive vibrissae (tactile sensory organs that vibrate) on each side of the upper lip (Vennen and Mitchell 2009). This is an evolutionary adaptation of wild rabbits to remain visually alert (on the horizon) to potential predators. In the mouth, the first structures that process food are the teeth. Rabbit uses its incisor teeth to seize and cut (type I chewing action) shoots of young and succulent plants (in nature), a process that is facilitated by both the high mobility of the lips and the presence of a long diastema (Davies and Davies 2003).

The lower incisors occlude just behind the upper primary incisors, and the wearing between the upper and lower incisors forms a sharp edge and compensates for the continuous growth of the teeth. The incisors are used to cut through the vegetation. The ingested food is then ground by the molars and premolars before being swallowed (Brewer and Cruise 1994; Varga 2014).

After cutting, the plant is ground through the premolar and molar teeth present in each arcade functioning as a whole (type II action, the main one) and using only one of the two hemimandibles at a time. Type III actions are those involving the formation of a bolus of food. Up to approximately 120 jaw movements may occur after eating fresh food. When the cecotrophs are ingested, type II actions do not occur, and the cecotrophs are swallowed intact (Davies and Davies 2003).

During mastication, rabbits' jaw movements are reported to be up to 120 per minute (O'Malley 2005). All the processes are accompanied by the continuous secretion of saliva, which contains amylase with activity similar to humans (Brewer 2006), but with a small quantity of lipase and urea (Davies and Davies 2003).

8.2 Stomach

After being swallowed, food passes to the stomach. The stomach itself represents approximately 15% of the volume of the gastrointestinal tract, and over 80% of the digesta are located in the stomach and cecum. In adult animals, the low pH of the stomach content (pH 1–2) sterilizes the ingesta before it passes to the small intestine (Varga 2014).

The fundic region of the stomach is the major secretory portion and presents low motor function, allowing the permanence of cecotrophs for 6–8 hours (Varga 2014). This permanence in the stomach allows bacterial fermentation within the cecotrophs, and by this time, the stomach pH rises to 3 due to the lactate buffer effect produced by these bacteria (Harcourt-Brown 2002b; Blas and Gidenne 2020).

The passage of the remaining material occurs for 3–6 hours. In this way, the stomach is always with content (feed and hair surrounded by fluid) that can reach 50% of food ingested 24 hours before (Kohles 2014).

In suckling rabbits, the stomach pH is higher, approximately 5–6.5, allowing for microbial inoculation of the hindgut. In these animals, the milk coagulates by enzymatic action similar to renin. The passage of this rennet to the intestine is slow, allowing a 24-hour suckling interval. Due to the action of an antibacterial "milk oil" (containing octanoic and decanoic acids), formed by the action of digestive enzymes on the doe's milk, there is no bacterial proliferation, which does not happen if a substitute or milk from another species is given, making them very susceptible to infections (Davies and Davies 2003; O'Malley 2005).

Young rabbits depend exclusively on milk in the first 10 days of life, slowly starting to ingest solid/fibrous food by day 15. From day 20, the feeding is predominantly solid, initiating cecotrophy, which is fully developed on day 30. When young rabbits begin cecotrophy, the "milk oil" present in the stomach does not act on the cecotrophs since they remain intact and surrounded by mucus (mucin) (Davies and Davies 2003).

With the decrease in milk intake and weaning, the production of the antibacterial compound decreases, and simultaneously, the pH, that is, 5–6.5, descends to 1–2, forming a new barrier to bacterial growth (antibodies of maternal origin also decrease rapidly at this point). It is the transition between the different local defense mechanisms that leads to a lower protective capacity of the animal, contributing to the emergence of enteropathies such as weaning colibacillosis, coccidiosis, and mucoid enteritis, among others (Davies and Davies 2003).

8.3 Small Intestine

The rabbit has a rapid intestinal transit where the smaller fibrous particles are retained in the cecum for fermentation (Varga 2014). There are different processes of intestinal motility called segmentation and peristalsis (Box 2).

Box 2 Intestinal Motility Processes

I. The segmentation, defined by periodic static contractions of the intestinal wall, with consequent mixing of the contents and with predominance in the duodenum.

II. Peristalsis involves annular contraction with progression of the intestinal contents. It is regulated by cholecystokinin, somatostatin, and vasoactive intestinal peptide, among other substances.

Peristaltic contractions occur slowly every 10–15 minutes and are not altered by the cycle of cecotrophy (Davies and Davies 2003). The retention of chyme is approximately 10–20 minutes in the jejunum and 30–60 minutes in the ileum

(Blas and Gidenne 2020; Carabaño et al. 2020). It is the motilin protein (as in humans and in contrast to other animal species) produced by enterochromaffin cells, one of the regulatory mechanisms of intestinal motility, that stimulates smooth muscle contraction by binding to specific receptors whose macrolide antibiotics are agonists. Its secretion is stimulated by fats and inhibited by the carbohydrates present in the intestinal contents. This regulation by motilin decreases gradually in the distal part of the small intestine, being absent in the cecum and reappearing in the large intestine and rectum (Davies and Davies 2003).

In the duodenum and jejunum, the simple carbohydrates and proteins of chyme are digested into monosaccharides and amino acids that are absorbed during their passage. This digestion includes the cecotrophs, which in this case are digested and therein are absorbed, namely, amino acids, volatile fatty acids, vitamins, and those derived from the digested bacteria. The lysis of these bacteria releases mainly amylase, which by itself also regulates the digestive process. However, the digestion of the protein of microbial origin intensifies with the addition of lysozyme in the colon (Davies and Davies 2003; Kohles 2014).

8.4 Large Intestine

Since the physiology of digestion in the stomach and small intestine of rabbits is similar to that of other monogastric mammals, the high digestion efficiency of the rabbit is explained by the large length of the digestive system and by the fact that it is a hindgut fermenter, relying on extensive microbial fermentation in the cecum for a significant nutrient supply (O'Malley 2005; Varga 2014).

In the hindgut, the separation of indigestible and potentially digestible matter takes place at the proximal colon, according to particle size. There, the large particles of indigestible fiber pass down the colon and are quickly eliminated as hard and dry fecal pellets. At the same place (proximal colon), small particles and fluid are directed to the cecum, where they are fermented. This fermentation is mainly of bacterial origin and releases volatile fatty acids (VFAs), microbial protein and vitamins (Varga 2014). VFAs are absorbed across the cecal epithelium and used as an energy source, accounting for up to 40% of the rabbits' maintenance energy requirement. The VFA proportion in the rabbit cecum varies according to the time of day, diet, and rabbit developmental stage. Usually, VFA in cecum contents is 60% to 70% acetic, 15% to 20% butyric, and 10% to 15% propionic acid (Campbell-Ward 2012; Meredith 2014). Rabbits fed with low-fiber content and higher highly fermentable carbohydrate content have an increased appendix, showing their buffer function, or even evidencing the lymphoid tissue present there, due to the change in cecal population induced by this type of feeding (Davies and Davies 2003).

8.4.1 Cecal Microbiota

The cecal content corresponds to approximately 40% of the total content in the digestive tract (Brewer 2006). The cecum is the primary site of fermentation, and

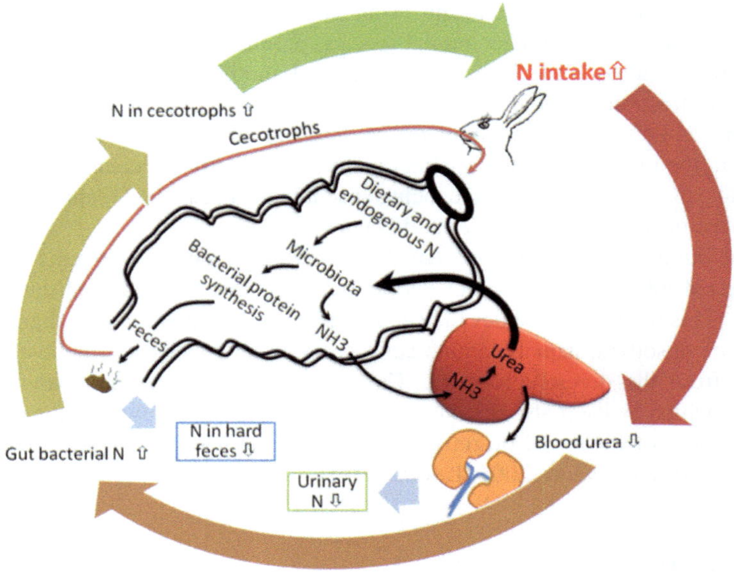

Fig. 1 Microbial nitrogen (N) recycling in growing rabbits. Modified from Xiao et al. (2015)

suitable cecal fermentation is only possible by the presence of large populations of bacteria (Crowley et al. 2017).

The cecum provides an anaerobic fermentation chamber for several bacteria with a diverse microbiota, such as *Alistipes*, *Ruminococcus*, *Akkermansia*, and *Subdoligranulum* (Bäuerl et al. 2014), and from the phyla Firmicutes (Velasco-Galilea et al. 2018) and Bacteroidetes (Montoro-Dasi et al. 2022). Nevertheless, as expected, feeding management or antibiotic use affects the cecal microbiota (Velasco-Galilea et al. 2020).

Lactobacillus and *E. coli* are usually absent but may be found in adult rabbits fed a high-carbohydrate, low-fiber diet (Varga 2014). Many nonpathogenic protozoan species, e.g., *Entamoeba cuniculi*, *Eutrichomastix*, *Enteromonas*, and *Retortamonas* spp., and yeasts are also present (Lelkes and Chang 1987; Owen 1992; Pakes and Gerrity 1994; Harkness and Wagner 1995).

The cecal microbiota breaks down, in a preferred order, ammonia (Fig. 1) (Xiao et al. 2015), urea, proteins, and enzymes from the small intestine and can also metabolize cellulose, xylan, and pectin (Carabaño et al. 2020). Substances that result from this metabolism are used as nutrients by microorganisms that use them to form their own protein and enzymatic structures, which will be subsequently digested as cecotrophs. The fermentation process also produces VFAs and vitamins, i.e. vit. B complex (e.g., niacin, riboflavin, pantothenate, cyanocobalamin), vit. C, and conversion of vit. K1 to K2.

8.4.2 Cecotrophy

The rabbit has the ability to produce two types of feces from the cecum: hard feces and soft feces called cecotrophs (Vennen and Mitchell 2009). Hard feces undergo a process similar to that of other herbivorous species, being well molded and excreted for the environment. Soft feces have different sizes, shapes, consistencies (higher water content), and odors than hard feces. At the chemical level, they also have a higher content of nitrogenous compounds, amino acids, and minerals, as well as the previously mentioned hydrosoluble vitamins, and a lower percentage of fibrous content when compared to hard feces (Carabaño et al. 1988; Fekete 1989).

Cecotrophs are produced primarily from fermentable material present in the cecal lumen and are partially fermented when adhered to the anus. Periodically, the fermented materials are excreted in the form of soft and moist cecal pellets, called cecotrophs, that are reingested by the animal directly from the anus. These reingested cecotrophs will be subjected to digestion a second time, and the nutrients released and produced in the fermentation process will pass through the small intestine and are absorbed (Varga 2014). The name of this behavior is cecotrophy.

Cecotrophs are produced approximately 4–8 h after feeding, and the behavior is exhibited during quiet periods, when the rabbit is in its burrow or in its cage (Harcourt-Brown 2002a, 2002b; Varga 2014). The production of cecotrophs varies inversely with the rabbit intake profile. When cecotrophy begins, ingestion ceases. Rabbits in cages and with normal photoperiods increase their intake between 15 and 17 h and remain elevated until 24 h. It decreases until 2 h in the morning, observing a new peak of intake at 6 h, which ends at 8 h. Thus, cecotrophy occurs between midnight and 2 h in the morning and returns at 8 h (Carabaño et al. 1988; Davies and Davies 2003; Carabaño et al. 2020).

Although the regulation of the daily profile of cecotrophy does not depend directly, for example, on satiety (the cyclic production of cecotrophs continues even if the animal does not eat food), cecotrophy is affected by energy and protein levels in the diet. If it is energy-deficient, a rabbit will consume all the cecotrophs produced (Meredith 2014). The intake of cecotrophs is most intense when rabbits are fed high levels of nondigestible fiber, whereas high protein levels reduce cecotrophy. Cecotrophs contain only approximately 50% of the crude fiber contained in hard feces. However, in situations of restriction of the protein level of feeding, cecotrophos maintain higher levels of protein compounds than hard feces (Davies and Davies 2003).

Cecotrophs are digested in the stomach and small intestine. The organization of the cecotroph, a soft pellet encapsulated in a gelatinous mucus coating, gives some protection from the low pH in the stomach while it is being stored in the fundic area (Harcourt-Brown 2002a; Fig. 2). Consequently, some fermentation occurs inside the pellet because of the bacteria present. These bacteria will be distributed by the lysozyme, secreted in the colon, and retained inside the cecotroph when it is formed, and its microbial protein will be available for further digestion and absorption in the small intestine, together with the amino acids and vitamins in the cecotroph (Varga 2014). Within the cecotroph, the amylase produced by the bacteria will digest

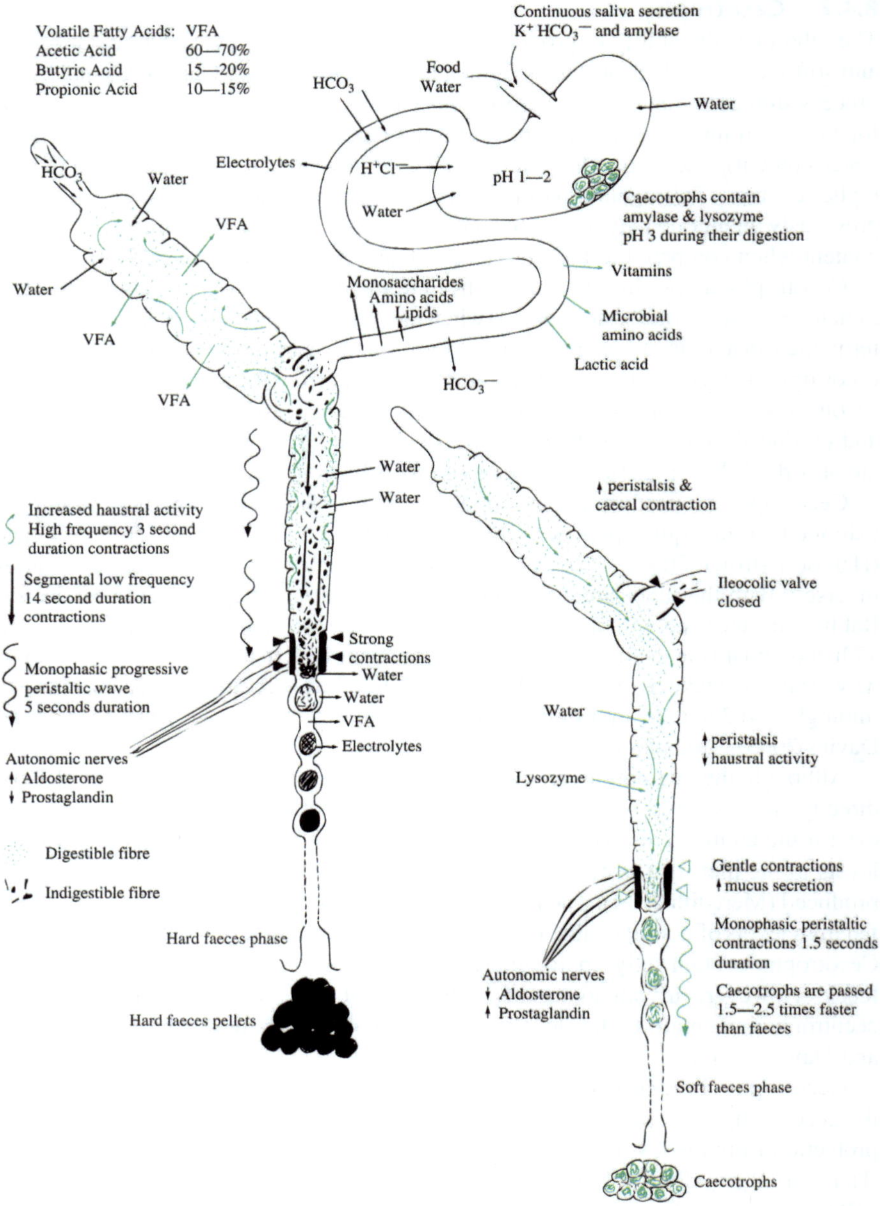

Fig. 2 Formation of hard feces and cecotrophs (soft feces) in rabbit. Adapted from Harcourt-Brown (2002a) with permission from Elsevier

carbohydrates, allowing for the degradation of glucose into carbon dioxide and lactic acid (Fekete 1989).

8.4.3 Regulation of Cecotrophy and Feces Production

One of the main factors affecting transit time is hindgut motility (Ehrlein et al. 1983). The excretion of feces and cecotrophs exhibits a circadian rhythm: the hard feces formation phase coincides with feeding activity, followed by a cecotroph production phase (Fioramonti and Ruckebusch 1976; Ehrlein et al. 1983; Varga 2014; Carabaño et al. 2020).

While hard feces are being produced, water secretion into the proximal colon helps in the mixing and separation process. The cecum and colon motility separates the large indigestible particles and the small particle, bacteria and water soluble intestinal material. The first fraction is rapidly transported through the proximal colon, *fusus coli* and distal colon and is excreted as hard feces, and the second fraction is directed backward to the cecum (Varga 2014).

During the hard feces phase, the cecal motility is higher, particularly during the mixing and separation process in the colon. After this phase, the motility pattern changes, and the cecal material is moved along the large colon. In the *fusus coli,* the material is molded into small pellets encapsulated in mucus (see Fig. 2). Ultimately, these soft pellets are excreted: it is the cecotroph excretion phase. This excretion takes place at least four hours after the feeding period, particularly in rest periods (Varga 2014).

The shifts in motility depend on the *fusus coli* pacemaker activity (Pairet et al. 1986), and the nature and direction of the peristaltic waves change with the different excretion phases (Ruckebusch and Fioramonti 1976). This organ, and its action potential producing capability, is affected by hormones, namely, aldosterone and prostaglandins. Consequently, the motility patterns change: aldosterone levels are high during the hard feces production phase and low during the cecotroph production phase; in this last phase, under the influence of prostaglandins, proximal colon motility is inhibited, and distal colon motility is enhanced (Pairet et al. 1986). This motility pattern (low motility in the proximal colon and high motility in the distal colon) coincides with the elimination of cecotrophs (Varga 2014).

The low water content of hard feces is the consequence of the high water, potassium, and sodium loss of the fecal pellets when they pass through the *fusus coli* (Snipes et al. 1982) because of the compression of the intestinal contents as a result of the strong contractions of the muscle wall of this organ (Varga 2014). These materials pass to the distal colon, where they are absorbed, and the fibrous indigestible material is eliminated as dry fecal pellets.

9 Conclusions

Several physiological particularities of rabbits are strongly related to their adaptation to survive and reproduce in the environment, such as the diurnal rhythm of behavioral and physiological processes. This information assumes relevance for clinical approaches and animal welfare evaluations.

Cecotrophy is one of the most relevant particularities of lagomorphs, improving feed efficiency and maintaining cecal and intestinal microbiota. Nonetheless, some other specificities related to agonistic behavior (e.g., does aggressivity) and temperature control (i.e., heat stress) can also pose challenges for rearing rabbits. Knowledge of these physiological features is crucial to attempt to secure adequate responses at the homeostasis and homeorhesis levels.

Acknowledgments The authors also thank Springer for the use of AJE Digital Editing. This work was supported by projects UIDP/CVT/00772/2020 (CECAV) and LA/P/0059/2020 (AL4AnimalS).

References

Alves JM, Carneiro M, Day JP, Welch JJ, Duckworth JA, Cox TE, Letnic M, Strive T, Ferrand N, Jiggins FM (2022) A single introduction of wild rabbits triggered the biological invasion of Australia. Proc Natl Acad Sci U S A 119:e2122734119. https://doi.org/10.1073/pnas.2122734119

Batchelor GR (1999) The laboratory rabbit. In: Poole TB, English P (eds) The UFAW handbook on the care and management of laboratory animals, 7th edn. Blackwell Science, Oxford, pp 395–409. ISBN: 978-0632051311

Bäuerl C, Collado MC, Zúñiga M, Blas E, Pérez Martínez G (2014) Changes in cecal microbiota and mucosal gene expression revealed new aspects of epizootic rabbit enteropathy. PLoS One 9(8):e105707. https://doi.org/10.1371/journal.pone.0105707

Blas E, Gidenne T (2020) Chapter 2 – Digestion of sugars and starch. In: De Blas C, Wiseman J (eds) Nutrition of the rabbit, 3rd edn. CABI, Oxfordshire, pp 21–40. ISBN: 978-1-78924-127-3. https://doi.org/10.1079/9781789241273.0021

Blumberg MS, Sokoloff G (1998) Thermoregulatory competence and behavioral expression in the young of altricial species--revisited. Dev Psychobiol 33:107–123. https://doi.org/10.1002/(sici)1098-2302(199809)33:2<107::aid-dev2>3.0.co;2-n

Brewer NR (2006) Biology of the rabbit. J Am Assoc Lab Anim Sci 45:8–24

Brewer NR, Cruise LJ (1994) Physiology. In: Manning PJ, Ringler DH, Newcomer CE (eds) The biology of the laboratory rabbit, 2nd edn. Academic Press, S. Diego, pp 63–70. ISBN: 9780124692350

Campbell-Ward ML (2012) Rabbits. Gastrointestinal physiology and nutrition. In: Quesenberry K, Carpenter JW (eds) Ferrets, rabbits, and rodents: clinical medicine and surgery, 3rd edn. Elsevier, St. Louis, MO, pp 183–192. ISBN 9781-416066217. https://doi.org/10.1016/B978-1-4160-6621-7.00014-2

Cannon B, Nedergaard J (2004) Brown adipose tissue: function and physiological significance. Physiol Rev 84:277–359. https://doi.org/10.1152/physrev.00015.2003

Carabaño R, Fraga MJ, Santoma G, de Blas JC (1988) Effect of diet on composition of cecal contents and on excretion and composition of soft and hard feces of rabbits. J Anim Sci 66:901–910. https://doi.org/10.2527/jas1988.664901x

Carabaño R, Piques J, Menoyo D, Badiola I (2020) The digestive system of the rabbit. In: De Blas C, Wiseman J (eds) Nutrition of the rabbit, 2nd edn. CABI, Oxfordshire, pp 1–20. ISBN: 978-1845936693

Cheeke PR (1987) Rabbit feeding and nutrition (Animal feeding and nutrition series). Academic Press, Orlando, FL. ISBN 0-12-17060 5–2

Cheeke PR, Amberg JW (1973) Comparative calcium excretion by rats and rabbits. J Anim Sci 37: 450–454. https://doi.org/10.2527/jas1973.372450x

Corbet GB (1986) Relationships and origins of the European lagomorphs. Mammal Rev 16:105–110. https://doi.org/10.1111/j.1365-2907.1986.tb00029.x

Cox RH (1978) Comparison of carotid artery mechanics in the rat, rabbit, and dog. Am J Phys 234: H280–H288. https://doi.org/10.1152/ajpheart.1978.234.3.H280

Crowell-Davis S (2021) Rabbit behavior. The veterinary clinics of North America. Exot Anim Pract 4:669–679. https://doi.org/10.1016/s1094-9194(17)30030-0

Crowley EJ, King JM, Wilkinson T, Worgan HJ, Huson KM, Rose MT, McEwan NR (2017) Comparison of the microbial population in rabbits and Guinea pigs by next generation sequencing. PLoS One 12:e0165779. https://doi.org/10.1371/journal.pone.0165779

Cruise JL, Nathan RB (1994) Chapter 3 – Anatomy. In: Manning PJ, Ringler DH, Newcomer CE (eds) The biology of the laboratory rabbit, 2nd edn. Academic Press, London, pp 47–61. ISBN: 978-0124692350. https://doi.org/10.1016/B978-0-12-469235-0.50009-9

Davies RR, Davies JA (2003) Rabbit gastrointestinal physiology. The veterinary clinics of North America. Exot Anim Pract 6:139–153. https://doi.org/10.1016/s1094-9194(02)00024-5

Ehrlein HJ, Reich H, Schwinger M (1983) Colonic motility and transit of digesta during hard and soft faeces formation in rabbits. J Physiol 338:75–86. https://doi.org/10.1113/jphysiol.1983.sp014661

Elliott S, Lord B (2014) Chapter 4: reproduction. In: Meredith A, Lord B (eds) BSAVA manual of rabbit medicine. British Small Animal Veterinary Association, Gloucester, pp 36–44. ISBN: 978-1905319497. https://doi.org/10.22233/9781910443217.4

Fekete S (1989) Recent findings and future perspectives of digestive physiology in rabbits: a review. Acta Vet Hung 37:265–279

Fioramonti J, Ruckebusch Y (1976) [Cecal motility in the rabbit. III. Duality of fecal excretion]. Annales de recherches veterinaires. Ann Vet Res 7:281–295

Flatt RE, Carpenter AB (1971) Identification of crystalline material in urine of rabbits. Am J Vet Res 32:655–658

Foote RH, Carney EW (2000) The rabbit as a model for reproductive and developmental toxicity studies. Reprod Toxicol 14:477–493. https://doi.org/10.1016/s0890-6238(00)00101-5

Fox RR (1974) Taxonomy and genetics. In: Weisbroth SH, Flatt RE, Kraus AL (eds) The biology of the laboratory rabbit, 1st edn. Academic Press, pp 1–22. ISBN: 9781483270319

Harcourt-Brown F (2002a) Biological characteristics of the domestic rabbit (*Oryctolagus cuniculi*). Textbook of rabbit medicine. Butterworth–Heinemann, Oxford, pp 1–18. ISBN: 9780750640022. https://doi.org/10.1016/B978-075064002-2.50004-7

Harcourt-Brown F (2002b) Digestive disorders. Textbook of rabbit medicine. Butterworth–Heinemann, Oxford, pp 249–291. ISBN: 9780750640022. https://doi.org/10.1016/B978-075064002-2.50013-8

Harkness JE, Wagner JE (1995) Biology and medicine of rabbits and rodents, 4th edn. Philadelphia, Williams & Wilkins. ISBN: 978-0683039191

Harkness JE, Turner PV, Vande Woude S, Wheler CL (2010) Harkness and Wagner's biology and medicine of rabbits and rodents. Ames, IA, Wiley-Blackwell. ISBN: 978-0-813-81531-2

Hillyer EV (1994) Pet rabbits. The veterinary clinics of North America. Small Anim Pract 24:25–65. https://doi.org/10.1016/s0195-5616(94)50002-0

Hrapkiewicz K, Colby L, Denison P (2013) Rabbits. Clinical laboratory animal medicine, an introduction, 4th edn. Wiley Blackwell, Ames, Iowa, pp 249–295. ISBN: 9781-1183-4510-8

Hull J, Hull D (1982) Behavioral thermoregulation in newborn rabbits. J Comp Physiol Psychol 96: 143–147. https://doi.org/10.1037/h0077857

IUCN - International Union for Conservation of Nature (2023) The IUCN Red List of Threatened Species.Version 2022-2. ISSN 2307-8235. https://www.iucnredlist.org/ (Downloaded on 26 October 2023)

Jenkins JR (2001) Rabbit behavior. The veterinary clinics of North America. Exotic Animal Practice 4:669–679. https://doi.org/10.1016/s1094-9194(17)30030-0

Jenkins JR (2008) Rabbit diagnostic testing. J Exot Pet Med 17:4–15. https://doi.org/10.1053/j.jepm.2007.12.003

Johnson-Delaney CA, Orosz SE (2011) Rabbit respiratory system: clinical anatomy, physiology and disease. The veterinary clinics of North America. Exot Anim Pract 14:257–2vi. https://doi.org/10.1016/j.cvex.2011.03.002

Kennedy A (1965) Urinary excretion of calcium by Normal rabbits. J Comp Pathol 75:69–74. https://doi.org/10.1016/0021-9975(65)90049-6
Kohles M (2014) Gastrointestinal anatomy and physiology of select exotic companion mammals. The veterinary clinics of North America. Exot Anim Pract 17:165–178. https://doi.org/10.1016/j.cvex.2014.01.010
Lebas F (1997) Reproduction. In: Lebas F, Coudert P, de Rochambeau H, Thébault RG (eds) The rabbit: husbandry, health, and production. Food and Agriculture Organization of the United Nations, Rome, pp 45–60. ISBN 92-5-103441-9
Lebas F (2000) Biología. In: Rosell JM (ed) Enfermedades del Conejo, vol I. Ediciones Mundi-Prensa, Madrid, pp 55–126
Lelkes L, Chang CL (1987) Microbial dysbiosis in rabbit mucoid enteropathy. Lab Anim Sci 37: 757–764
Lockley RM (1976) Private life of the rabbit, 2nd edn. Corgi, London. ISBN: 978-0552980012
Mancinelli E, Lord B (2014) Urogenital system and reproductive disease. In: Meredith A, Lord B (eds) BSAVA Manual of Rabbit Medicine. British Small Animal Veterinary Association, Gloucester, pp 191–204. ISBN: 978-1905319497. https://doi.org/10.22233/9781910443217.13
Matsuki T, Cohen MV, Downey JM (1990) Free-radical scavengers preserve wall motion in the xanthine oxidase-deficient rabbit heart. Coron Artery Dis 1:383–389
McBride E (1988) Rabbits and hares. Whittet Books, London. ISBN: 9780905483672
McNitt JI, Lukefahr SD, Cheeke PR, Patton NM (2013) Rabbit production, 9th edn. Oxfordshire. ISBN: 9781780640129, CABI. https://doi.org/10.1079/9781780640129.0000
Melillo A (2007) Rabbit clinical pathology. J Exot Pet Med 16:135–145. https://doi.org/10.1053/j.jepm.2007.06.002
Meredith A (2014) Biology, anatomy and physiology. In: Meredith A, Lord B (eds) BSAVA Manual of Rabbit Medicine. British Small Animal Veterinary Association, Gloucester, pp 1–12. https://doi.org/10.22233/9781910443217.1
Montoro-Dasi L, Lorenzo-Rebenaque L, Ramon-Moragues A, Pérez-Gracia MT, de Toro M, Marin C, Villagra A (2022) Antibiotic removal does not affect cecal microbiota balance and productive parameters in LP robust rabbit line. Front Vet Sci 9:1038218. https://doi.org/10.3389/fvets.2022.1038218
Morimoto M (2009) General physiology of rabbits. In: Houdebine L-M, Fan J (eds) Rabbit biotechnology. Springer, Dordrecht. https://doi.org/10.1007/978-90-481-2227-1_5
Morton D, Weisbrode SE, Wyder WE, Maurer JK, Capen CC (1986) Spermatid giant cells, tubular hypospermatogenesis, spermatogonial swelling, cytoplasmic vacuoles, and tubular dilatation in the testes of normal rabbits. Vet Pathol 23:176–183. https://doi.org/10.1177/030098588602300211
Mykytowycz R (1968) Territorial marking by rabbits. Sci Am 218:116–126. https://doi.org/10.1038/scientificamerican0568-116
Nowak RM (1999) Walker's mammals of the world, 6th edn. Baltimore, Johns Hopkins University Press
Oladimeji AM, Johnson TG, Metwally K, Farghly M, Mahrose KM (2022) Environmental heat stress in rabbits: implications and ameliorations. Int J Biometeorol 66:1–11. https://doi.org/10.1007/s00484-021-02191-0
O'Malley B (2005) Rabbits. Clinical anatomy and physiology of exotic species: structure and function of mammals, birds, reptiles, and amphibians. Elsevier Saunders, Edinburgh, pp 173–195. ISBN: 9780702027826
Orgebin-Crist MC (1965) Passage of spermatozoa labelled with thymidine-3-H through the ductus epididymidis of the rabbit. J Reprod Fertil 10:241–251. https://doi.org/10.1530/jrf.0.0100241
Owen DG (1992) Parasites of laboratory animals. Royal Society of Medicine Services Ltd., London, pp 7–48. Laboratory animal handbooks, no 12, ISBN: 1853151599
Pairet M, Bouyssou T, Ruckebusch Y (1986) Colonic formation of soft feces in rabbits - a role for endogenous prostaglandins. Am J Physiol 250:G302–G308. https://doi.org/10.1152/ajpgi.1986.250.3.G302

Pakes SP, Gerrity LW (1994) Protozoal diseases. In: Manning PJ, Ringler DH, Newcomer CE (eds) The biology of the laboratory rabbit, 2nd edn. Academic Press, Orlando, FL, pp 205–229. ISBN: 9781483288826

Pelletier M (2021) Morphological diversity, evolution and biogeography of early Pleistocene rabbits (Genus *Oryctolagus*). Palaeontology 64:817–838. https://doi.org/10.1111/pala.12575

Ruckebusch Y, Fioramonti J (1976) The Fusus coli of the rabbit as a pacemaker area. Experientia 32:1023–1024. https://doi.org/10.1007/BF01933949

Snipes RL, Clauss W, Weber A, Hörnicke H (1982) Structural and functional differences in various divisions of the rabbit colon. Cell Tissue Res 225:331–346. https://doi.org/10.1007/BF00214686

Sohn J, Couto MA (2012) Anatomy, physiology, and behavior. In: Suckow MA, Stevens KA, Wilson RP (eds) The laboratory rabbit, Guinea pig, hamster, and other rodents. Academic Press, pp 195–215. ISBN: 9780123809209. https://doi.org/10.1016/B978-0-12-380920-9.00008-0

Somerville AD, Sugiyama N (2021) Why were new world rabbits not domesticated? Anim Front 11:62–68. https://doi.org/10.1093/af/vfab026

Varga M (2014) Rabbit basic science. In: Varga M (ed) Textbook of rabbit medicine, 2nd edn. Butterworth-Heinemann, Oxford, pp 3–108. ISBN 9780702049798. https://doi.org/10.1016/B978-0-7020-4979-8.00001-7

Velasco-Galilea M, Piles M, Viñas M, Rafel O, González-Rodríguez O, Guivernau M, Sánchez JP (2018) Rabbit microbiota changes throughout the intestinal tract. Front Microbiol 9:2144. https://doi.org/10.3389/fmicb.2018.02144

Velasco-Galilea M, Guivernau M, Piles M, Viñas M, Rafel O, Sánchez A, Ramayo-Caldas Y, González-Rodríguez O, Sánchez JP (2020) Breeding farm, level of feeding and presence of antibiotics in the feed influence rabbit cecal microbiota. Animal Microbiome 2:40. https://doi.org/10.1186/s42523-020-00059-z

Vella D, Donnelly TM (2012) Rabbits. Basic anatomy, physiology, and husbandry. In: Quesenberry K, Carpenter JW (eds) Ferrets, rabbits, and rodents: clinical medicine and surgery, 3rd edn. Elsevier, St. Louis, MO, pp 157–173. ISBN 9781-416066217. https://doi.org/10.1016/B978-1-4160-6621-7.00012-9

Vennen KM, Mitchell MA (2009) Rabbits. In: Mitchell MA, Tully TN (eds) Manual of exotic pet practice. Saunders Elsevier, St. Louis, MO, pp 375–405. https://doi.org/10.1016/B978-141600119-5.50017-2

Xiao J, Metzler-Zebeli BU, Zebeli Q (2015) Gut function-enhancing properties and metabolic effects of dietary indigestible sugars in rodents and rabbits. Nutrients 7:8348–8365. https://doi.org/10.3390/nu7105397

Yagci A, Zik B, Uguz C, Altunbas K (2006) Histology and morphometry of white New Zealand rabbit skin. Indian Vet J 83:876–880

Yokoyama E (1979) Flow-volume curves of excised right and left rabbit lungs. J Appl Physiol 46:463–466. https://doi.org/10.1152/jappl.1979.46.3.463

Part II

Rabbit Production

Worldwide Rabbit Meat Trade

Laping Wu

Abstract

In the past 60 years, the global rabbit industry has been growing steadily. This chapter studies global rabbit meat production, trade and consumption. First, rabbit meat production and regional structure are introduced, as the basis of trade and consumption. Then, the global rabbit meat trade is studied in detail, including trade growth, regional structural changes, comparative advantages and competitiveness of major countries. Finally, a Gravity model is built to test major factors affecting rabbit meat trade and to explore the behind-the-trade scenes. The results show that: (1) Global rabbit industry had achieved great progress from 397 thousand tonnes (1961) to 893 thousand tonnes (2020); (2) Rabbit meat trade had been increasing for about 20 years from 1961 to 1979, then it fluctuated for another 20 years. However, since 2001 it has been stable with only small fluctuations, the average output being around 30 thousand tonnes; (3) The Gravity model results show that the rabbit meat trade is driven mainly by demand. In this way, some suggestions and policy implications are provided.

Keywords

Rabbit meat · Meat production · Meat trade · Comparative advantages · The Gravity model

This chapter is based on the author's plenary report on the global rabbit meat trade (Wu and Luefahr 2021) held at the 12th World Rabbit Congress.

L. Wu (✉)
College of Economics and Management, China Agricultural University, Beijing, China
e-mail: wulp@cau.edu.cn

J. Simões, J. M. Monteiro (eds.), *Veterinary Care of Farm Rabbits*,
https://doi.org/10.1007/978-3-031-44542-2_4

1 Introduction

In the past several decades, the rabbit industry has contributed much to farmer's income and rural economic growth in many countries. This is attributed to its favourable characteristics, such as flexible investment, easy starting and management, saving grain and land resources, producing high-quality meat, and creating jobs for rural people, especially women, younger or elderly people. In many developing countries, rabbit raising has rescued countless poor people from poverty. From the global perspective, rabbit meat output has kept steady, increasing for nearly 60 years. According to the statistics of the Food and Agriculture Organization of the United Nations (FAO), from 1961 to 1985 world rabbit meat was mainly produced in Europe, which had been accounting for more than 80% for about 25 years. However, in the following years, the rabbit industry in Asia increased quickly and became a major producing region. In 2020, the global rabbit meat output reached 0.89 million tonnes with European countries accounting for only 19.63%, but Asian countries dominated with a share of 67.49%. Meantime, in 2020, the global rabbit slaughtered and stocks reached 639.79 million heads and 192.59 million heads, respectively. The proportions of slaughtered rabbit and rabbit stocks in Asia accounted for 66.61% and 77.32%, on the contrast, the ratio in Europe were only 18.33% and 10.42%.

It is clear that rabbit meat-producing areas moved from Europe to Asia, but was this regional change of production accompanied by the restructure of rabbit meat trade? How did trade patterns evolve in the past several decades? Does this mean that rabbit meat consumption also moved from Europe to Asia? What are the influencing factors driving the structural change of the global rabbit trade? These questions are of great importance to global rabbit industry development. However, there are not many studies in this area. Smutka and Rosochatecka (2010) analyzed the global trade of rabbit meat, which showed that rabbit meat production and trade had been constantly increasing. The major producing countries of rabbit meat were Italy, France, China and Spain (Niedzwiadek 1994). The consuming areas included Italy, France, Spain, Belgium, Portugal and Malta, where per capita consumption ranged from 2.0 kg to 5.3 kg. Some research studies focused on a specific country or region. Adanguidi (2020) studied the profitability and competitiveness of rabbit value chains in Benin. Golban and Golban (2020) analyzed the competitiveness of rabbit meat produced in the Republic of Moldova. Szendro (2015), Szendrö and Bleyer (1999) studied rabbit meat production and trade in Hungary. Popescu-Miclosanu and Stanciu (2013) analyzed the development of both world and Romanian meat rabbit production and trade balance.

Compared with trade research, there are more studies about rabbit meat production and consumption. Cullere and Zotte (2018) showed that rabbit meat had many good characteristics and could be considered an ideal meat-producing animal, but the consumption in the world is not popular. McNitt et al. (2013) summarized rabbit production in major areas of the world and discussed about future of world rabbit production and rabbit research. Before the 1990s, related studies mainly focused on meat supply in order to meet higher demand (Owen 1981; Parkin 1972; Roberts

1980). Recently, more reports focused on production systems, technology and efficiency subjects. Trocino et al. (2019) evaluated the availability of research funds and scientific activities relating to rabbits in Italy, Europe and all over the world during the past 20 years. Khan et al. (2017) examined the production performance of indigenous rabbits under traditional and intensive systems in Pakistan. Lukefahr et al. (2004) reviewed the limitations of meat rabbit production in North America. From the perspective of rabbit meat consumption, some scientists focused on specific groups of people, including children and university students (Escriba-Perez et al. 2019; Gonzalez-Redondo et al. 2010), but most of this research was concentrated on only one country, including Italy, France, Hungary, Spain, the United States and Mexico (Chalah and Hajj 1996; Beal et al. 2004; Olivares et al. 2005; Bodnar and Horvath 2008; Dairo et al. 2012; Kowalska 2015; Petracci et al. 2018). Sanah et al. (2020) studied the motivation and obstacles to rabbit meat consumption in Algeria. Sugiyama et al. (2017) studied meat consumption in Mexico from the view of ecological and social contexts. Szendro et al. (2020) focused on the production methods and purchase forms in eight countries to study consumers' attitudes to consumption of rabbit meat.

There is a large body of multidiscipline research on rabbit science, which covers rabbit meat quality and nutrition, rabbit behaviour and welfare etc. Many researchers studied the impacts of different factors affecting rabbit meat quality, including cage and pen housing (Combes et al. 2010; Matics et al. 2019), limited feed intake (Gidenne et al. 2012; Chodova et al. 2019), and different feed additives (Dabbou et al. 2017; Wahyuni et al. 2018). Cavani et al. (2009) reviewed the advances in research on poultry and rabbit meat quality. Antipov et al. (2020) studied technology to increase the shelf-life of dietary-influenced rabbit meat in a supercooled-frozen state whilst maintaining organoleptic properties using a stabilized cooling system. Lehmann (1991) examined the social behaviour in young rabbits under semi-natural conditions. Verga et al. (2009) analyzed the welfare aspects of rabbit rearing and transport, whilst Nielsen et al. (2020) focused on the health and welfare of rabbits that were farmed in different production systems. There are also many research studies on rabbit industry competitiveness of major countries (Foster and Telford 1996; Luo and Wu 2019; Wu and Qin 2019). Therefore, there is much knowledge about the principal subjects inside rabbit science, but there is a lack of the research on rabbit meat trade.

The objective of this chapter is to study the trade growth of global rabbit meat in past decades and its influencing factors. We will focus on the development and trends of rabbit meat trade, regional structure and their determinants.

2 Global Rabbit Production and Regional Structure

The rabbit is a small animal species that needs less feed with a low feed conversion ratio. Generally, feed costs could be reduced through the utilization of homegrown or less expensive feedstuffs, labour could be shared amongst family members, and less expensive housing and equipment would be needed (Lukefahr 1999). Therefore, during the early period of economic development, many countries gave the rabbit industry a much higher priority in policy-making for poverty reduction.

From the development of global rabbit production (Table 1), it can be found that: (1) the rabbit industry has experienced rapid growth since 1961. Meat output increased from 397 thousand tonnes (1961) to 893 thousand tonnes (2020) with an annual growth rate of 1.38% (Fig. 1); (2) The period of rabbit meat production from 1961 to 2020 can be divided into four phases, *fast increase* (1961–1980), *adjustment/transition period* (1981–1990), *steady increase* (1991–2015) and *adjust stable period* (2016–2020). The growth rate during these four periods is 3.23%, 0.89%, 3.20% and − 6.75%, respectively; (3) meat output growth in the first period mainly took place in Europe with an average share of 85.99%, whilst the Asian countries only accounted for 7.87%.

After adjustment in the transition period from 1981 to 1990, the major producing area moved from Europe to Asia. Since 1991, rabbit meat output in Europe has been continuously decreasing, but that of the Asian countries has continuously increased. The shares of European rabbit meat decreased from 67.96% (1991) to 21.77% (2015), but during the same period, rabbit meat output in the Asian countries increased from 17.55% to 70.05%. Although rabbit meat production underwent a major adjustment in 2016–2018, rabbit meat production has stabilized at around 900 thousand tonnes in the last 3 years. In 2020, rabbit meat production in Asia and Europe was 603 thousand tonnes and 175 thousand tonnes, respectively.

From the perspective of major producing countries, during the 1960s and 1970s, France was the largest producer, and Italy ranked second. In the 1960s, these two countries produced more than 58.68% of global rabbit meat, and in the 1970s they still produced more than 45.93%. In 1979, Italy became the largest producer with a portion of 22.23% and for the first time surpassed France (20.93%). This trend continued until 1990. Since 1991, Asian countries stepped on the rabbit market stage, especially China, which in 1991 became the largest producer. It is noteworthy that China's portion increased from 17.12% (1991) to 51.09% (2020).

In the past 60 years, the amount of rabbit slaughtered (Table 2) was basically consistent with the change in rabbit meat yield. The number of rabbits slaughtered increased from 271.34 million (1961) to 639.79 million (2020), with an average annual growth rate of 1.44% (Fig. 2). The growth rates of rabbit slaughtered in the four rabbit industry development periods were 3.23%, 0.90%, 2.93% and − 5.51%, respectively. With the transfer of rabbit-producing areas, the share of rabbit slaughtered in Europe fell from 89.87% (1961) to 18.33% (2020), whilst share in Asia rose from 3.55% (1961) to 66.61% (2020).

In 2020, 426.15 million rabbits were slaughtered in Asia and 117.27 million in Europe. The number of rabbits slaughtered in Asia is about four times that in Europe. In recent years, the global rabbit slaughtered has stabilized at about 650 million.

The trend in the global rabbit stocks (Table 3) is also broadly in line with rabbit meat production and rabbit slaughtered. Rabbit stocks increased from 98.04 million in 1961 to 192.59 million in 2020, with an average annual growth rate of 1.15% (Fig. 3). However, different from rabbit meat production, the rabbit meat stock increased rapidly in the second period (1981–1990), reaching 380 million in 1990.

Table 1 Global rabbit meat production in major years from 1961 to 2020

Year	World	Europe		Asia		Africa		Americas	
		Quantity (1000 tonne)	Portion (%)	Quantity (1000 tonne)	Portion (%)	Quantity (1000 tonne)	Portion (%)	Quantity (1000 tonne)	Portion (%)
1961	397.06	359.41	90.52	15.66	3.94	11.73	2.96	10.26	2.58
1966	421.68	372.68	88.38	24.27	5.76	13.61	3.23	11.12	2.64
1971	537.66	470.70	87.54	40.84	7.59	15.09	2.81	11.04	2.05
1976	686.79	601.04	87.51	54.68	7.96	16.36	2.38	14.71	2.14
1981	723.11	613.03	84.78	60.58	8.38	31.45	4.35	18.06	2.50
1986	776.98	622.44	80.11	79.63	10.25	57.48	7.40	17.43	2.24
1991	630.83	428.73	67.96	110.69	17.55	68.70	10.89	22.70	3.60
1996	809.58	390.83	48.28	323.47	39.96	74.28	9.17	21.00	2.59
2001	961.71	363.61	37.81	501.97	52.20	75.01	7.80	21.13	2.20
2006	1055.02	307.47	29.14	652.08	61.81	76.70	7.27	18.77	1.78
2011	1290.05	306.28	23.74	885.29	68.62	80.93	6.27	17.55	1.36
2015	1342.91	292.33	21.77	940.77	70.05	92.80	6.91	17.02	1.27
2016	1181.73	294.23	24.90	783.26	66.28	88.67	7.50	15.56	1.32
2017	989.50	259.04	26.18	626.17	63.28	89.03	9.00	15.26	1.54
2018	901.48	173.36	19.23	619.85	68.76	92.87	10.30	15.39	1.71
2019	919.98	175.34	19.06	628.01	68.26	101.48	11.03	15.15	1.65
2020	893.63	175.42	19.63	603.12	67.49	99.93	11.18	15.16	1.70

Data source: FAO Statistics, http://www.fao.org/faostat/en/

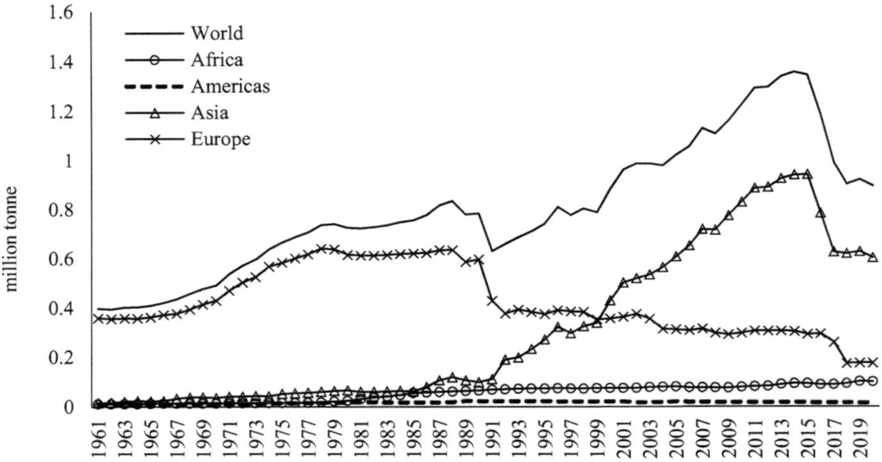

Fig. 1 Global rabbit meat outputs across continents (1961–2020). Data source: FAO Statistics, http://www.fao.org/faostat/en/

Then it decreased rapidly in the following years, and stabilized at about 200 million in recent years. From the regional perspective, with the transfer of rabbit production areas, the proportion of European rabbit stocks decreased from 78.13% in 1961 to 10.42% in 2020, whilst the share of Asia increased from 17.26% to 77.32%. In 2020, the number of rabbits in Asia and Europe will be 148.91 and 20.07 million, respectively. The number of rabbit stocks in Asia is seven times that of Europe.

In 2020, the top five countries for rabbit meat production are China, North Korea, Egypt, France and Italy (Fig. 4). They produced 0.74 million tonnes of rabbit meat, accounting for 85.09% of the global total. The share of top ten countries reached 89.23%. Therefore, the concentration rate of production was still high.

3 Rabbit Meat Trade and Structural Changes

Figure 5 shows the global rabbit meat export data across continents from 1961 to 2020. It is evident that rabbit meat exports had been increasing for about 20 years before 1980. However, since 1981, exports have been fluctuating for about 20 years at around 58.40 thousand tonnes. After 2001, it has been stable at the level of around 30 thousand tonnes with little fluctuation. In 2020 it was 26.70 thousand tonnes. Nevertheless, the destination of rabbit meat exports has been the Europe since 1961 with an average share of 95.26%. It is clear that before 1985 rabbit meat trade mainly took place between Asia and Europe, but since 1986, it has been traded mainly within the European countries, and with certain supplements from Asian countries (mainly China).

Concerning major trading countries, for many years, China, France and Hungary have been the major exporters. Since 2001, rabbit meat exports from Spain increased

Table 2 Global rabbit slaughtered in major years from 1961 to 2020

		Europe		Asia		Africa		Americas	
		Quantity	Portion	Quantity	Portion	Quantity	Portion	Quantity	Portion
Year	World	(million head)	(%)	(million head)	(%)	(million head)	(%)	(million head)	(%)
1961	271.34	243.86	89.87	9.63	3.55	10.08	3.72	7.77	2.86
1966	289.91	254.18	87.67	15.44	5.32	11.81	4.08	8.48	2.93
1971	362.97	314.73	86.71	26.74	7.37	13.19	3.63	8.32	2.29
1976	467.67	406.20	86.85	35.44	7.58	14.37	3.07	11.67	2.49
1981	493.32	411.94	83.50	39.77	8.06	27.02	5.48	14.60	2.96
1986	536.58	421.70	78.57	52.14	9.72	48.72	9.08	14.03	2.61
1991	458.69	308.39	67.23	73.70	12.20	58.19	10.24	18.42	3.43
1996	584.19	270.66	46.33	233.11	39.90	62.95	10.78	17.48	2.99
2001	691.51	246.91	35.71	362.92	52.48	63.67	9.21	18.02	2.61
2006	770.91	203.62	26.41	485.51	62.98	65.94	8.55	15.84	2.05
2011	864.18	185.41	21.45	592.49	68.56	71.72	8.30	14.57	1.69
2015	916.46	175.92	19.20	644.08	70.28	82.57	9.01	13.89	1.52
2016	802.47	167.46	20.87	543.24	67.70	78.71	9.81	13.07	1.63
2017	692.70	160.90	23.23	439.95	63.51	79.06	11.41	12.80	1.85
2018	646.26	115.91	17.93	434.64	67.25	82.64	12.79	13.08	2.02
2019	655.85	116.05	17.69	443.41	67.61	83.53	12.74	12.86	1.96
2020	639.79	117.27	18.33	426.15	66.61	83.50	13.05	12.87	2.01

Data source: FAO Statistics, http://www.fao.org/faostat/en/

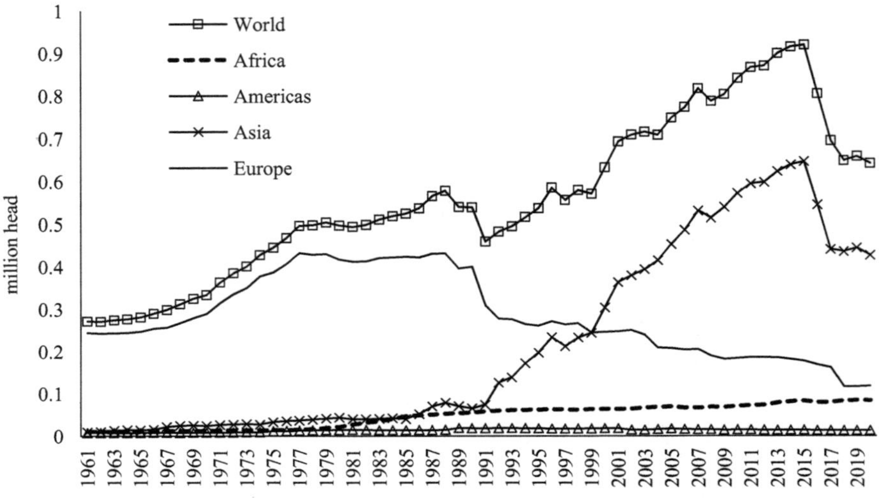

Fig. 2 Global rabbit slaughtered across continents (1961–2020). Data source: FAO Statistics, http://www.fao.org/faostat/en/

quickly. During the same period, the major importing countries included Germany, France and Italy. Figure 6 shows the net exports[1] of major countries since 1961.

It is clear that during the period of 1960s and 1970s, rabbit meat was mainly exported from China and Hungary to Italy, France and Germany. However, from the beginning of the 1980s, the exported rabbit meat from China started to decline, and since 1990s the exports from Hungary have also declined. Correspondingly, rabbit meat imports from France and Italy also decreased. After 2001, both exports and imports were at a low level and have been stable now for about 10 years. This situation shows that the self-sufficiency rate of major importing countries is increasing.

There were 43 countries exporting rabbit meat in 2021. The top 10 exporting and importing countries are listed in Table 4. It can be seen that the top 5 exporting countries were Spain, Hungary, France, China and Belgium. They exported 25,742 tonnes of rabbit meat, accounting for 80.97% of global total rabbit meat exports in 2021. The top 10 countries exported 30,561 tonnes, accounting for 96.12% in 2021. Except for China and Argentina, the other top 10 countries are all from Europe. Different from exports, the rabbit meat imports data are more scattered. In 2021, a total of 66 countries imported rabbit meat. The top 5 were Germany, Belgium, Italy, France and Portugal. Collectively, they imported 16,434 tonnes, accounting for 64.18% of global total imports. The share of the top 10 countries is 82.24%. Except

[1] Net exports are equal to exports minus imports. It is the sole contribution of one country to global market. If net exports are greater than 0, it means that this country exports more than imports, and vice versa.

Table 3 Global rabbit stocks in major years from 1961 to 2020

		Europe		Asia		Africa		Americas	
		Quantity	Portion	Quantity	Portion	Quantity	Portion	Quantity	Portion
Year	World	(million head)	(%)	(million head)	(%)	(million head)	(%)	(million head)	(%)
1961	98.04	76.60	78.13	16.92	17.26	2.78	2.84	1.74	1.78
1966	110.19	80.20	72.79	24.92	22.62	3.15	2.86	1.91	1.74
1971	138.62	99.58	71.83	33.72	24.33	3.41	2.46	1.91	1.38
1976	167.47	116.85	69.77	44.19	26.38	3.62	2.16	2.82	1.68
1981	189.03	125.51	66.40	55.06	29.13	4.71	2.49	3.75	1.98
1986	242.57	127.56	52.59	103.03	42.48	8.18	3.37	3.78	1.56
1991	324.10	145.51	44.90	162.73	50.21	11.26	3.48	4.59	1.42
1996	213.43	29.64	13.89	159.09	74.54	20.28	9.50	4.42	2.07
2001	254.85	28.43	11.16	210.58	82.63	11.19	4.39	4.64	1.82
2006	293.25	28.04	9.56	240.58	82.04	19.66	6.71	4.98	1.70
2011	302.04	28.62	9.48	252.90	83.73	14.92	4.94	5.60	1.85
2015	301.87	29.28	9.70	249.02	82.49	17.91	5.93	5.67	1.88
2016	253.69	29.60	11.67	200.88	79.18	17.49	6.90	5.71	2.25
2017	207.88	29.14	14.02	154.09	74.12	18.79	9.04	5.87	2.82
2018	198.04	20.93	10.57	152.76	77.13	18.44	9.31	5.91	2.98
2019	200.01	20.57	10.28	155.08	77.54	18.40	9.20	5.96	2.98
2020	192.59	20.07	10.42	148.91	77.32	17.70	9.19	5.91	3.07

Data source: FAO Statistics, http://www.fao.org/faostat/en/

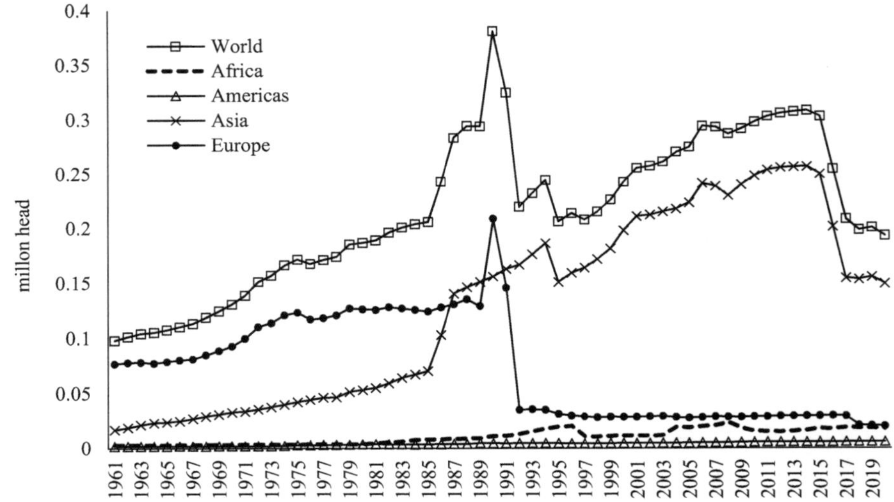

Fig. 3 Global rabbit stocks across continents (1961–2020). Data source: FAO Statistics, http://www.fao.org/faostat/en/

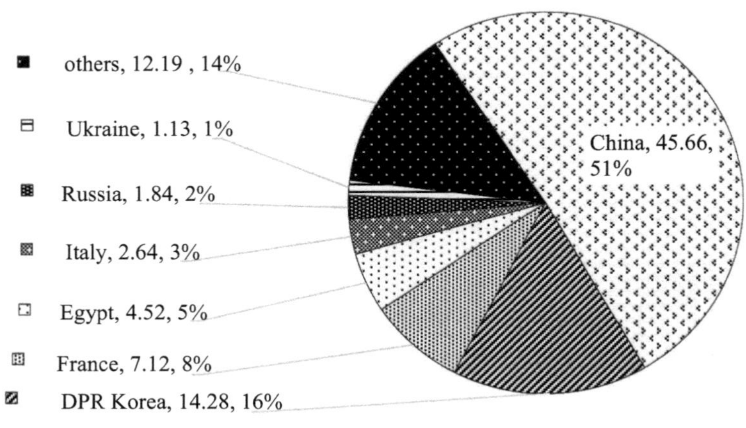

Fig. 4 Rabbit meat output and shares of main producing countries in 2020. Data source: FAO statistics, http://www.fao.org/faostat/en/. Unit: tonnes

for the United States, the other importing countries in the top 10 were from Europe, therefore Europe is still a traditional area of rabbit meat consumption.

There is an interesting phenomenon here, that is, one country both exports and imports rabbit meat. In economics, it is called *intra-industry trade*. From Table 4 it can be found that the intra-industry trade mainly took place in Belgium, France and Italy. To understand in more detail, we selected three countries (Belgium, France and Italy) to study their trade pattern (Table 5).

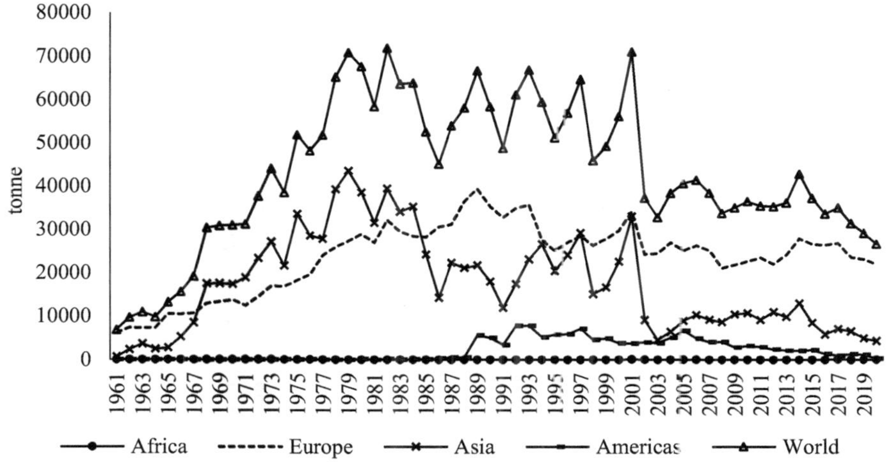

Fig. 5 Global rabbit meat exports across continents (1961–2020). Data source: FAO Statistics, http://www.fao.org/faostat/en/

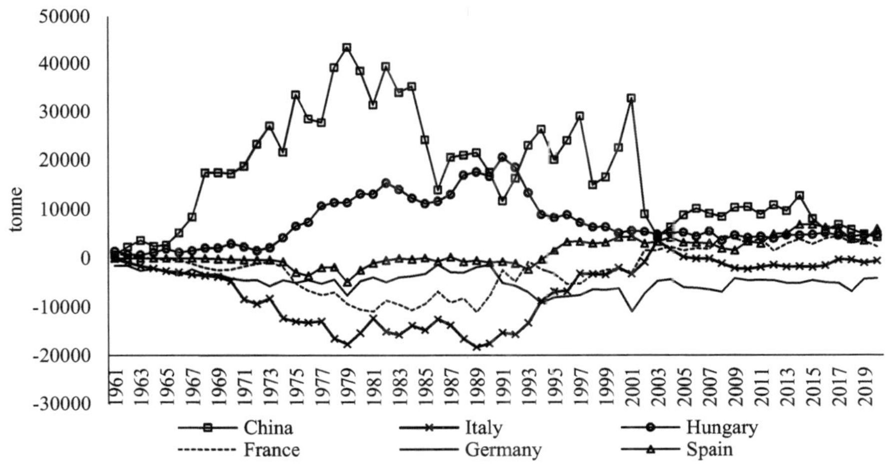

Fig. 6 Net export of rabbit meat in major countries (1961–2020). Data source: FAO Statistics, http://www.fao.org/faostat/en/

In 2021, Belgium exported rabbit meat to 23 countries, but mainly to 5 countries, which accounted for 88.78%. France exported to 46 countries or regions, but the top 5 and top 10 countries accounted for 69.47% and 92.87%, respectively. Italy exported to 27 countries, and 78.43% was exported to top 5 countries and 93.44% to top 10 countries. Compared with exports, the imports of these three countries were more concentrated in fewer countries. Belgium imported from 17 countries, France and Italy imported from 19 and 16 countries, respectively.

Table 4 Top 10 countries of rabbit meat export and import in 2021

	Export						Import				
	Country	Quantity (tonne)	Share (%)	Value (million USD)	Share (%)		Country	Quantity (tonne)	Share (%)	Value (million USD)	Share (%)
1	Spain	9475.01	29.80	33.74	23.66	**1**	Germany	5765	22.52	34.85	27.77
2	Hungary	4603.97	14.48	30.30	21.25	**2**	Belgium	3887	15.18	10.83	8.63
3	France	4356.54	13.70	19.57	13.73	**3**	Italy	2795	10.92	11.13	8.87
4	China	4026.67	12.66	17.49	12.27	**4**	France	2075	8.10	9.25	7.37
5	Belgium	3280.20	10.32	12.91	9.05	**5**	Portugal	1912	7.47	8.71	6.94
Top 5 sum		***25742.39***	***80.97***	***114.02***	***79.95***	***Top 5 sum***		***16,434***	***64.18***	***74.78***	***59.57***
6	Italy	1887.97	5.94	6.01	4.21	**6**	United States	1122	4.38	4.75	3.78
7	Poland	1158.20	3.64	4.91	3.44	**7**	Czech Republic	1028	4.01	5.61	4.47
8	Netherlands	875.35	2.75	6.49	4.55	**8**	Romania	927	3.62	2.92	2.33
9	Argentina	628.88	1.98	3.02	2.12	**9**	Netherlands	800	3.12	5.53	4.40
10	Portugal	268.81	0.85	0.96	0.67	**10**	Switzerland	745	2.91	6.26	4.99
Top10 sum		***30561.60***	***96.12***	***135.41***	***94.94***	***Top 10 sum***		***21,056***	***82.24***	***99.85***	***79.54***
Global total		***31794.07***	***100***	***142.62***	***100***	***Global total***		***25,604***	***100***	***125.52***	***100***

Data source: WITS database, http://wits.worldbank.org/WITS/

Note: Theoretically global exports are equal to global imports, however, there is difference due to statistical or other errors. The error rate of trade volume is 19.47%

Table 5 Intra-industry trade of Belgium, France and Italy in 2021

Belgium			France			Italy		
23 Export partners			*46 Export partners*			*27 Export partners*		
Top 10 Destination	Quantity (tonne)	Share (%)	Top 10 Destination	Quantity (tonne)	Share (%)	Top 10 Destination	Quantity (tonne)	Share (%)
Germany	1643	50.10	Italy	995	22.83	Germany	487	25.80
Netherlands	436	13.31	Belgium	598	13.72	Poland	414	21.94
Luxembourg	341	10.40	United States	550	12.63	Malta	267	14.12
France	248	7.57	Germany	461	10.58	China	208	11.03
United Kingdom	243	7.41	Spain	423	9.71	Romania	105	5.54
Italy	129	3.93	United Kingdom	314	7.22	Czech Republic	71	3.73
China	99	3.02	Netherlands	299	6.86	Belgium	69	3.66
Czech Republic	51	1.54	China	254	5.83	Hungary	56	2.97
Switzerland	37	1.11	Luxembourg	84	1.94	Greece	48	2.54
Spain	18	0.54	Portugal	68	1.56	Portugal	40	2.11
Sum of top 5	2912	88.78	Sum of top 5	3027	69.47	Sum of top 5	1481	78.43
Sum of top 10	3245	98.93	Sum of top 10	4046	92.87	Sum of top 10	1764	93.44
Total export	3280	100	Total export	4357	100	Total export	1888	100

Belgium			France			Italy		
17 Import partners			*19 Import partners*			*16 Import partners*		
Import origins	Quantity (tonne)	Share (%)	Import origins	Quantity (tonne)	Share (%)	Import origins	Quantity (tonne)	Share (%)
France	2435	62.65	Belgium	1474	71.06	Spain	1283	45.92
China	493	12.68	Spain	275	13.28	Poland	369	13.22
Spain	329	8.45	Netherlands	102	4.92	Hungary	324	11.58
Germany	184	4.73	Hungary	84	4.04	Germany	241	8.64

(continued)

Table 5 (continued)

Belgium			France			Italy		
17 Import partners			*19 Import partners*			*16 Import partners*		
Import origins	Quantity (tonne)	Share (%)	Import origins	Quantity (tonne)	Share (%)	Import origins	Quantity (tonne)	Share (%)
Argentina	181	4.65	United Kingdom	68	3.26	France	238	8.50
Netherlands	132	3.39	Italy	27	1.32	Slovenia	139	4.96
Uruguay	80	2.05	Germany	21	0.99	Finland	54	1.93
Poland	22	0.57	Poland	7	0.35	Netherlands	53	1.91
Austria	11	0.29	Austria	5	0.25	Belgium	34	1.20
Estonia	10	0.25	Slovenia	4	0.21	Argentina	28	1.02
Sum of top 5	3622	93.17	Sum of top 5	2003	96.56	Sum of top 5	2455	87.86
Sum of top 10	3876	99.72	Sum of top 10	2068	99.66	Sum of top 10	2763	98.88
Total import	3887	100	Total import	2075	100	Total import	2795	100

Data source: WITS database, http://wits.worldbank.org/WITS/

Theoretically, intra-industry trade may be due to the distance amongst countries or product diversity etc. The impacts of distance and borders between two countries will be tested later in the Gravity model. There are also some other explanations, such as differentiation of products. Rabbit meat is not homogeneous, which includes fresh or frozen meat, or different processing, and it may involve different parts (e.g. head and leg) etc. Some economists also explain this fact basing on the perspective of technology.

Table 5 shows that the trade concentration degrees[2] are much lower in France and Italy. Low concentration degree means more trade partners and lower risk. This explains why they can maintain a long-term trade position in the global rabbit meat market, without much fluctuation.

4 Rabbit Meat Consumption in Major Countries

Trade can be driven either by production or consumption. In following section, we will study the total rabbit meat consumption patterns in major countries. Firstly, total consumption in each country was calculated by the following method:

$$\text{Total consumption} = \text{total output} + \text{imports} - \text{exports}$$

Based on FAO statistics, the total and per capita consumption in major countries were calculated and illustrated in Table 6. There are two prominent countries: France and China. In France, both total consumption and per capita consumption of rabbit meat show a sharp decreasing trend since 1961. In contrast, both total and per capita consumption in China tended to increase continuously since 1961. These two countries show significant substitution. The per capita consumption in France decreased from 4.084 kg (1961) to 0.657 kg (2020). However, in China, during the same period it increased from 0.015 kg to 0.314 kg.

Like China, Spain also showed an increasing trend but with a low growth rate and slight fluctuations. However, in 2020, per capita consumption in Spain reached 1.088 kg, which was higher than other countries. In Italy, both total and per capita consumption did not change much with the exception of 1970s and 1980s. Germany also was stable compared with other countries. In 2020, the top countries of rabbit meat consumption included China, Spain, Italy and France. The total consumption of China in 2020 reached 452.34 thousand tonnes. Whereas in Spain, France and Italy, the consumption figures were 50.86, 42.86 and 26.93 thousand tonnes, respectively. In terms of per capita consumption, Spain ranked the first with 1.088 kg, followed by France with 0.657 kg.

From the above analysis, it can be found that: (1) with the regional change of production from Europe to Asia, consumption also basically developed the same

[2]Trade concentration degree, or concentration ratio, is usually used to show the extent of trade control of one country in global trade. For one country, low concentration ratio means more trade partners and low risks.

Table 6 Rabbit meat consumption (unit: tonne, kg) in major countries (1961–2020)

	France		Italy		Germany		Spain		China	
Year	Total (tonne)	Per capita (kg)	Total (tonne)	Per capita (kg)	Total (tonne)	Per capita (kg)	Total (tonne)	Per capita (kg)	Total (tonne)	Per capita (kg)
1961	188.97	4.084	49.03	0.980	17.78	0.241	20.56	0.669	9.86	0.015
1966	185.35	3.765	71.24	1.368	28.81	0.375	17.70	0.545	14.31	0.019
1971	181.67	3.551	100.03	1.856	48.24	0.612	25.85	0.754	18.73	0.022
1976	178.62	3.372	139.35	2.509	46.74	0.593	90.96	2.508	19.46	0.021
1981	154.12	2.848	191.31	3.386	39.88	0.511	76.85	2.024	23.96	0.024
1986	138.07	2.486	196.89	3.455	34.76	0.447	78.27	2.014	60.00	0.055
1991	94.76	1.665	49.42	0.866	40.46	0.509	78.69	2.002	96.24	0.081
1996	87.27	1.504	45.77	0.802	46.67	0.574	118.62	2.972	281.90	0.225
2001	76.19	1.283	46.04	0.810	47.24	0.580	107.13	2.593	373.04	0.287
2006	64.48	1.048	42.27	0.722	38.08	0.467	69.17	1.546	534.55	0.399
2011	52.30	0.827	53.34	0.895	39.68	0.491	61.17	1.299	722.01	0.525
2012	51.45	0.809	54.65	0.913	40.78	0.504	59.63	1.267	724.07	0.523
2013	49.18	0.770	56.89	0.946	41.49	0.511	58.16	1.239	761.15	0.547
2014	49.46	0.770	54.99	0.910	39.35	0.483	56.88	1.216	770.46	0.551
2015	49.30	0.765	56.83	0.938	38.30	0.468	47.21	1.011	775.35	0.551
2016	44.59	0.690	57.99	0.956	38.05	0.463	53.24	1.142	621.40	0.439
2017	41.20	0.635	29.01	0.478	37.63	0.455	51.31	1.100	461.76	0.325
2018	40.75	0.627	24.08	0.397	38.94	0.481	51.97	1.113	459.88	0.322
2019	42.60	0.654	27.62	0.456	36.83	0.461	53.98	1.155	452.89	0.316
2020	42.86	0.657	26.93	0.445	36.55	0.436	50.86	1.088	452.34	0.314

Data source: FAO Statistics, http://www.fao.org/faostat/en/

Total consumption is calculated by: output + imports – exports

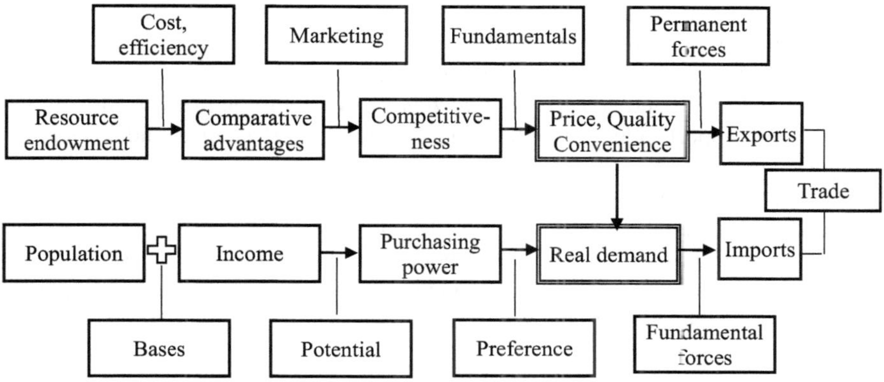

Fig. 7 International trade and its determinants

path; (2) consumption in Asia was gradually stimulated, especially in China and per capita consumption reached 0.314 kg in 2020, which is near the level of Germany (0.436 kg); (3) the long-distance trade of rabbit meat mainly involved frozen products, but now consumers mostly prefer fresh meat. This may be the main reason why trade between Europe and Asia decreased.

5 The Comparative Advantages and Competitiveness of Major Countries in the Rabbit Meat Trade

Theoretically, trade is determined by the *comparative advantages* of a country. The law of *comparative advantage* describes how, under free trade, a country will produce more of a good for which they have low costs or prices, so it can be exported to other countries (Dixit and Norman 1980). However, the *comparative advantages* are determined by *Resource endowment*, which refers to the amount of land, labour, capital and entrepreneurship that a country possesses for production exploitation. More resources mean low prices of resources and low production costs; therefore, countries with a large endowment of resources tend to be more prosperous and more competent than those with a small endowment.

Based on trade theory, the mechanism of exports and imports is as shown in the chart (Fig. 7). From above, we know that resource endowment determines comparative advantages. However, comparative advantages are just the potential for exports, only achieved through effective marketing to attract more buyers, can it become competitive. From the supply side (upper part of the Fig. 7), price, quality and convenience are the fundamentals of competitiveness. Besides this, there are also permanent factors that affect exports.

On the other hand, for imports (lower part of Fig. 7), population and income are the bases of demand. However, they are also the potential for purchasing or importing. Only by cultivating preferences of consumers, can potential purchasing power become real demand. Therefore, population, income and preference are the fundamental driving forces of imports.

Based on the above mechanism, from the sides of production and exports, we will focus on *cost*, *price* and *comparative advantage* of major countries. And from the demand and import side, we will focus on *population*, *income* and *preference change* in major countries.

Revealed comparative advantage (RCA) can measure and compare the advantages of different countries in global trade. It is the proportion of the export value of a good in one country's total good exports, divided by the share of the world's total export value of the same good in the world's total exports of all goods. The RCA is calculated by Eq. (1):

$$\mathrm{RCA}_{iC} = (X_{iC}/X_{it})/(X_{wC}/X_{wt}) \tag{1}$$

where RCA_{iC} represents the revealed comparative advantage index of good C in country i, X_{iC} is the export value of good C in country i, X_{it} is the total export value of all goods in the country i, X_{wC} is the total export value of good C in the world, X_{wt} is the total export value of all goods in the world. If $RCA_{iC} > 1$, it means that good C in country i has *revealed comparative advantage*, otherwise, it does not.

We calculated RCA in order to compare the advantages amongst major countries and study the changes in one country (Table 7). The results show that since 2000, Hungary and Argentina have been two strong competitors in the rabbit meat market, although for more than 10 years now, Argentina has exported very little, especially in 2020 and 2021. According to the FAO statistics, in 2021 Argentina exported 628.88 tonnes of rabbit meat and ranked in the top 9 in all exporting countries. Spain and Belgium were two other strong competitors. During the past two decades, their RCAs have been increasing continuously. As a traditional producer and exporter, France has maintained stable RCAs. However, China still accounted for largest share in world rabbit meat production and trade, although in the past 20 years, its advantages have been declining. The Netherlands also showed a decreasing trend in RCAs. In all, Hungary and Spain had strong comparative advantages with RCAs of 37.9 and 11.31 in 2020 with the following countries being Belgium and France (RCAs are 8.81 and 6.17, respectively), whilst the Netherlands and Argentina had lowest RCAs, and China was intermediate.

6 Driving Forces of the Global Rabbit Meat Trade: Based on Gravity Model

We have analyzed the impact mechanism of rabbit international trade above, and then a *Gravity model* will be applied to test the impacts of these factors on rabbit meat trade. Firstly, we need a brief introduction to the Gravity model.

The Gravity model of international trade is a model that, in its traditional form, predicts bilateral trade flows based on the economic sizes and distance between two

Table 7 Reveal Comparative Advantage Index (RCA) of main countries (2000–2020)

Year	Argentina	Belgium	China	France	Hungary	Netherlands	Spain
2000	29.40	1.41	8.44	3.50	28.48	3.81	4.75
2001	28.59	1.07	7.66	3.37	25.91	4.01	4.77
2002	32.81	1.68	2.35	4.67	37.23	3.40	5.34
2003	32.19	1.47	0.93	4.48	36.95	1.57	6.59
2004	39.70	2.11	0.98	4.68	28.90	1.38	7.04
2005	49.19	2.48	1.76	4.33	27.95	0.98	6.66
2006	35.82	4.24	1.84	4.40	22.64	1.37	7.02
2007	30.90	4.34	1.99	4.95	24.37	0.98	6.14
2008	33.67	5.21	2.24	6.06	19.48	0.22	4.34
2009	21.64	5.62	2.70	5.49	25.32	0.58	3.44
2010	25.59	8.99	2.39	5.45	25.28	0.83	6.64
2011	24.46	9.92	1.95	6.47	29.98	1.09	5.54
2012	19.20	8.98	2.24	5.58	32.86	1.02	8.83
2013	12.98	10.05	1.97	6.06	35.75	0.79	8.87
2014	11.63	8.99	2.33	5.65	31.03	0.64	9.24
2015	15.07	11.84	1.54	6.04	29.98	0.68	9.51
2016	16.29	9.28	1.27	6.93	31.25	0.76	10.51
2017	13.20	9.58	1.64	6.56	29.26	0.55	9.83
2018	15.27	9.53	1.53	6.11	30.01	0.72	9.30
2019	10.91	9.22	1.30	6.56	32.50	0.73	9.88
2020	0.57	8.81	1.14	6.17	37.90	0.61	11.31

Data source: FAO statistics, http://www.fao.org/faostat/en/, and WITS database, http://wits.worldbank.org/WITS/

countries, which was first introduced in economics by Walter Isard (1954). The general form of the Gravity model is given by Eq. (2):

$$T_{ij} = a_0 Y_i^{a_1} Y_j^{a_2} D_{ij}^{a_3} A_{ij}^{a_4} \tag{2}$$

where T_{ij} is the trade volume or value between country i and j, Y_i and Y_j are respectively GDPs of country i and j, D_{ij} is the distance between two countries, A_{ij} denotes trade policy and other factors and a_0, a_1, a_3, a_4 are coefficients.

Trade includes exports and imports, so we will build both export and import models, which are specified as follows:

$$\begin{aligned} lnEX_{ij} = a_0 + a_1\, ln\, \mathrm{GDP}_i + a_2\, ln\, \mathrm{GDP}_j + a_3\, ln\, \mathrm{POP}_i + a_4\, ln\, \mathrm{POP}_j \\ + a_5\, ln\, \mathrm{DIS}_{ij} + a_6\, ln\, \mathrm{WTO}_{ij} + a_7 \mathrm{BOR}_{ij} + a_8 \mathrm{Langu}_{ij} + \mu_{ij} \end{aligned} \tag{3}$$

$$\begin{aligned} lnIM_{ij} = b_0 + b_1\, ln\, \mathrm{GDP}_i + b_2\, ln\, \mathrm{GDP}_j + b_3\, ln\, \mathrm{POP}_i + b_4\, ln\, \mathrm{POP}_j \\ + b_5\, ln\, \mathrm{DIS}_{ij} + b_6\, ln\, \mathrm{WTO}_{ij} + b_7 BOR_{ij} + b_8 \mathrm{Langu}_{ij} + \sigma_{ij} \end{aligned} \tag{4}$$

$\mathrm{EX}_{ij}$and IM_{ij}represent exports and imports, respectively, from country i to country j. GDP *and* POP are Gross Domestic Product and population. DIS_{ij}stands for the distance between two countries. WTO_{ij} refers to the membership in GATT/WTO, if one country is a member of GATT/WTO, it will be 1, otherwise 0. BOR_{ij} denotes adjacency of two countries, if two countries are neighbours, it will be 1, otherwise 0. $Langu_{ij}$ represents the language, if two countries have common official or primary language, it will be 1, otherwise 0. $a_0 \sim a_8$ and $b_0 \sim b_8$ *are* coefficients to be estimated. μ_{ij} and σ_{ij} are disturbance.

To capture the general effects of the above factors on both exports and imports, the general trade model will also be estimated, so there are three models as follows: (1) model 1: *export determination model*, which is specified as Eq. 3; (2) model 2, *import determination model,* which is specified as Eq. 4 and (3) model 3, *general trade model* (total trade volume of exports and imports). The data in all these models cover the years from 1990 to 2018. In model 1, the top 6 exporting countries and their partners are selected as a sample, the sample number is 2259. In model 2, the top 6 importing countries and their partners are selected as a sample, the sample number is 1339. In model 3, these two samples are combined together, so the total sample is 3598.

Data used in this chapter come from different channels: rabbit meat production data was from the FAO statistics bank, the trade and WTO membership data were collected from the World Trade Organization, the data on distance between countries were from CEPII Database.[3] GDP and Population were from the World Bank, and other data were from related country government statistics. The models were estimated by econometric software STATA 14.

As introduced above, the major factors affecting trade have been determined, including: (1) resources endowment, which determines the production cost and prices; (2) distance between two countries, which determines the time to ship meat from exporters to importers. The distance is also related to marketing costs and convenience; (3) income and population, which determines purchasing power; (4) preferences and traditions. Similar preferences and traditions can cause more interaction and trade; and (5) trade policy. A free trade policy can improve trade.

To capture the impacts of all these factors, the Gravity model is set up by incorporating the following variables: (1) GDP of exporting country (GDP_i), to capture the resources endowment and production factors since the countries with richer resource endowment usually have the highest GDP level. For an importing country, GDP mainly reflects consumers' purchasing power, which is the potential for rabbit meat imports; (2) population of exporting country (POP_i) and importing country (POP_j) are included to capture the demand for rabbit meat; (3) distance between country i and j (DIS_{ij}), to denote the marketing costs. The short-distance transportation means fresher meat, which is the key factor attracting consumers. Adjacency (BOR_{ij}) is specified to capture the similarity of two countries in consumers' preferences and traditions, it also reflects fast transportation and convenience; (4) Common language. On one hand, language is a very important communication tool in trade. If two countries have a common official or primary language, it

[3] http://www.cepii.fr/CEPII/en/bdd_modele/bdd_modele.asp

is easier for them to trade. On the other hand, the same language also means that they might have similar habits. Though the same language may not always cause trade, compared with the multi-language countries, the same language may cause more trade. By employing the Gravity model, this hypothesis will be tested and (5) free trade or trade protection will cause different results. GATT/WTO membership is introduced to the model to control the variable of policy intervention. If one country is a WTO member, this also means it may trade more freely.

The Gravity model results are listed in Table 8. It can be found that most variables are significant. The GDP of the exporting country is negative and significant, which shows that when an exporting country has higher GDP, the rabbit meat trade will decrease. However, if an importing country has a higher GDP, the trade will increase. This shows that trade is mainly driven by demand. High GDP means high income and this causes more consumption. China is a typical case. It has high GDP level and GDP growth rate, but rabbit meat exports have been decreasing for many years. Since the model takes the double-log form, the coefficient of variable GDP_i is the *GDP elasticity of trade*. This means if GDP increases by 1% in an exporting country, the rabbit meat trade is predicted to decrease by 0.76%. Nevertheless, if importing countries increase their GDP by 1%, the imports will increase by an estimated 0.65 percent.

The population was significant in three models. It shows that when the country has a greater population, the exporting country will export a larger quantity of rabbit meat. However, in an importing country, there was an opposite effect. Therefore, if exporting countries have a higher population, rabbit meat imports will decline. This may be due to the low labour costs in countries with a higher population. Presently in many countries, rabbit raising is still labour-intensive compared with other sectors; therefore, a higher population means a strong comparative advantage in rabbit raising, when ultimately, they can meet their domestic demand for rabbit meat so as to reduce imports. This also shows that the demand for rabbit meat is driven mainly by income increase (GDP) instead of population growth. Moreover, in many countries, youngsters do not eat as much as their elder generation.

Luo and Wu (2019) compared the minimum monthly wages in main rabbit raising countries. During 2016 and 2018, the average minimum monthly wages were high in the Netherlands (1700 USD), Belgium (1695 USD), France (1622 USD) and Spain (846USD), but were low in Hungary (394 USD), Brazil (237USD) and China (230USD). Labour costs are one of the key factors affecting rabbit industry development.

The coefficient of distance variable (DIS) is negative and statistically significant suggesting that countries trade more rabbit meat from distant partners. This is in line with our knowledge and most related research findings. Meanwhile, the variable representing adjacency of two trade partners (BOR) has a positive and significant impact in all three models. This suggests that neighbouring countries are important for rabbit meat trade. Therefore, both distance and adjacency variables show that nearby location should be the key factor affecting rabbit meat trade.

From the regional perspective, the model results can explain the intra-industry trade of rabbit meat. In many countries, rabbit farm is far from the city. In some

Table 8 Results of the gravity model

Variable	Export model	Import model	General trade model
GDP_i: GDP of exporting country (current US$)	−0.76***	−0.84***	−0.76***
	(−9.34)	(−7.25)	(−11.69)
GDP_j: GDP of importing country (current US$)	0.67***	0.84***	0.65***
	(13.90)	(4.32)	(15.31)
POP_i: Population of exporting country (mn)	0.99***	1.36***	1.16***
	(11.78)	(11.50)	(16.81)
POP_j: Population of importing country (mn)	−0.13**	−0.80***	−0.20***
	(−2.24)	(−3.78)	(−3.71)
DIS_{ij}: Distance between country i and j (km)	−0.47***	−0.38***	−0.44***
	(−5.42)	(−4.76)	(−7.89)
WTO_i: Origin country i is GATT/WTO member, yes = 1, otherwise = 0	0.52	0.67	0.85***
	(1.38)	(1.30)	(2.83)
WTO_i: Destination country j is GATT/WTO member, yes = 1, otherwise = 0	0.16	−0.79	0.10
	(0.8096)	(−1.6331)	(0.59)
$Lang_{ij}$: Common official or primary language, yes = 1, otherwise = 0	0.61***	−0.76***	0.14
	(4.26)	(−3.40)	(1.16)
BOR_{ij}: Contiguity, yes = 1, otherwise = 0	1.59***	1.78***	1.56***
	(9.32)	(9.95)	(12.63)
Cons	12.21***	10.94**	11.71***
	(5.201)	(2.35)	(5.93)
N	2259	1339	3598
R-sq	0.37	0.25	0.33

*$p < 0.1$, **$p < 0.05$, ***$p < 0.01$; T value in parentheses

bordering and producing areas, rabbit meat was exported to the neighbouring country, but in other non-producing areas they possibly imported from neighbouring countries. In 2021, Belgium exported 2912 tonnes of rabbit meat to its neighbouring countries: Germany, Netherlands, Luxembourg, France and the United Kingdom, accounting for 88.78% of its total exports. At the same time, it imported 1960 tonnes from France, Spain, Germany and the Netherlands, which accounted for 79.22%. France is also the similar case, in 2021 it exported 2562 tonnes of rabbit meat to neighbouring countries: Italy, Belgium, Germany, Spain and Luxembourg, accounting for 58.78%, and imported 1749 tonnes from Belgium and Spain, accounting for about 84.34%. Italy mainly exported to Germany but imported from Spain, Hungary, Germany, France etc.

Language has a significant and positive effect on exports, but a negative effect on imports. The reason may be that it is easier for exporting countries to promote their products in countries with the same or similar language. Nevertheless, for the importing countries, what was pursued may have been low prices, instead of the same language. Furthermore, the concentration of trade may be another reason. We know from previous section that the concentration rate of rabbit meat exports was higher than that of imports. In 2021, a total of 43 countries exported rabbit meat, and the top 5 countries accounted for 80.97%; but the imports were more diversified. The importing countries in 2021 reached 66, and the top 5 countries only accounted for 64.18% of total exported rabbit meat in the global market. Diversified importing countries means more different languages, this ultimately caused the negative impacts on imports. In the general trade model, the language variable is not significant. This may be attributed to the offset of positive effect on rabbit meat exports and the negative effect on imports.

WTO membership generally is not significant, except for the exporting country in the general trade model. This may be because most trading countries have already been WTO members. The policies of rabbit meat trade in these countries are free and market-oriented and generally, there is no protection from governments.

7 Conclusions

In the past 60 years, the rabbit industry has achieved great progress globally. Before 2000, rabbit meat was mainly produced in Europe. In the past 20 years, rabbit meat production has moved largely to Asia. In 2000, rabbit meat output in Asia (428.66 thousand tonnes) first surpassed that of Europe (356.42 thousand tonnes). In 2020, Asia and Europe produced 67.49% and 19.63%, respectively, of global rabbit meat and the shares of Africa and America were 11.18% and 1.70%, respectively. In the meantime, changes in rabbit slaughtered and stock are basically consistent with rabbit meat production.

Different from production, the rabbit meat trade only increased for about 20 years from 1961 to 1979. After 1979, the global rabbit meat trade fluctuated around 60 thousand tonnes for another 20 years. Since 2001 it has been stable at the level of 30 thousand tonnes with only small fluctuations. Increasing production whilst showing stable and decreasing trade indicates that rabbit meat is gradually consumed locally within countries. The trade pattern currently is from Asia (mainly China) and Europe (Spain, Belgium and France) to European countries. In 2021, the top 5 destinations were Germany, Belgium, Italy, France and Portugal.

In regards to comparative advantage, Hungary is a strong competitor in the global market, whilst Spain and Belgium are gradually becoming two promising countries in the rabbit meat trade. China was once a strong exporter, but in the past decade the domestic demand has been increasing rapidly, whilst exports have been decreasing rapidly. France is a special case, which has been a key player in production, trade and consumption for about 50 years.

The Gravity model results show that rabbit meat trade is mainly driven by demand. The countries with high GDP tend to increase imports of rabbit meat, but decrease exports. As for production, countries with a higher population mean lower labour costs and higher comparative advantages in rabbit meat production and exports. Common language and close borders denote similar traditions and dietary habits. It also affects trade significantly.

Based on the above results, some suggestions and policy implications are provided: (1) Since rabbit meat is traded more locally or between neighbouring countries, rabbit farmers or processing companies should pay more attention to domestic consumers or neighbouring countries to exploit the market potential; (2) Experiences of France and Italy show that expanding multi-channels of exports can reduce risks and maintain a long-time stable position; therefore, traders should explore more markets so as to reduce trade concentration degree and avoid risks; (3) Population has significant negative effects on rabbit meat imports, which means the new generation consumes less rabbit meat. Governments should popularize the nutritional knowledge of rabbit meat to encourage people (especially young people) to consume more healthy rabbit meat because this food can reduce obesity, as well as the risk of heart diseases and other serious diseases.

Acknowledgements This research is funded by the China Rabbit Research System (CRRS, 2011–2022) of People's Republic of China, Many thanks go to chief scientist Professor Yinghe Qin and my colleagues at CRRS, and special thanks to professor Lukefahr Steven, Thierry Gidenne, Luiz Carlos Machado for their very valuable comments and suggestions.

References

Adanguidi J (2020) Analysis of the profitability and competitiveness of rabbit value chains in Benin. J Agric Sci (Toronto) 12:151–159. https://doi.org/10.5539/jas.v12n2p151

Antipov, A. V., Prantsuz, O. S., Beketov, S. V., & Klimov, V. A. (2020). Methods of storing rabbit meat to ensure functional nutrition. IOP Conference Series: Earth and Environmental Science 548, 072028 (072026 pp.)-072028 (072026 pp.). https://doi.org/10.1088/1755-1315/548/7/072028.

Beal MN, McLean-Meyinsse PE, Atkinson C (2004) An analysis of household consumption of rabbit meat in the Southern United States. J Food Distrib Res 35:24–29. https://doi.org/10.22004/ag.econ.27148

Bodnar, K., Horvath, J. (2008). Consumers' opinion about rabbit meat consumption in Hungary. In: Proceedings of the 9th World Rabbit Congress, Verona, Italy, 10–13 June 2008, pp 1519–1522.

Cavani C, Petracci M, Trocino A, Xiccato G (2009) Advances in research on poultry and rabbit meat quality. Ital J Anim Sci 8:741–750. https://doi.org/10.4081/ijas.2009.s2.741

Chalah T, Hajj E (1996) Rabbit production and consumption in Lebanon. World Rabbit Sci 4:69–74. https://doi.org/10.4995/wrs.1996.273

Chodova D, Tumova E, Volek Z (2019) The effect of limited feed intake on carcase yield and meat quality in early weaned rabbits. Ital J Anim Sci 18:381–388. https://doi.org/10.1080/1828051X.2018.1530961

Combes S, Postollec G, Cauquil L, Gidenne T (2010) Influence of cage or pen housing on carcass traits and meat quality of rabbit. Animal 4:295–302. https://doi.org/10.1017/S1751731109991030

Cullere M, Zotte AD (2018) Rabbit meat production and consumption: State of knowledge and future perspectives. Meat Sci 143:137–146. https://doi.org/10.1016/j.meatsci.2018.04.029

Dabbou S, Renna M, Lussiana C, Gai F, Rotolo L, Kovitvadhi A, Brugiapaglia A, Helal AN, Schiavone A, Zoccarato I, Gasco L (2017) Bilberry pomace in growing rabbit diets: effects on quality traits of hind leg meat. Ital J Anim Sci 16:371–379. https://doi.org/10.1080/1828051X.2017.1292413

Dairo FAS, Abi HM, Oluwatusin FM (2012) Social acceptability of rabbit meat and strategies for improving its consumption in Ekiti State Southwestern Nigeria. Livest Res Rural Dev 24:94. Article 94

Dixit A, Norman V (1980) Theory of international trade: a dual, general equilibrium approach. Cambridge University Press, Cambridge, pp 93–126. ISBN: 978–0521299695

Escriba-Perez C, Baviera-Puig A, Montero-Vicente L, Buitrago-Vera J (2019) Children's consumption of rabbit meat. World Rabbit Sci 27:113–122. https://doi.org/10.4995/wrs.2019.11991

Foster M, Telford R (1996) Structure of the Australian rabbit industry: a preliminary analysis. Australian Bureau of Agriculture and Resource Economics, Canberra, p 33

Gidenne T, Combes S, Fortun-Lamothe L (2012) Feed intake limitation strategies for the growing rabbit: effect on feeding behaviour, welfare, performance, digestive physiology and health: a review. Animal 6:1407–1419. https://doi.org/10.1017/S1751731112000389

Golban A, Golban R (2020) The competitiveness of rabbit meat produced in Republic of Moldova. Sci Papers Ser Manag Econom Eng Agric Rural Dev 20:255–259

Gonzalez-Redondo P, Mena Y, Fernandez-Cabanas VM (2010) Factors affecting rabbit meat consumption among Spanish University students. Ecol Food Nutr 49:298–315. https://doi.org/10.1080/03670244.2010.491053

Khan K, Khan S, Khan NA, Ahmad N (2017) Production Performance of indigenous rabbits under traditional and intensive production systems in northern Pakistan. J Anim Plant Sci 27:75–81

Kowalska D (2015) History of rabbit meat consumption in Poland. Wiadomości Zootechniczne *53*:45–49

Lehmann M (1991) Social-behavior in young domestic rabbits under seminatural conditions. Appl Anim Behav Sci 32:269–292. https://doi.org/10.1016/S0168-1591(05)80049-8

Lukefahr SD (1999) Small-scale rabbit meat production in the Western Hemisphere: back to basics? World Rabbit Sci 7:87–94. https://doi.org/10.4995/wrs.1999.384

Lukefahr SD, Cheeke PR, McNitt JI, Patton NM (2004) Limitations of intensive meat rabbit production in North America: A review. Can J Anim Sci 84:349–360. https://doi.org/10.4141/A04-002

Luo Y, Wu L (2019) Evolution of the international Competitiveness of China rabbit industry and its future development. J South China Agric Univ 24:201–214. https://doi.org/10.11841/j.issn.1007-4333.2019.12.22

Matics Z, Cullere M, Zotte AD, Szendro K, Szendro Z, Odermatt M, Atkari T, Radnai I, Nagy I, Gerencser Z (2019) Effect of cage and pen housing on the live performance, carcase, and meat quality traits of growing rabbits. Ital J Anim Sci 18:441–449. https://doi.org/10.1080/1828051X.2018.1532329

McNitt JI, Lukefahr SD, Cheeke PR, Patton NM (2013) Rabbit production worldwide (Chapter 3). In: McNitt JI, Lukefahr SD, Cheeke PR, Patton NM (eds) Rabbit production, 9th edn. CABI International Publishers, Wallingford, UK. ISBN: 9781780640129. https://doi.org/10.1079/9781780640129.0000

Niedzwiadek S (1994) World production and trade in rabbit meat. Biuletyn Informacyjny - Instytut Zootechniki 32:31–54

Nielsen SS, Alvarez J, Bicout DJ, Calistri P, Depner K, Drewe JA, Garin-Bastuji B, Gonzales Rojas JL, Gortázar Schmidt C, Michel V, Miranda Chueca MÁ, Roberts HC, Sihvonen LH, Spoolder H, Stahl K, Velarde Calvo A, Viltrop A, Buijs S, Edwards S et al (2020) Health and welfare of rabbits farmed in different production systems. EFSA J 18:e05944. https://doi.org/10.2903/j.efsa.2020.5944

Olivares R, Soriano R, Lopez M, Rivera J, Losada H (2005) Selling points and forms of consumption of rabbit meat in the Metropolitan Area of Mexico City. In: Proceedings 8th World Rabbit Congress, Septembre 7–10, 2004 - Puebla Mexico, pp 1146–1151

Owen JE (1981) Rabbit meat for the developing-countries. World Anim Rev:2–11

Parkin RJ (1972) Meat rabbit production. Agriculture 79:198

Petracci M, Soglia F, Baldi G, Balzani L, Mudalal S, Cavani C (2018) Technical note: estimation of real rabbit meat consumption in Italy. World Rabbit Sci 26:91–96. https://doi.org/10.4995/wrs.2018.7802

Popescu-Miclosanu E, Stanciu N (2013) Evolution of world and Romanian meat rabbit production and trade balance. Bull Univ Agric Sci Vet Med Cluj-Napoca Anim Sci Biotechnol 70:332–338. https://doi.org/10.15835/buasvmcn-asb:70:2:9646

Roberts S (1980) Rabbit breeding for plentiful meat supply. N Z J Agric 140:86–87

Sanah I, Becila S, Djeghim F, Boudjellal A (2020) Rabbit meat in the east of Algeria: motivation and obstacles to consumption. World Rabbit Sci 28:221–237. https://doi.org/10.4995/wrs.2020.13419

Smutka L, Rosochatecka E (2010) The development of agrarian foreign trade in rabbit meat. Acta Universitatis Bohemiae Meridionales 13:7–27

Sugiyama N, Azua RV, Galicia BR (2017) Faunal acquisition, maintenance, and consumption: how the Teotihuacanos got their meat. Archaeol Anthropol Sci 9:61–81. https://doi.org/10.1007/s12520-016-0387-z

Szendro K (2015) Rabbit meat production and foreign trade globally and in Hungary. Gazdalkodas 59:114–126. https://doi.org/10.22004/ag.econ.208275

Szendrö Z, Bleyer F (1999) The current situation in rabbit production in Hungary. World Rabbit Sci 7:209–216. https://doi.org/10.4995/wrs.1999.403

Szendro K, Szabo-Szentgroti E, Szigeti O (2020) Consumers' attitude to consumption of rabbit meat in eight countries depending on the production method and its purchase form. Foods 9:654. https://doi.org/10.3390/foods9050654

Trocino A, Cotozzolo E, Zomeno C, Petracci M, Xiccato G, Castellini C (2019) Rabbit production and science: the world and Italian scenarios from 1998 to 2018. Ital J Anim Sci 18:1361–1371. https://doi.org/10.1080/1828051X.2019.1662739

Verga M, Luzi F, Petracci M, Cavani C (2009) Welfare aspects in rabbit rearing and transport. Ital J Anim Sci 8:191–204. https://doi.org/10.4081/ijas.2009.s1.191

Wahyuni TH, Ginting N, Yunilas H, Mirwandono E, Siregar GA, Sinaga IG, Sembiring I (2018) The utilization of coconut waste fermented by Aspergillus niger and saccharomyces cerevisiae on meat quality of weaning males rex rabbit. IOP Conf Ser Earth Environ 122:012129. https://doi.org/10.1088/1755-1315/122/1/012129

Wu L, Luefahr SD (2021) Rabbit meat trade of major countries: regional patterns and influencing factors (invited paper). In: Proceedings 12th World Rabbit Congress - November 3–5 2021 - Nantes, France, Communication F-00, 15 pp. + presentation

Wu L, Qin Y (2019) China's rabbit industry development status in 2019 and outlook for 2020. Chinese J Anim Sci 55:152–156

Rabbit Farming: Industrial, Small-Scale, and Organic Housing Systems

Luiz Carlos Machado and João Simões

Abstract

Rabbit housing systems should be designed according to production purposes. These systems are crucial to rearing rabbits, considering ease of management, compliance with national legislation rules and animal welfare, and implementation of external and internal biosecurity measures, among other premises. The rabbit species is highly versatile to rear and can be well accommodated from basic (e.g., subsistence agriculture, familiar) to modern climatic-conditioned sheds. Currently, the design of large-scale rabbitries for meat purposes, as well as fixed and variable costs, are well defined at the entrepreneur level. For small-scale rabbit production, aspects related to the rearing objectives of the producer, expected gains, initial investment capital, types of products to be obtained, and possibilities to adapt farm facilities without high initial investment assume relevance. In both realities, edaphoclimatic characteristics of the local area, route accessibility, compliance with animal welfare requirements, and public health, including antimicrobial resistance issues, need to be taken into account for rabbitry implementation. This chapter aims to approach different housing systems for rearing rabbits at large- and small-scale levels throughout worldwide regions.

L. C. Machado (✉)
Instituto Federal de Minas Gerais (IFMG), Campus Bambuí, Bambuí, Minas Gerais, Brasil
e-mail: luiz.machado@ifmg.edu.br

J. Simões
Animal and Veterinary Research Centre (CECAV), Associate Laboratory for Animal and Veterinary Sciences (AL4AnimalS), Department of Veterinary Sciences, School of Agrarian and Veterinary Sciences, University of Trás-os-Montes and Alto Douro (UTAD), Vila Real, Portugal

J. Simões, J. M. Monteiro (eds.), *Veterinary Care of Farm Rabbits*,
https://doi.org/10.1007/978-3-031-44542-2_5

Keywords

Rabbit sheds · Small-scale production · Large-scale production · Meat rabbits

1 Introduction

Facilities should provide protection and comfort to animals and are a key factor in ensuring a higher level of comfort, enhancing the welfare of the environment. Despite the intrinsic highest level of animal welfare, its optimization can reflect greater productivity (Martínez-Paredes et al. 2018) and be related to functional longevity (Szendrő et al. 2013), according to the objectives of the farm.

The rabbit is the most versatile species in terms of different housing possibilities as well as products and services offered to humanity. Rabbits can be reared from very simple facilities, adapted to a model of subsistence agriculture, through organic production models, systems that consider small- and medium-scale production, and highly technical models with environmental control, such as those used today in modern European cuniculture (Cullere and Dalle Zotte 2018; EC-DGHFS 2018). All facilities should be designed according to the main objectives of the commercial plans and in accordance with the national legislation, climate of the place, type of animals, and housing systems, among other important points discussed in this chapter.

In the postwar period (1945–1959), the first efficient housing models for rabbits were defined. At that time, the greatest concerns were related to higher productivity in a smaller area and minimization of the number of health problems. Thus, traditional housing systems did not consider current relevant requirements, such as a greater degree of compliance with animal welfare precepts. Thus, new alternative systems have been proposed, seeking to offer a better quality of life for animals through measures such as environmental enrichment (Hansen and Berthelsen 2000; Luzi et al. 2003), greater social contact, and greater space per animal, reflecting a more appropriate behavioral range. The new housing systems seek to associate the simplicity of small-scale systems, such as pen housing, with the current prerequisite to ensure food security and mitigate environmental impact.

The main aim of this chapter is to describe the main housing systems for rabbits used worldwide according to the different production systems.

In 2021, the European Parliament pointed to the objective of replacing cages for several species, including rabbits, by 2027 (EPR 2021). In this sense, we must also consider that in the coming years, there will be major transformations in housing systems (Szendrő et al. 2019), which may impact the viability of each production system.

2 Location and Orientation of Sheds

The location of the sheds is one of the key factors for the success of the rabbitry and is one of the first decisions that the farmer should make. This decision will directly and indirectly impact items such as environmental comfort, ease of sales, ease of access, and other factors previously reported.

Aspects related to isolation of the local, orientation, as well as the build quality of the sheds or alternative facilities should be carefully considered so that the best cost–benefit ratio is sought. At this point, it should be noted that all of this information and decision-making may require time and dialog with the rabbit industry players. As with other livestock and poultry species (Fig. 1), the physical and socioeconomic interactions of the farm are major concerns (Ryschawy et al. 2019). Additionally, external biosecurity (Manuja et al. 2014) and climate change scenario (Ambazamkandi et al. 2015) issues need to be considered. The main factors to make a decision on local implementation are reported in Box 1.

Box 1 Desirable Premises for Locating Rabbitries

- If the purpose is meat production, the site should be strategic considering the distance from slaughterhouses and balanced food factories but preserving biosecurity distances between them.
- If the purpose is the production of pet animals, it should be close to urban centers so that it can facilitate the arrival of clients as well as the delivery of animals by land or even by air. The place should be noiseless, isolated, away from housing areas, busy roads that give rise to noise, and away from other rabbitries.
- Legislation should be accessed to verify any legal impediment regarding installation in rural, peripheral, and urban areas.
- Electric energy supply, safe access to water, feed, and technicians (e.g., inseminators and veterinarians) is needed.
- The total area and volume of the sheds must ensure adequate ventilation, hinder the spread of diseases, and comply with legal requirements.
- Easy access to charge and/or discharge animals, feed, manures, residues, equipment, and other materials.

Sheds are key structures (EFSA 2020) for controlling environmental factors and rabbits, ensuring farm biosecurity, optimizing rabbit management and production, and protecting rabbits, especially on extreme cold/hot days or rainy days (open sheds). Intensive and semi-intensive production systems are dependent on the sheds' characteristics. Modern industrial cuniculture is practiced, almost entirely, from the use of sheds, although there are initiatives implemented around the world that do not require the construction of this space.

It should also be considered that the construction of a shed is costly and requires previous studies to attest to the feasibility of the new development. In Europe,

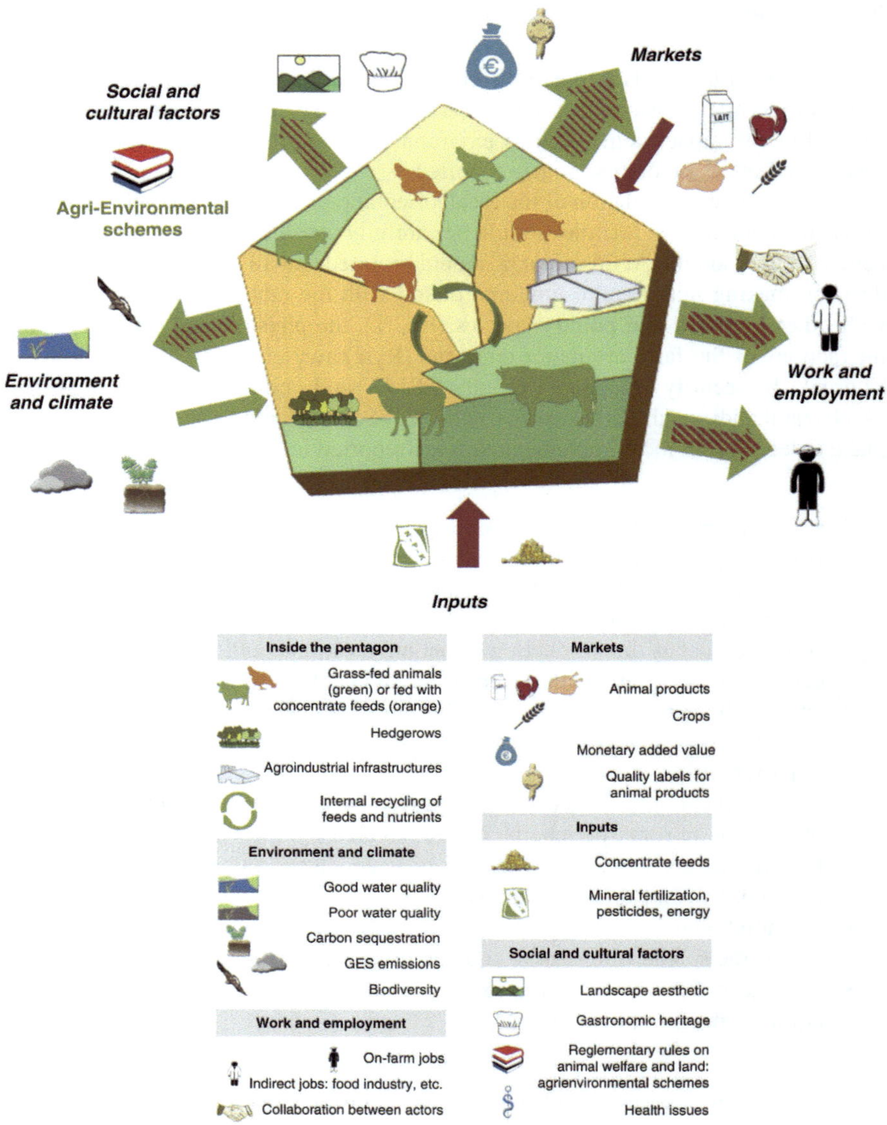

Fig. 1 Interactions between farm implementation and environmental or socioeconomic issues. Adapted from Ryschawy et al. (2019) with permission from Elsevier

currently (2023), a double shed for 1000 breeding does and 8500–9000 fattening rabbits in a 42-day semi-intensive production cycle can reach an initial investment varying from 450,000 to 700,000 euros. A total cost of 448 €/m^2 for two sheds (42 × 15 m = 630 m^2 each) was estimated by Mondin et al. (2021).

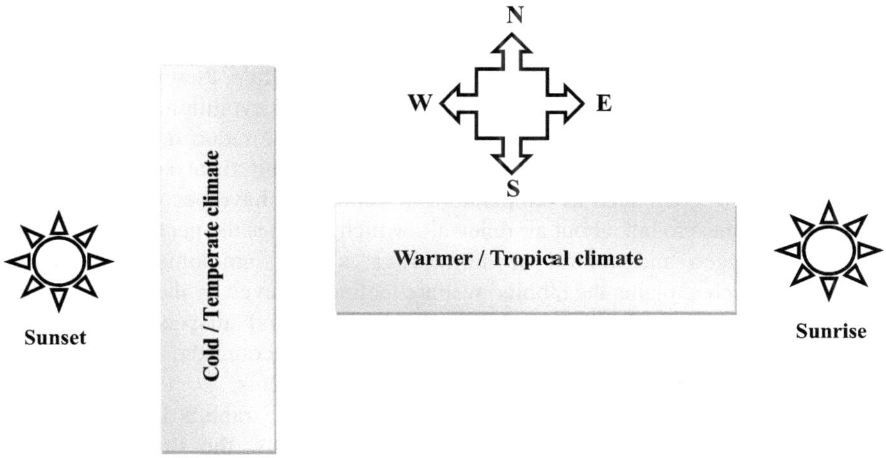

Fig. 2 Shed orientation

For small-scale production, other options are available. In addition, it may be crucial in some low-income countries that interested parties start taking advantage of some existing structures in their farm, which will reduce the initial costs of deployment. After a period of 1 year, and according to the success of the new venture, these beginners will be able to invest in new facilities.

The orientation of the sheds depends on the climate of the region where the rabbitry will be implemented. In countries with a warmer, tropical climate, where open sheds are used, it is common to use an orientation along the east–west axis to prevent direct sunstroke indoors (McNitt 2009; Ambazamkandi et al. 2015). For colder countries with temperate climates, the north–south orientation becomes more common (Fig. 2).

The magnetic axis of the earth is different from the geographical axis, and this difference should be considered for shed orientation. A compass can be useful, as well as other alternative methods, such as conducting an on-site study on summer solstice days or spring or autumn equinoxes, to check the best option for positioning. In addition, at high latitudes, even from the correct east–west positioning, there may be an incidence of sun inside the open shed from one of its sides. In this case, it is recommended to use longer eaves.

The sheds are better oriented when the dominant winds reach their longitudinal and orthogonal axes at an angle of 80–85°. Natural windbreaks, such as trees and shrubs well-placed around the sheds, can be useful in these situations. We emphasize that sun incidence and dominant winds contribute to controlling passive thermal comfort inside the sheds (Ferraz et al. 2019; Silva et al. 2021a, b).

3 Industrial Housing Systems

In Europe, the latest rabbitries have focused on animal welfare, thermal comfort, and air renewal, with strict control of harmful gases. With this evolution, they improved rabbits' life quality, achieved better productive results, reduced morbidity and mortality, and improved carcass yield. Nowadays, rabbit meat sector looks for excellence, and terms such as temperature or ventilation have become obsolete. It makes more sense to talk about air renewal—which resumes the necessity of correct levels of oxygen and control of harmful gases (e.g., ammonia) and thermal comfort—which explains the rabbits' welfare feeling achieved by the correct adjustment of temperature (heating or cooling), humidity, and airspeed (ventilation). Furthermore, industrial rabbitries are built taking into account the implementation of risk factor control measures and strict biosecurity systems.

Two main areas in rabbitries are required for rearing rabbits industrially: the multiplication (maternity) and flattening sectors. Usually, the sheds are paired, communicating by a central corridor, to ensure different environments from outside and to allow the implementation of biosecurity measures such as all-in and all-out systems and sanitation. Currently, the same shed can serve as maternity, followed by the displacement of rabbit does to the second shed at weaning time and fattening purposes by the use of polyvalent (dual-purpose) cages (Pérez-Fuentes et al. 2020). After removing the nest, each cage stores the 8–9 weaned rabbits until the end of their production cycle (60–70 days of age).

The structure of sheds varies depending on the type of housing system: 1) Natural environment. The shed is covered and naturally ventilated through side windows and upper openings (Silva et al., 2016). The width should be less than approximately 16 m to allow convective air flow; 2) Controlled environment (Borso et al. 2016). Air extractors are used to force ventilation by depression (Fig. 3). A disused alternative is overpressure with air injection ventilation.

Fig. 3 Sheds under controlled ventilation by air extractors (Portugal). Back view

Fig. 4 External (**a**) and indoor (**b**) views of two modern European sheds (Portugal), comporting 800 individual cages

For modern cuniculture, mainly practiced in European and Asian countries, there are already companies that offer services for the construction of sheds, which are assembled from prefabricated parts (Fig. 4). Several specific areas inside and outside the sheds should be designed as reported in Box 2. Outside sheds, a separation (i.e., hygienic barriers) promoting a buffer between clean and dirty zones to delimitate the farm perimeter is required for biosafety reasons (Fig. 5), i.e., to promote external biosecurity (Manuja et al. 2014).

Box 2 Rabbitry Structure Used in Industrial Cuniculture

- Breeding sector, containing multipurpose cages with folding nests.
- The fattening sector, containing multipurpose cages with folding nests, allows an increased area for rearing.
- Sector for rearing rabbit does for replacement.
- Zone of quarantine.

(continued)

Box 2 (continued)

- Farm vehicle disinfection area.
- Footbath or shoe disinfection area at the entrance of each shed.
- Disposal area of dead animals (freezer). In countries without specific regulations for freezer use, animals should be located outside the sanitary barrier.
- Area for cleaning and disinfection of material.
- Area for silos and/or feed warehouse.
- Area for administration, including veterinary storage area.
- Personal service area. Dressing room with sanitary facilities.
- Area and equipment to implement management plan of effluents: Ditch, septic tank, and solid effluent area.
- Area for transport/displacement of animals on the holding.

Fig. 5 Outside areas with a hygienic barrier (wire screen) to demarcate the limits of the farm (Portugal)

4 Small-Scale Housing Systems

4.1 Open Sheds

In subtropical and tropical regions, and contrary to temperate climatic areas, it is usual to operate with open sheds. The great advantage of sheds with natural ventilation is the fact that no electricity or other investments are needed for a higher level of climatization, reducing production costs (Silva et al., 2021).

In these regions, the sheds are most often structures assembled from masonry combined with wire screens. The side parts of the shed are closed, being built from bricks, and oriented toward the east or west directions (Fig. 6). The presence of a small sidewall, 30–50 cm tall, is important to stop small animals and rain. Wire screens are placed above this wall to protect the animals and allow the passage of air (Fig. 7) (Ferraz et al., 2021). In the breeding shed sector, these wire screens integrate curtains that are managed as needed.

Indoor, the roof is supported by a structure of steel, as preferable currently, instead of wood (Fig. 8). High ceiling height (distance between the base of the roof and the floor) values favor internal ventilation and gas exchange between animals and the environment. It is suggested that the ceiling height should be at least 3 m in countries with tropical climates, mainly in regions with high average annual temperatures (above 21 °C). Additionally, ceiling height should be directly proportional to the width of the shed.

Fig. 6 Lateral view of a masonry shed for rabbit fattening with small sidewall, wire screens, wooden pillars, and ceramic tile cover located in the tropical region (Brazil)

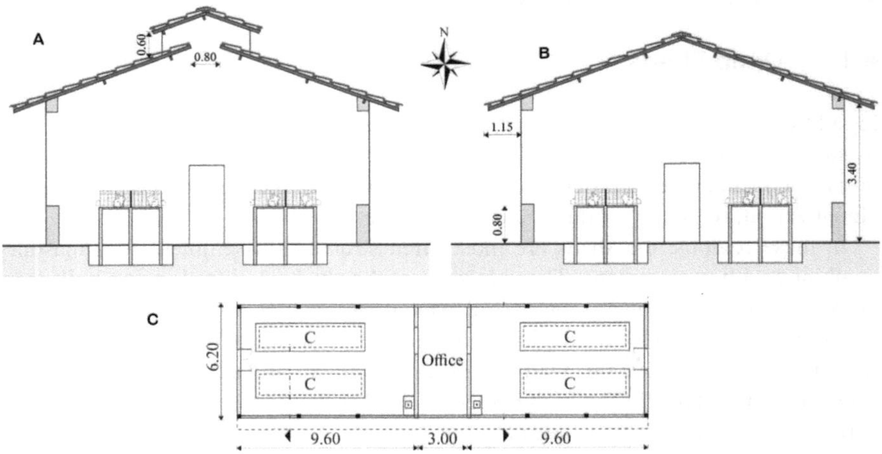

Fig. 7 Cross-section view of sheds with (**a**) and without (**b**) ridge vent and floor plan (**c**). Unit = meter or similar. Adapted from Ferraz et al. (2019)

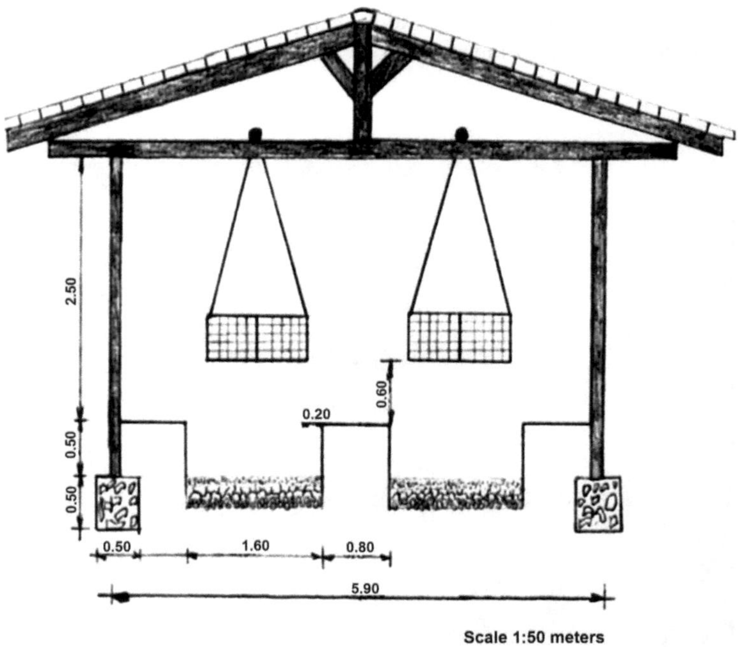

Fig. 8 Cross-section view of a shed with a central corridor and two sides, wood crate and ceramic tiles, cages hung with wire, and collecting fossa consisting of three layers, sand, gravel, and coal. Note that the ceiling height of 2.50 m is low for most tropical climates

Fig. 9 Inside view of a classical masonry shed in a tropical climate (Brazil)

The roof receives sunlight, acting as a thermal insulating barrier. There are several types of materials used for roofing. Polyurethane sandwich panels provide excellent thermal insulation, but ceramic tiles can be used. The use of white color outside the roof partially reflects the electromagnetic waves of different lengths, contributing to a decrease in the temperature inside the shed.

An internal corridor is positioned between two rows to facilitate people's transit and management (Fig. 9). The floor of these corridors can be concrete, although in more simplified facilities, wooden corridors or even beaten floors can be used.

Automatic cleaning systems are not used in these types of sheds. The most commonly used system in these places is that of a pit containing 80 cm depth (Fig. 10). The ditch contains three layers of 10 cm each from the bottom, sand, coal, and gravel, considering the allocation sequence from top to bottom (the gravel is in contact with the soil). This system is a biological filter and considerably favors the reduction of odors as well as effective control of the moisture of the rabbitry.

4.2 Breeding and Rearing Rabbits on the Floor of the Shed

Several studies have been conducted in recent years to improve animal welfare in housing systems, such as greater area per rabbit and more social interactions between rabbits to express natural behavior (see EFSA et al., 2020). Even in some countries (e.g., Brazil) where the activity is low and with diverse purposes (e.g., rabbit pet production), a growing concern on animal welfare monitoring is reported (Silva et al. 2022). A return to primordial rabbit rearing has been approached with some adaptations in some worldwide regions. For example, in the EU, organic rabbit

Fig. 10 Inside view of an experimental shed with a central corridor and waste-collecting pit

Fig. 11 Fattening rabbits in a floor collective system with wood shavings and aviculture equipment (feeder), indoor of a shed

farming allows the use of pure breeds and their crossbreeding (F1), and local breeds, even thermal stress (hot seasons), negatively influence doe performance (e.g., energy balance and prolificacy) (Dalle Zotte and Paci 2013). Furthermore, simplified housing systems can contribute to a significant reduction in the initial costs of small-scale rabbitries. Several pieces of equipment from other animal species, such as poultry, can be adapted for rabbit production to reduce cost in low-income regions where the poultry industry is relevant.

Floor rearing is mainly used for fattening rabbits (Fig. 11). A relatively large number of animals can be placed in a given area. Nonetheless, recent research has also aimed to develop solutions to adequately accommodate breeding rabbits in collective pens in Brazil.

On the floor, bedding material (e.g., wood shavings) is required to receive the excrement of the animals. The *bedding* should be changed every 30 days but varies according to animal density mainly in the presence of high humidity in the floor. Some researchers believe that the animals housed in this system make bed material intake, which contributes to a reduction in feed intake and therefore the daily weight gain of the animals, reflected in lighter animals at the end of the fattening period.

For breeding animals, which may remain in this system for years, problems arising from contact with the bedding, toxic gases, and feces may be more severe. An adequate ventilation level should be ensured, bed material should be periodically replaced, and rabbits should be monitored for respiratory problems.

Equipment for aviculture can be adapted in this accommodation system (see Fig. 11). Cup instead of nipple drinking fountains should be used to prevent leaking and bedding soaking. Although the size of the bay may vary, it is not recommended to use large areas that aim to house large groups of animals. It is preferable to divide the rabbits into small groups, suggesting approximately 3000 cm^2 per animal in the fattening phase and 6500 cm^2 per rabbit for the reproduction phase, in accordance with the EFSA (2020) recommendation. If placed on the floor, it is necessary that the nests are at a higher level to avoid excess moisture.

There are also alternatives to the floor collective system using rebar or screens that can be leaked and elevated. The great advantages of these last floor types are the low rabbit contact with feces and urine, excrement removal, and floor disinfection.

5 Alternatives to the Use of Sheds

Although sheds can optimize rabbit production, indoor environmental conditions, animal welfare, biosecurity, etc., it is necessary to consider that they require large initial investments and are more adequate for industrial production. In some countries, cuniculture is an incipient and high-risk activity, and large initial investments are not indicated (Machado and Ferreira 2014). In this context, for most cases, a greater investment would be indicated only from the first year of activity, when the farmer has a higher level of certainty of the profitability and accumulated experience. It should also be considered that for subsistence or small-scale cuniculture, the initial investment should be limited, and preferably, one can take advantage of the facilities and covered sites already available on the farm property.

5.1 Hut-Type Modules

Hut-type modules can be an alternative with less initial investment, as they do not require the construction of a large and complex structure such as industrial or masonry sheds. These structures can be purchased somewhere. They are modular structures that can easily be added over time, increasing the number of animals.

As the rabbits will be very close to the roof of the hut, which has a lower roof when compared to that of the shed, it is necessary that the modules are positioned as a place between trees offering shade to the covers.

It is noteworthy that this type of accommodation should not be used in regions of temperate climate because the intense cold or thermal amplitude inside the facilities would be a major impediment, reducing the productive performance of the animals as well as increasing the mortality of kits.

5.2 Outdoor Cages and Nurseries

In tropical climatic regions, where the winter temperature is not extreme and, in most zones, it is not negative, more simplified housing systems can be used. In these locations, the thermal amplitude usually varies between 8 and 30 °C, and the relative humidity varies between 55 and 85%. In temperate climatic regions of cold weather, it would become extremely difficult to keep animals efficient during the winter and even summer seasons, given the lack of a protective environment such as that provided by a shed.

Cage-based facilities and outdoor nurseries have the great advantage of reducing initial production costs, although they are only indicated for subsistence or small-scale farms. However, it should be noted that in some developing countries, considering the last decades of the twentieth century, there have already been farms of considerable size from this housing.

Rabbits need to be protected from rain, and a roof should be provided, even if rudimentary. The cages can be constructed of alternative material and must contain a small roof, and this facility is commonly called a nursery. In this type of housing system, cages or nurseries can be built in masonry, wood, or alternative materials such as bamboo. There is also the possibility of systems excavated in ravines.

Like hut-type modules, this housing model requires shade. It is necessary to provide trees around these modules to provide environmental protection against winds and other climate-dependent conditions. A shade screen can provide a lower incidence of light in animals to mitigate problems related to excess heat inside the cages. However, in regions of high rainfall season or low latitude near the equator, this housing system can pose constraints to the farmers' assessment and consequently rabbits' management. Furthermore, feces and urine may be carried by rainwater and reach other watercourses, which may cause widespread pathogens and may have high polluting potential to cause environmental impacts.

5.3 Delimited Fences

For versatile and familiar rabbit production, the simplest housing systems require few initial investments, provide more area per animal, and can allow full behavior of the rabbits. It should also be considered that these systems can provide a high level of animal welfare, and some are very close to those that were used in the early days of commercial rabbit farming. Nevertheless, some issues should be considered in this housing system, as reported in Box 3.

Box 3

- If male breeders remain close to females, the farmer loses control of reproduction. Growing rabbits at different ages are present.
- Rabbits can dig under the fence and exit the picket lines, causing damage in neighboring areas outside of the enclosure.
- If rabbit does compete for nests, the opposite litter can be eliminated by one of the females.
- It is essential that the picket has good fodder quality, which is the basis for feeding the rabbits.
- The production indexes will be median due to the exclusive use of grass. The interval between births should be low to improve the revenues for farmers.
- Health problems and sanitary measures are often more difficult to control and manage.

Rabbits can be reared in fenced areas with or without control of reproduction and balanced feeding. The housing system of delimited fences can be applied in subsistence cuniculture or in hunting farms, which produce animals with greater reactivity, ideal for repopulation of degraded areas. In this type of farm, there is already better control of sanitary, reproductive, and feeding management.

A small shelter to protect rabbits from sun, rain, and wind is needed. For the delimitation of the enclosure, galvanized wire mesh or even simpler materials, such as wood or bamboo, can be used. The wire mesh or other materials must be buried, approximately 50 cm, to prevent the animals from leaving the enclosure.

Organic farming can make use of the outdoor fenced model, where animals can be moved and displaced in an environment according to the availability of fodder.

6 Organic Rabbit Farming

Currently, there is great concern with animal production regarding the environment, animal welfare, and food safety. Considering this last issue, in the past decade, new systems, sometimes more rudimentary, have emerged when compared to conventional ones that provide more natural housing conditions and differentiated food to

Fig. 12 Organic rabbit production model. Adapted from Gidenne et al. (2020) with permission from Springer Nature

supply the market. Even if they have worse productive and reproductive indexes (Dalle Zotte and Paci 2013), these systems can add value to the products marketed. The standards for the production of organic rabbits are still incipient and are available in a few countries. See Gidenne et al. (2022) for all relevant aspects concerning organic production systems. One great challenge is to rear rabbits under grazing without feed supplementation and maintain performance (Martin et al. 2016; Legendre et al. 2019).

Countries such as France already have a commercial production of organic rabbits (Gidenne 2018; Gidenne et al. 2020). In this housing system, the rabbits must graze. Mobile fenced furniture allows nursery displacement according to the availability of fodder (Fig. 12).

A portion of the commercial feed can be offered but without the inclusion of antimicrobials, anticoccidials, or growth promoters. Moreover, an extended delivery interval, at least 73 days, and a high age for weaning, at least 42 days, are adopted (Gidenne et al. 2022). Only natural mounts should be carried out. All these requirements influence the profitability of the production system. Usually, organic production is an added value increasing the price of the product for a very specific market niche.

7 Group Housing Systems

The objectives of this new housing system are related to a greater demand for animal welfare by society, even though differences between different housing systems are not always evident (Trocino et al. 2022). The social contact between rabbits, environmental enrichment (Fetiveau et al. 2023), and area per animal is improved (Fetiveau et al. 2021) (see Fig. 11). In nature, rabbits are sociable and gregarious animals, where they live in small groups, with a well-defined hierarchy, which contributes to the stability of the group and the use of resources. On farms, this behavior is not evident, and adult rabbits are averse to social life (Dal Bosco et al. 2019).

In the first system of group housing, does and bucks were reared in the same pen, causing a loss of control of reproduction. Other systems already kept the pregnant does together, causing serious fights at peripartum due to their extreme aggressivity. Moreover, it was common to deliver one kit over the nest of another does, causing cannibalism, which raised the mortality of the kits to unreasonable levels and negatively impacted animal welfare.

The best housing system found to date seems to be the semigroup model, where rabbits remain part of the time in the collectivity and part of the time isolated (Huang et al., 2021). In some research studies carried out (Machado et al. 2019; Buijs et al. 2014), the does were isolated 3 days before delivery. Although the isolation time can be variable, it was suggested that regrouping doses and kits should occur after the kits complete at least 18 days of life. At this age, kits have reached a sufficient level of independence and resilience and do not get hurt due to the fights that occur between the mothers. It should be noted that these kits will mix over the next few days, and the rabbits have lost control of their litter, breastfeeding them indistinctly. Even if the rabbits already have a hierarchy defined in the period before segregation, there will be hierarchical disputes when placing the animals in collectivity again, which increases the frequency of fights that can result in injuries, accidents, elimination of animals, or even death. This agonistic behavior tends to be more intense if a previous socialization of the rabbit does not occur.

Although the hierarchy tends to stabilize the group and decrease disputes over time, the fights remain the biggest problem for the adoption of these systems in farms. In this sense, the best environmental enrichment of the pens seems to decrease the occurrence of fights. These new models of group housing imply a new management approach, the resolution of new agonistic behavioral challenges of group rabbits and health issues (e.g., disinfections, sanitary voids, and contacts).

The incidence of these problems can be reduced by introducing environmental enrichment in pens (Princz et al. 2009), such as the use of straw, which can help distract the attention of animals at the most critical moments. This housing system should provide escape routes as well as hiding for animals. The introduction of rabbits from the same litter into the collective system, avoiding the introduction of new breeders, can mitigate fights between them. Additionally, the selection of animals by temperament can be a new parameter to be considered.

Currently, there are already some new models of collective pens for rabbits, as happens in the system of cage colonies with removable divisions in Italian (Tufarelli et al. 2022) or German (Rauterberg et al. 2019) models of authentic pens whose welfare standards are regulated.

The simplest system of group housing seeks to house two rabbit does without dividing the environment (cage), where the physiological state of one cannot coincide with that of the other, i.e., while one is in the lactation phase, the other must be pregnant. These does will use the same nest space, which can be cleaned and disinfected a few days before delivery. This model has the advantage of optimizing the cage space for two rabbits without cannibalism or other aggressions. This space should be large and environmentally enriched to provide good housing conditions for the rabbits. As a disadvantage, if a negative pregnancy is detected in one of the rabbits, there will be a longer period of delay in the reproduction of the two animals.

8 Conclusions

Currently, two issues permeate animal production: the first is related to an increasing need for food production internally in each country (reduction of import dependence), and the second is related to the need for housing systems to be friendlier from the point of view of animal welfare.

Considering the first item, many countries suffer from the lack of available area for the production of animal feed, and thus, strategies related to vertical animal production emerge. The production of rabbits in this type of model was suggested in 2021 at the world Cuniculture Congress in Nantes, France.

The industrial production of meat rabbits is based on housing systems, providing an indoor microclimatic environment and optimizing production, sanitation, biosecurity, food biosafety, and other relevant issues.

Already considering the new guidelines of the European Parliament, the development of new collective housing systems is expected and will become more representative from 2027, the year limit indicated for rabbit farming to be cage-free. In this sense, the organic production of rabbits also grows.

Acknowledgments The authors thank Springer for the use of the AJE Digital Editing. This work was supported by projects UIDP/CVT/00772/2020 (CECAV) and LA/P/0059/2020 (AL4AnimalS), funded by the Portuguese Foundation for Science and Technology (FCT).

References

Ambazamkandi P, Thyagarajan G, Sambasivan S, Davis J, Shanmugam S, Joseph BA (2015) Shelter design for different livestock from a climate change perspective. In: Sejian V, Gaughan J, Baumgard L, Prasad C (eds) Climate change impact on livestock: adaptation and mitigation. Springer, New Delhi. https://doi.org/10.1007/978-81-322-2265-1_23

Borso F d, Chiumenti A, Mezzadri M, Teri F (2016) Noxious gases in rabbit housing systems: effects of cross and longitudinal ventilation. J Agric Eng 47:222–229. https://doi.org/10.4081/jae.2016.572

Buijs S, Hermans K, Maertens L, Van Caelenberg A, Tuyttens FAM (2014) Effects of semigsemi-grouping housing and floor type on pododermatitis, spinal deformation and bone quality in rabbit does. Animal 8:1728–1734. https://doi.org/10.1017/S1751731114001669

Cullere M, Dalle Zotte A (2018) Rabbit meat production and consumption: state of knowledge and future perspectives. Meat Sci 143:137–146. https://doi.org/10.1016/j.meatsci.2018.04.029

Dal Bosco A, Mugnai C, Martino M, Szendrő Z, Mattioli S, Cambiotti V, Cartoni Mancinelli A, Moscati L, Castellini C (2019) Housing rabbit does in a Combi system with removable walls: effect on Behaviour and reproductive performance. Animals 9:528. https://doi.org/10.3390/ani9080528

Dalle Zotte A, Paci G (2013) Influence of rabbit sire genetic origin, season of birth and parity order on doe and litter performance in an organic production system. Asian Australas J Anim Sci 26: 43–49. https://doi.org/10.5713/ajas.2012.12401

EC-DGHFS. European Commission, Directorate-General for Health and Food Safety (2018) Commercial rabbit farming in the European Union: overview report, Publications Office. https://data.europa.eu/doi/10.2772/62174

EFSA Panel on Animal Health and Welfare (AHAW), Nielsen SS, Alvarez J, Bicout DJ, Calistri P, Depner K, Drewe JA, Garin-Bastuji B, Gonzales Rojas JL, Gortázar Schmidt C, Michel V, Miranda Chueca MÁ, Roberts HC, Sihvonen LH, Spoolder H, Stahl K, Velarde Calvo A, Viltrop A, Buijs S, Edwards S, Winckler C (2020) Health and welfare of rabbits farmed in different production systems. EFSA J 18:e05944. https://doi.org/10.2903/j.efsa.2020.5944

EPR (2021) European Parliament resolution of 10 June 2021 on the European citizens' initiative 'end the cage age' (2021/2633(RSP)

Ferraz PFP, Ferraz GAS, Damasceno FA, de Moura RS, Silva MAJG, Rodrigues RL (2019) Spatial variability of enthalpy in rabbit house with and without ridge vent. Rev Bras Eng Agric Ambient 23:126–132. https://doi.org/10.1590/1807-1929/agriambi.v23n2p126-132

Fetiveau M, Savietto D, Gidenne T, Pujol S, Aymard P, Fortun-Lamothe L (2021) Effect of access to outdoor grazing and stocking density on space and pasture use, behaviour, reactivity, and growth traits of weaned rabbits. Animal 15(9):100334. https://doi.org/10.1016/j.animal.2021.100334

Fetiveau M, Savietto D, Fillon V, Bannelier C, Pujol S, Fortun-Lamothe L (2023) Effect of outdoor grazing-area access time and enrichment on space and pasture use, behaviour, health and growth traits of weaned rabbits. Animal 17:100724. https://doi.org/10.1016/j.animal.2023.100724

Gidenne T (2018) Organic rabbit farming: situation in France and Europe. 6th American rabbit congress, Goiania, Brazil, 28–30 Aug 2018

Gidenne T, Savietto D, Goby JP, Fortun-Lamothe L, Roinsard A (2020) A referencing system to analyse performances of French organic rabbit farms. Org Agric 10:125–129. https://doi.org/10.1007/s13165-020-00315-1

Gidenne T, Savietto D, Fortun-Lamothe L, Huang Y (2022) Cuniculture au pâturage et sous certification Agriculture Biologique en France: Fonctionnement des systèmes, performances et règlementation. INRAE Prod Anim 35:201–216. https://doi.org/10.20870/productions-animales.2022.35.3.7257

Hansen LT, Berthelsen H (2000) The effect of environmental enrichment on the behaviour of caged rabbits (*Oryctolagus cuniculus*). Appl Anim Behav Sci 68:163–178. https://doi.org/10.1016/s0168-1591(00)00093-9

Huang Y, Breda J, Savietto D, Debrusse AM, Combes S, Fortun-Lamothe L (2021) Part-time grouping of rabbit does in enriched housing: effects on performances, injury occurrence and enrichment use. Animal 15:100390. https://doi.org/10.1016/j.animal.2021.100390

Legendre H, Goby JP, Duprat A, Gidenne T, Martin G (2019) Herbage intake and growth of rabbits under different pasture type, herbage allowance and quality conditions in organic production. Animal 13:495–501. https://doi.org/10.1017/S1751731118001775

Luzi F, Ferrante V, Heinzl E, Verga M (2003) Effect of environmental enrichment on productive performance and welfare aspects in fattening rabbits. Ital J Anim Sci 2:438–440. https://doi.org/10.4081/ijas.2003.11676036

Machado LC, Ferreira WM (2014) Opinião: organização e estratégias da cunicultura Brasileira - buscando soluções. [Opinion: organization and strategies of Brazilian rabbitry – seeking solutions]. Rev Bras Cunicult 6:1–31

Machado LC, Martínez-Paredez E, Cervera C (2019) Performance of rabbit does housed in collective pens and individual cages. World Rabbit Sci 27:227–235. https://doi.org/10.4995/wrs.2019.11540

Manuja BK, Manuja A, Singh RK (2014) Globalization and livestock biosecurity. Agric Res (New Delhi, India) 3:22–31. https://doi.org/10.1007/s40003-014-0097-7

Martin G, Duprat A, Goby JP, Theau JP, Roinsard A, Descombes M, Legendre H, Gidenne T (2016) Herbage intake regulation and growth of rabbits raised on grasslands: back to basics and looking forward. Animal 10:1609–1618. https://doi.org/10.1017/S1751731116000598

Martínez-Paredes E, Ródenas L, Pascual JJ, Savietto D (2018) Early development and reproductive lifespan of rabbit females: implications of growth rate, rearing diet and body condition at first mating. Animal 12:2347–2355. https://doi.org/10.1017/S1751731118000162

McNitt JI (2009) Rabbit housing rabbit: hutches, cages, wires. Outhern university agricultural research and extension center, LA 7081, USA. Accessed from http://world-rabbit-science.com/Developping/Fichiers-pdf/McNitt-2009-rabbit-housing-manual.pdf. 22 Mar 2023

Mondin C, Trestini S, Trocino A, Di Martino G (2021) The economics of rabbit farming: a pilot study on the impact of different housing systems. Animals 11(11):3040. https://doi.org/10.3390/ani11113040

Pérez-Fuentes S, Muñoz-Silvestre A, Moreno-Grua E, Martínez-Paredes E, Viana D, Selva L, Villagrá A, Sanz-Tejero C, Pascual JJ, Cervera C, Corpa JM (2020) Effect of different housing systems (single and group penning) on the health and welfare of commercial female rabbits. Animal 14:1270–1277. https://doi.org/10.1017/S1751731119003379

Princz Z, Dalle Zotte A, Metzger S, Radnai I, Biró-Németh E, Orova Z, Szendrő Z (2009) Response of fattening rabbits reared under different housing conditions. 1. Live performance and health status. Livest Sci 121:86–91. https://doi.org/10.1016/j.livsci.2008.05.018

Rauterberg SL, Bill J, Kimm S, Kemper N, Fels M (2019) Evaluation of two different flooring designs for rabbit housing in accordance with German welfare regulations: soiling and mortality. Agriculture 9:257. https://doi.org/10.3390/agriculture9120257

Ryschawy J, Dumont B, Therond O, Donnars C, Hendrickson J, Benoit M, Duru M (2019) Review: an integrated graphical tool for analysing impacts and services provided by livestock farming. Animal 13:1760–1772. https://doi.org/10.1017/S1751731119000351

Silva KG, Borges TD, Costa LB, Sotomaior CS (2021a) Characteristics of meat, pet, and research rabbit farms in Brazil: an overview based on twelve farms. World Rabbit Sci 29:115–123. https://doi.org/10.4995/wrs.2021.14125

Silva MAJG, Ferraz PFP, Santos LMD, Ferraz GAES, Rossi G, Barbari M (2021b) Effect of the spatial distribution of the temperature and humidity index in a New Zealand white rabbit house on respiratory frequency and ear surface temperature. Animals 11:1657. https://doi.org/10.3390/ani11061657

Silva KG, Borges TD, Costa LB, Dalmau A, Sotomaior CS (2022) Rabbit welfare protocols under Brazilian conditions: the applicability of welfare protocols in rabbit farms for different purposes – first results. J Vet Behav 54:36–53. https://doi.org/10.1016/j.jveb.2022.06.002

Szendrő Z, Mikó A, Odermatt M, Gerencsér Z, Radnai I, Dezséry B, Garai E, Nagy I, Szendrő K, Matics Z (2013) Comparison of performance and welfare of single-caged and group-housed rabbit does. Animal 7:463–468. https://doi.org/10.1017/S1751731112001760

Szendrő Z, Trocino A, Hoy S, Xiccato G, Villagrá A, Maertens L (2019) A review of recent research outcomes on the housing of farmed domestic rabbits: reproducing does. World Rabbit Sci 27:1–14. https://doi.org/10.4995/wrs.2019.10599

Trocino A, Menegon F, Zomeño C, Pasqualin D, Cunial G, Xiccato G, Pirrone F, Bertotto D, Bortoletti M, Dorigo F, Lavazza A, Di Martino G (2022) A pilot study about on-farm assessment of health and welfare in rabbits kept in different housing systems. Front Vet Sci 9: 936643. https://doi.org/10.3389/fvets.2022.936643

Tufarelli V, Tateo A, Schiavitto M, Mazzei D, Calzaretti G, Laudadio V (2022) Evaluating productive performance, meat quality and oxidation products of Italian white breed rabbits under free-range and cage rearing system. Anim Biosci 35:884–891. https://doi.org/10.5713/ab.21.0327

Rabbit Farming: Indoor Microclimate

Luiz Carlos Machado and João Simões

Abstract

The surrounding environment plays a crucial role in livestock production. Several physical factors influence the efficiency of production, health, and animal welfare in small- and large-scale production systems. These interrelationships within and between the environment and animals assume a special relevance in high-producing meat rabbits reared on a large scale. A significant number of rabbits are produced as a close population developing a dynamic microclimate inside the sheds. Animal density, temperature, relative humidity, ventilation, and luminosity are the main factors accurately controlled to ensure the sustainability of the rabbitries. Most environmental monitoring is automated. In small-scale housing systems, rabbits are more dependent on natural conditions. Nonetheless, they can profit from technological advances in the rabbit industry to accomplish their function, mainly in impoverished rural areas, to provide food-derived products and services in local markets. This chapter highlights the main issues related to indoor environmental factors in both large- and small-scale rabbit production.

Supplementary Information The online version contains supplementary material available at https://doi.org/10.1007/978-3-031-44542-2_6. The videos can be accessed individually by clicking the DOI link in the accompanying figure caption or by scanning this link with the SN More Media App.

L. C. Machado (✉)
Instituto Federal de Minas Gerais (IFMG), Campus Bambuí, Bambuí, Minas Gerais, Brasil
e-mail: luiz.machado@ifmg.edu.br

J. Simões
Animal and Veterinary Research Centre (CECAV), Associate Laboratory for Animal and Veterinary Sciences (AL4AnimalS), Department of Veterinary Sciences, School of Agrarian and Veterinary Sciences, University of Trás-os-Montes and Alto Douro (UTAD), Vila Real, Portugal

J. Simões, J. M. Monteiro (eds.), *Veterinary Care of Farm Rabbits*,
https://doi.org/10.1007/978-3-031-44542-2_6

Keywords

Temperature · Relative humidity · Photoperiod · Shed microclimate

1 Introduction

The shed indoor microclimate is closely linked to adaptation or disruption of the physiological status of rabbits (Silva et al. 2021). The indoor environment refers to the environmental conditions to which the animals are subjected, significantly impacting the productive performance and longevity of rabbits. Aspects related to air temperature, hygrometry, ventilation, area/volume per animal/kg body weight (BW), and luminosity should be optimized. All these factors are related to the rabbit's welfare.

The indoor environment directly affects the performance and health of rabbits, e.g., thermal conform (Liang et al. 2022), lighting (Mousa-Balabel 2011), irritating gases and dust Calvet et al. 2011), but it also contributes, as a risk factor, to an overall dynamic indoor microbiota according to the housing production system (Faye et al. 1999; Bouquin et al. 2009). Pathogens, e.g., fungi (Martino et al. 2007; Miao et al. 2010) and sporulation of protozoan oocysts (Boyko et al. 2021), can develop under a specific indoor microclimate, which serves as a reservoir and source to infect rabbits. These features imply the periodical sanitation of the shed (e.g., all-in-all-out system).

In this chapter, we describe the main environmental factors that influence rabbit rearing and their general impact on performance and health.

2 Animal Density

Animal density is one of the relevant factors affecting rabbit behavior, performance, and health (El-Bayoumi et al. 2018; El-Sabrout 2018).

The density of rabbits within some enclosures, such as cages, parks, or pens, should be balanced (Fig. 1). If the animal density is low, the cost of accommodation per animal will be higher. On the other hand, high animal densities are related to a decrease in rabbit performance and an increase in the frequency of aggressivity (Trocino et al. 2015), especially in rabbits of advanced age, and may decrease the acceptable behavior level for the rabbitry, which negatively impacts animal welfare (Gerencsér et al. 2019).

Recommendations for rabbit density vary according to the regulation of each country, state, or province. As a general rule, 40–45 kg of live weight per m^2, corresponding to 15–18 fattening rabbits/m^2, is acceptable for industrial meat rabbit farming (Szendrő and Dalle Zotte 2011). Table 1 reports the indicator values of the rabbit density according to the productive phase.

Fig. 1 Animal density (8–9 rabbits per conventional cage; <16 rabbits per m^2). Early postweaned (left; Portugal) and late fattening (right; Brazil) stages. (Video 1) (► https://doi.org/10.1007/000-brt)

Table 1 Indicative values of animal density in industrial meat rabbitries. Adapted from González Redondo and Rodríguez (2007)

Production phase	Rabbits/m^2	Mean range cm^2/Rabbit	Minimum cm^2/Rabbit
Lactating does (individual cage with nest)	2.5–3.3	3000–4000[a]	<4 kg/BW: 2000 ≥4 kg/BW: 3000
Pregnant does (individual cage)	3.3–5.0	2000–3000	<4 kg/BW: 2000 ≥4 kg/BW: 3000
Fattening rabbits (up to 2.2 kg/BW)	15.0–17.0	600–700	400
Rabbits for replacement (individual cage)	4.0–5.0	2000–2500	<4 kg/BW: 800 ≥4 kg/BW: 1500
Bucks (individual cage)	2.5–3.3	3000–4000	<4 kg/BW: 2000 ≥4 kg/BW: 3000

[a] For cages with outside nest: 2500 cm^2/female

In countries where cuniculture is developing without national standardization, the indicative values of the animal density are 800, 4800 (including nest area) and 3200 cm^2 per fattening rabbits, does and bucks, respectively. These values are applied to the individual and collective cages, and other enclosures. They are slightly higher than those reported by González Redondo and Rodríguez (2007) for industrial meat rabbit production.

It is also important to highlight a future trend regarding animal welfare. The current trend is that the animal density will be lower than currently, which allows, in addition to greater free space, space for the introduction of environmental enrichment in the enclosure (see EFSA 2020). See Chapter "Welfare, Behavior, and Housing of Rabbits" for animal welfare housing.

3 Environmental Temperature

Air temperature is a major environmental factor that impacts the productive and reproductive performance of animals and is also very variable according to the various regions of the globe (Ragab et al. 2022). Rabbits are raised in very cold places, such as the European winter, until very hot places, such as summer, in tropical regions. Suitable installations are essential to avoid the adverse effects of extreme temperatures on the rabbit's homeostasis (i.e., thermal stress and dispense of net energy), considering excessive cold and excessive heat, providing thermal comfort (Liang et al. 2022). The large temperature variation throughout the day (thermal amplitude) can also negatively impact animal performance and animal health and welfare (Bakr et al. 2015). When sheds are not used, the thermal amplitude of the housing indoor environment is greater than the presence of rabbits (dissipation of body temperature).

Rabbits are animals that do not withstand high temperatures for long periods. This species is characterized by the absence of sweat glands, except inside the mouth (Yagci et al. 2006). In addition, the thick layer of hair makes dispersion difficult for heat dissipation, especially in poorly ventilated environments (Oladimeji et al. 2022). As physiological and behavioral mechanisms against thermal stress, rabbits perform dermal vasodilation, mainly in the ears, decrease food intake, increase respiratory frequency, and are less active during warmer hours, among other adaptations (Fayez et al. 1994).

By reducing feed consumption, the growing animal will reduce its weight gain capacity (Szendrő et al. 2018; Matics et al. 2021), have a greater predisposition to digestive syndrome and may increase mortality (Box 1). It was estimated that feed intake decreased by 30–40% when the air temperature increased from 20 to 30 °C (Lebas et al. 1986).

High temperatures also impair reproduction, as they reduce fertility in general, acting negatively on embryo implantation, spermatogenesis, and libido, among other aspects (Marai et al. 2002; Boni 2019). Additionally, air temperatures greater than 30 °C significantly affect kit survival (EFSA 2020).

Box 1 Effects of Temperature on Rabbit Production and Health

Low air temperature

- Increase in feed consumption.
- High incidence of respiratory syndrome.
- Increase in the abandonment of nests by rabbit does.

High air temperature

- Decrease in feed intake.
- Increase in embryonic losses before embryo implantation in rabbit does.
- Poor sexual behavior and low semen quality in bucks.

The comfort zone temperature for rabbits is estimated to be approximately 21 °C (Marai et al. 2002). Overall, the thermal comfort zone for the rabbit ranges between 15 and 22 °C and relative humidity (RH) of 60–70% (Lebas et al. 1986; Marai et al. 2002) and is higher for does than for bucks (Box 2; EFSA 2005; Verga et al. 2007). The thermoregulatory response of rabbits to air temperatures between 8 and 28 °C is considered normal. Outside this thermal amplitude, passive and active measures are required to mitigate excessive cold or heat. In general, rabbits can be raised in regions where the average annual temperature is up to 24 °C. Above that, stakeholders should evaluate the feasibility of rearing, given that performance will be lower and spending on strategies to mitigate thermal stress will be higher.

Box 2 Optimal Air Temperature according to Sector Production

- Rabbits does: 16–21 °C.
- Early weaned rabbits: 20–22 °C.
- Fattening rabbits: 15–20 °C.
- Bucks: 12–16 °C.

Low environmental temperatures are better supported by rabbits than high temperatures (Matics et al. 2021). Nonetheless, they lead to higher heat metabolic production activity by the animal to regulate its body temperature. This heat production negatively affects the conversion rate of feed to produce meat. Since the energy cost required to heat the environment through an external heat source is lower than the energy cost of feed, shed heating systems should be deployed in areas with significantly cold daily or seasonal periods.

In these situations, the complete insulation of the shed will also be a key part in the face of low external temperatures. It should be emphasized that the breeding of rabbits outside sheds (as presented in the following sections) becomes impractical in areas where the cold is very intense during the winter.

In the nest, low temperatures induce hypothermia in kids. This occurrence is also related to the thermal quality of the nest, which represents some of the causes that most impact the preweaning mortality rate. Inside the nest, the optimal temperature is approximately 38 °C at birth, which progressively decreases by 8 °C in the following 2 weeks according to the development of thermoregulatory control and the surrounding temperature (Hull et al. 1986). The rabbit does must have previously prepared the place, and the hairs coming from its skin surface are as important as a thermal insulator. The heat transferred from milk remains in the nest and contributes to the heating of the kids. Thus, a litter of adequate size allows the kids to agglomerate, reduce heat losses, and maintain their body temperature.

The comfort zone temperature can be achieved with passive and active systems.

3.1 Passive Thermal Control

To achieve a higher level of productivity, as in modern industrial cuniculture, the thermal insulation of pavilions becomes a fundamental item. This optimized insulation capability is also important to enhance energy savings.

Thermal insulation includes the insulation of the walls and roof, usually with sandwich panels (e.g., insulating compound between zinc-plated or other plates; Fig. 2) and/or the use of polyurethane plates of different thicknesses. It is a passive form of energy efficiency with a relatively high initial cost but with guaranteed economic return due to lower energy costs, especially in climates of extreme environmental temperatures or large daily thermal amplitude. The structural use of sandwich panels in walls and roofs can reduce the cooling energy cost by approximately 15 and 50%, respectively, to maintain a difference of 10 °C between external (30 °C) and internal (20 °C) environments when compared with structural masonries (Capra and Blumetto 2014).

Thermal bridges, i.e., areas where insulation is deficient, such as doors and structural elements without insulation, should be mitigated or even avoided.

In hot/tropical climates, much of the heat is transmitted indoors through the shed through the roof, and the influence of the walls is secondary.

3.2 Active Cooling Systems

In hot climate regions or in summer/hot seasons of most worldwide regions, it is necessary to cool the interior of the pavilion, ensuring ideal conditions for rabbits to

Fig. 2 Thermal insulation of a shed. Structure of the walls and roof with sandwich panels. Indoor view

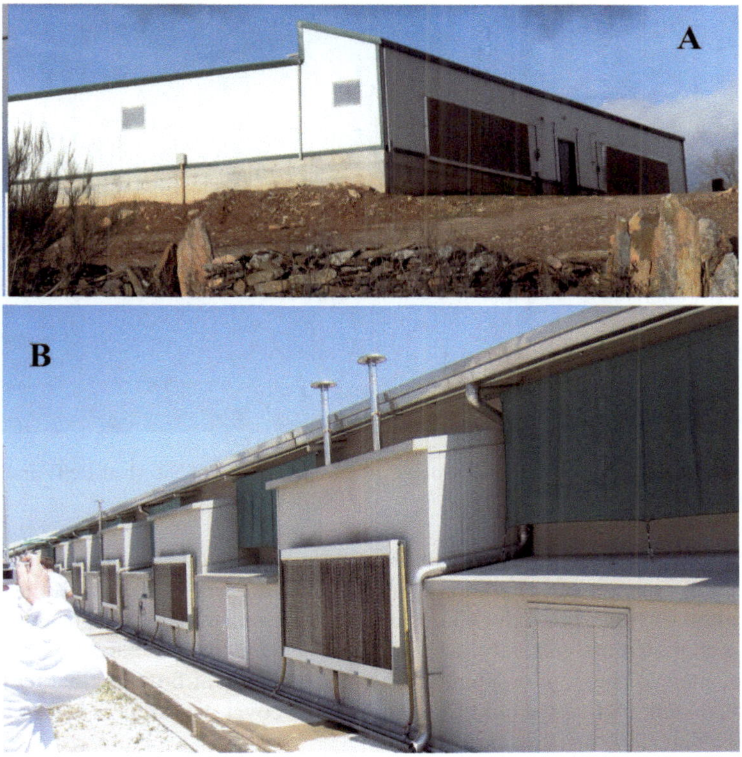

Fig. 3 Evaporative pads at the end (**a**) and lateral (**b**) of the shed

express all their productive and reproductive potentials (Zeferino et al. 2013; Dalmau et al. 2014; Mady et al. 2018).

A simple method is to wash the cover of the shed, which can reduce the inside temperature by 3–4 °C through water evaporation. This system, although it is easy to assemble, has the drawbacks of expanding a significant volume of water and being relatively ineffective.

Evaporative cooling is another method of cooling air by evaporating water inside the pavilion. As an effect, the RH of the air will be increased. Therefore, its efficiency is limited by the RH of hot air and the limit that is imposed as the maximum RH of cooled air. By norm, RH ranges between 80 and 70% (Lebas et al. 1986), a value that optimizes heat exchanges between the rabbit and the environment. When the temperature is very high (35 °C) and the RH is low, evaporative cooling can reduce the temperature by approximately 8–10 °C; however, when the RH of the air is high, only 2–3 °C is reduced. The most common evaporative cooling systems are the nebulization system and the wet panel system (Fig. 3).

Fig. 4 Block the air stream. Curtains down in the antecamera of the shed (left) and external semicover of the fiber panels (right)

A practical, efficient, and economical system, widely used in industrial cuniculture, consists of installing at one of the extremes of the shed a system of evaporation of water through cellulosic wall evaporative pads or, considering fiber as the material most currently used. In this system, the air extractors present at the opposite end (Fig. 3a) or laterally (Fig. 3b) create an airflow originating through the cooled wall. During the cold days or season, a cover or curtain is placed to block the air stream (Fig. 4).

In worldwide regions where small-scale cuniculture predominates, alternative measures for cooling can be implemented. There is low-cost evaporative air-cooling equipment (approximately 300€) that can efficiently reduce the temperature in small sheds. This air-cooling equipment, which creates airflow and water nebulization, must be placed above the level of the animals and may be associated with a thermostat that activates the system automatically. For these situations, water sprinklers on the roof, which are triggered automatically or by the farmer in the hottest periods of the day, can also be installed. The water can be collected and reused for this purpose.

3.3 Active Heating Systems

Considering the drawbacks caused by low winter temperatures in temperate regions, it is necessary to use heating systems in the shed, especially in the maternity sector.

There are several active heating systems that can be used, and you should look for systems with economical installation, reduced operating cost, easy and effective environmental temperature regulation of the shed and safety to use (Fig. 5).

Fig. 5 Gas heating equipment

Almond shell or wood shave burners, salamanders, coal kilns, hot water radiators, gas or diesel hot air generators, etc., are some of the shed environment heating systems that can be used.

Usually, in tropical regions, it is not necessary to heat the environment.

4 Relative Humidity

The RH can be defined as the percentage of vapor pressure relative to saturation vapor pressure in air (Kong and Singh 2011). Inside the rabbit shed, the environmental vapor comes from outside air, rabbits' breathing and excrements (feces and urine), drinking fountains, and washing water. RH alone has little influence on animal productivity, except when it is too high or too low.

Overall, the optimal RH varies between 55 and 65% (Liang et al. 2022). It can be approximately 60–70% in maternity, but a variation between 55 and 60% is acceptable in the fattening sector. RH has a strong impact on the heat exchange capacity between the animal and the environment.

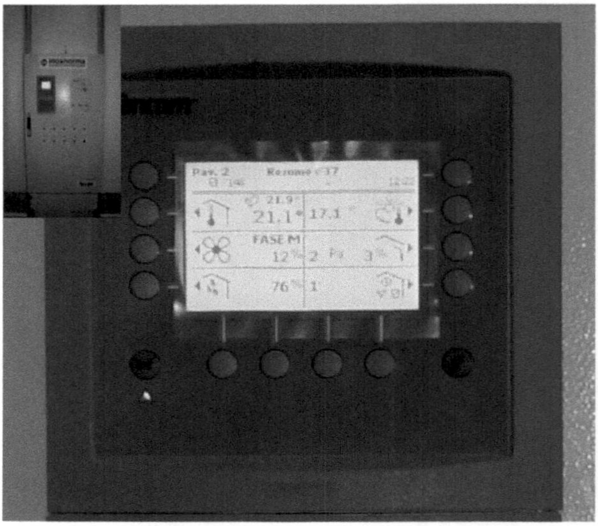

Fig. 6 Automatic control system of temperature, relative humidity, and ventilation with wireless (GSM) connection to smartphone app

Rabbits are the main source of moisture in the environment, as described in Box 3. Their excrement (manure) corresponds to only 10% of the total water that evaporates into the environment.

Box 3 Moisture Production (g H_2O/hour/animal) Increased by 10% from Manure

- Rabbit does (3.5 kg/BW): 8.6 g.
- Fattening rabbits (1.5 kg/BW): 4.4 g.
- Replacement females (2.5 kg/BW): 6.8 g.
- [Males (4 kg/BW): 9.9 g].

The RH excess in the shed (very high levels: >80%) creates conditions for microorganism (e.g., bacteria and fungi) development. Additionally, it hinders the thermoregulation of animals because in these conditions, the capacity of heat loss to the environment is limited. Moreover, condensation of water in cold surfaces, mainly in walls and cages, can occur during winter or morning.

On the other hand, very low levels (<55%) promote the occurrence of respiratory and digestive syndromes.

Both of these phenomena are intensified when there is an inadequate temperature. As a consequence, it is necessary to regulate these two environmental parameters (temperature and RH) together with ventilation and air velocity at the animal level (Fig. 6). Some remote controls for these automatic systems were designed to facilitate farmer work (Benhmad et al. 2022), including the incipient use of artificial intelligence (Zhang and Qian 2022).

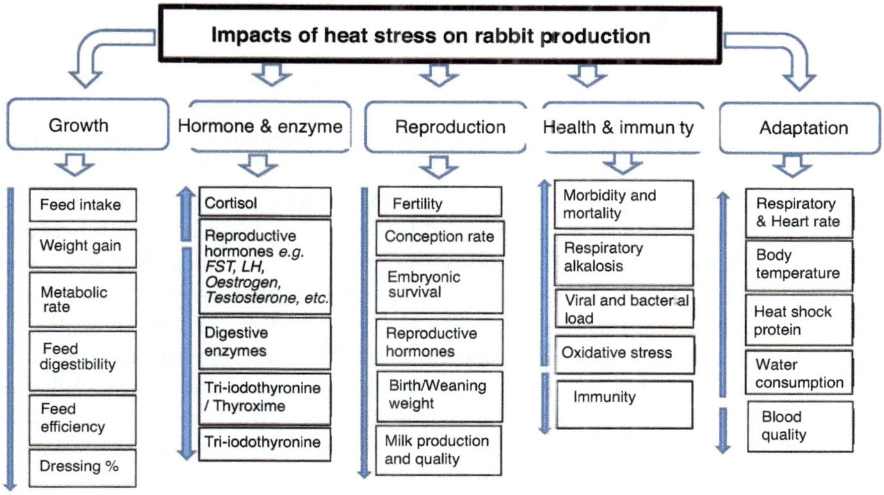

Fig. 7 Impacts of heat stress on rabbit production and health. Adapted from Oladimeji et al. (2022) with permission from Springer Nature

Rabbits are especially sensitive to air temperature > 26 °C and RH > 70% (Silva et al. 2021). The presence of high temperature associated with high RH hinders heat loss by evaporation, especially if there are no air flows, provoking a suffocation sensation in rabbits and leading to increased ammonia levels in the indoor environment and discomfort of the animals, negatively impacting animal welfare. RH excess also improves dust concentration, which may be associated with an increased presence of pathogens in the air. These conditions also trigger the process of irritation of the mucous membranes of the respiratory system.

High RH associated with low temperatures is also an undesirable situation. The existence of a heating system associated with adequate ventilation is recommended for RH decrease. This occurrence is related to the spread of dermatophytosis, contagious rhinitis, and enteritis.

Special attention, especially in low-scale productions, should be given to leakage drinking fountains. These leaks can significantly increase the RH of the environment and contribute to water waste and increased production costs.

A relevant occurrence related to both temperature and RH factors is environmental heat stress (HS) in rabbits. Heat stress is due to the imbalance between animal heat production and its environmental dissipation (Aggarwal and Upadhyay 2012). The combination of temperature and RH is usually measured in the environment using the temperature-humidity index (THI). The temperature-humidity index is categorized as lack of HS (< 27.8), moderate HS (27.8–28.9), severe HS (27.9–30.0), and very severe HS (≥30.0 units) using the formula THI = db °C-[(0.31–0.31RH) × (db °C-14)], where db is dry bulb temperature (Marai et al. 2001, 2002). HT has a significant impact on the production and health of rabbits (Aggarwal and Upadhyay 2012; Liang et al. 2022), as described in Fig. 7.

5 Ventilation

An adequate level of ventilation is the key to optimizing the environment within a shed. The main function of ventilation is to allow air renewal, improve gas exchange between the animal and the environment, and remove potentially toxic gases, mainly NH_3 and CO_2 but also CH_4, H_2S, N_2O, and dust (Calvet et al. 2011). The air renewal flow rate shall be between 1 and 4 m^3/h for each kilogram of live animal weight, with a speed of between 10 and 40 cm/sec. Table 2 describes the relationship between ventilation, temperature, and humidity reported by González Redondo and Rodríguez (2007). It should not be forgotten, however, that dynamic cooling and heating systems should include ventilation as a way of spreading cold or heat.

CO_2 must be properly controlled and close to 0.15% (1500 ppm) and should not exceed 0.3%. Each breeding and fattening rabbit produced 11 and 4.2 g CO_2/h and 55.9 and 10.1 NH_3 mg/h, respectively. When NH_3 exceeds 20 ppm, a disagreeable odor is detected and potentially irritates the eyes and the mucosa of nasal cavities. These persistent irritations cause local immunosuppression in rabbits and predispose them to respiratory syndrome. Currently, CO_2 and even NH_3 sensors strategically positioned indoors can be used to measure these gases. Dust particles with a diameter between 2.5 micron and 10 micrometers should not exceed 1.7 and 3.4 mg/m^3, respectively.

Adequate ventilation contributes to the control of the temperature, reduction of moisture produced by rabbits and evacuation of harmful gases and dust. The poor renewal of air has a negative impact on the health status of the animals and the productive indexes (Box 4).

Box 4 Negative Effects of Inadequate Ventilation

- High relative humidity level.
- Predisposition to disease occurrence.
- Worst conversion rate.
- Worse growth of fattening rabbits.

Table 2 Recommended values of temperature, relative humidity, and ventilation in maternity and fattening sectors (González Redondo and Rodríguez 2007)

	Sector	
Parameter	Maternity	Fattening
Temperature (°C)	16–20	19–22
Relative humidity (%)	60–70	
Ventilation flow (m^3/h/kg BW):		
• Shed temperature at 10–24 °C	2–6.5	1–5.5
• Shed temperature at 25–36 °C	7–10.5	6–9
Maximum air speed (m/sec)		
• Shed temperature at 10–24 °C	0.05–0.30	
• Shed temperature at 25–36 °C	0.30–0.60	

Fig. 8 Air extractors at different levels on one of the tops of the sheds

The air flows are more intense near the air apertures. When using side apertures, i.e., the windows are on one side and the extractors are on the other, they must be at the animal level to cause homogeneous ventilation at this location. In the tunnel system, the ventilation is longitudinal (Fig. 8) since the air inlet is on one of the tops and the extractors on the other. With this arrangement, it is possible to obtain higher air displacement speeds for the same ventilation flow when compared to lateral ventilation. This system, normally used in intensive production, allows strict control of the main environmental factors.

In sheds with natural environments, air movement is mainly caused by the difference in temperature between the exterior and interior of the pavilion, creating convection air flows. These air flows are essential for the removal of saturated air (rich in potentially toxic compounds) and therefore air renewal. When the temperature inside the shed is higher than outdoor, the hot air rises and exits through the top exits, allowing the air to flow through its entrance through the side windows. The use of some kind of forced ventilation is recommended, even if they are simple fans associated with climatization equipment.

For open sheds installed in tropical regions, a ceiling height, at least 3 m high, will be crucial for ensuring acceptable ventilation (see Chapter "Rabbit Farming: Industrial, Small-Scale, and Organic Housing Systems").

It should be noted that adequate ventilation should be planned according to the time choice of the local area and during shed construction. It is not desirable that it is in the highest part of a hill, which would favor excessive ventilation, or in a valley, which could impair the minimum level of ventilation. Additionally, natural barriers such as trees can be used in cases of excessive ventilation.

6 Luminosity

Luminosity refers to the intensity of light that animals receive, and the photoperiod refers to the amount of daily natural light. In nature, the photoperiod affects the reproductive physiology of rabbits (Szendrő et al. 2016). Lighting periods influence the circadian rhythm of rabbits, including feed intake and reproduction patterns. Peaks of activity were observed during night, between 20 and 8 h (Spain; Lat. 41° 53' N), in wild rabbits with insertion of an individual identifying microchip and under semi-freedom conditions in the autumn and winter seasons (Díez et al. 2005, 2013). A similar stay length in the feeding area (Díez et al. 2013) or feed intake pattern (Bellier et al. 1995) related to sunrise and sunset was observed. Additionally, maximum rabbit activity can be observed between January and June (Díez et al. 2013).

Periods of increasing luminosity stimulate reproduction. In species with a seasonal reproduction of "long days," such as rabbits, the decrease in melatonin levels in spring leads to increased levels of GnRH (gonadotropin-releasing hormone) produced by the hypothalamus. This increase will trigger the anterior pituitary to release greater amounts of gonadotropins, i.e., FSH (follicle stimulating hormone) and LH (luteinizing hormone), causing growth and follicular maturation in females (see Chapter "Reproduction in the Female Rabbit"). In the opposite direction, the decrease in photoperiod signals induces reproductive rest in does.

In farms located in high latitudes, where the influence of the seasons is more pronounced, artificial lighting can be offered to the animals to complete a total period of daily light. In closed sheds, luminosity is exclusively provided by artificial lighting, although a few apertures (e.g., cooling fans) are present.

The duration and light intensity affect rabbit activity, but some doubts persist regarding performance and animal welfare. In the maternity sector, a total of 16 h daily of white light with an intensity of 20–40 lux (5–6 W/m^2) and homogeneously distributed (e.g., a distance of 4 m between them) is recommended. This duration improves the fertility and prolificity of does. The 16:8 rhythm (light: darkness) reduces seasonal fluctuations in the normal duration of the day, and light rhythms with shorter durations (e.g., 14:10) are still adopted (Mousa-Balabel 2011).

An interesting management technique that can enhance fertility and prolificity is related to the maintenance of the total duration of the photoperiod up to 4–6 days before artificial insemination (AI), gradually decreasing (1 h/day) to 12 hours in the 4–6 days after AI (Quintela et al. 2004).

At the reproduction level, an tendency of increase in total newborn (11.3 vs. 10.6 kits; $P < 0.1$) and increase in liveborn kits (10.8 vs. 10.0 kits; $P < 0.05$) was observed when does were exposed to 150–200 lux compared to 10–20 lux, respectively (Matics et al. 2016). Additionally, suckling mortality (8.1 vs. 10.0% kits; $P < 0.01$) and litter size at day 35 (8.8 vs. 8.5 kits; $P < 0.05$) were advantageous for higher than lower lux exposure. Nevertheless, does showed a preference for cages with 10 lux.

Additionally, it was observed that blue light can improve the litter weight (+3.2% and body weight (+3.7%) of kits on the twenty-third day of life (Gerencsér et al.

2011) when compared with white light. Nevertheless, red light has a positive effect on the reproductive index (total litter size at birth × kindling rate × viability rate) and increases melatonin, FSH, and estrogens (on the 3rd day after kindling) of does as well as the performance of kits at weaning (Kalaba and Abdel-Khalek 2011). In addition to hormone improvement, red light modulates the gene expression of follicular development (Pan et al. 2022). A 16 L:8D photoperiod from 6 days before AI to 12 days post-AI with 80 lx red light was proposed for breeding purposes (Wu et al. 2021). Red light can affect the reproductive performance of female rabbits and the expression of key genes for follicular development.

In the fattening sector, the luminosity does not need to be constant, and twilight can be made available. The photoperiod in the fattening phase does not affect the daily intake of food or the average daily gain of live weight and can also be optimized for the use of only natural lighting, without any addition of artificial light.

In the tropics, especially in low-latitude regions, which do not suffer from the great influence of seasonality, it is not common to use artificial lighting in sheds. Nonetheless, there is an improvement in the prolificity of rabbits when at least 14 h light (e.g., 14 L:10D lighting program) is used (Theau-Clément et al. 1998). In open sheds, lighting at night can attract insects, which can be vectors of diseases such as myxomatosis, increasing the spread of diseases, especially in the summer season.

7 Noise

Rabbits are sensitive to noise from 0 to 20 decibels, particularly to the audio intensity (Marai and Rashwan 2004), which acts as a physical stressor (De la Fuente et al. 2007).

They are easily startled by sudden noise, presenting with nervous and behavioral alterations. The adverse influence of (loud) noise in different body systems of animals is documented (Kight and Swaddle 2011; Erbe et al. 2022). In rabbits, loud noise induces histopathological changes in kidney and liver tissues (Abouee Mehrizi et al. 2022).

Music environmental enrichment has been used in rabbitries for a long time. It was demonstrated that it can improve fecal cortisol concentration and, when removed, increase the heterophil:lymphocyte ratio in rabbits (Peveler and Hickman 2018).

8 Conclusions

Environmental factors play a key role in animal performance, health, and welfare in nature as well as in controlled closed environments. A microclimate is made in sheds that directly influence the rabbits, mainly through the temperature-humidity index and luminosity. Overall, the threshold ranges are defined. Additionally, these factors can create a favorable environment for the development of some pathogens and an exogenous cycle of protozoa.

Currently, several systems to control the indoor environment, mainly in industrial housing systems, are suitable to optimize rabbit production. New trends have arisen to optimize environmental factors toward an equilibrium between costs, performance, animal health, and animal welfare. Additionally, small-scale production can profit from these technical and scientific advancements.

Acknowledgments The authors thank Springer for the use of the AJE Digital Editing. This work was supported by projects UIDP/CVT/00772/2020 (CECAV) and LA/P/0059/2020 (AL4AnimalS), funded by the Portuguese Foundation for Science and Technology (FCT).

References

Abouee Mehrizi A, Rasoulzadeh Y, Saed Moucheshi S, Mehdipour A (2022) Histopathological changes caused by noise exposure in lung, heart, kidney, and liver tissues in New Zealand white rabbits. J Air Pollut Health 6:257–264. https://doi.org/10.18502/japh.v6i4.8584

Aggarwal A, Upadhyay RC (2012) Heat stress and animal productivity. Springer, New Delhi. https://doi.org/10.1007/978-81-322-0879-2

Bakr MH, Tusell L, Rafel O, Terré M, Sánchez JP, Piles M (2015) Lactating performance, water and feed consumption of rabbit does reared under a Mediterranean summer circadian cycle of temperature v. comfort temperature conditions. Animal 9:1203–1209. https://doi.org/10.1017/S1751731114003310

Bellier R, Gidenne T, Vernay M, Colin M (1995) In vivo study of circadian variations of the cecal fermentation pattern in postweaned and adult rabbits. J Anim Sci 73(1):128–135. https://doi.org/10.2527/1995.731128x

Benhmad T, Abdennour AB, Darghouthi A, Rhaimi CB (2022) Remote control of environmental parameters in rabbitry based on IoT. Internet of Things and Cyber-Physical Systems 2:111–119. https://doi.org/10.1016/j.iotcps.2022.06.003

Boni R (2019) Heat stress, a serious threat to reproductive function in animals and humans. Mol Reprod Dev 86(10):1307–1323. https://doi.org/10.1002/mrd.23123

Bouquin SL, Jobert JL, Larour G, Balaine L, Eono F, Boucher S, Huneau A, Michel V (2009) Risk factors for an acute expression of Epizootic Rabbit Enteropathy syndrome in rabbits after weaning in French kindling-to-finish farms. Livest Sci 125:283–290. https://doi.org/10.1016/j.livsci.2009.05.010

Boyko OO, Shendryk LI, Rudyk VS, Volovyk IA, Biben IA, Brygadyrenko VV (2021) Influence of temperature on sporulation of Eimeria arloingi and Eimeria perforans oocysts. Regul Mech Biosyst 12:369–373. https://doi.org/10.15421/022150

Calvet S, Cambra-López M, Estellés F, Torres AG (2011) Characterisation of the indoor environment and gas emissions in rabbit farms. World Rabbit Sci 19:49–61. https://doi.org/10.4995/wrs.2011.802

Capra G, Blumetto O (2014) Tecnología de producción de conejos para carne. Série técnica 216. INIA, Unidad de Comunicación y Transferencia de Tecnología, Montevideo. ISSN: 1688-9266

Dalmau A, Catanese B, Rafel O, Rodriguez P, Fuentes C, Llonch P et al (2014) Effect of high temperatures on breeding rabbit behaviour. Anim Prod Sci 55:1207–1214. https://doi.org/10.1071/AN13440

De la Fuente J, Díaz M, Ibáñez M, De Chavarri E (2007) Physiological response of rabbits to heat, cold, noise and mixing in the context of transport. Anim Welf 16(1):41–47. https://doi.org/10.1017/S0962728600030918

Díez C, Pérez JA, Prieto R, Alonso ME, Olmedo JA (2005) Activity patterns of wild rabbit (Oryctolagus cuniculus, L. 1758), under semi-freedom conditions, during autumn and winter. Wildl Biol Pract 1:41–46. https://doi.org/10.2461/wbp.2005.1.6

Díez C, Sánchez-Garzia C, Pérez JA, Bartolomé DJ, González V, Wheatley CJ, Alonso ME, Gauidoso VR (2013) Behavioural activity of wild rabbit (Oryctolagus Cuniculus) under semi-natural rearing systems: establishing a seasonal pattern. World Rabbit Sci 21:263–270. https://doi.org/10.4995/wrs.2013.1332

EFSA (European Food Safety Authority) (2005) The impact of the current housing and husbandry systems on the health and welfare of farmed domestic rabbits. EFSA J 267:1–31. https://doi.org/10.2903/j.efsa.2005.267

EFSA Panel on Animal Health and Welfare (AHAW), Nielsen SS, Alvarez J, Bicout DJ, Calistri P, Depner K, Drewe JA, Garin-Bastuji B, Gonzales Rojas JL, Gortázar Schmidt C, Michel V, Miranda Chueca MÁ, Roberts HC, Sihvonen LH, Spoolder H, Stahl K, Velarde Calvo A, Viltrop A, Buijs S, Edwards S et al (2020) Health and welfare of rabbits farmed in different production systems. EFSA J 18:e05944. https://doi.org/10.2903/j.efsa.2020.5944

El-Bayoumi KM, Abdel-Hamid TM, El-Tarabany MS, Sherief WRIA (2018) Influence of stocking density on growth performance traits, blood chemistry and the expression of HSP70 and IGF-I genes in New Zealand white rabbits. Slov Vet Res 55:201–210. https://doi.org/10.26873/SVR-646-2018

El-Sabrout K (2018) Effect of rearing system and season on behaviour, productive performance and carcass quality of rabbit: a review. J Anim Behav Biometeorol 6:102–108. https://doi.org/10.31893/2318-1265jabb.v6n4p102-108

Erbe C, Dent ML, Gannon WL, McCauley RD, Römer H, Southall BL, Stansbury AL, Stoeger AS, Thomas JA (2022) The effects of noise on animals. In: Erbe C, Thomas JA (eds) Exploring animal behavior through sound: Volume 1. Springer, Cham. https://doi.org/10.1007/978-3-030-97540-1_13

Faye B, Waltner-Toews D, McDermott J (1999) From 'ecopathology' to 'agroecosystem health'. Prev Vet Med 39:111–128. https://doi.org/10.1016/s0167-5877(98)00149-4

Fayez I, Marai M, Alnaimy A, Habeeb M (1994) Thermoregulation in rabbits. In: Baselga M, Marai IFM (eds) Rabbit production in hot climates. CIHEAM, Zaragoza, pp 33–41. (Cahiers options Méditerranéennes; n. 8). 1. International conference of rabbit production in hot climates, 1994/09/06-08, Cairo (Egypt)

Gerencsér Z, Matics Z, Nagy I, Szendrő Z (2011) Effect of light colour and reproductive rhythm on rabbit doe performance. World Rabbit Sci 19:161–170. https://doi.org/10.4995/wrs.2011.827

Gerencsér Z, Matics Z, Szabó RT, Kustos K, Mikó A, Nagy I, Odermatt M, Atkári T, Szendrő Z (2019) Aggressiveness, mating behaviour and lifespan of group housed rabbit does. Animals 9:708. https://doi.org/10.3390/ani9100708

González Redondo P, Rodríguez FPC (2007) Producción de conejos de aptitud cárnica (chapter 30). In: Rodríguez FPC, Redondo PG (eds) Sistemas ganaderos en el siglo XXI. Universidad de Sevilla, Sevilha, pp 443–461. ISBN 978-84-472-0929-3

Hull D, Hull J, Vinter J (1986) The preferred environmental temperature of newborn rabbits. Biol Neonate 50:323–330. https://doi.org/10.1159/000242616

Kalaba ZM, Abdel-Khalek AE (2011) Reproductive performance of rabbit does and productivity of their kits in response to colour of light. Asian J Anim Vet Adv 6:814–822. https://doi.org/10.3923/ajava.2011.814.822

Kight CR, Swaddle JP (2011) How and why environmental noise impacts animals: an integrative, mechanistic review. Ecol Lett 14(10):1052–1061. https://doi.org/10.1111/j.1461-0248.2011.01664.x

Kong F, Singh RP (2011) 2 - Chemical deterioration and physical instability of foods and beverages. In: Kilcast D, Subramaniam P (eds) Food and beverage stability and shelf life. Woodhead Publishing, Philadelphia, pp 29–62. ISBN 9781845697013. https://doi.org/10.1533/9780857092540.1.29

Lebas F, Coudert P, Rouvier R, de Rochambeau H (1986) The rabbit husbandry, health and production, Animal production and health series, vol 21. FAO, Rome, p 259. F3017

Liang ZL, Chen F, Park S, Balasubramanian B, Liu WC (2022) Impacts of heat stress on rabbit immune function, endocrine, blood biochemical changes, antioxidant capacity and production performance, and the potential mitigation strategies of nutritional intervention. Front Vet Sci 9: 906084. https://doi.org/10.3389/fvets.2022.906084

Mady EA, Karousa MM, El-laithy SM, Souad AA (2018) Effect of season on New Zealand White (NZW) rabbits' behavior and reproductive and productive performance. Benha Vet Med J 35: 274–284

Marai IFM, Rashwan AA (2004) Rabbits behavioural response to climatic and managerial conditions. Arch Anim Breed 47:469–482. https://doi.org/10.5194/aab-47-469-2004

Marai IF, Ayyat MS, Abd el-Monem, U. M. (2001) Growth performance and reproductive traits at first parity of New Zealand white female rabbits as affected by heat stress and its alleviation under Egyptian conditions. Trop Anim Health Prod 33:451–462. https://doi.org/10.1023/a:1012772311177

Marai IFM, Hareeb AAM, Gad AE (2002) Rabbits' productive, reproductive and physiological performance traits as affected by heat stress: a review. Livest Prod Sci 78:71–90. https://doi.org/10.1016/j.rvsc.2013.04.012

Martino PA, Luzi F, Verga M (2007) Fungal environment in different rabbit intensive farms. Ital J Anim Sci 6:774–776. https://doi.org/10.4081/ijas.2007.1s.774

Matics Z, Gerencsér Z, Radnai I, Kasza R, Szendrő ZS (2016) Effect of different light intensities on reproductive performance, nursing behaviour and preference of rabbit does. World Rabbit Sci 24:139–144. https://doi.org/10.4995/wrs.2016.4112

Matics Z, Gerencsér Z, Kasza R, Terhes K, Nagy I, Radnai I, Zotte AD, Cullere M, Szendrő Z (2021) Effect of ambient temperature on the productive and carcass traits of growing rabbits divergently selected for body fat content. Animal 15:100096. https://doi.org/10.1016/j.animal.2020.100096

Miao ZM, Chai TJ, Qi CH, Cai YM, Liu JB, Yuan W, Yao ML (2010) Composition and variability of airborne fungi in an enclosed rabbit house in China. Aerobiologia 26:135–140. https://doi.org/10.1007/s10453-009-9150-6

Mousa-Balabel TM (2011) Using light and melatonin in the management of New Zealand white rabbits. Open Vet J 1(1):1–6

Oladimeji AM, Johnson TG, Metwally K, Farghly M, Mahrose KM (2022) Environmental heat stress in rabbits: implications and ameliorations. Int J Biometeorol 66:1–11. https://doi.org/10.1007/s00484-021-02191-0

Pan X, Wang X, Shao L, Yang J, Qin F, Li J, Zhang X, Zhai P (2022) The effects of light colour on female rabbit reproductive performance and the expression of key genes in follicular development. J Anim Sci Technol 64:432–442. https://doi.org/10.5187/jast.2022.e31

Peveler JL, Hickman DL (2018) Effects of music enrichment on individually housed male New Zealand white rabbits. J Am Assoc Lab Anim Sci 57:695–697. https://doi.org/10.30802/AALAS-JAALAS-17-000153

Quintela LA, Peña AI, Vega MD, Gullón J, Prieto MC, Barrio M, Becerra JJ, Maseda F, Herradón PG (2004) Ovulation induction in rabbit does submitted to artificial insemination by adding buserelin to the seminal dose. Reprod Nutr Dev 44:79–88. https://doi.org/10.1051/rnd:2004015

Ragab M, Elkhaiat I, Younis H, Ahmed M, Helal M (2022) Genotype by heat conditions interaction effects on growth and litter traits in rabbits. Front Vet Sci 9:1018625. https://doi.org/10.3389/fvets.2022.1018625

Silva MAJG, Ferraz PFP, Santos LMD, Ferraz GAES, Rossi G, Barbari M (2021) Effect of the spatial distribution of the temperature and humidity index in a New Zealand white rabbit house on respiratory frequency and ear surface temperature. Animals 11:1657. https://doi.org/10.3390/ani11061657

Szendrő Z, Dalle Zotte A (2011) Effect of housing conditions on production and behaviour of growing meat rabbits: a review. Livest Sci 137:296–303. https://doi.org/10.1016/j.livsci.2010.11.012

Szendrő Z, Gerencsér Z, McNitt JI, Matics Z (2016) Effect of lighting on rabbits and its role in rabbit production: a review. Livest Sci 183:12–18. https://doi.org/10.1016/j.livsci.2015.11.012

Szendrő Z, Papp Z, Kustos K (2018) Effect of ambient temperature and restricted feeding on the production of rabbit does and their kits. Acta Agraria Kaposvariensis 22:1–17. https://doi.org/10.31914/aak.2272

Theau-Clément M, Castellini C, Maertens L, Boiti C (1998) Biostimulations applied to rabbit reproduction: theory and practice. World Rabbit Sci 6:179–184. https://doi.org/10.4995/wrs.1998.339

Trocino A, Filiou E, Tazzoli M, Birolo M, Zuffellato A, Xiccato G (2015) Effects of floor type, stocking density, slaughter age and gender on productive and qualitative traits of rabbits reared in collective pens. Animal 9:855–861. https://doi.org/10.1017/S1751731114003188

Verga M, Luzi F, Carenzi C (2007) Effects of husbandry and management systems on physiology and behaviour of farmed and laboratory rabbits. Horm Behav 52:122–129. https://doi.org/10.1016/j.yhbeh.2007.03.024

Wu Y, Zhao A, Qin Y (2021) Effect of lighting schedule, intensity, and colour on reproductive performance of rabbit does. World Rabbit Sci 29:59–54. https://doi.org/10.4995/wrs.2021.14623

Yagci A, Zik B, Uguz C, Altunbas K (2006) Histology and morphometry of white New Zealand rabbit skin. Indian Vet J 83:876–880

Zeferino CP, Komiyama CM, Fernandes S, Sartori JR, Teixeira PS, Moura AS (2013) Carcass and meat quality traits of rabbits under heat stress. Animal 7:518–523. https://doi.org/10.1017/S1751731112001838

Zhang H, Qian S (2022) Network architecture for intelligent identification of faults in rabbit farm environment monitoring based on a biological neural network model. Comput Intell Neurosci 2022:6377043. https://doi.org/10.1155/2022/6377043

Rabbit Farming: Cages and Related Equipment

Luiz Carlos Machado and João Simões

Abstract

Rabbits for production purposes are currently mainly reared in individual and collective confinement. Cages and related equipment provide protection to animals, make the management of large groups more suitable, such as band systems, and improve the implementation of sanitary measures to produce safe food. Nonetheless, the intensification of rabbit production, which is highly productive, has led to new challenges related to animal welfare and health. A significant technical evolution of cage design associated with automatization of feed and water supply has been observed in the last two decades. Currently, the traceability and safety of rabbit meat and rabbit derivative products provided to consumers are fully ensured. This chapter describes the main characteristics of cage-based housing in both large- and small-scale rabbit production.

Keywords

Polyvalent cages · Rabbit parks · Nests · Manure · Automatization

L. C. Machado (✉)
Instituto Federal de Minas Gerais (IFMG), Campus Bambui, Bambui, Minas Gerais, Brasil
e-mail: luiz.machado@ifmg.edu.br

J. Simões
Animal and Veterinary Research Centre (CECAV), Associate Laboratory for Animal and Veterinary Sciences (AL4AnimalS), Department of Veterinary Sciences, School of Agrarian and Veterinary Sciences, University of Trás-os-Montes and Alto Douro (UTAD), Vila Real, Portugal

J. Simões, J. M. Monteiro (eds.), *Veterinary Care of Farm Rabbits*,
https://doi.org/10.1007/978-3-031-44542-2_7

1 Introduction

In large-scale production systems, rabbits are housed indoors in sheds and confined to cages. These housing systems allow the optimization of the production process, providing normalized meat rabbit products as a source of animal protein to feed the human population. Furthermore, similar to poultry, domestic rabbits can also be reared at familiar or subsistence levels to provide food for rural populations. For example, rabbit meat is part of the Mediterranean diet (Petracci et al. 2018).

After a period of great industrialization, when several animal species (e.g., swine, poultry, and rabbits) reached a high level of production, a new societal demand for animal compassion and welfare was requested in both urban and rural areas of the world (see Cima 2013). In rabbit rearing, as well as in other species, scientific and technological advances toward economic and environmentally sustainable production have been made. From confined environments and high animal density housing to threats to animal well-being and health, a transitive innovation period is in progress toward more ethical and green animal production.

In the present chapter, a three-decade evolution in individual and collective housing systems is described, as well as the main equipment to sustain rabbit management. Both industrial and small-scale rabbit production are addressed. We believe that knowledge from both realities can be profitable for further research to meet the expectations of producers and consumers.

2 Cages

2.1 Cages for Large-Scale Housing Systems

Almost all the production of meat and pet rabbits is currently carried out in cages (EFSA 2020). They are usually made of a wire mesh (galvanized steel), allowing easy evacuation of excrement into the cesspool (pits) and suitable washing and disinfection. Cages need to be sufficiently resistant to pressure (animal weight) and high heat caused by the flame of a burner, which burns the hair adherents to the mesh and reduces the microbial load.

Although most of the cages used are metallic, they can be made of polycarbonate material. The floor of the commercial cages consists of a metal or plastic mesh, with sufficient aperture to allow the drop excrements. Excessive aperture can cause foot lesions such as pododermatites, mainly in rabbit does (see Chapter "Staphylococcosis in Rabbits and Rabbitries") (Masthoff and Hoy 2019).

Throughout recent decades, four individual types of cages (Box 1) have been used. The polyvalent (multipurpose) flat-deck model is most commonly used in Europe (Fig. 1) and is very common in countries where cuniculture is inceptive. In fact, currently, cages with removable nests are quite popular; the does can be moved from the original cage to another previously disinfected cage, and kits can be kept on to avoid transport stress (Kaźmierczak et al. 2019). This procedure contributes to decreasing the mortality rate of weaned rabbits in the fattening phase.

Fig. 1 Conventional polyvalent flat-deck (**a**) and California (**b**; Courtesy of Laping Wu) rabbit cages

Box 1 Type of Cages

I. *Flat-deck or single plan*. They are easier to handle and monitor. Effective ventilation and illumination are achieved. They represent greater convenience for the farmer, and the removal of manure is facilitated.

II. *California or staggered plans*. Two or three planes partially overlapping with metal plates collecting feces and urine.

(continued)

Box 1 (continued)

III. *Battery system or overlapping plans.* Two or three planes overlapped vertically with the collecting plates between each cage.

IV. *Collective parks.* They allow a greater number of animals to be housed in a larger area, also allowing a lower animal density. Able to provide an adequate environmental enrichment and natural behavior of the rabbits.

Most of the cages that house the rabbit does are individual. A removable nest can be coupled, becoming multipurpose cages. This characteristic allows sequential use of the cages for maternity and for fattening the rabbit's phase. The size of these cages should be sufficiently large to allow correct displacement, feeding, and rest (Box 2) of suckling does and kits up to weaning. At this time, the pregnant does leave the maternity cage to another previously disinfected cage, and weaned rabbits are collectively raised in multipurpose cages. Other options involve moving weaned kits to another shed to house them in (enriched) collective cages or parks.

Box 2 Size of Polyvalent Cages and Nests (Width × length × height)

- Maternity cage: 40–45 × 85–95 × 30–35 cm.
- Nest place (included in the maternity cage): 24–25 cm length.
- Cages for replacement, pregnant and nonlactating does: 30–38 × 40–43 × 33–35 cm.
- Mesh: 2.5–3 mm in diameter with rectangular conformation of 73 × 13 mm.

A major improvement in the housing of breeding rabbits was the introduction of second-level platform cages, i.e., enriched cages (Fig. 2). In these cages, the rabbit does have access to an elevated platform, which will provide more space and better

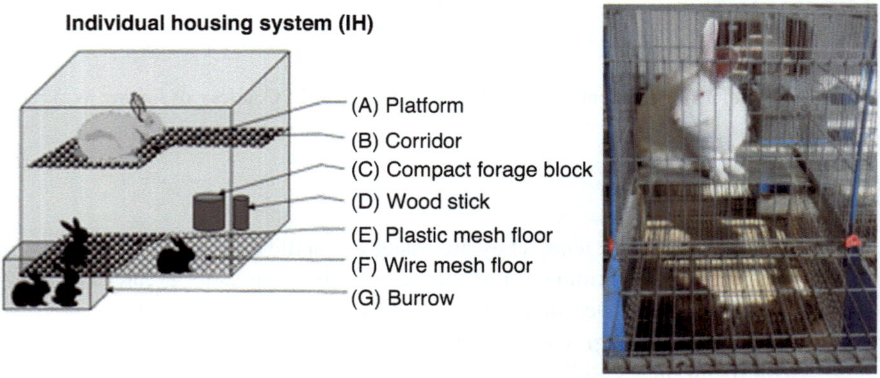

Fig. 2 Two-level platform cages from rabbit does at maternity. The schematic figure (left) was adapted from Huang et al. (2021)

Fig. 3 Indoor shed under all-in/all-out sanitation measures for 7 days

behavioral expression toward improved welfare (Hansen and Berthelsen 2000). The does can escape when the kits are leaving the nest and try numerous attempts to suckle. Additionally, these cages can also provide improvements in the skeletal system of the animal due to a greater number of possibilities for locomotion. Similar use of platforms was evaluated for flattening rabbits (Rauterberg et al. 2019). Other environmental enrichments can be done by the presence of wood sticks and gnawing blocks (Huang et al. 2021). Environmental enrichment mitigates rabbit restlessness, excessive grooming, bar-gnawing, and timidity behavior (Hansen and Berthelsen 2000) and improves species-typical rabbit behaviors (Coda et al. 2020).

The polyvalent cages are adapted to management systems working with bands or production cycles, i.e., a group of does are reared at the same physiologic phase (Izco and Martin 1992). The weaning time and fattening phase of offspring are also the same within this group. Procedures related to sanitation (all-in/all-out principles; Fig. 3), breeding, logistics related to purchases and rabbit sales, and other management will be possible or facilitated (see Chapter "Preventive Medicine and Rabbit Health Management").

The design of polyvalent cages needs to contemplate main practical aspects such as the easy insertion and removal of mobile cages and expansion of the useful area of the cage for growing rabbits in the fattening phase (Pérez-Fuentes et al. 2020). An intermediate wall between the main cage area and the nest area should be removed at weaning time. The nest is usually made in thermal material with floor fenestration, which was previously cleaned and disinfected, and placed on the nest area 3 days before kindling. Sterilized wood shavings were inserted where the does place their hair.

2.2 Cages for Small-Scale Housing Systems

Although they are widely used in modern European cuniculture, polyvalent commercial cages are not used in subsistence cuniculture and small-scale cuniculture, as well as in countries where this activity is still emerging. The lack of competitiveness in cuniculture is a major factor that makes it very difficult to purchase polyvalent cages (Machado and Ferreira 2014). In these conditions, galvanized wire cages, manufactured in common metalworks and not specialized in cuniculture, are suitable to make rabbit cages with a 20-year useful life.

For subsistence or even familiar rabbit production, where the initial investment needs to be low or residual, alternative materials to accommodate rabbits can be used, despite their poor disinfection. In these situations, wood, which can be associated with wire screens or even structures with bamboo, is usually freely acquired in rural tropical regions (Fig. 4). The floor of these cages must be 1 cm wide so that the rabbit excrement can be directed into the pits. Spaces smaller than 1 cm facilitate the accumulation of feces and the occurrence of foot traumatisms.

Fig. 4 Alternative simple pen (**a**, **b**, **d**; courtesy of Berilo de Souza Brum Junior, Instituto Federal de Educação Ciência e Tecnologia Farroupilha), and wooden and wire screen cage (**c**, **f**) used in subsistence or familiar rabbit production (Brazil)

These cages should be placed in strategic locations on the land to receive a lower incidence of rain, intense wind, and sunbeams.

The cages can also be constructed from masonry. In this situation, a second extra space inside the cage was used for feces and urine excretion. In the first days of accommodation, rabbits are placed in this temporally closed area to create habits for this purpose. This cage model can also be applied to cages dug into the ground. As an alternative, daily cleaning of the cage is needed.

2.3 Collective Cages, Pens, or Parks

In Europe, due to social pressure (see https://www.endthecageage.eu/), a transient period of moving rabbit production in cages to collective parks is planned. As an example, in 2016, Belgium started to ban rabbit cages to reach free-cage rabbit production by 2025 (Act n°. 2014024303 2014). For instance, the EU does not specifically regulate rabbit production concerning animal welfare (Ex-Post Evaluation Unit 2021), and only national legislation has been published in a few countries (Germany, the Netherlands, Sweden, and Austria).

Several studies have been conducted to solve this issue, proposing different collective cages (Fig. 5). One major problem of collective housing is related to does. Does partially housed in collective cages present higher stress, haptoglobin and cortisol serum levels, a greater rate of injured and culled females, and higher kit mortality than individually housed in higher and deeper cages or those with a platform (Pérez-Fuentes et al. 2020), which confirms the results previously obtained by Szendrő et al. (2013) for group-housed does. Housing suckling does in pairs seems to improve the welfare of their kits (Hube et al. 2023).

The use of different collective parks also evidences poor growth performance and carcass traits of fattening rabbits when compared with bicellular cages (housing two rabbits) (Xiccato et al. 2013). Nevertheless, rabbits can show less fear and incomplete behavior in collective parks (Trocino et al. 2013). Plastic tubes have been used for environmental enrichment (Fig. 6), but their effectiveness in providing better welfare to fattening rabbits in collective cages is still under discussion (Trocino et al. 2019).

More research is needed to optimize production and animal welfare in animals housed in collective parks at the industrial level, focusing on rabbit management systems and environmental factors (Trocino et al. 2022).

3 Nests

Maternal nest characteristics are determinants of kit development and survival mainly during the first 2 weeks (Hamilton et al. 1997). The main characteristics of nest buckets are mainly related to thermal insulation, easy cleanliness, and resistance to disinfectants (e.g., muriatic acid). Usually, in industrial housing systems, nest buckets are made in rigid plastic (e.g., PVC) material with dimensions of

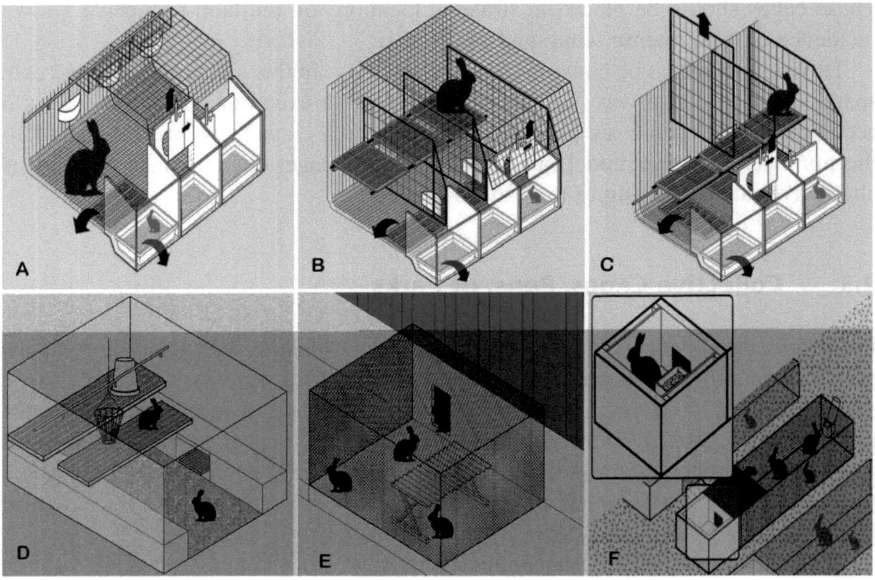

Legend: A- Polyvalent (dual-purpose) cages; B- Polyvalent enriched cage; C- Modular polyvalent elevated pen (park); D- Floor pen; E- Outdoor niche system; F- Organic housing system

Fig. 5 Evolution of individual to collective rabbit housing (EFSA, 2020). Legend: (**a**)—Polyvalent (dual-purpose) cages; (**b**)—Polyvalent enriched cage; (**c**)—Modular polyvalent elevated pen (park); (**d**)—Floor pen; (**e**)—Outdoor niche system; (**f**)—Organic housing system

approximately 40 (length) × 25 (width) × 10 (height) cm and fenestrated floors to prevent urine storage (Fig. 7).

In multipurpose cages, the nest is placed inside the allocated nest zone (see Fig. 5), while in more conventional cages, the nest remains outside. For both situations, the nest is placed approximately 10–12 cm below the floor of the mother cage (Fig. 8). This placement prevents the exit of the kits to the maternal cage in the first 10–15 days of life. The nest zone is separated from the remaining cage by a hatch that can be opened and closed (Fig. 9). The nest is placed 3 days before kindly and managed with sterilized wood shavings as described in this chapter.

For familiar or subsistence rabbit production purposes, some homemade nests based on wood, rigid plastic, or metal are suitable to accommodate kits (Fig. 10). Inside thermal conform needs be ensured, mainly on cold days. The does need adequate conditions to prepare the nest and ensure sufficient milk to nourish kits and to produce metabolic heat. In these cases, other insulating and absorbent materials, such as strip paper, can be used if wood shavings are not available.

Fig. 6 Plastic tubes (PVC; 20 cm in length and 15 cm in diameter; left) were used for environmental enrichment. Courtesy of Berilo de Souza Brum Junior, Instituto Federal de Educação Ciência e Tecnologia Farroupilha (right)

Fig. 7 Nest buckets and semiautomatic machines were used to clean, wash, and disinfect material for industrial production systems

Legend: A- Nest 12 cm above the floor of does cage; B- 5-day-old kits; 11-day-old kits. Courtesy of Teresa Tavares (B, C).

Fig. 8 Rabbit nests. Legend: (**a**)—Nest 12 cm above the floor of does cage; (**b**)—5-day-old kits; 11-day-old kits. Courtesy of Teresa Tavares (**b**, **c**)

4 Feeders

4.1 Feeders for Industrial Housing Systems

Approximately 70% of direct production costs are due to rabbit nutrition. Feeding management and efficiency depend primarily on the feeder system, which should be properly designed to reduce waste. Automatic feeding is currently used in industrial production, and the pelleted feed is distributed by dosing machines (Fig. 11). Automatic or manually supplied feeders should be able to host food only up to 48 hours after distribution, especially when rabbits are fed ad libitum. Pellet fermentation and the decrease in nutritional value are evident after this time. Restricted feed regimens have been used in growing rabbits (Martignon et al. 2022) and can mitigate waste.

The feeders can be made from plastic material and/or galvanized sheets (Fig. 12), which are materials that provide easy cleaning and disinfection. The bottom of these feeders can be drilled to evacuate fine particles (flour aspect). If the feeder is not drilled, it will be necessary to clean it weekly to remove excess fine particles, which tend to accumulate at the bottom of the equipment. Additionally, there must be an anti-waste edge that makes it difficult for rabbits to spill food.

4.2 Alternative Feeders for Small-Scale Rabbit Production

In small-scale rabbit production. Semiautomatic feeders can be used, which are supplied daily. This type of feeder allows storing a significant amount of feed in an upper feed recipient, which fills by gravity the opening used for feeding.

Fig. 9 Metallic structures of the nest to sustain nest buckets in the outside area of the cage (up) and movable metallic walls with gates to isolate the does from the nest (bottom) for conventional cages

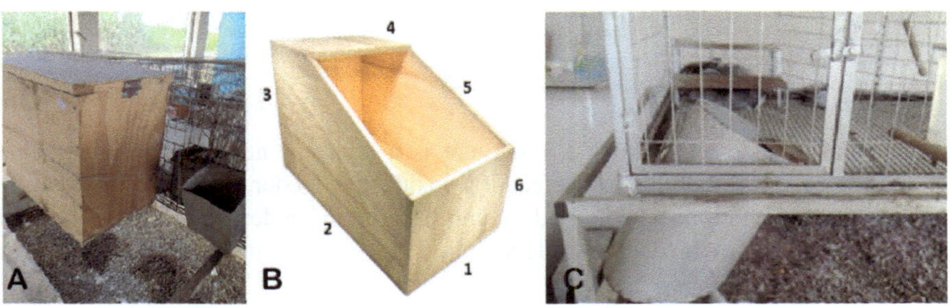

Size- 1: 25 cm, 2: 45 cm, 3: 30 cm, 4: 25 cm, 5: 15 cm, 6: 15 cm.

Fig. 10 Low-cost nests built with wood (**a** and **b**) or plastic materials (**c**) for small-scale rabbit production. Size—1: 25 cm, 2: 45 cm, 3: 30 cm, 4: 25 cm, 5: 15 cm, 6: 15 cm

Fig. 11 Automatic feed system in a rabbitry. Legend: Three outside silos containing pelletized feed for does, and early and finishing ("clean feed") phase for fattening rabbits and phase are transported (endless screws technology) inside to be distributed to feeders

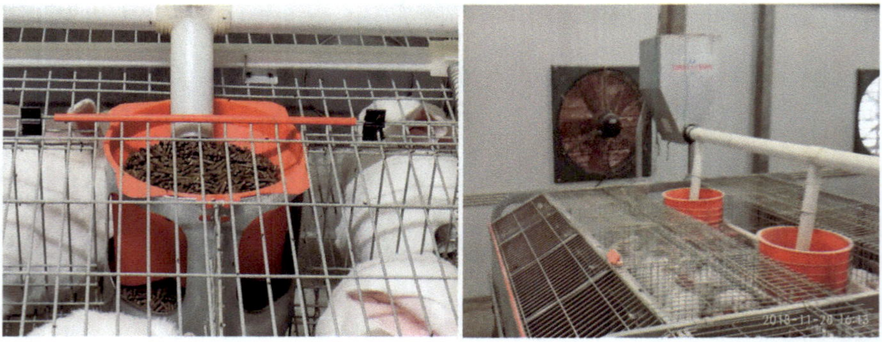

Fig. 12 Feeders used in industrial housing systems

Ceramic feeders can be used, but there is a constant need for cleaning. One limitation is the possibility of defecation inside the feeder, especially by young rabbits. An adaptation can be made using a ceramic feeder with a cross-shaped braided wire. The rabbits can introduce only their mouth to ingest the pelleted flour feed (Fig. 13).

Fig. 13 Ceramic feeders can also serve as drinkers (left) and with cross-shaped braided wire (right)

Fig. 14 Chlorinated water was stored in tanks (left) and two water-powered medicators (right) to supply different lines of production. Courtesy of José Manuel Monteiro

5 Drinkers

Rabbitry water can be supplied by public water systems and owner wells. The drinking water should be fresh, and its chlorination is crucial. Automatic drinkers are used in industrial housing systems. Water tanks are installed in sheds to ensure a continuous source and adequate pressure of water. Chemical and bacteriological control every 6 months is required by norm. Tanks, pipelines, and drinkers need to be regularly cleaned and disinfected (EFSA 2020). Additionally, water-powered medicators are accoupled to the water system for treatments (Fig. 14).

Automatic drinkers are made up of a low-pressure system, which must be adequate and controlled to facilitate the intake of water by rabbits. Water shedding from drinkers occurs if the water pressure is high. Despite the wastage, the moisture of the shed increased.

Fig. 15 Bowl (left) and nipple (right) drinkers

Nipple (pipe) or bowl drinkers can supply water to rabbits (Fig. 15). Nipple drinkers should be located approximately 18–22 cm from the cage floor. One disadvantage of this system is the greater waste of water, which can increase moisture in the farm. Bowl drinking systems can reduce water waste and must be installed approximately 14 cm from the cage floor. Their disadvantage is related to organic material accumulation, requiring frequent cleanliness. Nipple and bowl drinkers are triggered by suction and pressure exerted by the rabbits.

The temperature of the water is also an important issue. Warm water reduces its ingestion and consequently the feed intake. By norm, the water intake of rabbits is double the feed intake. In industrial production systems, the water tanks are inside the sheds, but in small-scale production, insulating protection may be useful for this purpose.

6 Resting Mats (Footrests)

Cages housing breeding rabbits, especially does, must be equipped with resting mats to avoid problems related to calluses formation and pododermatitis and contribute to improving their welfare in general (Rosell and de la Fuente 2013; Dalmau et al. 2020). These pads, based on soft and resistant rubber or plastic material, are composed of a mesh with sufficient aperture for suitable feces evacuation (Fig. 16). In small-scale production, PCV mesh squares have been used for this purpose.

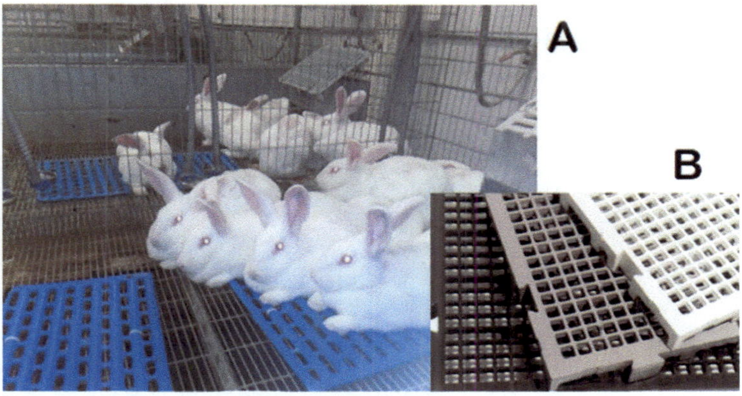

Fig. 16 Resting mats for rabbit does

7 Cleaning and Disinfection Equipment

Biosecurity programs are crucial for the implementation of rabbitries (see Chapter "Preventive Medicine and Rabbit Health Management"). One of the main internal biosecurity measures is related to the periodical cleanliness and disinfection of materials and equipment.

Disinfection reduces microbial load, as well as other biotic agents, and should be carried out whenever an installation, equipment, and utensils are used by other animals. Chlorine disinfectants or quaternary ammonium compounds are largely used to disinfect common materials used daily, such as buckets and small equipment sensitive to corrosion (e.g., metals), respectively. Cresols and phenolic substances are highly corrosives and should be limited to structural housing (e.g., corridors' floor). These last disinfectants are also irritating for animals and should be used in the absence of rabbits (see Fig. 3). General cleaning equipment, such as spatulas, brushes, and jet pumps, is needed.

Nests should be sanitized and disinfected after use (see Fig. 7). By norm, a rest period of 7 days is required in each production cycle. The cages can also be washed and disinfected periodically, mainly when the incidence of digestive diseases in growing animals is significant.

To eliminate hair from the mesh of cages, a flame thrower was used once or two times per week (Fig. 17). This is an important hygienic measure to break the transmission of fungi and bacteria between rabbits.

Fig. 17 Flame thrower burning hair adhered to the metal mesh

8 Rabbit Transport Equipment

The transport of rabbits inside the sheds, withdrawal of fattening rabbits to slaughterhouse, and replacement breeding females (in some rabbitries) are ensured by transport cages. At weaning, the pregnant does were transported to the other shed through a common corridor using a specific transport box, as reported in Fig. 18, minimizing their stress during transport.

At the time of withdrawal, fattening rabbits were manually placed in cages to form groups. It is recommended to use a specific exit hall with a pier to maximize the biosecurity of the shed.

In small-scale production, where the feeding system is not automatized, small wagons are adapted to transport feed and other materials.

Fig. 18 Transport box to move rabbits inside the sheds (left) and a pier for fattening rabbits' removal to slaughterhouse (right). Note the presence of a freezer (left photo; right quadrant top) to accommodate rabbits' cadavers and that the pier is near the silos (right photo). Both the pier and silos are located close to the main entrance of the farm to minimize truck movement inside the clean area

9 Manure Pits and Management

Rabbits' excrements fall into a pit beneath flat-deck cages. The pit characteristics depend on the system and the frequency of waste removal indoors in the shed. The inside removal of manure can be done manually or mechanically to an outside cesspool or similar.

The deep tanks, up to 5 m deep, have a large storage capacity, allowing the accumulation of waste for months or even up to 2–3 years, removed manually or with a tractor with a load shovel. This type of pit is in disuse regarding industrial production systems. When used, calcium superphosphate or calcium dihydrogen phosphate is periodically scattered over manure to mitigate odors, control gas emission (NH_3), and even control bacterial growth (Estellés et al. 2014). Additionally, insecticide and larvicide agents can be used. Semi-deep pits, approximately 1–2 m deep, require manure removal every 3–4 months. Manure treatment should be carried out in the shed during its stay on the farm.

The ammonia levels inside the shed are generally low and below acceptable values (Calvet et al. 2011). Only an increase in NH3 levels occurs at manure removal. Small-depth pits, 20–100 cm deep, with daily mechanical manure

Fig. 19 Manure pit system of a shed. Legend: Rabbits' excrements (left up), which are dragged by a mechanical drag (left down) to outside of the shed (right). Outside the shed, the manure is also mechanically transported into a cesspool for further processing

extraction were implemented in sheds, mainly during the last two decades (Fig. 19). An alternative is manure removal at the end of the fattening cycle.

Cleanliness with pressurized water and periodical disinfection is recommended, but in regions with water resource limitations, judicious water use is needed.

For small-scale cuniculture or family and subsistence farming, the use and cleaning of the site below the cages can be simplified. A layer of sawdust to receive waste is recommended. Small channels can be built with a slight inclination to drive urine directly to a collector.

Further processing of rabbit manure can be used as plant fertilizer (Li et al. 2022) or for biogas (methane) production (Zapata Martínez et al. 2019).

10 Conclusions

Overall, rabbit does remain currently housed in individual cages to ensure performance and prevent health problems derived from stress and aggressivity. Polyvalent cages with one level are prevalent, but enriched cages with platforms seem to be more appropriate. Additionally, bicellular cages comporting two does or even modular cages are promising.

At the technological level, a relevant automatization and instrument related to supplying feed, sanitation, and manure removal and treatment allow the management of industrial rabbit production. In the last 10 years, intense research on cage design has been performed to improve animal welfare. Nevertheless, some issues about the relation between rabbits' performance and being well need to be refined regarding the expectations of both producers and consumers.

Acknowledgments The authors thank Springer for the use of the AJE Digital Editing. This work was supported by projects UIDP/CVT/00772/2020 (CECAV) and LA/P/0059/2020 (AL4AnimalS), funded by the Portuguese Foundation for Science and Technology (FCT).

References

Act n°. 2014024303 (2014) Royal Decree concerning the welfare of rabbits in farms, 29/06/14. Arrêté royal relatif au bien-être des lapins dans les élevages, 29 juin 2014. Accessed from https://etaamb.openjustice.be/fr/arrete-royal-du-29-juin-2014_n2014024303 (05-03-2023)

Calvet S, Cambra-López M, Estellés F, Torres AG (2011) Characterisation of the indoor environment and gas emissions in rabbit farms. World Rabbit Sci 19:49–61. https://doi.org/10.4995/wrs.2011.802

Cima G (2013) Animal welfare continues improving with demands from buyers. J Am Vet Med Assoc 242:1314–1317. https://doi.org/10.2460/javma.242.10.1314

Coda KA, Fortman JD, García KD (2020) Behavioral effects of cage size and environmental enrichment in New Zealand White rabbits. J Am Assoc Lab Anim Sci 59:356–364. Advance online publication. https://doi.org/10.30802/AALAS-JAALAS-19-000136

Dalmau A, Moles X, Pallisera J (2020) Animal welfare assessment protocol for does, bucks, and kit rabbits reared for production. Front Vet Sci 7:445. https://doi.org/10.3389/fvets.2020.00445

EFSA Panel on Animal Health and Welfare (AHAW), Saxmose Nielsen S, Alvarez J, Bicout DJ, Calistri P, Depner K, Drewe JA, Garin-Bastuji B, Gonzales Rojas JL, Gortázar Schmidt C, Michel V, Miranda Chueca MÁ, Roberts HC, Sihvonen LH, Spoolder H, Stahl K, Velarde Calvo A, Viltrop A, Buijs S, Edwards S, Winckler C (2020) Health and welfare of rabbits farmed in different production systems. EFSA J 18:e05944. https://doi.org/10.2903/j.efsa.2020.5944

Estellés F, Cambra-López M, Jiménez-Belenguer A, Calvet S (2014) Evaluation of calcium superphosphate as an additive to reduce gas emissions from rabbit manure. World Rabbit Sci 22:279–286. https://doi.org/10.4995/wrs.2014.3223

Ex-Post Evaluation Unit (2021) Animal welfare on the farm – ex-post evaluation of the EU legislation: Prospects for animal welfare labelling at EU level. European Parliamentary Research Service, E 662.643 – June 202. Accessed from https://www.europarl.europa.eu/RegData/etudes/STUD/2021/662643/EPRS_STU(2021)662643_EN.pdf (05-03-2023)

Hamilton HH, Lukefahr SD, McNitt JI (1997) Maternal nest quality and its influence on litter survival and weaning performance in commercial rabbits. J Anim Sci 75:926–933. https://doi.org/10.2527/1997.754926x

Hansen LT, Berthelsen H (2000) The effect of environmental enrichment on the behaviour of caged rabbits (*Oryctolagus cuniculus*). Appl Anim Behav Sci 68:163–178 https://doi.org/10.1016/s0168-1591(00)00093-9

Huang Y, Breda J, Savietto D, Debrusse AM, Combes S, Fortun-Lamothe L (2021) Part-time grouping of rabbit does in enriched housing: effects on performances, injury occurrence and enrichment use. Animal 15:100390. https://doi.org/10.1016/j.animal.2021.100390

Hube D, Bill J, Knop ES, Herbrandt S, Kemper N, Fels M (2023) Physical injuries and hair corticosterone concentration in rabbit kits from single- and group-housed does kept on a commercial farm. Animals 13:196. https://doi.org/10.3390/ani13020196

Izco ML, Martin XI (1992) [Band management]. XVII Symposium de Cunicultura, Salamanca, Spain; September

Kaźmierczak S, Cwojdzińska A, Górecki MT (2019) Behaviour of domestic rabbits during 2 weeks after weaning. Arch Anim Breed 62:49–57. https://doi.org/10.5194/aab-62-49-2019

Li R, Hao H, Yang C, Wang L, Wang H (2022) Rabbit manure compost as a peat substitute for compound growing media: proportioning optimization according to physiochemical characteristics and seedling effects. Front Plant Sci 13:1008089. https://doi.org/10.3389/fpls.2022.1008089

Machado LC, Ferreira WM (2014) Opinião: organização e estratégias da cunicultura Brasileira - buscando soluções. [Opinion: organization and strategies of Brazilian rabbitry – seeking solutions]. Rev Bras de Cunicult 6:1–31

Martignon M, Burel C, Guinebretière M, Postollec G, Huonnic D, Boilletot E, Michel V, Gidenne T (2022) Feeding behaviour of the growing rabbit fed freely or restricted, and impact on performances and digestive organ. World Rabbit Sci 30:119–130. https://doi.org/10.4995/wrs.2022.14513

Masthoff T, Hoy S (2019) Investigations on the influence of floor design on dirtiness and foot pad lesions in growing rabbits. Animals 9:354. https://doi.org/10.3390/ani9060354

Pérez-Fuentes S, Muñoz-Silvestre A, Moreno-Grua E, Martínez-Paredes E, Viana D, Selva L, Villagrá A, Sanz-Tejero C, Pascual JJ, Cervera C, Corpa JM (2020) Effect of different housing systems (single and group penning) on the health and welfare of commercial female rabbits. Animal 14:1270–1277. https://doi.org/10.1017/S1751731119003379

Petracci M, Soglia F, Leroy F (2018) Rabbit meat in need of a hat-trick: from tradition to innovation (and back). Meat Sci 146:93–100. https://doi.org/10.1016/j.meatsci.2018.08.003

Rauterberg SL, Bill J, Kimm S, Kemper N, Fels M (2019) Effect of a new housing system on skin lesions, performance and soiling of fattening rabbits: a German case study. Animals 9:650. https://doi.org/10.3390/ani9090650

Rosell JM, de la Fuente LF (2013) Assessing ulcerative Pododermatitis of breeding rabbits. Animals 3:318–326. https://doi.org/10.3390/ani3020318

Szendrő Z, Mikó A, Odermatt M, Gerencsér Z, Radnai I, Dezséry B, Garai E, Nagy I, Szendrő K, Matics Z (2013) Comparison of performance and welfare of single-caged and group-housed rabbit does. Animal 7:463–468. https://doi.org/10.1017/S1751731112001760

Trocino A, Majolini D, Tazzoli M, Filiou E, Xiccato G (2013) Housing of growing rabbits in individual, bicellular and collective cages: fear level and behavioural patterns. Animal 7:633–639. https://doi.org/10.1017/S1751731112002029

Trocino A, Zomeño C, Filiou E, Birolo M, White P, Xiccato G (2019) The use of environmental enrichments affects performance and behavior of growing rabbits housed in collective pens. Animals 9:537. https://doi.org/10.3390/ani9080537

Trocino A, Menegon F, Zomeño C, Pasqualin D, Cunial G, Xiccato G, Pirrone F, Bertotto D, Bortoletti M, Dorigo F, Lavazza A, Di Martino G (2022) A pilot study about on-farm assessment of health and welfare in rabbits kept in different housing systems. Front Vet Sci 9: 936643. https://doi.org/10.3389/fvets.2022.936643

Xiccato G, Trocino A, Majolini D, Tazzoli M, Zuffellato A (2013) Housing of growing rabbits in individual, bicellular and collective cages: growth performance, carcass traits and meat quality. Animal 7:627–632. https://doi.org/10.1017/S175173111200198X

Zapata Martínez O, Serrano A, Aguilera Flores MM, Villa-Gómez DK (2019) Rabbit manure as a potential inoculum for anaerobic digestion. J Environ Sci Health A Toxic/Hazard Subst Environ Eng 54:943–950. https://doi.org/10.1080/10934529.2019.1610293

Welfare, Behavior, and Housing of Rabbits

Carmen Cano, Patricia Carulla, and Arantxa Villagrá

Abstract

Animal welfare is defined as the physical and mental state of an individual in relation to its environment. Currently, society's increasing concern about animal welfare on farms calls for regulation. For this, it is important to determine the optimal requirements in rabbit production from the three fundamental spheres of welfare based on the response of the animals. The three spheres of animal welfare are biological functioning and health, affective states or cognitive judgment, and natural living. In recent years, there has been a shift from avoiding negative experiences to exploring positive experiences for animals. In addition, the welfare of rabbits is greatly influenced by the housing conditions in which they are kept. Therefore, scientific research is looking for a balance between welfare and productivity through new housing designs, prioritizing the ethology of the species. Different types of flooring, cage sizes, densities, and group sizes are studied depending on the physiological and psychological needs of the productive stage of the animal. Additionally, highlighting the different environmental enrichments that can be included in the housing. This chapter also focuses on the current trend of implementing alternative systems where the aim is to improve animal welfare.

Keywords

Welfare · Breeding does · Fattening rabbits · Housing · Environmental enrichment · Cage design

C. Cano · A. Villagrá (✉)
Centro de Tecnología Animal - Instituto Valenciano de Investigaciones Agrarias (CITA-IVIA), Segorbe, Castellón, Spain
e-mail: villagra_ara@gva.es

P. Carulla
Institute of Animal Science and Technology, Universitat Politècnica de València, Valencia, Spain

J. Simões, J. M. Monteiro (eds.), *Veterinary Care of Farm Rabbits*,
https://doi.org/10.1007/978-3-031-44542-2_8

1 Introduction

Animal welfare is a requirement currently demanded by society, covering from the physical to the mental state (Brambell 1965). Animal welfare can be defined as a balance between animal and the environment, both internal and external (Vandenheede 2003). This balance refers to the stability of an organism that, through physiological and ethological regulation, fluctuates within tolerable limits for the organism (Broom and Johnson 1993). If any factor, both internal and external, exceeds these tolerable limits supported by the animal, this balance can be disturbed (Vandenheede 2003). This, together with the inability of the organism to return to this equilibrium with the environment, generates a stress response, and the welfare of the animal may be compromised (Broom and Johnson 1993). This basis is supported by a conceptual framework in which difficult or inadequate adaptation can lead to welfare problems in animals (Hemsworth and Coleman 2011). These alterations would become a stress agent for the animal, which can come from the environment, production densities and group sizes, production systems, housing, and nutrition (Lee 1993).

Therefore, especially in Europe, society and consumers are becoming increasingly demanding improved animal welfare in livestock production systems (Szendrő and McNitt 2012). This has promoted the publication of directives by the Council of the European Union (Directive 98/58/EC) for the protection of farm animals in different species. However, to date, in the case of rabbit farming, no regulations have been published by the European Union (Szendrö et al. 2019), despite having reached more than 18 drafts of the same. The lack of a European Union regulation for rabbit production has prompted several countries to elaborate their own legislation or production recommendations, such as the Netherlands (2006), Hungary (2009), Belgium (2014), and Germany (2014). In fact, a proposal for a protocol adapted from the Welfare Quality system to assess animal welfare in rabbit farms in Spain and Portugal has even been published (Dalmau et al. 2020). Despite the possible limitations that this protocol may have, its use has highlighted some problems in some facilities, such as the available spaces that rabbits have in the cages (Botelho et al. 2020) and the limited attention that has been paid to its possible effects on the behavior of rabbits.

One of the causes of this lack of specific regulation could be the existing disagreement on the evaluation of animal welfare. In recent years, there has been a shift from avoiding negative experiences to exploring positive experiences for animals. Recognizing that good welfare, a "good life," is not only about avoiding negative states but also about promoting positive experiences and emotional states (Edgar et al. 2013; Lawrence et al. 2019). Positive animal welfare and its assessment emphasize the resources valued by animals, positive emotions, and natural behaviors that animals are motivated to perform (Lawrence et al. 2019).

This chapter aims to summarize the behavioral and physiological needs of rabbits, both adult and young, as well as review the existing literature in regard to housing conditions for these animals.

2 Concepts on Animal Welfare

It is essential to know the indicators of animal welfare. These indicators can be grouped into three fundamental conceptions (represented by spheres) initially defined by Fraser et al. (1997). The three fundamental spheres are represented in Fig. 1 and are:

- Biological functioning and health, in which good health indicates the correct physiological functioning of the animal
- Affective states or cognitive judgment, considering how the animal feels as it experiences and perceives its environment
- Natural living, referring to the evolutionary adaptation undergone by the animal to its environment

For all the factors mentioned above, it is of great importance to determine the optimal requirements in rabbit production from the three spheres of welfare based on the response of the animals. This requires detailed knowledge of the physiology and behavior of rabbits (Szendrő and Dalle Zotte 2011; Szendrő and McNitt 2012).

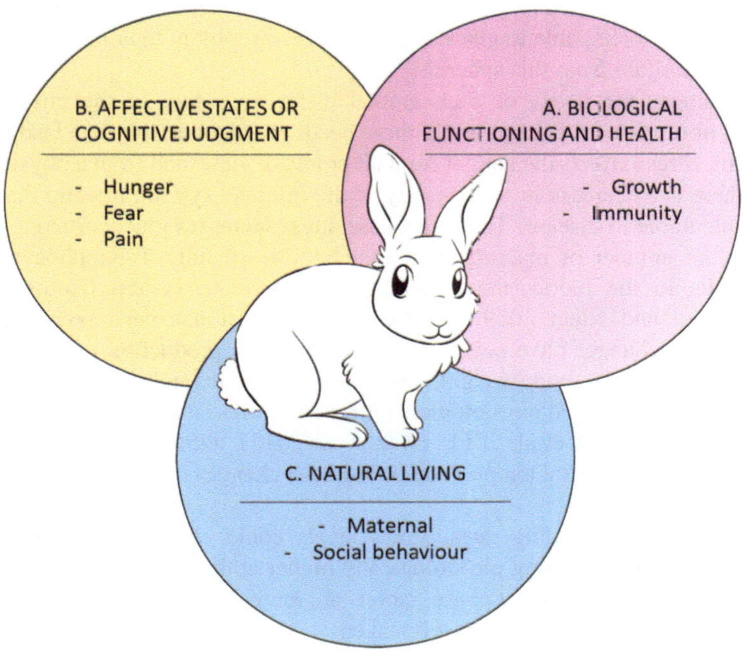

Fig. 1 The three key areas of animal welfare

2.1 Physiological Functioning and Health

Rabbits are prey animals, not predators, and have been domesticated for a relatively short time. Therefore, they remain very similar to their wild counterparts in terms of anatomy, physiology, behavior, psychology, and social requirements (Debra 2011).

Anatomically, rabbits have very powerful muscles in the hind limbs, but their skeleton represents a relatively low proportion of their body weight (Richardson 2000). This is very important in the correct handling of the animals on the farm. When a rabbit is handled incorrectly, it can kick with its hind legs and cause serious injuries or even a fractured spine (Keeble 2002). In daily handling on the farm, rabbits should be held firmly so that their body weight is supported at all times from below (Debra 2011), immobilizing their hind legs so that they cannot kick or jump, avoiding any type of injury.

In terms of physiology and as a generality, rabbits are herbivores and have continuous growth of teeth. Therefore, the composition of the diet is essential to maintain healthy teeth, without overgrowth, and adequate intestinal function (Debra 2011). In this sense, cecotrophs are a digestive strategy of the rabbit that allows it to take advantage of the nutrients resulting from cecal fermentation that have a high biological value (Davies and Davies 2003; Romero 2008) and, therefore, are important for their health. Cecotrophs, also called soft feces, are produced in the anus from partially fermented matter in the cecum and are rich in amino acids and vitamins, among others. The cecotrophs are ingested by the rabbit directly from the rectum and are swallowed whole without chewing (Davies and Davies 2003). For all the above mentioned, not being able to gnaw or perform cecotrophy in their daily life directly affects the welfare from this sphere.

In addition, the needs of each animal differ according to the physiological moment or reproductive stage, and if these needs are not met, they can lead to stress problems. Stress affects the animal from all points of view, and from a physiological view, there is a decrease in the capacity of the immune system, making the animal more vulnerable to disease. This is because stress increases glucocorticoids, which reduces the number of macrophages, lymphocytes (mainly T lymphocytes), and eosinophils in the blood, thus inhibiting the immune system (Romero 2008; Cunningham and Klein 2009). At the same time, it has been seen that these physiological changes have negative consequences on production parameters such as feed consumption, body weight, feed conversion rate, and meat quality (Téllez 2022). Therefore, a good physiological indicator of stress in rabbits is fecal or blood glucocorticoids (Buijs et al. 2011; Prola et al. 2013), although the most common technique is to take rectal temperature or to detect changes in productive parameters (Téllez 2022).

In the case of breeding does, stress levels could change depending on the reproductive moment, being physiologically higher at the time of parturition (Prola et al. 2013). Therefore, in that phase, rabbits are more susceptible to the surrounding environment, and special care must be taken.

One of the main problems affecting fattening rabbits is heat stress, among others. Rabbits are very sensitive to high environmental temperatures due to the presence of thick fur and the absence of sweat glands (Marzoni and Mori 1992). Heat stress negatively affects welfare and adaptation, food consumption and utilization, immunity, health status, and growth (Oladimeji et al. 2022).

2.2 Cognitive and Affective States

Traditionally, all studies have focused on ensuring good health and improving productive parameters, leaving the other two spheres of welfare. However, in recent decades, studies related to emotions and normal animal behavior have increased. In the case of the sphere of affective states or cognitive judgment, we study how the animal feels and perceives their environment.

Animals must be free of hunger, fear, pain, stress, anxiety, and any other prolonged and intense negative state. Furthermore, it has been shown that these negative affective states lead to increased morbidity and mortality due to persistent activation of the stress response (Walker et al. 2012). For a long time, it was thought that providing animals with a good life was only guaranteeing the absence of these negative states (Fraser et al. 1997; Paul et al. 2005). However, it is also important to ensure positive states such as comfort or pleasure (Carter 2001; Paul et al. 2005).

In humans, verbal information on emotional state has generally been considered the "gold standard." However, feelings and emotions cannot be measured directly in animals. For this purpose, it is possible to measure behavioral and physiological changes related to emotional responses to external stimuli. These tools are important but have limitations, such as the difficulty of interpretation, knowing if their feelings are pleasant or unpleasant and how much it has affected them emotionally (Paul et al. 2005).

One emotional state to which rabbits are predisposed is fear due to their condition as prey. Adequate management is necessary on farms to reduce this fear as much as possible. According to Lanier (2008), three conditions interact in this emotion: genetics, frequency of contact with humans, and the quality of that contact. Rabbits see humans as predators (Téllez 2022), but on farms, there are a great variety of handling tasks, such as inseminations, farrowing and birth controls, litter standardization (Trocino and Xiccato 2006), veterinary care and cage changes. In these moments, it is important for farmers to have adequate treatment to reduce fear and stress.

Animal contact with humans begins soon after birth, and early handling has been shown to help develop a positive relationship with the farmer and reduce fear levels (Trocino and Xiccato 2006). Methodical handling of younger rabbits during lactation habituates them to the farmer, making them less fearful and stressed, showing lower mortality while growing and higher fertility later (López et al. 2021). In contrast, handling pregnant or lactating does causes increased mortality in rabbits at birth and weaning compared to unmanipulated rabbits (Marzoni and Mori 1992).

In addition to the emotion of fear of humans, rabbits are hierarchical animals, so subordinate rabbits are constantly under fear and stress. Dominant rabbits have lower levels of aggression and experience more social interactions. In contrast, subordinate rabbits receive aggression from dominant rabbits and lack social support, which induces a negative affective state in these animals. Additionally, subordinates present a higher corticosteroid level and heart rate and significantly shorter lives (Walker et al. 2012).

The assessment of the affective and cognitive state of animals provides a basis for the future study of animal welfare (Mendl and Paul 2019). This plays an important role in the ethical concern for the quality of life of animals (Fraser et al. 1997).

2.3 Natural Living

The third sphere of animal welfare is natural living, which refers to the natural behavior of animals and the evolutionary adaptation they have suffered due to changes in their environment. As mentioned above, rabbits are a prey species and were also domesticated relatively recently (Debra 2011), so the behaviors observed in rabbits are related to this aspect.

Rabbits are very curious animals and exhibit daily exploratory behavior. This can be observed when a new object is introduced into the pen, and they show great interest, exploring it for long periods of time. In addition, upon an auditory stimulus, they stand on their hind legs to examine the situation. Rabbits quickly learn to differentiate sounds, which means that if food is always introduced into the room with the same bag, it provokes an immediate response in the animals. Additionally, if they have space, they run, jump, and scream, presenting play behavior (Ricca et al. 2018). Additionally, they present significant maternal behavior.

In nature, rabbits live in colonies in burrows and exhibit social and maternal behaviors (González-Redondo et al. 2015; Benedek et al. 2020), but they also exhibit an important hierarchy (Walker et al. 2012). Wild rabbits exhibit a set of behaviors during peripartum and litter rearing: digging a burrow, introducing dry plant material into, plucking hair to line the nest, calving in the burrow, and suckling the litter once a day. This maternal behavior must be taken into account in rabbit farms (González-Redondo et al. 2015).

The wild rabbit makes its nest inside the burrow, but in rabbit farming, it is replaced by a nest box that the rabbit farmer puts in the cage. On farms, the rabbits accept the nest box provided to make the nest with material (straw, wood shavings or litter) that the farmer puts inside and then they cover it with hair that they pull out from the abdomen and chest (González-Redondo et al. 2015). If they are deprived of these innate behaviors by not providing them with a nest, their stress level increases, and their welfare is damaged (Dalmau et al. 2020; Oliveira et al. 2022). Nest quality, measured by the abundance of hair and the degree to which it covers young rabbits, correlates positively with the number of weaned rabbits, while poorer quality nests have higher mortality and poorer growth (Szendrö et al. 1988; López et al. 2021). In general, nest quality improves with the order of parturition (Ross et al. 1956), but it is also important to regulate environmental conditions because if it is cold, it tends to deposit more hair in the nest (Szendrö et al. 1988).

Once the farmer makes the nest available to the doe 3–4 days before parturition, the maternal-filial relationship begins. From that moment on, the female is in charge of building the nest, making a series of visits to the nest that reaches its peak on the day of parturition and the following day (González-Redondo et al. 2015). At the time of parturition, females present a high level of physiological stress and are more susceptible to the surrounding environment, so special care must be taken with the nest (Prola et al. 2013). After parturition, visits decrease as the age of the suckling rabbits increases (Fernández-Carmona et al. 2005; González-Redondo et al. 2015).

The rabbits were born hairless, with closed eyes and retracted ear flaps. In addition, they have poor motor coordination and generate little endogenous heat, so they survive the first few days thanks to the nest temperature. They open their eyes at 10 days of age (Hudson and Distel 1982) and progress toward homeothermia as they grow.

Maternal care is highly efficient although the doe has limited contact with the rabbits during lactation, practically no care and little cleaning. Doe suckles the rabbits for the first time immediately after parturition and thereafter suckles regularly once a day, before the onset of the birthing period (González-Mariscal et al. 2013a). This brief once-daily suckling is an evolutionary defense mechanism to minimize the risk of being preyed upon (Lloyd and McCowan 1968). When the doe enters the nest, the suckling rabbits are awake and ready to initiate suckling. This behavior is due to synchronization triggered by food (Caba and González-Mariscal 2009; Morgado et al. 2011). Nursing lasts for a few minutes and abruptly ceases when the doe leaves the nest until the next day (González-Mariscal et al. 2013b).

At approximately 16–18 days of age, the young rabbits begin to leave the nest, recognize the territory, investigate the floor and walls, and interact with each other and with the mother. It is very common to observe grooming behaviors, scampering, climbing on the mother to access the drinking or feeding trough, etc. One way to improve the welfare of the doe during this period until weaning occurs is the use of hiding places or platforms that allow the female to isolate herself from the litter (González-Redondo et al. 2015).

Weaning is a stressful process for rabbits. In rabbit husbandry, it is done abruptly, separating the rabbits from the dam at 28–35 days of age. It is considered close to natural because the doe produces little milk, especially if she is simultaneously pregnant (Lincoln 1974; Hudson et al. 1996).

Currently, farms tend to use multipurpose cages so that the rabbits are fattened in the same cage in which they were born to reduce the stress of weaning. In this case, the dam is moved to a clean cage and room for the next birth (González-Redondo et al. 2015).

During the first weeks after weaning, Kimm et al. (2021) observed that rabbits preferred to huddle in the outdoor area and that elevated platforms favored resting and comfort behaviors.

In fattening, conventional cages (6–9 rabbits) promote specific behaviors (locomotor, social and exploratory). In addition, if locomotion is promoted by enlarging the space, it increases the strength and weight of the hind leg bones and may increase the proportion of the hind third (López et al. 2021). Additionally, Feng et al. (2022)

observed that with good environmental enrichment, rabbits increase feeding and drinking time, and exploratory behavior is further promoted. Environmental enrichments, such as wooden elements, help to express natural gnawing behavior and do not worsen productive parameters, while the platform broadens the repertoire of behaviors but worsens hygiene (López et al. 2021). However, being prey animals, when faced with a dangerous situation, they tend to flee, hide, or freeze and may even remain very still instead of screaming if they feel pain (Magnus 2005; Debra 2011).

As noted above, the welfare of rabbits is greatly influenced by the housing conditions in which they are kept. On the one hand, the nest box is a very important aspect to take into account in breeding rabbits, and on the other hand, the cage design will have a great influence on both flattering and breeding rabbits. The most important aspects of housing will be detailed below.

3 Housing

As in other species, the size of farms is increasing and, therefore, current production systems must be evaluated and investigated in search of improvements related to animal welfare, but always from a productive context, considering also the sustainability of such production.

In rabbit rearing, one of the strategies for specialization of production was the adoption of wire mesh as cage material, thus reducing the hours of handling and facilitating social contact between animals (Morisse and Maurice 1994). Currently, however, changes in housing design must be considered, taking into account factors such as comfort and the possibility of expression of species-specific behavior. Therefore, it is necessary to find a balance between productivity and animal welfare.

Some of the countries that have already developed their own legislation in rabbit farming have opted for the elimination of cage production through a gradual transition. A few farms have started to use parks for growing rabbits, where they are housed in larger groups. This system is usually the so-called all-in all-out system, and in this case, the walls of the cages, usually four cages, are removed to form a single space for fattening rabbits (32–36) (EFSA 2020). This alternative system has also been developed for the housing of does, such as those proposed by Maertens and Debie (2017). Park systems can have a wire mesh or plastic floor and be equipped with platforms (Zomeño et al. 2018).

The all-in all-out system is the so-called dual-purpose cage, in which does are moved to another cage after weaning, while their litter remains in the same cage until the end of fattening. Table 1 shows the typical conventional cages, enriched cages, and parks. They are all dual-purpose cages, except for the bicellular cage and the cage for young females, which are implemented in commercial farms (EFSA 2020).

It is important to consult the studies conducted and the results obtained in scientific research for future policy decisions, as the production system has a great effect on animal welfare but also on the economics of the farmer. Moreover, a new

Table 1 Housing sizes available on commercial rabbit farms for different rabbit production times

Housing systems	Width (cm)	Length (cm)	Height (cm)	Total available surface (cm^2)
Bicellular conventional cage for growing rabbits (1–2)	25.4	44	28	1200
Conventional cage for young or nonpregnant female and growing rabbits	38	43.5–66	28.41	1650–2510
Dual-purpose (all-in all-out system) conventional cage	38	87–102	32–39	3300–3900
Wider version of dual-purpose conventional cage	46	95–102	35	4370–4700
Dual-purpose enriched cage with wire mesh platform (width 20 cm)	38–46	95–102	60–65	4370–5600
Dual-purpose enriched cage with plastic-mesh platform	46–52.5	102	65–80	5600–6400
Dual-purpose elevated pen (park) for 4 does or for growing rabbits (32–36) with plastic-mesh platform (width 20–25 cm)	180–200	80–102	Open top	Total: 18,000–25,400 Per doe: 4500–6350 Per growing rabbit: 563–800

system may improve one aspect of welfare but worsen another, so a hierarchy in the importance of different welfare aspects would be useful.

For all these reasons, it is recommended to review in depth how the type of housing used in production systems can affect welfare and the expression of natural behavior and which modifications will improve both. Following this, the housing for breeding rabbits will be detailed, both conventional housing as the new alternative system and, finally, growing rabbits.

3.1 Breeding Does

As discussed above, in nature, the rabbit is a gregarious animal with a strong social instinct (Thurston et al. 2018). However, the most common housing system for breeding does is individual cage housing with wire mesh floor, whose surface area is between 3300 and 3900 cm^2 (Szendrö et al. 2019), and with the following measures: 83 × 38 × 32 cm (length × width × height; Prola et al. 2013). They have a feeder, drinker, and nest box.

In addition, individual housing in production systems limits the number of behaviors that animals can perform, one of which is affiliation behavior, which can lead to animal frustration (DiVincenti and Rehrig 2017).

Social contact is also a point to be taken into account (Thurston et al. 2018); in fact, some specialists have suggested that rabbits should be housed in groups for greater comfort and have living conditions similar to those of wild rabbits (Szendrő et al. 2016). Although in individual commercial cages they can interact with the does next to them through visual and olfactory contact, the animals have less space that impairs movement and have fewer opportunities for social interactions than in group housing (Szendrő et al. 2013). There are numerous studies comparing the rearing of breeding rabbits individually and in groups.

To improve the conditions and welfare of individually housed rabbits, numerous studies have been conducted in relation to environmental enrichment (Mikó et al. 2014; Prola et al. 2013; Shomer et al. 2001; Huls et al. 1991), cage size and comfort improvement (Cornale et al. 2016), or social contact (Dal Bosco et al. 2020), among others.

3.1.1 Environmental Enrichment

The main aim of environmental enrichment is to encourage normal species-specific behavior and minimize possible abusive behavior directed at oneself (Shomer et al. 2001). It also aims to avoid possible feelings of frustration and decrease stereotypies that might appear due to individual housing and low social contact (Thurston et al. 2018). Furthermore, according to Baumans (2005), one of the main reasons why animals play is the need to develop behavioral skills for their survival, so environmental enrichment could even be considered key to meeting these needs.

In general, the addition of enrichment devices can help to encourage a wide range of species-typical active behaviors, such as nest building (González-Redondo et al. 2015), exploration (Coda et al. 2020), gnawing behavior (through devices that can chew; Huls et al. 1991), and locomotor or visual behavior (Baumans 2005).

In addition to their effectiveness as an enrichment device, all devices that may be used for environmental enrichment must be reviewed for safety before use, and risk assessment must include both the characteristics of the device itself and those that it generates if it is misused or damaged, causing potential harm that needs to be attended by veterinarians (Shomer et al. 2001). Moreover, it is necessary to evaluate which these devices benefit the animal and its welfare through animal preference studies, as well as the evaluation of the use of these devices (Baumans 2005). In this way, the effect on welfare, on the development of species-typical behavior, and on the physiological parameters of the animal can be seen (Coda et al. 2020).

According to Shomer et al. (2001), environmental enrichment can be elements such as balls, chains, wooden blocks, PVC pipes, Nylabones®, and corrugated plastic tunnels. It may also include fodder, hay blocks, chewing sticks, as well as a retreat and surveillance area, for example, a platform (Baumans 2005).

On the other hand, toys have a limited time of attraction in animals, usually 1 day. To maintain a certain level of interest in the environmental enrichment offered, it is necessary to change the objects frequently (Baumans 2005) or to use objects of greater interest to the animal. Authors such as Harris et al. (2001) propose that this interest may be greater when the objects are food. They observed that the addition of environmental enrichment with food significantly increased the amount of time spent interacting with the object compared to animals whose enrichment provided to the cage was nonfood. In this case, Gugolek et al. (2008) found that it positively affected the productive results of does.

The most relevant enrichment elements for breeding rabbits housed in individual cages are explained in more detail below.

Nest Boxes

As mentioned above, the maternal-filial relationship in individual cages starts when the farmer offers to the doe a nest box and then the female starts building the nest (Fernández-Carmona et al. 2005; González-Redondo et al. 2015). Therefore, it is essential to provide rabbits with the nest box and the necessary material (Baumans 2005). This type of enrichment avoids frustration and allows the expression of the natural behavior of the species in the wild (Thurston et al. 2018).

Although infrequent, inadequate housing sometimes means that the doe has to build her nest on the floor of the cage or in a box whose entrance is permanently open without light protection (Schlolaut et al. 2013). This management system allows free access to the nest, but females cannot close it. In the opinion of Schlolaut et al. (2013), the permanent opening of the nest may be one of the reasons for the doe's behavioral alterations, such as nest building failures and problems when the doe gives birth to a litter. Females place litter outside the nest, or even leading to infanticide, but this opinion is not supported by scientific results.

Studies have shown that the use of nests with cat flaps (Baumans 2005) can improve the welfare of the female and this litter compared to nests that are always open. Females perform twice as many potentially agonistic actions when they have an ever-open nest compared to females who have a nest that can be closed (cat flap entrance), as well as having less stress, and show a decrease in mortality and problems with the litter (Baumans 2005).

It is essential to take into account the above to give reproducing does appropriate housing conditions and welfare.

Platforms

Platforms are seen as an environmental enrichment element, allowing animals to retreat or watch but also to jump and exercise. However, they have become popular because they also serve a space-enhancing function in a third (vertical) dimension, allowing for an increase in space without losing seats.

Despite the interest in such platforms, there are few scientific articles in which objective data can be found. Mikó et al. (2014) found that the introduction of the platform increased the weight of litter during lactation but reduced the number of visits of the does to the nest (as Selzer et al. 2004). On the other hand, Alfonso-Carrillo et al. (2014) observed that the introduction of a platform helps to improve feed conversion rates in kits between 3 and 21 days of age. These results could indicate that the introduction of a platform could contribute to improved milk production in reproducing does (Szendrő et al. 2016). However, different studies at ITAVI (Mirabito et al. 2005) showed no differences in litter size, lactation survival, or reproductive parameters. Similarly, Alfonso-Carrillo et al. (2014) did not find changes in the mortality and reproductive parameters of does, although they did find an increase in litter weight.

According to a recent review by López et al. (2020), in which they also collect data published in congresses, platforms do not seem to have a negative effect on productive parameters, and in addition, they increase the available space, provide more opportunity for movement and a greater repertoire of behaviors. However, platforms deteriorate the cage hygiene, health, and cleanliness of the animals, which is also supported by Szendrő and McNitt (2012).

Finally, in recent work (Pérez-Fuentes et al. 2020), the inclusion of a platform even led to an increase in the level of cortisol in the hair of the does, contrary to what would be expected with an enrichment element. Therefore, it can be deduced that not only the incorporation of a platform should be evaluated but also the characteristics that the platform should have (material, size, placement, etc.).

Footrests

The popularly known "sore hocks," which refers to injuries in the footpad caused mainly by the wire mesh floor of the cages, leads to poor welfare of the animals and, in addition, to economic damage for the rabbit farmers. The way to prevent or minimize this is with the use of footrests.

The rabbit research center of INRA (Institut National de la Recherche Agronomique) in France carried out a 10-month study with young breeding does housed individually in wire cages with and without footrests. In terms of production parameters, there was a 26% improvement in newborn kits alive in the cages with footrests compared to the cages without footrests. In addition, of the total deaths, 49% occurred in does without footrests, but only 27% occurred in does with footrests. In conclusion, the use of footrests is beneficial for improving the welfare of does, as well as prolonging their longevity and, therefore, increasing their cumulative production (Camps-Rabadà 2000).

It has been demonstrated that the use of footrests in wire cages reduces the incidence of sore hocks by more than 60% and up to 40% in the appearance of plantar hyperkeratosis. Furthermore, the introduction of this element in cages with a wire mesh floor without any covering helps to have a healing effect of 80% (Rosell and De la fuente 2009).

The presence of footrests on the wire mesh platforms is also important because the animals prefer the plastic platform, using it approximately twice as much (Mikó et al. 2014).

3.1.2 Cage Design and Comfort

Cage Size

Another important aspect in reproducing does housing is cage size. Studies such as Mikó et al. (2014) found that increasing cage size (102.5 × 38 × 61 cm and a platform of 28.5 × 38 cm) improved the weaning weights of sucking rabbits and reduced the incidence of sore hucks due to plastic footrests.

However, it is difficult to evaluate which factor (size, platform, or footrest) should be attributed to the observed changes. Munari et al. (2020a) compared cages of 60 × 53.3 × 35 cm (with external nest) with larger cages of 97 × 53.3 × 113 cm (with external nest and plastic platform). They found that does have a higher number of born alive in the larger cages when the breeding system was extensive (not when it was intensive or alternating both). However, the does had lower fertility, litter size at weaning, weaning weights of the litter and the suckling rabbits, and higher mortality during lactation with the larger cages.

On the other hand, following the same study, the results indicate that the larger cage meant a reduction in feeding and drinking time, leaving more time for other tasks such as self-grooming. According to these authors (Munari et al. 2020a), self-grooming is an improvement in reducing stereotypies that affect animal anxiety. However, animals housed in larger cages showed almost three times more corticosterone in saliva, which would not be expected in a less stressed animal.

On the other hand, Prola et al. (2013) conducted a study in which they used size cages of 113 × 46 × 46 × 46 cm and equipped with footrests covering 25% of the floor, and other more standard cages (83 × 38 × 32 cm) without footrests. It was observed that the larger housing with footrests produced a slight reduction in fecal corticosterone in the does. Using the same housing for male rabbits but shorter (90 and 60 cm, respectively), Cornale et al. (2016) did not observe changes in the level of fecal corticosterone.

Finally, Pérez-Fuentes et al. (2020) observed that cages provided with more area than others (5250 cm vs. 3920 or 4270 cm) showed lower hair cortisol and blood haptoglobin, which are indicators often associated with welfare improvement.

In summary, in relation to the available surface area, the results are sometimes contradictory, probably due to possible confounding effects and the different indicators used, but it does seem to be advisable to try to increase the available floor space with regard to conventional cages. For this purpose, appropriate platforms can be installed or the width of the cages can be increased, as in most studies, which seems to improve some welfare indices.

Cage Height

Another point to consider in the size of the cage is the height. One of the problems with this type of housing is that it limits both movement and resting possibilities due to the lack of space. Dresher (1996) reported in a study without statistical analysis an increase in spinal deformities with age in does housed individually in 40 cm high cages (40%) compared to does housed in an alternative system (where the dimensions are not defined). However, Buijs et al. (2014) also observed 38% spinal deformities in does housed in cages with a height of 60 cm, which are not reduced when they are housed in housing without height limitation. Although without statistical analysis, there appears to be an increase in spinal deformities, up to 70%, when the cage height is 32 cm (Dresher 1996). Based on these results, cages of at least 40 cm in height would prevent an increase in these deformities.

Regarding cage height, Alfonso-Carrillo et al. (2014) did not find differences in the productive parameters of does, although the repertoire of behaviors is greater and stereotyped behaviors are also greater (Munari et al. 2020b) when the height is greater than 35 cm. There is not enough information to know what the minimum height should be to ensure such improvements in behavior, although a height of 40 cm would be recommended.

Comfort and Pododermatitis

Regardless of the size and height of the cage, the floor material is also important to provide the animal with good comfort. Conventional cages have a wire mesh floor that, despite being the most effective for separating the animal's feces, often causes footpad problems (Rosell and de la Fuente 2009), as explained in the previous point.

Pododermatitis is a multifactorial chronic skin disease that appears mainly on the plantar surface of the hind legs and can cause pain leading to poor welfare (Ruchti et al. 2019). The prevalence of pododermatitis increases with wire mesh floors and the number of reproductive cycles in females (Buijs et al. 2014), as long periods of housing females on wire mesh floors complicate the treatment of this disease. In addition, tasks such as cleaning lesions and treatment with adjuvant products and disinfectants require animal handling and a working time that is not compatible with today's industrial farms (Camps-Rabadà 2000).

There is a 6.4% prevalence of pododermatitis in farms in the Iberian Peninsula, although the appearance of this disease is also associated with the use of energy-rich diets (Sánchez et al. 2012) due to the increased locomotion of does.

In addition to the use of footrests, another remedy to reduce the occurrence of this disease is the use of padded mats studied by Eady (2009), who managed to reduce the occurrence of pododermatitis by 3%. Even so, the main risk factors for the occurrence of this disease are the high relative humidity of the farm, the body weight of does, the number of kindlings and age, and finally the claw length (Ruchti et al. 2019).

From all the abovementioned factors, it can be concluded that both cage dimensions and design and the comfort provided to the reproducing doe can have an effect on animal welfare (Eady 2009; Rosell and de la Fuente 2013; Sánchez et al. 2012; Buijs et al. 2014; Ruchti et al. 2019), and its revision must take into account the mentioned factors.

3.1.3 Collective Housing

Wild rabbits live in colonies; they are social animals, which can contradict the way of life of reproducing doe on commercial farms. For this reason, promoted by growing public objections and the tendencies of some legislators against individual housing, which is predominant in today's livestock production, group housing systems for breeding does that are implementable on farms have been proposed (Ruis 2006).

Group Housing and Natural Behavior of Rabbits

Rabbits are gregarious animals, so it would be expected that stress would decrease when they are housed together. However, it should also be considered that the time of kindling is one of the most stressful moments of the productive cycle, radically changing behavior (Prola et al. 2013), and as they are hierarchical animals, they need to define their social ranking when grouped together. It is important to emphasize that domestic rabbits are still similar to wild rabbits, as most of the behaviors that have been reported in wild rabbits are currently observed in domestic rabbits (Marai and Rashwan 2003).

Returning to the time of kindling, wild doe elaborates the nest and enters only the nest to suckle her litter in a blind burrow independent from the rest of the animals, far from the common burrow (González-Redondo et al. 2015). Thus, group housing may be preferable from a natural behavioral developmental point of view and for ethical reasons, but it can lead to a large amount of agonistic behaviors that can cause severe injury among them, especially in peripartum (Rommers et al. 2014; Braconnier et al. 2020a). In fact, Szendrő et al. (2013) and Dal Bosco et al. (2020) observed that females with their litters preferred to remain confined, having less mortality of offspring in the individual cages.

In group housing, the most important thing is the establishment of the hierarchy between does. After grouping, the frequency of biting increases (Andrist et al. 2013; Munari et al. 2020b) until approximately the 6th day post-grouping, when hierarchies are already fully established (Munari et al. 2020b). The dominant female is quite clear in the hierarchy, while no differences between subordinate females are detected, and mortality is related to their rank order in that hierarchy (Gerencsér et al. 2019). Therefore, it is important to know the ethology of rabbits and to know that these peaks of aggressiveness occur until the hierarchy is established (Thurston et al. 2018).

Despite the fact that group housing could give does the option of having a life more similar to wild rabbits, aggression increases in captivity, with 52% of animals with notable injuries and between 13 and 39% with severe injuries (Rommers et al. 2014), resulting in worse welfare compared to individually housed does.

To decrease these agonistic behaviors, the improvement of environmental enrichment in parks is sought (López et al. 2020). Environmental enrichment ranges from objects that capture the attention of the animals, mentioned above in individual housing, to the design of parks, creating hiding places and escape possibilities that help to divert attention away from fighting (Rommers and De Greef 2018), thus reducing skin lesions (Rommers et al. 2014). Modifications of cages and parks with the use of platforms (Szendrő et al. 2016), PVC barriers that allow contact but prevent fights (Valuska and Mench 2013), removable walls, which are installed a few days before kindling and removed at the end of lactation (Dal Bosco et al. 2019), and the use of food items such as straw (Rommers et al. 2014) could be methods to reduce aggression.

In addition, some females with litter prefer to move away from the group, which is probably related to hierarchical status, so that those of lower social rank are more likely to develop this behavior (Dal Bosco et al. 2020). In addition, during breeding in group housing, females sometimes kill or steal the kits of other does to demonstrate their position in the hierarchy (Szendrő and McNitt 2012). Therefore, another modification of group housing is the installation of electronic nests with individual recognition, but they increase manpower and production costs (Hoy et al. 2006). In fact, does in group housing have more nest visit events of longer duration (more than 5 min) than does in individual housing, probably because they use it as a refuge from the rest of their mates (Rommers et al. 2012).

Despite this, reproducing does are not selective with their kits, as it has been observed that in group housing, they can suckle the kits of other females (González-Mariscal and Gallegos 2007; Zomeño et al. 2018).

In general, the main advantages of group housing are that it presents greater opportunities for social and locomotor behavior, but the disadvantages of this type of system are the increase in fights and injuries (Buijs et al. 2015), as well as high mortality and the slaughter of does due to reproduction problems or the presence of abscesses (Pérez-Fuentes et al. 2020).

Grouping Protocols

A large number of grouping protocols (Munari et al. 2020a, b; Braconnier et al. 2020a; López et al. 2020), housing types (Rommers et al. 2014; Hoy et al. 2006; Rommers and De Greef 2018), and the number of animals per group (Valuska and Mench 2013; Zomeño et al. 2017) have been studied to improve the disadvantages of group housing explained in the previous point.

Regarding grouping protocols, continuous or partial grouping is possible. In continuous grouping (does are permanently together, even during kindling and lactation), higher mortality rates of litter, more aggressions and injuries and, finally, more mortality in females have been observed compared to partial grouping (López et al. 2020). The latter consists of housing the breeding doe individually before kindling and regrouping for lactation. Several studies have been performed testing regroupings at 3 days (Dal Bosco et al. 2019), 12 days (Munari et al. 2020b; Braconnier et al. 2020a, b), 18 days (Buijs et al. 2016; Zomeño et al. 2017, 2018; Machado et al. 2019; Pérez-Fuentes et al. 2020; Munari et al. 2020b), 22 days (Munari et al. 2020b; Braconnier et al. 2020a, b), or 23 days postpartum (Rommers and De Greef 2018).

Munari et al. (2020b) observed that delaying the regrouping of females from 12 to 18 and 22 days postpartum did not reduce the number of aggressive interactions but did reduce severe aggression. However, at the productive level, there were no differences between the different grouping systems, but there were differences with individual housing, obtaining worse performances in the collective ones (as in the studies by Braconnier et al. 2020a, b). In addition, Machado et al. (2019) also found a drop in milk production in the regrouping of does at 18 days. This is in line with the findings of Rommers et al. (2012), who note that does visit the nest fewer times in group housing than does in individual housing, which could result in lower weaning weights.

Group Size

Different sizes of group does pens have been proposed. In the study by Rommers et al. (2014), four females with their litters were grouped in a pen of 1.0 × 1.5 × 0.6 m (0.225 m^2/does), while Valuska and Mench (2013) compared two types of housing for pairs, in small cages (0.76 × 1.22 × 0.61 m; 0.283 m^2/doe) and in a larger park (1.22 × 1.22 × 0.61 m; 0.454 m^2/doe), in both cases with PVC barriers. In this study, although there were no significant differences between the two types of housing, the authors indicate that fewer injuries occurred in those does housed in the larger enclosures due to the provision of more space per female (Valuska and Mench 2013).

The number of does in the group is also important. It has been observed that if pairs are socially compatible, they have better production levels than larger groups but worse levels than those housed individually (Zomeño et al. 2017). The main disadvantage of this type of grouping is when one member of the pair is much more dominant or aggressive than the other. Carter et al. (2016) even describe pseudogestation in the subordinate doe in the presence of an aggressive dominant doe in pair-housing.

The most common group size is four females, but this type of group housing presents worse data than pair-housing (Szendrő and McNitt 2012; Zomeño et al. 2017). Szendrő and McNitt (2012) used a group of four females and one male with an available space of 7.7 m^2 in their study, but they obtained a high mortality rate in kits, which was associated with maternal stress and aggressive behaviors such as scratching or biting. In addition, mating does not seem to be very effective in group systems (Gerencsér et al. 2019). These results are in line with those described by Buijs et al. (2016), who obtained results showing a higher number of agonistic behaviors in larger groups.

According to the literature, it seems to be demonstrated that continuous group housing systems for breeding females challenge animal welfare by increasing aggression and injuries between females and litters. In contrast, it has been observed that part-time group housing systems (also known as semigroups) might have potential but cannot yet be recommended on farms until the main problems of aggression and injuries between animals are solved, especially in reintroductions (Szendrö et al. 2019).

3.2 Fattening Rabbits

Housing for fattening rabbits can be very variable between countries, even within the same country. In European countries, the use of dual-purpose cages is becoming more common, and the use of enriched cages, where small groups of 4–5 rabbits are housed, has increased. Recently, rabbits have also been housed in pens (large groups of 32–36 rabbits) (EFSA 2020).

In general, the most common housing system is in collective cages, with groups of 7–9 rabbits, usually from the same litter, with the presence of males and females. It is important to provide a minimum surface area so that the animals can perform comfort activity, explore, mainly sniffing, digging, and gnawing (Trocino and Xiccato 2006), locomotor activity and social behavior (López et al. 2021).

Specific behaviors are performed using resources and structures or within a specific area (most rabbits try to divide their space into separate areas for feeding, resting, and excreting). These structures should allow the animals to control their environment (Baumans 2005) and may be included within the cage, such as tubes and platforms that provide shelter and vigilance possibilities.

Thus, it is important to study cage dimensions, platform material, pen housing, and enrichment systems to improve the animals' quality of life (Finzi et al. 2009).

3.2.1 Environmental Enrichments

As in the case of reproducing does, it has also been described in fattening rabbits that stress levels in group housing decrease when animals are provided with some kind of environmental enrichment (Buijs et al. 2011), so different types of environmental enrichment have been evaluated, and it seems to be related to a higher frequency of allo-grooming behavior, which could indicate a social behavior related to pheromonal olfactory stimulation and mutual recognition (Zucca et al. 2012). This should be considered when considering enrichments.

The effect of many environmental enrichment devices seems to be demonstrated because the repertoire of positive behaviors increases (Harris et al. 2001; Mastellone et al. 2019; Coda et al. 2020). However, the possible effect of these factors on stress levels when rabbits are subjected to a potentially stressful situation seems less clear (Coda et al. 2020).

Some environmental enrichments studied by different authors are detailed below.

Stainless Steel Rattles

Johnson et al. (2003) evaluated stainless steel rattles in individually housed rabbits. They found no significant differences in weight or blood parameters evaluated compared to the control group. However, they did note decreased use of the element over time, suggesting that enrichment elements should be changed periodically to maintain their attractiveness.

Mirrors

Another enrichment that has been tested is the presence of mirrors. These seemed to be able to modify the behavioral repertoire of isolated rabbits by acting on social perception and increasing self-grooming (Mastellone et al. 2019). Moreover, Musco et al. (2019) and Secci et al. (2020) note that their presence increases animal growth and may even improve rabbit carcass and meat characteristics, especially in small groups.

In general, it seems that the presence of mirrors offers advantages related to comfort and welfare. Therefore, it could be used as environmental enrichment for fattening rabbits and especially recommended for rabbits caged individually for prolonged periods of time in visual isolation from the rest of the animals (Dalle Zotte et al. 2009).

Food Environmental Enrichment

When environmental enrichment elements are related to food or contain food, they are used more than when they do not contain food (Harris et al. 2001). This can be of interest to encourage their use in certain conditions and, moreover, to maintain attention over time.

Thus, playing behavior is more frequent in cages enriched with straw or sand floor, and stereotypies are more frequent in cages without enrichment (Siloto et al. 2009). Some authors argue that rabbits should have straw or hay in addition to feed to satisfy their ethological needs, regardless of whether they are in more or less intensive housing (Lehmann 1990). According to the results of Szendrö et al. (2015), the availability of hay above the cage (or in special feeders) reduces growth rate (by 3%) but also ear lesions (by 60%) in rabbits between 5 and 12 weeks of age. These rabbits were fed granulated feed ad libitum at a density of 15 rabbits/m^2.

These results would be related to the fact that rabbits housed individually and fed at maintenance (to avoid fattening) interacted much more with hay (supplied in plastic feeders) than with hay sticks or wooden sticks and reduced abnormal behavior and stereotypies (Lidfors 1997). In this experiment, it should be noted that animals with hay or sticks were able to feel less hunger than those that were provided with a stick.

The most common practice is to place hay on top of the cage. However, rabbits consuming hay in elevated feeders placed on the cage wall prefer to consume it in the natural grazing posture (with all four feet on the ground), either by throwing the hay from the feeder to the ground with the front feet and consuming it once on the ground or by jumping to the feeder and consuming it there directly (Prebble et al. 2015).

Therefore, hay appears to be preferred by rabbits (Rottler et al. 2019), and it appears that supplementation with hay on branches (or branches from some trees) reduced boredom and the incidence of aggression. It is also more effective than a bucket of grass, gnawing sticks, or a box to reduce abnormal behavior and give individually housed rabbits some alternative occupation (Lidfors 1997).

However, most of the work is carried out on adult rabbits used in laboratories or on rabbits fed pet food, which differs from the conditions on farms. On the other hand, there are no data on hay consumption (due to the losses that occur and are difficult to avoid), nor what would be the best way to supply it (type of trough, particle size, etc.).

There are also feeds that can help animals confront potentially stressful situations and therefore improve their welfare in such conditions. Such is the case for chestnut tannins under heat stress conditions (Liu et al. 2012), where they relate their antioxidant properties to a reduction in cortisol.

Gnawing Sticks

Another type of environmental enrichment in fattening rabbits, for which there is considerable literature, is the use of sticks of any material. Thus, Dalle Zotte et al. (2009) state that sticks are neither an advantage nor a disadvantage in terms of meat quality, and the amount of wood eaten by each rabbit is very low (Zucca et al. 2012). However, the benefits from an animal welfare point of view may be high, since they provide an important incentive for the expression of the rabbit's exploratory behavior and reduce stereotypies, especially in small groups (Zucca et al. 2012).

Wooden sticks from different tree species have been tested: false acacia (Zucca et al. 2012), apple tree, willow (Mohammed and Nasr 2017) and Norway spruce (Jordan et al. 2008), among others. Mohammed and Nasr (2017) obtained significantly better performance and higher carcass weight in rabbits that had wooden sticks. Similarly, Princz et al. (2009) observed higher body weight at 11 weeks of age in animals housed with sticks. In contrast, Jordan et al. (2008) found no differences between the presence or absence of sticks.

At the behavioral level, stick has the greatest effect at the beginning of the growing period (Zucca et al. 2012) and continues to a lesser extent throughout the growing period, such as the reduction of abnormal behaviors (Mohammed and Nasr 2017) and aggressiveness (Princz et al. 2008, 2009; Szendrő and Dalle Zotte 2011).

Regarding the placement of the sticks, the safest and most effective seems to be to fix the sticks to the cage roof (Marin et al. 2018) or to the cage wall at the height of the rabbit's head (Szendrő and Dalle Zotte 2011).

Platforms

Platforms provide two levels inside the cage and do not seem to negatively affect the productive parameters, the level of feed intake or the rest of the animals (Postollec et al. 2008; Matics et al. 2018). In addition, it improves animal welfare, as animals prefer to spend more time on the platform than under it, as observed in the results of Finzi et al. (1998).

The use of the platform in fattening rabbits also has different benefits depending on where it is used; in the case of use in cages, they provide additional space (Postollec et al. 2008). However, in the case of use in large pens, it is an exercise element that, in the opinion of Postollec et al. (2008), reduces the average daily gain with respect to cages without platforms but could improve animal welfare. Martino et al. (2016) indicated that the use of platforms provides more possibilities of movement for rabbits and has no effect on carcass characteristics. Nevertheless, the quality of the meat may be affected, as increased activity influences the oxidative status, decreases the antioxidant content and impairs fat oxidation in rabbit meat. However, Combes et al. (2010) observed that carcass characteristics were improved in rabbits housed in pens with platforms compared to rabbits housed in conventional cages.

On the other hand, Trocino et al. (2019) observed that the use of these platforms in pens where rabbits were housed in high group sizes did not affect productivity. Although it increased the percentage of injured rabbits, it contributed to reduced aggressive behavior and improved resting activities.

Therefore, the authors conclude that this type of element may be suitable for increasing behavioral repertoire and opportunity for movement without negatively impacting productivity. In any case, the motivation of rabbits to use the platform is high since, when available, they use it approximately half of the time (Finzi et al. 1998), although only approximately 20% of rabbits use it, according to a study by Rottler et al. (2019).

Despite all of the above, cages with platforms in fattening rabbits also present hygienic challenges (Rauterberg et al. 2019a, b), as in the case of reproducing does, and therefore require further study.

Other Enrichment Types

One type of enrichment that has been tested is the provision of a shelter area (Uczay et al. 2015), but no behavioral advantages were obtained, and it decreased the feeding rate, so it did not positively influence the behavior of the rabbits. Another type of environmental enrichment is music, which decreases fecal cortisol levels (Peveler and Hickman 2018) and therefore the level of stress.

Other types of enrichment tested have been pinewood sticks, straw in a plastic container, compressed wood blocks or a combination of straw and pinewood sticks. The second and third were used for the longest time and with which the animals develop the most natural behaviors (Rommers et al. 2014).

3.2.2 Cage Design

Floor Type

Fattening rabbits in conventional farms are housed in wire mesh cages, similar to reproducing does. Conventional wire mesh cages are being strongly challenged, and alternatives are being sought, such as pens on the ground, since in general, rabbits in this system show less abnormal behavior (such as gnawing on the bars) than caged rabbits (Krohn et al. 1999). However, it should be noted that this factor may be more important in does that have to spend long periods of time in cages compared to fattening rabbits that spend only a few weeks on the farm and are much lighter animals.

According to González-Redondo (2009), inadequate flooring or inadequate hole diameter can lead to injuries to the hind legs of rabbits and even trauma. Tillmann et al. (2019) determined that among nonwire mesh floors, plastic floors with slats spaced 10 mm apart give better results than floors with 12 mm circular holes. This is mainly for reasons related to hygiene, as cleaning is easier (Masthoff and Hoy 2019). In addition, plastic flooring gives better results than wooden flooring in pens: higher final weight of rabbits and higher carcass quality (Trocino et al. 2015), less restriction of movement, less reactivity to the environment and even stress (Trocino et al. 2018), although these results may be due more to the distance between the wood slats in that work than to the material itself.

In addition, rabbit injuries are lower with plastic flooring, as proposed by Masthoff and Hoy (2019), which has a slat width of 5 mm and a slot width of 13 mm, although other authors (Princz et al. 2008) have found no differences for any behavioral pattern (eating, drinking, moving, resting, comfort, social, exploration) although growing rabbits have a preference for plastic flooring. Other authors have even found higher mortalities in housing with plastic flooring compared to cages with wire mesh flooring (Rauterberg et al. 2019a).

Another alternative that has been proposed is straw bedded pens. From an ethological point of view, housing rabbits in pens, especially in deep bedding, has a favorable effect on the frequency of social and stereotyped behaviors (Jekkel et al.

2010) but gives worse productive results due to straw ingestion and more direct contact with excreta (Dal Bosco et al. 2002). In fact, rabbits, given their tendency toward pododermatitis, should be kept on bedding materials with desirable physical properties, such as cotton (Wolf et al. 2020). Other causes that make deep bedding more unfavorable are the higher risk of coccidia contamination (higher mortality) and lower productivity and carcass quality (Szendrő and Dalle Zotte 2011; Windschnurer et al. 2019).

Some authors, such as Schoerwerth et al. (2019), argue that pen rearing of fattening rabbits is currently the preferable alternative husbandry system, as the animals experience a higher level of animal welfare that cannot be defined exclusively by animal health. However, when animals were confronted with a choice test in their thermoneutral zone, rabbits preferred wire mesh flooring over straw bedding (Morisse et al. 1999; Siloto et al. 2009). Rabbits kept at low temperature showed no preference (Siloto et al. 2009), probably due to the cleanliness and dryness of wire mesh flooring.

Therefore, the rearing of fattening rabbits in indoor pens should be done with wire mesh floor (Dal Bosco et al. 2002) or preferably with plastic slat flooring that allows the correct separation of the rabbits from their droppings (Szendrő and Dalle Zotte 2011), to be considered as a possible alternative housing system to improve rabbit welfare.

Cage Height

Currently, industrial cages have a height of 38–40 cm, but due to the growing interest in animal welfare, higher cages have been proposed. A cage height of 50–60 cm has been proposed, even reaching a height of 70 cm (modifying other cage characteristics) with the aim of allowing the rabbit to stand up on its hind legs with its ears erect (Negretti et al. 2010). Studies such as those of Martrenchar et al. (2001) and Olivas et al. (2013) showed the use of cage height by rabbits when this space is provided. In conventional cages, rabbits cannot perform behaviors such as standing on their hind legs, which is a typical vigilance behavior of the species. Therefore, the provision of increased cage height may be beneficial for the welfare of fattening animals (Olivas et al. 2013).

However, a study by Negretti et al. (2010) showed that 99% of rabbit behavior does not involve the use of a height greater than 40 cm, including the erect posture where rabbits are standing on their hind legs. This posture involves the use of 36–40 cm in height. The most common postures that imply the use of 30–35 cm in height are lying down, sitting, grooming, walking, standing on all four legs, eating, and drinking. Standing on their two extended hind legs is very unusual.

Other authors found no productive or behavioral differences between a cage height of 38 cm and a height of 45.7 cm (Stewart and Suckow 2016) or between 30 and 35 cm in height (Princz et al. 2008).

Considering that it is debatable to say whether or not an unusual posture is important in rabbit behavior, there are authors such as Princz et al. (2008) and Negretti et al. (2010) who argue for the use of standard cages because it does not negatively affect the behavior and comfort of the animals.

3.2.3 Animal Density and Group Size

Animal Density

Another aspect to consider in welfare issues in fattening rabbits is animal density, since this can lead to restricted movement, affect the biotic parameters of the body, and can significantly reduce the growth performance of rabbits (Rahman et al. 2018). The determination of the ideal density in fattening rabbits has been the subject of numerous studies, and an optimum density of 40 kg/m^2 is accepted for groups of 6–9 kits in conventional cages (López et al. 2021).

On the other hand, it has been shown that lowering animal density slightly improves feed intake and feed conversion rate, but if the density is reduced to 7.5 rabbits/m^2 compared to the usual density of 15 rabbits/m^2, this can lead to a negative economic balance on the farms (Verspecht et al. 2011).

There are numerous studies in relation to animal density in fattening rabbits. Trocino et al. (2015) worked with densities up to 16 rabbits/m^2 and observed that lesions tend to increase with density, just as the establishment of hierarchies can be altered (Vervaecke et al. 2010). Less lying behavior and more locomotor and sitting activity were observed at low densities (Zucca et al. 2012). There are higher yields, better daily feed intake, and better feed conversion rates, as well as lower stress parameters (Araújo et al. 2017; El-Tarabany et al. 2019, 2021)

Group Size

The other aspect related to animal density is group size. When the group size exceeds four or five rabbits per cage, there are more disadvantages (higher risk of contamination, diseases and mortality, higher rates of aggressions and injured rabbits) than advantages (higher possibility of locomotion and more social contact) (Szendrő and Dalle Zotte 2011). Thus, rabbits from smaller pens tend to weigh more than rabbits from larger pens (Postollec et al. 2006). Furthermore, the number and severity of injuries in larger groups (more than 16 animals per pen) is also higher (Bigler and Oester 1998; Szendrö et al. 2015).

Similarly, when fattening rabbits are housed in groups (Xiccato et al. 2013; Martrenchar et al. 2001), they show lower live weight and worse carcass characteristics compared to those in bicellular cages, regardless of animal density or pen size. In contrast, the skeletal development of animals improves in collective systems (Martrenchar et al. 2001), while meat quality is only slightly affected by the housing system (Xiccato et al. 2013).

Combes et al. (2010) showed that group size has an effect on carcass size when comparing large versus small pens (6, 10, and 61 rabbits housed at the same density; 15 rabbits/m^2), altering carcass weight, increasing reddish meat color and bone strength of the animal when rabbits were housed in large pens. Because large pens favor the locomotor activity of the rabbits, carcass and meat characteristics are altered. According to Dalle Zotte et al. (2009), when groups are smaller (2 rabbits per cage and 13 rabbits per pen), the differences in carcass and meat characteristics are diluted.

Furthermore, according to Martrenchar et al. (2001), rabbits jump more in pens than in cages (housed at the same density but in smaller groups and with less height) and walk significantly less on top of each other, which reduces the frequency of ear injuries. However, Bigler and Oester (1994) point out the possibility that pens with very large numbers of rabbits may present social problems at the end of the fattening period (from 80 days of life) as a consequence of the beginning of sexual development, which would be a problem in systems in which the animals are slaughtered very old.

According to the study of Drescher and Reiter (1996), the optimal group size among those studied (4, 8, 16, 32, and 64) was that of 16 individuals, since a higher percentage of rabbits in resting position and a lower percentage of aggressions were obtained, while the productive parameters were not adversely affected.

Some authors have gone further in the study of group housing and have evaluated the possibility of housing with outdoor access. Such is the case of D'Agata et al. (2009), who observed that rabbits with outdoor access had a higher frequency of escape and burrowing attempts and a lower frequency of gnawing. They relate this to the animals' greater ability to cope with stressors, as well as increased physical activity, greater hind limb development and, consequently, improved meat characteristics (higher slaughter weight, less water in the meat, and less cooking losses).

Finally, there are still young animals in individual housing, mainly replacement animals. In such housing, young animals tend to exhibit a greater number of abnormal behaviors and, therefore, a greater need for environmental enrichment (Katsarou et al. 2011). Some of these anomalous behaviors include bar gnawing and excessive grooming. One way to reduce the presence of these abnormal behaviors is the inclusion of hay as environmental enrichment (Berthelsen and Hansen 1999).

4 Conclusions

In general, it can be observed that some of the behavioral needs of female rabbits were already satisfied in traditional systems (for example, availability of building their nest). However, other needs such as gnawing, exploring or alert behaviors, have to be facilitated in the new housing systems, by providing the animals a proper environment. In addition, scientific information regarding group housing systems is clear regarding the problem of aggressiveness, so these systems must be deeply studied in order to be a real alternative to individual systems.

With regard to growing rabbits, group size and stocking density are the main challenges to solve, as abnormal behaviors and injuries have to be minimized, at the same time that group sizes have to be higher than traditionally. Furthermore, in this type of animals, environmental enrichment is crucial to improve housing conditions, as otherwise, the environment is too poor to satisfy their behavioral needs.

References

Alfonso-Carrillo C, Garcia-Rebollar P, De Bias C, Ibanez MA, Garcia-Ruiz AI (2014) Effect of late weaning and use of alternative cages on performance of does, suckling and fattening rabbits under extensive reproductive management. Livest Sci 167:425–434. https://doi.org/10.1016/j.livsci.2014.05.018

Andrist CA, Van den Borne BH, Bigler LM, Buchwalder T, Roth BA (2013) Epidemiologic survey in Swiss group-housed breeding rabbits: extent of lesions and potential risk factors. Prev Vet Med 108(2-3):218–224. https://doi.org/10.1016/j.prevetmed.2012.07.015

Araújo JP, Marques R, Pires P, Cantalapiedra J, Durão J, Cerqueira JL (2017) Desempenho produtivo de coelhos criados em jaulas sob o efeito de duas densidades. Rev Cienc Agrar 40: 521–530. https://doi.org/10.19084/RCA16196

Baumans V (2005) Environmental enrichment for laboratory rodents and rabbits: requirements of rodents, rabbits, and research. ILAR J 46(2):162–170. https://doi.org/10.1093/ilar.46.2.162

Belgium (2014) Arrêté Royal relatif au bien-être des lapins dans les élevages, Moniteur Belge [C – 2014/24303], 60861–60864

Benedek I, Altbäcker V, Zsolnai A, Molnár T (2020) Exploring the genetic background of the differences in nest-building behavior in European rabbit. Animals 10(9):1579. https://doi.org/10.3390/ani10091579

Berthelsen H, Hansen LT (1999) The effect of hay on the behavior of caged rabbits (Oryctolagus cuniculus). Anim Welf 8(2):149–157

Bigler L, Oester H (1994) The evaluation of animal welfare of housing systems for small and large groups of fattening rabbits. Berl munch tierarztl 107(5):150–156

Bigler L, Oester H (1998) Group housing for male rabbits. In: Proceedings of the 6th world rabbit congress. Vol 1: nutrition, wool & fur; vol 2: reproduction & reproduction physiology—genetics & selection—general physiology—ethology & welfare; vol 3: pathology & prophylaxis—growth & meat—management & production, pp b411–b415

Botelho N, Vieira-Pinto M, Batchelli P, Pallisera J, Dalmau A (2020) Testing an animal welfare assessment protocol for growing-rabbits reared for meat production based on the welfare quality approach. Animals 10(8). https://doi.org/10.3390/ani10081415

Braconnier M, Gomez Y, Gebhardt-Henrich SG (2020a) Different regrouping schedules in semi group-housed rabbit does: Effects on agonistic behavior, stress and lesions. Appl Anim Behav Sci 228:105024. https://doi.org/10.1016/j.applanim.2020.105024

Braconnier M, Munari C, Gomez Y, Gebhardt-Henrich SG (2020b) Grouping of breeding rabbit does at different time points: effects on fertility, mortality and weight. World Rabbit Sci 28(2): 73–80. https://doi.org/10.4995/wrs.2020.13107

Brambell R (1965) Report of the technical committee to enquire into the welfare of animals kept under intensive livestock husbandry systems. Parliament by the Secretary of State for Scotland and the Minister of Agriculture, Fisheries and Food by Command of her Majesty, pp 1–84

Broom D, Johnson K (1993) Stress and animal welfare. Kluwer Academic, London. ISBN: 0412395800

Buijs S, Keeling LJ, Tuyttens FAM (2011) Behavior and use of space in fattening rabbits as influenced by cage size and enrichment. Appl Anim Behav Sci 134(3):229–238. https://doi.org/10.1016/j.applanim.2011.06.008

Buijs S, Hermans K, Maertens L, Van Caelenberg A, Tuyttens FA (2014) Effects of semigroup housing and floor type on pododermatitis, spinal deformation and bone quality in rabbit does. Animal 8(10):1728–1734

Buijs S, Maertens L, Hermans K, Vangeyte J, Tuyttens FAM (2015) Behavior, wounds, weight loss and adrenal weight of rabbit does as affected by semigroup housing. Appl Anim Behav Sci 172: 44–51. https://doi.org/10.1016/j.applanim.2015.09.003

Buijs S, Vangeyte J, Tuyttens F (2016) Effects of communal rearing and group size on breeding rabbits' postgrouping behavior and its relation to ano-genital distance. Appl Anim Behav Sci 182:53–60. https://doi.org/10.1016/j.applanim.2016.06.005

Caba M, González-Mariscal G (2009) The rabbit pup, a natural model of nursing anticipatory activity. Eur J Neurosci 30:1697–1706. https://doi.org/10.1111/j.1460-9568.2009.06964.x

Camps-Rabadà J (2000) Comparación científica entre suelos de varilla, o los mismos con "relax", para control del "mal de patas". Lagorpha 112:36–38

Carter CS (2001) Is there a neurobiology of good welfare? In: Broom DM (ed) Coping with challenge. Welfare in animals including humans. Dahlem University Press, Berlin

Carter CL, Adams JK, Czarra JA, Coan PN (2016) An incidence of pseudopregnancy associated with the social enrichment of rabbits (Oryctolagus cuniculi). J Am Assoc Lab Anim 55(1): 98–99

Coda KA, Fortman JD, García KD (2020) Behavioral effects of cage size and environmental enrichment in New Zealand white rabbits. J Am Assoc Lab Anim 59(4):356–364. https://doi.org/10.30802/AALAS-JAALAS-19-000136

Combes S, Postollec G, Cauquil L, Gidenne T (2010) Influence of cage or pen housing on carcass traits and meat quality of rabbit. Animal 4(2):295–302. https://doi.org/10.1017/S1751731109991030

Cornale P, Macchi E, Renna M, Prola L, Perona G, Mimosi A (2016) Effect of cage type on fecal corticosterone concentration in buck rabbits during the reproductive cycle. J Appl Anim Welf Sci 19(1):90–96. https://doi.org/10.1080/10888705.2015.1072468

Council Directive 98/58/EC of 20 July 1998 concerning the protection of animals kept for farming purposes

Cunningham JG, Klein BG (2009) Fisiología veterinaria, 4tª edn. Elsevier, Barcelona, España

D'Agata M, Preziuso G, Russo C, Zotte AD, Mourvaki E, Paci G (2009) Effect of an outdoor rearing system on the welfare, growth performance, carcass and meat quality of a slow-growing rabbit population. Meat Sci 83(4). https://doi.org/10.1016/j.meatsci.2009.08.005

Dal Bosco A, Castellini C, Mugnai C (2002) Rearing rabbits on a wire net floor or straw litter: behavior, growth and meat qualitative traits. Livest Prod Sci 75(2):149–156. https://doi.org/10.1016/S0301-6226(01)00307-4

Dal Bosco A, Mugnai C, Martino M, Szendrő Z, Mattioli S, Cambiotti V, Cartoni Mancinelli A, Moscati L, Castellini C (2019) Housing rabbit does in a combi system with removable walls: Effect on behavior and reproductive performance. Animals 9(8):528. https://doi.org/10.3390/ani9080528

Dal Bosco A, Cartoni Mancinelli A, Hoy S, Martino M, Mattioli S, Cotozzolo E, Castellini C (2020) Assessing the preference of rabbit does to social contact or seclusion: Results of different investigations. Animals 10(2):286. https://doi.org/10.3390/ani10020286

Dalle Zotte A, Princz Z, Metzger S, Szabo A, Radnai I, Biro-Nemeth E, Orova Z, Szendrö Z (2009) Response of fattening rabbits reared under different housing conditions. 2. Carcass and meat quality. Livest Sci 122(1):39–47. https://doi.org/10.1016/j.livsci.2008.07.021

Dalmau A, Moles X, Pallisera J (2020) Animal welfare assessment protocol for does, bucks, and kit rabbits reared for production. Front Vet Sci 7:445. https://doi.org/10.3389/fvets.2020.00445

Davies RR, Davies JAE (2003) Rabbit gastrointestinal physiology. Vet Clin North Am Exot Anim Pract 6(1):139–153. https://doi.org/10.1016/s1094-9194(02)00024-5

Debra B (2011) Physical and psychological needs of rabbits: A rabbit is not a cat. Vet Nurse 2(4): 184–189. https://doi.org/10.12968/vetn.2011.2.4.184

DiVincenti L Jr, Rehrig A (2017) Social behavior of adult male New Zealand white rabbits housed in groups or pairs in the laboratory. J Appl Anim Welf Sci 20(1):86–94. https://doi.org/10.1080/10888705.2016.1247352

Drescher B, Reiter J (1996) The optimization of group size for fattening rabbits in group housing on grids made of artificial material. Berl Munch Tierarztl 10(8):304–308

Dresher B (1996) Deformations of vertebral column in breeding rabbits. In: Proc. 6th world rabbit congress, Toulouse, France, vol 2, pp 417–421

Eady SJ (2009) Evaluation of equipment for breeding does housed under Australian meat rabbit farming conditions. Anim Prod Sci 49(8):717–726. https://doi.org/10.1071/EA08232

Edgar JL, Mullan SM, Pritchard JC, McFarlane UJC, Main DCJ (2013) Toward a 'good life' for farm animals: Development of a resource tier framework to achieve positive welfare for laying hens. Animals 3(3):584–605. https://doi.org/10.3390/ani3030584

El-Tarabany MS, Ahmed-Farid OA, El-Tarabany AA (2019) Impact of space allowance on performance traits, brain neurotransmitters and blood antioxidant activity of New Zealand White rabbits. Prev Vet Med 163:44–50. https://doi.org/10.1016/j.prevetmed.2018.12.011

El-Tarabany MS, El-Tarabany AA, Ahmed-Farid OA (2021) Effect of cage density on growth rate, carcass traits, muscle amino acid profile, and antioxidant biomarkers of commercial rabbits. Trop Anim Health Prod 53(3):393. https://doi.org/10.1007/s11250-021-02842-9

European Food Safety Authority (EFSA) Panel on Animal Health and Welfare (AHAW), Saxmose Nielsen S, Alvarez J, Bicout DJ, Calistri P, Depner K, Winckler C (2020) Health and welfare of rabbits farmed in different production systems. EFSA J 18(1):e05944

Feng Y, Fan H, Liang X, Wang X, Gao G, Gun S (2022) Environmental enrichment changes rabbits' behavior, serum hormone level and further affects cecal microbiota. PeerJ 10:e13068. https://doi.org/10.7717/peerj.13068

Fernández-Carmona J, Solar A, Pascual JJ, Blas E, Cervera C (2005) The behavior of farm rabbit does around parturition and during lactation. World Rabbit Sci 13:253–277. https://doi.org/10.4995/wrs.2005.513

Finzi A, Margarit R, Calabrese A (1998) A two-floor cage for rabbit welfare. In: Proceedings of the 6th world rabbit congress, vols 1–3: vol 1: Nutrition, wool & fur; vol 2: Reproduction & reproduction physiology—genetics & selection—general physiology—ethology & welfare; vol 3: Pathology & prophylaxis—growth & meat—management & production, pp B423–B424

Finzi A, Negretti P, Macchioni P, González-Redondo P (2009) La cría en jaulas y las necesidades etológicas del conejo. In: XXXIV Symposium de cunicultura de la Asociación Española de Cunicultura, pp 141–147

Fraser D, Weary D, Pajor E, Milligan B (1997) A scientific conception of animal welfare that reflects ethical concerns. Anim Welf 6:187–205

Gerencsér Z, Matics Z, Szabó RT, Kustos K, Mikó A, Nagy I, Odermatt M, Atkári T, Szendrő Z (2019) Aggressiveness, mating behavior and lifespan of group housed rabbit does. Animals 9(10):708. https://doi.org/10.3390/ani9100708

Germany (2014) Verordnung zum Schutz landwirtschaftlicher Nutztiere und anderer zur Erzeugung tierischer Produkte gehaltener Tiere bei ihrer Haltung in der Fassung von 2014

González-Mariscal G, Gallegos JA (2007) New Zealand white rabbits show nonselective nursing in various types of nests. World Rabbit Sci 15(3):167–172. https://doi.org/10.4995/wrs.2007.592

González-Mariscal G, Toribio A, Gallegos JA, Serrano-Meneses MA (2013a) The characteristics of suckling stimulation determine the daily duration of mother-young contact and milk output in rabbits. Dev Psychobiol 55:809–817. https://doi.org/10.1002/dev.21071

González-Mariscal G, Lemus AC, Vega-Gonzalez A, Aguilar-Roblero R (2013b) Litter size determines circadian periodicity of nursing in rabbits. Chronobiol Int 30:711–718. https://doi.org/10.3109/07420528.2013.784769

González-Redondo P (2009) Injuries in cage-bred wild rabbits (Oryctolagus cuniculus) caused by the size of the orifices in the cage floor. Arq Bras Med Vet Zoo 61(5):1246–1250

González-Redondo P, Gonzalez-Mariscal G, Lopez M, Fernandez-Carmona J, Finzi A, Villagra A (2015) Maternal behavior and welfare of the domestic and wild rabbit doe and its litter. ITEA Inf Tec Econ Ag 111(4):326–347. https://doi.org/10.12706/itea.2015.021

Gugolek A, Lorek MO, Janiszewski P (2008) Effect of application of onion bulbs and birch twigs on production performances of rabbits. J Appl Anim Res 34(2):189–192. https://doi.org/10.1080/09712119.2008.9706969

Harris LD, Custer LB, Soranaka ET, Burge JR, Ruble GR (2001) Evaluation of objects and food for environmental enrichment of NZW rabbits. Contemp Top Lab Anim Sci 40(1):27–30

Hemsworth PH, Coleman GJ (2011) Human-livestock interactions: The stockperson and the productivity and welfare of intensively farmed animals, 2nd edn. CABI, Wallingford, pp 47–83

Hoy S, Ruis M, Szendrő Z (2006) Housing of rabbits—results of a European research network. Arch Geflugelkund 70(5):223–227

Hudson R, Distel H (1982) The pattern of behavior of rabbit pups in the nest. Behavior 79:255–271

Hudson R, Schaal B, Bilko Á, Altbäcker V (1996) Just three minutes a day: the behavior of young rabbits viewed in the context of limited maternal care. In: 6th world rabbit congress, Toulouse, vol 2, pp 395–403

Huls WL, Brooks DL, Bean-Knudsen D (1991) Response of adult New Zealand white rabbits to enrichment objects and paired housing. Lab Anim Sci 41(6):609–612

Hungary (2009) Ministry of Agriculture. Minimum requirements for the housing of rabbit on large-scale farms. Decree of the Ministry of Agriculture (32/1999./III. 31./and 178/2009./XII. 29./) (in Hung.)

Jekkel G, Milisits G, Nagy I (2010) Effect of alternative rearing methods on the behavior and on the growth and slaughter traits of growing rabbits. Archiv Tierzucht 53(2):205–215. https://doi.org/10.5194/aab-53-205-2010

Johnson CA, Pallozzi WA, Geiger L, Szumiloski JL, Castiglia L, Dahl NP, Destefano JA, Pratt SJ, Hall SJ, Beare CM, Gallagher M, Klein HJ (2003) The effect of an environmental enrichment device on individually caged rabbits in a safety assessment facility. Contemp Top Lab Anim Sci 42(5):27–30

Jordan D, Gorjanc G, Kermauner A, Stuhec I (2008) Wooden sticks as environmental enrichment: effect on fattening and carcass traits of individually housed growing rabbits. World Rabbit Sci 16(4):237–243. https://doi.org/10.4995/wrs.2008.619

Katsarou A, Tsironi A, Serafetinidou M, Voyazaki C, Baumans V, Kostomitsopoulos N (2011) Age-related behavior on individually caged rabbits. J Hellenic Vet Med Soc 62(1):21–28

Keeble E (2002) Nervous and musculoskeletal disorders. In: Meredith A, Redrobe S (eds) BSAVA manual of exotic pets, 4th edn. British Small Animal Veterinary Association, Gloucester, UK, pp 103–116

Kimm S, Rauterberg SL, Bill J, Stracke J, Kemper N, Fels M (2021) Use of space, active and resting behavior in fattening rabbits (Oryctolagus cuniculus) housed in a combi park system: A case study. Anim Welf 230:493–506. https://doi.org/10.7120/09627286.30.4.012

Krohn TC, Ritskes-Hoitinga J, Svendsen P (1999) The effects of feeding and housing on the behavior of the laboratory rabbit. Lab Anim 33(2):101–107. https://doi.org/10.1258/002367799780578327

Lanier JL (2008) El estrés y el miedo en procedimientos estándares agropecuarios. REDVET 9:1–13

Lawrence AB, Vigors B, Sandøe P (2019) What is so positive about positive animal welfare?—A critical review of the literature. Animals 9:783. https://doi.org/10.3390/ani9100783

Lee CN (1993) Environmental stress effects on bovine reproduction. Vet Clin North Am Food Anim Pract 9(2):263–273

Lehmann M (1990) Activity requirement for young domestic rabbits: raw fiber consumption and animal welfare. Schweiz Arch Tierh 132(7):375–381

Lidfors L (1997) Behavioral effects of environmental enrichment for individually caged rabbits. Appl Anim Behav Sci 52(1):157–169. https://doi.org/10.1016/S0168-1591(96)01141-0

Lincoln W (1974) Suckling: a time constant in the nursing behavior of the rabbit. Physiol Behav 13: 711–714. https://doi.org/10.1016/0031-9384(74)90247-9

Liu H, Zhou D, Tong J, Vaddella V (2012) Influence of chestnut tannins on welfare, carcass characteristics, meat quality, and lipid oxidation in rabbits under high ambient temperature. Meat Sci 90(1):164–169. https://doi.org/10.1016/j.meatsci.2011.06.019

Lloyd HG, McCowan D (1968) Some observations on the breeding burrows of the wild rabbits Oryctolagus cuniculus on the island of Skokholm. J Zool 156:540–549. https://doi.org/10.1111/j.1469-7998.1968.tb04376.x

López M, Cervera C, Pascual JJ (2020) Welfare and performance traits of meat rabbit females. A review. ITEA Inf Tec Econ Ag 116(2):131–149. https://doi.org/10.12706/itea.2020.002

López M, Cervera C, Pascual JJ (2021) Welfare and performance results in meat rabbits during their lactation and fattening. A review. ITEA Inf Tec Econ Ag 117(2):108–129. https://doi.org/10.12706/itea.2020.014

Machado LC, Martinez-Paredes E, Cervera C (2019) Performance of rabbit does housed in collective pens and individual cages. World Rabbit Sci 27(4):227–235. https://doi.org/10.4995/wrs.2019.11540

Maertens L, Debie Y (2017) Logement de lapines "part-time" en groupe: résultats dans un élevage équipé avec des parcs polyvalents. In: Proc. 17èmes Journ. Rech. Cunicole, Le Mans, France, pp 55–58

Magnus E (2005) Behavior of the pet rabbit: what is normal and why do problems develop? Clin Pract 27:531–535. https://doi.org/10.1136/inpract.27.10.531

Marai FM, Rashwan AA (2003) Rabbits behavior under modern commercial production conditions—a review. Arch Tierzucht 46(4):357–376. https://doi.org/10.5194/aab-46-357-2003

Marin C, Simarro-Catala L, Villagra A (2018) Technical note: assessment of best location of gnawing sticks in growing rabbit cages. World Rabbit Sci 26(3):249–254. https://doi.org/10.4995/wrs.2018.7547

Martino M, Mattioli S, Farkas P, Szendro Z, Dal Bosco A, Ruggeri S, Matics Z, Castellini C, Gerencser Z (2016) Carcass traits and meat quality of growing rabbits in pens with and without different multilevel platforms. World Rabbit Sci 24(2):129–138. https://doi.org/10.4995/wrs.2016.3922

Martrenchar A, Boilletot E, Cotte JP, Morisse JP (2001) Wire-floor pens as an alternative to metallic cages in fattening rabbits: Influence on some welfare traits. Anim Welf 10(2):153–161

Marzoni M, Mori B (1992) Factores estresantes y comportamiento del conejo. Conigliocultura 29(2):19–23

Mastellone V, Bovera F, Musco N, Panettieri V, Piccolo G, Scandurra A, Di Meo C, Attia YA, Lombardi P (2019) Mirrors improve rabbit natural behavior in a free-range breeding system. Animals 9(8):533. https://doi.org/10.3390/ani9080533

Masthoff T, Hoy S (2019) Investigations on the influence of floor design on dirtiness and foot pad lesions in growing rabbits. Animals 9(6):354. https://doi.org/10.3390/ani9060354

Matics Z, Farkas TP, Dal Bosco A, Szendrő Z, Filiou E, Nagy I, Odermatt M, Paci G, Gerencsér Z (2018) Comparison of pens without and with multilevel platforms for growing rabbits. Ital J Anim Sci 17(2):469–476. https://doi.org/10.1080/1828051X.2017.1363640

Mendl M, Paul ES (2019) Assessing affective states in animals. Ment Health Well Being Anim 328–344. https://doi.org/10.1079/9781786393401.0328

Mikó A, Matics Z, Gerencsér Z, Odermatt M, Radnai I, Nagy I, Szendrő Z (2014) Performance and welfare of rabbit does in various caging systems. Animal 8(7):1146–1152. https://doi.org/10.1017/S1751731114001244

Mirabito L, Galliot P, Souchet C, Dumont F, Thomeret F (2005) Logement collectif des lapines reproductrices: Conséquences zootechniques. In: 11émes Journées Recherche Cunicole, Paris, France, pp 53–56

Mohammed H, Nasr M (2017) Growth performance, carcass traits, behavior and welfare of New Zealand White rabbits housed in different enriched cages. Anim Prod Sci 57(8):1759–1766. https://doi.org/10.1071/AN15865

Morgado E, Juárez C, Melo AI, Domínguez B, Lehman MN, Escobar C, Meza E, Caba M (2011) Artificial feeding synchronizes behavioral, hormonal, metabolic and neural parameters in mother-deprived neonatal rabbit pups. Eur J Neurosci 34:1807–1816. https://doi.org/10.1111/j.1460-9568.2011.07898.x

Morisse JP, Maurice R (1994) Welfare and the intensive production of rabbits. Rev Sci Tech 13(1):131–152

Morisse JP, Boilletot E, Martrenchar A (1999) Preference testing in intensively kept meat production rabbits for straw on wire grid floor. Appl Anim Behav Sci 64(1):71–80. https://doi.org/10.1016/S0168-1591(99)00023-4

Munari C, Ponzio P, Macchi E, Elkhawagah AR, Tarantola M, Ponti G, Mugnai C (2020a) A multifactorial evaluation of different reproductive rhythms and housing systems for improving welfare in rabbit does. Appl Anim Behav Sci 230:105047. https://doi.org/10.1016/j.applanim.2020.105047

Munari C, Mugnai C, Braconnier M, Toscano MJ, Gebhardt-Henrich SG (2020b) Effect of different management protocols for grouping does on aggression and dominance hierarchies. Appl Anim Behav Sci 227:104999. https://doi.org/10.1016/j.applanim.2020.104999

Musco N, Lombardi P, Addeo NF, Secci G, Parisi G, Pero ME, Piccolo G, Nizza A, Bovera F (2019) Mirrors can affect growth rate, blood profile, carcass and meat traits and cecal microbial activity of rabbits reared in a "small group" free-range system. Animals 9(9):639. https://doi.org/10.3390/ani9090639

Negretti P, Bianconi G, Finzi A (2010) Postura y comportamiento del conejo para determinar la altura de la jaula en relación con el bienestar animal. ASESCU, 54

Oladimeji AM, Johnson TG, Metwally K, Farghly M, Khalid M (2022) Environmental heat stress in rabbits: implications and ameliorations. Int J Biometeorol 66:1–11. https://doi.org/10.1007/s00484-021-02191-0

Olivas I, Rodriguez-Latorre A, Estelles F, Calvet S, Villagra A (2013) Short communication: assessment of activity patterns of growing rabbits in a flux-controlled chamber. World Rabbit Sci 21(2):107–110. https://doi.org/10.4995/wrs.2013.1187

Oliveira ACF, Bernardi LM, Monteiro ALB, Silva KG, Weber SH, Borges TD, Dalmau A, Costa LB (2022) Induced stress and tactile stimulation applied to primiparous does and their consequences on maternal behavior, human-animal relationships, and future offspring's sexual disorders. Vet Res Commun 46(3):925–938. https://doi.org/10.1007/s11259-022-09920-9

Paul ES, Harding EJ, Mendl M (2005) Measuring emotional processes in animals: The utility of a cognitive approach. Neurosci Biobehav Rev 29(3):469–491. https://doi.org/10.1016/j.neubiorev.2005.01.002

Pérez-Fuentes S, Muñoz-Silvestre A, Moreno-Grua E, Martínez-Paredes E, Viana D, Selva L, Villagrá A, Sanz-Tejero C, Pascual JJ, Cervera C, Corpa JM (2020) Effect of different housing systems (single and group penning) on the health and welfare of commercial female rabbits. Animal 14(6):1270–1277. https://doi.org/10.1017/S1751731119003379

Peveler JL, Hickman DL (2018) Effects of music enrichment on individually housed male New Zealand white rabbits. J Am Assoc Lab Anim Sci 57(6):695–697. https://doi.org/10.30802/AALAS-JAALAS-17-000153

Postollec G, Boilletot E, Maurice R, Michel V (2006) The effect of housing system on the behavior and growth parameters of fattening rabbits. Anim Welf 15(2):105–111

Postollec G, Boilletot E, Maurice R, Michel V (2008) The effect of pen size and an enrichment structure (elevated platform) on the performances and the behavior of fattening rabbits. Anim Welf 17:53–59

Prebble JL, Langford FM, Shaw DJ, Meredith AL (2015) The effect of four different feeding regimes on rabbit behavior. Appl Anim Behav Sci 169:86–92. https://doi.org/10.1016/j.applanim.2015.05.003

Princz Z, Dalle Zotte A, Radnai I, Biro-Nemeth E, Matics Z, Gerencser Z, Nagy I, Szendro Z (2008) Behavior of growing rabbits under various housing conditions. Appl Anim Behav Sci 111(3):342–356. https://doi.org/10.1016/j.applanim.2007.06.013

Princz Z, Dalle Zotte A, Sz M, Radnai I, Biro-Nemeth E, Orova Z, Zs S (2009) Response of fattening rabbits reared under different housing conditions. 1. Live performance and health status. Livest Sci 121(1):86–91. https://doi.org/10.1016/j.livsci.2008.05.018

Prola L, Cornale P, Renna M, Macchi E, Perona G, Mimosi A (2013) Effect of breed, cage type, and reproductive phase on fecal corticosterone levels in doe rabbits. J Appl Anim Welf Sci 16(2):140–149. https://doi.org/10.1080/10888705.2013.768903

Rahman SU, Wang X, Yu L (2018) Observations on biotic parameters of Angora rabbit breed under controlled conditions in different housing systems. Vet World 11(1):88–92. https://doi.org/10.14202/vetworld.2018.88-92

Rauterberg SL, Bill J, Kimm S, Kemper N, Fels M (2019a) Effect of a new housing system on skin lesions, performance and soiling of fattening rabbits: a German case study. Animals 9(9):650. https://doi.org/10.3390/ani9090650

Rauterberg SL, Bill J, Kimm S, Kemper N, Fels M (2019b) Evaluation of two different flooring designs for rabbit housing in accordance with German Welfare Regulations: Soiling and mortality. Agriculture 9(12):257. https://doi.org/10.3390/agriculture9120257

Ricca M, Valenzuela S, Aedo P (2018) Alojamiento grupal de conejos como enriquecimiento ambiental. Promoviendo el bienestar en escenarios experimentales. SECAL 78:8–11

Richardson VCG (2000) Rabbits: health, husbandry, and diseases. Blackwell, Oxford, UK. ISBN 063205221-X

Romero C (2008) La importancia de la Cecotrofia en el conejo. ASESCU 156:53–56

Rommers J, De Greef KH (2018) Are combi parks just as useful as regular parks for fatteners for part-time group housing of rabbit does? World Rabbit Sci 26(4):299–305. https://doi.org/10.4995/wrs.2018.9587

Rommers J, Kemp B, Houwers HW, Gunnink H, de Jong IC (2012) Description of nestbox visits and suckling events in a group housing system for rabbit does as compared to individual cages. World Rabbit Sci 20(4):231–240. https://doi.org/10.4995/wrs.2012.1231

Rommers J, Bracke MBM, Reuvekamp B, Gunnink H, De Jung IC (2014) Cage enrichment: rabbit does prefer straw or a compressed wooden block. World Rabbit Sci 22(4):301–309. https://doi.org/10.4995/wrs.2014.1353

Rosell JM, de la Fuente LF (2009) Effect of footrests on the incidence of ulcerative pododermatitis in domestic rabbit does. Anim Welf 18(2):199–204

Rosell JM, de la Fuente LF (2013) Assessing ulcerative pododermatitis of breeding rabbits. Animals 3(2):318–326. https://doi.org/10.3390/ani3020318

Ross S, Denenberg VH, Sawin PB, Meyer P (1956) Changes in nest-building behavior in multiparous rabbits. Brit J Anim Behav 4:69–74. https://doi.org/10.1016/S0950-5601(56)80024-5

Rottler S, Schoerwerth A, Lee HW, Schmidt P, Erhard MH, Bergmann S (2019) Comparison of a pen husbandry system with a dual purpose cage system for fattening rabbits concerning ethological aspects for the evaluation of animal well-being. Berl Munch Tierarztl 132(11): 513–531

Ruchti S, Kratzer G, Furrer R, Hartnack S, Würbel H, Gebhardt-Henrich SG (2019) Progression and risk factors for pododermatitis in part-time group housed rabbit does in Switzerland. Prev Vet Med 166:56–64. https://doi.org/10.1016/j.prevetmed.2019.01.013

Ruis M (2006) Group housing of breeding does. In: Maertens L, Coudert P (eds) Recent advances in rabbit sciences. ILVO, 99, 105, Melle, Bélgica

Sánchez JP, de la Fuente LF, Rosell JM (2012) Health and body condition of lactating females on rabbit farms. J Anim Sci 90(7):2353–2361. https://doi.org/10.2527/jas.2011-4065

Schlolaut W, Hudson R, Roedel HG (2013) Impact of rearing management on health in domestic rabbits: a review. World Rabbit Sci 21(3):145–159. https://doi.org/10.4995/wrs.2013.1029

Schoerwerth A, Rottler S, Lee H-W, Schmidt P, Erhard MH, Bergmann S (2019) Comparative studies on the animal health of fattening rabbits kept in pen husbandry and a dual purpose cage system under practical conditions. Berl Munch Tierarztl 132(11):532–549

Secci G, Bovera F, Musco N, Husein Y, Parisi G (2020) Use of mirrors into free-range areas: effects on rabbit meat quality and storage stability. Livest Sci 239:104094. https://doi.org/10.1016/j.livsci.2020.104094

Selzer D, Lange K, Hoy S (2004) Frequency of nursing in domestic rabbits under different housing conditions. Appl Anim Behav Sci 87(3):317–324. https://doi.org/10.1016/j.applanim.2004.01.013

Shomer NH, Peikert S, Terwilliger G (2001) Enrichment-toy trauma in a New Zealand white rabbit. Contemp Top Lab Anim Sci 40(1):31–32

Siloto EV, Zeferino CP, Meira Tavares Moura ASA, Fernandes S, Sartori JR, de Siqueira ER (2009) Temperature and environmental enrichment on the welfare of growing rabbits. Ciencia Rural 39(2):528–533

Stewart KL, Suckow MA (2016) Effects of nominal differences in cage height and floor space on the wellbeing of rabbits. J Am Assoc Lab Anim Sci 55(2):168–171

Szendrő Z, Dalle Zotte A (2011) Effect of housing conditions on production and behavior of growing meat rabbits: a review. Livest Sci 137:296–303. https://doi.org/10.1016/j.livsci.2010.11.012

Szendrő Z, McNitt IJ (2012) Housing of rabbit does: Group and individual systems: A review. Livest Sci 150:1–10. https://doi.org/10.1016/j.livsci.2012.09.017

Szendrö Z, Kustos K, El-Din SS (1988) Hair pull of rabbits does and its relationship with their rearing ability. In: 4th world rabbit congress, Budapest, 10–14 October, vol III, pp 173–181

Szendrő Z, Mikó A, Odermatt M, Gerencsér Z, Radnai I, Dezséry B, Garai É, Nagy I, Szendrő K, Matics Z (2013) Comparison of performance and welfare of single-caged and group-housed rabbit does. Animal 7:463–468. https://doi.org/10.1017/S1751731112001760

Szendrö K, Szendrö Z, Matics Z, Zotte AD, Odermatt M, Radnai I, Gerencser Z (2015) Effect of genotype, housing system and hay supplementation on performance and ear lesions of growing rabbits. Livest Sci 174:105–112. https://doi.org/10.1016/j.livsci.2015.01.008

Szendrő Z, McNitt JI, Matics Z, Miko A, Gerencser Z (2016) Alternative and enriched housing systems for breeding does: a review. World Rabbit Sci 24(1):1–14. https://doi.org/10.4995/wrs.2016.3801

Szendrö ZS, Trocino A, Hoy ST, Xiccato G, Villagra A, Maertens L (2019) A review of recent research outcomes on the housing of farmed domestic rabbits: reproducing does. World Rabbit Sci 27(1):1–14. https://doi.org/10.4995/wrs.2019.10599

Téllez JA (2022) Etología y bienestar animal en la especie cunícula: termografía aplicada a la valoración del estrés en conejos. Tesis Doctoral Inédita, Universidad de Sevilla, Sevilla

The Netherlands (2006) Verordening welzijnsnormen konijnen (PPE). (Regulations for welfare standards of rabbits). Law of The Netherlands Government: nl/BWBR0019560

Thurston S, Burlingame L, Lester PA, Lofgren J (2018) Methods of pairing and pair maintenance of New Zealand white rabbits (Oryctolagus Cuniculus) via behavioral ethogram, monitoring, and interventions. J Vis Exp 133:57267. https://doi.org/10.3791/57267

Tillmann K, Windschnurer I, Gamper J, Hinney B, Rülicke T, Podesser BK, Troxler J, Plasenzotti R (2019) Welfare assessment in rabbits raised for meat and laboratory purposes in enclosures with two floor types: Perforated plastic with holes versus slats. Res Vet Sci 122:200–209. https://doi.org/10.1016/j.rvsc.2018.11.016

Trocino A, Xiccato G (2006) Animal welfare in reared rabbits: a review with emphasis on housing systems. World Rabbit Sci 14:77–93. https://doi.org/10.4995/wrs.2006.553

Trocino A, Filiou E, Tazzoli M, Birolo M, Zuffellato A, Xiccato G (2015) Effects of floor type, stocking density, slaughter age and gender on productive and qualitative traits of rabbits reared in collective pens. Animal 9(5):855–861. https://doi.org/10.1017/S1751731114003188

Trocino A, Filiou E, Zomeño C, Birolo M, Bertotto D, Xiccato G (2018) Behavior and reactivity of female and male rabbits housed in collective pens: effects of floor type and stocking density at different ages. World Rabbit Sci 26(2):135–147. https://doi.org/10.4995/wrs.2018.7747

Trocino A, Zomeño C, Filiou E, Birolo M, White P, Xiccato G (2019) The use of environmental enrichments affects performance and behavior of growing rabbits housed in collective pens. Animals 9(8):537. https://doi.org/10.3390/ani9080537

Uczay M, Rodrigues RB, Goncalves D, Roesler T (2015) Ethology of rabbits reared in cages with environmental enrichment. Braz J Hyg Anim Sanity 9(3):439–449. https://doi.org/10.5935/1981-2965.20150040

Valuska AJ, Mench JA (2013) Size does matter: The effect of enclosure size on aggression and affiliation between female New Zealand White rabbits during mixing. Appl Anim Behav Sci 149(1):72–76. https://doi.org/10.1016/j.applanim.2013.10.002

Vandenheede M (2003) Animal welfare: The contribution of ethology. Ann Med Vet 147:17–22

Verspecht A, Maertens L, Vanhonacker F, Tuyttens FAM, Van Huylenbroeck G, Verbeke W (2011) Economic impact of decreasing stocking densities in broiler rabbit production based on Belgian farm data. World Rabbit Sci 19(3):123–132. https://doi.org/10.4995/wrs.2011.849

Vervaecke H, De Bonte L, Maertens L, Tuyttens F, Stevens JMG, Lips D (2010) Development of hierarchy and rank effects in weaned growing rabbits (Oryctolagus cuniculus). World Rabbit Sci 18(3):139–149. https://doi.org/10.4995/wrs.2010.8229

Walker MD, Duggan G, Roulston N, Van Slack A, Mason G (2012) Negative affective states and their effects on morbidity, mortality and longevity. Anim Welf 21:497–509. https://doi.org/10.7120/09627286.21.4.497

Windschnurer I, Waiblinger S, Hanslik S, Klang A, Smajlhodzic F, Löwenstein M, Niebuhr K (2019) Effects of ground floor type on selected health-parameters and weight of rabbits reared in group pens. Animals 9(5):216. https://doi.org/10.3390/ani9050216

Wolf P, Speers R, Cappai MG (2020) Influence of different types of bedding material on the prevalence of pododermatitis in rabbits. Res Vet Sci 129:1–5. https://doi.org/10.1016/j.rvsc.2019.12.004

Xiccato G, Trocino A, Filiou E, Majolini D, Tazzoli M, Zuffellato A (2013) Bicellular cage vs. collective pen housing for rabbits: Growth performance, carcass and meat quality. Livest Sci 155(2):407–414. https://doi.org/10.1016/j.livsci.2013.05.013

Zomeño C, Birolo M, Zuffellato A, Xiccato G, Trocino A (2017) Aggressiveness in group-housed rabbit does: Influence of group size and pen characteristics. Appl Anim Behav Sci 194:79–85. https://doi.org/10.1016/j.applanim.2017.05.016

Zomeño C, Birolo M, Gratta F, Zufellato A, Xiccato G, Trocino A (2018) Effects of group housing system, pen floor type and lactation management on performance and behavior in rabbit does. Appl Anim Behav Sci 203:55–63. https://doi.org/10.1016/j.applanim.2018.03.002

Zucca D, Marelli SP, Redaelli V, Heinzl E, Cardile H, Ricci C, Verga M, Luzi F (2012) Effect of environmental enrichment and group size on behavior and live weight in growing rabbits. World Rabbit Sci 20(2):89–95. https://doi.org/10.4995/wrs.2012.1082

Welfare Protocols for Rabbits

Mariana Almeida and Severiano R. Silva

Abstract

Animal welfare assessments are widely used in a range of situations and animal species. In recent years, there has been a shift towards developing more standardized animal welfare assessment protocols. Some examples are the Welfare Quality® and the Animal Welfare Indicators (AWIN) projects. These projects have produced protocols aimed to be flexible and adaptable, allowing them to be used with different farm animal species. Nevertheless, due to a low economic impact or the limitations of its geographical distribution, the welfare assessment of rabbits in commercial production was left out of both projects, even though this species is primarily reared in intensive systems. Nevertheless, several rabbit welfare certifications are now being applied in countries like Spain, Portugal, France, and Italy. Still, there is a need for more research in indicator validation, feasibility, and repeatability. In this chapter, we analyze and compare each protocol proposal recently designed for rabbit production.

Keywords

Welfare · Protocols · Indicators · Rabbits · Does · Bucks · Kits

M. Almeida (✉) · S. R. Silva
Veterinary and Animal Research Centre (CECAV), Associate Laboratory of Animal and Veterinary Science (AL4AnimalS), University of Trás-os-Montes e Alto Douro, Vila Real, Portugal
e-mail: mdantas@utad.pt

J. Simões, J. M. Monteiro (eds.), *Veterinary Care of Farm Rabbits*,
https://doi.org/10.1007/978-3-031-44542-2_9

1 Introduction

Animal welfare assessments are widely used in a range of situations and animal species. Examples of protocols can be identified for research laboratory animals, companion animals, and in zoos, as well as farm animals, and are seen as an essential tool to ensure that animals are treated ethically and humanely (Webster 2016; Blokhuis and Veissier 2022). Animal welfare has been considered a crucial element in animal science for many years, and there is a long history of research for indicators as tools for its assessment (Webster 2005; Fraser 2008; EFSA 2012). Developing protocols took a step forward in assessing animal welfare, which has been particularly noticeable for the main farm animal species, where welfare assessment protocols have become more widespread in the last two decades, both in intensive and extensive systems (Sørensen and Fraser 2010; Stilwell et al. 2020; Battini et al. 2021).

In recent years, there has been a shift towards developing more standardized animal welfare assessment protocols. Some examples are the Welfare Quality® and the Animal Welfare Indicators (AWIN) projects, which were launched in 2004 and 2011, respectively, and aimed to develop a set of standardized indicators for assessing the welfare of different farm species (Blokhuis et al. 2010; Stilwell et al. 2020). Welfare Quality® was a European Union-funded research project to develop a standardized animal welfare assessment protocol for the European Union. The project involved more than 40 partners from 13 countries and was conducted over 4 years, from 2004 to 2008 (Blokhuis et al. 2010). The Welfare Quality® protocol was designed to apply to cattle, pigs, and poultry. It is intended to be used by farmers, animal welfare auditors, and other stakeholders in the livestock industry to assess and improve farm animal welfare. After its success, the European Union decided to fund the Animal Welfare Indicators (AWIN) project, which involved more than 20 partners from 12 countries and was conducted for 4 years, from 2011 to 2015. In addition, the AWIN protocols were designed to assess other species which had not been studied in the previous project, such as sheep, goats, horses, donkeys, and turkeys (Zanella 2016; Stilwell et al. 2020). These protocols are based on four principles linked to the Five Freedoms, which outline the basic needs of animals for a good quality of life (Blokhuis et al. 2010; García 2021). The protocols include both direct and indirect measures of animal welfare. However, emphasis is given to animal-based indicators, which are considered better suited to the animal experience than input measures or resource-based indicators (Richmond et al. 2017).

These protocols aimed to be flexible and adaptable, allowing them to be used with different farm animal species. Nevertheless, due to a low economic impact or the limitations of its geographical distribution, the welfare assessment of rabbits in commercial production was left out of both projects, even though these animals are primarily reared in intensive systems (Dalmau et al. 2020). Even so, some protocols have been developed based on existing ones and adding important welfare indicators for the rabbit species (Dalmau et al. 2020; da Silva et al. 2022). This chapter aims to review the current information on welfare assessment protocols for rabbits.

2 Welfare Protocols for Rabbits

Animal welfare protocols are a set of guidelines and regulations designed to ensure that farms meet the Five Freedoms (Blokhuis et al. 2010; García 2021) in various settings, including farms, laboratories, and zoos. The evolution of animal welfare protocols has been a gradual process, with an increasing awareness of the ethical considerations surrounding the treatment of animals leading to the development of more comprehensive and scientifically informed guidelines over time. According to Dalmau et al. (2020), the work of de Jong et al. (2011) would have been the first attempt at a standardized welfare assessment protocol for rabbits based on the Welfare Quality project. The Welfare Quality® project has significantly impacted how animal welfare is assessed and managed in the European livestock industry, and its protocols have been used as a model for similar initiatives in other parts of the world. The Welfare Quality® project developed a science-based animal welfare assessment protocol focusing on four key principles of animal welfare: good feeding, good housing, good health, and appropriate behavior. The assessment protocol includes a series of criteria to evaluate each of these principles (Table 1), such as the availability of fresh water and feed, the presence of environmental enrichment, and the absence of disease. Following these principles, de Jong et al. (2011) first developed observation and scoring methodologies for on-farm commercially housed rabbit assessment. Starting with a literature study, the authors compiled the most relevant parameters for rabbit welfare, considering the different criteria and principles used in Welfare Quality while taking advantage of the knowledge of a group of experts who also proposed other indicators. This protocol focuses on the welfare assessment of reproductive does and meat rabbits, although the authors clearly state that it can also be applied to does and bucks. This protocol was then validated over the following years in Spain, and eventually, some indicators were changed or removed.

Dalmau et al. (2020) present the results of applying a protocol based on de Jong et al. (2011) work. Following the previous research, the authors refined and implemented it in 30 Spanish farms. Even though the protocol would be applicable to assess the welfare of bucks, does, and kits, the authors stated that most of the farms where the assessments took place did not have any males (for reproductive purposes), which is the case for most European rabbit commercial farms (EFSA 2020). While still following the criteria and principles of Welfare Quality, the more recent protocol presents different indicators. For example, Dalmau et al. (2020) have replaced most of the indicators related to good feeding with an assessment of the kits' access to milk and the feeder, along with the feeders cleanliness (Table 1). Body Condition Score (BCS) was maintained. Most indicators for good housing were maintained, with some indicators changing to different protocol criteria, such as stocking density and group size or being "translated" into a more objective indicator. For example, the indicator "respiration rate" (de Jong et al. 2011) transformed into "panting" (Dalmau et al. 2020), which is more feasible to measure. While the criteria principle of good health had a multiplication of indicators, the contrary happened to Appropriate Behavior. This is possibly due to the lack of studies on rabbit behavior,

Table 1 Comparison of indicators in different protocols based on the Welfare Quality principles and criteria

		de Jong et al. (2011)	Dalmau et al. (2020)	Botelho et al. (2020)	da Silva et al. (2022)	
		Bucks, does, and meat rabbits	Bucks, does, and kits	Growing rabbits	Bucks, does, and kits	Growing rabbits
Good feeding	Absence of prolonged hunger	Body condition Emaciated rabbits at the slaughter plant Rabbits per feeder	Body condition Cleanliness of feeders Access to milk in kits Access to food in kits	Body condition Cleanliness of feeders Animals per feeder	Body condition Cleanliness of feeders Access to milk in kits Access to food in kits	Body condition Cleanliness of feeders Animals per feeder
	Absence of prolonged thirst	Drinking points per rabbit Functioning of drinkers Cleanliness of drinkers Height of drinkers	Drinking points per doe/buck Functioning of drinkers Cleanliness of drinkers Height of drinkers	Animals per drinker Working drinkers Cleanliness of drinkers	Drinking points per doe/buck Functioning of drinkers Cleanliness of drinkers Height of drinkers	Animals per drinker Working drinkers Cleanliness of drinkers
					Supplementary drinker with medical water	Supplementary drinker with medical water
Good housing	Comfort around resting	Fully stretched animals	Fully stretched animals		Fully stretched animals	
		Simultaneous resting in group housing Dust Stocking density and group size Presence of shelter and elevated platform Cage size	Wet animals Dirty animals Dust Presence of resting mat Presence of an elevated platform	Wet animals Dirty animals Dust	Wet animals Dirty animals Dust Presence of resting mat Presence of an	Wet animals Dirty animals Dust

		Size of elevated platform and shelter			elevated platform (kits > 21 days)	
		Light pattern Floor types and quality of littered floor	Light quality Quality of littered floor	Light quality	Light pattern and quality Quality of nest (kits < 8 days)	Light pattern and quality
	Thermal comfort	Temperature Respiration rate Red ears	Temperature Burning hair Panting Shivering	Temperature Burning hair	Temperature Burning hair Panting Shivering	Temperature Burning hair
	Ease of movement	Hopping, jumping, turning, running Lameness Length of the pen/cage, cage/pen height Stocking density	Free movement Height of the cage Stocking density	Height of the cage Stocking density	Free movement Height of the cage Stocking density	Height of the cage Stocking density
Good health	Absence of injuries	Skin injuries/wounds (meat rabbits and does) Pododermatitis (does) Number of toes and ear damage (meat rabbits) Trichophagy (meat rabbits)	Wounds on the body Wounds on the ears Fallen ears Pododermatitis Gait score Hairless areas Risk of injuries	Wounds on the body Wounds on the ears Fallen ears Gait score Hairless areas Risk of injuries	Wounds on the body Wounds on the ears Fallen ears Pododermatitis Gait score Hairless areas Risk of injuries Cage condition	Wounds on the body Wounds on the ears Fallen ears Gait score Hairless areas Cage condition
	Absence of diseases	Mortality and selection Technical performance Management procedures like all-in-all-out, age of breeding material bought, breeding	Mortality Culling Replacement Time between parturitions	Mortality Culling Coughing Sneezing Nasal discharge	Mortality Culling Replacement Time between parturitions	Mortality Culling Coughing Sneezing Nasal discharge

(continued)

Table 1 (continued)

		de Jong et al. (2011)	Dalmau et al. (2020)	Botelho et al. (2020)	da Silva et al. (2022)	
		Bucks, does, and meat rabbits	Bucks, does, and kits	Growing rabbits	Bucks, does, and kits	Growing rabbits
		Interval between parturition (does) Age at first insemination Cleanliness of facilities	Coughing Sneezing Nasal discharge Ocular discharge Dermatophytosis, dermatitis, abscesses Neck torsions Enteropathy Diarrhea Mange Cleanliness of facilities Age of weaning	Ocular discharge Dermatophytosis, dermatitis, abscesses Neck torsions Mange Cleanliness of facilities Flies presence	Coughing Sneezing Nasal discharge Ocular discharge Dermatophytosis, dermatitis, abscesses Neck torsions Enteropathy Diarrhea Mange Cleanliness of facilities Age of weaning (kits) Interval between parturition (does)	Ocular discharge Dermatophytosis, dermatitis, abscesses Neck torsions Mange Cleanliness of facilities
	Absence of pain induced by management	Mutilations for identification Tissue growth (ear marks) Artificial insemination (willingness of the doe is stimulated)	Flies presence Killing methods Mutilations for identification	Killing methods	Killing methods Mutilations for identification	Killing methods
Appropriate behavior	Social behavior	Injuries and wounds Social behavior Group housing (does)	Negative social behavior Isolated animals	Negative social behavior Isolated animals	Negative social behavior Isolated animals	Negative social behavior

		Group size (does, meat, rabbits) Absence of solid cage walls				
	Other behaviors	Abnormal behaviors Coat condition (meat rabbits) Kit mortality Individual or group housing (does) Group size (meat rabbits) Elevated platform present (dimensions) Cage height Gnawing materials (type, amount) Nesting material (type, amount) Nest quality	Abnormal behavior Enrichment material Nesting material Time of access to the nest	Abnormal behavior	Abnormal behavior Enrichment material Nesting material (does 24 h prior to parturition) Time of access to the nest (does)	Abnormal behavior
	Human–animal relationship	Human approach test (does and meat rabbits)	Human approach test Training of personnel Touching the kits	Training of personnel	Human approach test Touching the kits	
	Positive emotional state	Fear for novel objects (does and meat rabbits); Group hopping behavior in young rabbits Elevated platform or shelter				

compared to other species bred in commercial farms (Trocino et al. 2019) and the fact that behavioral observations are time-consuming. Overall, the topic of rabbit welfare and ethology has been neglected over the years, being one of the topics with less published research (Trocino et al. 2019).

In the same year of Dalmau et al. (2020) publication, another manuscript was published based on the Welfare Quality approach and de Jong et al. (2011) work, which was tested in farms in Portugal and Spain (Botelho et al. 2020). This work focused on growing rabbits, so some indicators were not measured, and others were added to reflect the reality of this production stage. Most criteria show a decrease in the number of measured indicators, mostly due to their lack of applicability (Botelho et al. 2020). The same is observed in a recent publication (da Silva et al. 2022), in which the authors tried to apply the work of Botelho et al. (2020) and Dalmau et al. (2020) to the Brazilian reality. The more prevalent differences of this work are the application to both bucks, does, kits and growing kits, along with the addition of an indicator "Supplementary drinker with medical water." Although the authors add this indicator, no explanation is found in the text to explain this alteration.

While most of the published works in the last decades on welfare assessment protocols are related to the Welfare Quality approach, not all follow the same exact specifications. Around the time of the Welfare Quality protocols development, Cerioli et al. (2011) worked on identifying health and welfare parameters in eight farms in Italy, following an objective welfare assessment and previously collected parameters (Broom 1993; Cerioli and Lavazza 2006). The assessed parameters belonged to four criteria: health, immunological, environmental, and productive. These parameters have different sampling frequencies and are presented on a scale of importance from low to high. The proposed protocol is one of the only to include productivity as a welfare indicator, which the authors justify with the need to identify the basal level of health and welfare of the animals to identify possible issues (Cerioli et al. 2011).

In 2021, the Ebene® Welfare Assessment Measures For Rabbits were presented, a protocol for does and growing rabbits (EURCAW-Poultry-SFA 2022). Although this protocol still follows the criteria and principles of the Welfare Quality project, the indicators were reduced when compared to other species. The protocol is presented very practically, with an indication of time consumption for each part and similar indicators for both reproducing does and growing rabbits. It provides simple visuals, making reading easier and more intuitive. Warin et al. (2021) then present the results of its application in eight farms and study the feasibility and repeatability of the indicators. The protocol is deemed feasible by these authors since the assessment takes less than an hour, around 40 ± 3 min for reproducing does and 43.5 ± 5 min for growing rabbits. Furthermore, the proposed indicators are, in their majority, repeatable, except for behavior indicators which had to be altered and are still to be tested for repeatability (Warin et al. 2021).

Trocino et al. (2022) present a protocol that uses the Welfare Quality criteria to identify possible indicators. It was one of the most recently published works in which different housing systems were subjected to the same evaluation. In order to evaluate both reproducing does and litters, and growing rabbits, Trocino et al. (2022)

propose two visits: pre-weaning and pre-slaughtering. Many indicators proposed by these authors can be found in the previously described literature, although some are different, such as "litter size," "animal genotype," and "NH_3, and CO_2 concentrations." Trocino et al. (2022) propose an approach without a behavioral assessment, basing the animals' fulfilling the criteria of "Appropriate Behavior" through resource and management-based indicators and animal-based measures (ABMs), such as injuries.

3 Applicability and Limitations

Most of the protocols described in this chapter are being perfected and tested by different organizations and countries. However, just like any other research field, there are limitations to their applicability that must be considered to better understand how to improve them and move forward.

Due to the rabbit's short production cycle, it is crucial to establish the best times for welfare assessment since indicators like the body condition score of the does fluctuate during different reproductive phases. de Jong et al. (2011) point to this limitation at the beginning of the document, providing a list of ideal moments to assess does, according to their reproductive cycle. The best option to assess the body condition score of does, for example, is around artificial insemination; however, it can also be assessed at the end of lactation, which allows for determining litter performance and the percentage of palpated positives, reducing does' handling (de Jong et al. 2011). In a recent protocol proposal, Trocino et al. (2022) suggest two assessments: the first pre-weaning, around 27–31 days after kindling, and the second pre-slaughtering, around 2–5 days before slaughter. On the first visit, the assessors would make recordings of the indicators for does and litters; on the second, the same would be done for growing rabbits. Although it is necessary to assess the animal's welfare when it is most at risk (de Jong et al. 2011), it is fair to say that this proves to be a near-impossible task for producers and assessors. Time is frequently pointed out as one of the most significant limitations of welfare assessment protocols. In this case, allied with the short production cycle and a large number of animals usually present at the facilities, it may prove challenging for everyone involved.

Most of the existing research is focused on applying the modified protocols; however, there is little focus on validating the results and the protocols themselves. There needs to be more research in indicator validation, feasibility, and repeatability, specifically for rabbits. Although most of these indicators were validated for other species, it is necessary to provide more information on repeatability and accuracy between assessors. Warin et al. (2021) describe an assessment carried out by two assessors and presenting their results and making a feasibility and repeatability (intra- and inter-assessor) study. Sanitary indicators presented high repeatability in growing rabbits, so most were not modified, in contrast to the behavioral indicators for both does and growing rabbits, which needed to be modified due to low repeatability (Warin et al. 2021). EURCAW-Poultry-SFA (2022) extensively reviews the indicators used across different protocols available at the moment, including their validity, feasibility, and reliability. This comparison will be helpful in the future to study and improve indicators.

Experts in the field have developed all of the before-mentioned protocols and followed years of research in rabbit welfare. However, recent versions of the welfare assessment protocols for rabbits present a lower number of indicators in common with each other and a lack or severe decrease in behavioral assessments. In some cases, no behavioral assessment is carried out (Cerioli et al. 2011; Botelho et al. 2020; Trocino et al. 2022). The authors justify that this is due to the difficulties in obtaining behavioral measurements, especially in cages with more than one individual (Botelho et al. 2020), or due to achieving a notion of the animal's behavior through resource and management-based indicators and ABMs (Trocino et al. 2022). However, it is essential to remember that one of EFSAs rabbit welfare assessments conclusions was that behavioral restrictions are one of the main concerns both in reproducing does and growing rabbits in conventional cages (EFSA 2020), which is still one of the most used housing systems in commercial farms. In the same document, EFSA also recommends that basic research on rabbit behavior is carried out since it is essential to ensure these species' welfare (EFSA 2020).

It is well known that promoting consistency between assessments is essential but also challenging. Frequently the auditors or assessors in the cited manuscripts are trained to meet the gold standard of the Welfare Quality assessment protocols. This is ideal since it is essential to promote consistency between assessments, which can only be obtained through credible and trained assessors (Blokhuis et al. 2010). However, since the Welfare Quality project did not include rabbits, both Botelho et al. (2020) and Dalmau et al. (2020) state that the assessors are trained according to the training procedure established in Welfare Quality (Blokhuis et al. 2010). In some cases, there is no mention of the assessor's training (Cerioli et al. 2011; Warin et al. 2021; Trocino et al. 2022). da Silva et al. (2022) mention training by the Institute of Agrifood Research and Technology (IRTA), which has been developing the Welfair® certification program and created welfare protocols for rabbits (Welfair®; https://www.animalwelfair.com/).

Several welfare certifications are now being applied in different countries. In Portugal and Spain, several certification companies are applying the Welfair® welfare protocols for rabbits (Welfair®; https://www.animalwelfair.com/). The EBENE protocols (ITAVI 2018) are being employed in France. According to the EURCAW-Poultry-SFA (2022), Italy is about to approve and apply a welfare assessment protocol for rabbits at the beginning of the present year (2023). The final draft was provided to EURCAW Poultry-SFA, which compares the different European protocols, focusing mainly on those created in Spain, France, and Italy. This protocol has a similar system to the ones presented by Botelho et al. (2020) and Dalmau et al. (2020) and is based on Classyfarm (https://www.classyfarm.it/) (EURCAW-Poultry-SFA 2022). Although there is an increased interest in the topic, and there have been efforts to create mechanisms to assess rabbit welfare properly, EFSA et al. (2020) recommend that "a validated welfare assessment protocol suitable for on-farm use should be standardized across the EU." However, this is a challenging task due to the diversity of rabbit farming systems.

4 Conclusions

Most welfare assessment protocols for rabbits have been developed using the Welfare Quality approach. Undoubtedly, this project has been widely adopted in Europe and is considered a benchmark for animal welfare assessment. In addition to the assessment protocol, the Welfare Quality® project also developed a set of guidelines for animal welfare improvements and a training program for auditors and farmers.

The increase in publications concerning rabbit welfare and the development of several welfare assessment protocols all over Europe is a clear sign of change. Although there are still limitations to these protocols, the challenges of rabbit production are not forgotten, and there is an interest in improving this species' welfare in commercial farms. The authors of the presented protocols propose indicators that cover a wide range of animal welfare domains, including physical health, behavior, and affective states. Like other species' indicators, they will need to be thoroughly validated and tested for reliability and repeatability. The following years should be focused on making sure that the most relevant indicators remain and working to create a merge between these protocols all over Europe.

References

Battini M, Renna M, Giammarino M, Battaglini L, Mattiello S (2021) Feasibility and reliability of the AWIN welfare assessment protocol for dairy goats in semi-extensive farming conditions. Front Vet Sci 8. https://doi.org/10.3389/fvets.2021.731927

Blokhuis HJ, Veissier I (2022) Animal welfare assessment. In: Knight A, Phillips C, Sparks P (eds) Routledge handbook of animal welfare. Routledge, Abingdon. https://doi.org/10.4324/9781003182351-4

Blokhuis HJ, Veissier I, Miele M, Jones B (2010) The Welfare Quality® Project and beyond: Safeguarding farm animal well-being. Acta Agric Scand A Anim Sci 60(3):129–140. https://doi.org/10.1080/09064702.2010.523480

Botelho N, Vieira-pinto M, Batchelli P, Pallisera J, Dalmau A (2020) Testing an animal welfare assessment protocol for growing-rabbits reared for meat production based on the welfare quality approach. Animals 10(8):1–25. https://doi.org/10.3390/ani10081415

Broom DM (1993) Animal welfare: its scientific measurement and current relevance to animal husbandry in Europe. In: Phillips C, Piggins D (eds) Farm animals and the environment. CAB, Wallingford

Cerioli M, Lavazza A (2006) Viral enteritis of rabbits. In: Coudert P (ed) Recent advances in rabbit sciences. Institute for Agricultural and Fisheries Research (ILVO), Animal Science Unit, Melle

Cerioli M, Brivio R, Tittarelli C, Grilli G, Lavazza A (2011) Identification of health and welfare parameters for rabbit production and definition of an evaluation score. J Agric Sci Technol 1: 500–507

Dalmau A, Moles X, Pallisera J (2020) Animal welfare assessment protocol for does, bucks, and kit rabbits reared for production. Front Vet Sci. https://doi.org/10.3389/fvets.2020.00445

da Silva KG, Borges TD, Costa LB, Dalmau A, Sotomaior CS (2022) Rabbit welfare protocols under Brazilian conditions: The applicability of welfare protocols in rabbit farms for different purposes—first results. J Vet Behav 54:36–53. https://doi.org/10.1016/j.jveb.2022.06.002
de Jong IC, Reuvekamp BFJ, Rommers JM (2011) A welfare assessment protocol for commercially housed rabbits—First step in the development of the protocol. Wageningen UR Livestock Research. https://edepot.wur.nl/197480
EFSA (2012) Statement on the use of animal-based measures to assess the welfare of animals. EFSA J 10(6). https://doi.org/10.2903/j.efsa.2012.2767
EFSA AHAW Panel (EFSA Panel on Animal Health and Welfare), Saxmose Nielsen S, Alvarez J, Bicout DJ, Calistri P, Depner K, Drewe JA, Garin-Bastuji B, Gonzales Rojas JL, Gortázar Schmidt C, Michel V, Miranda Chueca MÁ, Roberts HC, Sihvonen LH, Spoolder H, Stahl K, Velarde Calvo A, Viltrop A, Buijs S, Edwards S, Candiani D, Mosbach-Schulz O, Van der Stede Y, Winckler C (2020) Scientific Opinion on the health and welfare of rabbits farmed in different production systems. EFSA J 18(1):5944, 96 pp. https://doi.org/10.2903/j.efsa.2020.5944
EURCAW-Poultry-SFA (2022) List of welfare indicators and methods of assessment for rabbits on-farm. In: EURCAW-Poultry-SFA – 2022 – DL. 2.1.5 – 1/63
Fraser D (2008) Understanding animal welfare. Acta Veterinaria Scandinavica 50(1):1–7. https://doi.org/10.1186/1751-0147-50-S1-S1
García A (2021) Housing and rabbit welfare in breeding does. In: Argente M-J, De La Luz García Pardo M, Dalton KP (eds) Lagomorpha characteristics. IntechOpen, London. https://doi.org/10.5772/intechopen.91829
ITAVI (2018) Evaluer le Bien-être des Lapins en Maternité et en Croissance
Richmond SE, Wemelsfelder F, de Heredia IB, Ruiz R, Canali E, Dwyer CM (2017) Evaluation of animal-based indicators to be used in a welfare assessment protocol for sheep. Front Vet Sci 4: 1–13. https://doi.org/10.3389/fvets.2017.00210
Sørensen JT, Fraser D (2010) On-farm welfare assessment for regulatory purposes: Issues and possible solutions. Livestock Sci 131(1):1–7. https://doi.org/10.1016/j.livsci.2010.02.025
Stilwell G, Vieira A, Can E, Krug C, Saraiva S, Battini M, Mattiello S (2020) Ruminants' welfare assessment. In: Duarte AF, da Costa LL (eds) Advances in animal health, medicine and production. Springer, Cham. https://doi.org/10.1007/978-3-030-61981-7_1
Trocino A, Cotozzolo E, Zomeño C, Petracci M, Xiccato G, Castellini C (2019) Rabbit production and science: the world and Italian scenarios from 1998 to 2018. Ital J Anim Sci 18(1): 1361–1371. https://doi.org/10.1080/1828051X.2019.1662739
Trocino A, Menegon F, Zomeño C, Pasqualin D, Cunial G, Xiccato G, Pirrone F, Bertotto D, Bortoletti M, Dorigo F, Lavazza A, Di Martino G (2022) A pilot study about on-farm assessment of health and welfare in rabbits kept in different housing systems. Front Vet Sci 9. https://doi.org/10.3389/fvets.2022.936643
Warin L, Mika A, Souchet C, Bouvarel I (2021) Feasibility and repeatability of the Ebene® Welfare Assessment measures for rabbits. In: Proceedings 12th World Rabbit Congress, November 3–5 2021, Nantes, France, p 4
Webster J (2005) Evaluación y materialización del bienestar animal: de la teoría a la práctica. Revue Scientifique et Technique de l'OIE 24(2):723–734. https://doi.org/10.20506/rst.24.2.1602
Webster J (2016) Animal welfare: Freedoms, dominions and "A life worth living". Animals 6(6): 2–7. https://doi.org/10.3390/ani6060035
Zanella A (2016) AWIN—Animal Health and Welfare—FP7 Project. Impact 2016(1):15–17. https://doi.org/10.21820/23987073.2016.1.15

Rabbit Nutrition and Feeding

Victor Pinheiro and Thierry Gidenne

Abstract

This chapter intends to address the feeding of rabbits for the production of meat in conventional intensive systems (indoor breeding). Feed costs represent the main cost of rabbit production and thus sharply impact the profitability. The main components of the diet for rabbits (i.e., fiber, starch and protein) are characterized and their implications in the performances. Balanced nutrients are needed to reach the expected performances of genetically improved lines of rabbits. Digestive pathologies have a high incidence around the weaning, and we propose nutrition and feeding strategies (outside the use of antibiotics) to minimize these occurrences, namely the use of feed intake limitation programs, and good fibers balance in the diets between indigestible fiber and more digestible fiber. The nutritional needs of rabbits are now well known and if respected the animal can express fully its productive potential without compromising its health. Rabbit farming is a production with low environmental impact and reduced competition with human population. Local ingredients from short circuits and many coproducts are already included in their feed, which meets some of the challenges of current animal production.

Keywords

Rabbit farming · Diet formulation · Feeding strategies · Fiber

V. Pinheiro (✉)
Animal and Veterinary Research Centre (CECAV), University of Trás-os-Montes and Alto Douro, Vila Real, Portugal

Associate Laboratory for Animal and Veterinary Sciences (AL4AnimalS), Lisbon, Portugal
e-mail: vpinheir@utad.pt

T. Gidenne
GenPhySE, Université de Toulouse, INRA, INPT, ENVT, Castanet Tolosan, France

J. Simões, J. M. Monteiro (eds.), *Veterinary Care of Farm Rabbits*,
https://doi.org/10.1007/978-3-031-44542-2_10

1 Introduction

The rabbit is an herbivore well adapted to the use of fibrous feed, due to the presence of microorganisms in its lower digestive tract and because it practices cecotrophy. The cecal microbiota plays a fundamental role in the digestive physiology and health status.

Fibers are the main components of the rabbit's diet and are essential for several digestive processes, such as digestive transit and cecal fermentation. The rabbit requires a minimum intake of fibers to maintain adequate digestive process and reduce the risks of digestive pathologies, particularly around the weaning. However, a high fiber intake corresponds to a lower feed conversion, and thus to moderate productive performances. There is thus an antagonism between productivity and animal health security. According to the type of animal and its physiological state, rabbits have different nutritional needs, so their knowledge is necessary to apply an adequate feeding program. Although fresh forage or hay can be used in small-scale "family" production systems, in European conventional intensive rabbit farming a complete pelleted and balanced feed is provided to animals, to meet all nutritional needs at each production phase (gestation, lactation, growth). Below we detail the feeding and nutrition of the genetically improved rabbit bred in intensive farming systems.

2 Feeds Components, Characteristics, Digestibility, Effects, and Antinutritional Factors

2.1 The Fiber

Dietary fibers are made up mostly of carbohydrates with complex bonds, which have a structural function in plant cell wall. The concept of fibers in animal nutrition is generally defined as polysaccharides and other substances resistant to endogenous enzyme digestion that can be partially or totally hydrolyzed by enzymes produced by the gut microbiota. Similarly, fibers include cellulose, hemicelluloses, and pectic substances, but also other components such as oligosaccharides, gums, resistant starch, and inulin that have variable digestibility. Other noncarbohydrate components (polyphenols such as lignin, waxes, saponins, cutin, phytates, and resistant protein) are also part of fiber.

The biochemical characteristics of fibers are one of the main factors that affect digestive physiology, and the impact depends on many factors, such as molecular weight, monomer nature, and composition and types of bonds between sugars, etc.

The different fiber fractions can be determined by different techniques (Fig. 1), which allow the characterization and control of the quality of the raw materials necessary for the diet formulation.

Depending on the location in the plant cell, two main groups of dietary fiber are considered (Fig. 1): (1) water-soluble or insoluble plant cell cytoplasmic components (fructans, resistant starch and mannans) and (2) cell wall components,

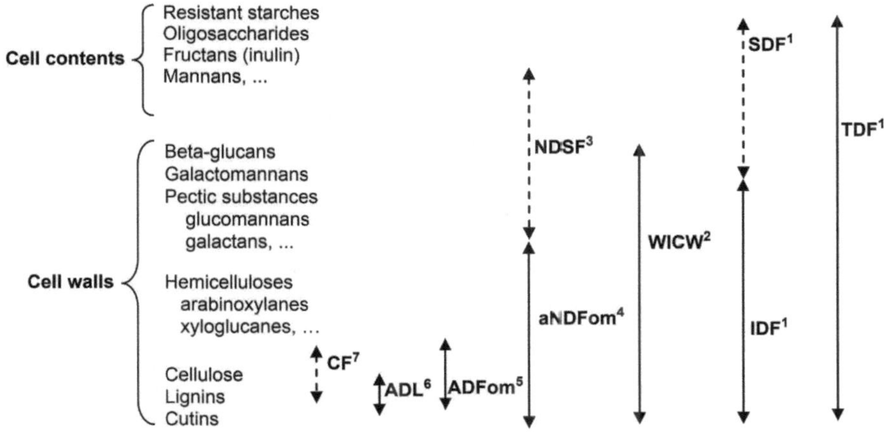

Fig. 1 Main fiber fractions and their quantification by some gravimetric methods used in animal feeding. (1) Total dietary fiber (TDF), insoluble dietary fiber; soluble dietary fiber is determined by difference (TDF–IDF) or directly; (2) water-insoluble cell wall; (3) neutral detergent-soluble fiber; (4) neutral detergent fiber assayed with a heat-stable amylase and expressed free of ash, (5) ADF expressed free of ash; (6) ADL; (7) crude fiber. Adapted from Gidenne et al. (2015) with permission from Elsevier

such as nonstarch polysaccharides (NSP; some β-glucans, arabinoxylans, part of pectic substances, etc.) and water-insoluble polymers (lignins, cellulose, hemicelluloses, and pectic substances).

Water-soluble fiber consists of nonstarch polysaccharides, namely pectic β-glucans, fructans, and gums. It is a more difficult fraction to determine and is rarely used in the characterization of diets.

From the point of view of the nutritional use of fiber, which is related to the effect on the health of rabbits, we can classify fiber as indigestible or digestible (components that are mostly fermentable in the ceco-colic segments).

The utilization of dietary fiber occurs essentially after the ileus and depends on the activity of cecal microorganisms. However, there is evidence that some fiber components can be degraded even in the ileum, resulting from cecotrophy, which in the rabbit is higher than other nonruminant species and whose extent can reach up to 43%, expressed in NDF (Gidenne et al. 2020a, b). The extent of fiber digestion is quite different according to the fraction (Table 1), ranging from 5% for cellulose to 90% for the most soluble fiber fractions. The digestibility of fiber is lower than that of protein or starch, which is reduced when levels of dietary fiber increase.

Dietary fiber is one of the most important factors to consider in rabbit diets, where accounting for 40–50% of the total diet (Gidenne et al. 2020a, b). Its effects can be observed in the productive performances of the growing and adult animals, in the health status, in the digestibility and use of the constituents of the diet, in the cecal fermentative characteristics, in the development of the caecum and in the digestive transit (Gidenne 2015).

Table 1 Whole-tract digestibility coefficients (g kg^{-1}) for some fiber fractions in the growing rabbit

	Range
Neutral detergent fiber (aNDFom)	100–600
Cellulose (ADFom − ADL)	50–250
Hemicelluloses (aNDFom − ADF)	200–600
Water-insoluble pectins	300–800
Uronic acids	300–850
Soluble fiber (TDF − aNDFom)	700–900
Lignins (ADL)	–150–150

Adapted from Gidenne et al. (2015)

Table 2 Fibers fraction in some raw materials for feed rabbit

Ingredients	Wheat straw	Wheat bran	Dehydrated alfalfa	Sugar-beet pulp	Sunflower meal	Soya bean hulls	Grape pomace
aNDFom	80	45	46	47	48	62	64
ADFom	54	11	34	22	32	44	54
ADL	16	3	8	2	11	2	34
NDSF	–	3	18	30	–	22	–
Crude Fiber	40	10	27	19	26	36	26

Adapted from Gidenne et al. (2015)
NDF, neutral detergent fiber assayed with a heat-stable amylase and expressed free of ash; ADF, acid detergent fiber expressed free of ash; ADL, lignin acid detergent; NDSF, neutral detergent-soluble fiber

Cellulose (found in alfalfa, straw, and cereal husks) and mainly hemicelluloses (wheat, rye, and oilseed husks) are partly digested by bacteria, also producing VFA with a decrease in cecal pH. Pectins (fruit pulp, beet pulp) are the most digestible cell wall elements (up to 75%) and together with hemicelluloses are the most digestible fibers. Also, fiber digestion is also dependent of the fiber origin, particle size, and previous treatment. When the lignin content (cereal straw, grape pomace, and sunflower seed husk) is increased, fiber is digested to a much lesser extent. However, together with cellulose, it is related to low incidence of digestive pathologies (lower mortality in fattening period). In return, high fiber intake stimulates the intestinal peristalsis and reduces the digesta transit time, particularly in the ceco-colic segments, and consequently, reduces the fermentation time.

The different raw materials incorporated in diets for rabbits present different fiber fractions (Table 2), which will influence the levels of incorporation in the formulation of the diets. In addition to the fiber content of the diet, the requirements are also related to the size of the fibrous particles, namely large particles (>0.3 mm), as this affects the digestive physiology of the rabbit and deserves attention in diet formulation (García et al. 2002). The particle size distribution varies with the incorporation of raw material.

An insufficient fiber intake causes an unfavorable intestinal environment, and slower digestive transit, which will be reflected in worse growth, and increased mortality and morbidity, especially after weaning.

2.2 The Starch

Starch is a major plant reserve carbohydrate found in nature as granules either in seeds, roots, or tubers. It is formed exclusively by glucose molecules, it constitutes an important source of energy for metabolism and may be digested mainly by intestinal enzymes (Blas 1986).

Starch digestibility depends on its concentration in the feed and on its biochemical structure according to its botanical origin. It depends also on the age of the animals and the technological treatments (Blas and Gidenne 2020). In adult rabbits, fecal starch digestion is very high (between 95 and 100%), and fecal excretion is close to zero (Gidenne and Perez 1993). For growing rabbits, starch that escapes intestinal digestion is quickly fermented by the cecal microbiota, resulting in the release of lactate and volatile fatty acids (VFAs) that are absorbed through the cecum wall. Several factors can influence starch digestion, notably: (1) the size and physical structure of the starch granules, (2) the particle size, (3) the amylopectin levels, (4) the presence of lipoproteins and tannins that can inhibit the action of α-amylase, and (5) the form of feed presentation.

In young rabbits (before 30 days old), amylolytic activity is low and the ability to digest starch is limited. Consequently, some starch may enter the cecum leading to the production of VFA by the bacteria.

In starches considered to be more resistant to enzymatic hydrolysis (corn, potato), the amount that escapes hydrolysis in the small intestine is greater and, consequently, the amount available for bacteria in the cecum increases. Starch from wheat, barley, cassava, and some legumes (chickpeas and peas) is more easily hydrolyzed than potato or corn starch. In addition to starch, small amounts of simple sugars (fructose, glucose), disaccharides (maltose, saccharose) and oligosaccharides (maltotriose, dextrin) may also be present in diets, which are easily hydrolyzed in the small intestine by endogenous enzymes (Blas and Gidenne 2020).

2.3 The Protein

Proteins are macromolecules made up of long chains of amino acids linked by peptide bonds. Essential amino acids for rabbits are: isoleucine, leucine, lysine, methionine, phenylalanine, threonine, tryptophan, and valine (Villamide et al. 2020). The protein supply comes from feed ingredients and from endogenous sources, from the microorganisms present in cecotrophes. Ingesting cecotrophes allows rabbits to use amino acids from microbial protein synthesis occurring in the cecum. Cecotrophy contributes to protein requirements for approximately one third of the total protein intake (García et al. 2005).

Legume seeds (soya, pea, etc.) are the main provider of proteins in rabbit feed. However, protein supply comes also from fiber sources such as alfalfa meal or sunflower meal. Dietary proteins are digested in the stomach by pepsin and then in the small intestine by pancreatic and intestinal peptidases and transformed into amino acids that are absorbed and used in cellular metabolism. The protein that escapes endogenous digestion is fermented in the cecum, leading to NH3 production, which in turn is incorporated into the bacteria. An increase in the concentration of protein in the diet corresponds to an increase in its digestibility, although it is also dependent on other factors, namely the ADL content, with which it presents a negative correlation (Villamide et al. 2020).

The use of dietary protein also depends on energy availability and the dietary protein/energy ratio must be considered, as energy is required for protein anabolism and catabolism. Low levels of protein (digestible) cause a decrease in feed efficiency and even delays in the growth of the young. Low energy levels promote the use of endogenous proteins as an energy source for bacteria and increase the production of NH3, with consequences on air quality and on the formation of nitrates and environmental contamination. Excess protein may favor the proliferation of proteolytic and alkalophilic bacteria, producing NH3 and increasing the pH value that may cause digestive, liver, and kidney disorders (the latter due to ammonia absorption). An excessive intake of crude protein (>18%) may favor diarrhea.

2.4 The Fat

Triglycerides are considered the true fats, as they represent the main form of stored fat in animals and plants. Dietary fat comes essentially from nuts and cereals. The use of fats of animal origin is restricted. Fat is a supplementary source of energy and should represent between 2 and 4% of the diet (Xiccato 2020). In breeding does, the addition of fat to the diet increases energy density and helps to attenuate the negative energy balance that occurs during lactation, improving their body condition (Maertens et al. 2006).

The digestibility of fat is generally high, but it depends on the level of incorporation and the origin. In adult animals, triglycerides are hydrolyzed and the fatty acids and glycerol are absorbed in the small intestine. In young rabbits there may be some preintestinal digestion. Some fat that escapes endogenous enzymatic digestion is then used by cecal microorganisms resulting in the production of VFA. Digestibility of fat as well as fiber increases with moderate incorporation of fat (up to 4%), which may result from decreased feed intake (Xiccato 2020). With higher incorporation levels (>6%), the effect on digestibility is adverse, due to the alteration of cecal fermentative activity (Falcão e Cunha et al. 2004).

The age of the rabbit does not greatly affect fat digestibility, it may remain constant or decrease slightly. At birth, the lapin has the ability to digest milk fat. However, it turns out that with age, the ability to digest fat, which in the young rabbits is mostly carried out by salivary lipase, is transferred to pancreatic lipase with the switch to solid feed (Gallois et al. 2008). In adult animals the ability to digest fat may increase slightly.

2.5 Antinutritional Factors in Rabbit Feed

Some plants develop mechanisms that protect them from being ingested, producing substances that have adverse physiological effects in animals, which are called antinutritional factors. These factors, which reduce the usability of diets or cause toxicity, can be classified according to their chemical nature or their effects with alkaloids, tannins, and trypsin inhibitors, being more frequent (Vikram et al. 2020). In some ingredients used in animal feed, due to genetic improvement, the content of antinutritional factors is reduced. Some of these factors are also destroyed by high temperatures, which happens, for example, with the trypsin inhibitors present in soya, which are inactivated by the manufacturing technology.

The importance of antinutritional factors is more worrying when the rabbit's diet uses local plants or highly lignified coproducts with alkaloids which can limit their incorporation. For example, mimosine present in the genus Leucaena leucocephala can decrease performance in rabbits (Ramteke et al. 2019), which limits its use to 5–10%.

The origin of antinutritional factors in the diets can also originate from contamination with weed seeds. In Europe, some legumes or cereals may contain seeds of Datura stramonium, which contain atropine (alkaloid) which is a molecule that can cause acute or chronic poisoning. However, rabbits are not affected, and they synthesize the atropine esterase enzyme (Piva and Piva 1995).

2.6 The Drinking Water

The quality of the water available to the rabbit is extremely important, as food intake is dependent on water consumption. Water intake is approximately 2–5 times greater than feed intake when solid feed is a complete pelleted diet.

Daily water consumption by rabbits varies according to age, season (temperature and humidity), the percentage of water incorporated in the feed, and the physiological phase. The amount of water ingested can vary from 100 to 300 ml day^{-1} in young

Table 3 Daily drinking water consumption in rabbits of medium size

Animal	Consumption (ml by kg body weight)
Young	200–300
Adult	300–1300
Doe with litter (0–14 days)	800–1500
Doe with litter (14–28 days)	1500–2500
Doe with litter (28–42 days)	2500–3800

Adapted from Bodnar and Bodnar (2020)

animals to 1–4 L day^{-1} in adult animals of medium breeds (Table 3). The physiological state of the rabbit also affects water consumption, with a strong increase in the lactation phase and depending on milk production. If we consider the mother and the litter as a whole, consumption may reach 4 L day^{-1}, close to weaning (Table 3). Water intake will also depend on the ambient temperature, being lower at moderate temperatures (within the animal's thermal comfort zone, between 15 and 20 °C) and increasing slightly when the temperature decreases (due to higher dry matter intake), and suffers a greater increase in hot temperatures (needs to regulate body temperature). Water intake is reduced if the animals have access to green fodder, but even in this situation, good quality water must be made available.

Water temperature will also influence your consumption. Intake is maximum when the water has a temperature close to 15–18 °C. For temperatures below 8 °C, the drastic reduction in consumption can even cause digestive disorders (Bodnar and Bodnar 2020); above 29 °C, the reduction in water intake compromises productive results, with greater relevance in breeding rabbits (Yassein et al. 2008).

Some authors suggest that the high mineral content of drinking water (generally referring to sodium chloride) may decrease the performance of the animals, depending on the effect of the type and amount of minerals and the tolerance of the animals and the ambient temperature. For rabbits, water with a relatively low salt content (limit value of approximately 3000 ppm) is recommended to avoid impairing the performance and health of the animals (Abdelsattar et al. 2020).

In practice, the rabbit must have water permanently and of good quality. The initial water may be of good quality, but contamination in tanks or pipes may later occur. To maintain good water quality, tanks, distribution pipes, and even drinking fountains must be kept clean.

The parameters to be evaluated in drinking water quality differ in different parts of the world, but there are general recommendations from the World Health Organization, which define some guidelines and concepts to consider to prevent risks and guarantee the quality of drinking water for human consumption. In Europe, the parameters to be considered in water analysis are framed by some legislation (Dir 80/778/EEC and Dir 98/83/EEC), in which 38 parameters are considered divided into three groups: (1) microbiological parameters, (2) organoleptic parameters (color, taste, odor, and turbidity), and (3) physical-chemical parameters (conductivity, pH, ionic composition). Analyses of the parameters of group 1 should be carried out regularly (at least twice a year). The analyses of groups 2 and 3 may be carried out less frequently. Pfost and Fulhage (2001) define the chemical composition shown in Table 4 for production animals.

Water quality is generally good if the source is public water supply, which is controlled and meets quality criteria necessary for human consumption. When the source is underground extraction, the water quality is not guaranteed, as it may be polluted (nitrates, pesticides) as a result of intensive agriculture (Fortun-Lamothe et al. 2009).

Table 4 Desired and potential levels of pollutants in livestock water supplies

Substance	Desired range	Problem range
Total bacteria per 100 ml	<200	>1,000,000
Fecal coliform per 100 ml	<1	>1 for young animals >10 for older animals
Fecal strep per 100 ml	<1	>3 for young animals >30 for older animals
pH	6.8–7.5	<5.5 or >8.5
Dissolved solids, milligrams per liter	<500	>3000
Total alkalinity, milligrams per liter	<400	>5000
Sulfate, milligrams per liter	<250	>2000
Phosphate, milligrams per liter	<1	Not established
Turbidity, Jackson units	<30	Not established

Adapted from Pfost and Fulhage (2001)

3 Nutrient Requirements in the Different Stages of Production and Diet Formulation

3.1 Nutrient Requirements

Throughout growth and in view of body development, needs will increase. In European intensive systems, fattening rabbits have an average daily weight gain between 35 and 55 g day^{-1}, which is maximum at the 5th or 7th week of life. If the animal is housed within the zone of thermal neutrality, the maintenance needs are proportional to the metabolic weight (the metabolic weight (MW) is an exponential function of the live weight (LW); MW = LW 0.75).

During the gestation period, the nutritional requirements for the formation of the uterine tissues of the doe and the developing embryos are small during the first 2 weeks of gestation but increase later in the third and are more important in the last third of gestation. A nutritional deficit implies a reduction in the weight of the pups at birth with a decrease in the protein content and mobilizable lipid reserves.

During the first 2 weeks of lactation, the rabbit doe can increase its body weight, but in general, afterwards, the rabbit doe loses weight. These changes in live weight reflect the balance between needs and intake capacity, which results in a negative energy balance during most of lactation.

Energy

In animal nutrition, energy needs can be expressed in different parameters referring to different requirements, considering their use in the body: gross energy (GE), digestible energy (DE), metabolizable energy (ME), and net energy (NE), where gross energy is the total energy present in the diet and net energy is the energy that is available for cellular metabolism, which in rabbits represents 35–40% of gross

Table 5 Digestible energy requirements for maintenance (DEm) and efficiency of utilization of DE intake (DEI) and body energy reserves

	Young rabbits	Pregnant does	Lactating does	Pregnant and lactating does	Nonreproducing does and bucks
DEm (kJ day^{-1} kg^{-1} LW$^{0.75}$	430	430	430	470	400
Efficiency of energy utilization					
Body retained energy (RE/DEI)	0.5	0.5			0.5
RE as protein (RE/DEI)	0.4				
RE as fat (RE/DEI)	0.65				

LW, live weight; RE, retained energy
Adapted from Xiccato and Trocino (2020)

energy (Xiccato and Trocino 2020). In rabbits, requirements are usually expressed in digestible energy.

The energy metabolism and consequently the energy needs of rabbits are influenced by several factors, the most important of which are (1) body size/weight, (2) physiological status and production level (maintenance, growth, pregnancy, and/or lactation), and (3) environmental conditions (air temperature and humidity). The Table 5 summarizes the average values indicated in the literature for maintenance energy needs and efficiency of energy use in rabbit a different physiological state (Parigi Bini and Xiccato 1986).

The growth rate is related to energy intake (when the energy-to-protein ratio is constant and balanced in the supply of amino acids). It is estimated that the maximum daily weight gain is achieved when the concentration of digestible energy in the diet is approximately 10–10.5 MJ kg^{-1}. An increase in the energy density of the diet and a consequent increase in energy intake will affect the composition of body tissues (protein and fat), and these changes are not linearly correlated.

The reproduction rhythm most used in intensive rabbit farming is mating on a fixed day after parturition (generally 11 days). In this case, during the cycle part of time, the female is in negative energy balance and utilizes body reserves, and in other phases, the rabbit is in positive balance and body energy recovery. The breeding system affects the energy balance of lactating does (Fig. 2).

Protein and Amino Acids

Protein needs should be expressed as digestible protein (DP). As animals have specific amino acid requirements, their requirements must also be considered (Table 6).

Protein needs vary according to the type of animal and the physiological state (Table 7)

The use of dietary protein will depend on energy and therefore it is important to pay attention to the digestible protein/digestible energy (DP/DE) ratio, which should

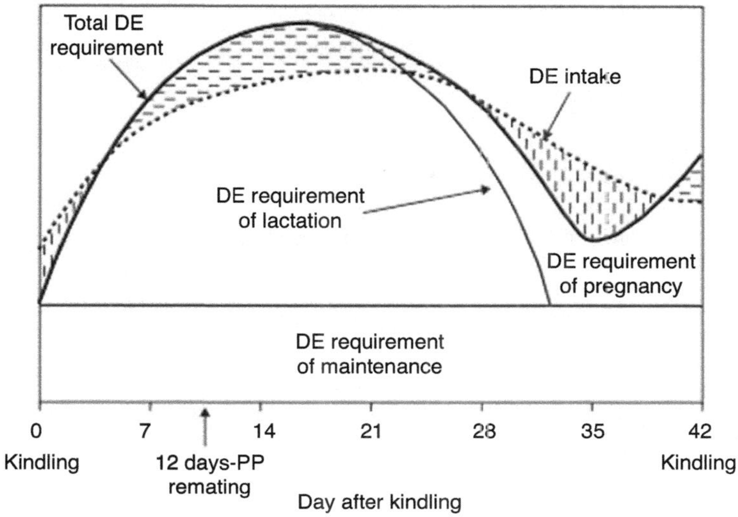

Fig. 2 Evolution of digestible energy (DE) requirement (full line) and DE intake (dotted line) in does submitted to 12 days postpartum insemination. Adapted from Xiccato (1996)

Table 6 Digestible protein requirements for maintenance (DPm) and the efficiency of utilization of DP intake (DPI) and body protein reserves (RPr)

	Growing rabbits	Pregnant and or lactating does	Nonreproducing does and bucks
DPm (g day^{-1} kg^{-1} LW0.75)	2.9	3.7	2.9
Efficiency of protein utilization			
Body retained protein (RP/DPI)	0.56		0.56
RP as fetuses (Pfetus/DPI)		0.44	
Milk protein from DP (Pmilk/DPI)		0.78	
Milk protein from body reserves (Pmilk/RPr)		0.60	

LW, live weight
Adapted from Xiccato and Trocino (2020)

Table 7 Crude protein requirements in rabbit diets

Stage	Crude protein (g kg^{-1} feed)
Maintenance	120
Breeding	170–180
Gestation	180
Lactating	160–170
Growing (young) rabbits	180
Fattening rabbits	150–160

Adapted from Xiccato and Trocino (2020)

Table 8 Amino acid requirements (g kg^{-1} feed) for rabbits

	Growing rabbits and nonreproductive does	Reproductive does
Methionine	5.4	6.3
Lysine	6.8	7.6–8
Threonine	–	5.8–6.4

Adapted from Xiccato and Trocino (2020)

Table 9 Fiber requirements (g kg^{-1} feed) in growing rabbits

Unit g kg^{-1} feed[a]	Postweaning 35–49 days old	End of fattening 49–77 days old	Rabbit does
Acid detergent fiber (ADFom)	>190	>170	>140
Acid detergent lignin	>55	>50	>30
Digestible fiber/ADFom[b]	<1.3	<1.3	
Cellulose (ADFom-ADL)	>130	>110	
Hemicelluloses (aNDFom-ADFom)	>120	>100	
Ratio lignins/cellulose	>0.4	>0.4	

Adapted from Gidenne et al. (2015)
[a]Unit: g kg^{-1} as-fed – 1 g kg^{-1} as-fed basis, corrected to a dry matter content of 900 g kg^{-1}
[b]Digestible fiber fractions = [hemicelluloses (NDF−ADF) + water-insoluble pectins]

be 10.5–11.0 g MJ^{-1} for young and male rabbits and 11.5–12.5 g MJ^{-1} for rabbit does (Xiccato and Trocino 2020).

The cecotrophy carried out by rabbits is an important source of amino acids, but it is not enough, especially in the most demanding conditions of growth or lactation. Research on the importance of amino acids in rabbit diets has focused mainly on essential amino acids, from which the values are shown in Table 8.

Fiber

Fiber is essential in rabbit diets, as it affects the digestive physiology and health of the animals, particularly for the growing rabbit. In rabbits, it is necessary to consider not only the concentration in dietary fiber but also the quality of the fibers. Thus, to reduce the risk of digestive troubles after weaning, for conventional rabbit breeding systems, one criterion is not sufficient for fiber recommendations.

Three key points must be taken into account. The first criterion to be controlled is the level of ADF that should be over 19% in a complete pelleted feed (Table 9). Second, the quality of lignocelluloses also plays a role in digestive health, and the minimum level of lignin should be 5% in feed. Third, the balance between the low-digested ADF and high-digested fiber fractions should be respected: the ratio DgF/ADF should be under 1.3 to avoid an unbalanced intake of highly fermentable polysaccharides (pectins, beta-glucans, etc.). Recent data about the role of the compound "soluble fiber" in rabbit digestion, revealed contrasting results (inadequate number animal, use of antibiotics, etc.) and appear at present not sufficiently consistent to deserved a supplementary recommendation (Table 9).

Table 10 Macromineral and micromineral requirements for rabbits

	Growing–fattening rabbits	Lactating does
Macrominerals ($g\ kg^{-1}$-fed)		
Calcium	4–10	7.5–15
Phosphorus	22–7	5–8
Sodium	2–3.5	2–3
Chloride	3–4.8	2.8–4.8
Potassium	6–14	6–17
Microminerals ($mg\ kg^{-1}$ fed)		
Iron	35–50	30–100
Copper	3–10	5–10
Zinc	25–60	30–70
Iodine	0.2–1.1	0.2–1.1
Manganese	5–8.5	2.5–15
Cobalt	0–0.25	0.1–1
Selenium	0–0.15	0.01–0.15

Adapted from Mateos et al. (2020)

Minerals

Macrominerals include ions that are needed in larger amounts (expressed g per day) and include calcium, phosphorus, sodium, magnesium, potassium, and chlorine. They are elements that make up the skeleton, intervene in the osmotic balance of body fluids and are essential for triggering action potentials in muscle and nerve cells. Microminerals or trace elements include elements whose daily needs are on the order of mg. Among the different microminerals of iron involved in the transport of oxygen, copper is a major component of metalloenzymes involved in energy utilization, and zinc is an ion of numerous enzymes involved in the biosynthesis of nucleic acids and in cell division processes. The requirements for macro- and microminerals are very variable (Table 10).

Vitamins

Vitamins are a group of complex organic compounds present in small amounts in natural foods that are essential for cellular metabolism and for life. Some vitamins can be synthesized in the body, but others their supply depends on intake. Vitamins are also classified based on their solubility in water (water soluble) or fat (fat soluble). In Table 11, we can see the vitamin requirements in growing rabbits and rabbit does.

3.2 Feed Formulation

Rabbit feed represents the highest production cost, ranging from 60 to 70% of the total production cost. One of the main objectives of formulating diets is to reduce the cost without compromising the nutritional needs and health of the animals. Ideally, we may formulate one feed for each physiological state of the rabbit. However, in the

Table 11 Vitamin requirements (mg kg^{-1} feed) of rabbits

	Growing–fattening rabbits	Lactating does
Vitamin A (IU)	0.58–10	0.01–12
Vitamin D (IU)	0.8–1	0.9–1
Vitamin E	20–50	30–50
Vitamin K	0–2	1–2
Niacin	31–180	31–50
Pyridoxine	0.5–39	0.5–2
Riboflavin	2–6	2–6
Folic acid	0.1–5	0.1–5
Pantothenic acid	10–20	10–20
Cyanocobalamin	0.01	0–0.01
Choline	100–1200	100–300
Biotin	0.2–10	0.01

Adapted from Mateos et al. (2020)

Table 12 Usual range of ingredients (g kg^{-1} feed) for rabbit feed

Ingredient	Amount
Cereals (barley, wheat)	100–200
Vegetable fats	5–30
Molasses	0–30
Beet, apple and citrus pulp, soy hulls	0–100
Cereal coproducts (wheat bran, maize gluten feed distillers)	150–350
Lucerne hay	150–300
Lignified fibrous (wheat straw, olive, and grape coproducts)	50–150
Proteins concentrates (Sunflower, soybean, rapeseed meal)	120–220

Adapted from De Blas and Mateos et al. (2020)

practice of European intensive rabbit farming, three different feeds are classically used: one for lactating does, one for the weaning period, and one for the fattening period. Feeds for doe may also be used in special seasons, such as during the summer. On small farms, only one feed can be used for all animals, more adapted to the needs of lactating rabbits, while the needs of other groups could be controlled by feed restriction programs.

The rabbit is an herbivore whose nutritional needs can be satisfied with raw materials of moderate cost and with little competition with human food. Table 12 reports the main ingredients used in the feed formulation for rabbits.

When formulating diets, it is necessary to know the chemical composition of the main ingredients to be included (Table 13). This composition must be known for each batch, as it varies with several factors (phenological development, conservation status, variety). Thus, with the ingredient composition, it is possible to meet the nutritional needs of rabbits (Tables 14 and 15) by calculating the right proportions of each raw material

Table 13 Total and digestible protein and amino acid content (g kg^{-1} dry matter) of the most important sources of protein in rabbit diets

	Total content				Apparent fecal digestible			
	CP	Lys	Met	Thr	CP	Lys	Met	Thr
Sunflower meal	393	14.2	8.6	14.3	335	11.3	7.9	11.0
Soybean meal	540	31.9	6.6	20.1	518	30.6	6.4	18.5
Full-fat soybean	442	26.5	5.6	17.0	405	25.0	5.2	15.0
Peas	283	17.4	2.9	9.4	231	15.4	2.5	6.8
Barley	121	4.2	2.0	4.1	83	2.5	1.4	2.2
Maize	119	4.2	2.3	4.1	79	2.7	1.8	1.7
Wheat	156	6.2	2.6	4.7	118	4.7	2.3	2.7
Wheat shorts	186	7.9	2.8	5.7	126	5.9	2.3	2.9
Wheat bran	160	6.9	3.2	5.9	90	3.3	1.9	3.1
Gluten feed	237	11.7	3.9	8.7	179	9.7	3.3	5.8
Soybean hulls	163	11.6	2.4	6.7	107	8.7	1.8	4.5
Lucerne hay	183	8.9	2.8	7.4	125	6.7	2.1	4.5

CP, crude protein
Adapted from Villamide et al. (2020)

4 Relationship Between Feeding and Health of the Rabbit

Nutrition and feeding strategies play a key role in rabbit breeding, not only to optimize production itself (e.g., meat, milk, fur) but also to prevent various pathologies through (1) the presence of toxic compounds in the feeds or utilization of unbalanced diets and (2) the presence of pathogenic agents (viruses, bacteria, parasites) in feeds or drinking water. This last aspect is not considered in this chapter since it does not cover nutrition directly but is rather a question of feeding management and hygiene. Similarly, the presence of undesirable pesticides in feed ingredients can impair rabbit health. Very few specific data are available for rabbits in production conditions, so this aspect is also not considered here; readers can consult more specialized books devoted to this aspect of animal feeding.

In this section, it is assumed that the daily minimum requirements are covered for the main individual components such as energy, protein and amino acids, minerals and vitamins, as recommended in other chapters. Nevertheless, it is generally difficult to provide all nutrients and energy exactly at the optimum level, and it is necessary to accept an excess or imbalance in some components to ensure that the minimum of other nutrients is met. We aimed to indicate the rules able to minimize the risk of disorders, and some give ideas on "acceptable" imbalances in feeding practice. In addition to imbalance problems, an absolute excess of ingredients such as some minerals (e.g., phosphorus) or vitamins (e.g., vitamin D) can be toxic independently of the health status of rabbits. The only question is when does a nutrient supplied above the recommended minimum or optimum become toxic?

Table 14 Level of nutrients to be included in the diets of intensively reared rabbits (kg^{-1} feed)

		Breeding does		Fattening rabbits		Mixed feed	
Nutrient	Unit	Mean	Range	Mean	Range	Mean	Range
Digestible energy	MJ	10.7		10.2		10.2	
Metabolizable energy	MJ	10.2		9.8		9.8	
NDF[a]	g	320	310–335	340	330–350	335	320–340
ADF	g	175	165–185	190	180–200	180	160–180
Crude fiber	g	145	140–150	155	150–160	150	145–165
ADL	g	*55*[b]		*50*		*55*	
Soluble NDF	g	Free		115		80	
Starch	g	170	160–180	150	140–160	160	150–170
Ether extract	g	45		Free		Free	
Crude protein[c]	g	175	165–185	150	142–160	159	154–162
Digestible proteind	g	128	115–140	104	100–110	111	108–113
Lysine (total/ Digestible)[d]	g	81/ 64		78/ 61		78/ 61	
Sulfur (total/ Digestible)[e]	g	63/ 48		52/ 40		59/ 45	
Threonine (total/ Digestible)[f]	g	67/ 46		62/ 43		65/ 45	
Calcium	g	105		60		100	
Phosphorus	g	60		40		57	
Sodium	g	23		22		22	
Chloride	g	29		28		28	

Adapted from De Blas and Mateos et al. (2020)
NDF, neutral detergent fiber; ADF, acid detergent fiber; ADL, acid detergent lignin
[a]The proportion of long fiber particles (>0.3 mm) should be >0.22 for breeding does and >0.205 for fattening rabbits
[b]Values in italics are provisional estimates
[c]The digestibility of crude protein and essential amino acids is expressed as fecal apparent digestibility
[d]Total amino acid requirements were calculated for a contribution of synthetic amino acids of 0.15
[e]Methionine should provide a minimum of 35% of the total sulfur amino acid requirements
[f]Maximal levels of 50 and 72 g kg^{-1} digestible and total threonine, respectively, are recommended for breeding does

Thereafter, we will consider health troubles (mainly digestive) linked to the balance of dietary components and the presence of nutrients in excess, mainly in relation to the initial composition of feed ingredients but also to the feeding strategy. We will consider feeding strategies (feed quantity and nutritional quality) for doe and for the growing rabbit. The last section will cover the health consequences of nonnutritional components that are frequently associated with feed ingredients (mycotoxins, etc.) and with water quality. First, we will present methods for health status assessment that are usable in rabbit breeding.

Table 15 Levels of microminerals and vitamins (kg^{-1} feed) to be included in diets of intensively reared rabbits

Nutrient	Unit	Breeding does	Fattening rabbits	Mixed feed
Cobalt	mg	0.3	0.3	0.3
Copper	mg	10	6	10
Iron	mg	50	30	45
Iodine	mg	1.1	0.4	1.0
Manganese	mg	15	8	12
Selenium	mg	0.05	0.05	0.05
Zinc	mg	60	35	60
Vitamin A	mlU	10	6	10
Vitamin D	mlU	0.9	0.9	0.9
Vitamin E	IU	50	15	40
Vitamin K3	mg	2	1	2
Vitamin B1	mg	1	0.8	1
Vitamin B2	mg	5	3	5
Vitamin B6	mg	1.5	0.5	1.5
Vitamin B12	μg	12	9	12
Folic Acid	mg	1.5	0.1	1.5
Niacin	mg	35	35	35
Pantothenic acid	mg	15	8	15
Biotin	μg	100	10	100
Choline	mg	200	100	200

Adapted from De Blas and Mateos (2020)

4.1 Digestive Troubles Related to Nutrient Imbalances

Digestive disorders are responsible for mortality around weaning but also for morbidity often characterized by growth depression and poor feed conversion. Fiber intake is the main nutrient to be controlled to reduce the risk of digestive disorders around weaning. It should be expressed in terms of the quantity or quality (type) of cell wall constituents (see Table 9). Starch intake has very little impact on the incidence of digestive disorders, as demonstrated by Gidenne et al. (2020a, b). The favorable effect of a high level of low-digestible fiber (lignocellulose or ADF) on digestive health would correspond to a stimulation of gut motility, particularly in the ceco-colic segment. Moreover, most results indicate that factors contributing to a long retention time (low fiber level, reducing the dietary particle size) contribute to modifying the cecal microbial balance and activity and may favor digestive troubles. It could be speculated that an excessively low cecal turnover rate of digesta leads to an insufficient supply of substrates available for the fibrolytic flora.

The digestible fiber "DgF" fraction also plays a role in digestive efficiency and digestive health since it is rapidly fermented (compared to ADF) in a delay compatible with the retention time of the ceco-colic segment (9–13 h; Gidenne 1997). The favorable effect of DgF on digestive health would originate from stimulated cecal fermentative activity and possibly from its moderate effect on the rate of passage.

Consequently, an excessively high incorporation of DgF, with respect to lignins and cellulose, should be avoided to minimize the risk of digestive troubles during fattening. It is thus recommended that the ratio DgF/ADF remain under 1.3 for diets having an ADF level over 15% (see Table 9).

Therefore, a balanced supply of low- and high-digested fiber fractions is required to reduce the risk of digestive trouble for the rabbit after weaning. When a sufficient supply of lignocellulose (at least 18%) is provided, it is advisable to replace some starch with digestible fiber fractions.

Furthermore, a substitution of protein by DgF also led to a significant improvement in the digestive health status of the growing rabbit, without significant impairment in growth performance (Gidenne et al. 2013).

4.2 Dietary Fiber for the Doe and Litters

The favorable effect of dietary fiber has also been analyzed in young rabbits during the weaning period (3–5 weeks old) in a large-scale study (six sites and three reproductive cycles) by Fortun-Lamothe et al. (2005). A lower mortality rate was reported for litters fed a diet rich in fiber or when fiber and lipids replaced starch. The substitution of starch for fiber has also been studied for rabbit doe diets using five iso-energetic diets (10.6 MJ digestible energy (DE) kg^{-1}) with increasing levels of NDF (from 278 to 371 g kg^{-1}) and fat (from 20 to 51 g kg^{-1}) at the expense of starch (decreasing from 237 to 117 g kg^{-1}) (De Blas et al. 1995). Some impairment in the performances of does was observed in those fed the highest levels of fiber. This might be explained by higher fermentation losses in the cecum, together with an insufficient uptake of glucose from the gut to meet the requirements for pregnancy and milk lactose synthesis. Conversely, negative effects of high dietary starch concentrations were also mentioned and were related to an increase in diarrhea mortality for the does. The combination of soluble fiber with n-6 fat reduced the number of does removed (culled + mortality) during the first four parturitions (Delgado et al. 2018), in part related to the increase in TDF digestibility.

4.3 Protein Level and Quality

We assumed that an excessive supply of protein substrates for the cecal microbiota will increase the prevalence of pathogenic species. Accordingly, reducing the protein content (21% vs. 18%; Chamorro et al. 2007) would reduce mortality and affect the ileal and/or cecal bacterial community. The reduction in dietary crude protein content led to a reduction in the frequency of detection of *Clostridium* spp. (Chamorro et al. 2007). The importance of the reduction in the ileal flow of protein (by using digestible sources or reduced protein levels) in reducing the mortality rate was supported by two other studies (Chamorro et al. 2007). Most feed manufacturers limit the dietary protein level in fattening diets because of the increased mortality rate when protein levels exceed 20 g kg^{-1}. Moreover, an excessive protein supply will probably become increasingly unusual in Europe because of increased dietary cost and, most importantly, because the European animal management strategy favors a reduction in nitrogen excretion to the environment through the use of low-protein diets (Maertens 1999).

4.4 Digestive Pathology of the Growing Rabbit and Feeding Program to Limit the Feed Intake

Since 2003, several authors have dealt with the relationship between postweaning intake level and digestive trouble incidence for the growing rabbit, including studies with experimental infections (either ERE or EPEC) or large-scale studies in the French network of rabbit experimental units (Gidenne et al. 2012).

Various short-term postweaning restriction strategies have been studied in rabbits in terms of the duration of the restriction period (1–5 weeks), the intensity of the intake limitation (90–40% of the voluntary intake), or the method (quantitative feed restriction, water restriction, limited time access to the feeder, etc.). Dietary restriction is one of the most effective nondrug ways to reduce the incidence of nonspecific enteropathy after weaning (Gidenne et al. 2012). We will focus here on the main results with respect to the impacts of transitory feed restriction after weaning on the digestive health of growing rabbits.

The favorable effect of a quantitative linear reduction in feed intake (ad libitum to 0.6 of ad libitum) on rabbit digestive health was first shown in 2003 in a multicentric experiment (six breeding units, 2000 rabbits per treatment; Gidenne et al. 2009): during feed restriction, the mortality and morbidity rates were sharply reduced (from 12 to 3.5% and from 12 to 6% for ad libitum + 0.9 ad libitum feeding level versus 0.7 + 0.6 ad libitum). Moreover, Boisot et al. (2003) also demonstrated a similar positive effect of feed restriction when rabbits were challenged with ERE inoculum. The favorable effect of limiting feed intake on the digestive health of young rabbits has been confirmed by several studies, as shown by a meta-analysis (Gidenne et al. 2012). However, the response of a restriction strategy to mortality after weaning is highly variable according to the breeding system (housing, feeding, weaning age, etc.)

In addition, feed restriction proportionally reduced the growth rate. Thereafter, returning to ad libitum feed intake led to compensatory growth and better feed efficiency (Gidenne et al. 2017). Over the whole fattening period, the live weight loss of the more restricted rabbits (0.6 ad libitum) was 7.7% compared to the control rabbits fed ad libitum from weaning.

Apart from the advantages for digestive health, a restriction strategy reduces feed costs, as feed conversion is improved during and particularly after a restriction period because of compensatory growth. Globally, the margin on the feed cost is estimated at 0.30 € per weaned rabbit. Drug consumption (estimated at approximately 0.10–0.15 € per rabbit in France) is also reduced. In conclusion, beneficial effects of a feed intake limitation after weaning have been proven in experiments and in field conditions. Postweaning intake limitation strategies are now widely practiced by rabbit breeders (over 90% of French professional breeders), in parallel with the development of new automatic feeding equipment, since beneficial effects have been obtained on resistance to digestive troubles and on feed efficiency. To reduce the risk of postweaning digestive troubles, it is recommended to limit the intake to less than 80% of the ad libitum level for 2 weeks after weaning (35 days). However, intake limitation strategies must be adapted to every breeding system, according to the aims of the farmer: improving health status, reducing feed costs, or standardizing performance.

4.5 Feeding and Reproductive Health of the Young Female

Like many farming animals, future breeding rabbit females must be prepared for their reproductive life, which often begins at approximately 19 weeks of age in conventional farming systems. Optimizing its preparation since weaning (or even birth) is key to the success of the rabbit's reproductive performance and longevity. To prepare young females for their future reproductive life, two factors must be taken into consideration: (1) the age at the first breeding or first artificial insemination (AI), in general, it is 19 weeks for selected European lines; and (2) body condition: overweight associated with a high fattening state (prior to first mating) should be prevented to avoid fertility problems at first AI (or mating). In practice, these two factors are often integrated into a single recommendation: young females must reach 80% of the adult weight (for their genotype) at the time of the first AI (or mating). However, the defect of this practice is related to the fact that live weight is a poor indicator of the physiological maturity of females.

If the young female is fed freely after weaning with a feed that is too energetic ("maternity" type), 80% of the adult weight is reached too quickly with excessive fattening. The farmer can then choose to limit the energy intake with several techniques: either by controlling the growth curve to reach a target weight (80% adult) at a fixed age by varying the dietary energy concentration or by applying a "quantitative" feed restriction (lower feed supply) for a limited period (with possible alternation of free intake periods) to reach the target weight at a fixed age at the first AI. The advantages of a quantitative feed restriction between 9 and 11 weeks of age and between 17 and 19 weeks of age (except for flushing before AI) are a reduction in the mortality of young females, a reduction in perinatal mortality of the litter at the beginning of their careers, and a more moderate mobilization of body reserves between the first AI and the birth and between the first birth and the second AI. However, applying a feed restriction strategy to control the growth of young rabbits can be complicated. In particular, care should be taken to avoid protein, vitamin, and mineral ingestion deficiencies during this key period of body development in young females.

One solution, faced with the problems of excess fattening or a difficult restriction strategy, is to use a low-energy feed (<10 MJ kg^{-1}) rich in fibers ($>22\%$ ADF) that is able to stimulate the ingestion capacity of the growing female. Thus, the daily feed intake of females is higher, leading to a stronger intake of protein and a better energy balance. The size of the litter at first birth and the weight of the rabbits at weaning would also be improved, but these results are to be confirmed. The most important aspect of using high-fiber feeds is related to the preparation of young females to better utilize their body reserves at the end of gestation and at the end of milking period, which can improve fertility at the second AI (with a breeding cycle of 6 weeks).

4.6 Feeding and Health of the Adult Reproducing Female

The prevalence of digestive disorders is lower in adult females than in growing rabbits. Its fiber requirements are usually covered by a feed containing at least 15% ADF and 4.0% lignin. Critical points mainly concern the energetic supply and digestible protein needs and the protein/energy ratio.

The breeding female must cover her needs for maintenance (renewal of tissues in neutral thermal conditions), thermoregulation, pregnancy, lactation, physical activity, etc. The adult female regulates her intake according to the dietary energy content, and the energy intake is proportional to the metabolic weight (MW), approximately 1100–1300 kJ DE kg^{-1} PM for a female in full lactation (2–3 weeks after kindling). However, in conventional breeding systems, the energy needs often exceed the intake capacity, and the female thus uses her body reserves to meet her needs. Even with high energy feeds (10.4–11.2 MJ DE kg^{-1}), the energetic balance could be negative at the end of the milking period. Along the successive reproductive cycle, the female thus gains and then loses body weight. The main difficulties are to avoid too large variations in body weight that impair the health of the female and of the litter.

Obviously, the reproductive rhythm strongly influences the energy balance of the female and thus its health status. For an intensive reproduction rhythm, the female is pregnant during lactation, which increases the requirements (milk + fetal growth). In the long term, this situation can become problematic because females do not have (postpartum or 32-day cycle) or few (35- or 42-day rhythm) periods with less production, during which they can rebuild their body reserves. This results in either an increase in female loss (culled or dead) or periods of infertility during which females restore their body reserves.

The need for digestible proteins must also be covered by the feed, and for breeding females, it is recommended to use a feed containing 17.5–19% crude protein and 12.5–13.8% digestible protein, corresponding to a DP/DE ratio of 11.5–12.5 g DP MJ^{-1} DE. The highest values are recommended for high-producing females in intensive rhythm. A decrease in the PD/ED ratio will reduce milk production and therefore the size and/or weight of the litter rather than the body condition of the female. In contrast, an increase in the dietary protein level (up to 21%) can increase milk production but may slightly reduce the number of weaned rabbits, including increased nest losses.

4.7 Problems Associated with Dietary Compounds Present at Toxic Levels

It is important for this chapter to consider maximum acceptable levels in diets. In effect, it is important that, during diet formulation, there is control over nutrient levels such that, even if analysis is not available, they are well below toxic levels. The main available information is given by Virag et al. (2008) for vitamin E, by Abdel-Khalek et al. (2008) for vitamins E and C in breeding does, Abd El-Rahim et al. (1996) for iron, and Guimarães and Motta (2000) and Ayyat and Marai (2000) for zinc. It must be noted that the maximum acceptable level is generally far higher than the recommended level, but with some noticeable exceptions, such as potassium, phosphorus, and vitamin D.

Mycotoxins are metabolites produced by certain fungi in the field on standing crops or during the harvesting of feedstuffs. Mold growth can also occur on stored grains or other raw materials because of nonhygienic storage conditions. These toxic substances may be contained within the spore or secreted into the substrate on which the fungi are growing. Most of these substances have a high degree of animal toxicity. Feeding rabbits naturally molded diets (mixed toxin contamination) is responsible for many problems, such as decreased feed intake, functional alteration of the liver and genital tract, and changes in blood constituents (Abdelhamid 1990). Mycotoxicosis appears in chronic and acute forms. The acute form is caused by the rapid ingestion of large amounts of toxins over a short period. For more details, see the review of Mézes and Balogh (2009).

Aflatoxins are naturally occurring toxins produced in grains and other feedstuffs both before and after harvest by toxigenic strains of the fungi *Aspergillus flavus* and *Aspergillus parasiticus*. Aflatoxin B1 (AFB1) is of primary concern because it is the most abundant and the most toxic. Acute or chronic aflatoxicosis may occur depending on the dietary concentration of toxins. Rabbits are extremely sensitive to aflatoxin. The acute, oral, single-dose median lethal dose is approximately 0.3 mg kg^{-1} body weight (Newberne and Butler 1969), among the lowest of any animal species. Moderate to severe death losses can be encountered with diets containing even low concentrations of toxin (<100 ppb) (Krishna et al., 1991). Signs of toxicity include hepatic lesions (Abdelhamid et al. 2002), anorexia, weight loss, and emaciation, followed by icterus in the terminal stages (Morisse et al. 1981). Acute aflatoxin poisoning (AFB1 daily doses >0.04 mg kg^{-1} body weight) causes a prolonged blood-clotting time, extensive liver damage, and death from liver failure (Clark et al. 1980).

Zearalenone (F-2 toxin) is an estrogenic substance that is frequently recovered from maize and other grains contaminated by *Fusarium graminearum* (Perez and Leuillet 1986). Zearalenone causes hypertrophic development of the genital tract of female rabbits (Abdelhamid et al. 1992). It can also affect components of the uterine tubal fluid known to be of critical importance during the early preimplantation period (Osborn et al. 1988). Zearalenone induces changes in blood serum enzyme activities. Low doses (10 μg kg^{-1}) result in significant increases in alkaline phosphatase (ALP) activity, while higher doses (100 μg kg^{-1}) lead to significant increases in the activity of aspartate aminotransferase, alanine aminotransferase, ALP, α-glutamyl transpeptidase, and lactate dehydrogenase, indicating possible liver toxicity due to chronic effects of the toxin (Conková et al. 2001). Levels of zearalenone in feed as low as 1–2 ppm can interfere with the normal reproductive activity of rabbits when fed for only a few days (1–2 weeks). This high sensitivity of rabbits to this mycotoxin could be related to the slow hepatic transformation of zearalenone mainly into α-zearalenol, a more uterotrophic metabolite (Pompa et al. 1986).

Another group of toxins produced by *Fusarium* species is trichothecenes: T-2 toxin and vomitoxin. T-2 toxin is produced by some strains of the fungus *Fusarium tricinctum*. It is relatively common in fibrous raw materials that have been harvested or stored in poor conditions. In affected rabbits, T-2 toxin causes marked feed refusal, lesions of the digestive tract and impairment of blood-clotting mechanisms (Gentry 1982; Fekete et al. 1989). Long-term (4–7 weeks) feeding of sublethal quantities of T-2 toxin (0.19 ppm) has been found to alter the ovarian activity of sexually mature female rabbits (Fekete and Huszenicza 1993).

Vomitoxin (4-deoxynivalenol) may be found in cereal grains. Contamination of rabbit feeds with this toxin results in feed refusal and vomiting. Adverse effects on fetal development have also been encountered in does. Khera et al. (1986) observed that a level of 0.00024 mg vomitoxin g^{-1} diet caused a 100% incidence of fetal resorption.

The nephrotoxins (ochratoxin and citrinin) have also been implicated in rabbit mycotoxicosis. Ochratoxin is produced by toxigenic strains of *Aspergillus ochraceus*. Galtier et al. (1977) examined the excretion of ochratoxin An in rabbit females after a single intravenous administration (1–4 mg kg^{-1} body weight) and demonstrated transfer of the toxin into the milk: the level in milk reached 1 ppm for the highest dose of administration. The actual toxicity for rabbits is unknown, but it can be pointed out that, in the abovementioned experiment, lactating does accommodate a single dose of 4 mg kg^{-1} body weight.

Citrinin is found in moldy cereals contaminated by various fungal species of *Aspergillus* and *Penicillium*. Ingestion of this toxin induces acute erosive gastritis and fluid diarrhea, with some rabbits dying less than 24 h after oral administration of a single 100–130 mg kg^{-1} body weight dose (Hanika et al. 1984). In rabbits, citrinin also causes renal damage with tubular dysfunction and necrosis similar to that found in other animal species (Hanika et al. 1984).

5 Feeding Management and Feed Availability

5.1 Feed Management and Programs

In growing rabbits, their needs vary, and the diets must be adapted to their requirements to be able to grow without compromising their health status. The young rabbit starts eating solid feed at approximately 18 days of age, and from 21 days onward, intake increases rapidly, and it can be weaned after 28 days. At this stage of life, the rabbit doe and the litter live together in the same cage, and in terms of feed management, there are three possible options: (a) feed adapted to the rabbit doe is distributed, which is high in digestible energy and low in fiber and may increase the risk of digestive disorders in the growing rabbit; (b) feed formulated for the growing rabbits is distributed, with more fiber, in general the reproductive performances are not much affected but may jeopardize the recovery of the female's body reserves; (c) using specialized equipment it is possible to carry out a differentiated distribution to the female and the litter, which from the point of view of feed management will be favorable, but more expensive in the acquisition of equipment.

After weaning and during the growth phase, two or three feeds should be provided, adapted to the rabbit's needs. In the beginning, there is a more fibrous feed (up to 45–50 days old) and then more energy and more protein. Feeding in the first 2 or 3 weeks of fattening should be restricted (see Sect. 4.3) and then ad libitum. The feed must be formulated to cover the nutritional requirements at each stage, but the raw materials must be identical to minimize the problems of adapting to the change in the feed.

The diet of the doe must be adapted to its age and physiological state. Future breeding rabbits must be fed in a controlled way to present a good body condition at the beginning of the breeding activity. The feed should be available in a restricted way (approximately 150–200 g day^{-1}) if it is more energetic or ad libitum if it is less energetic. If the feed is distributed in a restricted way, this feed management must be implemented ad libitum in the week before mating to increase receptivity and fertility (flushing) and in the last week of gestation to avoid compromising the development of the young rabbits.

The feeding of the rabbit doe must be carried out to maintain a good body condition and not compromise the productive results. During the rabbit's productive life, she may be simultaneously pregnant and lactating, so her energy and protein needs are high. Thus, the feed must be available ad libitum, and the intake increases with the course of lactation and can reach up to 450–500 g day^{-1} at the peak of milk production. If the rabbit is just pregnant, feeding should be restricted and passed ad libitum in the last 8–10 days before giving birth. When the rabbit is not pregnant or lactating, feeding should be restricted. During the entire production phase, we can maintain the chemical composition of the feed to be fed to the rabbit. However, with the increase in the ambient temperature, the voluntary intake of feed decreases, and we can provide a feed with greater energy density for the rabbit to maintain its performance.

The nutritional needs of breeding males are maintenance only. In insemination centers, a specific feed should be available, formulated according to needs, but enriched with some micronutrients (copper, zinc) that may improve semen quality. Alternatively, the feed of the breeding rabbits can be used, but it must be available in a rationed way (180–200 g day^{-1}).

The rabbit's diet should be based on pelleted feed. The form of feed presentation can affect the digestive use of diet components. With floury feeds, there is less growth, a reduction in feed intake and an increase in feed conversion. With floury feed, respiratory problems also increase due to the susceptibility to the ingestion of dust. Thus, concentrated feeds must be granulated, and the ideal dimensions for pellets are 2.5–5 mm in diameter and 5–10 mm in length (Blum 1989).

5.2 Dietary Soluble and Insoluble Fiber

Quickly fermentable fibers correspond mainly to *water-soluble polysaccharides* (β-glucans and fructans and some pectins) and oligosaccharides (degree of polymerization (DP > 15)) and to water-insoluble pectins. They are not digested in the small intestine but are rapidly fermented and highly digested in the hindgut. For instance, fructans from chicory roots (inulin-rich ingredient) are almost totally digested and stimulate cecal fermentation (Volek and Marounek 2011) without changing growth performance. According to Maertens et al. (2004), synthetic fructans would be approximately half-digested before the cecum, and they did not find a higher cecal VFA level but only a higher butyrate proportion. The addition of inulin to the diet sharply increased (+30%) the cecal VFA concentration but failed to reduce the mortality rate (21–77 days old) of the growing rabbits (Volek et al. 2007). Another way to analyze the role of quickly fermented polysaccharides is to determine the neutral detergent-soluble fiber (NDSF) residue (Hall et al. 1997), which corresponds to the polysaccharides soluble in the neutral detergent solution. Although the level of NDSF is moderate in rabbit feeds, a reduction in its level (12% vs. 8%) could be unfavorable for the digestive health of early-weaned rabbits (Gómez-Conde et al. 2009). Conversely, a higher level of NDSF may improve the mucosal morphology and functionality and its immune response (Gómez-Conde et al. 2007). However, the NDSF criteria remain difficult to analyze, and precision is relatively low for complete feeds with a low content of pectins or soluble fibers.

Accordingly, another approach is to estimate the content of the quickly fermentable fibers, also called "soluble fibers, SF," by the difference between the TDF (total dietary fiber) and the aNDFom, with the latter that must be corrected for its crude protein content. SF would thus be easier to handle in a routine laboratory for feed analysis. It would recover the part of TDF that comprises the nonstarch, non-NDF polysaccharides, including pectic substances, β-glucans, resistant starch, oligosaccharides, fructans, and gums. The soluble fiber level is generally increased in complete feed by supplying raw materials rich in pectins (beet pulp, citrus or apple pulp) or fructans (chicory pulp); thus, most of the studies in fact relate "pulp levels" to the performance of physiological data. Accordingly, the SF dietary level is positively related to the fecal digestibility of *"insoluble"* fiber fractions (NDF and ADF) and favors microbial activity with higher fermentation levels and lower pH, as reviewed by Trocino et al. (2013). As a consequence, the soluble fiber level is likely to affect the ileal and, especially, cecal microbiota (Gómez-Conde et al. 2007, 2009) by modifying the amount and type of substrate reaching the cecum. These changes in microbiota may also modify the immune response observed in young rabbits fed soluble/insoluble fermentable fibers. However, regardless of the advantages and disadvantages of the different methods and calculation procedures, the choice of the method to quantify SF will depend on the correlation with in vivo data collected in animals, particularly the impact on digestive health.

There is no clear global relationship between the soluble fibers, analyzed as TDF-NDF, and the postweaning mortality ($r^2 = -0.04$), although a tendency toward a reduction in mortality might be observed (Gidenne 2015). Even using the same set of studies, we did not observe any significant relationship between mortality and dietary SF. This lack of relationship seems logical since SF did not include the hemicellulose fractions that are in large amounts in rabbit feeds. Thus, it remains very risky to recommend an SF concentration in rabbit feeds to reduce the risk of digestive troubles. Nevertheless, it seems that over an SF level of 7%, the mortality rate decreases, but in fact, this level is generally reached in feeds that follow the current recommendations for ADF and DgF. Moreover, criteria for quickly fermentable fibers correspond to a lower amount of fiber residue than for DgF criteria, and due to a higher analytical error, this could add further imprecision in recommendations. Further studies are needed to elucidate the health response of rabbits to soluble fiber intake, with large-scale studies comparing the health of large groups of rabbits (over 100). The main problem is to obtain an analytical method that is sufficiently robust (Xiccato et al. 2012) and that could be used routinely in feed control laboratories.

6 Feed Perspectives in the Coming Years

Maintaining a minimal production level of food from animal origin seems necessary to meet the needs of a growing population and to compensate for the decrease in the area of agricultural crops, due to climate change. Conventional protein production is a challenge to meet the increased demand for animal feeding. As a consequence, there is a need to find alternative sources of proteins and energy. In a world undergoing constant climate change, there is pressure for production to take place based on short circuits and incorporating local resources. Coproducts from agricultural production and the agro-feed industry, such as olive pomace and leguminous straw, among others, can be incorporated in the rabbit feeds.

6.1 Agroindustry Coproducts in Rabbit Feed

Reducing the cost of feeding rabbits can be achieved by incorporating nonconventional ingredients, namely, agroindustry coproducts. These materials, grown locally, are also a potential source of phenolic compounds with antioxidant, antimicrobial, and antifungal effects. The use of these vegetable coproducts can also minimize environmental pollution. Due to their nutritional characteristics and abundance, wastes from agricultural production, particularly straws, hulls, and pods, are materials that can be incorporated into compound feeds for herbivores such as the rabbit.

Agroindustry coproducts (olive leaves, grape stalks) generally have high levels of lignin, with consequent negative effects on animal performance, which can limit their incorporation into diets. Several pretreatments (chemical, physical, and biological) have been used to improve the nutritional value of agricultural wastes. The application of white-rot fungi through solid-state fermentation may be a viable solution to increase the nutritional value of the materials, as it promotes changes in its lignin structure and facilitates access to structural polysaccharides (Ribeiro et al. 2012; Costa-Silva et al. 2022).

Some studies were carried out to investigate the possibility of using vegetable coproducts (olive leaf, olive cake, leguminous straws, grape stalks, among others) in rabbit feeding as alternatives to fibrous ingredients. The level of incorporation will depend on its composition, but it is generally limited to a maximum of 10–15% to avoid compromising productive performance.

The olive cake (*Olea europaea* L.) is a solid by-product of olive oil extraction composed essentially of a mixture of peels, pulps, and woody endocarps and is locally available in large quantities, especially in the Mediterranean region, although its availability is restricted to the olive harvest season. Olive cake is an important source of nondigestible fiber, with a high lignin content and a good balance between the different fiber fractions in the diet, which can prevent digestive problems in growing rabbits (Carraro et al. 2005). Its protein and energy content are low, but the fat content can be interesting (it can go up to 10–12% in DM), depending on the olive oil extraction method and with a balanced fatty acid profile. Among the works carried out, we can highlight those of Carraro et al. (2005), Dal Bosco et al. (2012), Dorbane et al. (2019) and Rupić et al. (2020), who highlight the possibility of using olive cake (10 and 20% incorporation in diets) without negatively affecting performance, digestibility, health, slaughter parameters, and meat quality.

6.2 Protein Sources in Rabbit Feed

The reduction of feed costs in animal and rabbit production can be achieved by incorporating nonconventional ingredients in the diet. In the Mediterranean area, there is some tradition of producing rainfed legumes (Alfalfa, Sainfoin, etc.), among which grains of the lupine species (*Lupinus* spp.) stand out, which can help to mitigate dependence on soy. Nutritionally, white lupine has a high protein content (more than 30%), a good fat content (on the order of 9–13%), a high concentration of polyunsaturated fatty acids (PUFAs), and substances that promote antioxidant activity (Gabriel-Santos et al. 2021). A disadvantage of some lupine varieties is the presence of antinutritional factors, which can reduce animal performance and compromise the suitability of these legumes as an alternative protein source (Volek et al. 2020) but can be overcome with the adoption of sweet varieties with low alkaloid content (Gresta et al. 2023), such as white lupine, the friendly variety.

Some studies (Volek and Marounek 2011; Volek 2016) have examined the possible use of white lupine (*Lupinus albus* cv. Amiga) to feed growing rabbits compared with diets with traditional protein sources (bagasse soybean or sunflower).

The level of incorporation in the diet varied between 60 and 250 g kg^{-1} and sought to evaluate the effect on zootechnical performance, diet digestibility, digestive health, and meat quality. The authors indicated that no adverse effect was observed by the inclusion of white lupine in the performances (final live weight, food intake, weight gain, and conversion index) in most of the works. However, Gabriel-Santos et al. (2021) already reported a lower average daily gain in live weight and a worse feed conversion index, with no change in intake and final live weight, in growing rabbits in which soybean meal was completely replaced by lupine (15%). Due to its fatty acid profile, the inclusion of white lupine in the diet can improve the fatty acid profile of meat, increasing its MUFA (monounsaturated fatty acids) content (Volek and Marounek 2011; Guedes et al. 2022). In some studies (Volek 2016; Strychalski et al. 2020), the origin of the protein did not affect the total apparent digestibility of crude protein and crude energy, but in the trial by Gabriel-Santos et al. (2021), a slight decrease in organic matter digestibility and an increase in protein digestibility with no effect on fiber digestibility were observed when soybean meal was replaced by *Lupinus albus* or *Lupinus luteus*.

Studies on the inclusion of white lupine in diets for breeding does are less frequent. Volek et al. (2014) studied the effect of its inclusion in the diet on milk composition and performance of does and their litter. No adverse effects were observed on the feed consumption and milk production of the does or the growth and mortality of the young rabbits

6.3 Insects in Rabbit Feed

Insects constitute a very diverse group of animals that belong to the phylum Arthropoda. More than 1900 species of insects that can be used in human feed have already been described in the literature (Kaczmarowski 2019). Insect meal can be incorporated into animal feed. The main constituent of insect meal is protein, which, depending on its stage of development, can range from 40 to 60% (Kowalska et al. 2019). The amino acid composition of this protein source is similar to that of mammalian proteins and has high digestibility, ranging from 77.9 to 98.9% (Bukkens 1997). The average fat content in insect meal ranges from 14.4 to 37.1%, being higher in larvae and pupae than in adults. Insect fat is relatively rich in unsaturated acids, including oleic, linoleic, and α-linolenic acid (Yang et al. 2006), which may make it interesting. The nutritional composition of insect meal can vary, depending on the stage of development, habitat, and substrates used to feed the insects (Rumpold and Schlüter 2013). Chitin, an indigestible component for most animals, can limit its use in feed, but its content depends on the species and the stage of development of the insects and can be removed by technological methods before its incorporation into diets (Yang et al. 2006).

Some works have studied the effect of using different sources of fat, such as animal fat or oilseeds (Xiccato 2020), on the performance, digestibility, health status and quality of rabbit meat. The use of insect meal has received more attention for birds and fish, but some work has also been carried out on rabbits. The purpose of incorporating insect meal into rabbit diets is to replace the protein materials or fat. The incorporation of insect meal as a protein source was studied by Carregal and Takahashi (1987) in a study that replaced soybean meal with silkworm larvae and did not observe negative effects in the measurements carried out. The use of insect meal as a source of fat was carried out by Gasco et al. (2019) and Dalle Zotte et al. (2018). Their work showed that black soldier fly (*Hermetia illucens*) and mealworm (*Tenebrio molitor*) oil can be used in rabbit feeding as a source of fat. The effect of replacing soybean oil with insect fat (from black soldier flies or mealworms) in diets for growing rabbits was studied by Gasco et al. (2019) on growth, digestibility, and meat quality and found no differences in the evaluated parameters between the experimental groups, with even an improvement in the fatty acid profile of the meat. Dalle Zotte et al. (2018) studied the effect of incorporating black soldier fly larvae (*Hermetia illucens* L.) or extruded flaxseed in the diet of growing rabbits. The authors concluded that fly fat could be an alternative source of lipids in the diet of growing rabbits, as the performance results were similar in both groups of animals.

7 Conclusions

The success of efficient and sustainable raising rabbits depends on knowledge needs to be able to formulate diets with balanced nutrients. The requirements of the rabbit in its different stages of production are known and must be followed so that the animal can produce. Fiber in the diet is fundamental, and energy and protein (also the energy/protein ratio) are also essential and are parameters to be considered when formulating diets.

The rabbit has a particular digestive physiology that will also influence the formulation of diets under the penalty of digestive disturbances of some severity, which can be minimized with adequate feed management. Traditionally, raw materials from agroindustry are already used in rabbit feeding. However, the search for alternative sources of protein and fat (protein crops of European origin, insects and others) must continue so that we have a real understanding of their value and possibility of use in rabbit nutrition.

References

Abdelhamid AM (1990) Effect of feeding rabbits on naturally moulded and mycotoxin-contaminated diet. Arch Anim Nutr 40:55–63

Abdelhamid AM, Kelada IP, Ali MM, El-Ayouty SA (1992) Influence of zearalenone on some metabolic physiological and pathological aspects of female rabbits at two different ages. Arch Anim Nutr 42:63–70

Abdelhamid AM, Ragab MA, El-Shaieb AF (2002) The use of tafla or aluminosilicate for alleviating toxic effects of aflatoxin-contaminated diets of growing rabbits. In: Proceedings of the 1st conference of animal and fish production, Mansoura, Egypt, pp 389–413

Abdel-Khalek AM, Selim NA, El-Medany SA, Nada SA (2008) Response of doe rabbits to dietary antioxidant vitamins E and C during pregnancy and lactation. In: Xiccato G, Trocino A,

Lukefahr SD (eds) Proceedings of the 9th world rabbit congress, Verona. Fondazione Iniziative Zooprofilattiche e Zootecniche, Brescia, Italy, pp 519–523 (on line in the WRSA web site, http://world-rabbit-science.com/)

Abd El-Rahim MI, El-Kerdawy Dawlat A, El-Kerdawy HM, Abdallah Fatma R (1996) Bioavailability of iron in growing rabbits fed excess levels of dietary iron, under Egyptian conditions. In: Lebas F (ed) Proceedings of the 6th world rabbit congress, Toulouse, vol 1. Association Française de Cuniculture, Lempdes, France, pp 51–57 (on line in the WRSA web site, http://world-rabbit-science.com/)

Abdelsattar MM, Hussein AM, El-Ati A, Saleem AM (2020) Impacts of saline water stress on livestock production: A review. SVU Int J Agricult Sci 2(1):1–12

Ayyat MS, Marai IFM (2000) Growth performance and carcass traits as affected by breed and dietary supplementation with different zinc levels, under Egyptian conditions. In: Becerril CM, Pro A (eds) Proceedings of the 8th world rabbit congress, Puebla, Colegio de Postgraduados, Montecillo, Spain, pp 83–88 (on line in the WRSA web site, http://world-rabbit-science.com/)

Blas E (1986) El almidón en la nutrición del conejo: utilización digestiva e implicaciones prácticas. Doctoral thesis, Universidad de Zaragoza, Spain

Blas E, Gidenne T (2020) Digestion of sugars and starch. In: De Blas C, Wiseman J (eds) The nutrition of the rabbit, 3nd edn. CAB, Oxon, UK, pp 21–40

Blum JC (1989) L´Alimentation des Animaux Monogastriques: Porc, Lapin Volailes, 2éme edn. INRA, Paris, p 282

Bodnar K, Bodnar G (2020) Drinking water supply in rabbit production: short review. Lucrari Stiintifice, I XXII(1):19–24

Boisot P, Licois D, Gidenne T (2003) Feed restriction reduces the sanitary impact of an experimental reproduction of epizootic rabbit enteropathy syndrome (ERE), in the growing rabbit. In: Bolet G (ed) Proceedings 10ème Journées Recherche Cunicole, Le Mans. ITAVI, Paris, France, pp 267–270

Bukkens SGF (1997) The nutritional value of edible insects. Ecol Food Nutr 36(2-4):287–319

Carraro L, Trocino A, Xiccato G (2005) Dietary supplementation with olive stone meal in growing rabbits. Ital J Anim Sci 4:88–90

Carregal RD, Takahashi R (1987) Use of silkworm (Bombyx mori L.) chrysalis meal as a replacement for soyabean meal in the feeding of growing rabbits. Rev Soc Bras Zoot 16:158–162

Chamorro S, Gomez Conde MS, Perez De Rozas AM, Badiola I, Carabaño R, de Blas JC (2007) Effect on digestion and performance of dietary protein content and increased substitution of lucerne hay with soya-bean protein concentrate in starter diets for young rabbits. Animal 1:651–659

Clark JD, Jain AV, Hatch RC, Mahaffey EA (1980) Experimentally induced chronic aflatoxicosis in rabbits. Am J Vet Res 41:1841–1845

Conková E, Laciaková A, Pástorová B, Seidel H, Kovác G (2001) The effect of zearalenone on some enzymatic parameters in rabbits. Toxicol Lett 121:145–149

Costa-Silva V, Pinheiro V, Alves A, Silva JA, Marques G, Lorenzo J, Rodrigues M, Ferreira J (2022) Effects of dietary incorporation of grape stalks untreated and fungi-treated in growing rabbits: A Preliminary Study. Animals 12:112, 15 pg

Dal Bosco A, Mourvaki E, Cardinali R, Servili M, Sebastiani B, Ruggeri S, Mattioli S, Taticchi A, Esposto S, Castellini C (2012) Effect of dietary supplementation with olive pomaces on the performance and meat quality of growing rabbits. Meat Sci 92(4):783–788

Dalle Zotte A, Cullere M, Martins C, Alves SP, Freire JPB, Falcão e Cunha L, Bessa RJB (2018) Incorporation of Black Soldier Fly (Hermetia illucens L.) larvae fat or extruded linseed in diets of growing rabbits and their effects on meat quality traits including detailed fatty acid composition. Meat Sci 146:50–58

De Blas C, Mateos GG (2020) Feed formulation. In: De Blas C, Wiseman J (eds) The nutrition of the rabbit, 3nd edn. CAB, Oxon, UK, pp 243–253

De Blas C, Taboada E, Mateos GG, Nicodemus N, Méndez J (1995) Effect of substitution of starch for fiber and fat in isoenergetic diets on nutrient digestibility and reproductive performance of rabbits. J Anim Sci 73:1131–1137

Delgado R, Abad-Guamán R, Nicodemus N, Villamide MJ, Ruíz-López N, Carabaño R, Menoyo D, García J (2018) Effect of level of soluble fiber and n-6/n-3 fatty acid ratio on performance of rabbit does and their litters. J Anim Sci 96:1084–1100

Dorbane Z, Kadi SA, Boudouma D, Gater-Belaid N, Bannelier C, Berchiche M, Gidenne T (2019) Nutritive value of two types of olive cake (Olea europaea L.) for growing rabbit. World Rabbit Sci 27:69–75

Falcão e Cunha L, Peres H, Freire JPB, Castro-Solla L (2004) Effects of alfalfa, wheat bran or beet pulp, with or without sunflower oil, on caecal fermentation and on digestibility in the rabbit. Anim Feed Sci Technol 117:131–149

Fekete S, Huszenicza G (1993) Effect of T-2 toxin on ovarian activity and some metabolic variables of rabbits. Lab Anim Sci 43:646–649

Fekete S, Tamas J, Vanyi A, Glavits R, Bata A (1989) Effect of T-2 toxin on feed intake, digestion and pathology of rabbits. Lab Anim Sci 39:603–606

Fortun-Lamothe L, Lacanal L, Boisot P, Jehl N, Arveux A, Hurtaud J, Perrin G (2005) Effects of level and origin of dietary energy on reproduction performance of the does and health status of the young. In: Bolet G (ed) Proceedings of 11ème J. Rech. Cunicoles, 29 and 30 Nov. 2005, Paris. ITAVI, Paris, France, pp 129–132

Fortun-Lamothe L, Combes S, Gidenne T (2009) Contribution of intensive rabbit breeding to sustainable development. A semi-quantitative analysis of the production in France. World Rabbit Sci 17(2):79–85

Gabriel-Santos SA, Almeida MB, Clossonm M, Guedes C, Barros A, Ferreira LM, Trindade HA, Pinheiro V (2021) Effect of total replacement of the soya bean meal by lupine seeds (L. albus and L. luteus) on performance and digestion characteristics of growing rabbits. Anim Feed Sci Technol 278:114996

Gallois M, Le Huërou-Luron I, Fortun-Lamothe L, Lallès JP, Gidenne T (2008) Adaptability of the digestive function according to age at weaning in the rabbit: I. Effect on feed intake and digestive functionality. Animal 2:525–535

Galtier P, Baradat C, Alvinerie M (1977) Etude de l'élimination d'ochratoxine A par le lait chez la lapine. Ann Nutr Aliment 31:911–918

García J, Nicodemus N, Carabaño R, de Blas JC (2002) Effect of inclusion of defatted seed meal in the diet on digestion and performance of growing rabbits. J Anim Sci 80:162–170

García AI, de Blas JC, Carabaño R (2005) Comparison of different methods for nitrogen and amino acid evaluation in rabbit diets. Anim Sci 80:169–178

Gasco L, Biasato I, Dabbou S, Schiavone A, Gai F (2019) Animals fed insect-based diets: State-of-the-art on digestibility, performance and product quality. Animals 9(4):170

Gentry PA (1982) The effect of administration of a single dose of T-2 toxin on blood coagulation in the rabbit. Can J Comp Med 46:414–419

Gidenne T (1997) Caeco-colic digestion in the growing rabbit: impact of nutritional factors and related disturbances. Livest Prod Sci 51:73–88

Gidenne T (2015) Dietary fibres in the nutrition of the growing rabbit and recommendations to preserve digestive health: a review. Animal 9:227–242

Gidenne T, Perez JM (1993) Effect of dietary starch origin on digestion in the rabbit. 1. Digestibility measurements from weaning to slaughter. Anim Feed Sci Technol 42:237–247

Gidenne T, Combes S, Feugier A, Jehl N, Arveux P, Boisot P, Briens C, Corrent E, Fortune H, Montessuy S, Verdelhan S (2009) Feed restriction strategy in the growing rabbit. 2. Impact on digestive health, growth and carcass characteristics. Animal 3:509–515

Gidenne T, Combes S, Fortun-Lamothe L (2012) Feed intake limitation strategies for the growing rabbit: effect on feeding behaviour, welfare, performance, digestive physiology and health: a review. Animal 6(9):1407–1419

Gidenne T, Kerdiles V, Jehl N, Arveuxe P, Eckenfelder B, Briens C, Stephan S, Fortune H, Montessuy S, Muraz G (2013) Protein replacement by digestible fibre in the diet of growing rabbits: 2-Impact on performances, digestive health and nitrogen output. Anim Feed Sci Technol 183:142–150

Gidenne T, Lebas F, Savietto D, Dorchies P, Duperray J, Davoust C, Fortun-Lamothe L (2015) Nutrition et alimentation. In: Gidenne T (ed) Le lapin. De la biologie à l'élevage, Quae éditions, pp 152–196

Gidenne T, Garreau H, Drouilhet L, Aubert C, Maertens L (2017) Improving feed efficiency in rabbit production, a review on nutritional, technico-economical, genetic and environmental aspects. Anim Feed Sci Technol 225:109–122

Gidenne T, Garcia J, Lebas F, Licois D (2020a) Nutrition and feeding strategy: impacts on health status. In: De Blas C, Wiseman J (eds) The nutrition of the rabbit, 3rd edn. CABI, Wallingford, UK, pp 193–221

Gidenne T, Carabano J, Garcia., De Blas, J.C. (2020b) Fibre digestion. In: De Blas C, Wiseman J (eds) The nutrition of the rabbit, 3nd edn. CAB, Oxon, UK, pp 69–88

Gómez-Conde MS, García J, Chamorro S, Eiras P, Rebollar PG, Pérez de Rozas A, Badiola I, de Blas C, Carabaño R (2007) Neutral detergent-soluble fiber improves gut barrier function in 25 d old weaned rabbits. J Anim Sci 85:3313–3321

Gómez-Conde MS, Pérez de Rozas A, Badiola I, Pérez-Alba L, de Blas C, Carabaño R, García J (2009) Effect of neutral detergent soluble fibre on digestion, intestinal microbiota and performance in twenty-five-day-old weaned rabbits. Livest Sci 125:192–198

Gresta F, Oteri M, Scordia D, Costale A, Armone R, Meineri G, Chiofalo B (2023) White lupin (Lupinus albus L.), an alternative legume for animal feeding in the Mediterranean area. Agriculture 13:434

Guedes CM, Almeida M, Closson M, Garcia-Santos S, Lorenzo JM, Domínguez R, Ferreira L, Trindade H, Silva S, Pinheiro V (2022) Effect of total replacement of the soya bean meal by lupine seeds and gender on meat quality and fatty acids profile of growing rabbits. Foods 11: 2411

Guimarães CS, Motta FW (2000) Bioavailability of dietary zinc sources for fattening rabbits. In: Blasco A (ed) Proceedings of the 7th world rabbit congress, Valencia, vol 3. Valencia University Publications, Valencia, Spain, pp 255–261 (on line in the WRSA web site, http://world-rabbit-science.com/)

Hall MB, Lewis BA, Van Soest PJ, Chase LE (1997) A simple method for estimation of neutral detergent-soluble fibre. J Sci Food Agricult 74:441–449

Hanika C, Carlton WW, Boon GD, Tuite J (1984) Citrinin mycotoxicosis in the rabbit: clinicopathological alterations. Food Chem Toxicol 22:999–1008

Kaczmarowski M (2019) Gatunki owadów zaliczone do zwierząt gospodarskich w Unii Europejsckiej. Higiena Żywności i Pasz 94(2):158–161

Khera KS, Whalen C, Angers G (1986) A teratology study on vomitoxin (4-deoxynivalenol) in rabbits. Food Chem Toxicol 24:421–424

Kowalska D (2019) Insects as a source of nutrients in animal feed. Sci Ann Pol Soc Anim Prod 15 (3):25–37

Maertens L (1999) Towards reduced feeding costs, dietary safety and minimal mineral excretion in rabbits: a review. World Rabbit Sci 7:65–74

Maertens L, Aerts JM, De Boever J (2004) Degradation of dietary oligofructose and inulin in the gastro-intestinal tract of the rabbit and the effects on caecal pH and volatile fatty acids. World Rabbit Sci 12:235–246

Maertens L, Lebas F, Szendrö Z (2006) Rabbit milk: a review of quantity, quality and non dietary affecting factors. World Rabbit Sci 14:205–230

Mateos GG, Rebollar PG, De Blas CA (2020) Minerals, vitamins and additives. In: De Blas C, Wiseman J (eds) The nutrition of the rabbit, 3nd edn. CAB, Oxon, UK, pp 126–158

Mézes M, Balogh K (2009) Mycotoxins in rabbit feed: a review. World Rabbit Sci 17:53–62

Morisse JP, Wyers M, Drouin P (1981) Aflatoxicose chronique chez le lapin. Essai de reproduction expérimentale. Recueil de Médecine Vétérinaire de l'Ecole d'Alfort 157:363–368
Newberne PM, Butler WH (1969) Acute and chronic effects of aflatoxin on the liver of domestic and laboratory animals: a review. Cancer Res 29(1):236–250
Osborn RG, Osweiler GD, Foley CW (1988) Effects of zearalenone on various components of rabbit uterine tubal fluid. Am J Vet Res 49:1382–1386
Parigi Bini R, Xiccato G (1986) Utilizzazione dell'energia e della proteina digeribile nel coniglio in accrescimento. Rivista di Coniglicoltura 23(4):54–56
Perez JM, Leuillet M (1986) Com position et valeur nutritive des céréales. Perspectives Agricoles 105:56–61
Pfost DL, Fulhage CD (2001) Water quality for livestock drinking, 4 p
Piva G, Piva A (1995) Anti-nutritional factors of *Datura* in feedstuffs. Nat Toxins 3(4):238–241
Pompa G, Montesissa C, Di Lauro FM, Fadini L (1986) The metabolism of zearalenone in subcellular fractions from rabbit and hen hepatocytes and its estrogenic activity in rabbits. Toxicology 42:69–75
Ramteke R, Doneria R, Gendley MK (2019) Antinutritional factors in feed and fodder used for livestock and poultry feeding. Acta Sci Nutr Health 3(5):39–48
Ribeiro L, Pinheiro V, Outor-Monteiro D, Mourão JL, Bezerra R, Dias A, Bennett R, Marques G, Rodrigues M (2012) Effects of the dietary incorporation of untreated and white-rot fungi (Ganoderma resinaceum Boud) pretreated olive leaves on growing rabbits. Anim Feed Sci Technol 173:244–251
Rumpold BA, Schlüter OK (2013) Nutritional composition and safety aspects of edible insects. Mol Nutr Food Res 57:802–823
Rupić V, Svjetlana LM, Cacić M, Romanović V, Cacić I (2020) Olive cake in the feeding of fattening rabbits: histopathological changes. Stočarstvo 73(1-2):21–34
Strychalski J, Gugołek A, Matusevicius P (2020) Usability of white lupin and pea seeds in the feeding of growing New Zealand white rabbits. Pol J Nat Sci 35(2):151–164
Trocino A, Fragkiadakis M, Majolini D, Tazzoli M, Radaelli G, Xiccato G (2013) Soluble fibre, starch and protein level in diets for growing rabbits: Effects on digestive efficiency and productive traits. Anim Feed Sci Technol 180:73–82
Vikram N, Katiyar SK, Singh CB, Husain R, Gangwar LK (2020) A review on anti-nutritional factors. Int J Curr Microbiol App Sci 9(5):1128–1137
Villamide MJ, Nicodemus N, Fraga MJ, Carabano R (2020) Protein digestion. In: De Blas C, Wiseman J (eds) The nutrition of the rabbit, 3nd edn. CAB, Oxon, UK, pp 41–57
Virag G, Eiben C, Tóth T, Schmidt J (2008) Colour and pH of rabbit meat and fat deposits as affected by the source and dose of dietary vitamin E supplementation. In: Xiccato G, Trocino A, Lukefahr SD (eds) Proceedings of the 9th world rabbit congress, Verona. Fondazione Iniziative Zooprofilattiche e Zootecniche, Brescia, Italy, pp 1467–1471
Volek Z (2016) White lupine is a suitable feed component in rabbit diets: A review. Slovak J Anim Sci 49(4):147–150
Volek Z, Marounek M (2011) Effect of feeding growing-fattening rabbits a diet supplemented with whole white lupin (Lupinus albus cv. Amiga) seeds on fatty acid composition and indexes related to human health in hind leg meat and perirenal fat. Meat Sci 87:40–45
Volek Z, Marounek M, Skrivanová V (2007) Effect of a starter diet supplementation with mannan oligosaccharide or inulin on health status, caecal metabolism, digestibility of nutrients and growth of early weaned rabbits. Animal 1:523–530
Volek Z, Marounek M, Volkov L, Kudrnova E (2014) Effect of diets containing whole white lupin seeds on rabbit doe milk yield and milk fatty acid composition as well as the growth and health of their litters. J Anim Sci 92(5):2041–2049
Volek Z, Uhlířová L, L., Zita, L. (2020) Narrow-leaved lupine seeds as a dietary protein source for fattening rabbits: a comparison with white lupine seeds. Animal 14(4):881–888
Xiccato G (1996) Nutrition of lactating does. In: Lebas F (ed) Proceedings of the 6th world rabbit congress, Toulouse. Association Française de Cuniculture, Lempdes, France, pp 29–50

Xiccato G (2020) Fat digestion. In: De Blas C, Wiseman J (eds) The nutrition of the rabbit, 3nd edn. CAB, Oxon, UK, pp 58–68

Xiccato G, Trocino A (2020) Energy and Protein metabolism and requirements. In: De Blas C, Wiseman J (eds) The nutrition of the rabbit, 3nd edn. CAB, Oxon, UK, pp 89–125

Xiccato G, Trocino A, Tazzoli M, Majolini D, Carabaño R, Villamide MJ, García J, Nicodemus N, Abad R, Blas E, Cervera C, Ródenas L, Martínez E, Falcão-e-Cunha L, Bengala Freire JP, Maertens L, Bannelier C, Segura M, Gidenne T (2012) European ring-test on the chemical analyses of total dietary fibre and soluble fibre of compound diets and raw materials for rabbits. In: Proceedings of the 10th world rabbit congress, 3–6 September 2012. Sharm El-Sheik, Egypt pp 453–471

Yang LF, Siriamornpun S, Li D (2006) Polyunsaturated fatty acid content of edible insects in Thailand. J Food Lipids 13(3):277–285

Yassein SA, Mahmoud KGM, Maghraby N, Ezzo OH (2008) Hot climate effects and their amelioration on some productive and reproductive traits in rabbit does. World Rabbit Sci 16(3):173–181

Reproduction in the Female Rabbit

Rosa M. Garcia-Garcia, Daniela Jordán-Rodriguez, Pedro L. Lorenzo, Pilar G. Rebollar, and María Arias-Alvarez

Abstract

The focus of this chapter is summarizing the most important aspects of reproductive physiology (folliculogenesis in the ovary, sexual behavior, ovulation, mating, embryo development, pregnancy, parturition, and lactation) as a basis to better understand and become successful in rabbit reproduction management in rabbit farms. Considering reproductive physiology, the current advances in reproductive and nutritional management to meet reproductive efficiency, welfare standards, consumer demand, and circular economy challenges will be described.

Keywords

Reproductive physiology · Folliculogenesis · Oocyte · Embryo · Ovulation · Pregnancy · Lactation · Artificial insemination · Reproductive rhythm · Reproductive outcome · Rabbit

R. M. Garcia-Garcia (✉) · D. Jordán-Rodriguez · P. L. Lorenzo
Department of Physiology, Veterinary Faculty, Complutense University of Madrid, Madrid, Spain
e-mail: rosa.garcia@vet.ucm.es

P. G. Rebollar
Department of Agrarian Production, Technical School of Agronomic, Alimentary and Biosystems Engineering, Technical University of Madrid, Madrid, Spain

M. Arias-Alvarez
Department of Animal Production, Veterinary Faculty, Complutense University of Madrid, Madrid, Spain

J. Simões, J. M. Monteiro (eds.), *Veterinary Care of Farm Rabbits*,
https://doi.org/10.1007/978-3-031-44542-2_11

1 General Considerations

The reproductive characteristics exhibited by female rabbits are unique. The estrous cycle is not well defined, and ovulation is usually induced by coitus or by the administration of gonadotrophin-releasing hormone (GnRH) after artificial insemination (AI), independent of positive estrogen feedback, as occurs in spontaneous ovulatory females (Rebollar et al. 1992). Rabbit females can be sexually receptive a short time after kindling, showing post-partum estrus. Thus, females can be simultaneously pregnant and lactating. Furthermore, rabbit females may show pseudopregnancy because ovulation can be induced by mechanical, hormonal (GnRH), or stressful stimuli without fertilization of oocytes. Parity order has an important impact on the reproductive performance of rabbit does, affecting live body weight, litter size, and litter weight.

The development of several management systems and the use of prolific genetic strains (Bolet et al. 2004; Castellini 2007) have improved the productivity of rabbit farms. The use of cycled production with defined reproductive rhythms (Maertens et al. 1995; Castellini 1996) and the spread of AI have greatly increased and homogenized rabbit production in farms from developed countries. However, females under intensive and semi-intensive reproductive rhythms show a negative energy balance (NEB), more critical in primiparous rabbits' dams (Fortun-Lamothe 2006; Arias-Álvarez et al. 2009a), which affects female receptivity and reproductive outcome. Reproductive strategies such as estrus synchronization methods using hormones or biostimulation methods and nutritional strategies have been used to increase reproductive performance and reduce NEB. In any case, the increasing attention of rabbit production actors such as farmers, consumers, genetic centers, technicians, etc., in the production system to respect animal welfare according to European rules (EU Directive 2010/63/EU, European Parliament, 2010) has promoted the development of "more precision management" and "less impactful strategies" to improve the rabbit's reproduction performance (Mattioli et al. 2021).

The key points of this chapter are reported in Box 1.

Box 1: Key Points for Rabbit Does' Reproduction

- Rabbit does are polyestrous, poly-ovulating, and polytocous females.
- Assembly of follicles and folliculogenesis starts in the postnatal period and over an extended period, in contrast to most mammals.
- Puberty usually starts at 10–11 weeks old, and sexual maturity arises at approximately 4–5 months in medium-size breeds.
- The estrous cycle is not well defined.
- Folliculogenesis occurs in continuous waves.
- Estrous behavior appears every 4–6 days.
- Sexual receptivity in rabbit females involves changes in behavior and vulva color and aspect.

(continued)

Box 1 (continued)

- In cases of non-pregnancy, new preovulatory follicles appear in the rabbit ovary 40–50 h after the LH peak and can be ovulated.
- Ovulation is induced by neurohormonal stimuli indepencently of positive estrogen feedback.
- Rabbit females can be sexually receptive a short time after kindling, showing post-partum estrus.
- Pseudopregnancy occurs in the absence of fertilization and lasts 17–18 days.
- A mucin coat surrounds rabbit embryos in the female reproductive tract.
- Rabbit does nurse their kits usually once or twice every 24-h period, during 3–4 min at night and/or early morning.
- Hormonal (equine chorionic gonadotropin, eCG) and non-hormonal methods (biostimulation) are used for estrus synchronization in rabbit does.
- Artificial insemination is extensively used in commercial rabbit farms.
- Administration of GnRH and its analogs is the usual method for ovulation induction in rabbit females.

2 Reproductive Physiology

2.1 General Characteristics of Female Rabbits

Rabbit females display estrous behavior every 4–6 days (Milligan 1982). They are poly-ovulating and polytocous females because they can ovulate several oocytes and have several offspring at each kindling. Fetuses implant in both uterine horns (that are completely separate), and more than one viable fetus is necessary for gestation to take place (Adams 1962).

Sexual maturity depends on breed and body weight. Body weight is more important than age for the beginning of reproductive function. Female rabbit achieves puberty when they reach 75% of mature body weight. Then adipose tissue releases leptin, which acts on follicular steroidogenesis and the secretion of pituitary gonadotrophins (Boiti 2004). Puberty occurs just after the maximum growth rate (Quesenberry et al. 2021). In females that are usually on farms, puberty starts at 10–11 weeks (Díaz 1987), and sexual maturity at 4–5 months. At this moment, the first mating usually occurs. Rabbit does can be fertile until 3 years of age, but fertility changes according to the breed (Table 1).

Table 1 Reproductive features of medium-sized rabbit does (i.e., New Zealand rabbit)

Reproductive parameters of rabbit female	
Puberty	10–11 weeks
Sexual maturity	4–6 months
Estrus cycle	Non-regular, several each 4–6 days
Gestation period	30–31 days
Litter size	8–12 kits
Normal birth weight	30–50 g
Weaning age	6–8 weeks (30 days in commercial farms)

2.2 Ovarian Physiology

2.2.1 Folliculogenesis

Follicle formation begins prenatally in the majority of species. However, in rabbits, the formation and assembly of primordial follicles occur entirely in the postnatal period. Oogenesis is initiated at birth, and most primordial follicles assemble at 2 weeks of life, whereas antral follicles are first observed during week 12 of life (Hutt et al., 2006).

The structural and functional unit of the ovary is the ovarian follicle, which governs oocyte growth and maturation with the support of granulosa cells and theca cells and vice versa. Folliculogenesis consists of the growth of follicles from the primordial follicle to the antral follicle, which contains a fully mature and developmentally competent oocyte that will be ovulated and fertilized. Folliculogenesis is regulated by the hypothalamus–hypophysis–ovarian (HHO) axis, where different hormones participate and have feedback control in this axis (Fig. 1). However, it is important to note that oocytes also have a crucial function in follicle growth due to the bidirectional signals with the surrounding cells, which can negatively or positively affect follicle development (van den Hurk and Zhao 2005). Cumulus cells are necessary for the growth and maturation of oocytes (Lorenzo et al. 1996a, b) and for transcriptional activity during folliculogenesis (Eppig 2001).

There are two main types of follicles: preantral follicles (primordial, primary, and secondary) and antral follicles (tertiary and preovulatory) with different characteristics (Table 2). Follicle growth to the antral stage involves granulosa cell proliferation, theca cell development, and formation of the antrum filled with follicular fluid. These events generate a steroidogenic environment.

The follicles develop to the preovulatory stage in waves that include recruitment and growth of follicles, the selection of the dominant follicles and their ovulation. Preantral follicles are non-gonadotropin responsive follicles and develop slowly. Early folliculogenesis is mainly controlled by paracrine and autocrine intrafollicular signaling, whereas the follicles of later stages are gonadotropin dependent. It has been demonstrated that different factors, such as bone morphogenetic proteins (BMPs), insulin-like growth factors (IGFs), epidermal growth factor (EGF), fibroblast growth factors (FGFs), growth hormone (GH), and activin, promote follicular growth by inducing the proliferation and differentiation of granulosa cells in the

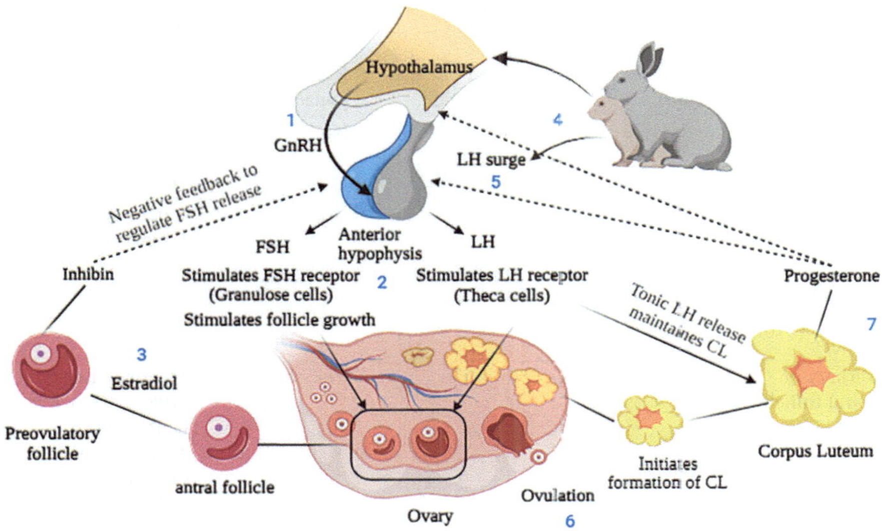

Fig. 1 Axis hypothalamus–hypophysis–ovary (HHO). Folliculogenesis is controlled by the HHO axis. First, the hypothalamus secretes GnRH (1), which acts in the anterior hypophysis, releasing gonadotropin hormones such as follicle-stimulating hormone (FSH) and luteinizing hormone (LH) (2). FSH promotes follicular growth by stimulating FSH receptors in granulosa cells. Both hormones play a role in the final follicular development of tertiary and antral follicles (gonadotropin-dependent phase). LH binds to its receptors in theca cells, where it promotes the synthesis of estrogen precursors (androstenedione). Granulosa cells synthetize 17-β-estradiol from the androstenedione produced by theca cells. Inhibin regulates FSH release by negative feedback. In rabbit does, the LH surge to induce ovulation is induced by a neurohormonal stimulus (i.e., coitus) (4 and 5). When ovulation occurs, the corpus luteum is formed in the ovary and is maintained by tonic levels of LH (6). The corpus luteum secretes progesterone, which has negative feedback in the hypothalamus and pituitary, regulating LH release

non-gonadotropin dependent phase (van den Hurk and Zhao 2005). Some studies in other species show a possible role of these gonadotrophins at basal level in follicular development and oocyte competence (Eppig 2001; Webb et al. 2003). In vitro studies suggest that many of these factors can produce similar actions in oocytes of rabbit does (Lorenzo et al. 1996a). FSH has an important role as a survival factor by stimulating the production of some growth factors, such as EGF and IGFs, and suppressing apoptosis-inducing genes (Smitz et al. 2001). In the absence of gonadotropins, GH promotes follicular development and the restart of meiosis in the cumulus–oocyte complex directly by binding to its receptors in the follicles and oocytes of rabbit does (Lorenzo et al. 1996a) or indirectly through IGF-I stimulation. As the antrum increases in size, growth factors, peptides, proteins, steroid hormones, energy metabolites, and other unknown substances accumulate in follicular fluid (Sutton et al. 2003). Antral follicles (tertiary and preovulatory) show receptors for gonadotrophins (Webb et al. 2003). At these stages, oocyte size increases, and granulosa and theca cells proliferate. Sensitivity to gonadotrophin shift and

Table 2 Developmental stages of rabbit follicles and their characteristics

Developmental stage	Oocyte diameter (μm)	Follicle diameter (μm)	Follicular fluid	Receptivity to gonadotropins	References
Primordial follicle	30	30–44	No	No	Pincus and Enzmann (1935); Hutt et al. (2006)
Primary follicle	60	100–120	No	No	Pincus and Enzmann (1935); Hutt et al. (2006)
Secondary or preantral follicle	80–104	200	No	No	Pincus and Enzmann (1935)
Tertiary or antral follicle	133–135	200–250	Yes	Yes	Fleming et al. (1984); Kranzfelder et al. (1984); Jelínkova et al. (1994); Hutt et al. (2006)
Preovulatory or Graff follicle	140–143	800–900	Yes	Yes	Kranzfelder et al. (1984); Jelínkova et al. (1994)

Adapted from Arias-Álvarez et al. (2007)

steroidogenesis promote theca cell differentiation and expansion of cumulus during ovulation (Lorenzo et al. 1997). Follicles that do not reach the size, therefore, cannot respond to gonadotrophin hormones and undergo atresia. If ovulation does not occur, preovulatory follicles also undergo atresia, and a new follicular wave starts (Baird and McNeilly 1981).

In rabbits, the mechanism of follicle recruitment and precise regulation of folliculogenesis in vivo is not fully understood (reviewed by Arias-Álvarez et al. 2007). Follicular growth in continuous waves with 5–10 follicles in each ovary is produced. Follicles may take 97 days to reach the preovulatory stage in rabbit females (Mariana et al. 1989). Follicular development has a cyclicity of approximately 10–12 days, starting with follicles that have a size of 450 μm and ending when they reach the preovulatory stage, with a size larger than 1.5 mm (Díaz 1987). Follicular recruitment is preceded by an acute increase in the gonadotrophin FSH, as in cyclic females (Fleming et al. 1984). FSH activates the aromatase function of granulosa cells to synthesize 17-β-estradiol, whose levels increase as follicle size increases (Lefèvre and Caillol 1978) (Fig. 1). Nevertheless, after recruitment, the selection and growth of a few follicles to the preovulatory stage is due to the progressive decrease in FSH sensitivity (Souza et al. 1996) caused by the negative feedback of increasing concentrations of 17-β-estradiol (Dufy-Barbe et al. 1978; Turckheim et al. 1983), inhibin (Pau et al. 2000), and follistatin. This produces the inhibition of GnRH release (Pau et al. 2000), and it indirectly inhibits FSH release (Webb et al. 2003; van den Hurk and Zhao 2005). In addition, dominant follicles show high levels of insulin growth factor I (IGF-I) (Yoshimura et al. 1994). Indeed, in vitro studies in rabbits have shown that IGF-I promotes 17-β-estradiol production, follicular development, and oocyte maturation (Yoshimura et al. 1996; Lorenzo et al. 1997). Synergic IGF and FSH increase the follicular response to gonadotrophins and aromatase activity (Webb et al. 2003) and the expression of LH receptors on theca cells (van den Hurk and Zhao 2005). LH is the predominant hormone in the final development of follicles because the LH receptors in dominant follicles promote final follicular growth by LH even when FSH levels are low (Smitz et al. 2001; Webb et al. 2003). In rabbit does, follicles in the final development stage have a size over 850 μm (Lefèvre and Caillol 1978; Molina et al. 1986). Antral follicles with a diameter greater than 600 μm which have not ovulated become atretic (Kranzfelder et al. 1984).

When ovulation occurs, the follicles become the corpus luteum. A luteal phase starts with high levels of progesterone that prevent ovulation.

2.2.2 Estrus Cycle and Sexual Receptivity

A certain rhythm can be identified in intervals because follicle development is continuously produced. Sexual receptivity is non-cyclic. Females are sexually receptive for 5–14 days, and then, there is a period of 1–2 days when the doe is not receptive (Vella and Donnelly 2012). If mating does not occur, the follicles degenerate, estrogen secretion decreases, and rabbits show a non-receptive period (Harcourt-Brown 2002). In the absence of coitus, waves of follicles develop and return at 7–10-day intervals under the tonic action of FSH (Boiti 2004). If mating occurs, genital somatosensory receptors are stimulated (Bakker and Baum 2000), and ovulation is triggered.

Sexual receptivity in female does can be assessed by behaviour and vulva characteristics. Some specific behaviors, such as lordosis, reverse bending or flattening of the back, and raising of the pelvis showing the perineum are indicative of a sexually receptive female (Quesenberry et al. 2021). In addition, sexual receptivity is accompanied by changes in vulvar color and aspect (Fig. 2). An enlarged, moist,

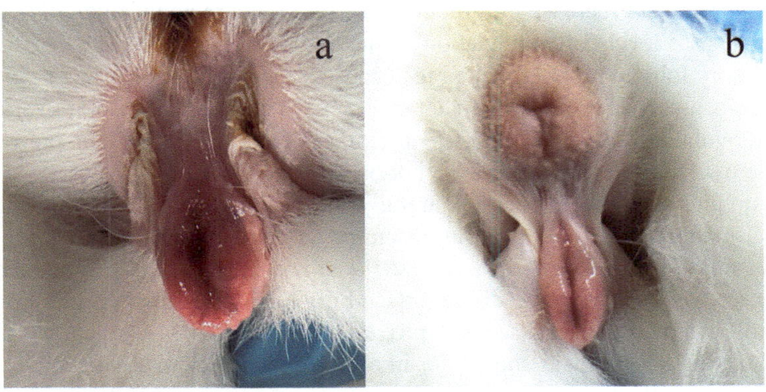

Fig. 2 Vulva color and aspect. (**a**) Enlarged, moist, reddish-purple vulva (receptive female); (**b**) Palish vulva (non-receptive female)

reddish-purple vulva indicates sexual receptivity (Rebollar et al. 1992; Maertens and Luzi 1995). However, estradiol levels and vaginal smears cannot predict sexual receptivity in this species (Rebollar et al. 1992). Non-receptive females usually show a particular behavior of running away, cornering, biting, and vocalizing (Quesenberry et al. 2021), and they exhibit white or purple vulva (Rebollar et al. 1992).

2.2.3 Ovulation, Mating, Fertilization, and Early Embryo Development

Rabbit is a reflex ovulatory species. This means that in rabbit does, ovulation is not produced by the negative feedback of estrogens (as occurs in spontaneous ovulatory species). This is triggered by a neurohormonal stimulus provoked by coitus that comprises mechanical stimulation and action of seminal plasma factors. Nervous stimulus by neural pathways, in which noradrenergic neurons and cholinergic neurons of the brainstem are major participants, elicits the pulsatile release of gonadotropin releasing hormone (GnRH) from GnRH neurons in the mediobasal hypothalamus (Pau and Spies 1986; Bakker and Baum 2000). Other factors in the seminal plasma, such as nerve growth factor (NGF) can also be involved in the ovulation process (Garcia-Garcia et al. 2020). It is important to note that in this species, certain factors, such as animal manipulation, nutrition, and the environment can generate auditory, olfactory, visual, and stressful stimuli that generate neuromodulatory signals that interfere with hormonal secretion in the HHO axis and can positively or negatively modify reproduction (Theau-Clément 2000).

As occurs in spontaneous ovulatory species, GnRH release is followed by an immediate preovulatory LH surge from the pituitary gland (Mills and Gerardot 1984; Pau et al. 2000), resulting in the final maturation of the oocyte (resumption of meiosis), ovulation, and corpus luteum formation. Ovulation occurs 11–12 h post-coitus (Díaz 1987; Bakker and Baum 2000) compared with 30 h in other reflex ovulators, such as cats and ferrets (Quesenberry et al. 2021). LH levels peak 1–2 h after mating and return to basal concentrations 4–5 h later. This coincides with the temporary increase in FSH levels (60–90 min) (Kaynard et al. 1990; Pau et al. 2000) and the consequent increase in follicular steroids (Lefèvre and Caillol 1978), which return to basal levels after ovulation. Osteen and Mills (1980) showed that new preovulatory follicles at 6 days emerge because a second FSH peak at 24–48 h post-coitum appears. However, Kranzfelder et al. (1984) reported that rabbit ovaries form new preovulatory follicles 40–50 h after the LH peak, which are prepared for possible ovulation if pregnancy does not occur.

The ovulation rate is higher in naturally mated female rabbits because they are always sexually receptive than in inseminated rabbits (Maranesi et al. 2018). Females normally ovulate 8–12 oocytes (Foote and Carney 2000) from each ovary and are recovered by the infundibulum (see Fig. 3). Corpus luteum formation in the ovary follows ovulation and secretes progesterone.

For mating, the female should be introduced into the male's cage. Receptive females are usually rapidly mounted. Coitus lasts only a few seconds, and males may attempt to copulate again within 2–3 min (Quesenberry et al. 2021). Males deposit semen into the anterior vagina, and the sperm pass to the uterus and to the isthmus.

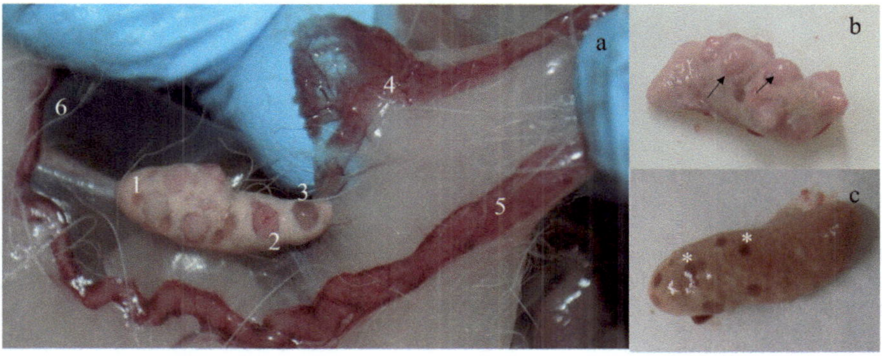

Fig. 3 Ovary and oviduct macroscopic morphology. (**a**) Ovary (bean shape) and oviduct of rabbit does are shown: (1) antral follicle, (2) corpus luteum with ovulation fossa, (3) luteinized follicle, (4) infundibulum, (5) ampulla and (6) isthmus; (**b**) ovary with corpus luteum (arrows show some of them); (**c**) ovary with small antral follicles (asterisks show some follicles)

Table 3 Reproductive events around ovulation and embryo development in rabbit. Embryo development times were adapted from Sultana et al. (2009) and Meng et al. (2013)

Peri-ovulatory events and embryo development		
Event	Time post-coitus (h)	Localization
Ovulation	11–12 h	Ovary
Fertilization	12–14 h	Oviduct (ampulla)
Zygote formation	18–20 h	Oviduct
2-cell embryo	24–26 h	Oviduct
32-cell embryo (morula stage)	54–56 h	Oviduct-uterus
Compact morula	64–66 h	Oviduct-uterus
Early blastocyst	76–78 h	Uterus
Blastocyst	84–86 h	Uterus
Expanded blastocyst	94–96 h	Uterus
Corpus luteum maintenance	17–18 days	Ovary

Afterwards, capacitated sperm travel to the ampulla of the oviduct, where fertilization occurs approximately 12–14 h after copulation or AI. Most mature oocytes are fertilized during the first hour after ovulation (Sultana et al. 2009), and their ability to be fertilized decreases at 8 h post-ovulation (Austin 1974). A high fertilization rate is physiologically produced after mating (Bolet and Theau-Clément 1994).

The first embryonic divisions occur at 18–20 h in the oviduct, where the embryo stays until 60–72 h after ovulation (Table 3). At this moment, embryos in the morula stage progress toward the uterus. Compaction occurs in the morula stage and coincides with the establishment of cellular junctions (Fischer et al. 2012). Blastocysts show a well-defined inner cell mass (ICM) surrounded by the trophoblast that gives rise to fetal organs and extraembryonic membranes and tissues (Fig. 4). Implantation in the blastocyst stage takes place around day 7 post-ovulation

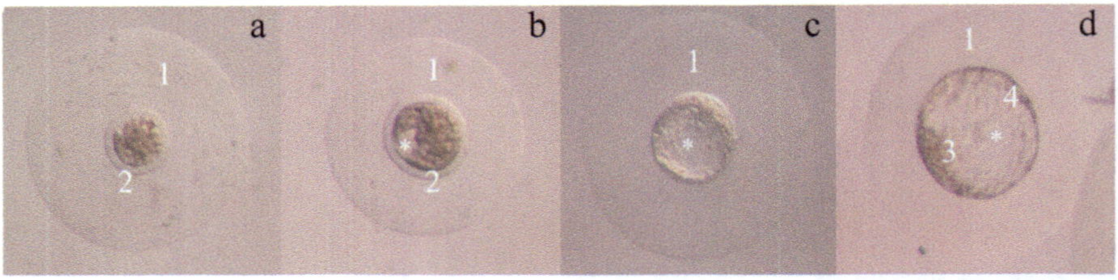

Fig. 4 Blastocyst developmental stages. (**a**) Morula; (**b**) early blastocyst; (**c**) blastocyst; and (**d**) expanded blastocyst. (1) Mucin coat, (2) zona pellucida; (3) ICM, inner cell mass; (4) trophoblast; asterisks show blastocoel cavity. Adapted from Arias Álvarez (2010)

(Orsini 1962). Maternal-to-zygotic transition (MZT) after fertilization occurs at the 8–16 cell stage in the rabbit (Christians et al. 1994). MZT involves switching control of embryo development from the maternal genome to the embryo genome.

Embryos are surrounded by a glycoprotein layer termed the zona pellucida (ZP), which is involved mainly in species-specific fertilization and early development. This shell is made from proteins (ZP1, ZP2, ZP3, and ZP4), and the rabbit ZP4 protein has a structural role and is implicated in fertility (Lamas-Toranzo et al. 2019). In addition to the zona pellucida, in vivo embryos present a mucin coat (Fig. 4), which is a layer of glycoprotein that accumulates during passage through the oviduct. It may protect against pathogens and ensure blastocyst expansion for implantation (Murakami and Imai 1996). At 72 h after ovulation, the thickness of the mucin coat is the greatest (approximately 110 μm) (Fischer et al. 1991). Blastocyst expansion and blastocele formation start at approximately 84 h post-ovulation, but they do not hatch as in other species. Instead, the zona pellucida disappears at approximately 96 h (Fischer et al., 1991), and the mucin layer dissolves at 140 h post-ovulation (Adams 1958). A new layer called neozone appears at this moment (Denker and Gerdes 1979). Before implantation at day 6, the embryo size is approximately 2–6 mm (Denker 1977), being one of the largest spherical blastocysts in mammals (Fischer et al. 2012).

2.3 Gestation and Parturition

Functional development of the corpus luteum and progesterone synthesis are necessary for uterine implantation and maintenance of gestation until delivery (Gadsby et al. 1983; Donnelly et al. 1991). Fetal resorption, mummification, retention, and abortion are common in this species. Fetal resorption is produced up to Day 21 of gestation, and then fetal death results in abortion (Quesenberry et al. 2021). Embryo loss usually occurs in the preimplantation phase (approximately 3%) or after embryo implantation, between days 8 and 17 of gestation (approximately 20%). In this species, 25% of fetuses can be lost without interfering with gestation. Mummified fetuses can remain in the uterus for extended periods. Rabbits have a discoid and hemochorial placenta independently for each fetus (Fig. 5).

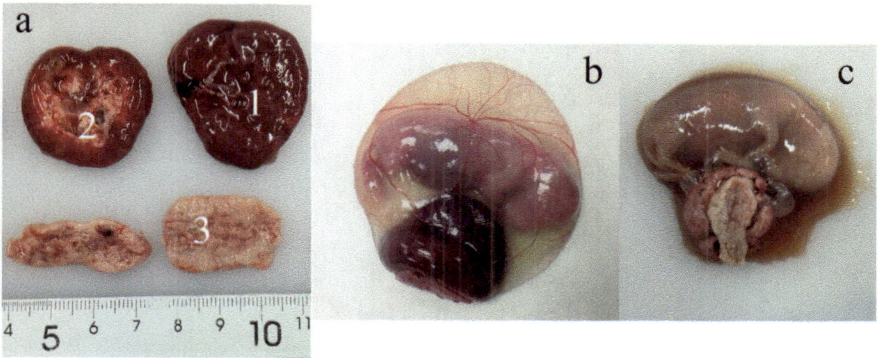

Fig. 5 Rabbit does' placenta and fetuses. (**a**) Hemochorial and discoidal placenta is shown. (1) The labyrinthine zone is proximal to the fetus, (2) the placental junctional zone is in contact with the maternal decidua, and (3) the decidua is in close contact with the uterus; (**b**) fetus 21 days old where the placenta and embryonic sac filled with amniotic fluid can be appreciated; (**c**) mummified fetus 21 days old

Fig. 6 Embryonic vesicles in uterine horns (14 days of pregnancy). Each asterisk indicates one fetus

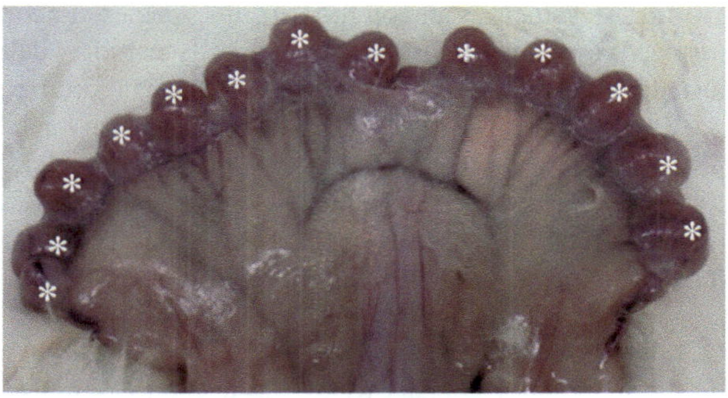

Gestation lasts 30–32 days with a range of 28–35 days. Pregnancy is detectable by palpation at 10 days of gestation, although diagnosis is more reliable on days 13 and 14 when embryonic vesicles are approximately 20 mm in size (Fig. 6). Palpation for pregnancy diagnosis is not recommended after day 20 of pregnancy to avoid the risk of pregnancy loss.

Before delivery (72 h before parturition), the nest is prepared for the female with different materials and with hair from her body (Canali et al. 1991; Hudson et al. 2000) (Fig. 7). Female plucks hair from chest, sides, and dewlap to make nest warmer and more comfortable. This hair usually becomes loose around delivery (5 days before parturition day) (Quesenberry et al. 2021) because of the high levels of testosterone and prolactin (González-Mariscal et al. 2007). Nests prevent the firstborn young from hypothermia while the rest of the litter is being born. Wood shavings or straw should be provided in order for rabbit doe to make the nest properly.

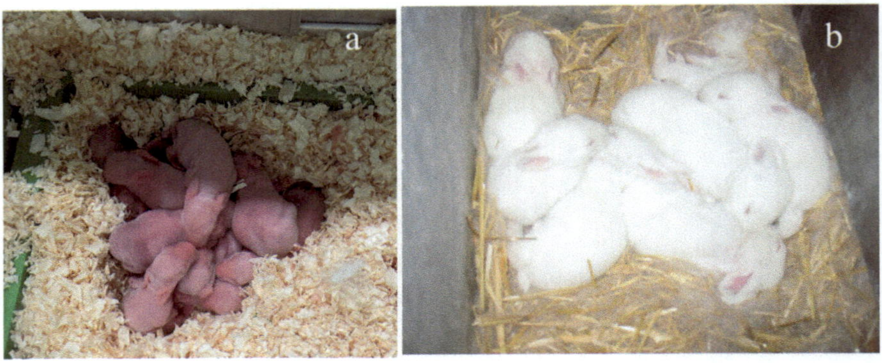

Fig. 7 Litter of newborn pups and kits. (**a**) A litter of naked newborn pups with closed eyes grouped in nests with wood shavings and doe's hair. (**b**) 14-day-old kits in nests with straw and doe hair

Parturition usually takes place early in the morning and lasts approximately 30 min. Dystocia is uncommon in this species. Each fetus is accompanied by its respective placenta, which rabbits usually eat after delivery. Cannibalism does not usually occur, although litter abandonment is more frequent.

The litter size is 8–10 kits but depends on parity and breed or strain (range 4–14). Primiparous does tend to produce smaller litters. Larger breeds have large litters (8–12 young), whereas small breeds produce smaller litters (4–5 kits). Rabbit young weighing 30–50 g at birth (López-Tello et al. 2017a) and are not fully mature with no hair and closed eyes (Fig. 7). Eyes are open at 10 days.

2.3.1 Pseudopregnancy

Pseudopregnancy appears when non-fertile mating takes place. It usually lasts from 15 to 19 days, and the common causes are copulation with an infertile male, fertilization failure, or early embryonic death.

Pseudopregnancy is characterized by corpus luteum development and progesterone synthesis. Ten to 12 days after coitus, the corpus luteum reaches its maximum size in pregnant and pseudopregnant does. The corpus luteum of the rabbit is dependent on estradiol by day 6 of pseudopregnancy (Osteen and Mills 1980). Plasma progesterone levels are high until day 12 post-copulation, and this causes changes in the uterus and mammary glands similar to those of pregnant does, but mammary involution usually occurs on day 16 (O'Malley 2005). From days 12 to 13, corpus luteum regression begins, and progesterone concentrations decrease, reaching basal levels between days 16 and 21 (Browning et al. 1980). Luteolytic action by prostaglandin F2α (PGF2α) secreted by the uterine endometrium spontaneously ends pseudopregnancy around days 17–18 (Lytton and Poyser 1982). Nesting behavior (including hair loosening) on day 17 post-coitus is indicative of pseudopregnancy.

2.4 Lactation

Rabbit females have five pairs of mammary glands in different regions (axilla, thorax, abdomen, and inguinal areas) (Fig. 8). Nursing of rabbit neonates is distinctive. Youngs have no preference for any one teat, and rabbit does nurse their kits a short time (3–4 min), usually once or twice a day, at night or early morning (Hudson and Distel 1989). Oxytocin and prolactin are secreted in high concentrations as a consequence of suckling, and therefore, milk production is also affected by maternal hormonal levels induced by the suction reflex (Freeman et al. 2000; Fernández-Pacheco et al. 2021a). Prolific rabbit females have similar pituitary responses for litters of 7–9 or 10–12 kits immediately after being exposed to a suckling stimulus (Fernández-Pacheco et al. 2021a). Plasma prolactin (PRL) concentrations are delayed until 1–5 minutes after suckling has ceased at the beginning of lactation, and in mid-lactation, prolactin reaches peak levels that plateau for 2–3 h (Fuchs et al. 1984). Dams and their litters have synchronized circadian rhythms, so the kits are able to anticipate their mother's arrival and increase their activity when she arrives, thereby preparing for the competitive situation that comes with suckling (Jilge et al. 2000).

Rabbit milk presents a high energy content (8.5 MJ kg^{-1}) and high percentages of protein (10–15%) and lipid content (12–15%). The lactose content is lower than that in other mammals (5% of dry matter) (Gidenne et al. 2010; Xiccato and Trocino 2010). Kits do not need to take colostrum because passive immunity comes from the placenta. According to Poigner et al. (2000), heterogeneous birth weights within the litter could play a major role in mortality rates during lactation.

Natural weaning usually occurs at 6–8 weeks, but weaning of litters in commercial rabbit farms usually takes place from days 28 to 35 of age. The minimum period for suckling to ensure litter survival is 25 days. If the doe dies at parturition, fostering for another lactating doe is produced. There is a mammary pheromone in milk (2-methyl-but-2-enal) essential for guiding kits to teats. Estradiol, progesterone, and prolactin regulate the secretion of an olfactory signal that triggers nipple-seeking behavior in kits, although its identity and relationship to mammary pheromone are not known (Melo and González-Mariscal, 2010).

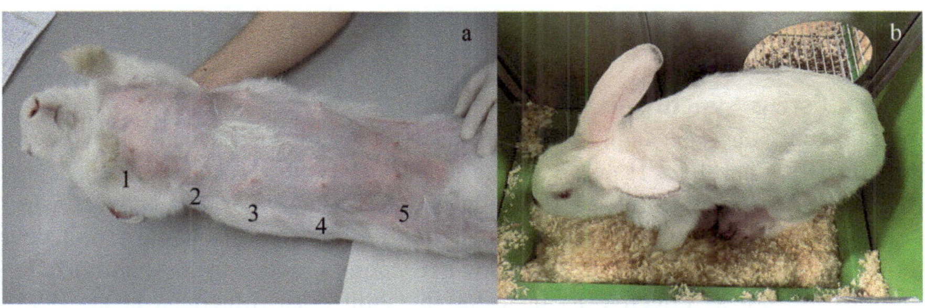

Fig. 8 Mammary gland chain and suckling rabbit does. (**a**) Mammary glands: (1) axillar, (2) thoracic, (3) abdominal, (4) and (5) inguinal; (**b**) rabbit doe nursing kits

At the end of pregnancy, a first wave of follicular development starts concomitantly with progesterone level decline on days 29 and 30 of pregnancy (Díaz et al. 1987). As stated in Sect. 2.2.1, waves of follicular development can overlap every 4–6 days because a new group of follicles can be recruited and grow before the previous wave enters atresia. In the post-partum period, these waves of follicles reach their maximum development around days 3 and 9; consequently, the sexual behavior of rabbit does is not the same during all post-partum periods (Ubilla and Rebollar 1995). If the female is mated in post-partum estrus or is inseminated, the doe can be pregnant and lactating simultaneously. Nevertheless, high prolactin levels provoke inefficient reproduction in this case. This physiological knowledge is used to manage female reproduction during the post-partum period, as shown in the following section.

3 Reproductive Management

Management of reproduction is essential in rabbit farms, and different strategies can be used to achieve successful reproductive management of females. In developed countries, rabbit reproduction is under the scheme of "single batch," and the females are handled simultaneously for different tasks (AI, palpation for pregnancy diagnosis, parturition, and kit weaning). Fertility is key in this well-organized scheme because non-pregnant females remain "empty" until the following AI and parturition interval increase (Theau-Clément et al. 2016). The use of AI combined with estrus synchronization protocols in rabbit does is common practice in rabbit farms (Rebollar et al. 2006a, 2008). Rabbit does are used for reproduction on modern farms for a maximum of 8–10 reproductive cycles. The replacement of females is correlated with the age at first mating, the interval between inseminations, and between parturitions. The average culling age for breeding does is 14.9 months and 6 parities (reproductive lifespan is approximately 1 year), but females in the first three parturitions show the highest risk of culling and mortality (Rosell and de la Fuente 2009). Low reproductive performance and poor body condition are the main causes of culling in productive systems.

3.1 Reproductive Strategies

3.1.1 Estrus Synchronization

As stated previously, lactating rabbit does can be inseminated because they do not show lactating anestrus. However, these females show low sexual receptivity that affects the fertility rate after AI (Ubilla and Rebollar 1995) and require estrus synchronization methods to increase their low fertility. Non-lactating rabbit does show high individual variability of sexual receptivity, and estrus synchronization methods are not as effective as in lactating rabbit does (Maertens et al. 1995). Hormonal and non-hormonal treatments can be used to synchronize rabbit estrus.

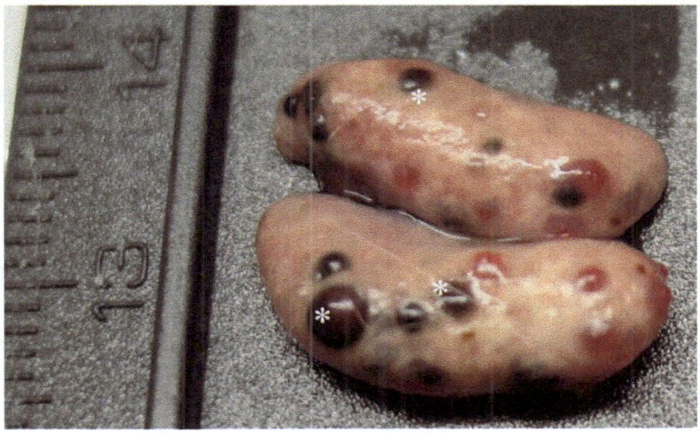

Fig. 9 Ovary showing hemorrhagic follicles (non-ruptured anovulatory follicles full of blood). Some representative hemorrhagic follicles are indicated (asterisks)

The most widely used hormone is equine chorionic gonadotropin (eCG). To increase the number of receptive does at AI time, a dose of 20–25 IU eCG is commonly used 48 h before insemination in lactating does at 11 days post-partum (semi-intensive rhythm). Administration of high doses of eCG (≥50 IU) provokes an elevated number of hemorrhagic and cystic follicles (Mehaisen et al. 2005) (Fig. 9). eCG promotes reproductive tract development and follicle maturation (Hu et al. 2019). In nulliparous rabbit does and during the first cycles of rabbit life, eCG improves follicle development (Rebollar et al. 2008), oocyte cytoplasmic maturation (Arias-Álvarez et al. 2013b), ovulation rate and embryo development (Theau-Clément et al. 2008; Arias-Álvarez et al. 2013b). However, this hormone induces progressive loss of reproductive efficiency in successive inseminations (Boiti et al. 1995; Rebollar et al. 2006a; Garcia-Garcia et al. 2009). The number of hemorrhagic follicles increases (Stradaioli et al. 1997), and an immune response is induced (Lebas et al. 1996). It has been postulated that the lower antioxidant defense found in embryos from eCG-treated mothers probably reduces the prolificacy and fertility rates observed in females when eCG administration is used repeatedly (Arias-Álvarez et al. 2013b).

To replace hormonal treatments for estrus synchronization, other non-hormonal methods have been implemented, such as biostimulation methods. These are more natural and less expensive procedures based on physiological events. Some biostimulation methods commonly used include a controlled light schedule and transient doe–litter separation.

The normal lighting schedule in rabbit farms consists of 16 h light:8 h dark. This large daily light period simulates the spring season, when rabbits are in favorable conditions (abundance of vegetation) (Quesenberry et al. 2021). The lighting program more successful is changing 8 h light to 16 h light 8 days before AI. This schedule increases the sexual receptivity of does and the kindling rate, but the weight of kits at weaning can be decreased (for review, Szendro et al. 2016).

Transient doe–litter separation consists of lactation control to prevent the negative effect of the hormone prolactin on the ovarian response. Nursing of the litter is usually controlled in a 24–48 h period of doe–litter separation (Bonanno et al. 2002;

Rebollar et al. 2008) with different regimes (i.e., implementing 10-min nursing of the litter at 24 h of mother-separation) (Rebollar et al. 2008). This method is as effective as eCG treatment, especially in the first AI cycles in females under intensive rhythm (AI at day 4 post-partum) or in primiparous does AI at day 11 post-partum (Rebollar et al. 2006a; Arias-Álvarez et al. 2010b). Later, separation becomes more difficult to perform because kits are older, and management is more complex. This type of biostimulation improves the sexual receptivity and fertility of females (Eiben et al. 2008; Rebollar et al. 2011) by stimulating ovarian function prior to AI. It promotes follicular growth and follicular steroidogenesis (Ubilla et al. 2000; Rebollar et al. 2008) and enhances oocyte quality (improves cytoplasmic maturation and reduces follicular atresia) (Arias-Álvarez et al. 2010b).

3.1.2 Artificial Insemination and Reproductive Rhythms

AI in rabbits is generally performed to improve breeding management in commercial farms (Fig. 10). Fertility rates are similar to those obtained with natural mating (Morrel 1995) but with greater advantages: (a) a cycle-based production system can be implemented, and more efficient breeding programs can be developed in rabbit farms; (b) sanitary status is controlled because animal contact (male–female) is avoided (Carluccio et al. 2004).

The success of insemination depends on female sexual receptivity, parturition order, lactation status, and pseudopregnancy (Castellini 1996; Theau-Clément 2007). Reproductive performance is higher in multiparous females (3–4 parturitions) and in receptive, non-lactating females. Primiparous rabbit females show the lowest reproductive performance and fertility rate depending on the reproductive rhythm applied (Arias-Álvarez et al. 2009b).

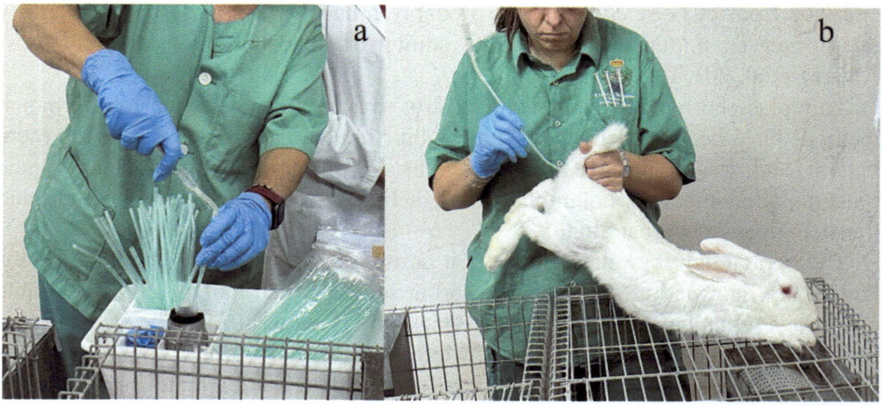

Fig. 10 Artificial insemination in rabbit does. (**a**) Preparing the artificial insemination (AI) dose (0.5 ml of pooled semen containing 10–20 million spermatozoa) to be administered by plastic cannulas or AI gun; (**b**) for AI, females should be slowly raised by the tail to expose the vulva, and the cannula should be introduced parallel to the spine in the vagina

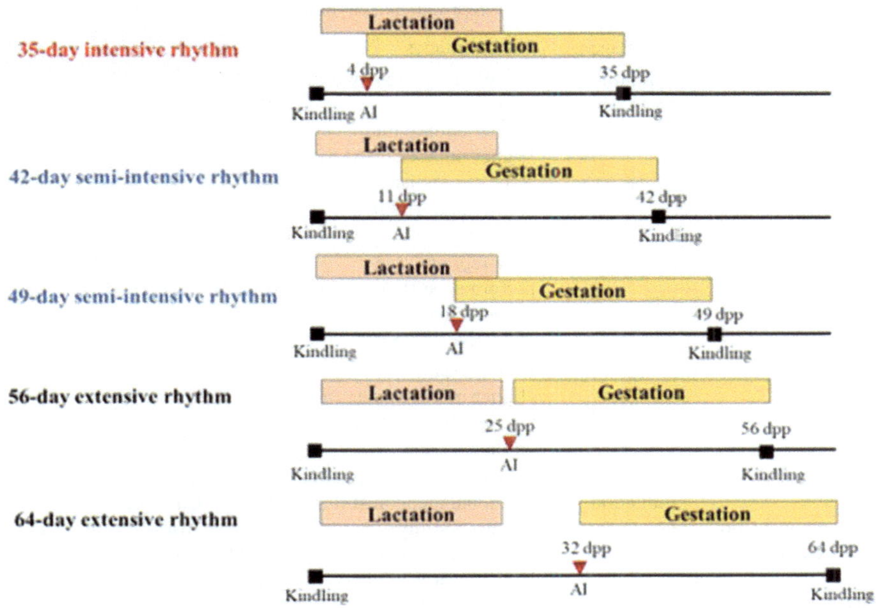

Fig. 11 Reproductive cycles in rabbit females at the farm. From Arias Álvarez (2010)

Rabbit reproduction is carried out under reproductive rhythms according to the AI regime and weaning time of the kits (Fig. 11). In commercial European farms, selected rabbit does are normally under 42-day semi-intensive rhythms (AI on Day 11 post-partum) or 49-day semi-extensive rhythms (AI on Day 18 post-partum). Another rhythm applied is extensive rhythm, where AIs are practiced after weaning (25 days or 32 days post-partum) because females show high receptivity immediately (Theau-Clément et al. 1990). As rabbit does exhibit high receptivity after kindling, it is also possible to inseminate rabbit does 4 days after parturition using AI in an intensive rhythm. Each of these rhythms shows advantages and disadvantages (see Table 4). Farmers should choose the best management for their farms. Longevity of the does and fertility rate are the contributing factors to the economic profitability of rabbit farms (Ludwiczak et al. 2021).

Continuous intensive reproductive rhythms can affect long-term reproductive performance (Cervera et al. 1993; Garcia-Garcia et al. 2009). Intensive production in rabbits diminishes reproductive efficiency, mainly of primiparous females, and decreases the lifespan of the doe (Arias-Álvarez et al. 2009a). This ultimately affects productive outcomes but also negatively impairs consumer perception of rabbit meat production (Trocino et al. 2019). In addition, infectious diseases such as mastitis increase the culling rate (Rosell and de la Fuente 2009). Therefore, this management is contrary to the European Community guidelines that tend to reduce the use of antibiotics and hormones in animal husbandry, including rabbit farming. This

Table 4 Advantages and disadvantages of reproductive rhythms used in rabbit farms

Rhythm	Advantages	Disadvantages	References
35-day intensive (AI 4 dpp and weaning at 25 days of lactation)	High number of kits per year	Lower energy reserves in does Higher culling rate Higher number of AI to achieve pregnancy Higher parturition interval	Cervera et al. (1993); Rebollar et al. (2006a, b)
		Low reproductive performance (low quality of the oocytes and endocrine response)	Rebollar et al. (2009)
		Low lifespan Higher mortality rate in young rabbits (25 and 35 days of age)	Garcia-Garcia et al. (2009)
42-day semi-intensive (AI on day 11 pp and weaning at 28–30 days of lactation)	High production rate	Low body condition and hormonal antagonist that impairs fertility Low conception rate High parturition interval Low reproductive lifespan	Theau-Clément and Roustan (1992); Castellini et al. (2003); Arias-Álvarez et al. (2009a); Fortun-Lamothe (1998, 2006); Castellini et al. (2010)
Extensive (AI at 30–32 dpp and weaning at 28–32 days of lactation)	Enhance nutritional status Higher fertility High reproductive lifespan Better welfare state Better oocyte quality	If females are not pregnant: Excessive fat Impaired reproduction	Castellini et al. (2003, 2010)

AI, artificial insemination

system seems to be less efficient, and semi-intensive and extensive rhythms are recommended.

Conversely, extensive reproductive rhythm can be adapted to enhance rabbit nutritional status, fertility, length of reproductive activity, and welfare (Castellini et al. 2003, 2010), as they eliminate the hormonal and energetic antagonisms between lactation and pregnancy (Castellini et al. 2003). In addition, this system reduces costs by avoiding repetitive tasks such as AI, nest control, pregnancy diagnosis, and weaning (Velasco et al. 2009) and compensates for the lower production efficiency because doe longevity is improved. Insemination after weaning (32 days post-partum) is the best management, at least for primiparous does (Rebollar et al. 2009; Arias-Álvarez et al. 2009a). It improves the estimated body composition, energy and protein body content, as well as the ovarian physiology of rabbit does (Arias-Álvarez et al. 2009a). As a consequence, fertility and prolificacy rates are substantially improved (Arias-Álvarez et al. 2009a). However, a lower number of parturitions are reached in the reproductive life of the females, and if these animals do not become pregnant at the moment of AI, they can show excessive fattening, which also impairs reproduction (Cardinali et al. 2008). Then, nutritional strategies should be applied to improve reproductive outcomes and avoid fattening (see Sect. 3.2). Additionally, farmers should consider applying this rhythm only at specific moments of reproductive life, such as the second insemination when NEB is more pronounced in the does.

To decrease the energy deficit of lactating rabbit does prior to AI, early weaning can be applied to reduce the lactation period (Xiccato et al. 2004). It takes place at 21–25 days, when young individuals are able to intake solid food. However, this approach does not improve ovarian physiology (follicle development and oocyte quality) (Sakr et al. 2010), fertility rate (Feugier and Fortun-Lamothe 2006) and kit mortality increases 3.6-fold in early weaning at 25 days compared to 35 days of age (Rebollar et al. 2009).

3.1.3 Ovulation Induction

As rabbit does are reflex ovulators, exogenous hormones such as GnRH and its analogs are essential to induce ovulation at AI time. However, estrous behavior and sexual receptivity in females, as in spontaneously ovulating species, are necessary.

GnRH and its analogs are routinely given by the intramuscular (i.m.) route. Usually, mating and GnRH administration induce the rapid release of plasma LH, as GnRH rapidly reaches the pituitary gland. Simultaneous AI and GnRH administration are the most common practices in rabbit farms. The ovulation rate is very successful, and nearly 100% of females can ovulate with the i.m. administration of synthetic analogs of GnRH. GnRH can be repeatedly administered without causing antibody formation. In addition, human gonadotrophin hormone (hCG) and LH can also be used to induce ovulation in rabbits (Hulot and Mariana 1988; Molina et al. 1991). The response to hormonal administration is affected by some factors, such as the age and body weight of females, nutritional and health status, and environmental conditions.

Several commercial GnRH analogs for triggering ovulation of rabbits are available (Table 4). The intramuscular route is the most common route of administration. Experienced farmers usually inject hormonal doses and perform AI on farms. There are some disadvantages that include a risk for hormone over- or underdosage, a time-consuming process, and a low sustainable practice because consumables such as needles and syringes are wasted in each AI. For rabbit females, it is an invasive and stressful operation with serious consequences on the longevity and welfare of does (Munari et al. 2019). Furthermore, it can affect ovarian physiology, including oocyte maturation (Zanagnolo et al. 1996). For all these reasons, research efforts have been targeted to replace the i.m. route by including GnRH analogs in the seminal dose at AI time to be absorbed by vaginal mucosa. Effective absorption of the analog depends on vaginal mucosa vascularization associated with sexual receptivity of

Table 5 Doses and fertility rates of different commercial preparations of synthetic analogs of gonadotrophin-releasing hormone (GnRH) administered intramuscularly and intravaginally

Synthetic analog GnRH	Intramuscular route		Intravaginal route		References
	Doses (μg)	Fertility (%)	Doses (μg)	Fertility (%)	
Buserelin	0.8	91.7	8	79.2	Quintela et al. (2004)
			12	87	
			16	87.5	
	1	78.6	5	74.8	Viudes-de-Castro et al. (2007)
	1	83.4	5	76	Vicente et al. (2008)
	1	84.0	5	65	Casares-Crespo et al. (2016)
	–	–	4	70	Casares-Crespo et al. (2018)
			5	85	
	1	87.5	10	100	Rebollar et al. (2012)
Lecirelin	5	80.0	5	80	Dal Bosco et al. (2014)
	5	40–60	7.5	50–100	Munari et al. (2019)
Gonadorelin	20	85.6	–	–	Quintela et al. (2009)
des-Gly10, D-Ala6]-LHRH ethylamide	–	–	25	91.1	Quintela et al. (2009)

LHRH, luteinizing hormone-releasing hormone

the female. In addition, extender semen diluent and proteases of the seminal plasma are important factors to be fully effective (Dal Bosco et al. 2014). However, higher doses are needed to induce ovulation when GnRH analogs are administered by the intravaginal route. For example, same fertility rate is reached by vaginal route using 15-fold times the buserelin dose used by the i.m. via (Quintela et al. 2004) (Table 5). Then, some approaches to avoid this problem have been studied, such as new GnRH analogs such as [des-Gly10, D-Ala6]-LHRH ethylamide, with successful results in ovulation induction rates and kindling rates (Quintela et al. 2009). In addition, to avoid enzyme degradation of the analog for the seminal plasma proteases, enzyme inhibitors and nanoencapsulation methods were used (Viudes-de-Castro et al. 2014; Casares-Crespo et al. 2016, 2018). Nevertheless, the results of studies are not convincing enough to replace i.m. route at the moment.

In recent years, searching for alternative molecules naturally present in the seminal plasma that participate in the process of ovulation has been the focus of considerable attention. An ovulation-inducing factor (OIF) in the seminal plasma, whose identity was not initially known, was implicated in the ovulation process in other reflex ovulatory species, such as camelids. Then, the first studies were performed using seminal plasma. Rabbit seminal plasma injected by i.m. via was unable to induce ovulation in rabbit females, although hemorrhagic follicles were found (Silva et al. 2011; Masdeu et al. 2012). In contrast, in group-housed females, ovulation induction was triggered using i.m. injection of rabbit seminal plasma but not in individually housed rabbits (Cervantes et al. 2015). Heterologous seminal plasma from llama and camel failed to provoke ovulation and corpus luteum formation in rabbit does i.m. injected (llama: Silva et al. 2011; camel: Masdeu et al. 2015).

In 2012, OIF was identified by biochemical studies as a neurotrophin called beta nerve growth factor (βNGF) (Ratto et al. 2012). In rabbit does, plasma βNGF concentrations arose after natural mating or AI with raw semen, probably synthesized by the uterus (Maranesi et al. 2018). Using heterologous β-NGF from mice via i.m. only a slight ovulation induction response was observed (17% of the rabbit does ovulate without catheter stimulation, and many ovaries showed anovulatory hemorrhagic follicles) (Garcia-Garcia et al. 2018). βNGF is a highly conserved protein among species, but some important differences between rabbits and other species in some sites where the receptors bind to the neurotrophin have been detected (Sanchez-Rodriguez et al. 2019). In this sense, the addition of homologous recombinant βNGF in the seminal dose induces ovulation in a dose-dependent manner in inseminated rabbit female does. Using 1 μg/ml recombinant rabbit NGF, 60% of females were able to ovulate (Sanchez-Rodriguez et al. 2019). In addition, neurotrophin shows a luteotrophic and follicle-stimulating effect, as in camelids (Carrasco et al. 2016; Silva et al. 2017). However, the ovulation rate is still not comparable to the usual protocol using GnRH by i.m. via. Protection of neurotrophin with microencapsulation of chitosan for better vaginal absorption also remains insufficient in rabbits, and the ovulation rate is approximately 50–60% (Quiroga et al. 2022a, b and unpublished data). Moreover, βNGF plays autocrine/paracrine roles in rabbit oocyte maturation and early embryonic development. In fact, evidence has also shown that βNGF directly affects oocyte maturation (Quiroga et al. 2022a) and early embryo development in in vitro culture (Pei 2010). Gene expression during embryo development of βNGF and its receptors varies according to the developmental stage of the embryo (Arias-Álvarez et al. 2022).

As mechanical stimuli seem to be very important in rabbit females, cannulas mimicking the mechanical effect are under study. Using a short and flexible cannula promotes ovulation in 64% of females, whereas a long and rigid cannula only provokes ovulation in 30% of does (Viudes-de-Castro et al. 2017). In addition, female sexual receptivity affects ovulation induction when a mechanical stimulus is applied by the vaginal route with a cannula.

Females synchronized with eCG show a higher ovulation rate than females that are not hormonally stimulated and those with unknown sexual receptivity (Garcia-Garcia et al. 2018; Sanchez-Rodriguez et al. 2019).

3.2 Nutritional Management Strategies

Some nutritional approaches have been proposed to improve the body condition and, indeed, the reproductive performance of rabbit females before or during pregnancy and during the post-partum period (lactation phase). Some strategies are focused on increasing the feed intake capacity, increasing the energy content of the diet, and supplementing the diet with natural compounds such as polyunsaturated fatty acids (PUFA n-3) or antioxidants.

Feed intake capacity has been improved when rabbits are reared with high fiber diets that have low digestible energy but enlarge the digestive tract (de Blas et al. 1986). It produces an improvement in the energy balance at the end of the first lactation and higher milk production (Pascual et al. 1999). However, improving feed intake capacity cannot efficiently meet the energy deficit and reproductive performance (Arias-Álvarez et al. 2009b, 2010a). Nevertheless, fertility rates at the first parturition can be improved when short flushing around the first AI is accompanied by fibrous diets (Martínez-Paredes et al. 2012).

The energy content of the diet can be increased using high-fat diets, high-starch diets, or glycogenic precursors such as propylene glycol (PG). A high-starch diet increased the live weight and the energy intake of primiparous rabbit does during lactation, but it did not improve the performance of their litters and could induce an excessive live weight of pregnant rabbit does (Pascual et al. 1999). As in other species, high-fat diet supply during rearing influenced the onset of puberty, follicular growth, and embryo gene expression in the very early stages of pregnancy, as well as hormonal responses to breeding and GnRH stimulation (Picone et al. 2011; Cordier et al. 2013). Glycogenic supplements such as PG in the drinking water of rabbit does for a short time (flushing 4 days before AI) were not enough to improve metabolic reserves, follicular development and oocyte quality or reproductive performance in primiparous lactating does (Sakr et al. 2012). In contrast, long-term flushing during lactation improved the body weight of the kits at weaning but not when PG was also provided during gestation, when kit mortality in the last days of the fattening period was increased (García-García et al. 2010). When PG is supplied from mid-pregnancy and during lactation, rabbit metabolic and endocrine status (serum glucose, insulin, and non-esterified fatty acids (NEFA) concentrations) and ovarian response (ovulation rate, follicular atresia, oocyte maturation, and number of blastocysts developed) were not enhanced. However, embryos from mothers supplemented with PG over the long term seem to adapt their "own metabolic pathways," providing a potential chance for embryo survival and successful implantation. Genes related to glucose metabolism, proliferation and differentiation, implantation capacity and fetal growth were upregulated, whereas lower oxidative stress and similar apoptosis-related transcripts were found in embryos from long-term supplemented dams (Arias-Álvarez et al. 2013a, b).

Other strategies to improve reproductive physiology in rabbit females are supplementation of the diet with natural compounds such as polyunsaturated fatty acids (PUFA n-3) or antioxidants. PUFA n-3 is mainly implicated in cholesterol metabolism, influencing prostaglandin F2α secretion and pregnancy losses. Long-term fed primiparous does with diets supplemented with docosahexaenoic acid (DHA; 40.0 g/kg total fatty acid methyl esters) and eicosapentaenoic acid (EPA; 33.9 g/kg total fatty acid methyl esters) showed a higher fertility rate than primiparous does fed a control diet (Rebollar et al. 2014). When PUFA n-3 is supplemented in the diet of primiparous rabbits, the lipid profile and the indicators of oxidative stress are altered, but endocrine and productive traits such as fertility rate, prolificacy, and newborn parameters are not improved (Rodriguez et al. 2017). However, long-term supplementation of PUFA n-3 in the diet from rearing and throughout two productive cycles enhanced plasma progesterone levels during pregnancy and fertility rate. In addition, the milk fatty acid profile improves and then newborn development (Rodríguez et al. 2018). However, DHA and EPA acids can produce high oxidative stress that can impair fertility potential (Guérin et al. 2001) and oocyte quality and early embryo development (Kitagawa et al. 2004), impairing reproductive success. New approaches using nanotechnology to protect PUFAs from oxidation or the addition of antioxidants in the diet have been challenged (Wang et al. 2021).

Supplementation with natural antioxidants can be an alternative to replace preventive antibiotics in livestock farming, according to "sustainable intensification," where safe and efficient methods to improve food production are accomplished. Natural antioxidant compounds are found in fruits, vegetables, various medicinal plants, and other dietary sources. They can be broadly divided into three groups: vitamins, carotenoids, and phenolic compounds. Natural antioxidants may enhance female fertility by upregulating the activities of endogenous antioxidants or gonadotropins (LH, FSH) and by downregulating or inhibiting oxidative stress or oxidative-induced damage in the female reproductive tract (Bhardwaj et al. 2021). Its effects are dependent on the compound and the dosage. Numerous studies have examined the effect of polyphenols on rabbit reproduction. For example, a polyphenol such as the flavonoid quercetin (30 mg/kg) improves follicular development and maintains oocyte competence in heat-stressed rabbits (Naseer et al. 2017). Other polyphenols, such as curcuminoids (5 g or 20 g/100 kg feed mixture), in the rabbit does diet increased folliculogenesis and the number of live-born and weaned pups and decreased the number of stillborn kits but did not affect weight gain, conception, or kindling rate. Furthermore, doe mortality decreased during the first reproductive cycle but not during the second cycle (Sirotkin et al. 2018).

In addition, different plant extracts or mixtures of additives have been challenged in the rabbit diet to enhance reproductive performance. One of the most studied is *Yucca schidigera*, which is rich in polyphenols such as resveratrol. This natural plant (5 or 20 g/100 kg feed mixture) positively influences rabbit reproductive performance, such as corpus luteum viability, ovarian folliculogenesis, and fertility rate, as well as the profile of reproductive hormones, stimulating progesterone and prostaglandin release (Földešiová et al. 2017). However, some adverse results have also been documented. Green tea (rich in epigallocatechin and resveratrol) supply (5 or 20 g/100 kg feed mixture) reduced ovarian function and fertility by changes in ovarian cell apoptosis, steroid hormone release, and blockade of the ovulation of large ovarian follicles (Baláži et al. 2019). Other natural supplements contain a mix of active compounds. The extract plant of Goji berry is rich in polysaccharides, carotenoids, phenylpropanoids, phenolics, and flavonoids. When it is included at 1% in the diet, the ovarian response and sexual receptivity increase (Andoni et al. 2021). Another plant used is *Moringa oleifera*. Seed extracts at 200 mg/kg improve reproductive performance by counteracting heat stress (Mutwedu et al. 2022). Brown seaweeds (Laminaria spp.) plus phenolic acid, hydroxycinnamic acids, tannins, and flavonoids from plant extracts promoted a lower survival rate and weaning weight of kits in the first reproductive cycle but not in the second cycle when they were fed a high dosage in the diet at 0.6% (Vizzarri et al. 2020). Vitamin E (α-tocopherol) improves rabbit oocyte developmental competence by reducing oxidative stress, cell damage, and apoptotic cascade activation in oocytes and cumulus cells (Arias-Álvarez et al. 2018). Therefore, a potential effect on reproductive outcome has been shown in this species.

Another strategy to improve the reproductive performance of females that fail to become pregnant after AI is moderate food adjustment during pregnancy. This management can be implemented in several pregnancy windows. When food adjustment is applied in the first 15 days of pregnancy, the viability and weight of the kits at birth are not impaired (Nafeaa et al. 2011). In addition, ad libitum refeeding in the last third of gestation, when fetal nutritional needs are higher, improves the energetic status of rabbit does before parturition (Manal et al. 2010). The compensatory food intake of the female doe at the end of gestation is able to attenuate changes in the rabbit's live body weight, body reserves, reproductive parameters, and offspring performance until weaning. However, some epigenetic changes in the liver and metabolic parameters of offspring were found (García-García et al. 2021a, b). Other regimes of food adjustment during different periods of gestation (0–7 days, 7–21 days, and 0–21 days) slightly affected the energy homeostasis and metabolism of the animals, and their general metabolic condition, productive performance, and welfare were not severely affected (Fernández-Pacheco et al. 2021b). However, when feed adjustment (50%) is settled in the first week of pregnancy where the preimplantation period is produced, fetal development is affected because placental functionality is impaired. This can be enhanced when the daily amount of feed provided increases to 60% of the doe's total voluntary intake (López-Tello et al. 2017a, b). Moderate maternal feed restriction (40% of their nutritional requirements) applied during pregnancy also has long-term consequences on daughters' ovarian physiology (F1) at the beginning of their reproductive life (Arias-Álvarez et al. 2019).

4 Conclusions

Rabbit females' specific reproductive characteristics are unique compared with those of other species regarding folliculogenesis, sexual cycle, and ovulation. Understanding these reproductive events is necessary to apply valuable strategies and precise management in commercial production systems to achieve maximum potential reproductive output of females, according to their physiological needs and animal welfare.

References

Adams CE (1958) Egg development in the rabbit: the influence of post-coital ligation of the uterine tube and of ovariectomy. J Endocrinol 16:283–293
Adams CE (1962) Studies on prenatal mortality in the rabbit, Oryctolagus cuniculus: the effect of transferring varying numbers of eggs. J Endocrinol 24:471–490
Andoni E, Curone G, Agradi S et al (2021) Effect of goji berry (Lycium barbarum) supplementation on reproductive performance of rabbit does. Animals 11. https://doi.org/10.3390/ani11061672
Arias Álvarez M (2010) Influencia de diferentes estrategias reproductivas y nutricionales sobre la fisiología ovárica en conejas primíparas. Complutense University of Madrid
Arias-Álvarez M, García-García RM, Rebollar PG, Lorenzo PL (2007) Desarrollo folicular en la coneja. ITEA 103:173–185
Arias-Álvarez M, García-García RM, Rebollar PG et al (2009a) Influence of metabolic status on oocyte quality and follicular characteristics at different postpartum periods in primiparous rabbit does. Theriogenology 72:612–623. https://doi.org/10.1016/j.theriogenology.2009.04.017
Arias-Álvarez M, García-García RM, Rebollar PG et al (2009b) Effects of a lignin-rich fibre diet on productive, reproductive and endocrine parameters in nulliparous rabbit does. Livest Sci 123: 107–115. https://doi.org/10.1016/j.livsci.2008.10.013
Arias-Álvarez M, García-García R, Rebollar P et al (2010a) Follicular, oocyte and embryo features related to metabolic status in primiparous lactating does fed with high-fibre rearing diets. Reprod Domest Anim 45:91–100. https://doi.org/10.1111/j.1439-0531.2009.01528.x
Arias-Álvarez M, García-García RM, Torres-Rovira L et al (2010b) Influence of hormonal and nonhormonal estrus synchronization methods on follicular and oocyte quality in primiparous lactating does at early postpartum period. Theriogenology 73:26–35. https://doi.org/10.1016/j.theriogenology.2009.07.017
Arias-Álvarez M, García-García RM, Lorenzo PL et al (2013a) Embryo gene expression in response to maternal supplementation with glycogenic precursors in the rabbit. Anim Reprod Sci 142:173–182. https://doi.org/10.1016/j.anireprosci.2013.10.001
Arias-Álvarez M, García-García RM, Rebollar PG et al (2013b) Ovarian response and embryo gene expression patterns after nonsuperovulatory gonadotropin stimulation in primiparous rabbits does. Theriogenology 79:323–330. https://doi.org/10.1016/j.theriogenology.2012.09.019
Arias-Álvarez M, García-García RM, López-Tello J et al (2018) a-Tocopherol modifies the expression of genes related to oxidative stress and apoptosis during in vitro maturation and enhances the developmental competence of rabbit oocytes. Reprod Fertil Dev 30:1728–1738
Arias-Álvarez M, García-García R, Millán P et al (2019) Consequences of maternal food restriction in rabbits during gestation on oocyte and embryo quality of female offspring at first insemination. In: Proceeding of the 15th International Congress of the Spanish Society for Animal Reproduction (AERA). NLM (Medline), p 123
Arias-Álvarez M, Pericuesta E, Gutiérrez-Adán A et al (2022) Gene expression patterns of NGF-TrKA-p75 system in cumulus-oocyte complexes and early embryos in rabbit. In: 16th International Congress of the Spanish Society for Animal Reproduction (AERA). León, España
Austin C (1974) Fertilization. In: Lash J, Whittaker J (eds) Concepts of development. Sinauer, Stanford, pp 48–75
Bakker J, Baum MJ (2000) Neuroendocrine regulation of GnRH release in induced ovulators. Front Neuroendocrinol 21:220–262. https://doi.org/10.1006/frne.2000.0198
Baláži A, Sirotkin AV, Földešiová M et al (2019) Green tea can suppress rabbit ovarian functions in vitro and in vivo. Theriogenology 127:72–79
Bhardwaj JK, Panchal H, Saraf P (2021) Ameliorating effects of natural antioxidant compounds on female infertility: a review. Reprod Sci 28:1227–1256. https://doi.org/10.1007/s43032-020-00312-5
Boiti C (2004) Underlying physiological mechanisms controlling the reproductive axis of rabbit does. In: Proceedings-8th World Rabbit Congress

Boiti C, Castellini C, Canali C et al (1995) Long term effect of PMSG on rabbit does reproductive performance. World Rabbit Sci 3:51–56

Bolet G, Theau-Clément M (1994) Fertilisation rate and preimplantation embryonic development in two rabbit strains of different fecundity, in purebreeding and crossbreeding. Anim Reprod Sci 36:153–162

Bolet G, Brun JM, Lechevestrier S et al (2004) Evaluation of the reproductive performance of eight rabbit breeds on experimental farms. Anim Res 53:59–65. https://doi.org/10.1051/animres:2003043

Bonanno A, di Grigoli A, Alabiso M, Boiti C (2002) Parity and number of repeated doe-litter-separation treatments affect differently the reproductive performances of lactating does. World Rabbit Sci 10:63–70

Browning JY, Landis Keyes P, Wolf RC (1980) Comparison of serum progesterone, 20a-dihydroprogesterone, and estradiol-17b in pregnant and pseudopregnant rabbits: evidence for postimplantation recognition of pregnancy. Biol Reprod 23:1014–1019

Canali E, Ferrante V, Todeschini R et al (1991) Rabbit nest construction and its relationship with litter development. Appl Anim Behav Sci 31:259–266. https://doi.org/10.1016/0168-1591(91)90010-U

Cardinali R, Dal Bosco A, Bonanno A et al (2008) Connection between body condition score, chemical characteristics of body and reproductive traits of rabbit does. Livest Sci 116:209–215. https://doi.org/10.1016/j.livsci.2007.10.004

Carluccio A, Robbe D, de Amicis I et al (2004) Artificial insemination in rabbits: laboratory and field trial with three different semen extenders. World Rabbit Sci 12:65–79

Carrasco R, Singh J, Adams GP (2016) The dynamics of trkA expression in the bovine ovary are associated with a luteotrophic effect of ovulation-inducing factor/nerve growth factor (OIF/NGF). Reprod Biol Endocrinol 14. https://doi.org/10.1186/s12958-016-0182-9

Casares-Crespo L, Vicente JS, Talaván AM, Viudes-de-Castro MP (2016) Does the inclusion of protease inhibitors in the insemination extender affect rabbit reproductive performance? Theriogenology 85:928–932. https://doi.org/10.1016/j.theriogenology.2015.10.044

Casares-Crespo L, Fernández-Serrano P, Viudes-de-Castro MP (2018) Protection of GnRH analogue by chitosan-dextran sulfate nanoparticles for intravaginal application in rabbit artificial insemination. Theriogenology 116:49–52. https://doi.org/10.1016/j.theriogenology.2018.05.008

Castellini C (1996) Recent advances in rabbit artificial insemination. In: 6th World Rabbit Congress, Toulouse

Castellini C (2007) Reproductive activity and welfare of rabbit does. Ital J Anim Sci 6:743–747

Castellini C, Dal Bosco A, Mugnai C (2003) Comparison of different reproduction protocols for rabbit does: Effect of litter size and mating interval. Livest Prod Sci 83:131–139. https://doi.org/10.1016/S0301-6226(03)00070-8

Castellini C, Dal Bosco A, Arias-Álvarez M et al (2010) The main factors affecting the reproductive performance of rabbit does: A review. Anim Reprod Sci 122:174–182. https://doi.org/10.1016/j.anireprosci.2010.10.003

Cervantes MP, Palomino JM, Adams GP (2015) In vivo imaging in the rabbit as a model for the study of ovulation-inducing factors. Lab Anim 49:1–9. https://doi.org/10.1177/0023677214547406

Cervera C, Fernandez-Carmona J, Viudes P, Blas E (1993) Effect of remating interval and diet on the performance of female rabbits and their litters. Anim Prod 56:399–405. https://doi.org/10.1017/S0003356100006450

Christians E, Rao VH, Renard JP (1994) Sequential acquisition of transcriptional control during early embryonic development in the rabbit. Dev Biol 164:160–172

Cordier AG, Leveille P, Dupont C et al (2013) Dietary lipid and cholesterol induce ovarian dysfunction and abnormal LH response to stimulation in rabbits. PLoS One 8:1–11. https://doi.org/10.1371/journal.pone.0063101ï

Dal Bosco A, Cardinali R, Brecchia G et al (2014) Induction of ovulation in rabbits by adding Lecirelin to the seminal dose: In vitro and in vivo effects of different excipients. Anim Reprod Sci 150:44–49. https://doi.org/10.1016/j.anireprosci.2014.08.009

de Blas JC, Santomá G, Carabaño R, Fraga M (1986) Fiber and starch levels in fattening rabbit diets. J Anim Sci 63:1897–1904

Denker HW (1977) Implantation. The role of proteinases, and blockage of implantation by proteinase inhibitors. In: Advances in anatomy, embryology, and cell biology. Springer, Berlin

Denker HW, Gerdes HJ (1979) The dynamic structure of rabbit blastocyst coverings. I. Transformation during regular preimplantation development. Anat Embryol (Berl) 157:15–34. https://doi.org/10.1007/BF00315639

Díaz P (1987) Actividad reproductiva de la coneja doméstica en torno a la pubertad. Polytechnic University of Madrid

Díaz et al (1987) https://tesis.biblioteca.upm.es/tesis/1377

Donnelly TM, Kelsey SF, Levine DM, Parker TS (1991) Control of variance in experimental studies of hyperlipidemia using the WHHL rabbit. J Lipid Res 32:1089–1098. https://doi.org/10.1016/s0022-2275(20)41972-8

Dufy-Barbe L, Dufy B, Vincent JD (1978) Serum gonadotropin levels in the ovariectomized rabbit: effect of acute and chronic administration of estradiol. Biol Reprod 18:118–119

Eiben C, Bonanno A, Gódor-Surmann K, Kustos K (2008) Effect of controlled nursing with one-day fasting on rabbit doe performance. Livest Sci 118:82–91. https://doi.org/10.1016/j.livsci.2008.01.007

Eppig JJ (2001) Oocyte control of ovarian follicular development and function in mammals. Reproduction 122:829–838

Fernández-Pacheco C, Millán P, Rodríguez M et al (2021a) Pituitary and ovarian hormones: Is their plasma concentration affected by litter size in primiparous lactating rabbit does? World Rabbit Sci 29:161–168. https://doi.org/10.4995/WRS.2021.15122

Fernández-Pacheco C, Millán P, Rodríguez M et al (2021b) Influence of different regimes of moderate maternal feed restriction during pregnancy of primiparous rabbit does on long-term metabolic energy homeostasis, productive performance and welfare. Animals 11. https://doi.org/10.3390/ani11092736

Feugier A, Fortun-Lamothe L (2006) Extensive reproductive rhythm and early weaning improve body condition and fertility of rabbit does. Anim Res 55:459–470. https://doi.org/10.1051/animres:2006025

Fischer B, Mootz U, Denker HW et al (1991) The dynamic structure of rabbit blastocyst coverings. III. Transformation of coverings under non-physiological developmental conditions. Anat Embryol 183:17–27

Fischer B, Chavatte-Palmer P, Viebahn C et al (2012) Rabbit as a reproductive model for human health. Reproduction 144:1–10

Fleming MW, Rhodes RC III, Dailey RA (1984) Compensatory responses after unilateral ovariectomy in rabbits. Biol Reprod 30:82–86

Földešiová M, Baláži A, Chrastinová Ľ et al (2017) Yucca schidigera can promote rabbit growth, fecundity, affect the release of hormones in vivo and in vitro, induce pathological changes in liver, and reduce ovarian resistance to benzene. Anim Reprod Sci 183:66–76. https://doi.org/10.1016/j.anireprosci.2017.06.001

Foote RH, Carney EW (2000) The rabbit as a model for reproductive and developmental toxicity studies. Reprod Toxicol 14:477–493

Fortun-Lamothe L (1998) Effects of pre-mating energy intake on reproductive performance of rabbit does. Anim Sci 66:263–269

Fortun-Lamothe L (2006) Energy balance and reproductive performance in rabbit does. Anim Reprod Sci 93:1–15. https://doi.org/10.1016/j.anireprosci.2005.06.009

Freeman ME, Kanyicska LA, Lerant A et al (2000) Prolactin: structure, function, and regulation of secretion. Physiol Rev 80:1523–1631

Fuchs A-R, Cubile L, Dawood MY, Jørgensen FS (1984) Release of oxytocin and prolactin by suckling in rabbits throughout lactation. Endocrinology 114(2):462–469. https://doi.org/10.1210/endo-114-2-462

Gadsby JE, Landis Keyes P, Ii CHB (1983) Control of corpus luteum function in the pregnant rabbit: role of estrogen and lack of a direct luteotropic role of the placenta. Endocrinology 113: 2255–2262

Garcia-Garcia RM, Arias-Alvarez M, Rebollar PG et al (2009) Influence of different reproductive rhythms on serum estradiol and testosterone levels, features of follicular population and atresia rate, and oocyte maturation in controlled suckling rabbits. Anim Reprod Sci 114:423–433. https://doi.org/10.1016/j.anireprosci.2008.10.007

García-García RM, Sark O, Velasco B et al (2010) La duración del suministro de propilenglicol a conejas primíparas afecta al peso y viabilidad de sus camadas. Boletín de Cunicultura 34(163): 39–46

Garcia-Garcia RM, Masdeu MDM, Sanchez Rodriguez A et al (2018) β-nerve growth factor identification in male rabbit genital tract and seminal plasma and its role in ovulation induction in rabbit does. Ital J Anim Sci 17:442–453. https://doi.org/10.1080/1828051X.2017.1382315

Garcia-Garcia RM, Arias-Alvarez M, Sanchez-Rodriguez A et al (2020) Role of nerve growth factor in the reproductive physiology of female rabbits: A review. Theriogenology 150:321–328. https://doi.org/10.1016/j.theriogenology.2020.01.070

García-García RM, Arias-Álvarez M, Millán P et al (2021a) Gestation food restriction and refeeding compensate maternal energy status and alleviate metabolic consequences in juvenile offspring in a rabbit model. Nutrients 13:1–18

García-García RM, Arias-Álvarez M, Rodríguez M et al (2021b) Effects of feed restriction during pregnancy on maternal reproductive outcome, foetal hepatic IGF gene expression and offspring performance in the rabbit. Animal 15. https://doi.org/10.1016/j.animal.2021.100382

Gidenne T, Lebas F, Fortun-Lamothe I (2010) Feeding behavior of rabbits. In: de Blas C, Wiseman J (eds) The nutrition of the rabbit, 2nd edn. CABI, Wallingford, pp 233–252

González-Mariscal G, McNitt JI, Lukefahr SD (2007) Maternal care of rabbits in the lab and on the farm: Endocrine regulation of behavior and productivity. Horm Behav 52:86–91. https://doi.org/10.1016/j.yhbeh.2007.03.028

Guérin P, Mouatassim SE, Ménézo Y (2001) Oxidative stress and protection against reactive oxygen species in the pre-implantation embryo and its surroundings. Hum Reprod Update 7: 175–189

Harcourt-Brown F (2002) Textbook of rabbit medicine, 1st edn. Butterworth-Heinemann, Oxford

Hu S, Liang X, Ren X et al (2019) Integrated analysis of mRNA and miRNA expression profiles in the ovary of Oryctolagus cuniculus in response to gonadotrophic stimulation. Front Endocrinol (Lausanne) 10:1–15. https://doi.org/10.3389/fendo.2019.00744

Hudson R, Distel H (1989) Temporal pattern of suckling in rabbit pups: a model of circadian synchrony between mother and young. In: Reppert S (ed) Research in perinatal medicine. Vol. IX Development of circadian rhythmicity and photoperiodism in mammals. Perinatology Press, Ithaca, pp 83–102

Hudson R, Shaal B, Martínez-Gómez M, Distel H (2000) Mother-young relations in the European rabbit: physiological and behavioral locks and keys. World Rabbit Sci 8:85–90

Hulot F, Mariana JC (1988) HCG-induced ovulation in two rabbit breeds: effects of dose, season and sexual behaviour. Livest Prod Sci 20:257–267

Hutt KJ, McLaughlin EA, Holland MK (2006) Primordial follicle activation and follicular development in the juvenile rabbit ovary. Cell Tissue Res 326:809–822. https://doi.org/10.1007/s00441-006-0223-3

Jelínkova L, Kubelka M, Motlfk J, Guerrier P (1994) Chromatin condensation and histone H1 kinase activity during growth and maturation of rabbit oocytes. Mol Reprod Dev 37:210–215

Jilge B, Kuhnt B, Landerer W, Rest S (2000) Circadian thermoregulation in suckling rabbit pups. J Biol Rhythms 15:329–335

Kaynard AH, Pau KYF, Hess DL, Spies HG (1990) Gonadotrophin-Releasing Hormone and Norepinephrine release from the rabbit mediobasal and anterior hypothalamus during the mating-induced luteinizing hormone surge. Endocrinology 127:1176–1185

Kitagawa Y, Suzuki K, Yoneda A, Watanabe T (2004) Effects of oxygen concentration and antioxidants on the in vitro developmental ability, production of reactive oxygen species (ROS), and DNA fragmentation in porcine embryos. Theriogenology 62:1186–1197. https://doi.org/10.1016/j.theriogenology.2004.01.011

Kranzfelder D, Korr H, Mestwerdt W, Maurer-Schultze B (1984) Follicle growth in the ovary of the rabbit after ovulation-inducing application of human chorionic gonadotropin. Cell Tissue Res 238:611–620

Lamas-Toranzo I, Fonseca Balvís N, Querejeta-Fernández A et al (2019) ZP4 confers structural properties to the zona pellucida essential for embryo development. Elife 8. https://doi.org/10.7554/eLife.48904.001

Lebas F, Theau-Clement M, Remy B et al (1996) Production of anti-PMSG antibodies and its relation to the productivity of rabbit does. World Rabbit Sci 4:57–62

Lefèvre B, Caillol M (1978) Relationship of estrous behaviour with follicular growth and sex steroid concentration in the follicular fluid in the domestic rabbit. Ann Biol Anim Biochim Biophys 18:1435–1441

López-Tello J, Arias-Álvarez M, Jiménez-Martínez MÁ et al (2017a) The effects of sildenafil citrate on feto-placental development and haemodynamics in a rabbit model of intrauterine growth restriction. Reprod Fertil Dev 29:1239–1248. https://doi.org/10.1071/RD15330

López-Tello J, Arias-Álvarez M, Jiménez-Martinez MA et al (2017b) Competition for materno-fetal resource partitioning in a rabbit model of undernourished pregnancy. PLoS One 12. https://doi.org/10.1371/journal.pone.0169194

Lorenzo PL, Rebollar PG, Illera MJ et al (1996a) Characterization of rabbit follicular oocytes and their ability to mature in vitro. Arch Zootec 45:25–35

Lorenzo PL, Rebollar PG, Illera MJ et al (1996b) Stimulatory effect of insulin-like growth factor I and epidermal growth factor on the maturation of rabbit oocytes in vitro. J Reprod Fertil 107: 109–117

Lorenzo PL, Illera JC, Silván G et al (1997) Steroid-level response to insulin-like growth factor-1 in oocytes matured in vitro. J Reprod Immunol 35(1):11–29

Ludwiczak A, Składanowska-Baryza J, Kuczyńska B et al (2021) Reproductive performance of hycole rabbit does, growth of kits and milk chemical composition during nine consecutive lactations under extensive rhythm. Animals 11. https://doi.org/10.3390/ani11092608

Lytton FDC, Poyser NL (1982) Prostaglandin production by the rabbit uterus and placenta in vitro. J Reprod Fertil 66:591–599

Maertens L, Luzi F (1995) Effect of diluent and storage time of rabbit semen on the fertility of does reared under two different lighting schedules. World Rabbit Sci 3:27–34

Maertens L, Luzi F, Grilli G (1995) Effects of PMSG induced oestrus on the performances of rabbit does: a review. World Rabbit Sci 3:191–199

Manal AF, Tony MA, Ezzo OH (2010) Feed restriction of pregnant nulliparous rabbit does: Consequences on reproductive performance and maternal behaviour. Anim Reprod Sci 120: 179–186. https://doi.org/10.1016/j.anireprosci.2010.03.010

Maranesi M, Petrucci L, Leonardi L et al (2018) New insights on a NGF-mediated pathway to induce ovulation in rabbits (Oryctolagus cuniculus). Biol Reprod 98:634–643. https://doi.org/10.1093/biolre/ioy041

Mariana JC, Hulot F, Drevin C et al (1989) Estimation de la durée moyenne de croissance d'ovarie de lapine agée de 20 semaines, dans deux souches. Archiv Biolo 100:47–63

Martínez-Paredes E, Ródenas L, Martínez-Vallespín B, Cervera C, Blas E, Brecchia G, Boiti C, Pascual JJ (2012) Effects of feeding programme on the performance and energy balance of nulliparous rabbit does. Animal 6:1086–1095

Masdeu M, García-García RM, Millán P et al (2012) Effect of rabbit seminal plasma in ovulating response. Reprod Fertil Dev 25:243. https://doi.org/10.1071/rdv25n1ab188

Masdeu M, Garcia-Garcia RM, Cardinali R et al (2015) Induction of ovulation in rabbit does using purified nerve growth factor and camel seminal plasma. Reprod Fertil Dev 27:224. https://doi.org/10.1071/RDv27n1Ab271

Mattioli S, Maranesi M, Castellini C et al (2021) Physiology and modulation factors of ovulation in rabbit reproduction management. World Rabbit Sci 29:221–229. https://doi.org/10.4995/wrs.2021.13184

Mehaisen GMK, Vicente JS, Lavara R, Viudes-De-Castro MP (2005) Effect of eCG dose and ovulation induction treatments on embryo recovery and in vitro development post-vitrification in two selected lines of rabbit does. Anim Reprod Sci 90:175–184. https://doi.org/10.1016/j.anireprosci.2005.01.015

Melo AI, González-Mariscal G (2010) Communication by olfactory signals in rabbits. Its role in reproduction. Vitam Horm 83:351–371

Meng Q, Polgar Z, Tancos Z et al (2013) Cloning of rabbits. In: Cibelli J, Wilmut I, Jaenisch R, Gurdon J, Lanza R, West M, Campbell KHS (eds) Principles of cloning, 2nd edn. Academic, New York, pp 227–244. https://doi.org/10.1016/B978-0-12-386541-0.00018-7

Milligan S (1982) Induced ovulation in mammals. In: Finn CA (ed) Oxford reviews of reproductive biology. Clarendon, Oxford, pp 1–46

Mills TM, Gerardot RJ (1984) Dissociation of copulation from ovulation in pregnant rabbits. Biol Reprod 30(5):1243–1252

Molina I, Pla M, García F (1986) Poblaciones de folículos antrales en función del comportamiento de monta en conejas: utilización de un método simpre para la medición de folículos. ITEA 66: 21–26

Molina I, Pla M, Vicente JS et al (1991) Induction of ovulation in rabbits with pure urinary luteinizing hormone and human chorionic gonadotrophin: comparison of oocyte and embryo quality. Hum Reprod 6:1449–1452. https://doi.org/10.1093/oxfordjournals.humrep.a137287

Morrel JM (1995) Artificial insemination in rabbits. Br Vet J 151(5):477–488

Munari C, Ponzio P, Alkhawagah AR et al (2019) Effects of an intravaginal GnRH analogue administration on rabbit reproductive parameters and welfare. Theriogenology 125:122–128. https://doi.org/10.1016/j.theriogenology.2018.10.024

Murakami H, Imai H (1996) Successful implantation of in vitro cultured rabbit embryos after uterine transfer: A role for mucin. Mol Reprod Dev 43:167–170. https://doi.org/10.1002/(SICI)1098-2795(199602)43:2<167::AID-MRD5>3.0.CO;2-P

Mutwedu VB, Nyongesa AW, Kitaa JM et al (2022) Effects of Moringa oleifera aqueous seed extracts on reproductive traits of heat-stressed New Zealand white female rabbits. Front Vet Sci 9:883976

Nafeaa A, Ahmed SAE, Fat Hallah S (2011) Effect of feed restriction during pregnancy on performance and productivity of New Zealand white rabbit does. Vet Med Int. https://doi.org/10.4061/2011/839737

Naseer Z, Ahmad E, Epikmen ET et al (2017) Quercetin supplemented diet improves follicular development, oocyte quality, and reduces ovarian apoptosis in rabbits during summer heat stress. Theriogenology 96:136–141. https://doi.org/10.1016/j.theriogenology.2017.03.029

O'Malley B (2005) Clinical anatomy and physiology of exotic species. Elsevier Saunders, Edinburgh

Orsini MW (1962) Study of ovo-implantation in the hamster, rat, mouse, guinea-pig and rabbit in cleared uterine tracts. J Reprod Fertil 3:288–293

Osteen KG, Mills TM (1980) Size and steroid content of rabbit follicles. Biol Reprod 22:1040–1046

Pascual JJ, Tolosa C, Cervera C et al (1999) Effect of diets with different digestible energy content on the performance of rabbit does. Anim Feed Sci Technol 81:105–117

Pau KY, Spies HG (1986) Estrogen-dependent effects of norepinephrine on hypothalamic Gonadotropin-Releasing Hormone release in rabbit. Brain Res 399:15–23

Pau K-YF, Pau CY, Berria M, Spies HG (2000) Ovarian influence on gonadotropin and prolactin release in mated rabbits. Endocrine 13(1):25–35

Pei Y (2010) Effect of nerve growth factor (NGF) on the development of preimplantation rabbit embryos in vitro. Vet Res Commun 34:11–18. https://doi.org/10.1007/s11259-009-9325-1

Picone O, Laigre P, Fortun-Lamothe L et al (2011) Hyperlipidic hypercholesterolemic diet in prepubertal rabbits affects gene expression in the embryo, restricts fetal growth and increases offspring susceptibility to obesity. Theriogenology 75:287–299. https://doi.org/10.1016/j.theriogenology.2010.08.015

Pincus G, Enzmann E (1935) The growth, maturation and atresia of ovarian eggs in the rabbit. J Physiol 61:351–382

Poigner J, Szendro ZS, Lévai A et al (2000) Effect of birth weight and litter size on growth and mortality in rabbits. World Rabbit Sci 8:17–22

Quesenberry KE, Orcutt CJ, Mans C, Carpenter JW (2021) Ferrets, rabbits and rodents: clinical medicine and surgery, 4th edn. WB Saunders, Philadelphia

Quintela L, Peña A, Vega M et al (2004) Reproduction nutrition development. Reprod Nutr Dev 44: 79–88. https://doi.org/10.1051/rnd:2004015ï

Quintela LA, Peña AI, Vega MD et al (2009) Reproductive performance of rabbit does artificially inseminated via intravaginal administration of [des-Gly 10, d-Ala6]-LHRH ethylamide as ovulation inductor. Reprod Domest Anim 44:829–833. https://doi.org/10.1111/j.1439-0531.2008.01095.x

Quiroga AC, Box-Catalán S, Rebollar PG et al (2022a) Effect of recombinant-rabbit nerve growth factor (NGF) on in vitro maturation of rabbit oocytes. In: 19th International Congress on Animal Reproduction (ICAR), Bologna, p 123

Quiroga AC, García-García RM, Arias-Álvarez M et al (2022b) Ovulation induction in rabbit does using microencapsulated rrbetaNGF. World Rabbit Sci 30:244

Ratto MH, Leduc YA, Valderrama XP et al (2012) The nerve of ovulation-inducing factor in semen. Proc Natl Acad Sci U S A 109:15042–15047. https://doi.org/10.1073/pnas.1206273109

Rebollar PG, Ubilla E, Alvariño JM et al (1992) Effect of the level of sexual receptivity on plasma estradiol and the ovulatory response during the postpartum period in the rabbit. Rev Esp Fisiol 48:13–17

Rebollar PG, Milanés A, Pereda N et al (2006a) Oestrus synchronisation of rabbit does at early post-partum by doe-litter separation or ECG injection: Reproductive parameters and endocrine profiles. Anim Reprod Sci 93:218–230. https://doi.org/10.1016/j.anireprosci.2005.06.032

Rebollar PG, Pereda N, Villarroel M et al (2006b) Oestrus synchronization of rabbit does at early post-partum by dam-litter separation or eCG injection: Effect on kit mortality and growth. Livest Sci 103:13–22. https://doi.org/10.1016/j.livsci.2005.12.008

Rebollar PG, Bonanno A, di Grigoli A et al (2008) Endocrine and ovarian response after a 2-day controlled suckling and eCG treatment in lactating rabbit does. Anim Reprod Sci 104:316–328. https://doi.org/10.1016/j.anireprosci.2007.02.018

Rebollar PG, Pérez-Cabal MA, Pereda N et al (2009) Effects of parity order and reproductive management on the efficiency of rabbit productive systems. Livest Sci 121:227–233. https://doi.org/10.1016/j.livsci.2008.06.018

Rebollar PG, Pereda N, Schwarz BF et al (2011) Effect of feed restriction or feeding high-fibre diet during the rearing period on body composition, serum parameters and productive performance of rabbit does. Anim Feed Sci Technol 163:67–76. https://doi.org/10.1016/j.anifeedsci.2010.10.005

Rebollar PG, Dal Bosco A, Millán P et al (2012) Ovulating induction methods in rabbit does: The pituitary and ovarian responses. Theriogenology 77:292–298. https://doi.org/10.1016/j.theriogenology.2011.07.041

Rebollar PG, García-García RM, Arias-Álvarez M et al (2014) Reproductive long-term effects, endocrine response and fatty acid profile of rabbit does fed diets supplemented with n-3 fatty acids. Anim Reprod Sci 146:202–209. https://doi.org/10.1016/j.anireprosci.2014.02.021

Rodriguez M, Garcia-Garcia RM, Arias-Alvarez M et al (2017) A diet supplemented with n-3 polyunsaturated fatty acids influences the metabolic and endocrine response of rabbit does and their offspring. J Anim Sci 95:2690–2700

Rodríguez M, García-García RM, Arias-Álvarez M et al (2018) Improvements in the conception rate, milk composition and embryo quality of rabbit does after dietary enrichment with n-3 polyunsaturated fatty acids. Animal 12:2080–2088. https://doi.org/10.1017/S1751731117003706

Rosell JM, de la Fuente LF (2009) Culling and mortality in breeding rabbits. Prev Vet Med 88:120–127. https://doi.org/10.1016/j.prevetmed.2008.08.003

Sakr OG, García-García RM, Arias-Álvarez M et al (2010) Body reserves and ovarian performance in primiparous lactating rabbit does submitted to early weaning as a strategy to decrease energy deficit. Anim Reprod Sci 121:294–300. https://doi.org/10.1016/j.anireprosci.2010.06.008

Sakr OG, García-García RM, Arias-Álvarez M et al (2012) Métodos de sincronización de celo en conejas primiparas lactantes a 25 días post-parto. Revista Complutense de Ciencias Veterinarias 6:6–13

Sanchez-Rodriguez A, Abad P, Arias-Alvarez M et al (2019) Recombinant rabbit beta nerve growth factor production and its biological effects on sperm and ovulation in rabbits. PLoS One 14: e0219780. https://doi.org/10.1371/journal.pone.0219780

Silva M, Niño A, Guerra M et al (2011) Is an ovulation-inducing factor (OIF) present in the seminal plasma of rabbits? Anim Reprod Sci 127:213–221. https://doi.org/10.1016/j.anireprosci.2011.08.004

Silva M, Urra F, Ulloa-Leal C, Ratto MH (2017) A comparative study of the effects of intramuscular administration of gonadorelin, mating and intrauterine infusion of either raw seminal plasma or seminal plasma purified β-NGF on luteal development in llamas. Reprod Domest Anim 52: 625–631. https://doi.org/10.1111/rda.12958

Sirotkin AV, Kadasi A, Stochmalova A et al (2018) Effect of turmeric on the viability, ovarian folliculogenesis, fecundity, ovarian hormones and response to luteinizing hormone of rabbits. Animal 12:1242–1249. https://doi.org/10.1017/S175173111700235X

Smitz J, Nogueira D, Albano C et al (2001) Improving in vitro maturation of oocytes in the human taking lessons from experiences in animal species. Reprod Domest Anim 36:11–17

Souza CJH, Campbell BH, Baird DT (1996) Follicular dynamics and ovarian steroid secretion in sheep during anoestrus. J Reprod Fertil 108:101–106

Stradaioli G, Verini-Supplizi A, Monaci M et al (1997) Effects of different doses of PMSG on ovarian response and in vitro embryo development in rabbits. World Rabbit Sci 5:143–148

Sultana F, Hatori M, Shimozawa N et al (2009) Continuous observation of rabbit preimplantation embryos in vitro by using a culture device connected to a microscope. J Am Assoc Lab Anim Sci 48:52–56

Sutton ML, Gilchrist RB, Thompson JG (2003) Effect of in-vivo and in vitro environments on the metabolism of the cumulus-oocyte complex and its influence on oocyte developmental capacity. Hum Reprod Update 9:35–48

Szendro Z, Gerencsér Z, McNitt JI, Matics Z (2016) Effect of lighting on rabbits and its role in rabbit production: A review. Livest Sci 183:12–18

Theau-Clément M (2000) Advances in biostimulation methods applied to rabbit reproduction. World Rabbit Sci 8:61–79

Theau-Clément M (2007) Preparation of the rabbit doe to insemination: a review. World Rabbit Sci 15:61–80

Theau-Clément M, Poujardieu B, Bellereaud J (1990) Influence des traitements lumineux, modes de reproduction et états physiologiques sur la productivité de lapines multipares. 5èmes Jour. Rech. Cunicole, 12-13 Décembre, 1990, Paris, France, Tome I: Comm. 7.

Theau-Clément M, Roustan A (1992) A study on relationships between receptivity and lactation in the doe, and their influence on reproductive performances. J Appl Rabbit Res 15:412–421

Theau-Clément M, Lebas F, Falières J (2008) Influence of different eCG doses on the rabbit doe ovary response, fertilizing aptitude and embryo development. World Rabbit Sci. https://doi.org/10.4995/wrs.2008.629

Theau-Clément M, Guardia S, Davoust C et al (2016) Performance and sustainability of two alternative rabbit breeding systems. World Rabbit Sci 24:253–265. https://doi.org/10.4995/wrs.2016.5154

Trocino A, Cotozzolo E, Zomeño C et al (2019) Rabbit production and science : the world and Italian scenarios from 1998 to 2018. Ital J Anim Sci 18:1361–1371. https://doi.org/10.1080/1828051X.2019.1662739

Turckheim M, Berger M, Jean-Faucher CH et al (1983) Changes in ovarian estrogens and in plasma gonadotropins in female rabbits from birth to adulthood. Acta Endocrinol 103:125–130

Ubilla E, Rebollar PG (1995) Influence of the postpartum day on plasma estradiol-17/3 levels, sexual behavior, and conception rate, in artificially inseminated lactating rabbits. Anim Reprod Sci 38:337–344

Ubilla E, Rebolla PG, Pazo D et al (2000) Pituitary and ovarian response to transient doe-litter separation in nursing rabbits. J Reprod Fertil 118:361–366

van den Hurk R, Zhao J (2005) Formation of mammalian oocytes and their growth, differentiation and maturation within ovarian follicles. Theriogenology 63:1717–1751. https://doi.org/10.1016/j.theriogenology.2004.08.005

Velasco B, Rebollar PG, Marco M, Costa R (2009) Cambios en la productividad anual de una explotación comercial por la extensificación del ritmo reproductivo. In: XXXIV Symposium de cunicultura de la Asociación Española de Cunicultura, pp 181–186

Vella D, Donnelly TM (2012) Basic anatomy, physiology and husbandry. In: Quesenberry K, Carpenter JW (eds) Ferrets, rabbits and rodents: clinical medicine and surgery, 3rd edn. Elsevier, Saunders, Philadelphia, pp 157–161

Vicente JS, Lavara R, Lavara F et al (2008) Rabbit reproductive performance after insemination with buserelin acetate extender. Livest Sci 115:153–157. https://doi.org/10.1016/j.livsci.2007.07.011

Viudes-de-Castro MP, Lavara R, Marco-Jiménez F et al (2007) Ovulation induced by mucosa vaginal absorption of buserelin and triptorelin in rabbit. Theriogenology 68:1031–1036. https://doi.org/10.1016/j.theriogenology.2007.08.004

Viudes-de-Castro MP, Mocé E, Lavara R et al (2014) Aminopeptidase activity in seminal plasma and effect of dilution rate on rabbit reproductive performance after insemination with an extender supplemented with buserelin acetate. Theriogenology 81:1223–1228

Viudes-de-Castro MP, Casares-Crespo L, Marco-Jiménez F, Vicente JS (2017) Efecto del estímulo físico de la cánula de inseminación sobre la inducción de la ovulación en coneja. In: AIDA, XVII Jornadas sobre Producción Animal, pp 380–382

Vizzarri F, Chiapparini S, Corino C et al (2020) Dietary supplementation with natural extracts mixture: effects on reproductive performances, blood biochemical and antioxidant parameters in rabbit does. Ann Anim Sci 20:565–578. https://doi.org/10.2478/aoas-2019-0084

Wang J, Han L, Wang D et al (2021) Stability and stabilization of omega-3 oils: A review. Trends Food Sci Technol 118:17–35. https://doi.org/10.1016/j.tifs.2021.09.018

Webb R, Nicholas B, Gong JG et al (2003) Mechanisms regulating follicular development and selection of the dominant follicle. Reprod Suppl 61:71–90

Xiccato G, Trocino A (2010) Energy and protein metabolism and requirements. In: de Blas C, Wiseman J (eds) The nutrition of the rabbit, 2nd edn. CABI, Wallingford, pp 83–118

Xiccato G, Trocino A, Sartori A, Queaque PI (2004) Effect of parity order and litter weaning age on the performance and body energy balance of rabbit does. Livest Prod Sci 85:239–251. https://doi.org/10.1016/S0301-6226(03)00125-8

Yoshimura Y, Iwashita M, Karube M et al (1994) Growth hormone stimulates follicular development by stimulating ovarian production of insulin-like growth factor-I. Endocrinology 135(3):887–894

Yoshimura Y, Nagamatsu S, Ando M et al (1996) Insulin-like growth factor binding protein-3 inhibits gonadotropin-induced ovulation, oocyte maturation, and steroidogenesis in rabbit ovary. Endocrinology 137(2):438–446

Zanagnolo V, Dharmarajan AM, Hesla J, Wallach EE (1996) Effects of a gonadotropin-releasing hormone analog on rabbit ovarian function. Endocrinology 137:5400–5406. https://doi.org/10.1210/endo.137.12.8940363

The Rabbit Male Reproduction

Cesare Castellini, Simona Mattioli, Alessandro Dal Bosco, and Giulia Collodel

Abstract

This chapter aims to furnish the state of the art on rabbit male reproduction focusing on factors affecting semen quality (all fractions: spermatozoa—SPZ, prostatic secretory granules—PSGs, and seminal plasma—SP), as genetic, diet, and management practices, and on innovative procedures for sperm analysis (motility assay in 2D *vs*. 3D) and cryopreservation (dilution and storage medium). Furthermore, insights on the use of rabbit as an animal model for human infertility, toxicology, or metabolic studies have been also done. Finally, a special focus on common limits (in vitro studies) and future prospective (mathematical models that mimic complex metabolic pathways of live systems) in rabbit reproduction studies has been also considered.

Keywords

Rabbit · Male · Spermatozoa · Seminal plasma · Prostatic secretory granules · Management · Diet · Genetic · CASA · Cryopreservation

C. Castellini · S. Mattioli (✉) · A. Dal Bosco
Department of Agricultural, Environmental and Food Science, University of Perugia, Perugia, Italy
e-mail: simona.mattioli@unipg.it

G. Collodel
Department of Molecular and Developmental Medicine, University of Siena, Policlinico Santa Maria alle Scotte, Siena, Italy

J. Simões, J. M. Monteiro (eds.), *Veterinary Care of Farm Rabbits*,
https://doi.org/10.1007/978-3-031-44542-2_12

1 Introduction

In the last 30 years, productivity of rabbit farms becomes more homogeneous thanks to the use of good management practices, such as artificial insemination (AI) (Rebollar et al. 1994; Castellini 1996). The diffusion of this technique permitted more controlled management and better planning. In addition, AI offers the same benefits for rabbit breeding as in other species in the control of genetic diversity, rapid upgrading of stock, the establishment of pregnancies in females, which refuse to mate, and avoidance of the spread of diseases. The AI, as well the optimization of reproductive performance, due to the management practices (i.e., reproductive rhythms, housing systems, etc.), dietary plans (i.e., ingredients, administration plan, etc.), and the genetic selection (i.e., commercial vs pure strains) are the main facts that assure high productivity on rabbitries. However, the success of the AI practice is mainly assured by an optimal semen quality that can be obtained using qualified rabbit strains and reliable semen manipulation.

The objective of this chapter is to furnish the state of the art on rabbit male reproduction focusing the attention on:

– Reproductive physiology
– Semen quality (deepening all semen fractions)
– Factors affecting semen quality
– Standard and innovative procedures for seminal analysis
– Innovation on dilution medium use and sperm cryopreservation
– Modelling for human infertility/toxicology/metabolic studies

The key points of this chapter are reported in Box 1.

Box 1: Key Points

- The rabbit male genital organs are constituted by three main portions: gland, tubular, and urogenital sections.
- The entire spermatogenesis in the rabbit lasts 55 days and is divided into eight stages, which occur in the testis and the final maturation, which takes place in the epididymis.
- Rabbit semen is a heterogeneous matrix constituted by spermatozoa (SPZ), prostatic secretory granules (PSGs), and many other molecules (protein, sugar, etc.) dispersed in seminal plasma (SP).
- The SP modulates sperm functions and deterioration through some complex mechanisms (proteins, e.g. β-Nerve Growth Factor-NGF, ATPase inhibitor, insulin-like growth factor-IFG, etc.).
- The SPZ membrane is mainly constituted by phospholipids (PLs) which moulted their functions (maturation, motility, capacitation, acrosome reaction).

(continued)

Box 1 (continued)

- The PSGs may be considered a sterols stock for the SPZ, indeed they are able to transfer sterols (cholesterol or desmosterol) from its membrane to head membrane of SPZ, delaying its capacitation.
- The semen quality depends on three macro-factors: genetic, diet, and management practices.
- *Genetic.* Cross-lines generally have better semen value of pure breeds with +37% of sperm concentration and +4% of sperm motility. Pure breeds (less selected; e.g., Silver of Champagne) have higher sperm count with respect to heavy ones (commercial lines; e.g., New Zealand White); however, it should be underlined that the repeatability of semen characteristics is very low.
- *Diets.* The dietary lipid and antioxidant contents strongly affected the sperm behaviours and its quality.
- *Management.* Semen collection procedures (collective rhythm, artificial vagina) and environmental traits (temperature, light, rearing system) affect the semen quality.
- Computer-assisted semen analysis (CASA) system allows the evaluation of several motility characteristics through the description of specific kinetic parameters.
- CASA system, for motility semen analysis, is studied in two dimensions (2D) by recording sperm swimming trajectories near a flat surface. However, 3D novel method and analysis of the trajectories of multiple free-swimming sperm could be more effective for motility evaluation.
- Medium extender is mandatory for the conservation of semen more than 2–3 h.
- Cryopreservation required the use of an optimal medium (e.g. TRIS-buffer, alcohols, dimethyl sulfoxide-DMSO, and amides) and a specific protocol; however, the fertilization ability of frozen semen is low.
- Rabbit is an excellent model for reproductive, toxicological, and metabolic studies. Moreover, rabbit semen may also represent a good model for in vitro studies. Semen is the more accessible reproductive endpoint, which can be easily analysed (without killing animals as in rodents), by well-established procedures.
- Mathematical models that represent complex metabolic pathways of live systems may be a useful predictive method for future advances.

2 The Reproductive Physiology of Rabbit Male

2.1 Male Genital Organs

Three main portions constitute the genital organs of rabbit male:

(i) The gland section constituted by two testes (where spermatogenesis occurred)
(ii) The tubular section, consisting of the spermatic lines, responsible for the collection and transport of sperm up to the urogenital tract. It is composed, on each side, by the epididymis (constituted by three segments: *caput*, *corpus*, and *cauda*), the vas deferens, and the vesicular gland
(iii) The urogenital sections, consisting of urethra, with the attached accessory glands (prostate and bulbourethral glands) and the corpus cavernosum, an erectile formation which joins the extra pelvic portion of the urethra for forming the penis, and male copulatory organ

The terminal portion of *ductus deferens* (ampulla) ends in the ventral wall of seminal vesicle. The central seminal vesicles open into the prostatic segment of the urethra immediately in front of the *colliculus seminalis*. The vesicular and prostate glands lie in the dorsal wall of the posterior part of the seminal vesicle, which they widely cover. The pair of vesicular gland ducts enter the urethra at either side of seminal *colliculus* and the prostatic ducts enter the urethra at either side or just posterior of them. The paraprostatic glands are minute finger-like extroflexions of the urethral lining imbedded in the outer part of its wall at either side of the seminal vesicle. The small-paired bulbourethral glands form a bilobed swelling in the dorsal wall of the urethra in which they are attached, immediately behind the prostate.

2.2 Spermatogenesis

The age at which rabbits reach puberty varies from 4 to 12 months in relation to the breed/strain, i.e. for the New Zealand White (NZW) strain, is around 8 months (Punnett and Bailey 1918). Male rabbits reach the sexual maturity slower than female, showing motile spermatozoa at 7–8 months. The process that brings to the production of mature spermatozoa consists of two main steps: i) the elaboration of cells (spermatogenesis) which occurred in the testis and takes approximately 55 days, and ii) the spermatozoa (SPZ) maturation which occurred in the epididymis and takes approximately 11 days for travelling through the epididymis from *caput* to *cauda* thanks to testosterone-induced contractions (Orgebin-Crist 1973).

The whole spermatozoa maturation includes eight specific stages (Fig. 1):

Stage 1. Type A (A) spermatogonia and Sertoli cells are present along the testis basement membrane. Two generations of primary spermatocytes are present: the "young" primary spermatocytes are in the *pre-leptotene* phase and are located near the basement membrane; the "old" primary spermatocytes are in *pachytene*

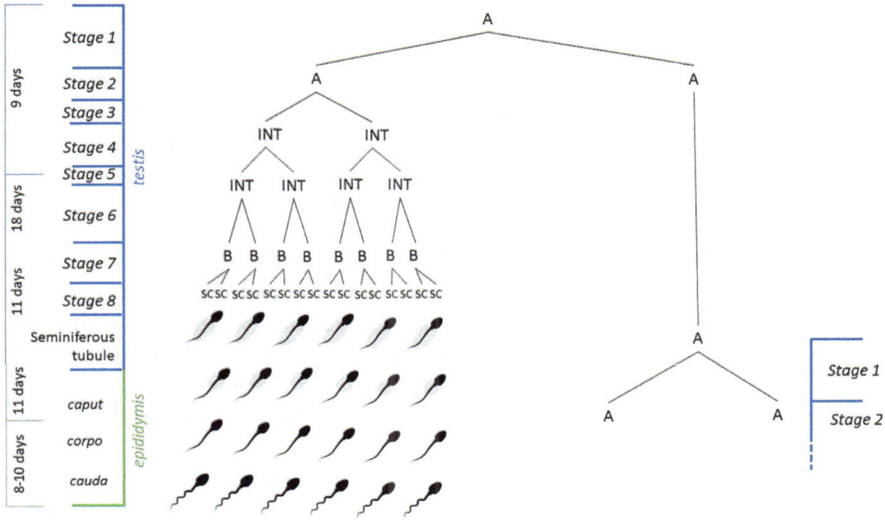

Fig. 1 Spermatogenesis in rabbit. A, type A spermatogonia; INT, intermediate-type spermatogonia; B, type B spermatogonia; SC, primary spermatocyte. The space given to each stage is proportional to its relative duration. Blue colour indicates process occurred in the testis whereas green colour indicates process occurred in epididymis. Sperm with grey shadow indicate immature SPZ with the cytoplasmic drop. This figure was created according to data retrieved from Kozma et al. (1974)

phase and are distributed between the basement membrane and the lumen of the tubule. The nuclei of the young primary spermatocytes are round with slightly staining chromatin, whereas the old primary spermatocytes have larger *pachytene* nuclei and the nuclear membrane of these cells is indistinct, finally the spermatids have round nuclei with a distinct nuclear membrane.

Stage 2. The characteristic feature of this stage is the elongation of the spermatid nuclei. In the beginning of Stage 2, the nuclei are slightly oval, and in the end, they have completely elongated. Two generations of primary spermatocytes are present. The appearance of the nuclear chromatin for both generations is similar to that described in Stage 1. Sertoli cells and A spermatogonia form the basement membrane. The number of A spermatogonia increased over the number present in Stage 1.

Stage 3. Sertoli cells and A spermatogonia continue to line the basement membrane. Again, two generations of primary spermatocytes are present. The nuclei of the young primary spermatocytes are still small and in the *zygotene* phase and those of the old primary spermatocytes are in the *pachytene* phase. The spermatids are elongated and constitute the Intermediate-Type (INT) Spermatogonia.

Stage 4. The composition of the cell population lining the basement membrane depends upon the extent to which this stage has progressed. During Stage 4, A spermatogonia start to divide themselves and as a result INT spermatogonia as well as A spermatogonia and Sertoli cells can be observed along the basement membrane. The INT spermatogonia have ovoid nuclei with the long axis parallel to the basement membrane. The chromatin of these nuclei is in the form of granules. In the early part of Stage 4, two generations of primary spermatocytes can be observed. The nuclei of the young generation are in the *pachytene* phase and the chromatin stains intensely. The nuclei of the older generation of primary spermatocytes are very large and are in the *diplotene* or diakinesis phase. At the beginning of Stage 4, these old primary spermatocytes start to undergo the first maturation division and give rise to secondary spermatocytes. The secondary spermatocytes have spherical nuclei with a distinct nuclear membrane and the chromatin is in the form of granules connected by filaments. Soon after formation, they undergo the second maturation division and form spermatids. The spermatids have spherical nuclei with a distinct nuclear membrane. The elongated spermatids from Stage 3 will develop in spermatozoa (Stage 4) since a new generation of spermatids is formed.

Stage 5. During this stage, the SPZ move towards the Sertoli cells. Many A spermatogonia divide to give rise to INT spermatogonia. In contrast to Stages1, 2, 3, and early Stage 4, where there are two generations of primary spermatocytes present, in Stages 5, 6, 7, and half of Stage 8 only one generation of primary spermatocytes is present. The chromatin of the nuclei of these cells starts to spread out and the nuclei move away from the basement membrane. The spermatids are spherical and have a distinct nuclear membrane.

Stage 6. Sertoli cells, A and INT spermatogonia line the basement membrane. Towards the end of this stage, the INT spermatogonia start to divide and form Type B (B) spermatogonia (Swierstra and Foote 1963). B spermatogonia have spherical nuclei and a distinct nuclear membrane. The chromatin of these cells is in the form of darkly staining granules. The spermatid nuclei are spherical and they establish contact with the Sertoli cells during this stage.

Stage 7. Sertoli cells and B spermatogonia line the basement membrane. The type of spermatogonia present depends on the stage progressed. INT spermatogonia can be observed in the early part of Stage 7. Soon these cells divide and give rise to B spermatogonia. In addition, a few A spermatogonia are also present throughout Stage 7. The *pachytene* nuclei of the primary spermatocytes have further increased in size and the chromatin is in the form of a dense network. The spermatid nuclei are spherical and reveals a nuclear membrane. Spermatozoa move towards the lumen.

Stage 8. The composition of the cell population lining the basement membrane depends upon the degree to which this stage has progressed. Sertoli cells and A spermatogonia are present at all times; B spermatogonia are present during the first half of Stage 8. During the second half, they undergo mitoses and produce primary spermatocytes. These newly formed primary spermatocytes in the *pre-leptotene* stage have spherical nuclei, which resemble B spermatogonia. The spermatid nuclei are spherical and reveal an acrosome preparation. The SPZ line the lumen.

The SPZ produced migrate from the lumen to the seminiferous tubules. It undergoes a series of changes in the acrosome and the glycoprotein layer: migration and reabsorption of the cytoplasmic drop present in the proximal part of the mid-piece; dehydration; loss of lipoproteins from the plasma membrane; adhesion of proteins from the secretions of the seminiferous epithelium; metabolic changes of DNA and lipids. The new structure is responsible for the appearance of the agglutinating properties of the sperm as well as for the development of its fertilizing aptitude and the ability to recognize the egg.

When the SPZ enter the epididymis, they are immobile, as they progress along the duct, they begin to move autonomously making a curvilinear trajectory, until they reach the *cauda* where they present a rectilinear progressive motility.

2.3 Rabbit Semen

Rabbit semen is a heterogeneous matrix (Fig. 2) constituted by SPZ, prostatic secretory granules (PSGs), and many other molecules (protein, sugar, etc.) dispersed in a fluid called seminal plasma (SP). All these factions are strictly inter-connected and participate in the egg fertilization process in different ways.

2.3.1 Seminal Plasma

The SP is a fluid mainly secreted by accessory glands, which provides the environment that ensures the success of fertilization. There is evidence that, reducing the SP concentration, the SPZ deteriorate (Dott et al. 1979), suggesting that it strongly modulates the SPZ behaviours. Castellini et al. (2000) demonstrated that increasing dilution rate of semen with SP (from one- to ten-fold) the sperm kinetic parameters worsen dramatically, due to sperm lipid membrane oxidation during storage (from 2 to 6 h). In rabbit SP, many protein constituents have been found, which modulate the protection, transport, capacitation, and motility of sperm within the female reproductive tract (Juyena and Stelletta 2012). Some of those most involved in the mechanisms of sperm modulation are the ATPase inhibitor (De Lamirande and Gagnon 1984) and insulin-like growth factor (IGF; Minelli et al. 2001), which mainly affect sperm motility or the ovulation inducing factor (OIF), a molecule that modulates the ovulation in rabbit does and also some sperm parameters (i.e. kinetic, capacitation, apoptosis; Castellini et al. 2010, 2022a; Dal Bosco et al. 2011).

In the last decades, one of the most studied seminal factors able to modulate sperm parameters is the nerve growth factor β (β-NGF), belonging to the OIF family (Adams et al. 2016). This protein factor is present in high concentration in the rabbit semen (0.2–150 μg/mL; Mattioli et al. 2021b; Castellini et al. 2010) and in other induced ovulation species (Maranesi et al. 2022). It is also present in different rabbit cell types: Leydig, Sertoli, germinal, and prostate cells (Maranesi et al. 2015). From a cellular viewpoint, the β-NGF exerts its own biological functions, binding two cell surface receptors: Tropomyosin receptor kinase A (TrKA), a high-affinity receptor, and p75 neurotrophin (p75NTR), a low-affinity receptor (Holgado-Madruga et al.

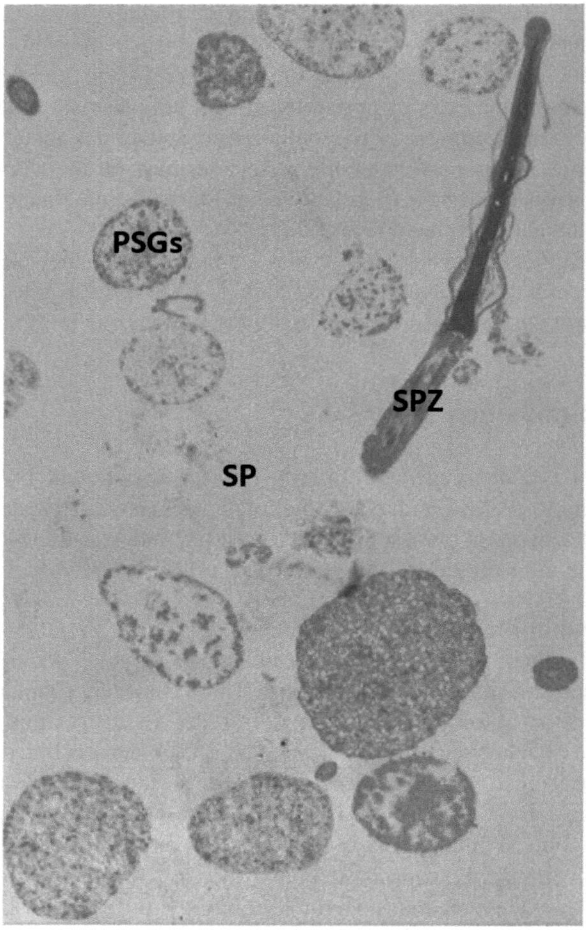

Original magnification: ×20 000; PSGs= Prostatic secretary granules; SPZ= spermatozoa; SP=seminal plasma.

Fig. 2 Transmission electron microscopy of rabbit sperm. Original magnification: ×20,000; PSGs, prostatic secretory granules; SPZ, spermatozoa; SP, seminal plasma

1997). Interactions among TrKA and p75NTR pathways are critical for the biological effects of β-NGF in the different cell types. These two receptors are differently distributed on sperm surface: TrKA is mainly located in the sperm head of the ejaculated sperm, whereas the p75NTR in the mid-piece (Fig. 3a). The different distribution also modulates their functions (Fig. 3b): Castellini et al. (2019b) in recent in vitro research demonstrated that NGF is able to enhance SPZ motility, liveability, and acrosome reaction, via TrKA, while the activation of

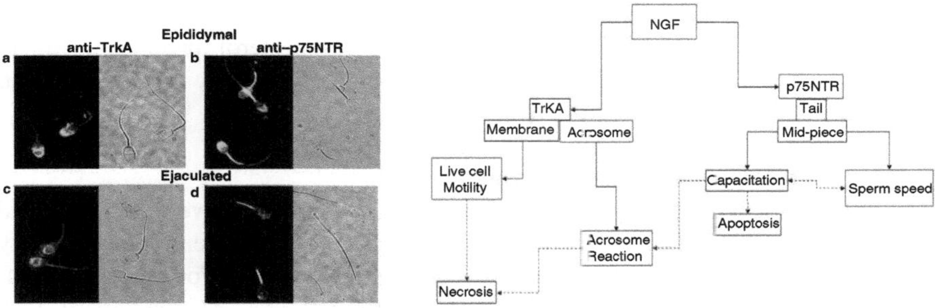

Fig. 3 Receptor distribution (**A**) on epididymal [(**a**) TrKA and (**b**) p75NTR] and ejaculated [(**c**) TrKA and (**d**) p75NTR] sperm and main NGF functions (**B**) associated to the NGF-receptors interaction on rabbit sperm. Modified from Castellini et al. (2019b)

p75NTR signalling is mainly involved in the regulation of apoptosis and sperm death.

2.3.2 Spermatozoa

The SPZ undergo maturation throughout spermatogenesis (see Sect. 2.2), during which it acquired the final morphology and functionality. Mature SPZ are costituted by three main districts: head, mid-piece, and tail. During maturation, the cytoplasm and nuclei are drastically reorganized and many components are created: acrosome is located in the apical portion of the head, it includes several hydrolytic enzymes involved in the fertilization process; the nucleus, located in the central part of the head, containing DNA linked to protein, to constitute the chromatin complex and the mitochondria, located in the mid-piece of SPZ, where is produced energy for the movement regulated by the tail tightening.

The daily SPZ production is almost 250 million, but about 30% is reabsorbed during the transit and storage, whereas 20% is eliminated with urine. Such reduction is modulated by processes with the aim to preserve only the "better" SPZ able to fertilize the egg. Most of SPZ is stored in the *cauda* of epididymis (1200–1500 million), followed by testis (130–180 million), *corpo* and *caput* of epididymis (150 million), and a small part in the deferens canal (50 million) (Prins and Zaneveld 1979).

Most of the head functions in SPZ are attributed to the specific composition of the lipid fraction, designated lipid rafts (Gadella et al. 2004). Similar to other mammalian SPZ, the lipid bilayer of the rabbit sperm membrane consists mainly of phospholipids (PLs) and cholesterol, at a molar ratio of 1:5 (Castellini et al. 2006). The rabbit sperm membrane PLs contain about 43% of long-chain polyunsaturated fatty acids (LC-PUFAs, more than 20 carbons atoms), especially of the n-6 series, with docosapentaenoic acid (C22:5n-6; DPA) being the most representative (Poulos et al. 1973; Castellini et al. 2022a). The LC-PUFAs in the sperm membrane derive from the essential linoleic acid (C18:2n-6; LA) and alpha-linolenic acid (C18:3n-3; ALA) resulting from the diet (see Sect. 2.4.2) and are converted into their LC derivatives by the liver (Cook and McMaster 2002) or in less way by testicular cells (Castellini et al. 2022a). Furthermore, the lipid bilayers are constituted by different sterols (cholesterol and desmosterol, a cholesterol precursor) which, together with the PLs fatty acids modulate the membrane fluidity and the sperm to the capacitation (Mourvaki et al. 2010a).

2.3.3 Prostatic Secretory Granules

The PSGs are a heterogeneous population of small granules and vesicles, dispersed in the seminal fluid. They were also called "prostasomes" and are produced mainly by prostate (Collodel et al. 2012a) for exocytosis (apocrine and/or holocrine pathway) or diacytosis (translocation of vesicles from the cell interior to the lumen). Two types of PSGs differing in size are present in the rabbit semen: large (2–4 μm) and small (0.03–0.2 μm) and have been isolated by gradient centrifugation and ultracentrifugation techniques, respectively (Castellini et al. 2006). Scanning electron microscopy revealed the presence of several clubbed protrusions and very small detaching particles on the surface of the bigger granules (Fig. 4).

Both PSGs population are delimited by a single or double membrane and contain electron-dense material, while others appear to be empty with a dense granular material attached to the inner membrane. Large granules constitute about 15–20% of the total PSGs, and several studies demonstrated that their concentration in semen might be higher than that of SPZ. However, their ratio varies depending on the collection rhythm of semen and the health status of rabbit (Castellini et al. 2006). The physiological PSGs concentration in seminal fluid ranges from 1 to 4 mg/mL estimated on protein basis (Castellini et al. 2012).

The PSGs have a relevant role in the SPZ activity. They may be considered a sterols stock for the SPZ; indeed, they are able to transfer cholesterol (or desmosterol) from its membrane to head membrane of SPZ, delaying its capacitation (Mourvaki et al. 2010a) and inhibit capacitation and acrosome reaction

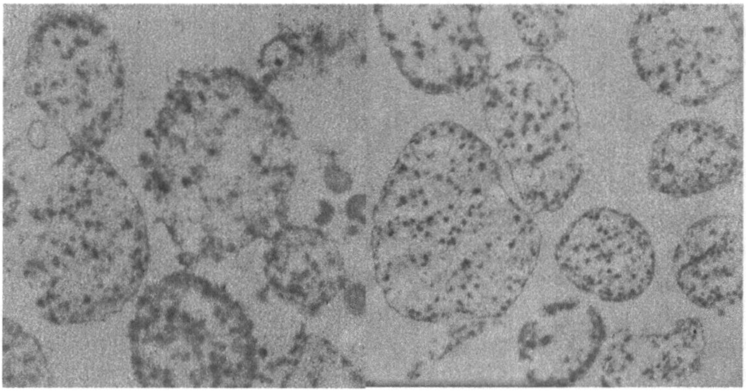

Fig. 4 Scanning electron microscopy of large prostatic secretory granules (PSGs). Original magnification ×11,500

(Castellini et al. 2007). Furthermore, PSGs are also rich in PLs (422 μg/10^9 spermatozoa; Castellini et al. 2006) which switches to the SPZ membrane in order to render it more stable (Nikolopoulou et al. 1986). Finally, also a minor concentration (like to that of SPZ, 22 ng/mL; Mourvaki et al. 2008) of lipid-soluble compounds such as Vitamin E and carotenes has been also found (Mourvaki et al. 2008; Castellini et al. 2007) suggesting that PSGs could be a potential adjuvant source of protection to vulnerable cells from related or not inflammatory-induced oxidative stress.

In the light of what reported, PSGs showed a prominent role in seminal fluids modulation: they interact with sperm to promote SPZ forward motility (Castellini 2013b). Probably, this last effect was related to the transfer of short chain FA which could be used by sperm mitochondria as a fuel for motility (via acetyl-CoA synthetase activity and the Krebs cycles; Rémésy et al. 1995). Furthermore, PSGs also are immunosuppressive (Kelly 1995), protective against reactive oxygen species (ROS; Mourvaki et al. 2010b), antibacterial (Carlsson et al. 2000), coagulant (Fernandez et al. 1997), and semen liquefaction and capacitance, reducing SPZ fertility (Lwaleed et al. 2004).

2.4 Factors Affecting Semen Quality

The quantity of semen produced by rabbit bucks depends on several factors: individual variability, genetic lines, age, photoperiod, temperature, diets, stress, or drugs. The principal of them may be grouped in three macro-factors: genetic, diet, and management practices.

2.4.1 Genetic

Several studies have been carried out to explore the accurate-enough genetic parameters (i.e. heritability or genetic correlation) for explaining the variability in semen production and sperm quality. However, it has been found relevant differences that are not related to the maternal or paternal rabbit lines (Piles et al. 2013). Based on some studies (Piles and Tusell 2012; Castellini 2007; Lavara et al. 2013) the crossbreed lines showed better semen condition of pure breeds with +37% of sperm concentration and +4% of sperm motility. At the same time, other studies demonstrated that less selected breeds showed higher sperm count with respect to commercial heavy lines; however, all studies concluded that the repeatability of such characteristics is very low. Furthermore, a higher heritability (h^2) was found only for sperm morphology traits (0.11–0.35) and sperm concentration (0.23) whereas lower values were recorded for ejaculate volume, kinetic parameters and status (from 0.05 to 0.18; García-Tomás et al. 2006). Accordingly, it was also demonstrated that genetic selection for improving male libido or semen characteristics was not very effective on ejaculate traits and is poorly correlated with male fertility and prolificacy.

Anyhow, differences between bucks may arise both from genetic and environmental factors: however, when bucks of a selected strain are reared and collected

Table 1 Semen traits (mean ± SEM) of different rabbit strains

Traits	NZW	Silver of Champagne[a]	California	Sign.
Ejaculate volume (μL)	744.2 ± 58.2	329.0 ± 102.9	1000.4 ± 28.9	**
Sperm concentration ($\times 10^6$/mL)	147.6 ± 9.3	51.7 ± 5.2	101.2 ± 8.7	**
Individual motility (%)	79.5 ± 3.7	42.4 ± 5.8	50.6 ± 6.0	**
VCL (μm/s)	384.4 ± 11.0	351.0 ± 40.03	251.1 ± 14.6	**
VSL (μm/s)	95.2 ± 3.6	96.5 ± 11.5	68.8 ± 4.4	**
VAP (μm/s)	192.2 ± 6.2	185.0 ± 20.4	133.5 ± 8.3	**
LIN (%)	25.0 ± 0.8	28.2 ± 1.6	26.6 ± 1.4	ns
ALH (μm)	3.0 ± 0.1	2.8 ± 0.3	2.1 ± 0.1	**
BCF (Hz)	29.6 ± 1.2	27.3 ± 2.3	18.4 ± 1.5	**

**$P < 0.01$

Sign., significance; VCL, curvilinear velocity; VSL, straight-line velocity; VAP, average path velocity; LIN, linearity; ALH, average SPZ track width; BCF, beat cross frame

[a]Fanti (1979)

according to a strict protocol, the variability is generally lower than what obtained in less controlled conditions (Panella et al. 1994). In this context, nowadays, the rabbit strain more used paternal lines are New Zealand White or synthetic lines derived from the latter because the rearing conditions and the semen parameters are better defined and standardized (Box 2 and Table 1).

Box 2: Particularities of Sperm Traits of New Zealand White and Pure Strains

(I) NZW rabbits showed better overall kinetic traits (motility, VLC, VSL, VAP, ALH, BCF) followed by Silver of Champagne and California.

(II) California showed higher sperm volume than NZW and Silver of Champagne.

(III) Silver of Champagne had a very low sperm concentration.

2.4.2 Diets

The diet represents one of the main environmental factors, more important, for obtaining high semen quality. It is reasonable to assume that improvements in the quantity and quality of semen are mainly affected by fat and antioxidants contents of diets, because these two components largely influence the sperm behaviours and semen traits as already specified in Sect. 2.3.2.

Over the past few years, several efforts have been made to increase the proportion of sperm LC-PUFA of n-3 series by supplementing animals with diets containing ALA (Mourvaki et al. 2010c; Castellini et al. 2013a, 2019a; Rodríguez et al. 2019; Collodel et al. 2020a) or n-3 LC-PUFA (Castellini et al. 2019a; Menchetti et al. 2020). Sperm cells of rabbit fed fish oil show a higher n-3 LC-PUFA content when compared to other tissues (blood, liver, and ovary; Mattioli et al. 2019; Rodríguez et al. 2019). Such an accumulation of n-3 PUFA in rabbit sperm is probably due to the demand of the SPZ, whose membrane consists mainly of PUFAs. Mourvaki et al. (2010b) found that the integration of 5% linseed into the rabbit diet improves sperm quality by modifying the sperm lipid composition, which means a reduction by half of LA and DPA, and a concomitant increase of ALA (+65%) and docosahexaenoic acid (C22:6n-3, DHA; + 83%). Furthermore, the authors also found a 70% reduction in sperm cholesterol. The relevant increase in DHA, mainly in the sperm tail, as well as the similar decrease in cholesterol, can influence sperm speed and the fluidity of their membranes improving the quality traits of semen (Mourvaki et al. 2010b).

Castellini et al. (2019a) demonstrated that supplementation of dietary linseed to rabbit bucks improves the motility rate and track speed of sperm (curvilinear velocity, VCL) and increases the blood testosterone concentration resulting in higher fertility. However, the positive effect on male rabbits is only found with concomitant intake of antioxidant compounds (vitamin E and/or vitamin C; Mattioli et al. 2021a, b). The high unsaturation level of sperm membrane renders these cells very susceptible to peroxidation, which could negatively affect the membrane structures, sperm physiology, and DNA integrity if not adequately protected. Many SP compounds protect against lipid oxidation (enzymes, albumin, tocopherols; Mann and Lutwak-Mann 1981); however, when an additional oxidative input is supplied (increase the PUFA profile of sperm membrane), a further dose of antioxidants is mandatory. In agreement, Gliozzi et al. (2009) demonstrated that the semen quality of rabbits fed a fish oil-enriched diet was strongly different if vitamin E was also provided, that is, rabbit fed fish oil diets (both with and without vitamin E) showed about fivefold higher values of DHA in semen PLs. Furthermore, in a recent paper, Mattioli et al. (2021a) demonstrated that also the testis and not only the semen was affected by the oxidant/antioxidant balance: a supplementary addition of dietary lipid source to rabbit bucks determined an increase of non-specific (malondialdehyde) and specific (isoprostanoids mainly derived from n-3 series, i.e. F2-isoprostanes and F4-neuroprostanes) oxidative products.

2.4.3 Management

Many management practices have an impact on rabbit semen quality, both referred to semen collection (collective rhythms, artificial vagina, etc.) both to animal handling (temperature, photoperiod, rearing system, etc.). The semen collection is currently done with an artificial vagina of different materials (e.g. glass, plastic, caoutchouc gum), maintained at a temperature around 42 °C. Considering the large differences in the artificial vagina models, one of the main aspects to take into consideration is the hygiene of collection, because a bacterial contamination during semen collection may negatively affect the semen quality (e.g. *Lactobacillus, Streptococcus, Staphylococcus*). For instance, it is reported that the plastic vagina is more effective than the glass one, because it is possible to change the plastic cover for every animal, whereas the glass vagina required a deep cleaning procedure to avoid cross contamination (Bredderman et al. 1964; Dal Bosco et al. 1996; Naughton et al. 2003), also reducing the economic costs ($125.00 for glass vagina, $6.00 for plastic vagina). Similarly, the caoutchouc gum often required the addition of a lubricating gel, which can interfere with semen characteristics.

Also, the collection rhythm affects the semen characteristics; in this regard, Boiti et al. (2005) reported that the best parameters have been obtained with one collection/animal twice a week, because an intensive collection rhythm reduced the semen volume and SPZ concentration. Furthermore, other studies demonstrated that also the rearing system influenced the semen characteristics: the extensive systems are friendlier and improving the animal welfare improves the sperm motility. Similarly, rearing bucks in group of three rabbits improved the volume concentration and motility, due to the testosterone effect, such as a good environmental temperature (between 15 and 23 °C), advanced the SPZ concentration and motility. Finally, also the experience of the technician may affect sperm parameters (mainly ejaculation volume).

2.5 Technical of Semen Analysis

The Computer-Assisted Semen Analysis (CASA) system is commonly used for assessment of sperm motility, viability, and morphology (Verstegen et al. 2002). For the last two practices, it is necessary to use fluorochromes coupled to optical (viability) or fluorescence (morphology) microscopy, thus using an approach similar to the manual one with the only difference that the image acquisition and cell count are computerized.

Conversely, for the kinetic parameters assessment, the use of the computer-assisted system is very useful and renders the evaluation more objective and sensible. The CASA system allows the evaluation of several kinetic characteristics thought the description of the movement in specific parameters: curvilinear velocity (VCL, mm/sec), average path velocity (VAP, mm/sec), straight-line velocity (VSL, mm/sec), linearity (% LIN=VSL/VCL), straightness (%STR=VSL/VAP), amplitude of lateral head displacement (ALH, average sperm track width, mm), beat cross frame (BCF, Hz). Furthermore, many other traits may be recorded: proportion of sperm with progressive/no-progressive motility or static, and the type of movement (rapid, medium, slow).

The standardization of semen evaluations was a prerequisite for the development of CASA analysis. It is generally agreed that changing the basal parameters (i.e. number of image acquired/sec, Hz, minimum and maximum range) the results drastically changed (Castellini et al. 2011). In rabbit, the setting parameters are standardized as reported in Box 3 and Table 2.

Box 3: Rabbits CASA System

(I) The minimum sperm size to analyse is 15 pixels^2.

(II) Suitable frame rate for rabbit sperm analysis is 100 frame/s.

Table 2 CASA set up parameters

Setting parameters	Values for rabbits
Min sperm area (pixels^2)	15
Max sperm area (pixels^2)	150
Min track length (frame)	30
Max sperm velocity (pixels)	40
Min VCL for motile sperm (μm/s)	25
Min VAP for motile sperm (μm/s)	30
Max VCL for slow sperm (μm/s)	15
Max VAP for slow sperm (μm/s)	20
Frame rates (frame/s)	100

VCL, curvilinear velocity; VAP, average path velocity

An important limiting factor in the use of CASA for semen analysis is that the sperm motility is studied in two dimensions (2D; Amann and Waberski 2014) by recording sperm swimming trajectories near a flat surface. However, spermatozoa swim in three-dimensions (3D) to find eggs in the female reproductive tract, with their speed being the main impediment tracking them under realistic conditions. Few studies describe a 3D method and analysis of the trajectories of multiple free-swimming sperm. Corkidi et al. (2008) using a piezoelectric device with a 70 Hz triangular waveform recorded the sperm motility in fish (sea urchin *Strongylocentrotus purpuratus*) and concluded that the 3D data obtained are significantly at variance with those of swimming sperm confined in 2D, emphasizing the use of 3D approach as more physiological in the sperm motility study (Fig. 5). However, today no simple tools, procedure, and software exist for recording sperm motility in 3D.

2.6 Sperm Refrigeration and Cryopreservation

Raw semen has a short life (6–8 h) in in vitro condition if not diluted in appropriate medium.

Currently, the AI practice is carried out within 2–3 h by semen collection, or, if the semen is well maintained in the dilution medium and refrigerated at 5 °C also after 24 h. For the dilution may be used medium with a well-defined organic component: i.e., Tyrode's albumin lactate pyruvate (TALP) or TRIS-buffers, added with citric acid, sugars (glucose and/or fructose), and antibiotic protection (i.e., Streptomycin, Penicillin). The dilution ratio is in a range of 1:5–1:10, and the insemination dose is generally of 0.2–0.5 μL (corresponding to about 10 million of sperm cells). Many compounds can be added to the standard buffer to further increase its nutritional characteristics: i.e., date palm pollen (Laghouati et al. 2021), vitamin E (Brecchia et al. 2010), egg yolk at a 10–20% of concentration (Samouilidis et al. 2001), skim milk at a final concentration of 8–10% (Sieme and Oldenhof 2015), etc. However, when the semen was stored for more than 24 h, it is suggested to use commercial medium.

Fig. 5 Trajectory of tracked sperm. (**a**) 2D representation of swimming on *x*,*y* axis; (**b**) 3D representation of free swimmer (with helical trajectory) on *x*,*y*,*z* axis. Modified from Corkidi et al. (2008) with permission from Elsevier

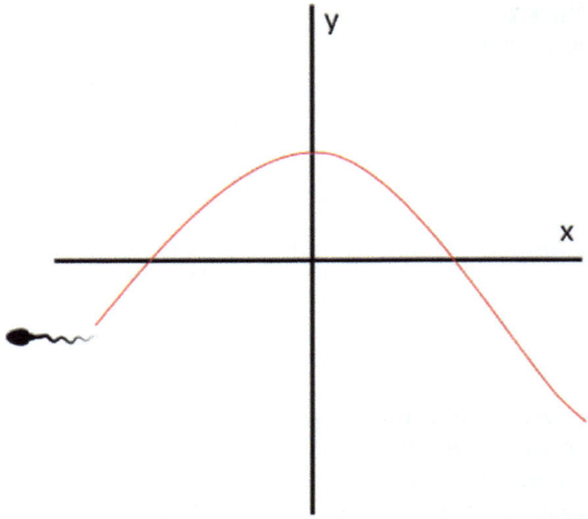

a)

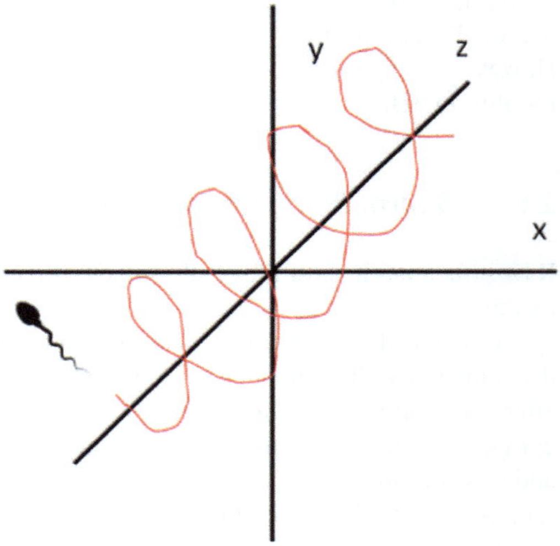

b)

Generally speaking, cryopreservation assures an easy transportation and conservation of semen. TRIS-based extenders are the base of the extenders used for rabbit sperm cryopreservation (Box 4; Table 3). Indeed, when several extenders were compared for rabbit sperm cryopreservation (Tris–citric acid–fructose, sodium

Table 3 Chemical composition of main medium used for rabbit semen dilution

Composition	TALP[a]	TRIS-buffer	TRIS-buffer for cryopreservation
NaCl (g)	5.690		
KCl (g)	0.231		
CaCl2 × 2H2O (g)	0.294		
MgCl2 × 6H2O (g)	0.081		
NaHPO4 (g)	0.040		
NaHCO3 (g)	2.090		
Na-pyruvate (g)	0.022		
D-Glucose (g)	0.900	1.250	1.250
Lactic acid (mL)	3.680		
Tris (g)	–	3.029	3.029
Citric acid (g)	–	1.676	1.676
HEPES (g)	2.380		
Streptomycin (IU)	75,000	75,000	75,000
G-Penicillin (IU)	166,200	166,200	166,200
DMSO (%)	–	–	5
Glycerol (%)			1.5
Total volume (L)	1	0.1	0.1
pH	7.40	7.14	7.14
Osmolarity (mOsm/kg)	296	299	299

[a]Tyrode's albumin lactate pyruvate

citrate, etc.), none of them offered better results than the TRIS-based extender (Mocé and Vicente 2009).

Box 4: Main Semen Dilutors for Rabbits Semen

(I) TALP and TRIS-buffer are the main medium used for AI in rabbit.

(II) TRIS-buffer for cryopreservation required additional ingredients (DMSO, glycerol, or amides).

However, conversely to other species, the cryopreservation in rabbits always showed many problems such as the fertility was often reduced and the prolificacy was lower than that of fresh semen. Indeed, it has been postulated that to the standard medium it is needed to add some molecules able to protect the sperm cells from damage: i.e., different amides, alcohols dimethyl sulfoxide (DMSO), glycerol, ethylene glycol, etc. (Kubovicova et al. 2022).

During the freezing, in the phase in which the temperature drops from 1 to −10 °C, crystals that are formed both inside and outside the cell, can damage the sperm membrane; in order to avoid these negative features, in addition to the agents described above, it is necessary to apply a freezing protocol which involves a gradual lowering of the temperature: generally, the semen is kept at 5 °C for 2–5 h before placing it on the nitrogen vapours for 10–15 min and successively preserved in liquid nitrogen.

However, the sperm yield following cryopreservation is lower than fresh semen, only 25–35% of the sperm cells remain intact and viable, the motility is reduced from 83 to 43% and the proportion of sperm with a damaged acrosome is 11 vs. 23% (Anger et al. 2003).

The addition of antibiotics in semen medium could and should be progressively reduced due to the problem of bacteria resistance to these compounds. It is widely acknowledged that the development of resistance by human bacterial pathogens to pharmaceutical antimicrobials due to their uncontrolled use (acquired antibiotic resistance) poses a major threat to global public health. Recent data suggest that over 1 million deaths per year are attributable to resistant bacterial infections (Jamrozik and Heriot 2022). Much of the debate about antibiotic resistance is related to the transfer of such resistance from livestock animals to humans, due to the preventive use of such drug in animal feeding (medicated feed) until now widely used to improve production performance. Such disproportionate use has led to a high persistence of resistance in the environment which, to be eliminated, is not sufficient to produce with the "antibiotic-free method", but to change the general habits of the use of antibiotics by citizens/doctors/producer/farmers/veterinaries, etc. (Jamrozik and Heriot 2022).

2.7 Rabbit as Animal Model for Human Studies

The renaissance of the laboratory rabbit as a reproductive model for human health is closely related to the growing evidence of periconceptional metabolic programming and its determining effects on offspring and adult health. Advantages of rabbit reproduction are the exact timing of fertilization and pregnancy stages, high cell numbers and yield in blastocysts, relatively late implantation at a time when gastrulation is already proceeding, detailed morphologic and molecular knowledge on gastrulation stages and a haemochorial placenta structured similarly to the human placenta (Fischer et al. 2012).

2.7.1 Reproductive Studies

The rabbit buck is an excellent model for studying the reproductive functions because mature sperm can be easily and continuously collected with an artificial vagina without having to sacrifice the animal (Foote and Carney 2000).

Rabbit has been a useful model for longitudinal, long-term studies on the effects of sub-acute endotoxaemia on sperm parameters. Intra-peritoneal inoculation of bacteria lipopolysaccharide (LPS) triggered a reversible inflammation-like status in buck rabbits which lasted 3 days, and subsequently, reproductive activity and sperm quality were progressively restored (Brecchia et al. 2010). Collodel et al. (2012b) induced an LPS inflammation in the same rabbit model, reducing sperm concentration and motility and increased ultrastructural alterations at 14–30 days after treatment.

2.7.2 Toxicology Studies

Most nanotoxicological research is performed in vitro and, consequently, cannot consider tissue distribution, accumulation in selected organs, clearance or diffusion across biological barriers such as the blood-brain barrier and blood-testis barrier.

Castellini et al. (2014) investigated the long-term effect of silver (Ag) nanoparticles (NPs) administered intravenously (0.6 mg/kg b.w.) on reproductive activity and sperm quality. The AgNPs reached the testes, compromising SPZ motility and speed, and acrosome and mitochondria shape and function. Successively it was demonstrated that the exposure to AgNPs produced a similar but more persistent effect than that of LPS on rabbit reproductive tissues: AgNPs administration triggered a proinflammatory response linked to oxidative thrust, worsening many sperm parameters. However, only anti-inflammatory treatment counteracted the negative effects of AgNPs, whereas vitamin E seemed to act as an adjuvant, attenuating the oxidative cascade. Similarly, Collodel et al. (2020b) confirmed the negative effect of AgNPs on rabbits reproductive tissues, underlining a progressive effect due to the synergic inflammatory/oxidative thrust that impairs the reproductive environment, affecting the physiological status of sperm, and DNA damage. An in vitro study reported the negative dose-dependent effect of ZnO (zinc oxide) NPs on rabbit SPZ motility and viability parameters (Halo et al. 2021).

2.7.3 Metabolic Studies

In the last five decades, the human sperm count linearly decreased in Western countries. Health problems, lifestyle, pollutants, and dietary behaviours are considered as the main risk factors (Sengupta et al. 2018).

In humans, diet is difficult to standardize, thus many researchers use an in vitro approach, which evaluates the effect on isolated cells or tissue and does not consider the effect on spermatogenesis. In vivo studies in animal models allow us to deeply understand the influence of different treatments, so, again the rabbit model has been used successfully.

For the first time, Mourvaki et al. (2010b) showed that the integration of flaxseed into the rabbit diet may improve sperm quality by modifying the sperm lipid composition and that, the sperm subfractions and the prostatic granules respond differently to the flaxseed-induced lipid manipulation (see Sect. 7.1.4.2). Collodel et al. (2014) tested chocolate and propolis-enriched diets observing a counterbalance in the negative effects on spermatogenesis as well as sperm motility and ultrastructure following LPS treatments and then the effect of different dietary sources of PUFA. An experimental rabbit model indicated that the development of metabolic syndrome produces detrimental effects on sperm quality and functionality by inducing metabolic disorders leading to alterations in testis and epididymis functions and evidence a role of hypertension as a new determinant of abnormal sperm morphology, in line with a previous human study (Marchiani et al. 2015).

Based on the ongoing debate about risk factors for subfertility associated with overweight and obesity in men, the effect of overweight on sperm quality parameters and fertility success in randomized controlled trial in a rabbit model was also studied (Marco-Jiménez and Vicente 2017).

2.7.4 Metabolic Studies

Rabbit semen may also represent a good model for in vitro studies. Semen is the more accessible reproductive endpoint, which can be easily analysed by well-established procedures and has been frequently used for metal reproductive toxicity studies.

Arsenic, cadmium, chromium, mercury, platinum, and vanadium compounds reduced sperm characteristics and probably male fertility, by inhibiting sperm motility and altering head membrane morphology, when they accumulate in seminal plasma in amounts higher than 1 M (Castellini et al. 2009). Rabbit spermatozoa model provided new information on the specific mechanisms of action involved in the in vitro toxic behaviour of fusariotoxins on male gametes (Tvrdá et al. 2022).

Moreover, in vitro studies of rabbit sperm also highlighted the effect of β-NGF receptors (TrKA or p75NTR) on sperm functions. It is known that β-NGF affects kinetic and other physiological traits of rabbit sperm. Considering that some seminal disorders in humans have been correlated with a lower β-NGF concentration and no studies have been done on the possible involvement of β-NGF receptors, these findings also provide new insights on human fertility (Castellini et al. 2019b). Recently, Castellini et al. (2022b) demonstrated that the concentration of p75NTR on ejaculated rabbit sperm modulates many semen outcomes (motility, apoptosis, viability) through NGF interaction affecting the senescence of sperm.

2.8 Future Prospective

In vitro studies determined only a partial understanding of the biological metabolic process that may occur in a live organism. At the same time, the in vivo ones are much more complex and have scarce sustainability considering the great attention of citizens and the public on the use of animals in experimental research. In this context, a future prospective may be to develop mathematical models based on computer software that mimics complex metabolic pathways of live systems. Recently, Mattioli et al. (2022) develop a predictive model of the ROS–PUFA–antioxidant system in rabbit, through the use of a mathematic tool named VENSIM model. This can be used to estimate a priori the effect of dietary treatments (including inputs like: antioxidants intake, PUFA) on the redox homeostasis of the rabbit, and in turn on the animal's health. The authors investigated the full body metabolism taking into consideration the blood outputs; however, in this view, the model can assist in assessing the effect of oxidative challenges and dietary interventions (bioactive lipids, Vitamin E, Vitamin C, selenium, etc.) also on specific organs/apparatus (i.e. reproductive system, muscular, etc.; Fig. 6). This approach can also be useful for estimating the range of values analysed, within which it may be possible to intervene to solve the problem or above which the pathogenicity process (chronic pathogenesis) begins.

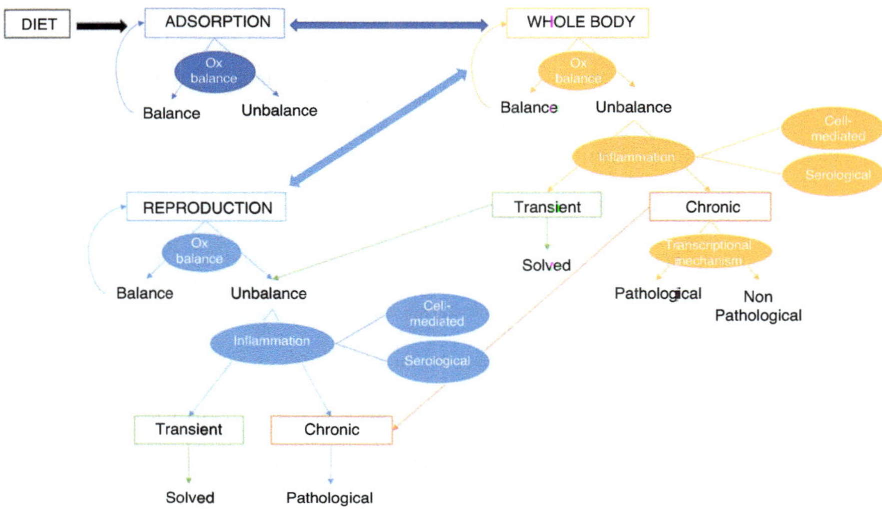

Fig. 6 Conceptual map of the application of modelling in the full body system. Dark blue patterns (arrow, shape, and ellipse) identify the first step of dietary compounds digestion and absorption. Yellow patterns identify the whole-body (blood) compounds destiny. Light blue patterns identify the compounds destiny in reproductive tissue. Ox balance: oxidative balance. When a molecule is introduced by the diets, it is adsorbed and can cause a modification of the oxidative balance, if such modification is solved, the system is in equilibrium (physiological status), on the contrary, a series of compensatory mechanisms are triggered by the body organism, which can bring to a stabilization (transient damage is solved) or to a chronic alteration (pathological or non-pathological). Such damage affects firstly the whole-body system, and subsequently a specific tissue (e.g. reproductive)

3 Conclusions

Rabbit semen quality is the main factor determining the success of AI practice. However, many considerations should be done in this regard:

(i) Semen quality may be modulated by many extrinsic (diet and management practices) or intrinsic (genetic, individual variability) factors
(ii) Motility assessment should forecast the use of innovative and more specific techniques (i.e. in 2D or 3D)
(iii) Reproductive outcomes in rabbit bucks may be predicted by the application of mathematical models that represents complex metabolic pathways of live systems

Many studies are in progress to deepen all the cited aspects with the aim to improve the semen characteristics and more in general the reproductive parameters of rabbit bucks.

References

Adams GP, Ratto MH, Silva ME, Carrasco RA (2016) Ovulation-inducing factor (OIF/NGF) in seminal plasma: a review and update. Reprod Domest Anim 51:4–17

Amann RP, Waberski D (2014) Computer-assisted sperm analysis (CASA): Capabilities and potential developments. Theriogenology 81(1):5–17

Anger JT, Gilbert BR, Goldstein MARC (2003) Cryopreservation of sperm: indications, methods and results. J Urol 170(4):1079–1084

Boiti C, Castellini C, Besenfelder U, Theau-Clément M, Liguori L, Renieri T, Pizzi F (2005) Guidelines for the handling of rabbit bucks and semen. World Rabbit Sci 13(2):71–91

Brecchia G, Cardinali R, Mourvaki E, Collodel G, Moretti E, Dal Bosco A, Castellini C (2010) Short-and long-term effects of lipopolysaccharide-induced inflammation on rabbit sperm quality. Anim Reprod Sci 118(2–4):310–316

Bredderman PJ, Foote RH, Yassen AM (1964) An improved artificial vagina for collecting rabbit semen. Reproduction 7(3):401–403

Carlsson L, Påhlson C, Bergquist M, Ronquist G, Stridsberg M (2000) Antibacterial activity of human prostasomes. Prostate 44(4):279–286

Castellini C (1996, July) Recent advances in rabbit artificial insemination. In: 6th World Rabbit Congress, Toulouse, vol 2, pp 13–26

Castellini C (2007) Reproductive activity and welfare of rabbit does. Ital J Anim Sci 6(suppl 1): 743–747

Castellini C, Lattaioli P, Moroni M, Minelli A (2000) Effect of seminal plasma on the characteristics and fertility of rabbit spermatozoa. Anim Reprod Sci 63(3–4):275–282

Castellini C, Cardinali R, Dal Bosco A, Minelli A, Camici O (2006) Lipid composition of the main fractions of rabbit semen. Theriogenology 65(4):703–712

Castellini C, Mourvaki E, Dal Bosco A, Galli F (2007) Vitamin E biochemistry and function: a case study in male rabbit. Reprod Domest Anim 42(3):248–256

Castellini C, Mourvaki E, Sartini B, Cardinali R, Moretti E, Collodel G, Fortaner S, Sabbioni E, Renieri T (2009) In vitro toxic effects of metal compounds on kinetic traits and ultrastructure of rabbit spermatozoa. Reprod Toxicol 27(1):46–54

Castellini C, Dal Bosco A, Arias-Álvarez M, Lorenzo PL, Cardinali R, Rebollar PG (2010) The main factors affecting the reproductive performance of rabbit does: A review. Anim Reprod Sci 122(3–4):174–182

Castellini C, Dal Bosco A, Ruggeri S, Collodel G (2011) What is the best frame rate for evaluation of sperm motility in different species by computer-assisted sperm analysis? Fertil Steril 96(1): 24–27

Castellini C, Mourvaki E, Cardinali R, Collodel G, Lasagna E, Del Vecchio MT, Dal Bosco A (2012) Secretion patterns and effect of prostate-derived granules on the sperm acrosome reaction of rabbit buck. Theriogenology 78(4):715–723

Castellini C, Brecchia G, Mattioli S, Ruggeri S, Cardinali R (2013a) Effect of dietary flaxseed on semen characteristics and testos-terone serum level of rabbit bucks. Reprod Domest Anim 48 (Suppl. 1):120–120

Castellini C, Mattioli S, Ruggeri S, Dal Bosco A, Collodel G (2013b) The time-dependent effects of prostate granules and seminal plasma on the capacitation, acrosome reaction, and motility of rabbit sperm. Anim Reprod Sci 140(1–2):97–102. https://doi.org/10.1016/j.anireprosci.2013.05.002

Castellini C, Ruggeri S, Mattioli S, Bernardini G, Macchioni L, Moretti E, Collodel G (2014) Long-term effects of silver nanoparticles on reproductive activity of rabbit buck. Syst Biol Reprod Med 60(3):143–150

Castellini C, Mattioli S, Signorini C, Cotozzolo E, Noto D, Moretti E, Brecchia G, Dal Bosco A, Belmonte G, Durand T, De Felice C, Collodel G (2019a) Effect of dietary n-3 source on rabbit male reproduction. Oxid Med Cell Longev. https://doi.org/10.1155/2019/3279670

Castellini C, Mattioli S, Dal Bosco A, Collodel G, Pistilli A, Stabile AM, Macchioni L, Mancuso F, Luca G, Rende M (2019b) In vitro effect of nerve growth factor on the main traits of rabbit sperm. Reprod Biol Endocrinol 17(1):1–11

Castellini C, Mattioli S, Moretti E, Cotozzolo E, Perini F, Dal Bosco A, Signorini C, Noto D, Belmonte G, Lasagna E, Brecchia G, Collodel G (2022a) Expression of genes and localization of enzymes involved in polyunsaturated fatty acid synthesis in rabbit testis and epididymis. Sci Rep 12(1):1–11

Castellini C, Mattioli S, Cotozzolo E, Pistilli A, Rende M, Bartolini D, Di Sante G, Menchetti L, Dal Bosco A, Stabile AM (2022b) The effect of interaction NGF/p75NTR in sperm cells: a rabbit model. Cells 11(6):1035

Collodel G, Cardinali R, Moretti E, Mattioli S, Ruggeri S, Castellini C (2012a) Role of rabbit prostate granules on sperm viability and acrosome reaction evaluated with different methods. Theriogenology 77(5):1021–1026

Collodel G, Castellini C, del Vecchio MT, Cardinali R, Geminiani M, Rossi B et al (2012b) Effect of a bacterial lipopolysaccharide treatment on rabbit testis and ejaculated sperm. Reprod Domest Anim 47(3):372–378

Collodel G, Moretti E, Del Vecchio MT, Biagi M, Cardinali R, Mazzi L, Brecchia G, Maranesi M, Manca D, Castellini C (2014) Effect of chocolate and Propolfenol on rabbit spermatogenesis and sperm quality following bacterial lipopolysaccharide treatment. Syst Biol Reprod Med 60(4):217–226

Collodel G, Castellini C, Lee JCY, Signorini C (2020a) Relevance of fatty acids to sperm maturation and quality. Oxid Med Cell Longev. https://doi.org/10.1155/2020/7038124

Collodel G, Simona M, Elena M, Daniela C, Lucia M, Anna Ida F, Laura M, Gabriele B, Cesare C (2020b) Oxidative and/or inflammatory thrust induced by silver nanoparticles in rabbits: Effect of vitamin E or NSAID administration on semen parameters. Mediators Inflamm. https://doi.org/10.1155/2020/6664062

Cook HW, McMaster CR (2002) Fatty acid desaturation and chain elongation in eukaryotes. In: Vance DE, Vance JE (eds) New comprehensive biochemistry, vol 36. Elsevier, Amsterdam, pp 181–204

Corkidi G, Taboada B, Wood CD, Guerrero A, Darszon A (2008) Tracking sperm in three-dimensions. Biochem Biophys Res Commun 373(1):125–129

Dal Bosco A, Scuota S, Castellini C, Cenci T (1996) Study of an artificial vagina to reduce the microbial contamination of rabbit semen. World Rabbit Sci 4(4)

Dal Bosco A, Rebollar PG, Boiti C, Zerani M, Castellini C (2011) Ovulation induction in rabbit does: current knowledge and perspectives. Anim Reprod Sci 129(3–4):106–117. https://doi.org/10.1016/j.anireprosci.2011.11.007

De Lamirande EVE, Gagnon C (1984) Origin of a motility inhibitor within the male reproductive tract. J Androl 5(4):269–276

Dott HM, Harrison RAP, Foster GCA (1979) The maintenance of motility and the surface properties of epididymal spermatozoa from bull, rabbit and ram in homologous seminal and epididymal plasma. Reproduction 55(1):113–124

Fanti A (1979) Silver rabbit of Champagne [France]. Rivista di Coniglicoltura

Fernandez JA, Heeb MJ, Radtke KP, Griffin MJ (1997) Potent blood coagulant activity of human semen due to prostasome-bound tissue factor. Biol Reprod 56(3):757–763

Fischer B, Chavatte-Palmer P, Viebahn C, Navarrete Santos A, Duranthon V (2012) Rabbit as a reproductive model for human health. Reproduction 144(1):1

Foote RH, Carney EW (2000) The rabbit as a model for reproductive and developmental toxicity studies. Reprod Toxicol 14(6):477–493

Gadella BM, Tsai PS, Boerke A, Brewis IA (2004) Sperm head membrane reorganisation during capacitation. Int J Dev Biol 52(5–6):473–480

García-Tomás M, Sánchez J, Rafel O, Ramon J, Piles M (2006) Heterosis, direct and maternal genetic effects on semen quality traits of rabbits. Livest Sci 100(2–3):111–120

Gliozzi TM, Zaniboni L, Maldjian A, Luzi F, Maertens L, Cerolini S (2009) Quality and lipid composition of spermatozoa in rabbits fed DHA and vitamin E rich diets. Theriogenology 71(6): 910–919

Halo M Jr, Bułka K, Antos PA, Greń A, Slanina T, Ondruška Ľ, Tokárová K, Massányi M, Formicki G, Halo M, Massányi P (2021) The effect of ZnO nanoparticles on rabbit spermatozoa motility and viability parameters in vitro. Saudi J Biol Sci 28(12):7450–7454

Holgado-Madruga M, Moscatello DK, Emlet DR, Dieterich R, Wong AJ (1997) Grb2-associated binder-1 mediates phosphatidylinositol 3-kinase activation and the promotion of cell survival by nerve growth factor. Proc Natl Acad Sci USA 94(23):12419–12424

Jamrozik E, Heriot GS (2022) Ethics and antibiotic resistance. Br Med Bull 141(1):4

Juyena NS, Stelletta C (2012) Seminal plasma: an essential attribute to spermatozoa. J Androl 33(4):536–551

Kelly RW (1995) Contraception: Immunosuppressive mechanisms in semen: Implications for contraception. Hum Reprod 10(7):1686–1693

Kozma C, Macklin W, Cummins LM, Mauer R (1974) Chapter 3—Anatomy, physiology, and biochemistry of the rabbit. In: Weisbroth SH, Flatt RE, Kraus AL (eds) The biology of the laboratory rabbit, 1st edn. Academic, New York, pp 49–72. ISBN: 9780127421506. https://doi.org/10.1016/B978-0-12-742150-6.50008-6

Kubovicova E, Makarevich AV, Balazi A, Vasicek J, Chrenek P (2022) Factors affecting rabbit sperm cryopreservation: a mini-review. Zygote (Cambridge, England) 30(1):1–8

Laghouati A, Belabbas R, Castellini C, Mattioli S, Dal Bosco A, Benberkane A, Iguer-Ouada M (2021) Impact of Algerian date palm pollen aqueous extract on epididymal and ejaculated rabbit sperm motility during in vitro incubation. Ital J Anim Sci 20(1):717–727

Lavara R, Vicente JS, Baselga M (2013) Genetic variation in head morphometry of rabbit sperm. Theriogenology 80(4):313–318

Lwaleed BA, Greenfield R, Stewart A, Birch B, Cooper AJ (2004) Seminal clotting and fibrinolytic balance: a possible physiological role in the male reproductive system. Thromb Haemost 92(10):752–766

Mann T, Lutwak-Mann C (1981) Biochemistry of seminal plasma and male accessory fluids; application to andrological problems. In: Mann T, Lutwak-Mann C (eds) Male reproductive function and semen. Springer, London, pp 269–336

Maranesi M, Zerani M, Leonardi L, Pistilli A, Arruda-Alencar J, Stabile AM, Rende M, Castellini C, Petrucci L, Parillo F, Moura A, Boiti C (2015) Gene expression and localization of NGF and its cognate receptors NTRK 1 and NGFR in the sex organs of male rabbits. Reprod Domest Anim 50(6):918–925

Maranesi M, Zerani M, Silva M (2022) Editorial: new insights on seminal factors signaling female reproduction in mammals. Front Vet Sci 9:893379

Marchiani S, Vignozzi L, Filippi S, Gurrieri B, Comeglio P, Morelli A, Danza G, Bartolucci G, Maggi M, Baldi E (2015) Metabolic syndrome-associated sperm alterations in an experimental rabbit model: relation with metabolic profile, testis and epididymis gene expression and effect of tamoxifen treatment. Mol Cell Endocrinol 401:12–24

Marco-Jiménez F, Vicente JS (2017) Overweight in young males reduce fertility in rabbit model. PLoS One 12(7):e0180679

Mattioli S, Dal Bosco A, Maranesi M, Petrucci L, Rebollar PG, Castellini C (2019) Dietary fish oil and flaxseed for rabbit does: fatty acids distribution and Δ6-desaturase enzyme expression of different tissues. Animal 13(9):1934–1942

Mattioli S, Collodel G, Signorini C, Cotozzolo E, Noto D, Cerretani D, Micheli L, Fiaschi AI, Brecchia G, Menchetti L, Moretti E, Oger C, De Felice C, Castellini C (2021a) Tissue antioxidant status and lipid peroxidation are related to dietary intake of n-3 polyunsaturated acids: A rabbit model. Antioxidants 10(5):681

Mattioli S, Maranesi M, Castellini C, Dal Bosco A, Lorenzo PL, Rebollar PG, García-García RM (2021b) Physiology and modulation factors of ovulation in rabbit reproduction management. World Rabbit Sci 29(4):221–229

Mattioli S, Dimauro C, Cesarani A, Dal Bosco A, Bartolini D, Galli F, Migni A, Sebastiani B, Signorini C, Oger C, Collodel G, Castellini C (2022) A dynamic model for estimating the interaction of ROS–PUFA–antioxidants in rabbit. Antioxidants 11(3):531

Menchetti L, Barbato O, Sforna M, Vigo D, Mattioli S, Curone G, Tecilla M, Riva F, Brecchia G (2020) Effects of diets enriched in linseed and fish oil on the expression pattern of toll-like receptors 4 and proinflammatory cytokines on gonadal axis and reproductive organs in rabbit buck. Oxid Med Cell Longev. https://doi.org/10.1155/2020/4327470

Minelli A, Moroni M, Castellini C (2001) Isolation and purification of the IGF-I protein complex from rabbit seminal plasma: Effects on sperm motility and viability. J Exp Zool 290(3):279–290

Mocé E, Vicente JS (2009) Rabbit sperm cryopreservation: a review. Anim Reprod Sci 110(1–2): 1–24

Mourvaki E, Collodel G, Moretti E, Cosci I, Castellini C (2008) Distribution of alpha-, gamma (+ beta)-and delta-tocopherol in the seminal plasma, spermatozoa and seminal vesicles of rabbit. Andrologia 40(5):282–285

Mourvaki E, Cardinali R, Dal Bosco A, Castellini C (2010a) In vitro antioxidant activity of the prostatic secretory granules in rabbit semen after exposure to organic peroxides. Reprod Biol Endocrinol 8(1):1–7

Mourvaki E, Cardinali R, Roberti R, Dal Bosco A, Castellini C (2010b) Desmosterol, the main sterol in rabbit semen: distribution among semen subfractions and its role in the in vitro spermatozoa acrosome reaction and motility. Asian J Androl 12(6):862

Mourvaki E, Cardinali R, Dal Bosco A, Corazzi L, Castellini C (2010c) Effects of flaxseed dietary supplementation on sperm quality and on lipid composition of sperm subfractions and prostatic granules in rabbit. Theriogenology 73(5):629–637

Naughton CK, Nelson DR, Thomas AJ Jr (2003) Development of an inexpensive artificial vagina for semen collection from rabbits. J Androl 24(5):712–715

Nikolopoulou M, Soucek DA, Vary JC (1986) Lipid composition of the membrane released after an in vitro acrosome reaction of epididymal boar sperm. Lipids 21(9):566–570

Orgebin-Crist MC (1973) Maturation of spermatozoa in the rabbit epididymis: effect of castration and testosterone replacement. J Exp Zool 185(3):301–309

Panella F, Castellini C, Facchin E (1994) Heritability of some male reproductive traits in rabbit. In: Baselga M, Marai IFM (eds) Rabbit production in hot climates. Cahiers Options Mediterraneennes (CIHEAM), Zaragoza

Piles M, Tusell L (2012) Genetic correlation between growth and female and male contributions to fertility in rabbit. J Anim Breed Genet 129(4):298–305

Piles M, Tusell L, Lavara García R, Baselga Izquierdo M (2013) Breeding programmes to improve male reproductive performance and efficiency of insemination dose production in paternal lines: feasibility and limitations. World Rabbit Sci 21(2):61–75

Poulos A, Darin-Bennett A, White IG (1973) The phospholipid-bound fatty acids and aldehydes of mammalian spermatozoa. Comp Biochem Physiol B Comp Biochem 46(3):541–549

Prins GS, Zaneveld LJ (1979) Distribution of spermatozoa in the rabbit vas deferens. Biol Reprod 21(1):181–185

Punnett RC, Bailey PG (1918) Genetic studies in poultry. J Genet 7:203–213

Rebollar PG, Alvariño JMR, Ubilla E (1994) Grouping of rabbit reproduction management by means of artificial insemination. World Rabbit Sci 2(3)

Rémésy C, Demigne C, Morand C (1995) Metabolism of short-chain fatty acids. In: Cummings JH, Rombeau JL, Sakata T (eds) Physiological and clinical aspects of short-chain fatty acids. Cambridge University Press, Cambridge, pp 171–190

Rodríguez M, Rebollar PG, Mattioli S, Castellini C (2019) n-3 PUFA sources (precursor/products): A review of current knowledge on rabbit. Animals 9(10):806

Samouilidis S, Saoulidis K, Foukos A, Ypsilantis P, Demertzis A, Sbiraki AP (2001) Freezing of rabbit semen after the addition of glycerol and dimethylsulfoxide in the extender. J Hellenic Vet Med Soc 52(4):299–302

Sengupta P, Borges E Jr, Dutta S, Krajewska-Kulak E (2018) Decline in sperm count in European men during the past 50 years. Hum Exp Toxicol 37(3):247–255

Sieme H, Oldenhof H (2015) Cryopreservation of semen from domestic livestock. In: Wolkers WF, Oldenhof H (eds) Cryopreservation and freeze-drying protocols. Springer, New York, pp 277–287

Swierstra EE, Foote RH (1963) Cytology and kinetics of spermatogenesis in the rabbit. J Reprod Fertil 5(3):309–322

Tvrdá E, Greifová H, Ďuračka M, Ondruška Ľ, Halenár M, Kolesárová A (2022) Comparative analysis of the detrimental in vitro effects of three fusariotoxins on the selected structural and functional characteristics of rabbit spermatozoa. Drug Chem Toxicol 45(6):2519–2527

Verstegen J, Iguer-Ouada M, Onclin K (2002) Computer assisted semen analyzers in andrology research and veterinary practice. Theriogenology 57(1):149–179

Assisted Reproductive Technologies

Rosa M. Garcia-Garcia, Daniela Jordán-Rodríguez, Silvia Gimeno-Martos, Pilar G. Rebollar, Pedro L. Lorenzo, and María Arias-Alvarez

Abstract

Rabbit is a valuable animal model in biomedical research due to many practical advantages, including reproductive ones. Assisted reproductive techniques (ARTs) are very useful tools to facilitate the progression of these studies in rabbit model. However, the majority of ARTs are not well developed in this species, and some problems remain unsolved. A brief overview of current ARTs for using rabbits as animal model for biomedical studies is addressed in this chapter.

Keywords

Oocyte · Embryo · In vitro oocyte maturation · Sperm capacitation · In vitro fertilization · ICSI · In vitro embryo culture · Embryo freezing · Embryo transfer · Gene editing · Assisted reproductive technology

R. M. Garcia-Garcia (✉) · D. Jordán-Rodríguez · P. L. Lorenzo
Department of Physiology, Veterinary Faculty, Complutense University of Madrid, Madrid, Spain
e-mail: rosa.garcia@vet.ucm.es

S. Gimeno-Martos
Department of Physiology, Veterinary Faculty, Complutense University of Madrid, Madrid, Spain

Department of Biochemistry, Veterinary Faculty, University of Zaragoza, Zaragoza, Spain

P. G. Rebollar
Department of Agrarian Production, Technical School of Agronomic, Alimentary and Biosystems Engineering, Technical University of Madrid, Madrid, Spain

M. Arias-Alvarez
Department of Animal Production, Veterinary Faculty, Complutense University of Madrid, Madrid, Spain

J. Simões, J. M. Monteiro (eds.), *Veterinary Care of Farm Rabbits*,
https://doi.org/10.1007/978-3-031-44542-2_13

1 Introduction

Rabbit (*Oryctolagus cuniculus*) is a valuable animal model for different approaches, including reproductive studies (Fischer et al. 2012; Lopez-Tello et al. 2019). In the XIX century, it was used to develop the rabies vaccine by Louis Pasteur (Esteves et al. 2018). For many years, the rabbit has been the preferred species for polyclonal antibody production, and over 40% of the total polyclonal antibody used in research labs and medical clinics has been produced in this species. In the late 1970s, rabbit models contributed to the development of statins prescribed annually for millions of patients worldwide for the treatment of atherosclerosis, since rabbits were used for the study of molecular and cellular mechanisms of this disease (Shiomi 2020). Rabbit model can offer accurate answers for any given biomedical question rather than mouse models that reflects a "use what is available and affordable" approach rather than a "use what is physiologically accurate and translational" approach (Xu et al. 2021). Then, this species is progressively becoming a translational model fulfilling the gap between rodent and larger animal models with exceptional advantages for biomedical research (Fan and Watanabe 2003; Ogonuki et al. 2005).

Rabbits have unique characteristics (Püschel et al. 2010). Unlike pigs and non-human primates, rabbits are relatively inexpensive and can be easily adapted to existing research facility infrastructures. They are easy to breed and handle and are recognized by the scientific and regulatory communities as a well-established laboratory-friendly model species. Compared to rodents, rabbits are phylogenetically closer to humans (Li et al. 1990; Graur et al. 1996) and share biochemical and physiological features with humans. Additionally, the relatively large size of rabbits makes it much easier to perform surgical procedures, collect serial blood samples, and perform tissue and organ biopsies than in a mouse. In addition, some reproductive characteristics, such as ovulation induction (that make available accurate timing of pregnancy), short pregnancy, and placental development (very similar) to humans, allow the rabbit to be considered an animal model to study the physiological processes associated with reproduction and to other research, such as eye, cardiovascular diseases, diabetes, or obesity (Fischer et al. 2012; Lopez-Tello et al. 2019) (Box 1). They are living bioreactors (i.e., milk therapeutic protein expression can be measured) (Meng et al. 2013).

Box 1: Specific Reproductive Features of Rabbit Females Useful to Be Chosen as Animal Model

- The timing of ovulation and oocyte age is predictable and controllable.
- Easy application of superovulation protocols.
- Oocytes show clear ooplasm and are not fragile.
- Embryo size is high enough for easy manipulation.
- Embryo transmigration between both uterine horns is not possible.
- Short gestational period (30–32 days).

(continued)

Box 1 (continued)

- Females produce large litters.
- Hemochorial placentation.
- Milk measurement is easy to perform.
- High reproductive performance.

Assisted reproductive technologies (ARTs) include in vitro embryo production (IVP), intracytoplasmic sperm injection (ICSI), embryo cryopreservation, embryo transfer, gene editing and cloning. Recently, gene-edited rabbit models have allowed a great opportunity in biomedical research due to their characteristics, such as their relatively large size and long lifespan compared to mouse models, and because rabbit models can mimic several diseases better than their mouse counterparts (Xu et al. 2021). ARTs are useful tools in cloning studies, basic research, and genetic progress. The development of suitable in vitro oocyte maturation (IVM), in vitro fertilization (IVF), and in vitro embryo culture (IVC) techniques in this species is crucial to produce somatic cell nuclear transfer (SCNT) embryos for animal breeding, preservation of genetic resources, developmental biology studies and as a model system for other animals and human reproduction. Therefore, numerous pioneering achievements in early reproductive physiological events such as fertilization or techniques such as in vivo and in vitro embryo culture and in vitro sperm capacitation were first studied in rabbits. However, the efficiency of ARTs in this species is poor and limits their practical application. This is due to the difficulty of achieving optimal rates of in vitro oocyte maturation and fertilization, among others. Currently, there is limited information about the cellular and molecular mechanisms controlling oocyte maturation and early embryo development. Therefore, some challenges should be faced before the wide application of ARTs in biomedical research and for commercial use (Cao et al. 2022). The objective of this chapter is to highlight the current state of the art of ARTs in rabbit species.

The key points of this chapter are reported in Box 2.

Box 2: Key Points

- Rabbits are a classic animal model in biomedicine studies.
- Assisted reproductive technologies are not well developed in rabbit species and have limitations.
- A variety of superovulation protocols can be used to maximize in vivo embryo production.
- Embryos can be recovered by flushing the reproductive tract postmortem or in vivo by laparoscopic techniques.
- In vitro embryo production involves in vitro oocyte maturation (IVM), in vitro fertilization (IVF), and in vitro culture of embryos (IVC).

(continued)

Box 2 (continued)

- IVM of oocytes can be successfully achieved in several media, but oocytes show poorer quality than in vivo-matured oocytes.
- IVF procedures are not adequately standardized.
- ICSI procedures can be used in this species.
- Low capacitation of spermatozoa in the IVF technique leads to low fertilization and poor cleavage and blastocyst rates.
- Several culture media can support the development of rabbit embryos.
- Embryo transfer can be successfully developed, and the morula stage is the preferred embryo stage for embryo transfer.
- Different media, times, and devices are used for embryo freezing.
- Somatic cell nuclear transfer (SCNT) is not efficient in rabbit species.
- Precise modification of the rabbit genome by gene targeting is currently an efficient technique and is opening a new field in biomedical research.

2 In Vivo Oocyte and Embryo Production

In vivo recovered oocytes and embryos are high-quality specimens compared with those produced in vitro (Arias-Álvarez et al. 2017). Therefore, in vivo embryos can be directly transferred to recipient females or used in vitro as controls in ART techniques.

To obtain in vivo oocytes or embryos, donor females should be sexually mature (5–12 months). Even if rabbit females can ovulate a high number of oocytes, superovulation protocols can be used to obtain the maximum number of embryos per donor female. A total of 20–50 embryos per female can be recovered. However, ovarian stimulation treatments provoke an increase in anovulatory follicles, non-survival of embryos after embryo transfer or freezing (Carney and Foote 1990), and even alterations in the expression of genes associated with the acquisition of oocyte maturation and competence, that also affect in vivo embryo viability (Vicente et al. 2022). Ovarian super stimulation treatments include the use of hormones. Some of the described protocols employ (1) equine chorionic gonadotropin (eCG) hormone at higher doses than those used in estrous synchronization protocols (>25 IU; i.e., 100–150 IU) combined with human chorionic gonadotropin (hCG) administration at insemination time, and (2) a combination of gonadotropin follicle stimulating hormone (FSH) and luteinizing hormone (LH). Protocols using FSH (Folltropin-V) every 12/24 h for 2–3 days to stimulate follicular development and hCG with LH action on day 4 have been described (Kane 2019). A recent protocol with long-acting FSH corifollitropin alfa combined with hCG promoted successful ovarian stimulation and survival rate in rabbits (Viudes-de-Castro et al. 2019). Treatment with eCG + hCG is usually less effective than FSH treatment.

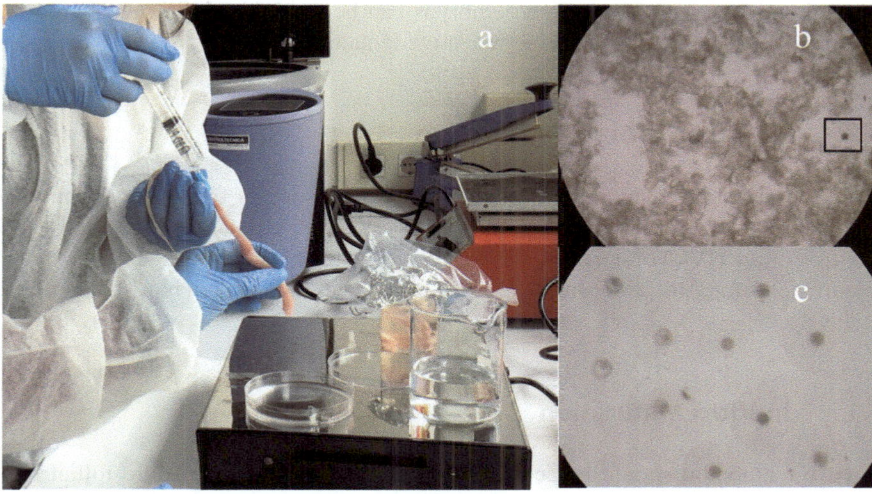

Fig. 1 Postmortem recovery of rabbit embryos. (**a**) Flushing of the uterine horn with warmed phosphate buffered saline (PBS) using a needle and syringe. (**b**) Petri dish under stereoscopic microscopy at low augmentation for search embryos. Uterine cells and an embryo (square) are shown; embryos are searched and picked up with a pipette under a stereoscopic microscope; (**c**) embryos are transferred to a 35 mm Petri dish after washing several times in PBS supplemented with bovine serum albumin (BSA) or polyvinylpyrrolidone (PVP). See large mucin coat surrounding in vivo produced embryos. Modified from Arias-Álvarez (2009)

Embryos can be recovered in rabbit females by flushing the reproductive tract. Laparoscopy techniques can be used. This is a less invasive technique than laparotomy, but anesthesia of animals, an experienced technician, and laparoscopic equipment are needed. The procedure usually lasts 20 min, and 50–75% of embryos can be recovered (Mehaisen et al. 2004). Other technical approaches consist of reproductive tract flushing *postmortem*. Once females are humanely euthanized, the reproductive tract is recovered and flushed. Efficiency for *postmortem* flushing is high as well (80–100%). The oviduct should be flushed to recover embryos at earlier stages (until morulae stage), whereas day 3 embryos may be in the oviduct or uterus or both. For recovering later embryo stages, flushing of the uterus is necessary. In both procedures, embryos are flushed from the oviduct or the uterotubal junction through the uterus in a tempered buffer such as phosphate buffer PBS supplemented with 0.2% bovine serum albumin (BSA), 0.1% polyvinylpyrrolidone (PVP) or polyvinyl alcohol (PVA). Embryos are recovered in a Petri dish plate, searched under a stereoscopic microscope, washed several times in the same medium that is used for embryo recovery, and usually transferred to a culture dish or frozen (Fig. 1). Embryos should be handled in a small fluid volume with a suitable pipette. Washes should include a macromolecule such as BSA to be properly handled, avoiding stickiness to handling pipettes, culture dish surfaces, and air bubbles. Once selected according to International Embryo Transfer Society (IETS) rules depending on their morphology (i.e., shape, color, zona pellucida intactness, accordance to chronological age, etc.), recovered embryos can be used for experimental studies, cryopreserved and/or transferred to receptive females.

3 In Vitro Embryo Production

In vitro embryo production includes the recovery of oocytes from ovaries, in vitro nuclear and cytoplasmic maturation of oocytes (IVM), in vitro fertilization (IVF) and in vitro culture (IVC) of the presumptive zygotes to the blastocyst stage. Some of the requirements for IVM, IVF, and IVC to the blastocyst stage remain non-standardized in this species.

3.1 In Vitro Maturation

In vitro maturation should be accomplished in oocytes from immature follicles to become competent. Meiotic resumption and cytoplasmic maturation with relocation of organelles such as mitochondria, rearrangement of cytoskeletal elements and cortical granule reaction are promoted with IVM. Recovery of fully grown cumulus–oocyte complexes (COCs) is normally performed by aspiration of antral follicles larger than 1 mm present in the ovarian surface (Jelínkova et al. 1994) under a stereoscopic microscope (Fig. 2). Then, selected compact COCs are placed in maturation medium to progress from germinal vesicles to metaphase II of meiosis and to induce changes in the cytoplasm necessary for fertilization and early embryonic development. In this species, nuclear chromatin condensation starts in growing oocytes before meiosis begins (Jelínkova et al. 1994). Then, oocytes exhibit germinal vesicles with bivalent chromosomes before the LH peak, which promotes shorter maturation times than in other species. There is no consensus about the best medium and times for IVM in rabbit oocytes. Some authors have used hormones such as

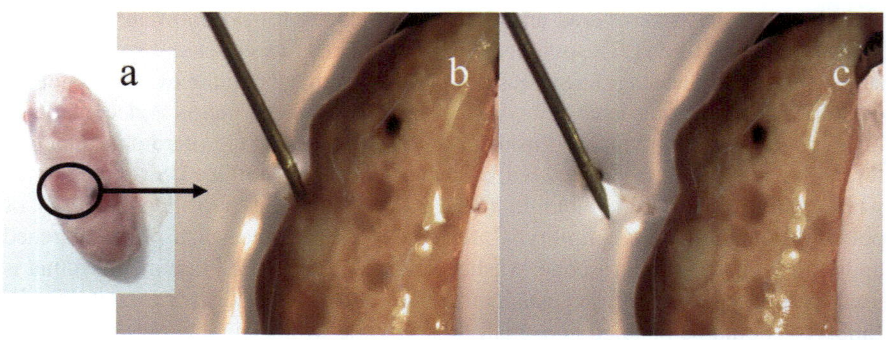

Fig. 2 Oocyte recovery from aspiration of antral follicles in the surface of rabbit ovary. (**a**) Follicle >1 mm to be aspirated; (**b**) rupture of the follicle wall by means of the needle; (**c**) aspiration of the content of the follicle, including the cumulus–oocyte complex (COC) and the follicular fluid. Modified From Arias-Álvarez et al. (2009)

FSH, LH, and estradiol (Ju et al. 2002) or amino acids such as glutamine and proline (Chung et al. 1974). The use of tissue culture medium (TCM 199) supplemented with glutamine, pyruvate, antibiotics, and growth factors such as epidermal growth factor (EGF) and insulin growth factor (IGF 1) for 16–18 h under controlled conditions such as 38.5 °C, 5% CO_2, and 90% humidity resulted in high maturation rates in rabbit oocytes (90% nuclear maturation rate and 70% cytoplasmatic maturation rate in terms of cortical granule migration) (Lorenzo et al. 1997; Arias-Álvarez et al. 2010a). Other factors, such as growth hormone (GH), can be included in the maturation medium in the absence of gonadotropins. GH promotes follicular development and restart of meiosis in COCs directly by binding to these receptors in follicles and oocytes of rabbit does or indirectly through IGF-I stimulation (Lorenzo et al. 1997; Rebollar et al. 2008). In addition, leptin directly stimulates rabbit oocyte nuclear in vitro maturation by activating intracellular MAPK and STAT3 pathways (Arias-Álvarez et al. 2010b).

High rates of nuclear and cytoplasmic maturation can be achieved in this species in vitro, but they are always lower than those in oocytes obtained in vivo. However, blastocysts from oocytes grown in vitro had reduced oxygen consumption and number of cells compared with those from ovulated oocytes (Sugimoto et al. 2012). In fact, the gene expression profile of in vivo-matured rabbit COCs is different from that of IVM COCs. These changes in gene expression are clearly accompanied by stage-specific cytoplasmic remodeling of mitochondria, a lower rate of apoptosis of in vivo-matured cumulus cells (CCs), and higher early embryo development, indicating that oocyte quality is compromised during IVM (Arias-Álvarez et al. 2017). Oxidative stress could be a key factor determining low developmental competence. In vivo-matured COCs are an adequate model in which to identify biomarkers to evaluate the developmental competence of oocytes and to understand the molecular mechanisms underlying the communication between oocytes and CCs to improve IVM systems. In this sense, for example, the supplementation of the IVM medium with α-tocopherol improves rabbit oocyte quality in different ways: it modifies mRNA transcripts, which reflect reduced oxidative stress (SOD2 and CAT), cell damage and apoptotic cascade activation (TP53 and CASP) in oocytes and CCs; it decreases the apoptotic rate in CCs; it reduces the rate of cytoplasmically immature oocytes; and it increases oocyte developmental competence because early embryo development in vitro was improved (Arias-Álvarez et al. 2018).

3.2 In Vitro Fertilization and ICSI

IVF was described very early in rabbit species (Chang 1959) and studied extensively, but despite the importance of rabbits as laboratory animal models, this process is not well standardized with IVM oocytes for optimal results in the IVP system. One of the reasons is the inefficiency of in vitro sperm capacitation and the few studies that have deeply investigated the pathways and molecules involved in rabbit sperm capacitation.

In 1963, it was demonstrated that capacitation could be induced in vitro (Yanagimachi and Chang 1963). Capacitation is a combination of a complex sequence of biochemical and structural events necessary for spermatozoa to acquire the competence to fertilize (Chang 1951; Austin 1952), such as an increase in intracellular Ca^{++} and cAMP, bicarbonate fluctuation (Visconti et al. 2011; Naresh and Atreja 2015), loss of cholesterol content, glycoprotein redistribution and phospholipid asymmetry in the sperm plasma membrane, and an increase in phospholipase activity (Gadella and Harrison 2000), acrosome reaction, protein tyrosine phosphorylation (PTP), and hyperactivation (Fraser 2010). The capacitation process seems to be programmed in mammals according to the time of egg availability in the oviduct to ensure the probabilities that an ovulated egg would meet spermatozoa in the best functional state (Giojalas et al. 2004). This is extended in spontaneous ovulatory species but short in induced ovulators, such as rabbits. In this species, the level of acrosome-reacted spermatozoa is low for approximately 8–10 h, after which it rises. At 16 h post-ejaculation, reacted spermatozoa reach a maximum, and then it declines (Giojalas et al. 2004). This timing of capacitation in rabbit sperm matches the time at which rabbit eggs are in the oviduct (approximately 10–18 h post-ejaculation) (Giojalas et al. 2004). This also agrees with the findings that spermatozoa recovered at 16 h post-copulation from the rabbit uterus have higher fertilizing potential than those recovered after 12 or 24 h (Chang 1955; Harper 1973).

IVF consists of the selection and capacitation of spermatozoa and co-incubation with oocytes. As well as in IVM, there is also no agreement on the best protocol, medium, and timing for in vitro rabbit sperm selection and capacitation. Semen from several bucks may be combined to avoid inter-male variability for IVF. First, protocols for the selection of spermatozoa involve the migration-sedimentation swim-up technique for 20 min (Hauser et al. 1992; Guidobaldi et al. 2008) or, routinely, sperm selection by centrifugation-resuspension (Saez Lancellotti et al. 2010; Arias-Álvarez et al. 2018). In this last swim-up procedure, which includes double centrifugation, the sperm sample is resuspended. Then, in a tube with the sperm sample in the bottom, medium (free of cells) is added on the top of the tube. The tube is inclined 45 degrees and incubated for 20 min in capacitating conditions, allowing motile spermatozoa to swim-up into the medium, where they are collected for the capacitation protocol (Brackett et al. 1982; Zeng et al. 1999; Arias-Álvarez et al. 2018) (Fig. 3).

For sperm capacitation, several protocols have been used. First one was a protocol with defined medium (DM) and high ionic strength (HIS) buffer with ejaculated semen (Brackett and Oliphant 1975; Brackett et al. 1982) or epididymal sperm (Hosoi et al. 1981; Niwa et al. 1983) with 12–22 or 10–10.5 h capacitating incubation and co-incubation with the oocyte for 6 or 1–5 h, respectively. The total cleavage rates in the mentioned conditions were 44% and 73.7%, respectively. The same medium with different shorter sperm capacitation times (6–8 h) and the co-incubation of IVM with sperm for 6–8 h improved the cleavage rate to 91.6% (Zeng et al. 1999). The supplementation of capacitation medium with lysophosphatidyl choline (LPC) before IVF experiments to induce acrosome

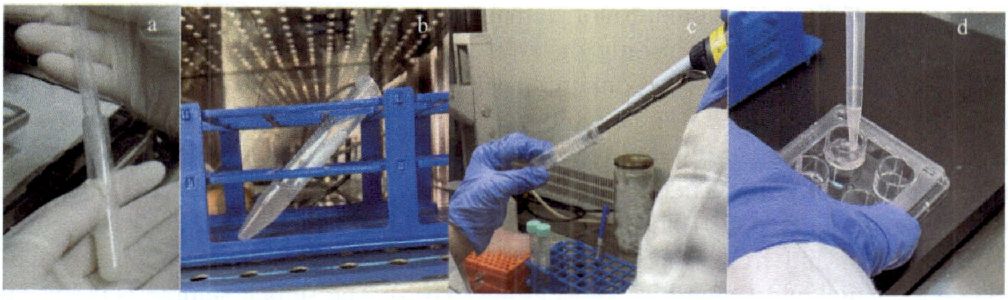

Fig. 3 Standard rabbit spermatozoa preparation for subsequent IVF: (**a**) semen was collected using an artificial vagina and pooled white ejaculates free from gel and sediments to avoid individual differences; (**b**, **c**) sperm selection by swim-up technique in a falcon tube inclined 45 degrees; (**d**) for the induction of in vitro capacitation, aliquots of swim-up-selected spermatozoa were incubated in four-well Petri dishes under capacitation conditions with 5% CO_2, maximum humidity, and 38.5 °C in the incubator

reaction (Kim et al. 1989) did not improve rates of cleavage compared to mentioned previous works of Brackett group (Brackett and Oliphant 1975; Brackett et al. 1982). Another medium studied in in vitro rabbit sperm capacitation for 6 h has been Tyrode medium supplemented with sodium bicarbonate, sodium pyruvate, sodium lactate, fatty acid-free BSA and heparin and the co-incubation of IVM with sperm for 4 h (Arias-Álvarez et al. 2018). However, the cleavage and blastocyst rates were lower in IVM oocytes than in vivo-matured oocytes. Using Tyrode's albumin lactate pyruvate (TALP) medium, most of the spermatozoa were capacitated after 8 h and acrosome-reacted after 12 h (Castellini et al. 2020). Biggers, Whitten, and Whittingham medium (BWW) (Biggers et al. 1971) is another medium used in rabbit sperm capacitation for 16 h (Giojalas et al. 2004; Guidobaldi et al. 2008; Saez Lancellotti et al. 2010). This medium and timing in the in vitro rabbit sperm capacitation showed an increase in phospho-tyrosine proteins (Saez Lancellotti et al. 2010, 2013). However, to the best of our knowledge, there are no IVF experiments in rabbits with these protocols and this medium in the literature.

In addition, few media have been proven as fertilization media for co-incubation with oocytes, such as DM (Brackett et al. 1982; Zeng et al. 1999) and Tyrode supplemented with different additives (albumin, lactate, pyruvate, heparin, hypotaurine, adrenalin) (Viudes-De-Castro et al. 2005; Arias-Álvarez et al. 2017) and different co-incubation times (from 4 to 16 h).

Intracytoplasmic sperm injection (ICSI) is a potent tool to be used in the field of ARTs because it provides opportunities for studying the basic mechanisms of fertilization and early embryo development. ICSI consists of fertilizing a metaphase II (MII) oocyte by the direct injection of a single spermatozoon into the ooplasm, in which the acrosome and membrane remain intact. It could be used for biodiversity conservation, transgenic production (using sperm as vector to introduce genes) or to solve fertilization problems in IVF systems of several species. In rabbits, this technology has considerably supported the gene editing technique (Li et al. 2010a).

For the ICSI procedure, donor females previously superovulated are used. Oocytes are collected by flushing the oviducts with an appropriate medium, such as HEPES-buffered TCM199 containing 10% FCS (Li et al. 2010a, b), PBS+0.1% PVA (Deng and Yang 2001), or Dulbecco PBS (Daniel and Renard 2010). In this technique, it is necessary to remove the cells from the COCs by repetitive pipetting (Deng and Yang 2001; Li et al. 2010a, b) and by using hyaluronidase (Daniel and Renard 2010; Li et al. 2010a) or not (Deng and Yang 2001).

Both fresh and frozen sperm support the cleavage and development of ICSI oocytes (Daniel and Renard 2010; Li et al. 2010b). The semen sample is collected using an artificial vagina (Bredderman et al. 1964; Sugimoto et al. 2012). Daniel and Renard (2010) described a protocol mixing fresh sperm with a semen extender suitable for artificial insemination in rabbits at a 1:4 dilution and stored in the dark for at least 24 h. Then, the post-sedimentation pellet was centrifuged, diluted 1: 2 with D-PBS, and mixed with 10% PVP in D-PBS at a concentration of 1:20. Other authors washed it with PBS 3–4 times by centrifugation and resuspended the pellet in 10% PVP (Deng and Yang 2001) or directly centrifuged and mixed the pellet with mDM medium (Wang et al. 2001) to make a 30-min swim-up. Then, the upper 100 μl of medium was taken and washed twice and diluted to a 1×10^6 spermatozoa/ml (Li et al. 2010a, b). For the use of frozen sperm, it is necessary to thaw the straws, wash sperm with 1 mL of semen extender by centrifuging, and resuspend the pellet in 1 mL of D-PBS; with this suspension, a new dilution (1:3) is made with 10% PVP (Daniel and Renard 2010).

Then, when the samples are processed, it is necessary to prepare the micromanipulator and microdrops in plates. The micromanipulator has two different pipettes that need to be mounted: one is the holding pipette with a suction pump, which will hold the oocytes; the second one is the injection pipette, which will inject the sperm into the oocyte, and it is attached to the microinjector (Daniel and Renard 2010). Different micro drops must be made. The injection microdrop, where oocytes are placed, can be either PBS+0.1%PVA (Deng and Yang 2001) or HM199S medium (Daniel and Renard 2010), and sperm microdrops have either 8–10% PVP (Li et al. 2010a, b; Deng and Yang 2001; Daniel and Renard 2010). An additional microdrop supplemented with 10% PVP to rinse the injection pipette can be prepared (Daniel and Renard 2010). The microdrops are covered with mineral oil.

In the injection microdrop, several oocytes are placed (7–15) (Deng and Yang 2001; Daniel and Renard 2010). First, the spermatozoon is aspirated tail first with the injection pipette, and the pipette is moved into a rinse microdrop (Daniel and Renard 2010) or directly into the injection microdrop (Deng and Yang 2001; Li et al. 2010a, b). In this injection microdrop, the oocyte is held with the holding pipette at the right position (the first polar body at 6 o'clock (Li et al. 2010a, b) or 12 o'clock (Daniel and Renard 2010)). The spermatozoon must be placed near the tip of the injection pipette. For the injection, the injection pipette should be pushed forward perpendicularly to the polar body until the spermatozoon is in contact with the oocyte plasma membrane. At this moment, the plasma membrane must be broken with repeated aspiration and expulsion with the microinjector. When the injection pipette is inside, the injected spermatozoon should be expelled in the center of the oocyte with a minimum amount of medium (Daniel and Renard 2010). Once the pipette has entered, some authors report that they aspirate a small amount of cytoplasm to ensure that the sperm has entered into the cytoplasm (Deng and Yang 2001). Several hours after ICSI (6–7 h), the sperm nucleus undergoes decondensation and forms a pronucleus. Then, two pronuclei and a second polar body can be observed in activated embryos.

In rabbits as well as in other species (Keefer 1989), activation of the oocyte in the ICSI procedure is not necessary. Indeed, a high percentage of fertilization (78%) and blastocyst development in vitro (39%) can be achieved when ICSI, without additional activation treatment, is used in in vivo-matured oocytes. The birth of live offspring has also been reported (Deng and Yang 2001). The rates of cleavage and blastocyst formation of ICSI of in vivo oocytes are significantly increased when artificial activation is applied following ICSI. The most effective method to activate rabbit oocytes for ICSI is using one pulse of electrical stimulation prior to sperm injection (Zhou et al. 2013). Double activation with ionomycin calcium I0634 may increase freeze–thawed oocyte developmental potential. After activation, oocyte cleavage development may be faster than that of oocytes without activation (Wang et al. 2009). Using IVM matured oocytes, the fertilization rate after ICSI is usually similar to that IVF (Lan et al. 2008). However, blastocyst rates are significantly higher when in vivo-matured oocytes are fertilized by ICSI than when oocytes are IVM (65.7 vs. 25.5%).

After embryo transfer, offspring were obtained from ICSI oocytes (Deng and Yang 2001; Li et al. 2010b; Zhou et al. 2013). Spermatozoa derived from infertile rabbits can successfully fertilize oocytes and produce offspring by ICSI as well. Some factors can affect embryo viability after ICSI, such as oocyte aging because of delayed sperm injection. This leads to failure of male pronucleus formation and a lower fertilization rate (Zheng et al. 2004) or lower rabbit embryo development after SCNT (Cervera and García-Ximénez 2003).

3.3 In Vitro Culture

At the end of the 1960s, Cornell University developed a semi-defined medium supplemented with 1.5% BSA called modified F10 to study the requirements for the development of one-cell rabbit embryos to blastocysts (Kane 2019). In contrast to other species, early cleavage block in rabbit embryos was not observed (Kane 1972). The embryonic genome activation (EGA) phase is usually a critical point in embryo culture because embryos shift from maternal transcripts synthesized and stored throughout oogenesis to their own mRNA synthesis resulting from the activity of the zygotic genome. However, in rabbits, EGA seems to be progressive (Brunet-Simon et al. 2001), and some culture media can support development without arrest (Kane 2019). However, some authors have reported that EGA occurs at the 8–16 cell stage (Telford et al. 1990).

Generally, presumptive zygotes from IVF or in vivo recovered zygotes are cultured to the blastocyst stage for 72–84 h. IVP embryos are not able to form the

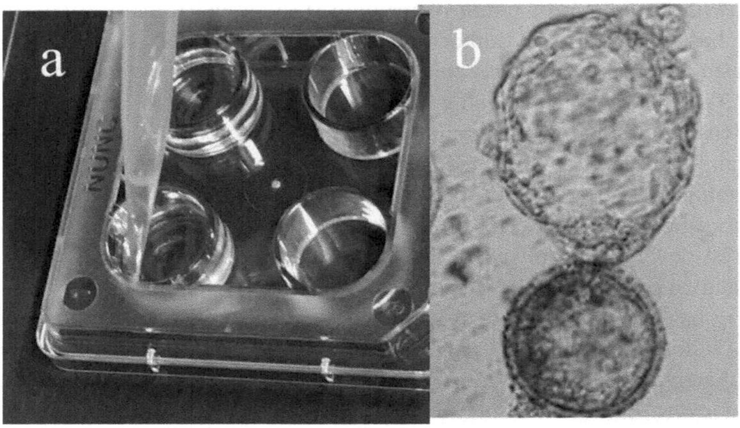

Fig. 4 (**a**) Microdrop covered by mineral oil; (**b**) in vitro cultured rabbit blastocyst where the zona pellucida is broken and part of the embryo cellular mass has gone out. Note the absence of mucin coat

mucin coat, and non-real hatching is observed, although the pellucida zone can be ruptured in vitro (Fig. 4). Hatched embryos expand to 500–700 μm diameter (Kane 2019). Implantation of cultured transfer embryos is compromised by the lack of mucin coat, but it could be previously transferred to the oviduct to be cultured and then recovered and transferred to the uterus of recipient females (Murakami and Imai 1996). Several culture media have been described, such as TCM-199, modified F10, Ham's F10, DMEM, and SOF, supplemented with 15–20% FBS or 1–2% BSA. Amino acids, citrate, and vitamins such as inositol are important for blastocyst formation in rabbit embryos because of their great growth until the blastocyst stage (Kane 2019). The phosphatidyl inositol signaling system is important for rabbit blastocyst growth (Fahy and Kane 1993). Mineral oils are usually used to cover microdrops where embryos are cultured. However, there is no consensus about the IVC medium to be used in rabbit species.

Embryos are cultured under a controlled atmosphere (38.5 °C, 5% CO2 and 5% O2, 90% humidity). However, blastocyst size after expansion is lower than in in vivo embryos around implantation (400–600 μm vs. 5 mm) (Denker 2000). Some studies have revealed that implantation genes and apoptotic genes, such as the Bax gene, are differentially expressed between in vivo and in vitro embryos (Henderson et al. 2014).

According to Kane (2019), blastocysts can be recorded as early blastocysts (when the blastocoel is just visible), blastocysts (when the blastocoel is at least half the size of the embryo), and expanding blastocysts (when the embryo is expanded).

IVC is an important tool for rabbit nuclear transfer. It is essential to be optimized to obtain repeatable cloning efficiency. New bioengineering tools, such as extracellular matrix (ECM) hydrogels and coatings from endometrial cells under ovarian hormone stimulation, are being developed for the in vitro culture of rabbit embryos as a promising tool (Campo et al. 2019).

4 Embryo Transfer

Embryo transfer consists of the transfer of preimplantation embryos to the uterus of pseudopregnant recipient females by laparotomy or laparoscopy. In this species, embryo transfer is mainly important for ARTs and cloning. One advantage is that rabbit doe shows two separate uterine horns, and each horn has its own cervix. There is no transmigration of embryos, as in other species, and then, experimental embryos can be transferred into the same doe at the same time in different horns. Moreover, the small size of rabbits, short gestation, and easy management make rabbits an ideal species for embryo transfer at a lower cost than other species.

In vivo or in vitro embryos can be transferred to recipient dams, usually from maternal lines (Marco-Jiménez et al. 2013). Non-surgical embryo transfer is not available in rabbits, and laparoscopic surgery is usually in use (for review, see Garcia-Dominguez et al. 2019) (Fig. 5). Recipient females are synchronized with female donors. First, females can be induced to ovulate by i.m. GnRH administration (22 h later than donor), but they are not inseminated or mated, and then oocytes are not fertilized (they will be reabsorbed). Pseudopregnant females show functional corpora lutea that can maintain gestation when embryos are transferred. However, in superovulation protocols, hemorrhagic follicles can appear, and females will not be suitable as recipients (Fig. 5). Embryos can be transferred to the oviduct at different times after ovulation induction depending on the embryo stage. For example, morulae are transferred at 48 h to the uterus (Vicente and Garciá-Ximénez 1993) or at 60 h for the blastocyst stage (Mehaisen et al. 2006) by surgical techniques. Proper synchronization of the age of the embryo with the recipient uterus is an essential point. At least five embryos are placed in each oviduct or horn (range 13–15 embryos) to promote gestation. Overcrowding of the uterus by the transfer of too many embryos can reduce progeny survival (Hafez 1964; Besenfelder and Brem 1993). Morula seems to be the more suitable stage for rabbit embryo recovery and transfer because these embryos can be successfully vitrified and laparoscopically transferred at the uterine horns (Garcia-Dominguez et al. 2019). Survival rates vary from 65.3 to 45.9% for compacted morula and non-compacted morula of in vivo-obtained embryos (Garcia-Dominguez et al. 2019).

Transfer of in vivo-obtained embryos is more successful than in vitro-produced embryo transfer. As stated before, in vitro embryos fail to develop a mucin coat, and higher failures of pregnancy are produced when they are transferred into recipients (Joung et al. 2004). However, the birth of normal young from IVF embryos transferred to recipients was reported early (Seidel et al. 1976).

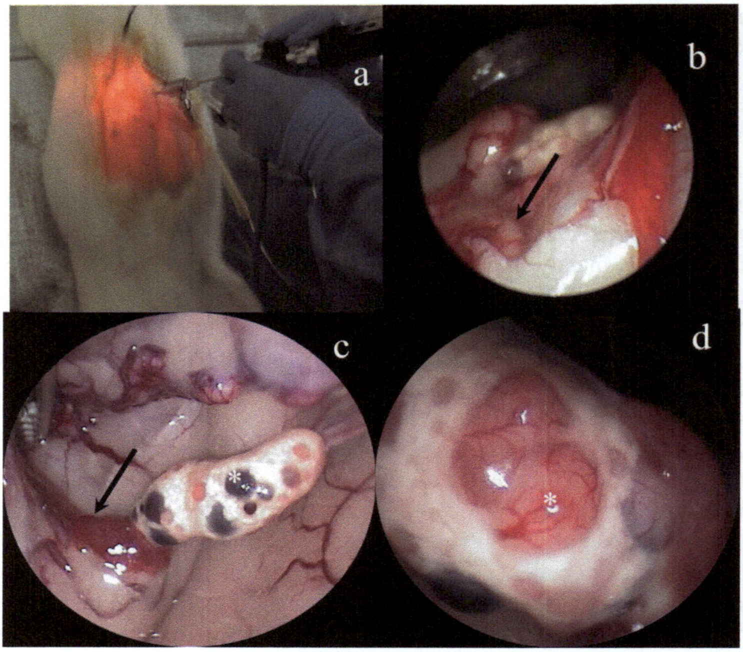

Fig. 5 Endoscopic procedure. (**a**) Rabbit female in recumbency position head down at 45 degrees and first endoscopic trocar inserted; (**b**) view of the ovary and oviduct just after the entrance of the trocar into the abdominal cavity. Arrow indicates the oviduct; (**c**) ovary showing an hemorrhagic follicle (asterisk) from a female not suitable for embryo transfer. Arrow indicates the infundibulum where the catheter should be inserted to place the embryos; (**d**) ovary showing a corpus luteum (asterisk) from a female suitable as recipient for embryo transfer

5 Embryo Freezing

Cryopreservation of rabbit embryos was a procedure early developed (Bank and Maurer 1974). This technique is a useful tool for the long-term preservation of genetic material in biobanks, rederivation of lines, ART, and cloning experiments. However, cryopreservation programs are not used regularly in rabbit species. Some problems derived from manipulation, such as oxidative stress, impair successful freezing. The genotypes of the embryo and the genetics of the recipient female affect the efficiency of the procedure (Vicente and Garciá-Ximénez 1993) as well as the housing and feeding conditions (Bruyère et al. 2013).

Rabbit embryos should be cryopreserved beyond the eight-cell stage because early stages are highly sensitive to the cryopreservation process (Lin et al. 2011). However, cryopreserved oocytes transferred are able to generate live pups (Jiménez-Trigos et al. 2014). Compacted morulae are the most suitable stage for successful cryopreservation (Garcia-Dominguez et al. 2019). The live-born rate (number of living pups/number of frozen-thawed embryos transferred) largely varies in the literature (35–38%, (Teixeira et al. 2018); 35–40% (Marco-Jiménez et al. 2016); 46–65% (Garcia-Dominguez et al. 2019). In addition, rabbits produced by vitrification and embryo transfer provoked some epigenetic changes, such as higher body weight in the first 4 weeks of age than control progeny (Lavara et al. 2015).

Conventional slow freezing and vitrification are the methods for embryo cryopreservation. In rabbit embryos, vitrification seems to be the best technique for cryopreservation, as the zona pellucida and mucin coat are frequently damaged during conventional freezing/thawing (Kasai et al. 1992). Implantation and birth rates improve for rabbit vitrified morula compared to the slow freezing method (Saenz-de-Juano et al. 2014). The vitrification technique promotes an amorphous state during cooling, avoiding intracellular ice crystal formation, as occurs in slow cooling. For vitrification, different protocols, vitrification media, combinations of cryoprotectants (CPA) and concentrations, and devices have been successfully employed. Media for embryo cryopreservation are commonly supplemented with SFB or BSA, although defined media use macromolecules such as hyaluronic acid, PVP, PVA, and Ficoll is also described (Teixeira et al. 2018). Vitrification protocols include one or two steps with higher solute concentrations of CPA than slow freezing, shorter equilibration times in CPA solution, and minimum volume surrounding embryos. The most commonly used combination of cryoprotectants is dimethyl sulfoxide (DMSO)+ ethylene glycol (ETG) or ETG+glycerol, but numerous combinations and equilibration times can be found. Embryos are allocated in different devices, such as open pull straws (OPS), cryotops, and French mini straws (0.25 ml), etc. (López-Béjar and López-Gatius 2002; Marco-Jiménez et al. 2013, 2016; Teixeira et al. 2018) and then directly plunged into liquid nitrogen. After storage in liquid nitrogen, vitrified embryos are rapidly warmed, and the vitrification solution is removed in buffer medium supplemented with sucrose. After several whases of embryos in buffer medium for embryo re-expansion, embryos with mucin coat and pellucida zona are transferred or cultured in vitro.

6 Gene Editing, Transgenesis, and Cloning

Successful nuclear transfer (NT) introducing embryonic cells from preimplantational embryos into enucleated oocytes was pioneering in rabbit species in the 1980–1990s (Stice and Robl 1988; Collas and Robl 1990). However, cloning is relatively more difficult in this species, and the first clone was not produced until 2002 (Chesné et al. 2002). Genome manipulation was delayed for many years until the discovery of genome editing techniques. Currently, the rabbit model has attracted many researchers to investigate several diseases because the completion of the rabbit genome project and the availability of genetic engineering technology such as gene editing precisely allow modification of the specific gene sequence. Then, the rabbit has again returned to at the forefront of advanced biomedical research.

Transgenesis in rabbits consists of the microinjection of exogenous DNA into pronuclear stage oocytes to be transmitted through the germ line. Embryo transfer of the transgenic embryos to recipient females should be performed later. However, random insertion of exogenous genetic material limited the technique. In addition, pluripotent stem cell line-mediated transgenesis and somatic cell nuclear transfer (SCNT)-mediated transgenesis are inefficient in rabbit species. Generation of competent pluripotent stem cell lines is difficult in rabbits, contrary to mice (Tancos et al. 2012). In addition, SCNT, by which a nucleus from a differentiated somatic cell is transferred into an enucleated oocyte to obtain a cloned individual (Wilmut et al. 2015), is also a complex and inefficient technique in rabbit species. This is probably due to the abnormal reprogramming of somatic cells in vitro (Du et al. 2009; Cao et al. 2022). Rabbit clones were generated using cultured adult fibroblasts, cumulus cells and genetically manipulated mesenchymal stem cells (Meng et al. 2013). Many studies use in vivo recovered oocytes to improve success because they are more competent. In fact, oocyte age and nuclear donor cell type influence the developmental rate of rabbit NT embryos in vitro (Cervera and García-Ximénez 2003; Du et al. 2009). In addition, electrically mediated membrane fusion is more effective for embryonic development than nuclear injection in cloned rabbit for NT (Du et al. 2009). The most successful protocol for somatic cell cloning is the use of two electrical pulses followed by a short chemical treatment to decrease meiosis promoting factor (MPF) activity (Meng et al. 2013).

However, new gene editing nuclease elements such as clustered regularly interspaced short palindromic repeats-CRISPR-associated nuclease 9 (CRISPR/Cas9), zinc-finger nucleases (ZFNs), and transcription activator-like endonucleases (TALENs) systems that cut the genome in a specific targeted position are designed and synthesized for generating the targeted genetic changes in rabbits. This comprises the injection of gene editing elements (e.g., Cas9 protein or its coding DNA, mRNA, guide RNA, and donor DNA) into pronuclear stage embryos followed by in vitro validation, embryo transfer (bypassing the need for germline transmitting stem cells), and genotyping of the founders. Gene targeting technology avoids the problems of transgenesis because DNA can be targeted and integrated into a specific locus by homologous recombination in pluripotent stem cells. Direct injection of the engineered endonucleases together with or without donor DNA into pronuclear stage embryos using CRISPR/Cas9 has proven very efficient for cloning. For the first time in 2014, the CRISPR/Cas9 system was used to develop gene KO in rabbit development models of hypercholesterolemia (Yang et al. 2014). Currently, CRISPR/Cas9 is the technique of choice, and the expansion of generated KO rabbits via these technologies is being used worldwide.

7 Conclusions

Many pioneering achievements in reproductive biology come from rabbits because this species plays a central role as a biomedical model for many diseases and for their use in biomedicine (antibody production, bioreactors, etc.). However, many ARTs are not sufficiently successful. Standardizing and optimizing the media and protocols are still needed. Advancements in ARTs in this species would expand the use of rabbits as an undeniable biomedical resource.

References

Arias-Álvarez M (2009) Influencia de diferentes estrategias reproductivas y nutricionales sobre la fisiologia ovarica en conejas primíparas. PhD thesis. Universidad Complutense de Madrid, Madrid, Spain; 266 pp. ISBN: 978-84-693-2400-4

Arias-Álvarez M, García-García RM, Rebollar PG et al (2009) Effects of a lignin-rich fibre diet on productive, reproductive and endocrine parameters in nulliparous rabbit does. Livest Sci 123: 107–115. https://doi.org/10.1016/j.livsci.2008.10.013

Arias-Álvarez M, García-García RM, Torres-Rovira L et al (2010a) Influence of hormonal and nonhormonal estrus synchronization methods on follicular and oocyte quality in primiparous lactating does at early postpartum period. Theriogenology 73:26–35. https://doi.org/10.1016/j.theriogenology.2009.07.017

Arias-Álvarez M, García-García RM, Torres-Rovira L et al (2010b) Influence of leptin on in vitro maturation and steroidogenic secretion of cumulus-oocyte complexes through JAK2/STAT3 and MEK 1/2 pathways in the rabbit model. Reproduction 139. https://doi.org/10.1530/REP-09-0309

Arias-Álvarez M, García-García RM, López-Tello J et al (2017) In vivo and in vitro maturation of rabbit oocytes differently affects the gene expression profile, mitochondrial distribution, apoptosis and early embryo development. Reprod Fertil Dev 29:1667–1679. https://doi.org/10.1071/RD15553

Arias-Álvarez M, Garciá-Garciá RM, López-Tello J et al (2018) α-Tocopherol modifies the expression of genes related to oxidative stress and apoptosis during in vitro maturation and enhances the developmental competence of rabbit oocytes. Reprod Fertil Dev 30:1728–1738. https://doi.org/10.1071/RD17525

Austin CR (1952) The 'capacitation' of the mammalian sperm. Nature 170:326–326. https://doi.org/10.1038/170326a0

Bank H, Maurer RR (1974) Survival of frozen rabbit embryos. Exp Cell Res 89:188–196. https://doi.org/10.1016/0014-4827(74)90201-8

Besenfelder U, Brem G (1993) Laparoscopic embryo transfer in rabbits. J Reprod Fertil 99:53–56. https://doi.org/10.1530/jrf.0.0990053

Biggers JD, Whitten WK, Whittingham DG (1971) The culture of mouse embryos in vitro. In: Daniel JC (ed) Methods in mammalian embryology. WH Freeman, San Francisco, pp 86–116

Brackett BG, Oliphant G (1975) Capacitation of rabbit spermatozoa in vitro. Biol Reprod 12:260–274

Brackett BG, Bousquet D, Dressel MA (1982) In vitro sperm capacitation and in vitro fertilization with normal development in the rabbit. J Androl 3:402–411. https://doi.org/10.1002/j.1939-4640.1982.tb00710.x

Bredderman PJ, Foote RH, Yassen AM (1964) An improved artificial vagina for collecting rabbit semen. J Reprod Fertil 7:401–403

Brunet-Simon A, Henrion G, Renard JP, Duranthon V (2001) Onset of zygotic transcription and maternal transcript legacy in the rabbit embryo. Mol Reprod Dev 58:127–136. https://doi.org/10.1002/1098-2795(200102)58:2<127::aid-mrd1>3.0.co;2-a

Bruyère P, Baudot A, Joly T et al (2013) A chemically defined medium for rabbit embryo cryopreservation. PLoS One 8. https://doi.org/10.1371/journal.pone.0071547

Campo H, García-Domínguez X, López-Martínez S et al (2019) Tissue-specific decellularized endometrial substratum mimicking different physiological conditions influences in vitro embryo development in a rabbit model. Acta Biomater 89:126–138. https://doi.org/10.1016/j.actbio.2019.03.004

Cao W, Zhao J, Qu P, Liu E (2022) Current progress and prospects in rabbit cloning. Cell Reprog 24:63–70

Carney EW, Foote RH (1990) Effects of superovulation, embryo recovery, culture system and embryo transfer on development of rabbit embryos in vivo and in vitro. J Reprod Fertil 89:543–551

Castellini C, Mattioli S, Dal Bosco A et al (2020) Nerve growth factor receptor role on rabbit sperm storage. Theriogenology 153:54–61. https://doi.org/10.1016/j.theriogenology.2020.04.042

Cervera RP, García-Ximénez F (2003) Oocyte age and nuclear donor cell type affect the technical efficiency of somatic cloning in rabbits. Zygote 11:151–158. https://doi.org/10.1017/S0967199403002181

Chang MC (1951) Fertilizing capacity of spermatozoa deposited into the fallopian tubes. Nature 168:697–698. https://doi.org/10.1038/168697b0

Chang MC (1955) Development of fertilizing capacity of rabbit spermatozoa in the uterus. Nature 175:1036–1037. https://doi.org/10.1038/1751036a0

Chang MC (1959) Fertilization of rabbit ova in vitro. Nature 184:466–467. https://doi.org/10.1038/184466a0

Chesné P, Adenot PG, Viglietta C et al (2002) Cloned rabbits produced by nuclear transfer from adult somatic cells. Nat Biotechnol 20:366–369

Chung OS, Bae I, Cho WK (1974) The maturation in vitro of the rabbit oocytes. Yonsei Med J 15: 11–16

Collas P, Robl JM (1990) Factors affecting the efficiency of nuclear transplantation in the rabbit embryo. Biol Reprod 43:877–884

Daniel N, Renard JP (2010) Artificial fertilization in rabbits by intracytoplasmic sperm injection (ICSI). Cold Spring Harb Protoc 5. https://doi.org/10.1101/pdb.prot5359

Deng M, Yang X (2001) Full term development of rabbit oocytes fertilized by intracytoplasmic sperm injection. Mol Reprod Dev 59:38–43. https://doi.org/10.1002/mrd.1005

Denker HW (2000) Structural dynamics and function of early embryonic coats. Cells Tissues Organs 166:180–207. https://doi.org/10.1159/000016732

Du F, Xu J, Zhang J et al (2009) Beneficial effect of young oocytes for rabbit somatic cell nuclear transfer. Cloning Stem Cells 11:131–140. https://doi.org/10.1089/clo.2008.0042

Esteves PJ, Abrantes J, Baldauf HM et al (2018) The wide utility of rabbits as models of human diseases. Exp Mol Med 50:1–10. https://doi.org/10.1038/s12276-018-0094-1

Fahy MM, Kane MT (1993) Incorporation of [3H]inositol into phosphoinositides and inositol phosphates by rabbit blastocysts. Mol Reprod Dev 34:391–395. https://doi.org/10.1002/mrd.1080340407

Fan J, Watanabe T (2003) Inflammatory reactions in the pathogenesis of atherosclerosis. J Atheroscler Thromb 10:63–71

Fischer B, Chavatte-Palmer P, Viebahn C et al (2012) Rabbit as a reproductive model for human health. Reproduction 144:1–10

Fraser LR (2010) The "switching on" of mammalian spermatozoa: molecular events involved in promotion and regulation of capacitation. Mol Reprod Dev 77:197–208. https://doi.org/10.1002/mrd.21124

Gadella BM, Harrison RAP (2000) The capacitating agent bicarbonate induces protein kinase A-dependent changes in phospholipid transbilayer behavior in the sperm plasma membrane. Development 127:2407–2420

Garcia-Dominguez X, Marco-Jimenez F, Viudes-de-Castro MP, Vicente JS (2019) Minimally invasive embryo transfer and embryo vitrification at the optimal embryo stage in rabbit model. J Vis Exp 2019:1–9. https://doi.org/10.3791/58055

Giojalas LC, Rovasio RA, Fabro G et al (2004) Timing of sperm capacitation appears to be programmed according to egg availability in the female genital tract. Fertil Steril 82:247–249. https://doi.org/10.1016/j.fertnstert.2003

Graur D, Duret L, Gouy M (1996) Phylogenetic position of the order Lagomorpha (rabbits, hares and allies). Nature 379:333–335

Guidobaldi HA, Teves ME, Uñates DR et al (2008) Progesterone from the cumulus cells is the sperm chemoattractant secreted by the rabbit oocyte cumulus complex. PLoS One 3. https://doi.org/10.1371/journal.pone.0003040

Hafez ES (1964) Effects of over-crowding in utero on implantation and fetal development in the rabbit. J Exp Zool 156:269–287. https://doi.org/10.1002/jez.1401560304

Harper MJK (1973) Relationship between sperm transport and penetration of eggs in the rabbit oviduct. Biol Reprod 8:441–450

Hauser R, Homonnai G, Paz GF et al (1992) Migration sedimentation technique as a predictive test for the fertilizing capacity of spermatozoa in an in-vitro fertilization programme. Int J Androl 15:498–503. https://doi.org/10.1111/j.1365-2605.1992.tb01143.x

Henderson GRW, Brahmasani SR, Yelisetti UM et al (2014) Candidate gene expression patterns in rabbit preimplantation embryos developed in vivo and in vitro. J Assist Reprod Genet 31:899–911. https://doi.org/10.1007/s10815-014-0233-0

Hosoi Y, Niwa K, Hatanaka S, Iritani A (1981) Fertilization in vitro of rabbit eggs by epididymal spermatozoa capacitated in a chemically defined medium. Biol Reprod 24:637–642

Jelínkova L, Kubelka M, Motlfk J, Guerrier P (1994) Chromatin condensation and histone H1 kinase activity during growth and maturation of rabbit oocytes. Mol Reprod Dev 37:210–215

Jiménez-Trigos E, Vicente JS, Marco-Jiménez F (2014) First pregnancy and live birth from vitrified rabbit oocytes after intraoviductal transfer and in vivo fertilization. Theriogenology 82:599–604. https://doi.org/10.1016/j.theriogenology.2014.05.029

Joung SY, Kim HJ, Choi WS et al (2004) Effects of transferring in vitro-cultured rabbit embryos to recipient oviducts on mucin coat deposition, implantation and development. Zygote 12:215–219. https://doi.org/10.1017/S0967199404002795

Ju JC, Chen TH, Tseng JK et al (2002) Cytoskeletal patterns, in vitro maturation and parthenogenetic development of rabbit GV oocytes. Asian Aust J Anim Sci 15:1695–1701

Kane MT (1972) Energy substrates and culture of single cell rabbit ova to blastocysts. Nature 238: 468–469

Kane MT (2019) Culture of preimplantation rabbit embryos. In: Herrick J (ed) Comparative embryo culture. Methods in molecular biology. Springer, New York, pp 63–91

Kasai M, Hamaguchi Y, Zhu SE et al (1992) High survival of rabbit morulae after vitrification in an ethylene glycol-based solution by a simple method. Biol Reprod 46(6):1042–1046

Keefer CL (1989) Fertilization by sperm injection in the rabbit. Gamete Res 22:59–69. https://doi.org/10.1002/mrd.1120220107

Kim CK, Im KS, Zheng X, Foote RH (1989) In vitro capacitation and fertilizing ability of ejaculated rabbit sperm treated with lysophosphatidylcholine. Gamete Res 22:131–141

Lan C, Xiaohui D, Qingzhao F et al (2008) Developmental potential of oocytes fertilized by conventional in vitro fertilization (IVF) or intracytoplasmic sperm injection (ICSI) after cryopreservation and mesometrial autotransplantation of rabbit ovarian tissue. Animal 2:1371–1376. https://doi.org/10.1017/S1751731108002528

Lavara R, Baselga M, Marco-Jiménez F, Vicente JS (2015) Embryo vitrification in rabbits: Consequences for progeny growth. Theriogenology 84:674–680. https://doi.org/10.1016/j.theriogenology.2015.04.025

Li W-H, Gouy M, Sharp PM et al (1990) Molecular phylogeny of Rodentia, Lagomorpha, Primates, Artiodactyla, and Carnivora and molecular clocks (mammalian phylogeny/DNA sequence trees/branching dates). Proc Natl Acad Sci USA 87:6703–6707

Li Q, Hou J, Wang S et al (2010a) Production of transgenic rabbit embryos through intracytoplasmic sperm injection. Zygote 18:301–307. https://doi.org/10.1017/S0967199410000250

Li QY, Hou J, Chen YF, An XR (2010b) Full-term development of rabbit embryos produced by ICSI with sperm frozen in liquid nitrogen without cryoprotectants. Reprod Domest Anim 45: 717–722. https://doi.org/10.1111/j.1439-0531.2009.01340.x

Lin TA, Chen CH, Sung LY et al (2011) Open-pulled straw vitrification differentiates cryotolerance of in vitro cultured rabbit embryos at the eight-cell stage. Theriogenology 75:760–768. https://doi.org/10.1016/j.theriogenology.2010.10.018

López-Béjar M, López-Gatius F (2002) Nonequilibrium cryopreservation of rabbit embryos using a modified (sealed) open pulled straw procedure. Theriogenology 58:1541–1552. https://doi.org/10.1016/S0093-691X(02)01045-2

Lopez-Tello J, Arias-Alvarez M, Gonzalez-Bulnes A, Sferuzzi-Perri AN (2019) Models of Intrauterine growth restriction and fetal programming in rabbits. Mol Reprod Dev 86:1781–1809

Lorenzo PL, Illera JC, Silván G et al (1997) Steroid-level response to insulin-like growth factor-1 in oocytes matured in vitro. J Reprod Immunol 35(1):11–29

Marco-Jiménez F, Lavara R, Jiménez-Trigos E, Vicente JS (2013) In vivo development of vitrified rabbit embryos: Effects of vitrification device, recipient genotype, and asynchrony. Theriogenology 79:1124–1129. https://doi.org/10.1016/j.theriogenology.2013.02.008

Marco-Jiménez F, Jiménez-Trigos E, Almela-Miralles V, Vicente JS (2016) Development of cheaper embryo vitrification device using the minimum volume method. PLoS One 11:1–9. https://doi.org/10.1371/journal.pone.0148661

Mehaisen GMK, Vicente JS, Lavara R (2004) In vivo embryo recovery rate by laparoscopic technique from rabbit does selected for growth rate. Reprod Domest Anim 39:347–351. https://doi.org/10.1111/j.1439-0531.2004.00526.x

Mehaisen GMK, Viudes-De-Castro MP, Vicente JS, Lavara R (2006) In vitro and in vivo viability of vitrified and non-vitrified embryos derived from eCG and FSH treatment in rabbit does. Theriogenology 65:1279–1291. https://doi.org/10.1016/j.theriogenology.2005.08.007

Meng Q, Polgar Z, Tancos Z et al (2013) Cloning of rabbits. In: Cibelli J, Wilmut I, Jaenisch R, Gurdon J, Lanza R, West M, Campbell KHS (eds) Principles of cloning, 2nd edn. Academic, New York, pp 227–244. https://doi.org/10.1016/B978-0-12-386541-0.00018-7

Murakami H, Imai H (1996) Successful implantation of in vitro cultured rabbit embryos after uterine transfer: A role for mucin. Mol Reprod Dev 43:167–170. https://doi.org/10.1002/(SICI)1098-2795(199602)43:2<167::AID-MRD5>3.0.CO;2-P

Naresh S, Atreja SK (2015) The protein tyrosine phosphorylation during in vitro capacitation and cryopreservation of mammalian spermatozoa. Cryobiology 70:211–216. https://doi.org/10.1016/j.cryobiol.2015.03.008

Niwa K, Hosoi Y, Ohara K, Iritani A (1983) Fertilization in vitro of rabbit eggs with or without follicular cells by epididymal spermatozoa capacitated in a chemically defined medium. Anim Reprod Sci 6:143–149

Ogonuki N, Inoue K, Miki H et al (2005) Differential development of rabbit embryos following microinsemination with sperm and spermatids. Mol Reprod Dev 72:411–417. https://doi.org/10.1002/mrd.20363

Püschel B, Daniel N, Bitzer E et al (2010) The rabbit (Oryctolagus cuniculus): A model for mammalian reproduction and early embryology. Cold Spring Harb Protoc 5. https://doi.org/10.1101/pdb.emo139

Rebollar PG, Bonanno A, Di Grigoli A et al (2008) Endocrine and ovarian response after a 2-day controlled suckling and eCG treatment in lactating rabbit does. Anim Reprod Sci 104:316–328. https://doi.org/10.1016/j.anireprosci.2007.02.018

Saenz-de-Juano M, Marco-Jimenez F, Viudes-de-Castro M et al (2014) Direct comparison of the effects of slow freezing and vitrification on late blastocyst gene expression, development, implantation and offspring of rabbit morulae. Reprod Domest Anim 49:505–511. https://doi.org/10.1111/rda.12320

Saez Lancellotti TE, Boarelli PV, Monclus MA et al (2010) Hypercholesterolemia impaired sperm functionality in rabbits. PLoS One 5. https://doi.org/10.1371/journal.pone.0013457

Saez Lancellotti TE, Boarelli PV, Romero AA et al (2013) Semen quality and sperm function loss by hypercholesterolemic diet was recovered by addition of olive oil to diet in rabbit. PLoS One 8. https://doi.org/10.1371/journal.pone.0052386

Seidel GE, Bowen RA, Kane MT (1976) In vitro fertilization, culture, and transfer of rabbit ova. Fertil Steril 27(7):861–870

Shiomi M (2020) The history of the WHHL rabbit, an animal model of familial hypercholesterolemia (II)-Contribution to the development and validation of the therapeutics for hypercholesterolemia and atherosclerosis. J Atheroscler Thromb 27:119–131. https://doi.org/10.5551/jat.RV17038-2

Stice SL, Robl JM (1988) Nuclear reprogramming in nuclear transplant rabbit embryos. Biol Reprod 39:657–664

Sugimoto H, Kida Y, Miyamoto Y et al (2012) Growth and development of rabbit oocytes in vitro: Effect of fetal bovine serum concentration on culture medium. Theriogenology 78:1040–1047. https://doi.org/10.1016/j.theriogenology.2012.04.007

Tancos Z, Nemes C, Polgar Z et al (2012) Generation of rabbit pluripotent stem cell lines. Theriogenology 78:1774–1786

Teixeira M, Commin L, Gavin-Plagne L et al (2018) Rapid cooling of rabbit embryos in a synthetic medium. Cryobiology 85:113–119. https://doi.org/10.1016/j.cryobiol.2018.07.006

Telford NA, Watson AJ, Schultz GA (1990) Transition from maternal to embryonic control in early mammalian development: a comparison of several species. Mol Reprod Dev 26:90

Vicente JS, Garciá-Ximénez F (1993) Effect of recipient doe genotype on survival rate at birth of frozen rabbit embryos. Reprod Nutr Dev 33:229–234

Vicente JS, Marco-Jiménez F, Pérez-García M et al (2022) Oocyte quality and in vivo embryo survival after ovarian stimulation in nulliparous and multiparous rabbit does. Theriogenology 189:53–58. https://doi.org/10.1016/j.theriogenology.2022.06.003

Visconti PE, Krapf D, De La Vega-Beltrán JL et al (2011) Ion channels, phosphorylation and mammalian sperm capacitation. Asian J Androl 13:395–405

Viudes-De-Castro MP, Mocé E, Vicente JS et al (2005) In vitro Evaluation of in vivo Fertilizing Ability of Frozen Rabbit Semen. Reprod Domest Anim 40:136–140

Viudes-de-Castro MP, Marco-Jiménez F, Más Pellicer A et al (2019) A single injection of corifollitropin alfa supplemented with human chorionic gonadotropin increases follicular recruitment and transferable embryos in the rabbit. Reprod Domest Anim 54:696–701. https://doi.org/10.1111/rda.13411

Wang HJ, Lin AX, Zhang ZC, Chen YF (2001) Expression of porcine growth hormone gene in transgenic rabbits as reported by green fluorescent protein. Anim Biotechnol 12:101–110. https://doi.org/10.1081/ABIO-100108336

Wang HL, Sui HS, Liu Y et al (2009) Dynamic changes of germinal vesicle chromatin configuration and transcriptional activity during maturation of rabbit follicles. Fertil Steril 91:1589–1594. https://doi.org/10.1016/j.fertnstert.2008.10.071

Wilmut I, Bai Y, Taylor J (2015) Somatic cell nuclear transfer: Origins, the present position and future opportunities. Philos Trans Roy Soc B Biol Sci 370. https://doi.org/10.1098/rstb.2014.0366

Xu J, Zhang J, Yang D et al (2021) Gene editing in rabbits: unique opportunities for translational biomedical research. Front Genet 12:1–12. https://doi.org/10.3389/fgene.2021.642444

Yanagimachi R, Chang MC (1963) Fertilization of hamster eggs in vitro. Nature 200:281–282. https://doi.org/10.1038/200281b0

Yang D, Xu J, Zhu T et al (2014) Effective gene targeting in rabbits using RNA-guided Cas9 nucleases. J Mol Cell Biol 6:97–99. https://doi.org/10.1093/jmcb/mjt047

Zeng SM, Zhu SE, Wang YS et al (1999) An efficient method for in vitro fertilization in rabbits. Anim Biotechnol 10:15–23. https://doi.org/10.1080/10495399909525918

Zheng Y-L, Jiang M-X, Zhang Y-L et al (2004) Effects of oocyte age, cumulus cells and injection methods on in vitro development of intracytoplasmic sperm injection rabbit embryos. Zygote 12:75–80. https://doi.org/10.1017/S0967199404002643

Zhou X, Yin M, Jiang W et al (2013) Electrical activation of rabbit oocytes increases fertilization and embryo development by intracytoplasmic sperm injection using sperm from deceased male. J Assist Reprod Genet 30:1605–1610. https://doi.org/10.1007/s10815-013-0113-z

Rabbit Meat Processing and Safety

Pramila Umaraw, Akhilesh K. Verma, V. P. Singh, Pavan Kumar, N. Mehta, and Devendra Kumar

Abstract

Rabbit meat is a highly nutritious and healthy meat. It is a lean meat with a low-fat (1.8–8.8%) content and less saturated fatty acids than other meats. It is also associated with low cholesterol (47 mg/100 g), high desirable unsaturated fatty acids (approximately 60% of total fat), better digestibility, and low allergenicity. It is rich in high biological value proteins, lean, and low in saturated fatty acids and cholesterol, which makes it an ideal meat for the current scenario where the meat sector is thriving to overcome its high carbon and water footprints. The present chapter exemplifies the quality characteristics of rabbit meat and elaborates its production and consumption and current scenario along with its processing, viz., preslaughter operations (such as loading and unloading of animals, transport of animals, lairage, and stunning), slaughtering, dressing, and cut-up parts. It also discusses the application of advanced technologies such as irradiation, sonication, high-pressure processing (HPP), and superchilling in rabbit meat processing. It also attempts to explore the strengths, weaknesses, opportunities, and threats (SWOT) of the rabbit meat/processing industry, as rabbit meat has the potential to become a sustainable meat source for future generations.

P. Umaraw (✉) · A. K. Verma · V. P. Singh
Department of Livestock Products Technology, College of Veterinary and Animal Sciences, Sardar Vallabhbhai Patel University of Agriculture and Technology, Meerut, Uttar Pradesh, India
e-mail: pramila.vet@svpuat.edu.in

P. Kumar · N. Mehta
Department of Livestock Products Technology, College of Veterinary Science, Guru Angad Dev Veterinary and Animal Sciences University, Ludhiana, Punjab, India

D. Kumar
Division of Livestock Products Technology, Indian Veterinary Research Institute, Bareilly, Uttar Pradesh, India

J. Simões, J. M. Monteiro (eds.), *Veterinary Care of Farm Rabbits*,
https://doi.org/10.1007/978-3-031-44542-2_14

Keywords

Production · Processing · Advanced technologies · SWOT of rabbit meat

1 Introduction

Meat has been part of the human diet for a long time. It is nutrient dense and provides most of the required essential components, such as high protein, minerals, and vitamins. It has been a major component of diet in Europe, America, South America, Russia, Australia, etc. Traditionally, when day-to-day working required more physical activities, the amount or quantity of meat was the major factor addressing human needs. However, with industrialization, modernization, and now artificial intelligence, the level of physical activity has decreased drastically, leading to changes in diet composition and requirements. It is now the quality of meat that determines human needs. Earlier high-fat diets were desired and were considered delicacies, as fat enhances the flavor and palatability of food and provides abundant energy. Currently, "fat" and red meat are regarded as culprits of most lifestyle diseases, such as cancers (Khan et al. 2010), cardiovascular diseases (Mota et al. 2019), and Alzheimer's disease (Amelianchik et al. 2022). Lifestyle diseases or non-communicable diseases (NCDs) are linked with consumption of high quantity and inferior quality fat, injudicious consumption of alcohol and tobacco, and lack of physical activities (Bennett et al. 2018).

The meat industry, with allegations and correlations to lifestyle diseases and cancers, has shifted toward the production of lean carcasses and products with functional components (Mota et al. 2019). Other than the production of lean carcasses, alternative meat sources have also been explored for the sustainability of the meat sector. Alternative meat sources, such as meat from deer, antelopes, rabbits, and hares, which are prevalent in localized areas and underutilized, have received much attention (Verma et al. 2017). The present chapter discusses in detail one such alternative meat source that has traditionally been localized in niche areas of Europe, China, Korea, and Africa. The rabbit meat industry has the potential to become a sustainable meat source for future generations.

2 Rabbit Meat

Rabbit meat has been part of the regular diet in countries such as France, Spain, and Italy in the Mediterranean region and China and Korea in the Eastern region and Africa. It is rich in high biological value proteins, lean, and low in saturated fatty acids and cholesterol, which makes it an ideal meat for the current scenario where the meat sector is thriving to overcome its high carbon and water footprints (Li et al. 2018). Its similarity to chicken meat and lower arable land and potable water dependency make it an ideal sustainable meat source. It is a monogastric animal with the characteristic behavior of cecotrophy, which enables it to utilize low-quality forages. It thus does not compete with humans for food; rather it converts inedible forages into edible meat (Cullere and Dalle Zotte 2018). It is extremely prolific and has a short gestation period and life cycle with a good feed conversion ratio. Rabbit meat is currently also regarded as functional meat due to its nutritional characteristics (Dalle Zotte and Szendrő 2011). Rabbit keeping or cuniculture can be used as a tool for uplifting poverty and generating employment in rural populations. The rabbit industry has also contributed significantly to the economic and social upliftment of the rural population in several countries (Wu 2022).

3 Current Scenario

Presently, the Republic of China is the largest producer of rabbit meat, followed by Korea. According to the continental distribution, major rabbit production is concentrated in Asia, with China and Korea being the major players producing three-quarters of total production. Europe, which was once the major hub of rabbit meat production and consumption, is now the second largest producer of rabbit meat. Africa has also adopted cuniculture and is utilizing it for socioeconomic development and is presently the third largest producer of rabbit meat (Cullere and Dalle Zotte 2018). It is expected that the market for rabbit meat will increase in the future. According to some market research analysts, the consumption of rabbit meat would increase at a CAGR of +2.3% for 2017–2025, which is estimated to be approximately 1.8 million tonnes by 2025 (Index Box Market survey; Euromeat.com).

The major market for rabbit meat consumption is China, consuming approximately 62% of the rabbit meat produced, followed by North Korea and Egypt (Euromeat news.com). Rabbit meat trade is also restricted to Asian regions (China, North Korea), Africa (Egypt), Mediterranean regions (Algeria, Cyprus, Egypt, France, Italy, Malta, Portugal, and Spain), and European countries (Belgium, Czech Republic, Germany, and Luxembourg). Li et al. (2018) reported that China, being the largest producer of rabbit meat, also dominates the export market of rabbit meat. During 2010–2016, China exported 5800–11,000 tonnes per year of global rabbit meat export. China, one of the first countries to export rabbit meat, presently exports approximately 27–31% of rabbit meat to the world, mainly from Shandong, Jilin, and Shanxi provinces. The major importing countries are Belgium, Germany, the Russian Federation, and Portugal (Li et al. 2018; Cullere and Dalle Zotte 2018). Rabbits are presently the second most farmed animal species in the European Union and the fourth most farmed animal species in the world (Voslarova et al. 2018). The market categories of rabbits popular in European countries are given in Table 1.

4 Quality Characteristics of Rabbit Meat

The dressing percentage of rabbit meat varies with breed. A dressing percentage of 56–58% was reported in meat breeds, which increases with larger breeds (Pla et al. 1998; Dalle Zotte et al. 2015). Rabbit meat is highly valued for its nutritional and

Table 1 Market category of rabbits

S. no.	Category name	Description
On the basis of age		
1.	Very young rabbit	Bunnies can be weaned as soon as they are out of the nest box and eating the doe's feed or eating feed from a creep feeder. Weaning time will vary between breeds but can range from 4 to 5 weeks
2.	Young rabbit	Less than 12 weeks of age
3.	Young mature rabbit	12 weeks to 6 months of age
4.	Mature rabbit	More than 6 months of age
5.	Open doe	More than 6 months of age, unfertilized mature rabbit
On the basis of production system		
1.	Free range	Rabbits are raised in heated and either ventilated or open-sided growing houses with access to the outdoors
2.	Pastured/ pasture-raised	Rabbits are raised outdoors utilizing movable enclosures located on grass after 3 weeks
3.	Conventional	Rabbits are raised in cages indoors
4.	Organic	Production methods conform to the legislation of the importing country concerning organic production

Source: UNECE Standard Rabbit Meat Carcasses and Cuts 2013 Edition

dietary properties; it is a lean meat with a low-fat (1.8–8.8%) content and less saturated fatty acids than other meats. It is also associated with low cholesterol (47 mg/100 g), high desirable unsaturated fatty acids (approximately 60% of total fat), better digestibility, and low allergenicity (Lan et al. 2016). Rabbit meat has a desirable fatty acid composition with approximately 35.6–37.9% saturated fatty acids (SFAs), 28.7–31.0% monounsaturated fatty acids (MUFAs), and 27.5–35.3% polyunsaturated fatty acids (PUFAs), which varies with feed. It is a good source of n-6 fatty acids, especially linoleic acid (C18:2 n-6), which constitutes approximately 22% of the total fat and is largely dependent on diet. The correlation of the α-linolenic acid content of food and rabbit meat is high, with $R^2 = 0.532$ (Hernández and Dalle Zotte 2020). It is also a good source of high biological value proteins (18.6–22.4%) with functional components such as lecithin; essential amino acids, viz., lysine (2.1 g/100 g), threonine (2 g/100 g), leucine (1.7 g/100 g), and phenylalanine (1 g/100 g); and minerals (1.2–1.3 mg/100 g), and the water-soluble vitamin B complex is especially rich in vitamin B12 (8.7–11.9%) (Dalle Zotte and Cullere 2019; Hernández and Dalle Zotte 2020). The sodium content of rabbit meat is also naturally low (37–47 mg/100 g meat), which makes it more desirable for people suffering from hypertension, high blood pressure, etc.

Cambero et al. (1991) reported that fat from meat of New Zealand white and commercial hybrid HYLA have comparatively higher palmitic acid, linoleic acid, and myristic acid than other species meats, and factors such as age, sex, breed, and feed can influence the lipid profile of meat. Similarly, Pla et al. (1998) observed that the fat percent in the longissimus muscle of rabbits is lower than that found in the breast muscle of chickens, which may be as low as 1% (extracted fat). In a study on different lines of rabbits, Pla et al. (1998) reported that the genetic origin of rabbits dominated the tenderness in the loin meat in rabbits more than the body weight. Between sexes, the meat of female rabbits was more acidic and retained more water than that of male rabbits. The adipose tissue content was also higher in female rabbits. The effect of genotype on the carcass characteristics and meat quality of Pannon Large and Hungarian Giant rabbit composition was contemplated by Dalle Zotte et al. (2015). It was observed that only fatty acids such as C12:0, C17:1, and C20:1 n-9 were affected by the genotype, and the effect was more significant on the Pannon Large breed than on the Hungarian Giant. The fatty acid composition of rabbit meat can be easily modulated, as rabbit meat is monogastric.

5 Rabbit Meat Processing

Rabbit meat has traditionally been used fresh with small-scale processing and preparation of carcasses. Most rabbit meat delicacies involve the utilization of whole carcasses owing to their small size, which is sufficient for small families and/or small meals. Thus, major carcass preparation was performed by expert butchers or rabbit rearers. With advancement and mechanization, cuniculture evolved, and the processing of rabbit meat has also developed. Small-scale farms and processing units are now available for processing rabbit meat. Modern processing involves preslaughter operations, such as loading and unloading of animals, transport of animals, lairage, and stunning, slaughtering, evisceration, chilling, and packaging (Cavani and Petracci 2004). Further processing, such as deboning, portioning, formulation, curing, smoking, etc., to develop ready-to-eat or ready-to-cook products has also now come up (Składanowska-Baryza and Stanisz 2019). In light of rabbit meat as a functional food and convenience, the demand for rabbit meat products such as marinated rabbit, sausages, hamburgers, and emulsified rabbit meat products such as patties and nuggets is also increasing.

Rabbit is comparatively less prone to textural and physical alterations by production factors such as PSE and DFD, as observed in pigs, cattle, and/or poultry (Cavani et al. 2009). Nevertheless, long journey duration, heat, long lairage period, etc. have been reported to affect yield and meat quality and increase mortality rates (Składanowska-Baryza and Stanisz 2019). Moreover, animal welfare has become one of the most important extrinsic cues influencing the demand for meat, especially rabbit meat, as rabbit is also categorized as a pet animal in many places. Modernization of rabbit meat processing similar to poultry processing is required to meet the future demand of rabbit meat.

6 Preslaughter Care and Handling

In intensive cuniculture, great care is required for transferring rabbits from their cages to the crates. Rabbits are generally caught by hand, and animals of the desired weight are collected and placed in crates to be transported. Animals to be delivered to processing plants should be collected during cool hours. Fasting is generally followed in food animals to avoid carcass contamination during carcass processing. In large animals, fasting for approximately 12 hours with ad libitum water is practiced to attain the desired carcass yield and meat quality (ultimate pH, restricted undesired microbial proliferation) and to minimize carcass contamination while maintaining animal welfare. A 12 h fasting in rabbits leads to 3–6% body weight loss, which might reach 8–12% in 36–48 h (Cavani and Petracci, 2004). Because rabbits are small in size, high body weight losses are not desirable, and rabbits are fasted for only 6–9 h at maximum. Animals to be transported are fasted before transportation if they are destined for short journey periods (1–4 h). Long transport (>7 h) has been shown to negatively affect meat quality and increase mortality.

According to Verga et al. (2009), critical factors for the transportation of rabbits are travel time, temperature, and relative humidity. In a study on the effect of seasons (summers and winters) on rabbit meat quality, Maпa et al. (2006) observed that seasons significantly influenced meat quality in terms of color, pH, water holding capacity, and WB texture parameters. Meat quality is adversely affected by an increase in travel time over 4 h, environmental temperatures beyond 18–20 °C, and 70–75% relative humidity. Crates with solid floors are preferred for transportation because they increase the number of crates carried on a truck and avoid soiling of rabbits in lower crates.

7 Transport

Plastic or wire mesh cages or crates are used for transportation of rabbits from the farm to the processing plant. These have generally solid floor and wired sides and tops. The top of the creates contains a sliding or lifting door for the transfer of animals within and out of the crate (Fig. 1).

Fig. 1 Crates used for transportation of rabbits (Courtesy of Julián Gullón Álvarez, Cogal S. Coop., Spain)

Fig. 2 Transportation of rabbit from farms to processing units (Courtesy of Julián Gullón Álvarez, Cogal S. Coop., Spain)

A stocking density of 40 kg/m^2 is desirable for housing fattening rabbits, which corresponds to 15 rabbits/m^2, while for transportation, a stocking density of approximately 75.5 kg/m^2 is used. Lambertini et al. (2006) reported that stocking density during transportation (1, 2, and 4 h studied) did not affect the weight loss or slaughter values. Transportation itself is a stress for animals; thus, stocking density and conditions during traveling should always be considered before transportation (Fig. 2). Muscle pH obtained after 24 h of slaughter is regarded as a standard for evaluating the quality of meat carcass and is generally denoted as pH 24. Long transportation generally results in increased pH 24 and ultimate pH and indirectly increases the water holding capacity of meat. Mazzone et al. (2010) observed more than double the serum corticosterone level (14.88 ng/mL) in rabbits transported for slaughtering than that before transportation, which was 6.23 ng/mL. Similarly, Fazio et al. (2015) observed an increase in rectal temperature, blood glucose, lactate, total proteins, and packed cell volume in rabbits transported for 2 and 6 h, but these secondary stress biomarkers decreased after 24 h of rest from transportation. The glucose level was 20 mg/dL lower in animals transported than in farms, and lactate, creatine phosphokinase, and corticosterone levels were more than double (Liste et al. 2009).

Transportation of rabbits for 7 h has been reported to affect the textural parameters measured by compression (Maria et al. 2006). The mortality rate of rabbits during transportation in Italy is lower than that of poultry and is approximately 0.09% (Petracci et al. 2010). In a study on transport-related mortality and the effect of batch size, external air temperature, weather, and transport distance on rabbits in the Czech Republic, Voslarova et al. (2018) reported that smaller batches and favorable or ambient temperatures evinced lower mortality. For a distance of less than 50 km, the mortality rate was 0.02%, which increased to 0.29% for distances over 400 km.

8 Loading and Unloading

In other food animals, loading and unloading are critical points concerning animal welfare. Use of electric goads, cane sticks, etc. for directing animals many times results in injuries to animals affecting the meat quality. Catching and hauling poultry also sometimes increases bruises on skin due to struggle. Such conditions are minimal with rabbits, but improper handling may still cause bruises on the thoracic area, loin, and legs (Składanowska-Baryza and Stanisz 2019). Due to the presence of fur, the bruises are visible only after the removal of pelt or skin. Mazzone et al. (2010) contemplated the effect of smooth and rough handling on rabbit meat quality and observed no significant difference in the slaughter live weight (g), weight losses (%), chilling losses (%), and dressing percentage (%) of rabbits loaded either smoothly or roughly. Rabbits are mostly loaded in a multirack system in trucks. Rabbits are either removed from their cages and placed on the crates to be transported or sometimes are taken near the trucks in their cages and are transferred directly to the loading crate on trucks. The design and size of crates may vary in different places, but EFSA recommends a stocking density of 0.06 m^2 per rabbit. A standard crate generally falls in the range of 110 × 50 × 22 cm to 110 × 60 × 30 cm and can house a maximum of 14–16 rabbits in one crate depending on the size of the rabbits (Verga et al. 2009).

Marıa et al. (2006) observed that the position of rabbits in the truck, such as in the bottom rack, top rack, or middle rack, did not greatly affect the eating quality of meat obtained from such rabbits after giving a lairage time of 3 h. Similarly, no effect was observed in the meat quality of rabbits placed at different positions in a multifloor cage rolling stand (Liste et al. 2009). However, it was observed that stress blood parameters such as creatine kinase were significantly higher in animals kept at the bottom than in those on top of the multifloor cage rolling stand, while their corticosterone level was lowest (Liste et al. 2009).

9 Lairage

On arrival, the crates are unloaded from the truck and placed in the lairage where animals are given rest to mitigate the stress effects of travel. Rabbits after transportation are received at the lairage where they are unloaded and placed for a few hours before slaughter. Rabbit lairages are generally a roofed reception area to protect animals from adverse weather conditions such as extreme sunlight and rain. Some modern abattoirs may have specially designed lairages with forced ventilation and water mist to reduce the stress of travel. A period of 6 h in proper lairage is sufficient to lower the stress level of transported rabbits (Maria et al. 2005). Liste et al. (2009) observed lower corticosterone (ng/dL), glucose (mg/dL), and lactate (mg/dL) levels in rabbits provided with 8 h of rest in lairage than in those given 2 h of rest. Even the presence of bruising on carcasses was significantly lower in animals provided longer resting periods (8 h) in lairage. The longer duration of rest in lairage slightly affected the meat quality, pH 24, and color, but meat texture measured by compression and

Warner-Bratzler analyses revealed a significant effect of lairage time (Liste et al. 2009).

10 Stunning

Stunning of animals before slaughter has been practiced for a long time to prevent unnecessary pain and suffering to animals slaughtered for food purposes. Rabbits were previously hunted, but with advancement, the rearing of rabbits as meat animals and the welfare of food animals have led to the development of various techniques and processes for stunning rabbits. Manual stunning or cervical dislocation of the head is usually performed at small backyard farms, but in modernized plants with high thorough output, standard stunning operations are followed. Stunning of rabbits can be done by electrical (head only), mechanical (penetrative captive bolt, nonpenetrative captive bolt, or percussive blow), or gaseous stunning. In the European Union, concussion, captive bolt pistol, electronarcosis, and carbon dioxide for gaseous stunning are allowed for stunning of rabbits by EU Council Directive 93/119/EC (Rota Nodari et al. 2009). Mostly for small groups of rabbits, the captive bolt method is preferred, while for larger stocks, electronarcosis is widely used. Induction of stunning in rabbits is characterized by breath cessation, increased salivation, and motor activity, while the corneal reflex is considered for monitoring return to sensibility (Rota Nodari et al. 2009). Electric shock is given at the frontal sinus by using V-shaped electrodes that usually work at a voltage of 49 and 250 Hz frequency applied for approximately 2 s (Lafuente and López 2014; Składanowska-Baryza et al. 2020). Stunning of food animals must be immediately followed by sticking. Poststunning animals should be promptly sticked by cutting blood vessels of the left side of the throat or by severing jugular and carotid arteries below the neck followed by exsanguination. The head is removed thereafter. Stunning in ritual slaughtering is exempted.

11 Dressing

Dressing of rabbits varies according to the stock. Rabbits, being small animals, are generally dressed on the board in small farms or at retail butchers. However, with the establishment of cuniculture, modern slaughterhouses have emerged with online processing of rabbit carcasses. Line dressing operations increase the convenience, hygiene, efficiency, and thorough output of the plant. In this process, the rabbits received at the farm, after proper rest, are hoisted live on the rails, which move forward to the stunning section. After stunning, animals are immediately exsanguinated by sticking or by dislocation of the head in ritual slaughtering. Rest dressing operations are performed online, where the animals are hung at eye level (Fig. 3). After cleaning and chilling, carcasses are packed as whole carcasses or as cut-up parts.

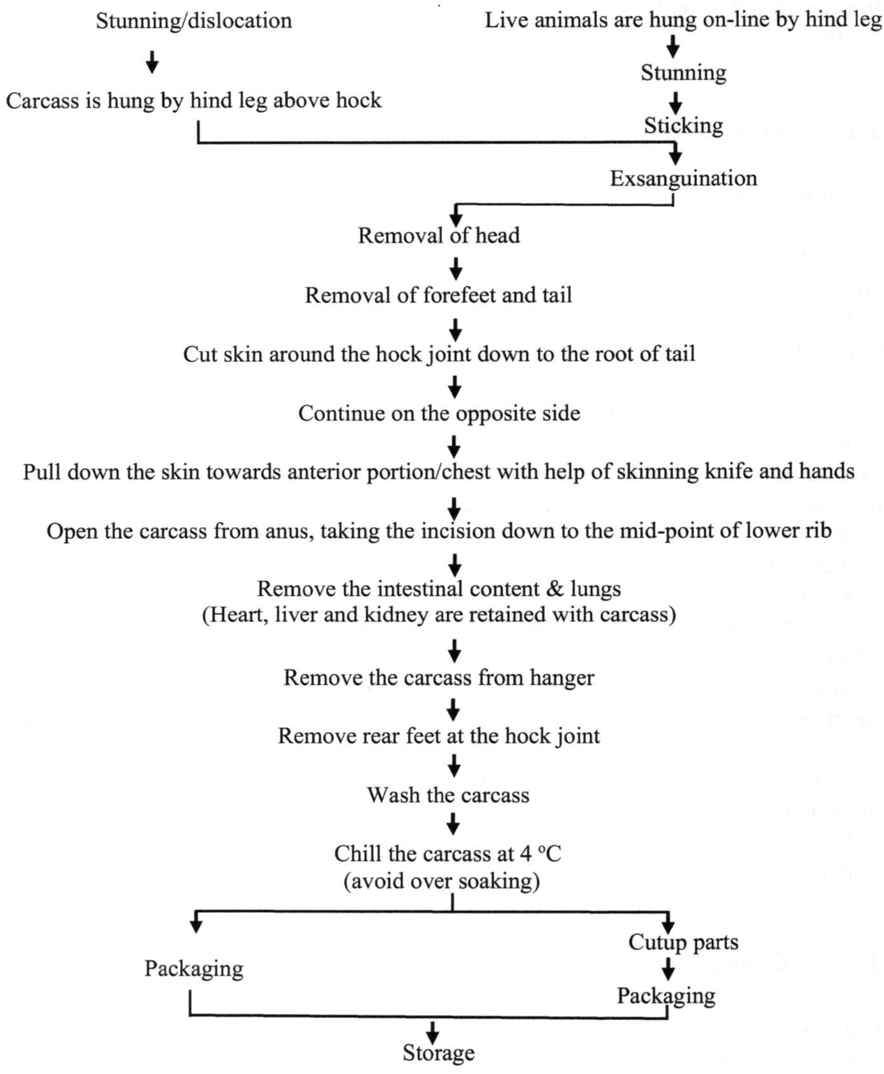

Fig. 3 Flowchart of rabbit carcass dressing in an abattoir

Rabbit carcasses are cut either as anatomical divisions or as traditionally done at the joints (Blasco et al. 1991). The anatomical division and technological divisions were identified and presented by Blasco et al. (1991) while defining terminology for rabbit carcass and meat that is widely accepted and adopted (Fig. 4). According to them, four cut points were identified and called:

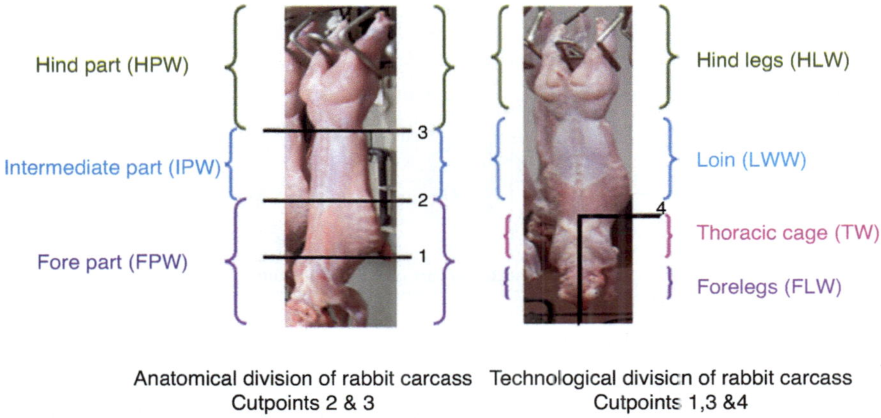

Fig. 4 Rabbit cuts: anatomical and technological cut points

- Cut point 1: Division of carcass into anterior portion by making a cut between the seventh and eighth thoracic vertebrae prolonging the cut toward the ventral side while cutting the thoracic wall, separating the rib cage, shoulders, and forelegs from the carcass.
- Cut point 2: Division of carcass by making a cut between the last thoracic and first lumbar vertebrae taking it down on the ventral side along the 12th rib for cutting the thoracic wall, separating the posterior part containing the loin, saddle, and hind legs.
- Cut point 3: Division of the hind portion or separating entirely the hind leg portion by cutting between the sixth and seventh lumbar vertebrae transversally across the abdominal cavity.
- Cut point 4: Halving of the anterior portion obtained by cut point 1. It contains the foreleg and the thoracic muscles.

Cut points 2 and 3 are anatomical divisions, while cut points 1, 3, and 4 can be used for technological division (Fig. 4). The parts produced by the technological division are also classified as first retail cuts comprising forelegs, loins, and hind legs, while thoracic cages are classified as second retail cuts.

Cut-up Parts

The cut-up parts of a rabbit carcass consist of seven parts: two forelegs, two rib pieces, one loin, and two hind legs based on the technological division or joints. Based on the anatomical divisions, the rabbit carcass is cut into an anterior part by making a cut at the last rib, and the left posterior section is further divided into the anterior saddle and posterior hind legs by making a cut at the last lumbar vertebrae (Fig. 5).

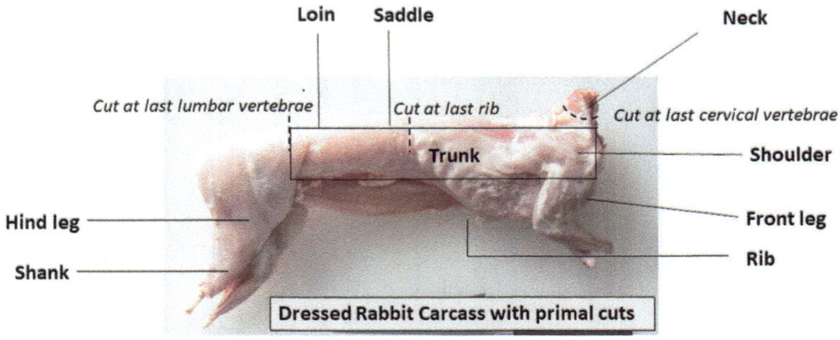

Fig. 5 Rabbit carcass showing commercial cuts

According to the UNECE Standard for Rabbit Meat – Carcasses and Cuts (2013), the popular categories or classes into which rabbit carcass and cuts are available in market are as follows:

- **Full carcass**: A full carcass of rabbit consists of the whole skeletal musculature along with the head and limbs, including the hock joint and knee joints. According to various national standards, a full carcass may or may not have a diaphragm, heart, lungs, kidneys, and kidney fats.
- **Carcass**: The term carcass is used for the rabbit dressed carcass without the head and tail removed from the carcass at the sacrococcygeal junction. During dressing of the carcass, the tender loin is removed, but the diaphragm, heart, lungs, and kidney may or may not be removed. Kidneys if included are devoid of kidney fat.
- **Side**: Sides are two equal portions of carcass obtained by longitudinal halving of the rabbit carcass without neck for which a cut is made centrally through the thoracic, lumbar, and sacral vertebrae. The neck should be removed before halving at the junction of the last cervical and first thoracic vertebrae. Sides contain respective kidneys with fat removed. The spinal cord and diaphragm are also removed during halving.
- **Trunk**: The portion of rabbit carcass obtained by removing the ventral portion at the sixth lumbar vertebrae, i.e., removal of hind legs. The trunk may or may not contain the head, neck, heart, and lungs, while kidney fat is partially or completely removed. Sometimes "trunk without legs" is also sold or available for sale in the market. These are the trunk portion as mentioned above, but the forelegs are removed. Trunk without legs generally does not include the head, but the neck may be retained or removed.
- **Forequarter**: The anterior portion of the rabbit carcass was obtained by separating the carcass at the end of the thoracic vertebrae and first lumbar vertebrae. All 11 ribs are specifically retained (the 12th rib is included in the hindquarter portion), while the diaphragm and spinal cord are removed.
- **Hindquarter**: The posterior part of the carcass obtained by cutting the carcass at the last thoracic and first lumbar vertebrae. It retains the last rib and kidney with

kidney fats removed. The tail and spinal cord were removed during preparation of the hindquarter.

- **Foreleg**: The forelegs are obtained by halving the forequarter cutting the backbone across until the eighth or ninth thoracic vertebrae.
- **Foreshank**: This is the portion of the forequarter containing the radius, ulna carpus, and humerus along with the musculature.
- **Hind shank**: This is the portion of the hindquarter containing the tibia, tarsus, and calcaneal bones. The heel muscles and leg tendon may or may not be retained.
- Leg pair and saddle, leg pair, saddle, loin, quarters set, five-piece cut-up, boneless rack, boneless loin, striploin, backstrap, tenderloin, eye of short loin, rabbit bag, rabbit roll, and rabbit steak are some other popular cut-up and dressed parts of rabbit specified by the regulatory authorities of the European Union, Kenya, etc. Edible byproducts such as skeleton, head, neck, kidney, liver, heart, testes, and stomach have also been specified in the guidelines (UNECE Standards rabbit carcass and cuts, 2013; Rabbit meat specification – Kenya 2017).

12 Advanced Technologies for Processing and Safety

The application of modern technologies and interventions can play a pivotal role in boosting the rabbit meat and meat processing industries (Fig. 6). These can help reduce the cost of production and energy consumption, leading toward sustainable meat production (Smetana et al. 2019). Novel technologies such as irradiation and high-pressure processing (HPP) have already been approved by the regulatory authorities of various countries along with standard guidelines for the implementation and labeling of such products. However, its application, guidelines, standards, and labeling requirements for rabbit meat are meager, which might be attributed to limited studies on the application of modern technologies to rabbit meat. Some recent studies exploring the application of novel technologies to rabbit meat have been discussed.

12.1 Irradiation

Irradiation is a process of exposing food products such as meat to ionizing gamma radiation under specific conditions to reduce microbial load, maintain hygiene, and extend shelf life without exposing food products to high temperatures. Thus, it is sometimes also known as “cold sterilization.” Irradiation can be used for preservation, sterilization, and decontamination. It inactivates microbial DNA by ionization. Badr (2004) investigated the effect of varying doses of gamma irradiation (0, 1.5, 3 kGy) on rabbit meat stored under refrigeration conditions. Irradiation of rabbit meat with a 1.5 kGy dose increased the shelf life of meat by 6 days, while irradiation with a 3 kGy dose extended the shelf life to 21 days compared to 6 days without irradiation. The enhanced shelf life was attributed to the reduced microbial proliferation (aerobic/mesophilic/psychrophilic bacteria and yeast and mold counts).

Fig. 6 Modernized rabbit processing plant (Courtesy of Julián Gullón Álvarez, Cogal S. Coop., Spain)

Bacteria such as *S. aureus*, *L. monocytogenes*, and *E. faecalis* were reduced at 1.5 kGy and further decreased at higher doses, but *Salmonella* was completely eliminated by higher doses without sensorially affecting the product characteristics.

12.2 Sonication

Sonication or application of ultrasound waves for processing meat and meat products has emerged as an alternative processing technology. These are sound waves with frequencies higher than 20 kHz. Sound waves when traveling in a medium cause compression and depression of particles depending on the energy. Sonication has been explored for microbial deactivation in meat and meat products. Microbial inactivation is primarily caused by acoustic cavitation, which is a characteristic of sonication. Carrillo-Lopez et al. (2021) explored the effect of ultrasonication at postmortem in rabbit carcasses and concluded that the application of a high-intensity ultrasound (HIU) frequency of 40 kHz with an intensity of 9.6 W/cm^2 for 20 min increased the tenderness and water holding capacity of rabbit carcasses. Reyes-Villagrana et al. (2020) observed that the application of high-intensity ultrasound sonication on rabbit carcasses evinces effects depending on the muscle type. The HIU of 24 kHz frequency, power 12 W/cm^2 when applied for 15 min on vacuum-packed rabbit carcass (weight 1.78 ± 0.17 kg), stored for 7 days at 2 °C increased the tenderness (shear force), lightness (L^*), and redness (a^*) values of longissimus dorsi muscle but showed no effect on the semimembranosus-semitendinosus muscles of the same carcass. Ultrasound and high-intensity ultrasound (HIU) technology is being widely explored in meat processing, such as curing and marination, as it conserves time, accelerates the salting process, and improves the product characteristics (Alarcon-Rojo et al. 2019). Ochoa-Montes et al. (2017) observed that application of power ultrasound of 40 kHz/110 W in acidic marination of rabbit meat slices improved salt penetration and decreased the immersion time. Similarly, Gómez-Salazar et al. (2018) also reported that sonicating rabbit slices at a frequency of 40 kHz and power of 110 W for 120 m at 4 °C increased the meat color and decreased the immersion time.

12.3 High-Pressure Processing

High-pressure processing is one of the most popular and successful nonthermal advanced processing technologies. Approximately 25–30% of HPP applications are performed on meat and meat product processing (Bolumar et al. 2021). HPP employs pressure to inactivate microorganisms without affecting the physico-chemical, nutritional, and sensory characteristics of a food item (Verma et al. 2021). It has come as a noninvasive, consumer-friendly, and safe process for preventing the growth of spoilage and pathogenic microorganisms in meat and meat products (Bolumar et al. 2021). Cao et al. (2012) contemplated processing of rabbit meat under high pressure for developing gels or gel-type rabbit meat products. Successful gel formation was achieved by applying HPP treatment at 400 MPa for 10 min, and it was concluded that myosin was mainly responsible for gel formation. Xue et al. (2017) utilized high-pressure processing to improve the product characteristics of emulsion-type rabbit meat sausages. It was reported that a high pressure of 200 MPa for 3 min induced better structural stability owing to a homogenous and more interconnected three-dimensional structure. The high pressure applied improved the hydrogen bonding and nonspecific associations, which led to higher water retention. The emulsion produced evinced lower cooking losses and higher product yield with better juiciness. The authors suggested the use of moderate high-pressure treatments such as 200 MPa for 9 and 15 min as well as 300 MPa for 3 and 9 min at 25 °C for developing designer rabbit meat products with low salt and fat contents.

12.4 Superchilling

Superchilling is an advanced chilling process in which the product is stored at a temperature 1–2 °C below its initial freezing point (Kumar et al. 2020). The process provides the beneficial effects of fast freezing, increases its storage life by approximately two- to fourfold, maintains food quality for longer times, and reduces the freeze/thaw energy requirements (Zhou et al. 2010; Kaale et al. 2011). The application of superchilling to rabbit meat was explored by Lan et al. (2016). Superchilling rabbit meat at −2.5 °C and −4 °C increased the shelf life of rabbit meat from 6 days in traditional chilling to 20 days and 36 days, respectively. The quality of meat was also better maintained along with a reduced degree of deterioration. Wang et al. (2018) compared lipid and protein oxidation in rabbit meat stored under refrigeration and superchilled conditions and reported that superchilled temperature restrained lipid, protein, and myoglobin oxidation better than refrigeration storage.

13 SWOT Analysis and Conclusions

The rabbit meat industry is still in its infancy and requires combined efforts from all stakeholders, such as breeders, producers, processors, and regulatory authorities, to evolve it. Here, an attempt has been made to bring out the strengths, weaknesses, opportunities, and threats (SWOT) of the rabbit meat processing industry to boost the rabbit meat/processing industry (Fig. 7). The revival of traditional cuisine and an increasing quest for healthy, functional, and minimally processed products have driven the search for alternative and/or traditional meat sources.

Fig. 7 Strength, weakness, opportunities, and threats (SWOT) of rabbit meat industry

With increasing globalization, the local delicacies that were once restricted to specific regions have now crossed boundaries and reached the plates of different cultures and regions. Rabbit delicacies such as Lapin á la Kriek from Belgium; Coniglio all'ischitana from Ischia, Italy; Fenkata from Malta; Civet de lapin and Lapin à la crème from France; Conejo al ajillo and Conejo con arroz de Calasparra from Spain; Sate kelinci (rabbit satay) from Java; and Kouneli stifado from Skiathos, Greece, are being introduced to other countries through globalization and tourism. Consequently, it would enhance the demand as well as acceptability of rabbit meat among consumers. Rabbit meat is still largely purchased fresh and cooked according to regional cuisine, which is also one of the limiting factors for its popularization, as skills and knowledge for such preparations are limited. The development of convenient and ready-to-cook or ready-to-eat meals based on rabbit meat can increase its popularity among young consumers.

The rabbit meat industry and researchers need to focus on standardizing traditional delicacies as well as developing convenient rabbit meat products. Another factor restricting the rabbit meat sector is the lack of uniform regulations on animal welfare, products, hygienic processing, marketing, labeling, and microbial safety. Consumers are currently becoming stringent in their food choices. They not only consider the taste and flavor of the product but are equally concerned about how the food product has been obtained. Consumer confidence in rabbit meat can be achieved by proper labeling, regulations, transparency, and traceability. Rabbit meat is a wholesome, nutritional meat with high protein and low fat, cholesterol, and sodium and high desirable unsaturated fatty acids. Rabbit production can also play an important role in the rural economy and sustainable development.

References

Alarcon-Rojo AD, Carrillo-Lopez LM, Reyes-Villagrana R, Huerta-Jiménez M, Garcia-Galicia IA (2019) Ultrasound and meat quality: A review. Ultrason Sonochem 55:369–382

Amelianchik A, Sweetland-Martin L, Norris EH (2022) The effect of dietary fat consumption on Alzheimer's disease pathogenesis in mouse models. Transl Psychiatry 12(1):293

Badr HM (2004) Use of irradiation to control foodborne pathogens and extend the refrigerated market life of rabbit meat. Meat Sci 67(4):541–548

Bennett JE, Stevens GA, Mathers CD, Bonita R, Rehm J, Kruk ME et al (2018) NCD Countdown 2030: worldwide trends in non-communicable disease mortality and progress towards Sustainable Development Goal target 3.4. Lancet 392(10152):1072–1088

Blasco A, Ouhayoun J, Masoero G (1991) Status of rabbit meat and carcass: Criteria and terminology. In: Options Mediterraneennes Serie A: Seminaires Mediterraneens (CIHEAM). CIHEAM, Zaragoza

Bolumar T, Orlien V, Sikes A, Aganovic K, Bak KH, Guyon C et al (2021) High-pressure processing of meat: Molecular impacts and industrial applications. Compr Rev Food Sci Food Saf 20(1):332–368

Cambero MI, de la Hoz L, Sanz B, Ordóñez JA (1991) Lipid and fatty acid composition of rabbit meat: Part 1.—Apolar fraction. Meat Sci 29(2):153–166

Cao Y, Xia T, Zhou G, Xu X (2012) The mechanism of high pressure-induced gels of rabbit myosin. Innovative Food Sci Emerg Technol 16:41–46

Carrillo-Lopez LM, Robledo D, Martínez V, Huerta-Jimenez M, Titulaer M, Alarcon-Rojo AD et al (2021) Post-mortem ultrasound and freezing of rabbit meat: Effects on the physicochemical quality and weight loss. Ultrason Sonochem 79:105766

Cavani C, Petracci M (2004) Rabbit meat processing and traceability. In Proceedings of the 8th World Rabbit Congress, pp 1318–1336

Cavani C, Petracci M, Trocino A, Xiccato G (2009) Advances in research on poultry and rabbit meat quality. Ital J Anim Sci 8(sup2):741–750

Cullere M, Dalle Zotte A (2018) Rabbit meat production and consumption: State of knowledge and future perspectives. Meat Sci 143:137–146

Dalle Zotte A, Cullere M (2019) Carcass traits and meat quality of rabbit, hare, Guinea pig and capybara. In: More than Beef, Pork and Chicken–The Production, Processing, and Quality Traits of Other Sources of Meat for Human Diet. Springer, Cham, pp 167–210

Dalle Zotte A, Szendrő Z (2011) The role of rabbit meat as functional food. Meat Sci 88(3):319–331

Dalle Zotte A, Szendrő K, Gerencsér Z, Szendrő Z, Cullere M, Odermatt M et al (2015) Effect of genotype, housing system and hay supplementation on carcass traits and meat quality of growing rabbits. Meat Sci 110:126–134

Fazio F, Casella S, Giudice E, Giannetto C, Piccione G (2015) Evaluation of secondary stress biomarkers during road transport in rabbit. Livest Sci 173:106–110

Gómez-Salazar JA, Ochoa-Montes DA, Cerón-García A, Ozuna C, Sosa-Morales ME (2018) Effect of acid marination assisted by power ultrasound on the quality of rabbit meat. J Food Qual 2018: 1–6

Hernández P, Dalle Zotte A (2020) Influence of diet on Rabbit meat quality. Nutrition of the Rabbit. CABI Publishing, Wallingford, pp 172–192

Kaale LD, Eikevik TM, Rustad T, Kolsaker K (2011) Superchilling of food: A review. J Food Eng 107(2):141–146

Khan N, Afaq F, Mukhtar H (2010) Lifestyle as risk factor for cancer: Evidence from human studies. Cancer Lett 293(2):133–143

Kumar P, Verma AK, Umaraw P, Kumar D, Mehta N (2020) Superchilling preserves freshness: The novel technology has emerged to improve long term storage life. Fleischwirtschaft International: Journal for Meat Production and Meat Processing 2:45–49

Lafuente R, López M (2014) Effect of electrical and mechanical stunning on bleeding, instrumental properties and sensory meat quality in rabbits. Meat Sci 98(2):247–254

Lambertini L, Vignola G, Badiani A, Zaghini G, Formigoni A (2006) The effect of journey time and stocking density during transport on carcass and meat quality in rabbits. Meat Sci 72(4): 641–646

Lan Y, Shang Y, Song Y, Dong Q (2016) Changes in the quality of superchilled rabbit meat stored at different temperatures. Meat Sci 117:173–181

Li S, Zeng W, Li R, Hoffman LC, He Z, Sun Q, Li H (2018) Rabbit meat production and processing in China. Meat Sci 145:320–328

Liste G, Villarroel M, Chacón G, Sañudo C, Olleta JL, García-Belenguer S et al (2009) Effect of lairage duration on rabbit welfare and meat quality. Meat Sci 82(1):71–76

Marıa GA, Buil T, Liste G, Villarroel M, Sañudo C, Olleta JL (2006) Effects of transport time and season on aspects of rabbit meat quality. Meat Sci 72(4):773–777

Mazzone G, Vignola G, Giammarco M, Manetta AC, Lambertini L (2010) Effects of loading methods on rabbit welfare and meat quality. Meat Sci 85(1):33–39

Mota JDO, Boué G, Guillou S, Pierre F, Membré JM (2019) Estimation of the burden of disease attributable to red meat consumption in France: influence on colorectal cancer and cardiovascular diseases. Food Chem Toxicol 130:174–186

Ochoa-Montes DA, Sosa-Morales ME, Cerón-García A, Gómez-Salazar JA (2017) Effect of power ultrasound on sodium chloride diffusion during marinating rabbit meat. In: In 2017 ASABE Annual International Meeting. American Society of Agricultural and Biological Engineers, p 1

Petracci M, Bianchi M, Biguzzi G, Cavani C (2010) Preslaughter risk factors associated with mortality and bruising in rabbits. World Rabbit Sci 18(4):219

Pla M, Guerrero L, Guardia D, Oliver MA, Blasco A (1998) Carcass characteristics and meat quality of rabbit lines selected for different objectives: I. Between lines comparison. Livest Prod Sci 54(2):115–123

Reyes-Villagrana RA, Huerta-Jimenez M, Salas-Carrazco JL, Carrillo-Lopez LM, Alarcon-Rojo AD, Sanchez-Vega R, Garcia-Galicia IA (2020) High-intensity ultrasonication of rabbit carcasses: a first glance into a small-scale model to improve meat quality traits. Ital J Anim Sci 19(1):544–550

Rota Nodari S, Lavazza A, Candotti P (2009) Rabbit welfare during electrical stunning and slaughter at a commercial abattoir. World Rabbit Sci 17(3):163–167

Składanowska-Baryza J, Stanisz M (2019) Pre-slaughter handling implications on rabbit carcass and meat quality–A review. Ann Anim Sci 19(4):875–885

Składanowska-Baryza J, Ludwiczak A, Pruszyńska-Oszmałek E, Kołodziejski P, Stanisz M (2020) Effect of two different stunning methods on the quality traits of rabbit meat. Animals 10(4):700

Smetana S, Terjung N, Aganovic K, Alahakoon AU, Oey I, Heinz V (2019) Emerging technologies of meat processing. In: Sustainable meat production and processing. Academic Press, pp 181–205

United Nations Economic Commission for Europe (UNECE) standard rabbit meat Carcasses and Cuts (2013). Accessed at https://unece.org/DAM/trade/agr/standard/meat/e/Rabbit_Meat-2013_E.pdf

Verga M, Luzi F, Petracci M, Cavani C (2009) Welfare aspects in rabbit rearing and transport. Ital J Anim Sci 8(Sup 1):191–204

Verma AK, Umaraw P, Singh VP, Kumar P, Mehta N, Kumar D (2017) Composition of unconventional meats. Fleischwirtschaft International: Journal for Meat Production and Meat Processing 3:12–21

Verma AK, Umaraw P, Singh VP, Kumar P, Mehta N, Kumar D (2021) Advantages of high pressure processing. Fleischwirtschaft International: Journal for Meat Production and Meat Processing 2:66–70

Voslarova E, Vecerek V, Bedanova I, Vecerkova L (2018) Mortality in rabbits transported for slaughter. Anim Sci J 89(6):931–936

Wang Z, He Z, Gan X, Li H (2018) Interrelationship among ferrous myoglobin, lipid and protein oxidations in rabbit meat during refrigerated and superchilled storage. Meat Sci 146:131–139

Wu L (2022) Rabbit meat trade of major countries: regional pattern and driving forces. World Rabbit Sci 30:69–82. https://doi.org/10.4995/wrs.2022.13390

Xue S, Wang H, Yang H, Yu X, Bai Y, Tendu AA et al (2017) Effects of high-pressure treatments on water characteristics and juiciness of rabbit meat sausages: Role of microstructure and chemical interactions. Innovative Food Sci Emerg Technol 41:150–159

Zhou GH, Xu XL, Liu Y (2010) Preservation technologies for fresh meat–A review. Meat Sci 86(1):119–128

Assessing Rabbit Body and Carcass Traits with Non-invasive and Non-destructive Image Techniques

Severiano R. Silva and Tamás Donkó

Abstract

The possibility of measuring body or carcass composition in animals is crucial for carrying out performance tests, carcass classification, and selection programs. The advances observed with non-invasive techniques, particularly imaging, have allowed for obtaining accurate, objective phenotypic data in the meat-producing species. This chapter will discuss the use of several imaging techniques, such as real-time ultrasonography (RTU), computed tomography (CT), magnetic resonance imaging (MRI), and dual-energy X-ray absorptiometry (DXA), to assess rabbit body and carcass traits. The preference for a specific imaging technique depends on the target to be evaluated and on technical and practical aspects such as accuracy, reliability, cost, portability, speed, and ease of use. Among the above techniques, CT and RTU are the ones that have been most used in rabbits. CT is more accurate than RTU, but the latter can be used more widely, even under field conditions. Therefore, a dual RTU and CT approach may be meritorious for a rabbit selection program.

Keywords

Rabbit · Carcass · Ultrasound · Computed tomography · MRI · DXA

S. R. Silva (✉)
Veterinary and Animal Research Centre (CECAV), Associate Laboratory of Animal and Veterinary Science (AL4AnimalS), University of Trás-os-Montes e Alto Douro, Vila Real, Portugal
e-mail: ssilva@utad.pt

T. Donkó
Department of Animal Science, Hungarian University of Agriculture and Life Sciences, Kaposvár, Hungary

Medicopus Nonprofit Ltd., Kaposvár, Hungary

J. Simões, J. M. Monteiro (eds.), *Veterinary Care of Farm Rabbits*,
https://doi.org/10.1007/978-3-031-44542-2_15

1 Introduction

Over the past decade, there has been a steady decline in rabbit meat consumption worldwide. According to the Food and Agriculture Organization of the United Nations (FAO 2020), global rabbit meat production decreased from 1.2 million tons in 2000 to 0.9 million tons in 2020. The most significant world and EU rabbit meat producers are China and Italy, respectively. Notably, the largest decrease of 41% was observed in Europe during this period.

Rabbit meat is considered a healthy and sustainable alternative to other types of meat, as it is low in fat and high in protein (Siddiqui et al. 2023; Trocino et al. 2019; Petrescu and Petrescu-Mag 2018). It is also a popular meat in many cuisines worldwide, particularly in Mediterranean and Latin American cuisine (Dalle Zotte and Szendrő 2011; Szendrő et al. 2020). In addition to its nutritional benefits, rabbit meat production has a relatively low environmental impact compared to other types of meat production (Cullere and Dalle Zotte 2018; Kumar et al. 2022).

Rabbits have a high reproductive rate and excellent feed efficiency and can be raised in relatively small spaces, making them a good choice for intensive as well as for small-scale and backyard farming (Trocino et al. 2019; Siddiqui et al. 2023). Moreover, the rabbit has also been used as an excellent experimental model to study meat and carcass traits since it allows one to perform experiments with much lower cost than other meat species (Hernández et al. 2006).

Despite its many attributes, rabbit meat production faces significant challenges to increasing consumption, particularly in some Western countries (Trocino et al. 2019). These challenges include concerns about animal welfare, lack of consumer awareness, economic crises, and the perception of rabbits as pets, which can increase consumer disaffection (Petracci et al. 2018; Magalhães et al. 2022). However, various efforts are being made to address these challenges. For instance, certification schemes have been developed to ensure high animal welfare standards in rabbit production. Additionally, rabbit meat has been promoted as a sustainable and healthy alternative to other types of meat (Petrescu and Petrescu-Mag 2018; Crovato et al. 2022). With such efforts in place, it is hoped that the perception of rabbit meat can be improved and its consumption can increase.

Furthermore, efforts have been made to improve growth, carcass, and meat quality in rabbit meat production through careful breeding and selecting rabbits with desirable traits (Blasco et al. 2018). However, the heritability of meat quality traits in rabbits varies depending on the trait. For example, growth rate and carcass traits generally have high to moderate heritability, with estimates ranging from 0.2 to 0.6 (Blasco et al. 2018). Therefore, assessing in vivo carcass traits with non-invasive image techniques can be an efficient way to obtain detailed information about the characteristics of the carcass, such as lean meat percentage and fat distribution (Silva and Cadavez 2012). Image techniques such as real-time ultrasonography, X-ray computed tomography, magnetic resonance imaging, and dual-energy X-ray absorptiometry have been used to assess carcass traits in ruminants and pigs (Silva et al. 2020; Scholz et al. 2021; Delgado-Pando et al. 2021; Leighton et al. 2021) but with much less expression in rabbits.

This chapter overviews the main imaging techniques for assessing body and carcass traits in rabbit production.

2 Image Techniques to Assess Body and Carcass Traits in Rabbits

Applying non-invasive and non-destructive image techniques in rabbit production has several advantages. Firstly, it enables accurate and objective measurements of carcass traits, improving the efficiency and effectiveness of selection and breeding programs. Secondly, it reduces the need for postmortem measurements, which can be time-consuming, labor-intensive, and expensive. Thirdly, it minimizes the wastage of valuable carcass material, as the animal can be returned to production after the non-invasive scan. Lastly, these techniques can provide valuable insights into the effects of nutrition, management, and environment on carcass traits. However, the image techniques used should be easily applicable, affordable, and suitable for a wide range of weights, body frames, and ages with minimal animal disturbance. This chapter will discuss the use of real-time ultrasonography (RTU), computed tomography (CT), magnetic resonance imaging (MRI), and dual-energy X-ray absorptiometry (DXA) to assess rabbit body and carcass traits.

2.1 Real-Time Ultrasonography

The challenge of knowing the carcass traits of a living animal has been pursued relentlessly since the 1950s (Silva and Stouffer 2019). Several techniques have been tested throughout this time, as evidenced by the numerous scientific articles found on the subject in the leading animal science journals (Silva and Stouffer 2019). The techniques that employ ultrasound are among the most successful in assessing carcass and meat traits in the main meat production species (Scholz et al. 2015; Silva and Stouffer 2019); nevertheless, only a few studies have been conducted using real-time ultrasonography on rabbits (Table 1).

Some of these studies are related to the body condition and fat depots prediction, especially perirenal fat (Pascual et al. 2000; Pascual et al. 2002; Pascual et al. 2004; Dal Bosco et al. 2003; Castellini et al. 2006; Quevedo et al. 2006). In these works, attention is paid to the scapular fat (SFT) and the perirenal fat thickness (PFT) since, in rabbits, the ability to assess the thickness of the subcutaneous fat is of little value given its reduced size (typically less than 1 mm), which drastically reduced the accuracy and precision of that measurement (Dal Bosco et al. 2003; Silva and Cadavez 2012).

Monitoring the variation in fat deposits in female rabbits throughout the reproductive cycle is an important objective for the in vivo use of RTU (Pascual et al. 2004; Quevedo et al. 2006). These works present results that show the suitability of the RTU to fulfill that objective and with sufficient rigor to be pointed out as an advantageous alternative to serial slaughter. In their pioneer study, Pascual et al.

Table 1. Summary of the studies to predict carcass composition, dressing percentage, monitoring body condition, and fat depots in vivo of rabbit using RTU

Reference	Aim	Breed	n	BW (kg)	Measurement point	Scanner	Probe	Image analysis	Measure	Lean	Bone	Fat	DoP
Silva et al. (2009)	CComp	NZ x C	69		6th and 7th LV	Aloka 500-V	7.5 MHz linear	ImageJ	LMA	0.88	0.84	0.66	
	CComp								LMW	0.78	0.72	0.46	
Silva et al. (2012)	CComp	NZ x C	63	2.09 ± 0.063	6th and 7th LV	Aloka 500-V	7.5 MHz linear	ImageJ	Lmvol	0.89	0.85	0.64	
	CComp								LMD	0.76	0.70	0.67	
Valencia-Garcia et al. (2022)	CComp	NZ x C	38	1.33 ± 0.291	6th and 7th LV	Mindray DP Vet-50®	7.5 MHz linear	Scanner software	LMA	0.80	0.57		
	CComp								LMW	0.76	0.40		
	CComp								LMD	0.84	0.76		
Lenoir and Morien (2016)	DoP	Seven breed groups	647	2.69 ± 0.260	2nd and 3rd LV	Hitachi EUB-555	7.5 MHz linear		LMA				0.74
Pascual et al. (2000)	Mbc	NZ x C	42	3.92 ± 0.42	Six different sites	UltraScan 900	5 MHz sectorial	VHGlobal V2	PFT			0.69 to 0.95	
Pascual et al. (2002)	Mbc	NZ x C	64		Six different sites	UltraScan 900	5 MHz sectorial	VHGlobal V2	PFT			0.07 to 0.66	
Dal Bosco et al. (2003)	Mbc	NZ	100	4.1 ± 0.495	2nd and 3rd LV	Aloka 500-V			SFT, PFT				

Pascual et al. (2004)	Mbc	NZ x C	60	4.57 ± 0.49	8th and 9th TV	Toshiba SSA-320A	5–7 MHz linear	Scanner software	PFT			0.72 to 0.84#	
Castellini et al. (2006)	Mbc	NZ	50	4.1 to 4.3	2nd and 3rd LV	Aloka 500-V			PFT				
Quevedo et al. (2006)	Mbc	Crossbred	166		Six different sites	Toshiba SSA-320A	5–7 MHz linear	Scanner software	PFT				
Yonkova et al. (2012)	Mbc	NZ	18			CHISON 600 VET	7 MHz convex	Thermoprinter	IFT, SFT, PFT				
Silva et al. (2021)	PfatD	NZ x C	42	4.5 ± 0.57		Aloka 500-V	7.5 MHz linear	ImageJ	PFT			0.91	
	PfatD								PFA			0.92	

CComp prediction of carcass composition, *DoP* dressing-out percentage, *Mbc* monitoring body condition, *PfatD* predicting fat depots, *NZ* New Zealand, *NZ x C* New Zealand x Californian, *LV* lumbar vertebra, *TV* thoracic vertebra, *IFT* inguinal fat thickness, *SFT* scapular fat thickness, *Lmvol* LM volume, *LMA* LM area, *LMD* LM thickness, *LMW* LM width, *PFA* perirenal fat area, *PFT* perirenal fat thick, # perirenal fat weight. The numbers in the lean, bone, fat, and DoP columns are correlation values

(2000) showed that RTU was suitable for fat deposit evaluation. They proposed a method based on the thickness measurement of the perirenal fat at the eighth–ninth thoracic vertebrae level. With that, these authors accurately predict carcass perirenal fat weight ($r^2 = 0.95$; $n = 42$) and total fat weight ($r^2 = 0.93$; $n = 42$). This result is in line with that observed by Silva et al. (2021), who reported a correlation between measurements obtained in perirenal fat and body fat ($r > 0.9$, $P < 0.01$). TUR was also used to estimate dissectible fat depots to indicate good health (Yonkova et al. 2012) and to analyze the variability of growth hormone gene associated with ultrasound imaging of LM muscle and perirenal fat (Amalianingsih and Brahmantiyo 2014).

Another group of works is related to predicting carcass composition using RTU measurements obtained in longissimus thoracis et lumborum (Silva et al. 2009, 2012; Valencia-Garcia et al. 2022).

The rabbit valuation carcass criteria are centered on the lean meat amount and yield and the ratio between lean meat and bone (Lukefahr and Ozimba 1991; Lambertini et al. 1996). However, several reports point out that that ratio is related to muscularity or cutability attributes of carcasses that can be improved by selection (Lukefahr et al. 1982; Lukefahr et al. 1983).

On the other hand, in rabbits, good results were achieved in studies conducted with morphometric measurements in vivo to predict the percentage of muscle and muscularity in the carcass (Lukefahr et al. 1982; Lukefahr and Ozimba 1991; Michalik et al. 2006). For example, Lukefahr and Ozimba (1991) point out that the loin cut weight was predicted accurately using body weight and the loin width ($r^2 = 0.797$).

This aspect deserves attention when using in vivo measures with RTU that can be potentially useful for application in genetic improvement programs or economic carcass evaluation (Silva et al. 2009).

Several works show that it is possible to estimate carcass composition in vivo from RTU measurements (Silva et al. 2009, 2012 Valencia-Garcia et al. (2022). In general, the results reported by this work show that measurements obtained from RTU images were able to explain a large amount of the variation of lean (r between 0.76 and 0.89), bone (r between 0.40 and 0.85), and fat (r between 0.46 and 0.67).

On the other hand, to predict the muscularity of the loin and leg, the research conducted shows that the RTU has the potential to estimate those carcass traits, but it needs new approaches to become a routine procedure in the in vivo evaluation of rabbits (Silva et al. 2007, 2008a, 2008b, 2009). There are practical difficulties, such as the need to shave the hair and the RTU image analysis, that have to be faster and more automated.

In this type of work, the most commonly used procedure involves restraining the rabbits, shaving the hair at the anatomical reading point, placing a gel to ensure acoustic contact between the skin and the probe, and then analyzing the images (Fig. 1).

Also, after RTU image analysis of multiple scanning images taken in the lumbar region, it was determined in vivo that the volume of the LM was significantly

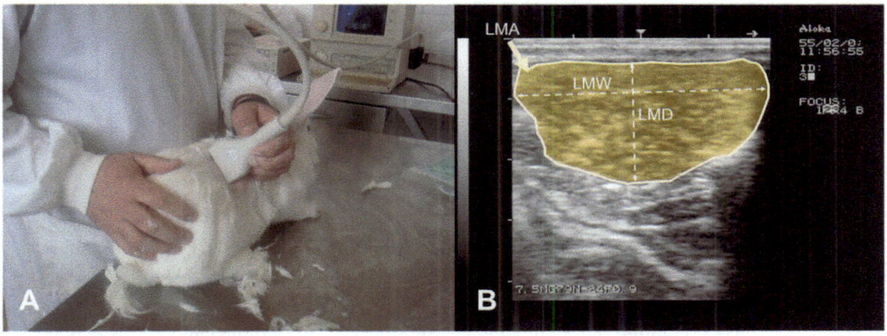

Fig. 1 Image illustrating the acquisition procedure of an RTU image with a linear probe placed over lumbar region where the fur was clipped close to the skin (**a**) and an example of RTU image showing the representations of longissimus thoracis et lumborum muscle area (LMA), depth (LMD), and width (LMW) measurements (**b**)

correlated with the volume measured postmortem ($r = 0.9$). The RTU volume was calculated from area measurements using Cavalieri's principle (Silva et al. 2008a).

As previously mentioned, muscularity is an important rabbit carcass trait, and Silva et al. (2009) verified that it was possible to use in vivo measurements of RTU to accurately predict loin muscularity (r between 0.76 and 0.81; $P < 0.001$) (Silva et al. 2009). On the contrary, a study focused on the leg could not estimate leg muscularity (Silva et al. 2008b).

These results show that, nevertheless, the use of a 7.5 MHz probe, which improves the resolution of the images and image analysis by software such as Image J, allows to improve the accuracy of the LTL muscle of loin measurements, and improvements are needed for the use of RTU for hind leg muscularity determination.

In any case, the results found in the various works encourage the use of in vivo RTU to predict rabbit carcass traits, and further research will be necessary for RTU image acquisition and image analysis but also for other issues such as rabbit restraint or procedure for shaving the fur together to the skin to improve the practicability of this technique in rabbits.

2.2 Computed Tomography

In the 1960s, Hounsfield was working at Electric and Musical Industries (EMI) Central Research Laboratories when he began developing the idea of using X-rays to create detailed images of the body. EMI supported Hounsfield's work and provided him with the resources and funding to develop the first CT scanner. Beatles that signed with EMI indirectly played a role in its development through its research and development efforts (Weiskirchen 2022). In 1979, Hounsfield and Allan Cormack, who developed the mathematical algorithms necessary to reconstruct the CT images

Table 2. Summary of Europe network of CT capability for animal science research

CT scanner type	Institution	City	Country
Mobile	Danish Technological Institute/DMRI	Taastrup	Denmark
	Institut National de la Recherche Agronomique (INRA) Aniscans	Saint-Gilles	France
	Scotland's Rural College (SRUC)	Penicuik, Midlothian	United Kingdom
	Institute of Agrifood Research and Technology (IRTA)	Monells, Girona	Spain
	IFIP – Institut du Porc	Le Rheu	France
Fixed	MRI, Institute of Safety and Quality of Meat	Kulmbach	Germany
	School of Veterinary Medicine Aristotle University	Thessaloniki	Greece
	Medicopus Nonprofit Ltd.	Kaposvár	Hungary
	Norsvin SA	Hamar	Norway
	Universidade de Trás-os-Montes e Alto Douro (UTAD)	Vila Real	Portugal
	Institute of Agrifood Research and Technology (IRTA)	Monells, Girona	Spain
	Scotland's Rural College (SRUC)	Penicuik, Midlothian	United Kingdom

from the X-ray data, won the Nobel Prize in Physiology or Medicine for that achievement (Seeram 2006). The development of CT revolutionized medical imaging and has had a significant impact on medicine but also veterinary and animal science. The first research study concerning animal body composition was conducted in Norway in the early 1980s, focusing on pigs (Skjervold et al. 1981).

In the last decade, despite its complexity and cost, CT has gained importance for the knowledge of body and carcass composition of the meat species, particularly in sheep and pigs (Bünger et al. 2014; Scholz et al. 2015) but also in chickens (e.g., Bentsen and Sehested 1989), turkeys (e.g., Dewez et al. 2018), cattle (e.g., Font-i-Furnols et al. 2021), and rabbits (e.g., Matics et al. 2020). Much of the research and practical application that has been carried out with these species have been conducted in Europe, where several countries have research facilities with both mobile and fixed CT. In 2015, a directory was published to highlight the Europe network of CT capability for animal science research (Daumas et al. 2015). Table 2 summarizes that network by institution and country. There are also CT research activities related to animal science in countries on other continents, such as New Zealand (Johnsona et al. 2020), Australia (Gardner et al. 2021), and Brazil (Geraldo et al. 2020).

Most research on rabbits has been carried out in the CT laboratory at Kaposvar (MATE Kaposvar Campus and Medicopus Nonprofit Ltd., former Kaposvár University). Since the 1990s of the last century, this laboratory has been carrying out

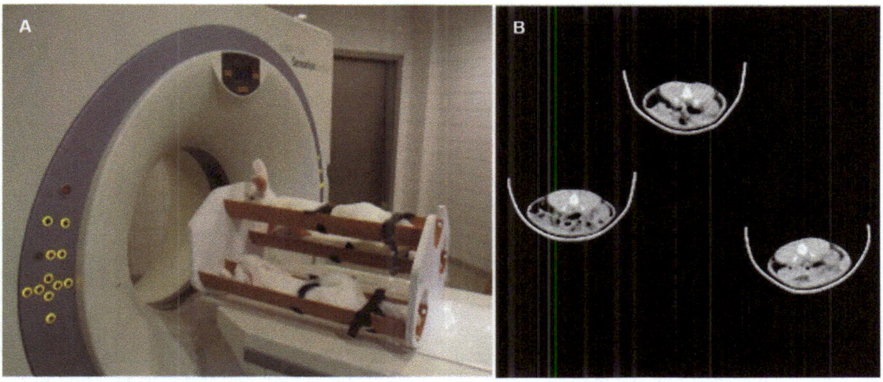

Fig. 2 Position of the three rabbits in a container during CT examination (**a**) (Kovács et al. 2013) and CT scans of the three rabbits (**b**) (Szendrő et al. 2005)

carcass trait selection in Pannon rabbit lines based on CT measurements (Matics et al. 2014).

Typically, CT scanning procedures are conducted following a fasting period of at least 8 hours to improve image quality by minimizing food residue in the gastrointestinal tract. The rabbits are immobilized using belts and placed in a stretched position while lying flat inside specialized plastic containers during the examination (Romvári et al. 1996; Matics et al. 2014). The animals are secured to ensure they do not move, and their legs are separated from their bodies. No anesthesia is used in the process (Matics et al. 2014). On the contrary, anesthetics are used for pregnant does instead of fixing straps to reduce the risk of abortion. Initially, rabbits were examined one at a time during CT examinations (Romvári et al. 1996). However, it was soon learned that multiple rabbits could be examined simultaneously. As a result, a holder was created to examine three rabbits at once (Fig. 2), significantly reducing the time required for CT scanning (Romvári et al. 1996; Matics et al. 2014).

The basic operating principle of a CT scanner involves using X-rays and advanced computer algorithms to generate detailed cross-sectional images of the body. The CT scanner emits a series of narrow X-ray beams that pass through the body at various angles. As the X-ray beams pass through the body, they are absorbed differently by different types of tissues based on their density and thickness, and the amount of X-rays that are transmitted is quantified in the detectors in the CT scanner; the resulting data is fed into a computer (Du et al. 2019). The computer uses mathematical algorithms to reconstruct detailed, consecutive two-dimensional body images based on the information gathered from the X-rays. The visualization can be 3D, because each of the pixels has a third dimension, which comes from the slice thickness. This process is called image reconstruction, and it allows us to visualize the body's internal structures in great detail (Du et al. 2019). Over time, the acquisition process, image reconstruction, and post-processing techniques have become faster and capable of producing higher-resolution images. Whereas in the

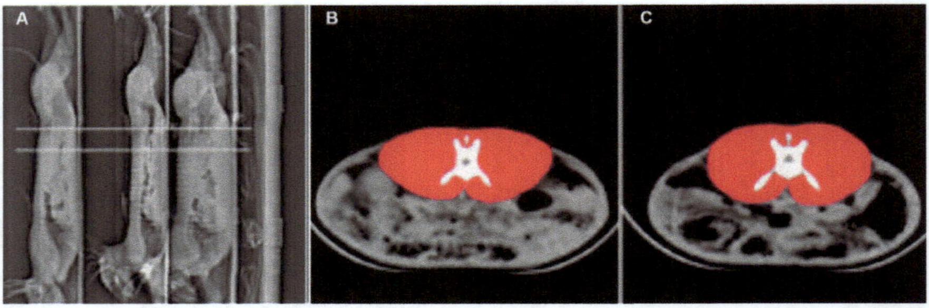

Fig. 3 The anatomical points between the second and third vertebrae and between the fourth and fifth lumbar vertebrae marked on the lateral view of the tomogram (**a**) and the surface of m. longissimus dorsi that is measured on the junction of the second and third (**b**) and that of the fourth and fifth (**c**) lumbar vertebrae (Szendrő et al. 2005)

late 1990s it took up to a minute to reconstruct a slice of an image, now it takes a few minutes to capture a whole series of images and reconstruct them.

In work with rabbits, three typical scanning methods have been used. In the beginning, a widely used and straightforward method is based on fixed anatomical points. In CT work linked to the breeding program of the Pannon White rabbits, examinations are carried out at two points between the second and third vertebrae and between the fourth and fifth lumbar vertebrae (Fig. 3). At these points, measurements of the area of the cross section of the LM are performed by manual segmentation. The average of the two areas was included in the selection program. This procedure has also been performed in other species, such as sheep, arguing that it makes the scanning process faster if it significantly compromises accuracy (Macfarlane et al. 2006). However, scanning time is not a limiting factor in more modern equipment (Silva et al. 2020).

Another approach is based on serial scans of the animal's entire body. In this case, the information collected is expected to correlate with whole-body animals' composition. At a time when technology did not yet allow for full overlap of scans, 20 scans are taken from the fifth cervical vertebra to the head of the femur (Romvári et al. 1996). In this way, different-sized animals can be compared.

The thigh muscle volume (TMV) can also be pointed out as an example. CT scans were taken from the iliac crest of the os ilium and the patella to determine the TMV. Depending on the length of the hind legs, 11–12 overlapping scans with 10 mm slice thickness are obtained, and the voxel frequency related to the HU range associated with the muscle (typically between +20 and +140 of the HU scale) is determined on each scan, and its sum allows estimating the TMV (Matics et al. 2014). In this case, the images are evaluated semi-automatically by thresholding the volume of the muscle voxels after manual selection of the slices.

Since 2004, the TMV has been measured, and it was observed that it is positively correlated with average daily gain. On the contrary, it shows a negative genetic correlation with the litter weight at day 21 (Matics et al. 2014; Gyovai et al. 2012). Furthermore, it has been observed in different works that the magnitude of maternal effects is generally reduced for CT muscle traits (Nagy et al. 2010; Nagy et al. 2013). Nevertheless, the TMV measurement has an impact on rabbit meat production. In a balanced work supported by experimental results and genetic evaluations, Matics et al. (2014) point to an improvement in the dressing-out percentage of rabbits and indicate CT-based selection as an efficient approach to achieve this result. These authors also point out that due to the CT-based selection, the estimated genetic trend for the thigh muscle volume is about 5 cm^3 per year, representing 10 tons more of thigh muscle produced per year at Hungarian slaughterhouses.

The segmentation of CT images of live animals or carcasses is usually done by simple thresholding of Hounsfield units (HU) from which the characteristic voxels of the different tissues are distinguished (Font-i-Furnols et al. 2009). With the aim of improving the ability to estimate loin weight, a segmentation technique based on CT images of rabbits was proposed by Matics et al. (2020). The region of interest was between the fifth pair of ribs and the cut in which the pelvic bone appears for the first time, as the loin muscle touches various organs and soft tissues with HU values very similar to LM. Consequently, the limits of the LM could not be sharp, making it difficult to use the thresholding method. However, after several steps, an automated post-processing pipeline was developed for in vivo prediction of rabbits' LM mass using morphological operators in the overlapping 2 mm scans. Linear regression was stable and could explain 73.7% of the variation of LM muscle.

The weight or volume of carcass cuts or carcass tissues is closely related to meat quality in various livestock species, and several CT studies to predict (determine) those traits and use this technique as a gold standard (Bünger et al. 2011; Scholz et al. 2015; Gardner et al. 2021). To estimate the weights of carcass cuts in live animals with images and CT, it is necessary to segment the region of interest and perform a regression between the histogram of radiodensity values and the reference weights per dissection. In this way, a predictive model is obtained that will be applied to the scans of non-dissected animals. However, this procedure can incorporate biases in the dissection process, limiting its rigor and accuracy. To overcome these problems, Csóka et al. (2022) conducted a study to develop a general application methodology to estimate the weight of several slices in CT images in rabbits that facilitate automation and reduce costs with the need for specific solutions for each species. The basic idea was the non-linear registration of multi-atlases derived from image analysis of MRI images of the human brain. The proposed approach was evaluated and presented r^2 values 12% higher than previous techniques and methods used in rabbit breeding selection programs. The advantage of this method is that it is less sensitive to small variations in the position of the rabbits during the CT scan.

Selection programs based on maximizing the growth rate and meat yield of the paternal line result in a reduction in body fat percentage, which can have detrimental consequences for reproductive performance. The amount of total body fat can be determined by CT scan and incorporated into breeding programs, as demonstrated by Kasza et al. (2020) in a two-way selection experiment.

2.3 Magnetic Resonance Imaging

Magnetic resonance imaging (MRI) has been identified as the imaging technique used in animal science as the most accurate for determining the body and carcass composition of meat-producing animals (Scholz et al. 2015). This technique exhibits several qualities that are not observed in either RTU or CT. The MRI allows an excellent contrast among or within the various tissues and organs, making it possible to measure its volume very precisely (Doneva 2020). This is undoubtedly why it has been identified as a reliable alternative to dissection (Monziols et al. 2006) and can also be used as a gold standard for other techniques (Weigand et al. 2020).

Despite these merits, MRI has significant constraints, such as those related to the need for a Faraday cage around the device; the high costs of the equipment, its operation, and maintenance; and the scanning time (Weigand et al. 2020). Nonetheless, several centers in the EU use MRI for animal body and carcass composition (Daumas et al. 2015). In this work, three centers conducting research in meat species (IRSTEA, Rennes, France; LMU, Munich, Germany, and the former Institute of Diagnostic Imaging and Radiation Oncology, Kaposvár, Hungary) are identified. The MRI has been used primarily on swine but also in poultry, turkey, fish, and sheep (Daumas et al. 2015). However, for rabbits, this technique has been little used. An example is the work by Kövér et al. (1996), which shows the suitability of MRI for the determination of kidney fat and shows significant correlations between the in vivo kidney fat volume and post-slaughter weight of the fat surrounding the kidney ($r > 0.90$). In another work using MRI to estimate the muscle volume of the hind part, the volume of the perirenal fat, and the total fat content of the empty body in 87 rabbits of 4 genotypes, Köver et al. (1998) reported correlation values between 0.77 and 0.94 for 16-week rabbits, which allowed them to conclude that MRI is an excellent method to determine volumes of fat and muscle.

2.4 Dual-Energy X-Ray Absorptiometry

Dual-energy X-ray absorptiometry (DXA) was introduced 35 years ago to measure bone mineral density. Since then, it has been the widely accepted technique for measurements of body composition to provide information for clinicians, nutritionists, and physiologists (Glüer 2017). Over the years, this technique has been the target of continuous technological innovations. Improvements opened DXA to new applications, offering the possibility of evaluating several musculoskeletal parameters other than bone mineral density (Messina et al. 2020). Due to DXA capabilities and high accuracy, it has been repeatedly used in the in vivo determination of body composition and to predict carcass and meat characteristics in meat-producing species (Scholz et al. 2015; Kušec et al. 2016; Pomar et al. 2017). Unfortunately, despite the results obtained in the main meat-producing species, no work was found in rabbits using DXA to predict carcass traits. However, there are multiple examples with DXA in which the rabbit is used as an animal model in the medical field (e.g., Adji et al. 2022; Özkan et al. 2023).

3 Conclusions

The use of imaging tools in animal husbandry dates back decades, yet most people first think of their use in diagnostics and reproductive biology. This chapter shows that these tools can also be used to determine the traits of rabbits in vivo, thus helping breeders in their selection work or the postmortem examinations that are useful for meat producers/meat scientists.

References

Adji AV, Pedersen AØ, Agyekum AK (2022) Obesity in pet rabbits: A narrative review. Journal of Exotic Pet Medicine 41:30–37. https://doi.org/10.1053/j.jepm.2022.02.003

Amalianingsih TI, Brahmantiyo B (2014) The variability of growth hormone gene associated with ultrasound imaging of longissimus dorsi muscle and perirenal fat in rabbits. Media Peternakan 37(1):1–7. https://doi.org/10.5398/medpet.2014.37.1.1

Bentsen HB, Sehested E (1989) Computerised tomography of chickens. Br Poult Sci 30(3): 575–589. https://doi.org/10.1080/00071668908417181

Blasco A, Nagy I, Hernández P (2018) Genetics of growth, carcass and meat quality in rabbits. Meat Sci 145:178–185. https://doi.org/10.1016/j.meatsci.2018.06.030

Bünger L, Glasbey CA, Simm G, Conington J, Macfarlane JM, McLean KA et al (2011) Use of X-ray computed tomography (CT) in UK sheep production and breeding. INTECH Open Access Publisher, Rijeka, pp 329–331

Bünger, L.; Moore, K.; McLean, K.; Kongsro, J.; Lambe, N. (2014). Integrating computed tomography into commercial sheep breeding in the UK: cost and value. In Farm Animal Imaging III; Maltin, C., Craigie, C., Bünger, J., Eds.; SRUC: Copenhagen, Denmark, pp. 22–27

Castellini C, Dal Bosco A, Cardinali R (2006) Long term effect of post-weaning rhythm on the body fat and performance of rabbit doe. Reprod Nutr Dev 46(2):195–204. https://doi.org/10.1051/rnd:2006009

Crovato S, Pinto A, Di Martino G, Mascarello G, Rizzoli V, Marcolin S, Ravarotto L (2022) Purchasing habits, sustainability perceptions, and welfare concerns of Italian consumers regarding rabbit meat. Foods 11(9):1205. https://doi.org/10.3390/foods11091205

Csóka Á, Kovács G, Ács V, Matics Z, Gerencsér Z, Szendrö Z et al (2022) A general technique for the estimation of farm animal body part weights from CT scans and its applications in a rabbit breeding program. Comput Electron Agric 196:106865. https://doi.org/10.1016/j.compag.2022.106865

Cullere M, Dalle Zotte A (2018) Rabbit meat production and consumption: state of knowledge and future perspectives. Meat Sci 143:137–146. https://doi.org/10.1016/j.meatsci.2018.04.029

Dal Bosco A, Castellini C, Mugnai C (2003) Evaluation of body condition in pregnant rabbit does by ultrasound scanner. Ital J Anim Sci 2(sup1):480–482

Dalle Zotte A, Szendrő Z (2011) The role of rabbit meat as functional food. Meat Sci 88(3): 319–331. https://doi.org/10.1016/j.meatsci.2011.02.017

Daumas G, Donkó T, Maltin C, Bünger L (2015) Imaging facilities (CT & MRI) in EU for measuring body composition. SRUC, Edinburgh, UK, pp 5–48

Delgado-Pando G, Allen P, Troy DJ, McDonnell CK (2021) Objective carcass measurement technologies: latest developments and future trends. Trends Food Sci Technol 111:771–782. https://doi.org/10.1016/j.tifs.2020.12.016

Dewez MF, Etourneau P, Lecompte F, Briere S, Froment P (2018) Use of computed tomography to determine body composition of heavy strain Turkey hens (Meleagris gallopavo) from rearing to early laying. Poult Sci 97(11):4093–4106. https://doi.org/10.3382/ps/pey267

Doneva M (2020) Mathematical models for magnetic resonance imaging reconstruction: an overview of the approaches, problems, and future research areas. IEEE Signal Process Mag 37(1): 24–32. https://doi.org/10.1109/MSP.2019.2936964

Du Z, Hu Y, Ali Buttar N, Mahmood A (2019) X-ray computed tomography for quality inspection of agricultural products: A review. Food Sci Nutr 7(10):3146–3160. https://doi.org/10.1002/fsn3.1179

FAOSTAT (2020) The Statistics Division of the FAO. Available online: http://www.fao.org/faostat/en/#data. Accessed 2 April 2023

Font-i-Furnols MF, Teran MF, Gispert M (2009) Estimation of lean meat content in pig carcasses using X-ray computed tomography and PLS regression. Chemom Intell Lab Syst 98(1):31–37. https://doi.org/10.1016/j.chemolab.2009.04.009

Font-i-Furnols M, Terré M, Brun A, Vidal M, Bach A (2021) Prediction of tissue composition of live dairy calves and carcasses by computed tomography. Livest Sci 243:104371. https://doi.org/10.1016/j.livsci.2020.104371

Gardner GE, Apps R, McColl R, Craigie CR (2021) Objective measurement technologies for transforming the Australian & New Zealand livestock industries. Meat Sci 179:108556. https://doi.org/10.1016/j.meatsci.2021.108556

Geraldo AT, Bartholazzi A, David CMG, de Freitas ACB, Antunes F, Rua MAS et al (2020) Body tissue development of Santa Inês breed sheep using computed tomography. Anim Prod Sci 61(2):191–200. https://doi.org/10.1071/AN20084

Glüer CC (2017) 30 years of DXA technology innovations. Bone 104:7–12. https://doi.org/10.1016/j.bone.2017.05.020

Gyovai P, Nagy I, Gerencsér Z, Matics Z, Radnai I, Donkó T et al (2012) Genetic parameters for litter weight, average daily gain and thigh muscle volume measured by in vivo computer tomography technique in Pannon white rabbits. Livest Sci 144(1–2):119–123. https://doi.org/10.1016/j.livsci.2011.11.006

Hernández P, Ariño B, Grimal A, Blasco A (2006) Comparison of carcass and meat characteristics of three rabbit lines selected for litter size or growth rate. Meat Sci 73:645–650. https://doi.org/10.1016/j.meatsci.2006.03.007

Johnsona PL, Juengela JL, Bain WE (2020) Predicting internal adipose from selected computed tomography images in sheep. New Zealand Journal of Animal Science and Production 80:113–116

Kasza R, Donkó T, Matics Z, Nagy I, Csóka Á, Kovács G, Gerencsér Z, Dalle Zotte A, Cullere M, Szendrő Z (2020) Rabbit lines divergently selected for total body fat content: correlated responses on growth performance and carcass traits. Animals 10(10101815):1–10. https://doi.org/10.3390/ani10101815

Kovács G, Donkó T, Emri M, Opposits G, Repa I (2013) Gabor-filter based automatic removal of troughs from CT images. Farm Animal Imaging. Kaposvár, The Rural Centre, Ingliston Newbridge, UK, pp 80–84

Kövér G, Sorensen P, Szendrö Z, Milisits G (1996) Resonance tomography. In Proceedings of the 6th world rabbit congress, Vol. 3, Toulouse

Köver GY, Szendrö ZS, Romvari R, Jensen JF, Sorensen P, Milisits G (1998) In vivo measurement of body parts and fat deposition in rabbits by MRI. World Rabbit Sci 6(2):231–235. https://doi.org/10.4995/wrs.1998.349

Kumar P, Abubakar AA, Verma AK, Umaraw P, Adewale Ahmed M, Mehta N et al (2022) New insights in improving sustainability in meat production: opportunities and challenges. Crit Rev Food Sci Nutr 1-29:1. https://doi.org/10.1080/10408398.2022.2096562

Kušec G, Scholz AM, Baulain U, Kušec ID, Bernau M (2016) Non-invasive techniques for exact phenotypic assessment of carcass composition and tissue growth in domestic animals. Acta Argiculturae Slovenica 5:12–17

Lambertini L, Bergoglio G, Masoero G, Gramenzi A (1996) Comparison between Provisal and Hyla rabbit strains. I. Slaughtering performances and muscle composition. In proceedings of the 6th world rabbit congress, Toulouse. France 3:195–199

Leighton PL, Marcoux M, Wei X, Soladoye P, Dugan ME, Juarez M, Prieto N (2021) Prediction of carcass composition and meat and fat quality using sensing technologies: A review. Meat Muscle Biol 5(3): 12951, 1–21. https://doi.org/10.22175/mmb.12951

Lenoir G, Morien F (2016) Estimation of genetic parameters for carcass traits evaluated by in vivo real-time ultrasonography in meat rabbit breeding. In: Proceedings of the 11th world rabbit congress, Qingdao, China, 15–18 June

Lukefahr SD, Ozimba CE (1991) Prediction of carcass merit from live body measurements in rabbits of four breed-types. Livest Prod Sci 29:323–334. https://doi.org/10.1016/0301-6226(91)90107-2

Lukefahr SD, Hohenboken WD, Cheeke PR, Patton NM, Kennick WH (1982) Carcass and meat characteristics of Flemish Giant and New Zealand white purebred and terminal-cross rabbits. J Anim Sci 54:1169–1174. https://doi.org/10.2527/jas1982.5461169x

Lukefahr SD, Hohenboken WD, Cheeke PR, Patton NM (1983) Appraisal of nine genetic groups of rabbits for carcass and lean yield traits. J Anim Sci 57:899–907. https://doi.org/10.2527/jas1983.574899x

Macfarlane JM, Lewis RM, Emmans GC, Young MJ, Simm G (2006) Predicting carcass composition of terminal sire sheep using X-ray computed tomography. Anim Sci 82(3):289–300. https://doi.org/10.1079/ASC200647

Magalhães LC, Costa RB, de Camargo GMF (2022) Consumption of rabbit meat in Brazil: potential and limitations. Meat Sci 191:108873. https://doi.org/10.1016/j.meatsci.2022.108873

Matics ZS, Nagy I, Gerencsér ZS, Radnai I, Gyovai P, Donkó T, Dalle Zotte A, Curik I, Szendrő ZS (2014) Pannon breeding program in rabbit at Kaposvár university. World Rabbit Sci 22(4): 287–300. https://doi.org/10.4995/wrs.2014.1511

Matics Z, Kovács G, Csóka Á, Ács V, Kasza R, Petneházy Ö et al (2020) Automated estimation of loin muscle mass in living rabbits using computed tomography. Acta Universitatis Agriculturae et Silviculturae Mendelianae Brunensis 68(1):63–71. https://doi.org/10.11118/actaun202068010063

Messina C, Usuelli FG, Maccario C, Di Silvestri CA, Gitto S, Cortese MC et al (2020) Precision of bone mineral density measurements around total ankle replacement using dual energy X-ray absorptiometry. J Clin Densitom 23(4):656–663. https://doi.org/10.1016/j.jocd.2019.01.006

Michalik D, Lewczuk A, Wilkiewicz-Wawro E, Brzozowski W (2006) Prediction of the meat content of the carcass and valuable carcass parts in French lop rabbits using some traits measured in vivo and post mortem. Czeh J Anim Sci 51:406–415. https://doi.org/10.4141/CJAS08060

Monziols M, Collewet G, Bonneau M, Mariette F, Davenel A, Kouba M (2006) Quantification of muscle, subcutaneous fat and intermuscular fat in pig carcasses and cuts by magnetic resonance imaging. Meat Sci 72(1):146–154. https://doi.org/10.1016/j.meatsci.2005.06.018

Nagy I, Gyovai P, Radnai I, Matics ZS, Gerencser ZS, Donkó T, Szendrő ZS (2010) Genetic parameters of growth in vivo CT based and slaughter traits in Pannon white rabbits. In 9th World Congress on Genetics Applied to Livestock Science, Leipzig, Germany, pp 1–6

Nagy I, Gyovai P, Radnai I, Nagyné Kiszlinger H, Farkas J, Szendrő Z (2013) Genetic parameters, genetic trends and inbreeding depression of growth and carcass traits in Pannon terminal line rabbits. Archives Animal Breeding 56(1):191–199. https://doi.org/10.7482/0003-9438-56-018

Özkan E, Şenel E, Bereket MC, Önger ME (2023) The effect of shock waves on mineralization and regeneration of distraction zone in osteoporotic rabbits. Ann Med 55(1):1346–1354. https://doi.org/10.1080/07853890.2023.2192958

Pascual JJ, Castella F, Cervera C, Blas E, Fernández-Carmona J (2000) The use of ultrasound measurement of perirenal fat thickness to estimate changes in body condition of young female rabbits. Anim Sci 70(3):435–442. https://doi.org/10.1017/S135772980005178X

Pascual JJ, Motta W, Cervera C, Quevedo F, Blas E, Fernández-Carmona J (2002) Effect of dietary energy source on the performance and perirenal fat thickness evolution of primiparous rabbit does. Anim Sci 75(2):267–279. https://doi.org/10.1017/S1357729800053029

Pascual JJ, Blanco J, Piquer O, Quevedo F, Cervera C (2004) Ultrasound measurements of perirenal fat thickness to estimate the body condition of reproducing rabbit does in different physiological states. World Rabbit Sci 12(1):7–21. https://doi.org/10.4995/wrs.2004.584

Petracci M, Soglia F, Leroy F (2018) Rabbit meat in need of a hat-trick: from tradition to innovation (and back). Meat Sci 146:93–100. https://doi.org/10.1016/j.meatsci.2018.08.003

Petrescu DC, Petrescu-Mag RM (2018) Consumer behaviour related to rabbit meat as functional food. World Rabbit Sci 26(4):321–333. https://doi.org/10.4995/wrs.2018.10435

Pomar C, Kipper M, Marcoux M (2017) Use of dual-energy x-ray absorptiometry in non-ruminant nutrition research. Rev Bras Zootec 46:621–629. https://doi.org/10.1590/S1806-92902017000700010

Quevedo F, Cervera C, Blas E, Baselga M, Pascual JJ (2006) Long-term effect of selection for litter size and feeding programme on the performance of reproductive rabbit does 1. Pregnancy of multiparous does. Animal Science 82:739–750. https://doi.org/10.1079/ASC200688

Romvári R, Milisits G, Szendrö Z, Sørensen P (1996) Non-invasive method to study the body composition of rabbits by X-ray computerised tomography. World Rabbit Sci 4:219–224. https://doi.org/10.4995/wrs.1996.298

Scholz AM, Bünger L, Kongsro J, Baulain U, Mitchell AD (2015) Non-invasive methods for the determination of body and carcass composition in livestock: dual-energy X-ray absorptiometry, computed tomography, magnetic resonance imaging and ultrasound: invited review. Animal 9(7):1250–1264. https://doi.org/10.1017/S1751731115000336

Scholz AM, Kusec G, Mitchell AD, Baulain U (2021) Tracing the inside of pigs non-invasively: recent developments. IntechOpen. https://doi.org/10.5772/intechopen.101740

Seeram E (2006) Nobel prize for CT and MRI pioneers: historical article. Radiographer: The Official Journal of the Australian Institute of Radiography 53(1):4–7. https://doi.org/10.1148/radiol.14140706

Siddiqui SA, Gerini F, Ikram A, Saeed F, Feng X, Chen Y (2023) Rabbit meat—production, consumption and consumers' attitudes and behavior. Sustainability 15(3):2008. https://doi.org/10.3390/su15032008

Silva SR, Cadavez VP (2012) Real-time ultrasound (RTU) imaging methods for quality control of meats. In: Computer vision technology in the food and beverage industries. Woodhead Publishing, pp 277–329. https://doi.org/10.1533/9780857095770.3.277

Silva SR, Stouffer JR (2019) Looking under the hide of animals. The history of ultrasound to assess carcass composition and meat quality in farm animals. História da Ciência e Ensino 20:523–535. https://doi.org/10.23925/2178-2911.2019v20espp523-535

Silva SR, Guedes CM, Mourão J, Pinheiro V (2007) Rabbit carcass chemical composition predicted by real time ultrasonography. In XXXII Symposium de ASESCU, 5 and 6 of June, Vila Real, Portugal, pp 9–12

Silva SR, Mourão JL, Guedes CM, Pio A, Pinheiro V (2008a) In vivo rabbit carcass composition and longissimus Dorsi muscle volume prediction by real time ultrasonography. Meat Quality & Safety, 9th World Rabbit Congress, 10–13 June, Verona, Italy, pp 1449–1453

Silva SR, Guedes CM, Mourão JL, Venâncio C, Pinheiro V (2008b) Estimation of rabbit hind leg muscle weight and muscularity by real time ultrasonography. Meat Quality & Safety, 9th World Rabbit Congress, 10–13 June, Verona, Italy, pp 1443–1447

Silva SR, Guedes CM, Mourão JL, Pio A, Pinheiro VM (2009) The value of in vivo real time ultrasonography in assessing loin muscularity and carcass composition of rabbits. Meat Sci 81: 357–363. https://doi.org/10.1016/j.meatsci.2008.08.013

Silva SR, Jorge AM, Mourão JL, Guedes CM, Pinheiro V (2012) Real-time ultrasound to predict rabbit carcass composition and volume of longissimus dorsi muscle. Rev Bras Zootec 41:2463–2466. https://doi.org/10.1590/S1516-35982012001200012

Silva S, Guedes C, Rodrigues S, Teixeira A (2020) Non-destructive imaging and spectroscopic techniques for assessment of carcass and meat quality in sheep and goats: A review. Foods 9(8): 1074. https://doi.org/10.3390/foods9081074

Silva SR, Guedes CM, Almeida M, Mourão LM, Pinheiro V (2021) Prediction of rabbit body fat deposits from perirenal fat measurements obtained with real-time ultrasonography. In Proceedings of the 12th world rabbit congress, Nantes (France) - November 3–5

Skjervold H, Grønseth K, Vangen O, Evensen A (1981) In vivo estimation of body composition by computerized tomography. Z Tierzuecht Zuechtungsbiol 98(1–4):77–79. https://doi.org/10.1111/j.1439-0388.1981.tb00330.x

Szendrő Z, Romvári R, Nagy I, Andrassy-Baka G, Metzger S, Radnai I, . . ., Horn P (2005) Selection of Pannon White rabbits based on computerised tomography. In Proceedings of the 8th World Rabbit Congress, September 7–10, 2004, Pueblo, Mexico. World Rabbit Science Association (WRSA), pp 175–180

Szendrő K, Szabó-Szentgróti E, Szigeti O (2020) Consumers' attitude to consumption of rabbit meat in eight countries depending on the production method and its purchase form. Foods 9(5): 654. https://doi.org/10.3390/foods9050654

Trocino A, Cotozzolo E, Zomeño C, Petracci M, Xiccato G, Castellini C (2019) Rabbit production and science: the world and Italian scenarios from 1998 to 2018. Ital J Anim Sci 18(1): 1361–1371. https://doi.org/10.1080/1828051X.2019.1662739

Valencia-Garcia CG, Arbez-Abnal TA, Portillo-Salgado R, García-Herrera RA, Cigarroa-Vazquez FA, Herrera-Camacho J et al (2022) Assessment of carcass tissue composition in growing rabbits using real time ultrasonography. Trop Subtrop Agroecosystems 25(2). https://doi.org/10.56369/tsaes.3982

Weigand A, Schweizer H, Aline Knob D, Scholz AM (2020) Phenotyping of the visceral adipose tissue using dual energy X-ray absorptiometry (DXA) and magnetic resonance imaging (MRI) in pigs. Animals 10(7):1165. https://doi.org/10.3390/ani10071165

Weiskirchen R (2022) The Beatles in life sciences: facts and fictions. Biochem Mol Biol Educ 50(3):334–344. https://doi.org/10.1002/bmb.21617

Yonkova P, Rusenov A, Kanakov D, Zapryanova D, Vachkova E, Serbest A et al (2012) Ultrasound imaging, biochemical blood analyses, and weight investigations of dissectible fat depots in New Zealand white rabbits. Turkish Journal of Veterinary & Animal Sciences 36(6): 635–641. https://doi.org/10.3906/vet-1105-17

Rabbit Skins and Furs: Production, Properties, and Potentiality as Raw Materials for Tanning and Other Industries

A. I. Nasr and E. A. Taha

Abstract

Rabbit skin is a by-product of rabbit meat production. For years, rabbit skins have not received adequate attention compared to the skins and hides of other animal species, despite their unique characteristics. In this regard, this chapter aims to discuss all matters related to rabbit skins, furs, and leathers. While rabbit skins have distinctive characteristics in terms of softness, low thickness, and distinguishing appearance, they compete with the skins or hides of other types in the areas of its industrial usages, although they are not used sufficiently. The main reasons for the limited industrial use of rabbit skin are improper skinning, small surface area, and the difficulty of assembling skins into larger pieces. However, with rabbit domestication, the application of correct skinning methods, proper handling of skins, and use of modern tanning processes may lead to an increase in the marketing of rabbit skins.

Keywords

Fur · Leather · Pelt · Properties · Tanning · Wool

1 Introduction

Rabbit animals are highly adaptable and can live and reproduce in diverse environmental conditions, making them a vital resource for pastoral people. There are many rabbit breeds that are raised mainly either for meat production or for using their hair fibers. Consequently, under the prevailing rearing conditions, three more

A. I. Nasr (✉) · E. A. Taha
Department of Wool Production & Technology, Animal & Poultry Production Division Desert Research Centre, Cairo, Egypt

J. Simões, J. M. Monteiro (eds.), *Veterinary Care of Farm Rabbits*,
https://doi.org/10.1007/978-3-031-44542-2_16

by-products are produced from their skins, the fur, the pelt, and the shorn hair, with no particular production constraint. The usages of these by-products are usually limited to fur garments, pelts for shorn hair (wool removed from the skin), and skins for use as fertilizer (Lebas et al. 1997). In some developing countries, rabbit skins are not exploited due to lack of interest in skinning, which damages them and thus becomes unusable for any industrial purposes (Taha et al. 2017).

Therefore, due to the lack of information on the uses of rabbit skins, furs, and leathers, this chapter will examine all aspects of rabbit skins, including their production, properties, and marketing.

2 Global and Regional Production of Rabbit Skins and Wool

There are no specific international records assigned for rabbit skin numbers or marketing values. Thus, the closest accurate to the scale of rabbit skin production is to rely on statistics of the slaughtered animals. According to the data of the Food and Agriculture Organization of the United Nations (FAO) during the past four decades from 1960 to 2020, there is a proportional relationship between the total number of rabbits and the number of slaughtered and thus the number of produced pelts. However, the resulting amount of pelts may be affected by a number of factors such as the low demand for rabbit's meat and thus the numbers of the resulting pelts due to the negative effects of the economic crisis all over the world, rearing of rabbits as pets, as well as the gradual change in lifestyle and eating habits for consumers (Trocino et al. 2019).

About two-thirds of the world's rabbit production is located in Asia (68%), where China and Korea are the two largest rabbit producers in the world (49 and 17%, respectively). The remaining third is mostly distributed in Europe (19%) and Africa (11%). Egypt is the largest African producer with about 9%, while Italy is the largest European producer with about 2.5% of the total world rabbit's production (FAO 2020). These data may enhance the potentiality of China, Korea, and Egypt to be promising countries in rabbit's pelt production. Unfortunately, there are no documented reports in this regard.

In fact, it is difficult to predict how each country will exploit the rabbit skins resulting from its meat industry. For example, although Egypt produces about 8% of the world rabbit meat production, most of produced rabbit skins are wasted and are not well utilized for reasons related mainly to the wrong skinning method used (Taha et al. 2017). On the other hand, Das et al. (2014) reported that France produces about 70 million rabbit fur skins and 56% of it is used for making different products like caps, jackets, etc. Poland and Russia utilize all the pelts whatever they produce for making different products. Australia exports fur skin produced in their country. The main importers of raw rabbit skins are Korea and the Philippines, and soon after processing, they export rabbit pelts to developed countries like the USA, Italy, and Germany.

With respect to rabbit wool production, Schlink and Liu (2003) explained that Angora production is truly an international industry. Angora rabbit-producing countries are not normally Angora wool producer, as well as they are not usually significant consumers of finished articles. For example, France is not a large consumer of Angora wool, and for a long time, French Angora production was developed mainly for the export market. The situation has continued to this day with the majority of fiber from producer countries such as China and Chile who export the raw fibers to be processed in Europe, Japan, and Korea. The main retail markets are in North America, Western Europe, and Japan. China is slowly increasing its Angora manufacturing capacity. India is an exception in the Angora industry, being a producer, processor, and consumer of Angora fibers.

3 Histological Construction of Rabbit Skin

Rabbit skins are thin, but they are structurally similar to other mammalian skins, with limited differences. Rabbit skin is a heterogeneous material covered with hair fibers (Ibrahim et al. 2017). Typically to other mammals, rabbit skins contain three main layers, epidermis, dermis, and hypodermis, in addition to the presence of a fourth layer of the "pellicle layer" (Ibrahim et al. 2017; Taha et al. 2017). Figure 1 shows the vertical section of rabbit skin.

3.1 Epidermis Layer

It is the upper thin layer of the skin that contains corneous and granulosum layers (Ibrahim et al. 2017). The thickness of the epidermis does not exceed 1% of the total thickness of the skin (BASF 2007).

3.2 Dermis Layer

It is the main layer of the skin that consists of both papillary and reticular layers. The papillary layer is the upper part of the dermis that contains hair follicles and sebaceous glands. Reticular layer locates below of the papillary layer and mainly contains the elastic fibers and dense collagen fibers in bundle form (Brewer 2006; Ibrahim et al. 2017). The thickness of the dermis is about 85% of the total skin thickness (BASF 2007).

3.3 Hypodermis and Pellicle Layers

They are the lower two layers of the skin, and their thickness is the rest of the total skin thickness (BASF 2007). Hypodermis layer contains blood vessels and fat, whereas pellicle layer consists of a subcutaneous skeletal muscle (Cruise and Brewer 1994; Ibrahim et al. 2017); thence, both layers should be removed when preparing the skins for tanning (Covington 2009).

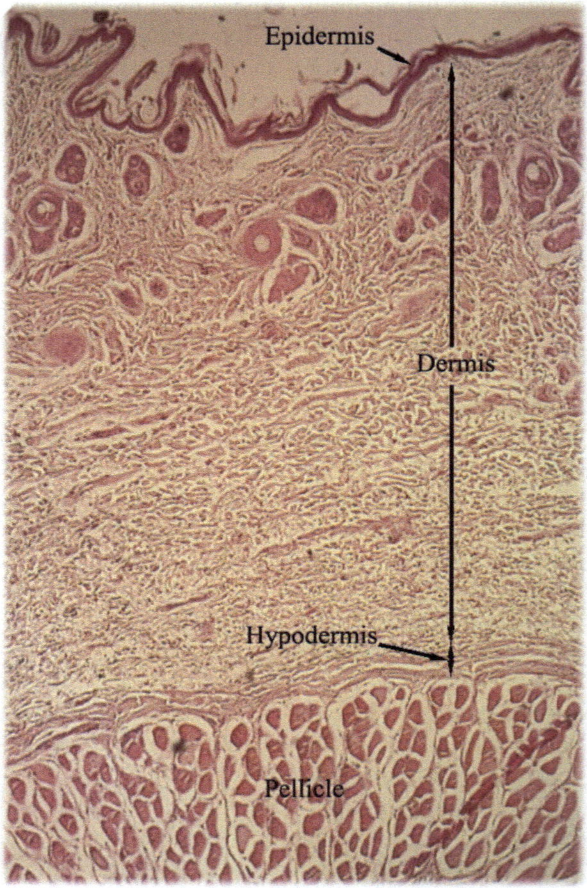

Fig. 1 Vertical section of rabbit skin

3.4 Follicular Construction of Rabbit Skins

Hair follicles are current in all body areas, and they arrange in follicular groups that contain two types of follicles – the primary and the secondary hair follicles—and both types are surrounded by distinguished connective tissues. The arrangement of hair follicles varies markedly among different rabbit breeds (Ibrahim et al. 2017; Lucy et al. 2019). In most of domesticated breeds, i.e., New Zealand white, California, and Rex, follicular groups contain one to three primary follicles accompanied with clusters of two to four secondary follicles lying to one side of each primary follicle (Taha et al. 2016). In other breeds, i.e., Angora, White Giant, and wild rabbit breeds, the majority of follicular groups consist of one central primary follicle that is surrounded by about 8 to 12 secondary follicles (Ibrahim et al. 2017; Lucy et al. 2019).

Primary follicles are rooted deeply in the dermis layer and are associated with accessory structures such as sebaceous and sweat glands and arrector pili muscle (Taha et al. 2016). They emerge primary hair fibers with thicker diameter and taller length than the secondary fibers that are produced by secondary follicles (Monteiro-Riviere and Stinson 1998). Moreover, primary follicles are larger in diameter and dimensions than secondary follicles, which are located superficially in the dermis. Secondary follicles may associate with sebaceous glands, but they lack sweat glands and arrector pili muscles (Oznurlu et al. 2009).

Sweat glands in rabbit skin excrete a serous secretion and are not functional in evacuating latent body heat. Therefore, sweating has a negligible effect on rabbit's thermal stability due to the presence of fur coat that envelops rabbit skin (Lebas et al. 1997).

The ratio of secondary to primary follicles, also known as the S/P ratio, depends on many factors, including breed, sex, nutrition, mineral intake, and season (Nasr et al. 2020; Taha et al. 2020). Meanwhile, it affects the fur traits. The higher S/P ratio associates higher coverage of the skin, more softness of fibers, lower prickling, and better body insulation (Taha et al. 2016, 2017). Meanwhile, S/P ratio varies among the different regions of rabbit's body where the highest ratio was in the back region and less in the ear region (Ibrahim et al. 2017). The lowest S/P ratio in ear skin causes sparse and limited hair growth on the posterior aspect and the distal parts of the anterior aspect to provide less coverage that may help more heat radiation from ear's skin during hot weather.

Hair growth is a highly regulated cyclical process (Soma et al. 1998). It goes through three specific stages of growing, regression, and resting phases (Stenn and Paus 2001). The first stage of the hair growth cycle is the anagen phase, in which hair follicles grow rapidly by cell proliferation and fibers reach their longest length. The second stage is the catagen phase, in which cell proliferation slows gradually until pausing follicular growth. The follicles then regress by apoptosis and enter their resting phase, the telogen phase, until the beginning of the next growth cycle (Soma et al. 1998; Ding et al. 2019; Lucy et al. 2019).

Rabbit skin has three main horizontal blood vessel networks: the sub-epidermal plexus, the dermal plexus, and the hypodermal plexus. The sub-epidermal plexus that consisted of fine vessels situated just beneath the epidermis that is a superficial network of fine vessels. The dermal plexus in the middle level of the skin. The hypodermal network that lies between the dermis and hypodermis (Geoffrey et al. 1966). During the active phase of hair growth, a continuous and very extensive network of fine vessels surrounds hair follicles. This network extends from the sub-epidermal plexus to the bases of follicles and embeds in the connective tissue sheaths, which surround the follicles to increase blood supply to the follicles. When follicles start their resting phase, they start to regress and locate their bases at a more superficial level in the dermis; subsequently, the blood supply to the follicles declines.

4 Chemical Composition of Skins

Rabbit skins are somewhat similar in their chemical composition to other mammals. Thus, it is essentially composed of water, minerals, natural fat, protein material, and carbohydrates (Gutterres 2003). The difference is usually in the fat content, while the values of other analyses are close to each other (Table 1).

5 Utilization of Rabbit Skins

5.1 Production of Rabbit Furs and Leathers

Using raw rabbit skins for different manufacturing requires tanning and finishing to be suitable for the desired purpose. But before referring to matters related to the tanning of rabbit skins, it should be clarified about what may affect the quality of raw skins. Generally, rabbit skins should be taken care of starting from the animal raising, slaughtering, skinning, and preserving the skins. Therefore, in order to improve and maximize raw skins' quality, some precautions must be taken into account, which can be summarized as follows:

- ***Seasonal molts in rabbits***
  Lebas et al. (1997) and Schlink and Liu (2003) reported that the summer coat in rabbits is usually the least valuable, as the fibers are thin and short, so the resulting skins are of less prized if they are used as furs. As for the winter coat, it is more stable and highly prized due to the greater contracting of collagen fibers in the dermis layer during this period, which makes the skins stronger and softer. Thus, these skins are often the most used for producing furs.
- ***Animal age and weight***
  Animal weight and dimensions are usually associated with its age, with reservations about other influences such as production and feeding systems. The infant rabbits have skins with too small surface area and a light hair coat that makes their skins unsuitable for using. In general, for most rabbit breeds, at the beginning of maturity stage (when animal weight reaches 1.7–1.9 kg), rabbit skin surface area, hair length, and coat density improve. Thence, it considers the first stage for using rabbit skins (Lebas et al. 1997; Taha et al. 2016).

Table 1 Chemical composition of rabbit skins vs. other animal species

Item	Cattle (BASF 2007)	Goat (BASF 2007)	Sheep (BASF 2007)	Rabbit (Liu et al. 2019)	Rabbit (Souza et al. 2016)
Moisture	65%	65%	65%	76.75%	64.37%
Protein	33%	33%	33%	14.56%	29.62%
Fat	2–6%	2–10%	5–30%	5.84%	5.96%
Ash	0.5%	0.5%	0.5%	1.35%	2.08%

- ***Animal health***
  Caring of animal health and avoiding any physiological or pathological disorders are reflected in the activity of the skin follicles, which make the hair fibers grow properly and result in usable skins, especially as fur (Lebas et al. 1997). Moreover, paying attention to the periodic external hygiene of animals would generally improve skin quality.
- ***Slaughtering and skinning***
  The methods of killing rabbits differ by countries and/or religious beliefs. Previous studies indicated that it may be done by removing the head directly, slaughtering using a sharp knife in the neck area for bleeding (Halal method), or stunning animal before slaughtering by different methods in order to minimize the suffering and pain of animals (Składanowska-Baryza and Stanisz 2019).

  Apart from the method of ending animal life, the bleeding should be done quickly and well as much as possible to reduce blood remaining in the skin vessels and thus decrease skin decomposition by microbial activity. Also, it is preferable to hang the animal from its hind legs after slaughter to avoid the fur from blood staining and remain clean (PNS 2022).

  On the other hand, skinning method is done differently based on the degree of interest of rabbit skins. In some countries with limited interest in using rabbit skins, such as Egypt, skinning is done in faulty way by cutting the skins from the abdomen area into two pieces, which damages them making it unusable (Taha et al. 2017). In standard conditions illustrated by Lebas et al. (1997), Składanowska-Baryza and Stanisz (2019), and PNS (2022), rabbit skins are flayed to obtain the maximum possible surface area, taking into account the avoidance of mutilation, knife marks, or grease, meat, or bloodstains. Skinning is done usually by making first cut at the hind feet and then making transverse incision between the two hind feet passing through thighs. Thereafter, the skin is pulled off toward the head until it is completely removed.
- ***Skin preservation***
  Skin preservation must be started after skinning immediately, during the first 15 minutes, to avoid bacterial decomposition and hair fall. Thus, skins must not be left in a pile for even a short time. Lebas et al. (1997) reviewed a standard method for preserving rabbit skins after slaughtering. The skins are turned inside out that the fleshy side is facing outward. Thereafter, it framed on inside a thick stainless or galvanized wire. In turn, skins are cooled off quickly and dried out to prevent the destructive enzymatic activity that attacks the hair root. Mostly, the preservation of rabbit skins is done by drying to decrease costs and the negative environmental impact when using salts. However, when the ambient temperature rises in hot regions, it is preferable to use salting to avoid any decomposition. After preservation, the well-dried rabbit pelts are usually placed in piles according to their grades, as well as insecticides, mainly naphthalene, are applied between each layer of skins to avoid any damage that may be caused by insects in the future until tanning process.

5.1.1 Tanning Rabbit Skins

The tanneries receive the rabbit pelts as classified piles in different grades according to their condition (Lebas et al. 1997). Actually, there are no standard methods for grading rabbit pelts, unlike other animal species, i.e., cattle, sheep, and goats. But due to the purpose of the final use of rabbit pelts, their size, color, and condition all play an important role in grading it. In general, the highest grade is the one that has the largest size and free from any apparent faults (i.e., cuts, holes, or scratches); thereafter, the pelt grade decreases according to the deterioration in these parameters. Moreover, pelts with a uniform hair coat and a white color are the highest prized as they can be dyed in various colors without restrictions. Figure 2 shows fur color variations due to rabbit breeds.

Although rabbit pelts could be sorted and graded according to the previous points, the classification into grades is considered realistically an agreement between producers and tanners. The classification is usually in four grades, with the first being the highest value while the last representing the rejected pelts as done in other animal species (BASF 2007).

The tanning process aims to remove non-collagenous components of the skin, leaving the collagen resistant to physical, chemical, and biological factors after reacting with tanning materials (Dutta 2008; Covington 2009). Figure 3 shows the difference between untanned and tanned rabbit furs. After completing tanning steps, the flesh side becomes clean and smooth, while the fur becomes flexible and therefore suitable for use.

In any case, the rabbit pelts, whatever their rank, have the same composition like other animal species. However, the purpose of their utilization affects the method of preparing and tanning. Accordingly, for fur production, pelts with good hair cover

Fig. 2 Fur color variations due to rabbit breeds

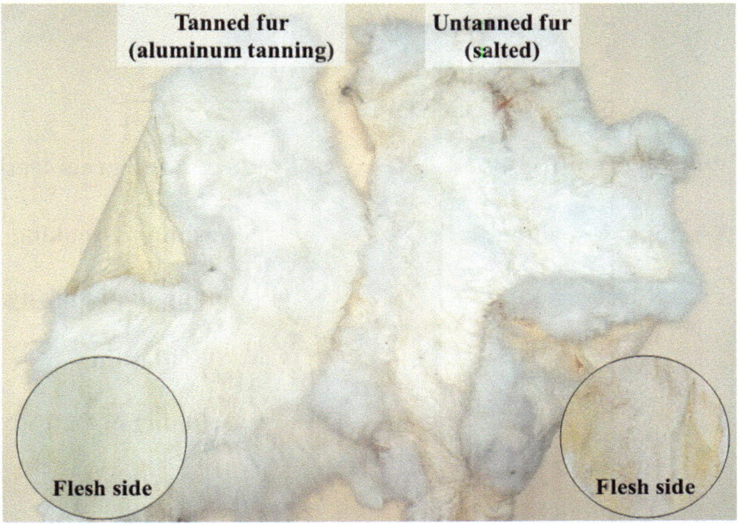

Fig. 3 Difference between untanned and tanned furs

are preferred. Skins with poor hair cover can be used to produce full-grain leather or nappa (leather without hair). Therefore, unhairing is an optional step in preparing skins for tanning, depending on the desired end product.

BASF (2007), Dutta (2008), and Covington (2009) explained the basics of tanning stages and steps for tanning animal skins, while some other studies have applied some of these steps on rabbit skins to suit the purpose of leather production, whether it is furs (Nasr et al. 2020; Taha et al. 2020; Maryati et al. 2021) or nappa leathers (Zapletal et al. 2020; Héctor-Julio et al. 2021). Figure 4 shows the preparation and tanning steps according to the production purpose of rabbit skins, while Table 2 summarizes the objective of each step and how to perform it.

5.1.2 Properties of Tanned Rabbit Furs and Leathers

After tanning, the tanned leather or fur acquired characteristics differ according to tanning materials used, as well as fatliquors, oils, or greases applied. Table 3 shows the physical properties of tanned rabbit leathers and furs as mentioned in literature. Chrome-tanned leathers are usually of high quality compared to other tannages (BASF 2007; Dutta 2008). Kanuri et al. (2019) found that chrome-tanned leather had higher tensile strength, tear strength, elongation, and shrinkage temperature than the corresponding vegetable-tanned leathers, whereas the latter are the thickest. Also, the percentage of adding fatliquors or oils affects the properties of the finished leathers or furs, as the elasticity of collagen fibers increases with the increase in the amount of added oils (Maryati et al. 2021). Also, from previous studies as shown in Table 3, there are other factors that may affect the characteristics of tanned rabbit leathers and furs, i.e., animal age, breed, sex, and breeding season; however, the quality of tanned leathers was mostly suitable for manufacturing in different usages

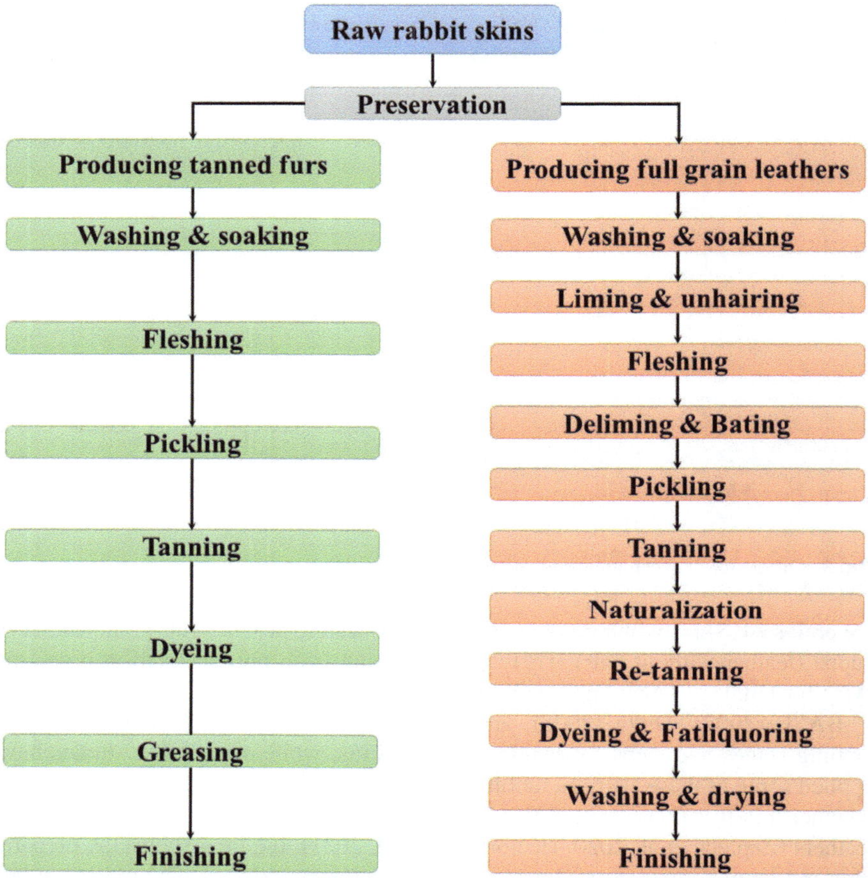

Fig. 4 Tanning flowchart for producing rabbits' furs and leathers

such as garments, gloves, footwear, fancy products, and lining. Additionally, Nasr et al. (2020) referred that reinforcing tanned furs with textile padding may increase their utility for most leather manufacturing purposes.

The other important point is the coloring effect of tanning materials that should be taken into account especially in the furs, as their hair color is changed if it was light. Chrome-tanned leathers acquire a color that tends to be light blue or light green, while vegetable-tanned leathers vary in colors among light yellow, orange, or brown, depending on the nature of plant extract that is used for tanning. Contrarily, aluminum and some of synthetic tannages produce wet white leathers, and thus it is more suitable for tanning white rabbit furs to save their hair color (Lebas et al. 1997). Figure 5 shows the difference between chrome- and aluminum-tanned New Zealand rabbit furs. Chrome-tanned fur has blue color in flesh side vs. white

Table 2 The procedures for tanning rabbit skins

Step	Objective	Chemicals / Conditions	Procedure	Remarks
Washing and soaking	– Remove salts, dirt, and blood – Restoring the natural swollen condition of the skin	– Water	Pelts immerse in soaking pits, mechanical drum, or paddle with water until restoring the natural swollen condition of the skin and removal of dirt, soluble proteins, and curing agents	– If stirring is applied by mechanical drums, it should be at low speed to avoid skin damages
Liming and unhairing	– Removal of hair and epidermis layer – Loosening up the collagen fiber texture – Partial saponification of the natural grease	– Water – Alkaline (Na_2S) – Lime (pH = 13)	Pelts are dipped in the bath with light mechanical stirring until the hair is completely removed	– If stirring is applied by mechanical drums, it should be at low speed to avoid skin damages – This step is not applied if producing furs
Fleshing	– Removal of pellicle membrane, meats, and fats from flesh side mechanically to increase the penetration of chemicals in the next steps into the skin layers	No chemical	Pelts are fleshed mechanically by fleshing machine or by hand using special fleshing knife	– It needs to be done delicately and carefully to avoid skin damage
Deliming and bating	– Removal of lime by conversion into readily soluble salts – Loosening and peptizing fiber texture of the skin and elimination of alkali swelling	– Water – Deliming agent – Pancreas protease enzyme (pH = 7–8)	Pelts are dipped in the bath with mechanical stirring and then washed until the desired pH is reached	– This step is not applied if producing furs
Pickling	– Acidify the pelts to a certain pH before tanning	– Water – Salt (NaCl) – Acid (H_2SO_4) (Bé = 8–10) (pH = 2–4)	Pelts are dipped firstly in bath of water and salt for 15 minutes with mechanical	

(continued)

Table 2 (continued)

Step	Objective	Chemicals / Conditions	Procedure	Remarks
			stirring and then acid is added gradually. The step is ended when the skin pH reaches the required level	
Tanning and basifying	– Converting skin substance to irreversible stabilization material and resist to putrefaction	– There are various tanning materials: mineral (Cr, Al, and Zr), vegetable (mimosa), aldehydes, and synthetic – Basifying with alkaline ($NaHCO_3$) (pH = 3.8–4.2)	The tanning agent is added to the same pickling bath, and mechanical stirring is done for at least a day. After tanning, the leathers are basified by adding an alkaline to reach the desired pH, and then washed to remove excess salts	– When tanning furs, the hair fibers may be affected by the color of the tanning agent
Naturalization	– Removal of free acids present in mineral-tanned leather to facilitate chemical penetration in the next steps	– Water – Alkaline ($NaHCO_3$) (pH = 4.5–5.5)	Leathers are dipped in the bath with mechanical stirring until the desired pH is reached and then washed	This step is not applied if producing furs
Re-tanning	– Reacting other type of tanning agent with collagen fibers to give leathers other characteristics	– Water – Tanning agent	Leathers are dipped in the bath with mechanical stirring for about 2 hours	This step is not applied if producing furs
Dyeing	– To give leathers the final color according to the purpose of use	– Dye	Dye is added to the previous bath	
Fatliquoring	– To give leathers desirable softness according to the purpose of use	– Fatliquor or oil	When producing full-grain leathers, fatliquor is added to the previous	

(continued)

Table 2 (continued)

Step	Objective	Chemicals / Conditions	Procedure	Remarks
			bath When producing furs, it is taken out and hung to dry. After then, oil is applied on flesh side	
Washing and drying	– Getting rid of excess water and moisture – To be ready for finishing	Water	Leathers are washed in water, set out, and then hung till dried	It is enough to dry furs without washing after lubricating it in previous step
Finishing	– Giving leathers or furs a pleasing appearance – Removing excess grease	There are different chemicals that can be applied by spraying on the leather surface such as pigments, waxes, acrylic resins, polyurethane resins, etc.	Different treatments could be applied: – Spray leather surface to give it its final appearance (color: gloss – insensitive to water) – Ironing, buffing, or burnishing grain side – Beating to increase softness – Stirring in a tub with absorbents to remove excess grease – Combing hair	Furs are usually finished by beating, combing hair, and removing excess grease

color in aluminum-tanned fur. Additionally, hair fibers tended to be blue or gray in chrome-tanned fur, while they appear bright white in aluminum-tanned fur.

Hong et al. (2016) noted that the use of traditional tannages which are often used in tanning rabbit skins such as chromium and aldehydes, as well as industrial dyes, causes harmful environmental effects on the health of consumers. Therefore, it has been suggested to use eco-friendly tanning methods, such as chrome-free and aldehyde-free methods. These methods include using combinations of vegetable extracts and aluminum salts, safe organic agents, or mixtures of aluminum and zirconium salts. In fact, it is difficult to apply the eco-friendly tanning on rabbit fur as using only one tanning agent because it gives poor properties to the tanned furs. For example, traditional aluminum tanning is a reverse tanning, meaning that tanning material could be lost from collagen fibers when leathers are washed. This

Table 3 Physical properties of tanned rabbit leathers and furs as mentioned in the literature

Author	Objective	Physical properties				Conclusion
		Thickness (mm)	Tensile strength (N/mm^2)	Elongation (%)	Tearing strength (N/mm)	
Maryati et al. (2021)	Evaluate the effect of fatliquor level on the physical quality of Indonesian rabbit fur leather	–	5% (9.60) 10% (31.9) 15% (17.5) 20% (17.7) 25% (10.50)	5% (129.47) 10% (173.54) 15% (175.44) 20% (154.03) 25% (81.48)	5% (32.43) 10% (9.35) 15% (24.05) 20% (12.50) 25% (36.82)	Fatliquoring rabbit furs with 20% is the optimal concentration to produce the highest quality
Héctor-Julio et al. (2021)	Evaluate tanned furs of New Zealand rabbits submitted to vegetable tanning of guanacaste tree fruit extract	Parallel (0.67) Perpendicular (0.65)	Parallel (20.2) Perpendicular (25.1)	Elongation at 50 N Parallel (4.8) Perpendicular (4.3)	Parallel (5.8) Perpendicular (22.5)	Tanned furs with prepared extract were partially in accordance with the international standards
Nasr et al. (2020)	Investigate the effect of breed and season on rabbit fur characteristics	Summer (0.87) Winter (0.91) NZW (0.90) Gabaly (0.87) Crossbred (0.88)	Summer (13.83) Winter (13.69) NZW (13.64) Gabaly (13.80) Crossbred (14.01)	Summer (45.86) Winter (43.03) NZW (44.07) Gabaly (45.41) Crossbred (44.09)	Summer (23.86) Winter (21.91) NZW (23.28) Gabaly (23.46) Crossbred (21.15)	The influence of season and breed is negligible, and tanned rabbit furs are suitable for manufacturing leather garments
Taha et al. (2020)	Investigate the effect of sex and seasonal changes on New Zealand white rabbit fur characteristics	Summer (0.90) Winter (0.91) Male (0.91) Female (0.89)	Summer (13.41) Winter (13.82) Male (14.11) Female (13.11)	Summer (43.95) Winter (44.16) Male (45.86) Female (42.01)	Summer (25.93) Winter (21.29) Male (23.73) Female (22.76)	Tanned furs of New Zealand rabbits could be used in garment leather manufacturing after reinforcing with textile padding

Zapletal et al. (2020)	Investigate the possibility of using tanned rabbit broiler leathers with combination tanning in the production of glove and footwear leathers	0.44–0.61	13–16	55–62	–	Tanned leathers can be used for making gloves and shoe uppers
Kanuri et al. (2019)	Evaluate tanned leathers of local Kenyan rabbits submitted to chrome and vegetable (mimosa) tannages	Chrome (0.7) Vegetable (0.8)	Chrome (17.09) Vegetable (15.93)	Chrome (50.0) Vegetable (36.5)	Chrome (28.2) Vegetable (20.5)	Both tannages produced leathers suitable for lining and fancy products
Maryati et al. (2019)	Evaluate tanned leathers of local Indonesian rabbits submitted to vegetable tanning of Kepok banana bunches	Mimosa (1.03) Kepok (1.17)	Mimosa (23.96) Kepok (30.2)	Mimosa (173.68) Kepok (174.12)	Mimosa (4.91) Kepok (12.02)	Leathers tanned with bunches of Kepok banana were of higher quality than mimosa-tanned leathers
Pancapalaga and Nitiharjo (2019)	Evaluate tanned leathers of local Indonesian rabbits submitted to vegetable tanning of mangrove	–	35.02–35.63	25.28–29.80	–	Leathers met the consumer acceptance standards for footwear
Taha et al. (2017)	Evaluate the properties of tanned rabbit furs from three different breeds for manufacturing purposes' suitability	Gabaly (0.96) NZW (1.18) Rex (0.82)	Gabaly (13.98) NZW (13.27) Rex (12.62)	Gabaly (43.60) NZW (49.60) Rex (50.40)	Gabaly (20.05) NZW (21.18) Rex (16.11)	The three rabbit breeds are suitable for garments and linings manufacture. The qualities of rabbit fur, in descending order, are Gabaly and then NZW followed by Rex tanned furs
Souza et al. (2016)	Evaluate tanned leathers of New Zealand white rabbits submitted to combination tannage	0.69–0.87	7.98–10.84	48.6–56.71	21.07–23.50	Rabbit leathers have not enough strength for clothing manufacture but may be used for the manufacture of other associated articles
Hoch et al. (2009)	Evaluate the action of different tanning agents in the	Chrome (1.02)	Chrome (11.24)	Chrome (62.85)	Chrome (22.70)	

(continued)

Table 3 (continued)

Author	Objective	Physical properties				Conclusion
		Thickness (mm)	Tensile strength (N/mm^2)	Elongation (%)	Tearing strength (N/mm)	
	processing of rabbit hides over skin resistance	Clarotan-x8 (1.06) Synthetic (0.92) Weibull (1.07)	Clarotan-x8 (10.25) Synthetic (11.63) Weibull (9.86)	Clarotan-x8 (70.28) Synthetic (44.25) Weibull (52.08)	Clarotan-x8 (25.94) Synthetic (19.49) Weibull (21.58)	Tanned leathers met the acceptance standards for clothing

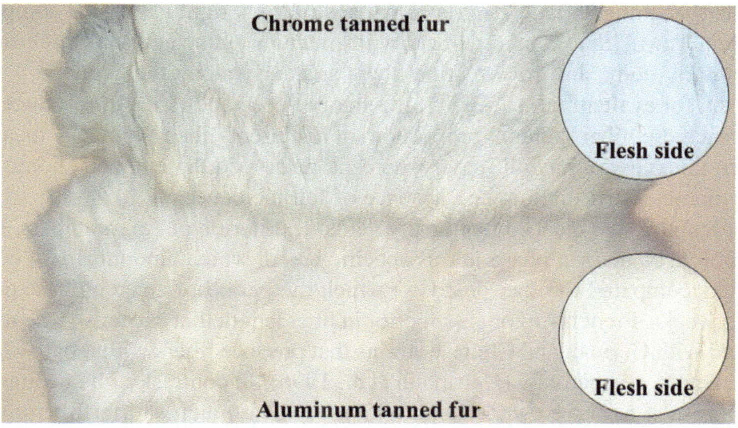

Fig. 5 Difference between chrome- and aluminum-tanned New Zealand furs

can lead to poor wash fastness, difficult storage, and matted hair fibers. Also, zirconium tanning is also a reverse tanning process, and it is additionally acidic and polymerizes in the tanning acid medium, damaging the hair fibers. Therefore, the harmful effect of using an eco-friendly tanning method on the hair fibers' properties should be limited as possible, especially the color.

5.2 Production of Rabbit Fibers

In order to achieve the suitability for textile manufactures, rabbit's fur fibers should have a set of special criteria, such as appropriate length, softness, tensile strength, and lack of felting. Although there have been some studies on fur fibers from meat and dual-purpose breeds, Angora rabbits are still the most popular and widely used for producing specialized fibers for fine yarns. This is because Angora rabbits have undergone multiple selection processes and continuous genetic improvement by breeders in different regions of the world. This continuous improvement produced genetic lines with different geographic origins such as the German, French, and Chinese Angoras. Continuous genetic improvement programs for these genetic lines led to among lines variation in fiber traits. Moreover, the characteristics of the fibers show differences between individuals that may be less determined whenever the improvement program is successful, to show greater symmetry between individuals. Variations in the characteristics of the fibers appear on the body of the same animal in different parts of the body and are also affected by the age of the animal (Schlolaut 1985; Taha et al. 2016), the effect of sex (Taha et al. 2020), the season of the year (Nasr et al. 2020; Taha et al. 2020), the quality of nutrition (Lebas 2000; Taha et al. 2016), and the efficiency of caring systems (Lebas et al. 1997).

Due to the variation in fiber morphology within the same fleece, three distinguishable grades of fiber were identified in Angora rabbit pelage (Vrillon and

Thebault 1988): the "bristle" that is composed of thick tight fibers that compose the guard hair, "awn" that contains fibers with medium diameters with thick tight top and thin curly body, and "down" fine fibers and curly along the fiber.

Criteria for evaluating the quality of Angora fleece and other fur producer breed fleeces include; fiber length, proportion of different fiber's grades, mean fiber diameter, thickness of medulla cavity, percentage of medullated fibers, homogeneity of fiber diameter distribution, and absence of felt incidence.

Fiber length is the most economically valuable criterion of Angora fibers. It refers to the prolonged anagen phase in this specific breed, which amounts to 14 consecutive weeks compared to other breeds in which the period of anagen phase does not exceed 5 weeks. French Angora is superior in fiber length that exceeds 19 centimeters compared with German and Chinese strains that produce fibers with length of 13 and 12 centimeters, respectively (Herrmann et al. 1996). In contrast, non-specific breeds of rabbits, such as the New Zealand White, produce much shorter fibers, with an average length of 2.2 to 2.3 millimeters at 4 and 6 weeks of age, respectively (Taha et al. 2016).

Producing finer fibers is highly correlated to the density of hair follicles in the skin dermis (Moore et al. 1998) and thinner medulla cavity within the fiber shaft (Herrmann et al. 1996). Highly medullated fibers are brittle and are more prone to breakage and less dyeable due to the presence of minor air chambers within the medulla that cause pale appearance of the end products. The incidence of prickle associates higher proportions of thick fibers particularly those with diameters above 30 micrometers (Naylor et al. 1992). In Angora (Herrmann et al. 1996) and in dual-purpose breeds of rabbits as well (Tao 1994; Ibrahim et al. 2017; Taha et al. 2020), fine fibers possess the majority of fur fiber's population. Meanwhile, fur fibers possess high homogeneity of fiber diameter distribution; therefore, it has a negligible prickling effect (Taha et al. 2020).

Angora rabbit hair is prized for its fineness, softness, lightness, and good insulating properties. It is widely used in the production of high-quality yarns for knitting garments and fashion wear (Herrmann et al. 1996). For its marked fineness, it is also used in lady's underwear, shawls, gloves, hosiery, and hats. Moreover, for its electrostatic, thermal, and moisture absorption properties, fabricated products from this type of fibers showed excellent thermal insulation properties and high efficiency in reducing muscle pains. Blending Angora fibers with cotton resulted in maintaining the comfort-related properties of the end product (Oglakcioglu et al. 2009). For its unique properties, Angora fibers are widely used in medical and thermal underwear, nightwear, blankets, and under-blanket sheets (Hunter 2020).

5.3 Production of Value-Added Proteinaceous Materials from Rabbit Skins

Animal hides or skins are materials with high protein content, especially collagen. Therefore, they are utilized to give additional value to this agro-industrial by-product by using in various purposes of pharmaceutical, food, and glue industries. In any case, the skins used for the production of value-added protein materials do not need to be from the first grade, but rejected skins or untanned tannery wastes such as cuts and trimmings can be utilized in this regard.

5.3.1 Collagen

Rabbit skin is a good source of type I collagen, which is mostly insoluble in its native form (Martínez-Ortiz et al. 2015). To maximize the yield, soluble collagen can be extracted first using acetic acid, followed by insoluble collagen extraction using the pepsin enzyme. The extracted collagen with high properties can be used in the cosmetology and pharmaceutical industries. Recently, Matei et al. (2020) demonstrated that electrospun non-active nanofibers based on collagen can be produced from rabbit skins using nanotechnology.

5.3.2 Gelatin

Gelatin is a type of cooked collagen that is typically used in food industry as an ingredient in gelatinous desserts and other uses. Gelatin is conventionally prepared from skins by soaking it in hot water (65 °C) for 6 hours, and then it is filtered and dried. Thus, rabbit gelatin had nearly all of the characteristic absorption peaks of collagen from rabbit skin (Yu et al. 2016). Gelatin characteristics are affected by collagen source and preparation conditions. The gelatin extracted from rabbit skins has a higher gel strength than that extracted from the skins of cows and pigs (Liu et al. 2019). Meanwhile, pretreatment of collagen with acids affects gelatinization. Ma et al. (2018) found that both the gelatin yield and gel strength were improved via acid pretreatment for 5 min with 1% HCl, as a result of the destruction of collagen covalent bonds and the formation of gelatinous collagen in a new form through new hydrogen bonds, where the folds of its surface become smaller. Applying microwave-assisted extraction for short-term (5–30 minutes) enhances gel strength due to increasing the powerful triple helix-like structures and the large amount of high-molecular-weight subunits, compared to a conventional extraction method (Liu et al. 2019). In terms of the nutritional value of rabbit skin gelatin, it contains high levels of glycine (23%), proline (10.9%), and hydroxyproline (10.4%) of the total amino acid content that are needed for human health. Previous investigations on rabbit skins have shown that the extracted gelatin is suitable for using in the food industry, such as in jelly candy and yogurt manufacture (Pancapalaga et al. 2021 and Nurwantoro et al. 2022). Unfortunately, other known uses of gelatin, such as manufacturing of capsules or tablets in the pharmaceutical industry, paper manufacturing, and a binder material for making the head of a match and sandpaper, have not yet been studied on rabbit skins (Schrieber and Gareis 2007).

5.3.3 Glue

Animal glues are natural polymers derived from collagen. Varied glue properties may be exhibited depending on animal source and preparation method. When treating insoluble collagen in skins with acids, alkalis, or hot water, it slowly becomes soluble and thus turned into glue (Schellmann 2007; Ebnesajjad 2009). There is no accurate scientific information available about the preparation methods of rabbit skin glue and its properties, despite the availability of various commercial products in the market.

Few investigations studied the properties of commercial rabbit skin glues. Tsetsekou et al. (2018) found that rabbit skin glue has the lowest bond strength to wood of several adhesive materials, including casein, fish glue, and two synthetic polymers, but it is less reversible after being subjected to relative humidity and temperature cycles. On the other hand, Schellmann (2007) reviewed the properties of animal glues and reported that commercial rabbit skin glues have a high molecular weight, gel/bloom strength, degree of helicity, and viscosity, with a pH ranging from 5 to 7.5. However, compared to other animal glues, rabbit skin glue has the lowest mechanical strength, the highest elasticity, and the highest stability in fluctuating relative humidity due to its high fat level of around 5% (compared to 1% for other animal glues). Therefore, rabbit skin glue could be occasionally used such as any other animal glues in traditional woodworking, gilding, packing, and painting techniques (Ebnesajjad 2009; Emblem and Hardwidge 2012; Ghezzi 2015).

6 Conclusions

Rabbit skins and furs, as well as their wool, are natural materials with multiple benefits. It produces distinctive crafts not matched by any products from other animal species. Despite its uniqueness, it has not yet received sufficient attention and optimal exploitation. Undoubtedly, genetic improvement of hair fiber characteristics of rabbit breeds and the application of modern technologies for leather tanning and fiber manufacturing will contribute to increasing its economic return.

Perhaps the development of expertise in developing countries during the next phase will significantly reduce the current waste in rabbit skins and fibers. Finally, the work of more applied studies on rabbit skins and fibers will help effectively to maximize their future benefits.

References

BASF (2007) Pocket book for leather technologist, 4th edn. BASF, Ludwigshafen

Brewer N (2006) Biology of the rabbit. J Am Assoc Lab Anim Sci 45(1):8–24

Covington AD (2009) Tanning chemistry, the science of leather. RSC Publishing, Cambridge, London

Cruise LJ, Brewer NR (1994) The biology of the laboratory rabbit. In: Manning PJ, Ringler DH, Newcomer CE (eds) The biology of the laboratory rabbit. Academic Press, San Diego, p 483

Das SK, Chakurkar EB, Singh NP (2014) Rabbit as an alternative source of meat production. Technical bulletin no. 45, ICAR research

Ding H, Zhao H, Cheng G, Yang Y, Wang X, Zhao X, Qi Y, Huang D (2019) Analyses of histological and transcriptome differences in the skin of short-hair and long-hair rabbits. BMC Genomics 20:140

Dutta S (2008) An introduction to the principles of leather manufacture, 4th edn. Indian Leather Technology Association, India

Ebnesajjad S (2009) Characteristics of adhesive materials. Chapter 5, adhesives technology handbook, 2nd edn. William Andrew publishing, pp 63–135, isbn 9780815515333
Emblem A, Hardwidge M (2012) Adhesives for packaging. In: Emblem A, Emblem H (eds) Packaging technology. Woodhead Publishing, Cambridge, pp 381–394
Geoffrey D, Thorburn MB, Brian H, Casey MB, Molyneux GS (1966) Distribution of blood flow within the skin of the rabbit with particular reference to hair growth. Circ Res XVM:650–659
Gutterres M (2003) Effect of tanning on the dermal microstructure. Revista do Couro Abqtic, Estância Velha 26:56–59
Héctor-Julio P, Astrid-Yeritza A, Dayanna-Angélica P, Mónica-María P, Ana-Milena S, Cristian-Jahir M (2021) Tannin extract from guanacaste tree fruit (*Enterolobium cyclocarpum*) as a tanning agent for European rabbit skin (*Oryctolagus cuniculus*). Biotecnología en el Sector Agropecuario y Agroindustrial 19(1):180–190
Herrmann S, Worthmann G, Worthmann FJ (1996) Characteristics of angora rabbit fiber 1– the influence of rabbit origin on fiber and medulla diameter in angora wool. World Rabbit Sci 4(3): 149–153
Hoch ALV, do Prado M, de Souza Franco MLR, Scapinello C, Franco N, Gasparino E (2009) Action of different tanning agents used on rabbit hide tanning: physical-mechanic tests. Maringá 31(4):411–415
Hunter L (2020). Mohair, cashmere and other animal hair fibres. In Handbook of natural fibres: Volume 2: Processing and Applications (The Textile Institute Book Series, Volume 2) 2nd Edition. Edi. Ryszard M. Kozlowski and Maria Mackiewicz-Talarczyk
Ibrahim RS, Hussein AA, Jabbar AI (2017) Comparative microscopically study of the skin in local and wild rabbits. Kufa J Vete Med Sci 8(1):151–156
Kanuri WD, Onyuka A, Tanui R (2019) An investigation on the properties of rabbit leather from different tannages. Int J Sci Res Publ 9(4):12–17
Lebas F (2000) Vitamins in rabbit nutrition: literature review and recommendations. World Rabbit Sci 8(4):185–192
Lebas F, Coudert P, de Rochambeau H, Thebault RG (1997) The rabbit husbandry, health and production. FAO, Rome
Liu T, Dai H, Ma L, Yu Y, Tang M, Li Y, Hu W, Feng X, Zhang Y (2019) Structure of Hyla rabbit skin gelatin as affected by microwave-assisted extraction. Int J Food Prop 22(1):1594–1607
Lucy KM, Ashok N, Maya S, Karthiayini K, Indu VR, Sreeranjini AR (2019) Histomorphological studies on distribution of hair follicles in different breeds of rabbits. Indian J Vet Anat 31(1): 47–48
Ma M, Ma L, Yu W, Zhang X, Shen Y, Zhang Y (2018) Research on rapid gelatinization of rabbit skin collagen as effect of acid treatment. Food Hydrocoll 77:945–951
Martínez-Ortiz MA, Hernández-Fuentes AD, Pimentel-González DJ, Campos-Montiel RG, Vargas-Torres A, Aguirre-Álvarez G (2015) Extraction and characterization of collagen from rabbit skin: partial characterization. CyTA – J Food 13(2):253–258
Maryati T, Pertiwiningrum A, Bachruddin Z (2019) The effects of tanning with Kepok banana (*Musa Paradisiaca L*) bunch on the physical quality of rabbit skin. Int J Recent Technol Eng 8(4):12564–12567
Maryati T, Nugroho T, Sundari DSHC (2021) Effect of fatliquor level on the physical quality of Indonesian rabbit fur leather. Earth Environ Sci 902:012021. https://doi.org/10.1088/1755-1315/902/1/012021
Monteiro-Riviere NA, Stinson AW (1998) Integument. In: Textbook of veterinary histology, (Eds. Delmann, H.D. and Eurell, J.A.) Lippincot Williams and Wilkins, Baltimore, Maryland, USA, pp 303–332. isbn: 978-0-7817-4148-4
Moore GPM, Jackson N, Isaacs K, Brown G (1998) Pattern and morphogenesis in skin. J. Theory Biol 191:87–94
Nasr AI, Taha EA, Naglaa SB, Essa DG (2020) Seasonal variations in furs of Gabaly and New Zealand white rabbits and their crossbred under Egyptian semi-arid conditions. World Rabbit Sci 28:49–57

Naylor GRS, Veitch CJ, Mayfield RJ, Kettlewell R (1992) Fabric evoked prickle of some wool fabrics. Text Res J 62:487–493

Nurwantoro HA, Legowo AM, Mulyani S, Quna TR, Sutaryo (2022) The functional properties of rabbit skin gelatin compared to commercial gelatin and its application in jelly candy. Jurnal Ilmu dan Teknologi Hasil Ternak 17(1):1–9

Oglakcioglu N, Celik P, Ute TB, Marmarali A, Kadoglu H (2009) Thermal comfort properties of angora rabbit/cotton fiber. Blended Knitted Fabrics Textile Res J 79(10):888–894. https://doi.org/10.1177/0040517508099396

Oznurlu Y, Celik I, Sur E, Telatar T, Ozparlak H (2009) Comparative skin histology of the New Zealand white and angora rabbits: histometrical and immunohistochemical evaluations. J Anim Vet Adv 8(9):1694–1701

Pancapalaga W, Nitiharjo S (2019) The quality of rabbit hide tanned by mangrove (Rhizophora mucronata). Revista de Pielărie Încălţăminte 19(4):189–194

Pancapalaga W, Khotimah K, Hartati ES (2021) Evaluation of rabbit skin gelatin and preparation of yogurt using rabbit skin gelatin. B Life Environ Sci 58(2):83–89

PNS (2022) Rabbit – code of practice – slaughtering. Philippine National Standard PNS/BAFS 327

Schellmann NC (2007) Animal glues: a review of their key properties relevant to conservation. Stud Conserv 52(Sup1):55–66. https://doi.org/10.1179/sic.2007.52.Supplement-1.55

Schlink AC, Liu SM (2003) Angora rabbits a potential new industry for Australia. A report for the Rural Industries Research and Development Corporation. CSIRO Livestock Industries. Publication no. 03/014. Project no. CSA-19A. isbn 0642585822

Schlolaut W (1985) A compendium of rabbit production appropriate for conditions in developing countries. Schriftenreihe der GTZ (Germany, F.R.), no. 168, 169. Eschborn, 1985

Schrieber R, Gareis H (2007) Gelatine Handbook: Theory and Industrial Practice. 82007 WILEY-VCH Verlag GmbH & Co. KGaA, Weinheim. isbn: 978-3-527-31548-2

Składanowska-Baryza J, Stanisz M (2019) Pre-slaughter handling implications on rabbit carcass and meat quality – a review. Ann Anim Sci 19:875. https://doi.org/10.2478/aoas-2019-0041

Soma TM, Suzuki OJ, Takahashi T, Hibino T (1998) Analysis of apoptotic cell death in human hair follicles in vivo and in vitro. J Invest Dermatol 111:948–954

Souza MLR, Hoch AL, Gasparino E, Scapinello C, Mesquita Dourado D, Claudino da Silva SC, Lala B (2016) Compositional analysis and physicochemical and mechanical testing of tanned rabbit skins. World Rabbit Sci 24:233–238

Stenn KS, Paus R (2001) Controls of hair follicle cycling. Physiol Rev 81:449–494

Taha EA, Samia AH, Badawey NS (2016) Studies of some skin and its coat characteristics in relation to feed additives of growing rabbits. Egyptian J Rabbit Sci 26(1):105–119

Taha EA, Samia AH, Nasr AI (2017) Evaluating skin quality of some rabbit breeds under Egyptian conditions. World Rabbit Sci 25:193–200

Taha EA, Nasr AI, Essa DG, Kassem ST, El-Shemy KA (2020) Effect of sex and seasonal changes on New Zealand rabbit fur under Egyptian semi-arid conditions. Int J Environ Agric Biotechnol 5(3):489–499

Tao YR (1994) Studies on the quality of rex rabbit fur. World Rabbit Sci 2(1):21–24

Trocino A, Cotozzolo E, Zomeño C, Petracci M, Xiccato G, Castellini C (2019) Rabbit production and science: the world and Italian scenarios from 1998 to 2018. Ital J Anim Sci 18(1): 1361–1371. https://doi.org/10.1080/1828051X.2019.1662739

Vrillon JL, Thebault RG (1988) Evaluation of angora wool, standardization of testing methods. 4th World Rabbit Congress, 10-14 October 1988. Budapest Hungary 1:79–82

Yu W, Wang X, Ma L, Li H, He Z, Zhang Y (2016) Preparation, characterization and structure of rabbit (Hyla rabbit) skin gelatine. Food Sci Technol 51(3):574–580

Zapletal P, Bierowiec-Widórek K, Czerniejewska-Surma B, Maj D (2020) The suitability of using broiler rabbit leathers in gloves and footwear manufacturing. World Rabbit Sci 28:239–250

Part III

Rabbit Herd and Health Management and Diseases

Preventive Medicine and Rabbit Health Management

Filipe Fontes Pinto, Tomás Rodrigues Magalhães, Ricardo Marcos, Joana Abrantes, Mariana Marrana, José Manuel Monteiro, João Simões, and Nuno Alegria

F. F. Pinto (✉)
HIPRA, Malveira, Portugal

UMIB – Unit for Multidisciplinary Research in Biomedicine, ICBAS – School of Medicine and Biomedical Sciences, University of Porto, Porto, Portugal

CIBIO, Centro de Investigação em Biodiversidade e Recursos Genéticos, InBIO Laboratório Associado, Campus de Vairão, Universidade do Porto, Vairão, Portugal

BIOPOLIS Program in Genomics, Biodiversity and Land Planning, CIBIO, Vairão, Portugal
e-mail: filipe.pinto@hipra.com

T. R. Magalhães · J. Simões
Department of Veterinary Sciences, University of Trás-os-Montes and Alto Douro, Vila Real, Portugal

Animal and Veterinary Research Centre (CECAV), University of Trás-os-Montes and Alto Douro, Vila Real, Portugal

R. Marcos
ICBAS - School of Medicine and Biomedical Sciences, University of Porto, Porto, Portugal

J. Abrantes
CIBIO, Centro de Investigação em Biodiversidade e Recursos Genéticos, InBIO Laboratório Associado, Campus de Vairão, Universidade do Porto, Vairão, Portugal

BIOPOLIS Program in Genomics, Biodiversity and Land Planning, CIBIO, Vairão, Portugal

Departamento de Biologia, Faculdade de Ciências, Universidade do Porto, Porto, Portugal

M. Marrana
World Organization for Animal Health (WOAH, founded as OIE), Paris, France

J. M. Monteiro
José Azevedo Monteiro, Lda., Vila Boa do Bispo, Portugal

N. Alegria
Department of Veterinary Sciences, University of Trás-os-Montes and Alto Douro, Vila Real, Portugal

J. Simões, J. M. Monteiro (eds.), *Veterinary Care of Farm Rabbits*,
https://doi.org/10.1007/978-3-031-44542-2_17

Abstract

The success of any animal production system is strictly related to the best practices in preventive medicine regarding rabbit welfare, biosecurity, prophylaxis, and technical-economic management. These four pillars are essential to reduce the risk of introducing, establishing, and spreading animal diseases, contributing to farms' success from a One Health perspective. Biosecurity programs must be implemented in each rabbitry, according to the specific epidemiological conditions, available resources, and production objectives, but the main goal is always the same: ensure the best rabbit health status by reducing the entry of pathogens into the farm, their potential transmission among animals, and disease development. Thus, a protocol based on systematic hygienic and sanitary measures involving animal management; appropriate infrastructures, equipment, and procedures; and responsible conduct of farm personnel and visitors, in association with adequate prophylaxis protocols, is always highly recommended. In this chapter, the currently recommended best practices in biosecurity, preventive medicine, and rabbit husbandry will be explored in detail.

Keywords

Biosecurity · Preventive medicine · Disease prevention · Herd health · Rabbit semi-intensive systems

1 Introduction

Infectious livestock diseases are a threat to global animal health, and their control is crucial to support the One Health concept, comply with international trade agreements, sustain economic growth, and promote sustainable livelihoods and feed security (Tomley and Shirley 2009; WOAH 2022).

Veterinarians have the ideal set of skills for the prevention of new and emerging infectious diseases at the human-animal-environment interface (WOAH 2022). In addition, there has been a remarkable evolution in the work methodologies of farm animal veterinarians, which has changed this practice, as there is now a greater concern to invest in preventive medicine than just in the therapeutic approach to animals with already established disease (Medeiros et al. 2022).

The World Organisation for Animal Health (WOAH, founded as OIE) Terrestrial Animal Health Code defines biosecurity as a set of preventive measures that aim to reduce the risk of introducing, establishing, and spreading animal diseases, infections, or infestations (WOAH 2022). These measures are essential to achieve high levels of animal health, feed safety, and wildlife preservation (Huber et al. 2022). Regarding the One Health concept, biosecurity planning is the key to preventing the spread of zoonoses, i.e., diseases that spread from animals to humans, as well as diseases from humans to other animals (Huber et al. 2022; Renault et al. 2022).

Biosecurity protocols are implemented to avoid the entrance of pathogens and to reduce the spread of infectious agents that may already be present on farms, minimizing their negative effects and enhancing animal health status (Huneau-Salahün et al. 2015; Nielsen et al. 2020). In rabbitries, successful biosecurity plans need high observation skills from producers and efficient cooperation with veterinarians (Rosell 2020). This synergy allows for early diagnoses, particularly of infectious diseases such as myxomatosis, rabbit hemorrhagic disease (RHD), encephalitozoonosis, and pasteurellosis, and once detected acts more quickly, improving the prognosis while protecting other animals.

In fact, a well-executed biosecurity program requires understanding which actions need to be done, planning how they will be performed, and then putting the plan into action fully and consistently. However, there is no "one-size-fits-all" program for rabbitries since programs must consider all unique aspects of the farm, such as management, type of production system, structural limitations, isolation, surrounding environment, staff, treatments, and prophylactic plans usually employed. In addition, biosecurity plans must also focus on rabbitry worker health by applying preventive measures to avoid zoonotic infections caused by ringworms, skin mites, hepatitis E virus, *Encephalitozoon* spp., *Salmonella* spp., and *Escherichia coli* (Beaucournu and Launay 1977; Nodari et al. 2009; Lhomme et al. 2013; Swennes et al. 2013; Burt et al. 2016; Kylie et al. 2017b; Magalhaes et al. 2022).

The implementation of a complete "all-in/all-out" approach in a closed-cycle rabbit production system can be logistically challenging in some rabbitries (Huneau-Salahün et al. 2015). Therefore, deworming protocols and specific vaccination programs must integrate all biosecurity protocols to further prevent against the main diseases that affect lagomorphs.

In livestock, preventive medicine comprises four main pillars: (1) animal welfare, (2) biosecurity, (3) prophylaxis, and (4) technical-economic management (Fig. 1). It is well known that rabbitry productivity is significantly affected by rabbit health (Huneau-Salahün et al. 2015). However, as well as other species, the cost/benefit relationships greatly influence the implementation of biosecurity measures on farms and need to be adequately accessed (Jarvis and Valdes-Donoso 2018; Osawe et al. 2022).

The farm rabbit's welfare is addressed specifically in Chapters "Welfare, Behavior and Housing of Rabbits" and "Welfare Protocols for Rabbits". Therefore, this chapter covers the concepts of biosecurity and prophylaxis applied to rabbitries and briefly describes the key indicators used to perform efficient technical-economic management.

2 Biosecurity Definition and Scope

Livestock biosecurity prevents the exit and entrance of pathogenic agents to contribute to a better health status of farm animals (Maes et al. 2008; Lambert et al. 2012; Zhang et al. 2013; Gillespie et al. 2015; Postma et al. 2015, 2016a, 2016b, 2017;

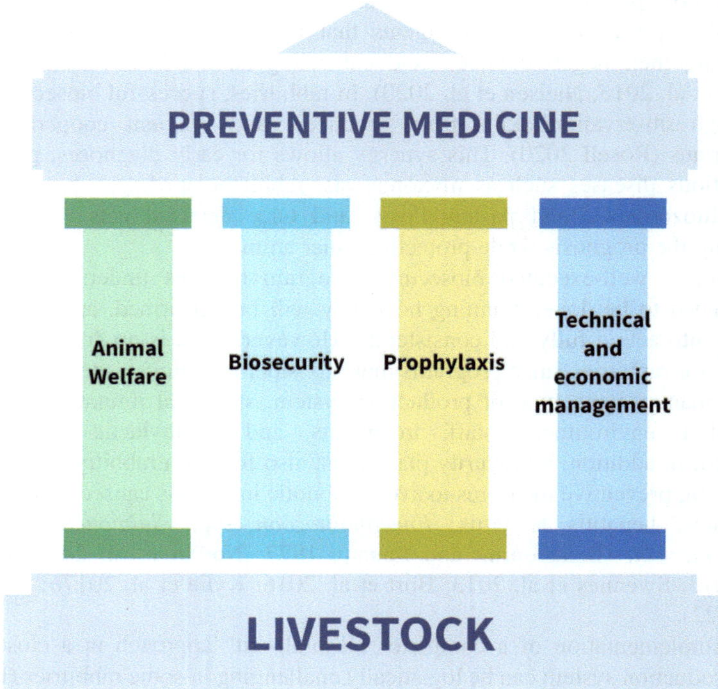

Fig. 1 The four pillars of preventive medicine applied to livestock and poultry farms

Huber et al. 2022). Biosecurity programs are designed to reach a high state of animal health and avoid economic or sanitary breakdowns in the farm, which is critical to the success of any livestock farm. Therefore, to ensure feed and water safety, reduce environmental pollution, and prevent the transmission of zoonoses and other infectious diseases, it is important to prioritize sanitary measures in a One Health approach in relation to economic benefits. It should be noted that these programs can be applied at different levels, including regional and national levels, and also for an individual or group of animals, such as a herd or a flock (Dewulf and Van Immerseel 2019).

Rabbit farm biosecurity programs must focus on the two notifiable rabbit diseases on the WOAH list, which, in addition to being associated with high rates of morbidity and mortality, are quite difficult to eradicate when they enter a farm: rabbit hemorrhagic disease virus (RHDV) and myxomatosis (WOAH 2023). However, in addition to preventing the entry of the infectious agent, it is important to prevent its exit and potential transmission to the surrounding environment, which could put other farms and wild animals at risk. Therefore, each farm must adapt the biosecurity program to its specific conditions in terms of infrastructure, location, and productive management to achieve the highest effectiveness possible (Nielsen et al. 2020).

These biosecurity programs are a multilevel and multistakeholder process, and their success depends on all parts involved, both inside, such as farmers, veterinarians, and other farm staff, and outside the farm, such as people associated with genetics, slaughterhouses, commercial areas, and the feed industry (Gordh and MckKirdy 2014; Macleod 2015; Reed and Curzon 2015; Mato-Amboage et al. 2019; Black and Bartlett 2020). Moreover, it is mandatory that all these plans comply with national and community legislation, as well as international agreements.

3 External and Internal Biosecurity

During livestock production, external and internal biosecurity aim to provide the ideal conditions for maximizing the rabbits' immune system and preventing the spread of pathogenic organisms inside and outside the farm (Fig. 2).

Internal biosecurity comprises all measures that aim to protect the health of animals (e.g., vaccination and deworming), prevent the spread of pathogenic agents among farm animals, and eliminate pathogens through good health management, isolation, treatment, or even elimination of sick animals (Filippitzi et al. 2018).

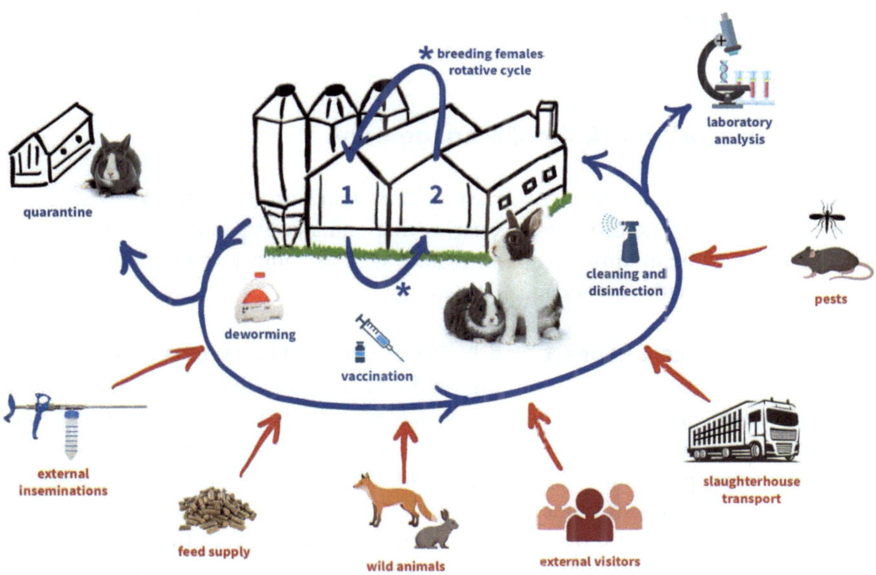

Fig. 2 Illustration of some of the main critical points of external (red circuit) and internal (blue) biosecurity in rabbit farms

External biosecurity comprises all measures that prevent the entrance (bioexclusion) and exit (biocontainment) of infectious agents from the farm. Potential direct and indirect transmission routes are avoided (Murphy et al. 2017). These measures are related to the actions that connect the farm to their external environment, such as the maintenance of the farm's infrastructures and organization, restrictions to the entry of animals and people, hygiene of transport vehicles, practices that ensure feed safety, and all strategies to exclude the possible transmission routes, either direct or indirect, with the outside (Murphy et al. 2017).

Usually, external biosecurity is better understood and implemented by farmers than internal biosecurity. Since many rabbitries have already experienced outbreaks of myxomatosis or RHD at a given time, there is now greater knowledge and awareness regarding the importance of external biosecurity measures to prevent the entrance of these causative agents. In fact, many competent authorities around the world currently rely on external biosecurity measures to evaluate the sanitary conditions of livestock farms.

4 Biosecurity Principles and Recommendations

Considering these two types of biosecurity (internal and external), the authors chose to explore, in this chapter, all the principles and recommended measures, grouped as follows: (1) biosecurity in animal management; (2) biosecurity regarding farm personnel, equipment, and procedures; (3) biosecurity in the rabbitry environment; and (4) biosecurity in disease prevention and treatment. These principles can be applied to different types of rabbitries and must be followed when planning individual programs.

4.1 Biosecurity in Animal Management

In rabbitries, biosecurity protocols must ensure the health of rabbits at all stages of animal production. Therefore, all scenarios that enhance the transmission of infectious agents must be minimized or eliminated, such as contact with sick animals and people as well as with vectors and fomites.

4.1.1 Avoidance of Direct and Indirect Contact Between Animals

Direct and indirect contact between high-risk and low-risk animals must be systematically avoided. This avoidance allows the prevention of clinically ill animals, including new arrivals and rabbits without adequate vaccination and deworming protocols, from potentially transmitting infectious agents to more susceptible animals (Dewulf and Van Immerseel 2019).

For that purpose, one of the best biosecurity measures is the effective separation of all animal areas, neutralizing possible sources of infection, whether internal or external to the farm (Racicot et al. 2011). Many rabbitries apply a rotative production cycle (deepened further described), separating the fattening rabbits, which are often unvaccinated, from the breeding females, protecting the latter group as they are associated with a higher farm value. Currently, in large-scale (industrial) rabbit production, rabbits are housed in double sheds separated by a central corridor (see Chapter "Rabbit Farming: Industrial, Small-Scale and Organic Housing Systems"). This facility design allows the physical separation of rabbit does (breeding females) and weaned rabbits.

Moreover, properly separated facilities on the farm are important to ensure the quarantine of newly introduced animals (Dewulf and Van Immerseel 2019). According to the Code of Practice of the Canadian National Farm Animal Care Council (NFACC), the quarantine period must consist of a minimum of 14 days, during which period the animals must be observed daily to detect possible signs of infectious diseases. Most of the worldwide semi-intensive rabbitries produce on-farm their owner replacement females from grandparent lines or purchase 1-day kits for this purpose. Nevertheless, purchasing animals or animal products (e.g., semen for artificial insemination) from other farms should always be considered as a potential route for the introduction of infectious agents. This risk should be minimized by avoiding animals or animal products from different origins and from farms with poor sanitary conditions (Kylie et al. 2017a; Rossi et al. 2017). Therefore, animals must always come from a farm with an equal or higher hygienic status (Dewulf and Van Immerseel 2019). In addition, due to the productive impact inherent to their activity, genetic centers have to guarantee high biosecurity standards in their facilities and good insemination practices when visiting customers' farms (Rosell 2020).

4.1.2 Reduction of Microbial Load

It is reasonable to understand that it is not possible to achieve a "sterile environment" in the real context of livestock farming. In addition, even if it were theoretical, it would not be ideal since it would limit the ability of these animals to develop a more robust immune system, which is important to protect them against numerous threats. Therefore, the aim of a biosecurity program is to reduce the presence of microorganisms to the minimum level that do not cause disease, which can be challenging since it is difficult to reconcile the rabbits' productive cycles with the recommended "all-in/all-out" systems (Huneau-Salaun et al. 2015).

In any case, to ensure the fulfillment of this objective, some hygienic and sanitary measures must be applied, such as (1) minimal animal handling necessary; (2) use of single-use personal protective equipment (PPE), such as gloves, masks, and footwear protection; (3) good hand hygiene after contact with any animal, especially those whose health status is uncertain or suspect; and (4) careful planning of procedures so that healthy animals are handled prior to suspicious or even sick animals.

4.1.3 Adaptation to High Animal Density Rabbitries

Another impactful aspect is animal density, since a greater number of animals are often associated with more intensive production systems, which entails increased biosecurity risks (Fig. 3). As animal-to-animal contact is more frequent and for a

Fig. 3 Intensive rabbitry production poses a high animal density

longer period, the risk of transmission of infectious agents is much higher, increasing the pathogen pressure on those animals and compromising their immune system, which tends to perpetuate the infection cycles. Thus, animals reared in those regimes tend to have a weaker immune system and a higher physiological demand, being more sensitive to environmental changes and more prone to immune system breakdowns and infections.

Prolonged hunger and thirst are major sources of stress and suffering in those rabbits; thus, a high number of fattening animals per cage should be avoided, as it limits access to feeders and drinkers (Nielsen et al. 2020). Furthermore, a high density of fattening rabbits also predisposes them to aggression and ear injuries, musculoskeletal problems, thermal stress, lack of rest, and fear (Nielsen et al. 2020; Rommers and Meijerhof 1998). Therefore, it is recommended that a maximum of 8 fattening rabbits should be placed per cage, and the maximum density should not exceed 17 rabbits per square meter. Concerning breeding females, the Institute of Agrifood Research and Technology (IRTA) protocol on welfare in farm rabbits defines 2500 cm^2 and 3500 cm^2 per doe as a minimum and an ideal value, respectively (Dalmau et al. 2020). See Chapter "Welfare, Behavior and Housing of Rabbits" (housing-animal welfare).

Finally, a greater number of animals require more personnel and more demanding procedures, such as more frequent handling for animal movement and transport and more visits, whether for feed delivery, animal health care, or structure maintenance, which carries a higher risk of pathogens entering the rabbitry.

4.1.4 Proper Supply and Control of Feed and Water

Feed and water, in addition to being essential for the nutrition and hydration of rabbits, must also be seen as potential sources of pathogen entry on the farm.

Feed companies are expected to have the highest sanitary standards, both during manufacturing and transport. In rabbitries, silos that store the feed must be well sealed to prevent contamination (Fig. 4), as well as regularly fumigated and disinfected. Moreover, supplier verification programs must be implemented to control and reduce the risk of feed contamination, and samples from new batches must be stored for subsequent control analysis, if necessary (Stewart et al. 2020).

The quality of the water supplied on a farm has a large impact on the rabbits' health. If the physical, chemical, and microbiological properties of water are not properly controlled, they may carry pathogenic agents and have negative effects on therapy by decreasing the solubility of drugs administered to animals, for example (El Allaoui et al. 2016; Dewulf and Van Immerseel 2019).

A wide range of water contaminants exist, either chemical (organic and inorganic) or biological (bacteria, fungi, parasites, and viruses). Control of water safety must be performed at different points, either at the water source, farm entrance, or different drinking fountains, as shown in Fig. 5. Water safety testing must be carried out regularly: at least twice a year (before and after extended rainstorm periods) and whenever animals change their water consumption habits or there are changes in water properties (e.g., changes in odor, taste, clarity).

Fig. 4 Tightly sealed silos are commonly used by rabbitries to store rabbit feed

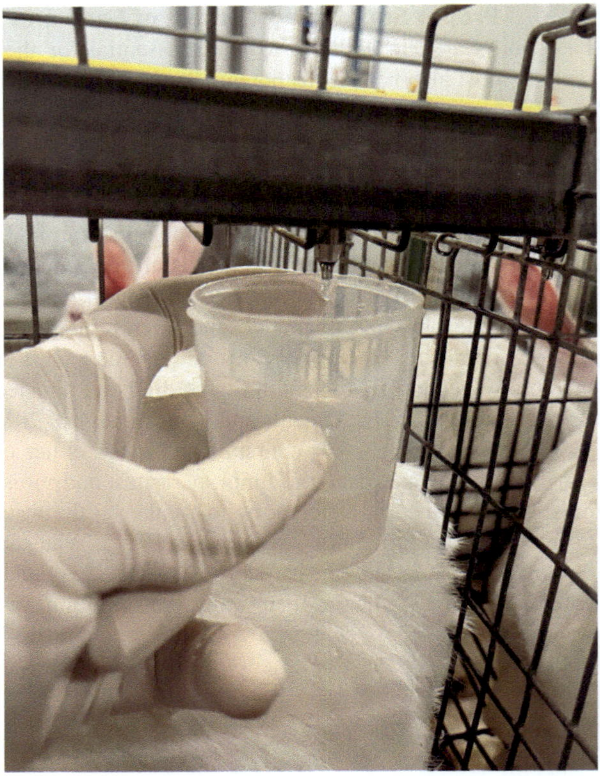

Fig. 5 Water sampling for analysis from a rabbit drinking system

Depending on the country, different water quality criteria must be met, with the presence of coliform bacteria (genera *Escherichia*, *Enterobacter*, and *Klebsiella*) being considered an important water-poor quality indicator, according to the World Health Organization (WHO) (WHO 2022).

To ensure that the water supplied to rabbits on a farm is free of contaminants and suitable for drinking, it must undergo a treatment process, which may include physical removal of contaminants and disinfection with ultraviolet light (Betancourt and Rose 2004), chlorine-based disinfectants (e.g., chlorine dioxide), ozone, or hydrogen peroxide. Moreover, as a protection measure, residual water treatment must be carried out daily against partial contamination by microorganisms (e.g., through chloramines) (WHO 2022).

4.2 Biosecurity in Animal Management

Any person, whether personnel or visitors, should be considered a potential mechanical vector capable of introducing infectious agents into the farm (Stewart et al. 2020). Consequently, the number of people with access to animal areas should be highly limited, all employees must receive biosecurity formal training, and rabbitries should keep a minimum of two workers to maintain biosecurity standards if one is unable to achieve those tasks (Rosell 2020).

4.2.1 Area Separation and Protective Equipment

Farms based on efficient biosecurity programs must have differentiated areas (clean versus dirty areas) separated by doors, walls, changing rooms, lines marked on the floor, and/or desks for disinfection of external materials, as shown in Fig. 6 (Alarcon et al. 2021; Filippitzi et al. 2018). Changing rooms are especially important to ensure the correct implementation of decontamination measures. These spaces should allow workers and visitors to remove their own clothes and shoes and perform hand hygiene using an aseptic technique before moving to the clean area, which is usually separated by a barrier (e.g., bench) and where they can change into the farm-specific clothes and shoes that are properly sanitized (Pritchard et al. 2005; Dewulf 2014; Backhans et al. 2015; Anderson et al. 2018).

Despite the standard procedure described, some rabbitries adopt even stricter biosecurity measures, requiring people to shower and/or use PPE, as shown in Fig. 7. In fact, PPE must be used in all operations involving the loading of animals or

Fig. 6 Clear separation between clean and dirty areas

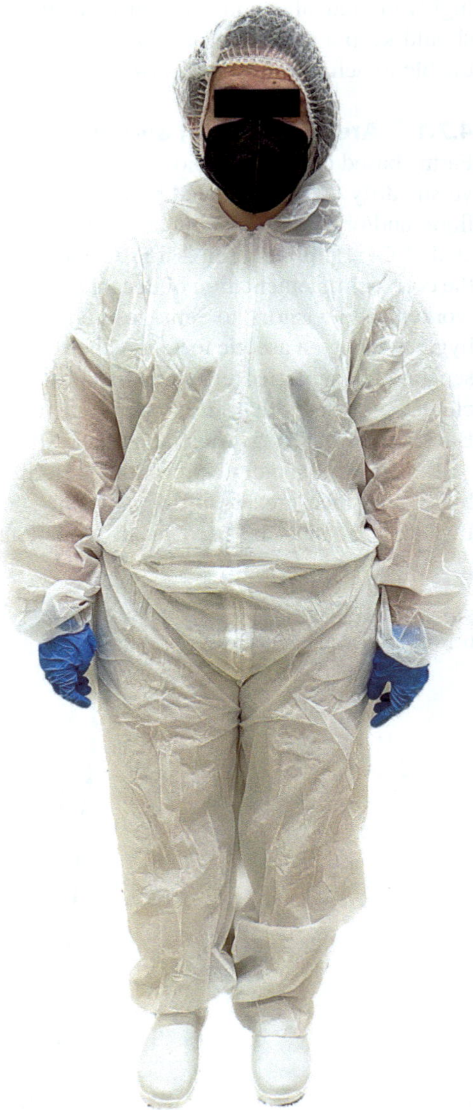

Fig. 7 External visitors should use single-use personal protective equipment (PPE) when entering clean areas

cadavers and similar procedures, which should always be supervised by the respective farmer to avoid incorrect or negligent behaviors (Graham et al. 2008; Millman et al. 2017). Additionally, the use of an FFP2 mask is highly recommended, as humans can be potential vectors of microorganisms (viruses or bacteria) transported in their upper airways (Lee et al. 2019; Bartoszko et al. 2020; Sommerstein et al. 2020).

Strict compliance with these preventive measures must be carried out, particularly on rabbit farms that do not vaccinate fattening rabbits, and the procedures on these animals must always be done after all the daily activities have been performed in the maternity building/ward.

4.2.2 Registration of Personnel and Visitors

An important procedure that is frequently overlooked and disregarded is the registration of all people who visit the farm (Garforth et al. 2013). This simple practice allows tracking all people or vehicles that could have been a possible source of infection in case of an outbreak of an infectious disease, identifying the possible source, and helping to contain the spread (Millman et al. 2017). It is recommended that all visits by people from outside the farm, such as commercial professionals, inseminators, staff involved in carcass collection, and personnel responsible for feed supply, be scheduled for the beginning of the week, preferably on Mondays, since it is unlikely that these persons may have interacted with another farm in the previous 24 to 48 hours (Alarcon et al. 2021). In line with this, when people visit a rabbit farm with an active infectious disease, they should avoid visiting another farm within the next 48 hours.

4.2.3 Sanitary Care and Traceability in Daily Procedures

Cleaning and disinfecting are essential procedures to break the infection cycle between consecutive production turnovers to eliminate or at least significantly reduce the burden of pathogenic microorganisms on surfaces and in the air. In rabbitries, hair burning is also a key sanitary procedure and must be followed by many other procedures, such as dry cleaning and organic material removal; soaking of all surfaces to release remaining organic material; cleaning through a high-pressure washer machine with water at 98°C degrees; drying the cages and surfaces to avoid disinfectant dilution in the following step; disinfection of cages and surfaces; drying the puddles of remaining disinfectant to avoid contact with rabbits; and finally collecting surface samples to test the efficiency of the procedures (Boucher and Nouaille 2013b; Dewulf and Van Immerseel 2019).

All procedures involving the transport of live animals, as well as the handling and removal of manure and carcasses, are associated with a high risk of contamination. This potential transmission can occur directly through infected animals, carcasses, and their secretions and excretions or indirectly from fomites, rodents, or other vectors (Dewulf and Van Immerseel 2019). To avoid this, animal movement, carcass removal, feed and waste management, and the passage of employees and visitors should take place in dirty areas, away from clean areas, where safe, hygienic, and disinfected transportation must be performed (Stewart et al. 2020).

Slaughterhouse trucks are one of the main foci of potential transmission of infectious diseases, representing a mechanical vector for several pathological agents to livestock animals (Ni et al. 2016; Rossi et al. 2017). To minimize this risk, transport companies must always ensure truck cleaning and disinfection before entering a new farm. These processes must be adjusted to weather conditions, as certain disinfectants (e.g., aldehyde-derived disinfectants) are inactivated at

Fig. 8 Disinfection arches are commonly applied to prevent the entry of pathogenic microorganisms through road vehicles

temperatures below 10 °C, for example (Pritchard et al. 2005; Dewulf 2014; Backhans et al. 2015; Ni et al. 2016). Additionally, to prevent the entry of pathogenic microorganisms through vehicles, all rabbitries must have a disinfection arch (Fig. 8) and wheel bath at the entrance (Kim et al. 2017; Rossi et al. 2017). Once an animal is loaded into the truck, it can never follow the reverse circuit (Pritchard et al. 2005; Backhans et al. 2015; Stewart et al. 2020).

Two other important measures are the use of single-use and disposable needles in all procedures involving animals, whether vaccinations, deworming, or other treatments, and careful nest management. In fact, although the standardization of nests is a routine and common practice in rabbitries, it should be avoided if disease is present, in which case adjustments are made in terms of organization and proximity of cages to prevent the spread of pathogens (Rosell and de la Fuente 2018).

Finally, in relation to the management of animal excretions, such as feces and urine, these must be collected and processed by specialized companies or properly used in agriculture. In this last scenario, they should be applied at a safe distance from the farm to avoid microorganisms and pests (Spencer and Guan 2004; Graham et al. 2008; Leibler et al. 2017).

4.3 Biosecurity in the Rabbitry Environment

Risk factors related to the environment are of great importance in animal farms. Accommodation is the main source of physical, chemical, and microbiological aggressions, and low temperatures, high humidity, and airspeed values have a negative impact on rabbit welfare (Rosell 2020). In fact, many health problems can be avoided with proper management.

4.3.1 Risk Level Assessment

The spread of a pathogen in a given environment depends on the route of transmission, which can be considered high or low risk. This difference must be taken into account when designing a successful biosecurity plan, focusing first on routes with higher transmission risk and then successively lower ones, as shown in Fig. 9 (Dewulf and Van Immerseel 2019). Moreover, the risk of pathogenic spread has also been associated with its occurrence rate, i.e., a low-risk route, such as handling animals for examination, nest care, prophylaxis procedures, and transport, can have increased its risk if such practices occur more frequently.

Nevertheless, some pathogens require increased attention and should therefore be prioritized when planning these biosecurity programs, especially those that most affect farm rabbits and that are more likely to spread through multiple routes, such as RHD, myxomatosis, ringworm, colibacillosis, and intestinal coccidiosis.

Furthermore, close proximity to slaughterhouses and other farms, especially those with less strict sanitary plans, is an example of a higher risk condition. Several

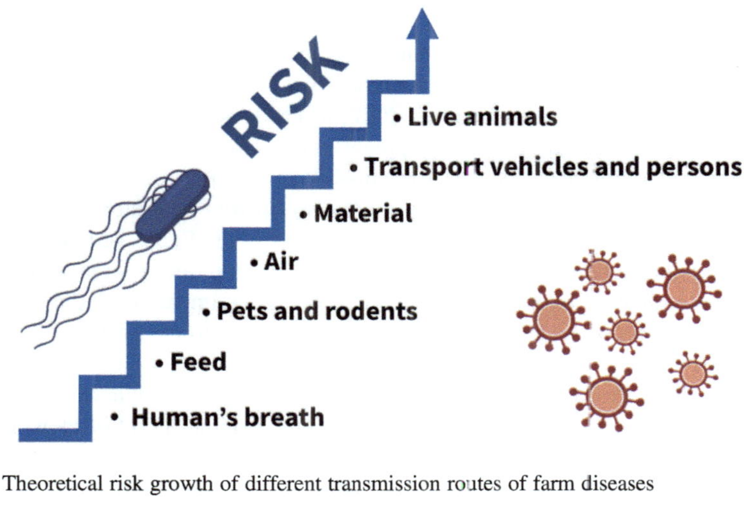

Fig. 9 Theoretical risk growth of different transmission routes of farm diseases

preventive measures have been recommended, such as air filtration equipment and tree plantation around the farm, to prevent the transmission of airborne pathogens and to act as an effective biological barrier, respectively (Dewulf and Van Immerseel 2019; Alarcon et al. 2021).

4.3.2 Infrastructure Organization

Many successful rabbitries have their infrastructures organized in at least two buildings, with polyvalent cages to separate breeding females from the fattening rabbits. In this setting, and considering a 42-day does reproduction cycle, does enter a 7-day prepartum in building number 1, which has already been sanitized. After giving birth, does stay with their kits until the weaning day, approximately at 35 days of age. At this point, does are sent to building number 2, which held the previous litter's fattening rabbits, loaded to the slaughterhouse at an average of 68 days old. Through this rotative cycle system applied in which building (Fig. 10), it is possible to perform 9 days of fallow period, including cleaning and disinfection measures of the "all-in/all-out" systems, before introducing the does. This practice, which allows for more effective cleaning and disinfection, has been associated with a reduction in the mortality rates of several diseases, such as epizootic rabbit enteropathy (Licois et al. 2006; Nielsen et al. 2020).

It is important to note that infrastructures must be designed to protect rabbits from other predators but also from wild rabbits, which are important reservoirs of multiple infectious agents (Brown et al. 2015; Hinney et al. 2016; Napoli et al. 2021). To restrict farm access from wild animals, dogs, and cats, it is critical to implement physical barriers, such as buried fences surrounding the entire farm, as shown in Fig. 11 (Bertagnoli et al. 2006). This delimitation also serves to define a clean area (buffer zone) outside the sheds, and even clear areas can be strictly defined as all zones, including equipment, in direct contact with animals (Alarcon et al. 2021). At this point, the area delimitation according to the level of biosecurity risks should be very well defined. As an example, in broiler production, a risk-based weighted scoring system was implemented in many industrial farms (Van Limbergen et al. 2018). Red (high disease risk external areas), orange (medium risk service areas), and green (clean and highly secure access-restricted area) zones were defined to access hazard risks (pigs, Scollo et al. 2022; broilers, Schreuder et al. 2023).

The farm infrastructure and equipment must also protect the animals from environmental outbreaks, such as fires and flash flooding due to heavy rain, or even contamination. In addition, they must also prevent heat stroke events inside the farm and avoid temperatures exceeding 30 °C, which affects the normal physiology of rabbits, potentially weakening the rabbits' immune system and favoring disease dissemination (Rosell 2020). In addition, the perimeter around the air intake fans must also be protected, avoiding the proximity of trucks, cars, or even other persons (Fig. 12).

Within the various infrastructures of a rabbitry, the area where carcasses are stored is one of the most important from a hygienic-sanitary point of view. Freezers must be in an indoor isolated area within the dirty area to reduce the risk of spreading infectious agents (Filippitzi et al. 2018; Dewulf and Van Immerseel 2019; OIE

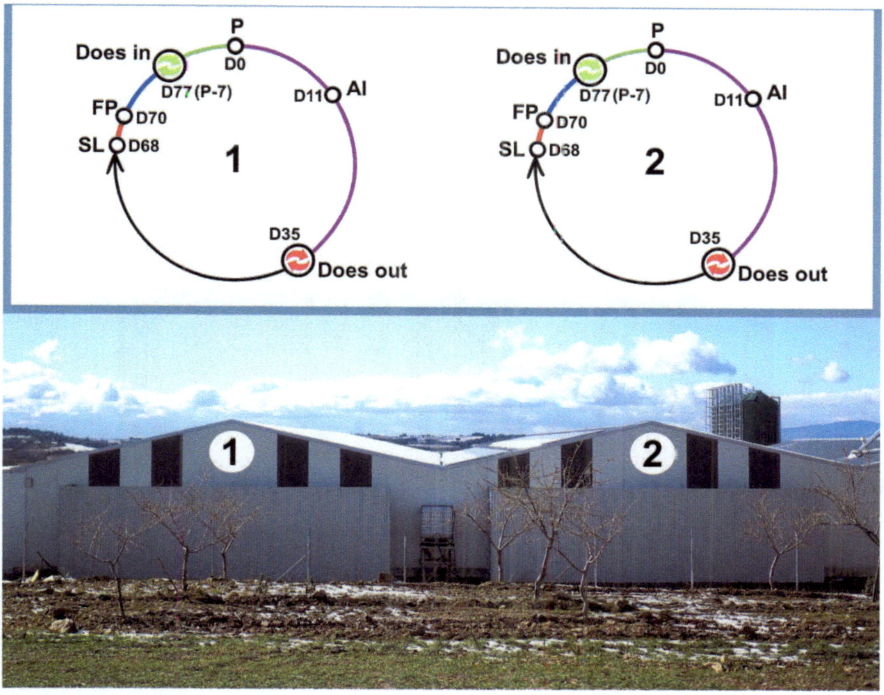

Fig. 10 In rabbitries with high biosecurity standards, a minimum of two buildings exist to perform rotative cycle systems. In this setting, a cycle occurs with an average of 84 days in each building. Does are programmed to 42 days of reproduction cycle and enter in building number 1 at 7-day prepartum (green line). After partum, does stay 35 days with their kits (purple line) and are inseminated (AI) at the 11-day postpartum. After weaning, does are moved out to building number 2, and fattening rabbits stay in building number 2 until they're 68 days old (black line), where they are loaded to the slaughterhouse (SL). At this point, cleaning and disinfection procedures are carried out in 2 days (red line) with a posterior fallow period (FP) of 7 days (blue line)

2019). Moreover, the temperature of the freezers must be low enough to avoid odor diffusion, which can attract parasites and pests, and the storage capacity should accommodate a high number of carcasses to reduce the frequency of carcass removal (Dewulf and Van Immerseel 2019; OIE 2019).

4.3.3 Pest and Parasite Control

Another aspect to consider when talking about biosecurity related to a farm's environment is the control of pests and parasites. In fact, several animals, such as birds, rats, dogs, and cats, can introduce pathogenic agents into rabbitries or may act as reservoirs and vectors for agent dissemination within these production areas (Dewulf and Van Immerseel 2019). Therefore, all farms should design a program to prevent and control all pests and animals capable of transmitting pathogens. First, it is essential to limit the growth of plants and the accumulation of organic waste in

Fig. 11 Implementing 30–40 cm buried fences (red line) around the entire farm is effective in preventing the entrance of wild animals, dogs, and cats

Fig. 12 Physical restriction of the perimeter of the air intake fans (yellow arrows) through fences

the vicinity of the farm, as they can attract and contribute to the local maintenance of certain pests. For example, rabbit feed, which is highly attractive for pests, must be stored in tightly sealed silos, as previously shown in Fig. 4.

As in livestock farming operations, rodents are one of the most significant pests in rabbitries, namely, the black rat (*Rattus rattus*), the brown rat (*Rattus norvegicus*), and the mouse (*Mus musculus*). Thus, all personnel must be aware of the sounds, odor, excrement, and burrows from these animals, and if they are identified during the day, appropriate measures must be taken immediately, as it is likely to be a major infestation (Dewulf and Van Immerseel 2019).

Rodents can endanger public health and lead to economic losses, as they are known to be effective reservoirs and mechanical vectors of several zoonoses, such as pasteurellosis, toxoplasmosis, leptospirosis, salmonellosis, and *Clostridium difficile* infection (Webster and Macdonald 1995; Lapuz et al. 2012; Luis et al. 2013; Lovera et al. 2017; de Oliveira et al. 2018; Rothenburger et al. 2018; Krijger et al. 2019; Jahan et al. 2021). Additionally, rodents can have other harmful effects on a rabbitry, such as causing damage to infrastructures and equipment, stressing the animals, and eating or contaminating the rabbits' feed (Dewulf and Van Immerseel 2019). Therefore, to avoid these situations with a negative productive impact, the farm's infrastructure must be inspected annually or more frequently if a possible infestation is suspected, and traps must be placed outside to eliminate pests.

Regarding parasites, it is known that rabbit health and consequently farm productivity can be seriously affected by the presence of ectoparasites. Mosquitoes, fleas, and ticks are important biological and mechanical vectors and can even cause injuries during their hematophagy process. Some of these parasites spread vector-borne diseases, several of which are zoonotic. Moreover, flies act as a mechanical vector for the transmission of *E. coli*, an infectious agent associated with high mortality in weaned rabbits (Ahmad et al. 2007; Solans et al. 2019), and hematophagous ectoparasites have been associated with the spread of the two main rabbit viral diseases, RHD and myxomatosis (Fenner et al. 1952; Abrantes et al. 2012). Therefore, the use of mosquito nets and pesticides in pit wetlands and daily removal of feces are key measures to control ectoparasites (Rosell 2002).

Other domestic animals, such as dogs and cats, are often found as cohabitants in rabbit farms, but biosecurity programs must restrict their access to clean areas and always avoid their contact with rabbits. Moreover, despite their presence helping to ward off pests, it is known that their use for this purpose is inappropriate, as these animals represent a potential vector for the transmission of pathogenic agents and can act as reservoirs within the farm environment (Dewulf and Van Immerseel 2019).

4.4 Biosecurity in Disease Prevention and Treatment

To guarantee good general health for all rabbits on the farm, it is essential to design a robust disease management strategy, including crucial internal measures, such as methods that allow systematic screening and early diagnosis of diseases; isolation of affected and suspected animals; registration and monitoring of all cases of disease; and adequate vaccination protocols to improve rabbits' immunity. Through the implementation of these measures, it is possible to characterize the health status of the farm and to establish an effective preventive protocol.

4.4.1 Approach to Sick Animals

Adequate biosecurity practices must include suitable management of sick animals, providing the therapeutic care they need to recover. From a productive point of view, the treatment of these animals aims to enhance their productive capacity and reduce economic losses, eliminating potential disease outbreaks and limiting the spread of pathogens within that environment.

As soon as a sick animal, or animal group, is identified, it must be isolated from the rest and placed in quarantine to prevent contagion through its excretions and secretions (Dewulf 2014). Once in this "nursery" area, rabbits must not return to the clean area, and their treatment must be performed after handling healthy animals and using disposable PPE to avoid iatrogenic transmission of pathogens (Backhans et al. 2015; Dewulf and Van Immerseel 2019). In the event that the animal dies before establishing the diagnosis, the veterinarian responsible for the rabbitry must perform a necropsy to analyze potential lesions in internal organs and collect samples for laboratory analysis to reach a definitive diagnosis (Rosell 2000).

Unfortunately, many rabbitries do not have available insulated nursery rooms due to their costs. Usually, ill and apparently healthy rabbits are separated, and the first ones are housed close to the air extraction fans to mitigate the circulation of potential pathogens within the shed if contagious disease is suspected. Other operational biosecurity measures, such as restricted and planned personal movements (e.g., health animals are first visited), take place. When a disease outbreak occurs, the affected shed or even the farm can remain in quarantine.

4.4.2 Proper Antibiotic Use in Rabbit Farming

Since infections caused by bacteria, such as *Pasteurella multocida*, *Bordetella* spp., *E. coli*, *Clostridium spiroforme*, and *Staphylococcus aureus*, are common in rabbits, antibiotic prescriptions may often be required in a rabbitry (Boucher and Nouaille 2013c). However, the use of antimicrobials should be limited and applied only when they are proven infectious, keeping in mind that their utility should be grounded in evidence-based medicine. Ideally, a bacteriological culture and antibiogram should be performed to choose the most appropriate drug and reduce antibiotic resistance (Ungemach et al. 2006).

Given the conditions in which farm rabbits are housed, antibiotics are usually administered orally, through water or feed, to all animals in the same group, except for breeding females, whose parental and individual treatment is preferably recommended (Boucher and Nouaille 2013c).

In line with Regulation 2019/6 of the European Parliament and of the Council of 11 December 2018 on veterinary medicinal products, these biosecurity programs are also essential to promote the development of increasingly drug-free production systems in the livestock industry. Thus, from a One Health perspective, the use of antimicrobials in farm rabbits should be reduced to what is strictly necessary and only prescribed when there is medical evidence of benefit, which is crucial to decrease the occurrence of antimicrobial resistance, especially associated with drugs of higher public health value (Rosell 2020).

5 Prophylaxis: Deworming and Vaccination

Deworming and vaccination are two essential prophylactic measures to keep farm rabbits healthy and increase the efficiency of their immune systems.

Deworming protocols prevent immunity disturbance while eliminating parasites and avoiding dietary energy losses and should be applied before starting any vaccination protocol to maximize their immune response. Currently, special attention should be given to the pinworm *Passalurus ambiguus*, an intestinal worm with the highest prevalence in rabbits that affects the small intestine, cecum, and colon (see Chapter "Major Endoparasitoses in Rabbits: Oxyuriosis and Metacestodosis"). This oxyurid releases eggs through animal feces without usually producing any clinical signs, although weight loss, anorexia, perianal pruritus, fecal impaction, self-trauma, and even rectal prolapse have already been reported with higher parasitic loads (Hess and Tater 2012). From the authors' experience, oral administration every 3 months of albendazole or fenbendazole at a dose of 20 mg/kg or flubendazole at a dose of 10 mg/kg to breeding rabbits is an effective form of parasite control in production rabbits.

Vaccination is one of the greatest tools of preventive medicine since it stimulates the rabbit immune system and is highly efficient in reducing infection and excretion of infectious agents. Myxomatosis and RHD are the two rabbit diseases notifiable to WOAH, for which vaccination is highly recommended in rabbitries. The most widely used commercial vaccines are live vaccines against myxomatosis (homologous and heterologous) and inactivated vaccines against classic RHDV (GI.1) and the new serotype RHDV2 (GI.2).

The vaccination protocols must be customized to each farm according to various risk factors, susceptibilities, and other epidemiological features. For example, additional vaccines are being applied in other countries, such as autogenous vaccines against pathogenic *E. coli* (type 0103), *S. aureus*, and *Pasteurella* spp. However, in general, many farms free of parasites and active viral infections successfully apply the prophylactic protocol listed in Table 1.

Sanitary practices must be employed before and during vaccination, such as checking the temperature of vaccines upon arrival, storing vaccines at the temperature recommended by the manufacturer, using only one needle per animal, and only administering vaccines within 2 hours of opening the vial. The equipment used for

Table 1 General prophylactic protocol applied in rabbitries

Agent	Primovaccination	Subsequent vaccinations	References
Myxomatosis	First vaccine at 4 weeks of age and booster vaccine at 10 weeks of age	Every 3.5 to 4 months	Boucher and Nouaille (2013a)
Rabbit hemorrhagic disease	First vaccine at 5 to 6 weeks of age and booster vaccine at 10 to 11 weeks of age	Every 6 months	Boucher and Nouaille (2013d)

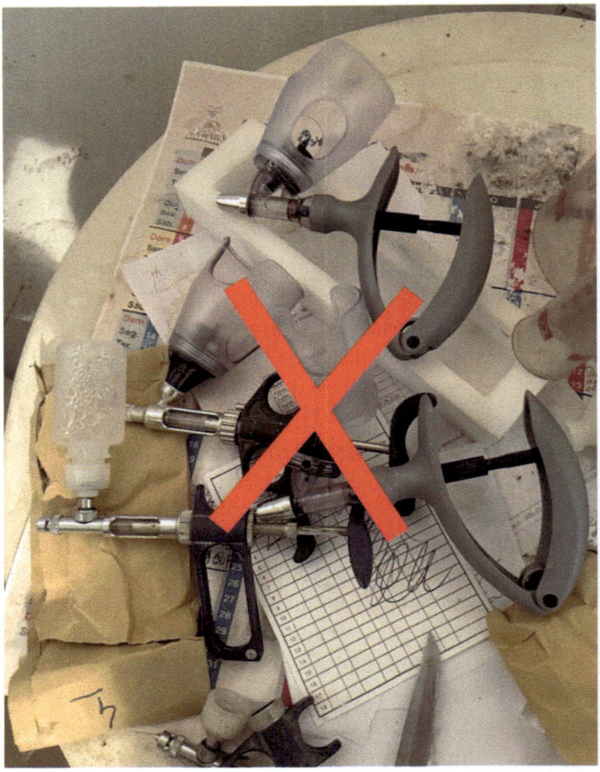

Fig. 13 Improper storage of material used in rabbit vaccinations may be a source of infection

vaccination must also be thoroughly cleaned and kept in a secure place, avoiding bad storage, as illustrated in Fig. 13.

6 Technical-Economic Management

As depicted in Fig. 1 at the beginning of this chapter, technical-economic management is an important pillar of preventive medicine, as it allows for the evaluation of the business performance in rabbit farming at technical, productive, and economic levels. Through the zootechnical data and registration of health indicators, this management can detect which points are critical and need to be optimized to prevent production breakdowns.

To calculate the productivity of a rabbitry, in addition to the financial income, it is crucial to evaluate the economic costs spent on the production system. Therefore, bioeconomic models have been proposed to facilitate this operation, considering age and reproductive status distribution, feed and water requirements, and expenses related to rabbit health care, infrastructure maintenance, and routine procedures (Krupová et al. 2020).

Despite the scarce literature on this subject, there is a growing demand for solutions that ensure the well-being and good health of rabbits without increasing maintenance costs (Mondin et al. 2021). Therefore, for proper technical-economic management, farmers need to rely on farm data related to fertility, growth curves, average weights, total numbers of does and weaned rabbits, and mortality rates (Krupová et al. 2020). It is recommended that all these parameters should be constantly recorded and evaluated, as well as the various farm expenses already referenced, so that the viability of the rabbitry is assured.

Another aspect that is frequently discussed in rabbit farming is whether the costs of treatment outweigh the benefits. From an economic point of view, all costs inherent to the treatments prescribed in a rabbitry should be evaluated based on the expected animal therapeutic response (Boucher and Nouaille 2013c). It should be noted that the treatment of an animal is not only viable and profitable when it allows the recovery of the individual productive capacity of that animal but also when it prevents an outbreak in rabbitry that could affect others with serious economic consequences.

In fact, wise investment in biosecurity measures has been associated with a higher economic return (Casal et al. 2007; Valeeva et al. 2011; Laanen et al. 2014; Dewulf and Van Immerseel 2019). However, the economic balance can still be improved by giving special focus to preventive measures, as it has been reported that several rabbitries still overuse prophylactic antimicrobials to compensate for the inadequate implementation of safety programs regarding cleaning and disinfection procedures, for example (Murphy et al. 2017).

7 Conclusions

Biosecurity in rabbitries should focus on farm rabbit health and welfare, environmental and public health protection, and economic benefits, all under a One Health perspective.

Preserving the highest animal health status through well-developed biosecurity protocols is a success factor to achieve more efficient production. None of this is possible without the effort, constant observation, control and analysis, and prompt response of rabbit producers aligned with the assistant veterinarian to detect any change in the health of farm rabbits. For this reason, all employees must receive biosecurity formal training based on the various methodologies and evaluation criteria recommended throughout this chapter.

Therefore, to achieve optimal results in a rabbitry, it is crucial to properly comply with deworming and vaccination protocols, as well as effective and systematic pest control, always in association with good nutrition and appropriate animal welfare measures.

As a final note, the authors would like to highlight the importance of designing and implementing adequate biosecurity protocols in rabbit production to maximize the productive and economic potential. Remember, prevention is always better than treatment, and good preventive management starts with the adoption of a complete, systematic, and well-established biosecurity practice.

Acknowledgments The authors thank Springer for the use of the AJE Digital Editing. The Portuguese Foundation for Science and Technology (FCT) supported the PhD fellowship of Filipe Fontes Pinto (2022.10138.BD) and the Assistant Researcher grant of Joana Abrantes (CEECIND/00078/2017). This work was also supported by projects UIDP/CVT/00772/2020 (CECAV) and LA/P/0059/2020 (AL4AnimalS).

References

Abrantes J, van der Loo W, Le Pendu J, Esteves PJ (2012) Rabbit haemorrhagic disease (RHD) and rabbit haemorrhagic disease virus (RHDV): a review. Vet Res 43:12. https://doi.org/10.1186/1297-9716-43-12

Ahmad A, Nagaraja TG, Zurek L (2007) Transmission of Escherichia coli O157:H7 to cattle by house flies. Prev Vet Med 80:74–81. https://doi.org/10.1016/j.prevetmed.2007.01.006

Alarcon LV, Alberto AA, Mateu E (2021) Biosecurity in pig farms: a review. Porc Health Manag 7: 5. https://doi.org/10.1186/s40813-020-00181-z

Anderson AV, Fitzgerald C, Baker K, Stika R, Linhares D, Holtkamp DJ (2018) Comparison of shower-in and shower-in plus bench entry protocols for prevention of environmental contamination due to personnel entry in a commercial swine facility. J Swine Health Prod 26:192–199

Backhans A, Sjolund M, Lindberg A, Emanuelson U (2015) Biosecurity level and health management practices in 60 Swedish farrow-to-finish herds. Acta Vet Scand 57:14. https://doi.org/10.1186/s13028-015-0103-5

Bartoszko JJ, Farooqi MAM, Alhazzani W, Loeb M (2020) Medical masks vs N95 respirators for preventing COVID-19 in healthcare workers: a systematic review and meta-analysis of randomized trials. Influenza Other Respir Viruses 14:365–373. https://doi.org/10.1111/irv.12745

Beaucournu JC, Launay H (1977) European rabbit and zoonosis. Med Mal Infect 7:495–501. https://doi.org/10.1016/s0399-077x(77)80117-0

Bertagnoli S, Messud-Petit F, Marlier D (2006) Myxomatosis. Recent advances in rabbit sciences. In: Maertens L, Coudert P (eds) . Institute for Agricultural and Fisheries Research (ILVO), Belgium, pp 139–145

Betancourt WQ, Rose JB (2004) Drinking water treatment processes for removal of cryptosporidium and giardia. Vet Parasitol 126:219–234. https://doi.org/10.1016/j.vetpar.2004.09.002

Black R, Bartlett DMF (2020) Biosecurity frameworks for cross-border movement of invasive alien species. Environ Sci Pol 105:113–119. https://doi.org/10.1016/j.envsci.2019.12.011

Boucher S, Nouaille L (2013a) Myxomatose. Maladies des lapins, Boucher S, Nouaille L (eds.) 3e édition. Éditions France Agricole, pp 136–147

Boucher S, Nouaille L (2013b) Nettoyage, désinfection et vide sanitaire. Maladies des lapins, Boucher S, Nouaille L (eds.), 3e édition. Éditions France Agricole, pp 321–328

Boucher S, Nouaille L (2013c) Utilisation des médicaments en élevage cunicole: aspects pratiques. Maladies des lapins, Boucher S, Nouaille L (eds.), 3e édition. Éditions France Agricole, pp 308–320

Boucher S, Nouaille L (2013d) VHD ou maladie hémorragique virale. Maladies des lapins, Boucher S, Nouaille L (eds.), 3e édition. Éditions France Agricole, pp 128–135

Brown VR, Adney DR, Bielefeldt-Ohmann H, Gordy PW, Felix TA, Olea-Popelka FJ, Bowen RA (2015) Pathogenesis and immune responses of *Francisella tularensis* strains in wild-caught cottontail rabbits (*Sylvilagus* spp.). J Wildl Dis 51:564–575. https://doi.org/10.7589/2015-02-030

Burt SA, Veltman J, Hakze-van der Honing R, Schmitt H, van der Poel WHM (2016) Hepatitis E virus in farmed rabbits, wild rabbits and petting farm rabbits in The Netherlands. Food Environ Virol 8:227–229. https://doi.org/10.1007/s12560-016-9239-3

Casal J, De Manuel A, Mateu E, Martin M (2007) Biosecurity measures on swine farms in Spain: perceptions by farmers and their relationship to current on-farm measures. Prev Vet Med 82: 138–150. https://doi.org/10.1016/j.prevetmed.2007.05.012

Dalmau A, Moles X, Pallisera J (2020) Animal welfare assessment protocol for does, bucks, and kit rabbits reared for production. Front Vet Sci 7:445. https://doi.org/10.3389/fvets.2020.00445

de Oliveira CA, Gabardo MD, Guedes RMC, Poncet F, Blanc DS, Lobato FCF, Silva ROS (2018) Rodents are carriers of Clostridioides difficile strains similar to those isolated from piglets. Anaerobe 51:61–63. https://doi.org/10.1016/j.anaerobe.2018.04.006

Dewulf J (2014) An online risk-based biosecurity scoring system for pig farms. Vet Ireland J 4:426–429

Dewulf J, Van Immerseel F (eds) (2019) Biosecurity in animal production and veterinary medicine: from principles to practice. Cabi, Oxfortshire. 978-1-78924-568-4. https://doi.org/10.1079/9781789245684.0000

El Allaoui A, Filali FR, Derouich A (2016) Water quality in Turkey farms in Khemisset (Morocco) and potential risk factors. Revue Scientifique et Technique-Office International des Epizooties 35:957–964. https://doi.org/10.20506/rst.35.3.2581

Fenner F, Day MF, Woodroofe GM (1952) The mechanism of the transmission of myxomatosis in the European rabbit (*Oryctolagus cuniculus*) by the mosquito *Aedes aegypti*. Aust J Exp Biol Med Sci 30:139–152. https://doi.org/10.1038/icb.1952.13

Filippitzi ME, Brinch Kruse A, Postma M, Sarrazin S, Maes D, Alban L, Nielsen LR, Dewulf J (2018) Review of transmission routes of 24 infectious diseases preventable by biosecurity measures and comparison of the implementation of these measures in pig herds in six European countries. Transbound Emerg Dis 65:381–398. https://doi.org/10.1111/tbed.12758

Garforth CJ, Bailey AP, Tranter RB (2013) Farmers' attitudes to disease risk management in England: a comparative analysis of sheep and pig farmers. Prev Vet Med 110:456–466. https://doi.org/10.1016/j.prevetmed.2013.02.018

Gillespie AV, Grove-White DH, Williams HJ (2015) Husbandry, health and biosecurity of the smallholder and pet pig population in England. Vet Rec 177:47. https://doi.org/10.1136/vr.102759

Gordh G, MckKirdy S (eds) (2014) The handbook of plant biosecurity. Springer, Dordrecht. https://doi.org/10.1007/978-94-007-7365-3. isbn:978-94-007-7365-3

Graham JP, Leibler JH, Price LB, Otte JM, Pfeiffer DU, Tiensin T, Silbergeld EK (2008) The animal-human interface and infectious disease in industrial food animal production: rethinking biosecurity and biocontainment. Public Health Rep 123:282–299. https://doi.org/10.1177/003335490812300309

Hess L, Tater K (2012) Chapter 18 - Dermatologic diseases. In Quesenberry KE, Carpenter JW (eds) Ferrets, rabbits, and rodents, 3rd edn. W.B. Saunders, Philadelphia, pp 232–244. https://doi.org/10.1016/B978-1-4160-6621-7.00018-X. isbn: 9781416066217

Hinney B, Sak B, Joachim A, Kvac M (2016) More than a rabbit's tale - Encephalitozoon spp. in wild mammals and birds. Int J Parasitol Parasites Wildl 5:76–87. https://doi.org/10.1016/j.ijppaw.2016.01.001

Huber N, Andraud M, Sassu EL, Prigge C, Zoche-Golob V, Käsbohrer A, D'Angelantonio D, Viltrop A, Żmudzki J, Jones H, Smith RP, Tobias T, Burow E (2022) What is a biosecurity measure? A definition proposal for animal production and linked processing operations. One Health 15:100433. https://doi.org/10.1016/j.onehlt.2022.100433

Huneau-Salahün A, Bougeard S, Balaine L, Eono F, Le Bouquin S, Chauvin C (2015) Husbandry factors and health conditions influencing the productivity of French rabbit farms. World Rabbit Sci 23:27–37. https://doi.org/10.4995/wrs.2015.3076

Jahan NA, Lindsey LL, Kipp EJ, Reinschmidt A, Heins BJ, Runck AM, Larsen PA (2021) Nanopore-based surveillance of zoonotic bacterial pathogens in farm-dwelling Peridomestic rodents. Pathogens 10:1183. https://doi.org/10.3390/pathogens10091183

Jarvis LS, Valdes-Donoso P (2018) A selective review of the economic analysis of animal health management. J Agric Econ 69:201–225. https://doi.org/10.1111/1477-9552.12131

Kim Y, Yang M, Goyal SM, Cheeran MCJ, Torremorell M (2017) Evaluation of biosecurity measures to prevent indirect transmission of porcine epidemic diarrhea virus. BMC Vet Res 13:89. https://doi.org/10.1186/s12917-017-1017-4

Krijger IM, Meerburg BG, Harmanus C, Burt SA (2019) *Clostridium difficile* in wild rodents and insectivores in The Netherlands. Lett Appl Microbiol 69:35–40. https://doi.org/10.1111/lam.13159

Krupová Z, Wolfová M, Krupa E, Volek Z (2020) Economic values of rabbit traits in different production systems. Animal 14:1943–1951. https://doi.org/10.1017/S1751731120000683

Kylie J, Brash M, Whiteman A, Tapscott B, Slavic D, Weese JS, Turner PV (2017a) Biosecurity practices and causes of enteritis on Ontario meat rabbit farms. Can Vet J 58:571–578

Kylie J, McEwen SA, Boerlin P, Reid-Smith RJ, Weese JS, Turner PV (2017b) Prevalence of antimicrobial resistance in fecal *Escherichia coli* and *Salmonella enterica* in Canadian commercial meat, companion, laboratory, and shelter rabbits (*Oryctolagus cuniculus*) and its association with routine antimicrobial use in commercial meat rabbits. Prev Vet Med 147:53–57. https://doi.org/10.1016/j.prevetmed.2017.09.004

Laanen M, Maes D, Hendriksen C, Gelaude P, De Vliegher S, Rosseel Y, Dewulf J (2014) Pig, cattle and poultry farmers with a known interest in research have comparable perspectives on disease prevention and on-farm biosecurity. Prev Vet Med 115:1–9. https://doi.org/10.1016/j.prevetmed.2014.03.015

Lambert ME, Arsenault J, Poljak Z, D'Allaire S (2012) Epidemiological investigations in regard to porcine reproductive and respiratory syndrome (PRRS) in Quebec, Canada. Part 2: prevalence and risk factors in breeding sites. Prev Vet Med 104(1–2):84–93. https://doi.org/10.1016/j.prevetmed.2011.11.002

Lapuz R, Umali DV, Suzuki T, Shirota K, Katoh H (2012) Comparison of the prevalence of salmonella infection in layer hens from commercial layer farms with high and Low rodent densities. Avian Dis 56:29–34. https://doi.org/10.1637/9704-030711-Reg.1

Lee J, Yoo D, Ryu S, Ham S, Lee K, Yeo M, Min K, Yoon C (2019) Quantity, size distribution, and characteristics of cough-generated aerosol produced by patients with an upper respiratory tract infection. Aerosol Air Qual Res 19:840–853. https://doi.org/10.4209/aaqr.2018.01.0031

Leibler JH, Dalton K, Pekosz A, Gray GC, Silbergeld EK (2017) Epizootics in industrial livestock production: preventable gaps in biosecurity and biocontainment. Zoonoses Public Health 64:137–145. https://doi.org/10.1111/zph.12292

Lhomme S, Dubois M, Abravanel F, Top S, Bertagnoli S, Guerin JL, Izopet J (2013) Risk of zoonotic transmission of HEV from rabbits. J Clin Virol 58:357–362. https://doi.org/10.1016/j.jcv.2013.02.006

Licois D, Coudert P, Marlier D (2006) Epizootic rabbit enteropathy. In: Maertens L, Coudert P (eds) Recent advances in rabbit sciences, Institute for Agricultural and Fisheries Research (ILVO), Belgium, pp 163–170

Lovera R, Fernández MS, Jacob J, Lucero N, Morici G, Brihuega B, Farace MI, Caracostantogolo J, Cavia R (2017) Intrinsic and extrinsic factors related to pathogen infection in wild small mammals in intensive milk cattle and swine production systems. PLoS Negl Trop Dis 11:e0005722. https://doi.org/10.1371/journal.pntd.0005722

Luis AD, Hayman DT, O'Shea TJ, Cryan PM, Gilbert AT, Pulliam JR, Mills JN, Timonin ME, Willis CK, Cunningham AA, Fooks AR, Rupprecht CE, Wood JL, Webb CT (2013) A comparison of bats and rodents as reservoirs of zoonotic viruses: are bats special? Proc Biol Sci 280:20122753. https://doi.org/10.1098/rspb.2012.2753

Macleod A (2015) In: Jarrad F, Low Chow S, Mengersen K (eds) The relationship between biosecurity surveillance and risk analysis. Biosecurity surveillance: quantitative approaches.

CABI, Wallingford, pp 109–122. isbn: 9781780643595. https://doi.org/10.1079/9781780643595.0109

Maes D, Segales J, Meyns T, Sibila M, Pieters M, Haesebrouck F (2008) Control of *mycoplasma hyopneumoniae* infections in pigs. Vet Microbiol 126:297–309. https://doi.org/10.1016/j.vetmic.2007.09.008

Magalhaes TR, Pinto FF, Queiroga FL (2022) A multidisciplinary review about *Encephalitozoon cuniculi* in a one health perspective. Parasitol Res 121:2463–2479. https://doi.org/10.1007/s00436-022-07562-z

Mato-Amboage R, Pitchford JW, Touza J (2019) Public-private partnerships for biosecurity: an opportunity for risk sharing. J Agric Econ 70:771–788. https://doi.org/10.1111/1477-9552.12315

Medeiros I, Fernandez-Novo A, Astiz S, Simões J (2022) Historical evolution of cattle management and herd health of dairy farms in OECD countries. Vet Sci 9:125. https://doi.org/10.3390/vetsci9030125

Millman C, Christley R, Rigby D, Dennis D, O'Brien SJ, Williams N (2017) "Catch 22": biosecurity awareness, interpretation and practice amongst poultry catchers. Prev Vet Med 141:22–32. https://doi.org/10.1016/j.prevetmed.2017.04.002

Mondin C, Trestini S, Trocino A, Di Martino G (2021) The economics of rabbit farming: a pilot study on the impact of different housing systems. Animals 11:3040. https://doi.org/10.3390/ani11113040

Murphy D, Ricci A, Auce Z, Beechinor JG, Bergendahl H, Breathnach R, Bureš J, Duarte Da Silva JP, Hederová J, Hekman P, Ibrahim C, Kozhuharov E, Kulcsár G, Lander Persson E, Lenhardsson JM, Mačiulskis P, Malemis I, Markus-Cizelj L, Michaelidou-Patsia A et al (2017) EMA and EFSA joint scientific opinion on measures to reduce the need to use antimicrobial agents in animal husbandry in the European Union, and the resulting impacts on food safety (RONAFA). EFSA J 15:e04666. https://doi.org/10.2903/j.efsa.2017.4666

Napoli E, Remesar S, Gaglio G, Giannetto S, Spadola F, Díaz P, Morrondo P, Brianti E (2021) Ectoparasites of wild rabbit (*Oryctolagus cuniculus*) in southern Italy. Vet Parasitol Reg Stud Rep 24:100555. https://doi.org/10.1016/j.vprsr.2021.100555

Ni L, Zheng WC, Zhang Q, Cao W, Li BM (2016) Application of slightly acidic electrolyzed water for decontamination of stainless steel surfaces in animal transport vehicles. Prev Vet Med 133:42–51. https://doi.org/10.1016/j.prevetmed.2016.09.010

Nielsen SS, Alvarez J, Bicout DJ, Calistri P, Depner K, Drewe JA, Garin-Bastuji B, Gonzales Rojas JL, Gortázar Schmidt C, Michel V, Miranda Chueca MÁ, Roberts HC, Sihvonen LH, Spoolder H, Stahl K, Velarde Calvo A, Viltrop A, Buijs S, Edwards S et al (2020) Health and welfare of rabbits farmed in different production systems. EFSA J 18:e05944. https://doi.org/10.2903/j.efsa.2020.5944

Nodari SR, Lavazza A, Candotti P (2009) Technical note: rabbit welfare during electrical stunning and slaughter at a commercial abattoir. World Rabbit Sci 17:163–167

OIE (2019) Disposal of dead animals. Terrestrial Animal Health Code, 28th Ed., Vol. I, pp 190–195. isbn 978-92-95108-85-1

Osawe OW, Läpple D, Mee JF (2022) Economic analysis of biosecurity adoption in dairy farming: evidence from Ireland. J Anim Sci 100(9):skac218. https://doi.org/10.1093/jas/skac218

Postma M, Stärk KD, Sjölund M, Backhans A, Beilage EG, Lösken S, Belloc C, Collineau L, Iten D, Visschers V, Nielsen EO, Dewulf J, MINAPIG consortium (2015) Alternatives to the use of antimicrobial agents in pig production: a multi-country expert-ranking of perceived effectiveness, feasibility and return on investment. Prev Vet Med 118:457–466. https://doi.org/10.1016/j.prevetmed.2015.01.010

Postma M, Backhans A, Collineau L, Loesken S, Sjölund M, Belloc C, Emanuelson U, Grosse Beilage E, Nielsen EO, Stärk KDC, Dewulf J, MINAPIG consortium (2016a) Evaluation of the relationship between the biosecurity status, production parameters, herd characteristics and antimicrobial usage in farrow-to-finish pig production in four EU countries. Porc Health Manag 2:9. https://doi.org/10.1186/s40813-016-0028-z

Postma M, Speksnijder DC, Jaarsma ADC, Verheij TJM, Wagenaar JA, Dewulf J (2016b) Opinions of veterinarians on antimicrobial use in farm animals in Flanders and The Netherlands. Vet Rec 179:68. https://doi.org/10.1136/vr.103618

Postma M, Vanderhaeghen W, Sarrazin S, Maes D, Dewulf J (2017) Reducing antimicrobial usage in pig production without jeopardizing production parameters. Zoonoses Public Health 64:63–74. https://doi.org/10.1111/zph.12283

Pritchard G, Dennis I, Waddilove J (2005) Biosecurity: reducing disease risks to pig breeding herds. In Pract 27:230–237. https://doi.org/10.1136/inpract.27.5.230

Racicot M, Venne D, Durivage A, Vaillancourt JP (2011) Description of 44 biosecurity errors while entering and exiting poultry barns based on video surveillance in Quebec, Canada. Prev Vet Med 100:193–199. https://doi.org/10.1016/j.prevetmed.2011.04.011

Reed MS, Curzon R (2015) Stakeholder mapping for the governance of biosecurity: a literature review. J Integr Environ Sci 12:15–38. https://doi.org/10.1080/1943815x.2014.975723

Renault V, Humblet MF, Saegerman C (2022) Biosecurity concept: origins, evolution and perspectives. Animals 12:63. https://doi.org/10.3390/ani12010063

Rommers J, Meijerhof R (1998) Effect of group size on performance, bone strength and skin lesions of meat rabbits housed under commercial conditions. World Rabbit Sci 6:299–302. https://doi.org/10.4995/wrs.1998.360

Rosell JM (2000) Enfermedades del Conejo. Madrid, Spain, Mundi-Prensa Libros SA. isbn: 9788471149077

Rosell JM (2002) Profilaxia em explorações de cunicultura. Proceedings II Jornadas Internacionais de Cunicultura 11 e 12 de Outubro de 2002, Vila Real, Portugal

Rosell JM (2020) Bioseguridad en granjas de conejos. Real Academia de Ciencias Veterinarias de España (RACVE). Accessed from https://www.asic-wrsa.it/documenti/tecnica/BiosecuridadConejosRosell.pdf. 16 March 2023

Rosell JM, de la Fuente LF (2018) Mastitis on rabbit farms: prevalence and risk factors. Animals 8:98. https://doi.org/10.3390/ani8060098

Rossi G, Smith RL, Pongolini S, Bolzoni L (2017) Modelling farm-to-farm disease transmission through personnel movements: from visits to contacts, and back. Sci Rep 7:2375. https://doi.org/10.1038/s41598-017-02567-6

Rothenburger JL, Rousseau JD, Weese JS, Jardine CM (2018) Livestock-associated methicillin-resistant *Staphylococcus aureus* and *Clostridium difficile* in wild Norway rats (*Rattus norvegicus*) from Ontario swine farms. Can J Vet Res 82:66–69

Schreuder J, Simitopoulou M, Angastiniotis K, Ferrari P, Wolthuis-Fillerup M, Kefalas G, Papasolomontos S (2023) Development and implementation of a risk assessment tool for broiler farm biosecurity and a health intervention plan in The Netherlands, Greece, and Cyprus. Poult Sci 102:102394. https://doi.org/10.1016/j.psj.2022.102394

Scollo A, Levallois P, Fourichon C, Motta A, Mannelli A, Lombardo F, Ferrari P (2022) Monitoring means and results of biosecurity in pig fattening farms: systematic assessment of measures in place and exploration of biomarkers of interest. Animals 12(19):2655. https://doi.org/10.3390/ani12192655

Solans L, Arnal JL, Sanz C, Benito A, Chacon G, Alzuguren O, Fernandez AB (2019) Rabbit enteropathies on commercial farms in the Iberian Peninsula: etiological agents identified in 2018–2019. Animals 9:1142. https://doi.org/10.3390/ani9121142

Sommerstein R, Fux CA, Vuichard-Gysin D, Abbas M, Marschall J, Balmelli C, Troillet N, Harbarth S, Schlegel M, Widmer A, Swissnoso (2020) Risk of SARS-CoV-2 transmission by aerosols, the rational use of masks, and protection of healthcare workers from COVID-19. Antimicrob Resist Infect Control 9:100. https://doi.org/10.1186/s13756-020-00763-0

Spencer JL, Guan J (2004) Public health implications related to spread of pathogens in manure from livestock and poultry operations. Methods Mol Biol (Clifton, N.J.) 268:503–515. https://doi.org/10.1385/1-59259-766-1:503

Stewart SC, Dritz SS, Woodworth JC, Paulk C, Jones CK (2020) A review of strategies to impact swine feed biosecurity. Anim Health Res Rev 21:61–68. https://doi.org/10.1017/s146625231900015x

Swennes AG, Buckley EM, Madden CM, Byrd CP, Donocoff RS, Rodriguez L, Parry NM, Fox JG (2013) Enteropathogenic Escherichia coli prevalence in laboratory rabbits. Vet Microbiol 163: 395–398. https://doi.org/10.1016/j.vetmic.2013.01.006

Tomley FM, Shirley MW (2009) Livestock infectious diseases and zoonoses. Philos Trans R Soc B-Biol Sci 364:2637–2642. https://doi.org/10.1098/rstb.2009.0133

Ungemach FR, Müller-Bahrdt D, Abraham G (2006) Guidelines for prudent use of antimicrobials and their implications on antibiotic usage in veterinary medicine. Int J Med Microbiol 296 (Suppl 41):33–38. https://doi.org/10.1016/j.ijmm.2006.01.059

Valeeva NI, van Asseldonk M, Backus GBC (2011) Perceived risk and strategy efficacy as motivators of risk management strategy adoption to prevent animal diseases in pig farming. Prev Vet Med 102:284–295. https://doi.org/10.1016/j.prevetmed.2011.08.005

Van Limbergen T, Dewulf J, Klinkenberg M, Ducatelle R, Gelaude P, Méndez J, Heinola K, Papasolomontos S, Szeleszczuk P, Maes D, PROHEALTH consortium (2018) Scoring biosecurity in European conventional broiler production. Poult Sci 97:74–83. https://doi.org/10.3382/ps/pex296

Webster JP, Macdonald DW (1995) Parasites of wild brown-rats (*Rattus norvegicus*) on UK farms. Parasitology 111:247–255. https://doi.org/10.1017/s0031182000081804

WHO (2022) Guidelines for drinking-water quality Guidelines for drinking-water quality: fourth edition incorporating the first and second addenda Geneva: World Health Organization; 2022. ISBN: 978-92-4-004506-4

WOAH (2022) Disease prevention and control (section 4). Terrestrial Animal Health Code. World Organisation for Animal Health (WOAH - OIE). Accessed from https://www.woah.org/en/what-we-do/standards/codes-and-manuals/terrestrial-code-online-access/. 16 March 2023

WOAH (2023) Notifiable diseases list of lagomorphs. World Organisation for Animal Health (WOAH, founded as OIE). Accessed from https://www.woah.org/en/what-we-do/animal-health-and-welfare/animal-diseases/?_tax_animal=terrestrials%2Clagomorph. 16 March 2023

Zhang YH, Li CS, Liu CC, Chen KZ (2013) Prevention of losses for hog farmers in China: insurance, on-farm biosecurity practices, and vaccination. Res Vet Sci 95:819–824. https://doi.org/10.1016/j.rvsc.2013.06.002

Administration of Antimicrobials and Resistance in Food-Producing Rabbits: Dietary Supplements as Potential Alternatives

Carla Miranda, Gilberto Igrejas, and Patrícia Poeta

Abstract

The food-producing rabbits, the fourth alternative of animal protein sources for human consumption, are an industry of great importance in the world, in particular in the largest rabbit meat producers like China, North Korea, Egypt, and Italy.

C. Miranda
Microbiology and Antibiotic Resistance Team (MicroART), University of Trás-os-Montes and Alto Douro (UTAD), Vila Real, Portugal

Department of Sciences, University Institute of Health Sciences (IUCS), CESPU, CRL, Gandra, Portugal

One Health - Toxicology Research Unit (1H-TOXRUN), IUCS, CESPU, CRL, Gandra, Portugal

Associated Laboratory for Green Chemistry (LAQV-REQUIMTE), University NOVA of Lisbon, Caparica, Portugal

G. Igrejas
Associated Laboratory for Green Chemistry (LAQV-REQUIMTE), University NOVA of Lisbon, Caparica, Portugal

Department of Genetics and Biotechnology, UTAD, Vila Real, Portugal

Functional Genomics and Proteomics Unit, UTAD, Vila Real, Portugal

P. Poeta (✉)
Microbiology and Antibiotic Resistance Team (MicroART), University of Trás-os-Montes and Alto Douro (UTAD), Vila Real, Portugal

Associated Laboratory for Green Chemistry (LAQV-REQUIMTE), University NOVA of Lisbon, Caparica, Portugal

Department of Veterinary Sciences, UTAD, Vila Real, Portugal

Veterinary and Animal Research Centre (CECAV), UTAD, Vila Real, Portugal

Associate Laboratory for Animal and Veterinary Science (AL4AnimalS), Vila Real, Portugal
e-mail: ppoeta@utad.pt

J. Simões, J. M. Monteiro (eds.), *Veterinary Care of Farm Rabbits*,
https://doi.org/10.1007/978-3-031-44542-2_18

However, the antimicrobial use in this sector is highly relative to other food-producing animals because of a lack of welfare legislation and enteric disease outbreaks in the farm rabbits, leading to high economic losses. Over the years, several strategies were implemented to improve welfare, health, and food safety in farmed rabbits and, consequently, reduce the cost of production and antimicrobial use. Very few antibiotics are approved for use in rabbits, although antibiotic-resistant bacteria detected in food-producing rabbits or their products have been reported. However, the knowledge of antimicrobial resistance in rabbits is scarce. Alternatives to mitigate antibiotic usage, such as diet supplementation with probiotics, organic acids, and postbiotics, have been tested in the rabbit industry, although with sometimes contradictory results. For these reasons, this review aimed to summarize the trend of antimicrobial use in farm rabbits and the role of antimicrobial resistance in these animals, as well as potential diet alternatives to antibiotic use.

Keywords

Antibiotics · Antimicrobial resistance · Livestock · Rabbits · Rabbit meat

1 Introduction

Nowadays, rabbits (*Oryctolagus cuniculus*) are classified as the fourth alternative of animal protein sources for human consumption following ruminants, pigs, and poultry production. In 2019, the total production of rabbit meat worldwide reached 883,94 tons, according to the FAO Statistical Database (FAO 2020). For these data, there are ten countries that more contribute to meat-producing rabbits: China, North Korea, Egypt, Italy, Russia, Ukraine, Algeria, Sierra Leone, Mexico, and Peru (FAO 2020). China is the largest rabbit meat producer in the world, contributing more than 80% of the total Asian production that increased from 690,000 to 849,150 tons between 2010 and 2016 (Li et al. 2018). After China, the European Union (EU) is the second largest rabbit meat producer, in which Spain, France, and Italy represent 83% of European rabbit meat production (Cullere and Dalle Zotte 2018; EFSA Panel on Animal Health and Welfare (AHAW) et al. 2020). In 2018, the population of food-producing rabbits in 31 European countries was estimated at 192,000 tons (EMA 2020).

The rabbit meat producers show great diversity, in which this production is developed on large or small scale with different housing systems (Fig. 1). In the EU, there are approximately 4500 commercial rabbit farms and 161,000 backyard farms (EFSA Panel on Animal Health and Welfare (AHAW) et al. 2020). Based on the report of the Directorate-General for Health and Food Safety, in the EU, 66% of farmed rabbits for meat consumption originated from commercial farms through approved slaughterhouses, and 34% originated from backyard farms by direct or local sales (HFSDG 2018).

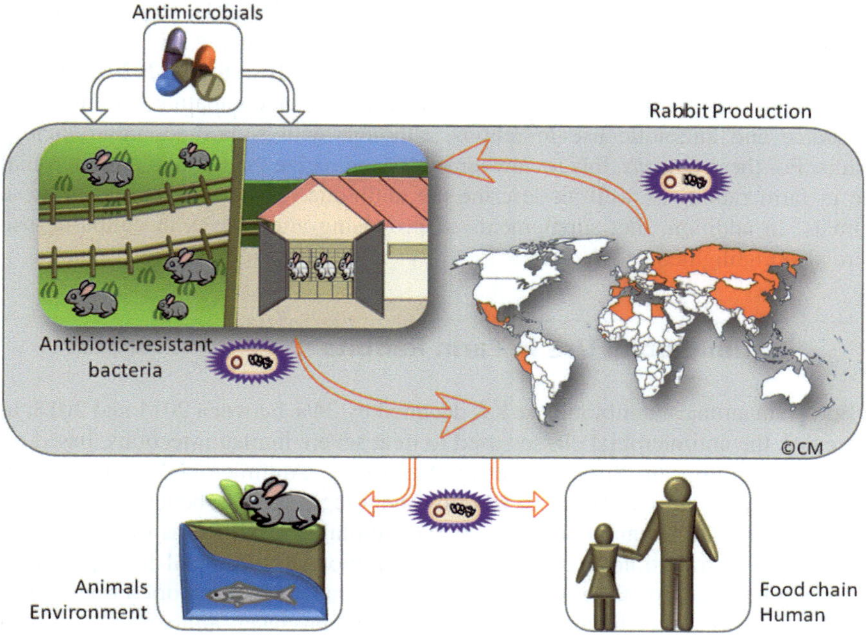

Fig. 1 Dissemination of antimicrobial-resistant bacteria among major rabbit-producing countries. Consequently, the administrated antibiotics increase this spread by food chain or direct contact with humans and the environmental ecosystem

With the intensification of this sector, the economic sustainability and the welfare of rabbit production are a growing concern for producers and consumers. In the last decades, there is a trend of continuous decrease in the consumption of rabbit meat in the EU. This reduction is mainly associated with the consumers' changing habits toward convenience foods relative to consumer price expectations and price competition with other animal meats and their increasing perception of the rabbit as a pet animal (HFSDG 2018). This global decrease in the rabbit meat market price may also be attributed to inadequate strategies for its product promotion (Li et al. 2018). Animal welfare is considered one of the key factors for consumers on the quality of meat and its derived products, orienting the meat choices and preferences of consumers. Additionally, this consumer's attitude is more evident for the rabbits since these animals are increasingly being considered pet animals and not food-producing animals (Cullere and Dalle Zotte 2018).

In this regard, several measures have been implemented for the protection of farm rabbits, although there is no specific European legislation. These strategies comprise improving good practices on farms and their housing conditions and mitigating the cost of production, according to the Commission services on rabbit farming based on the Council Directive 98/58/EC and the European Parliament standards in 2017. All strategies allow health and food safety, also contributing to the reduction of the use of antimicrobials (EP 2017; HFSDG 2018; Mondin et al. 2021). However, scarce data are available on the use of antimicrobials in food-producing rabbits relative to other livestock production species (Agnoletti et al. 2018; EMA 2020). Dietary supplements have emerged as potential alternatives in combating antimicrobial resistance and antibiotic use in rabbits, although with sometimes contradictory results. For these reasons, this review aimed to summarize the trend of antimicrobial use in farm rabbits as well as describe the antibiotic resistance reported in these animals. In addition, diet supplements as promising alternatives to antibiotic use were also highlighted.

2 Antimicrobial Use in Farm Rabbits

In European animals, antibiotic use has declined by 34% between 2011 and 2018, in particular, the antimicrobial classes used to treat severe human infections, based on the annual European Surveillance of Veterinary Antimicrobial Consumption (ESVAC) report (EMA 2020). In addition, new legislation is being implemented in the EU in 2022 (Regulation (EU) 2019/6), prohibiting all forms of routine and prophylactic farm antibiotic use and the importation of all animal food produced with antimicrobial growth promoters (AGPs). Around 99% of imported rabbit meat into the EU was from China, the major rabbit meat producer, although there was a decrease of up to 45% between 2014 and 2016 (HFSDG 2018).

The implementation of biosecurity measures has been the cornerstone for obtaining low rates of morbidity, mortality, and infectious diseases and, consequently, mitigating the use of antimicrobial substances (HFSDG 2018). In rabbit production, the information available about antimicrobials is very limited (Agnoletti et al. 2018). In a majority of used references, rabbits are not included in the livestock population or food-producing animals, such as pigs, bovines, sheep, goats, and poultry, since their production is lower in comparison to them. In addition, several times rabbits are grouped into other minor food-producing animal clusters like fish and/or horses without individual data or clustered into companion animal species, including rabbits, exotics, and horses (FDA 2019; EMA 2020). Overall, studies highlighted a serious lack of information on rabbit production. China leads, being the country with the most research funding, following France, Germany, and Italy. Moreover, Italy, Spain, and France stand among the first five publishing countries. The scientific studies are focused in particular on meat quality, nutrition, and feeding, and the remaining reported welfare, breeding, genetics, and ethology (EFSA 2005; Trocino et al. 2019). Additionally, some countries have recently implemented a system for monitoring antibiotic use in rabbits (NethMap-MARAN 2020), and the antibiotic usage level remained unknown until then. According to the World Organisation for Animal Health (OIE) report, the contributions of rabbits at risk for exposure to the antimicrobial quantities were negligible in the 60 analyzed countries, corresponding to 0.18% (756,000 tons) of the total animal biomass in 2014 (OIE 2017).

Table 1 Common antibiotic agents used in rabbits (Allen et al. 2005)

Classes	Antibiotic	Route of administration
Cephalosporins	Cefazolin, ceftiofur, cephalexin, cephalothin	Parenteral
Sulfonamides	Trimethoprim-sulfa combinations	Parenteral
	Sulfadimethoxine, sulfamerazine, sulfamethazine, sulfaquinoxaline	Oral
Amphenicols	Chloramphenicol	Parenteral, oral, and otic drops
Tetracyclines	Doxycycline Tetracycline (chlortetracycline, oxytetracycline)	Oral
Macrolide	Tylosin	Parenteral and oral
Aminoglycosides	Amikacin, gentamicin	Parenteral (nephrotoxic)
	Neomycin	Oral
Nitroimidazoles	Metronidazole	Oral
Penicillins	Penicillin G benzathine or procaine	Parenteral
Polymyxins	Colistin	Parenteral
Fluoroquinolones	Marbofloxacin	Oral
	Enrofloxacin	Parenteral, oral, and otic drops
	Ciprofloxacin	Oral and otic drops

In concordance with several studies, rabbit intensive production showed the highest administrated antimicrobial rates in comparison to other livestock industries, which may be associated with the high incidence of intestinal discords or enteritis verified in these animals, particularly in the weaning period (Drigo et al. 2015; Agnoletti et al. 2018; ANSES 2020). Although enteritis is the main health concern in rabbits requiring an antibiotic prescription, there is the risk of diarrhea, dysbiosis, or enteritis effects caused by some antibiotics administrated, in particular penicillins, lincosamides, and cephalosporins. So, few antibiotics can be used safely in rabbits, and most are not approved for them (Kylie et al. 2017). In this line, quinolones, ansamycins like rifamycin, β-lactams, and fluoroquinolones are rarely used in rabbit production (Table 1) (Drigo et al. 2015; Agnoletti et al. 2018).

In the Netherlands, antibiotic sales decreased by 70% for farm animals, between 2009 and 2019, resulting from efforts of the authorities, livestock sectors, and veterinarians. Since 2016, antibiotic use in rabbits started to be monitored, demonstrating substantial fluctuations up to 2020, according to the annual NethMap/MARAN 2020 report (NethMap-MARAN 2020). In addition, antibiotic use was high without improvement since 2016 in the rabbit farming sector, showing an increase of 7.2% between 2019 and 2020 (SDa 2021). Considering three categories of antibiotics, first-, second-, and third-choice antibiotics, in farmed rabbits were administrated about 70–85% of first-choice antibiotics (e.g., macrolides/lincosamides, tetracyclines, pleuromutilins, and trimethoprim-sulfonamide combination), 15–25% of second-choice antibiotics (e.g., aminoglycosides and penicillins), and up to 5% of third-choice antibiotics (e.g., fluoroquinolones and polymyxins). This last antibiotic option has not been recorded, as well as colistin administration in the rabbit farming sector in 2020 (NethMap-MARAN 2020; SDa 2021). Colistin as AGPs and colistin products' sales in livestock animals were banned in several countries because it is used as the last recourse for infections caused by multidrug-resistant (MDR) microorganisms in humans (Miranda et al. 2022).

In France, antibiotic sales for rabbits increased between 1999 and 2004, stabled between 2004 and 2007, and decreased between 2007 and 2018. In 2018 was sold 28 tons, less than 11% relative to 2017 and 60% relative to 2011. In concordance, the level of rabbit exposure to antimicrobials also decreased by 34% and 42% between 1999 and 2011, respectively. However, the exposure to antimicrobials increased by 2% for rabbits between 2017 and 2018, corresponding at 6% of total antimicrobial sales for all animals in 2018. Based on antimicrobial classes, the rabbit exposure decreased for all classes, in particular for macrolides (−48%), aminoglycosides (−47%), tetracyclines (−44%), polymyxins (−42%), and sulfonamides (−37%), between 2011 and 2018. The rabbit exposure also decreased for fluoroquinolones (−28%), trimethoprim (−15%), sulfonamides (−15%), and polymyxins (−13%) but increased for tetracyclines (29%) between 2017 and 2018. In addition, the administration of tetracyclines, trimethoprim and sulfonamides, aminoglycosides, pleuromutilins, and polymyxins (colistin and bacitracin) in food-producing rabbits was predominant in 2018 (ANSES 2020). Moreover, no quinolone administration is allowed in French rabbit production, which disagrees with Italian rabbit production where it was observed an increase in quinolone prescription and a reduction in tetracycline usage during 2010–2015 (Agnoletti et al. 2018).

An Italian study showed that 76% of rabbit farmers use antimicrobials for the treatment of diseases and 78% of the farmers can reduce antimicrobial use between 20% and 30%. Although a majority of farmers reported that the quality of rabbit meat products can be affected by high antimicrobial use, they showed a low level of awareness relative to negative implications on public health. Key factors for a significant reduction of these use in rabbits have been described as an improvement of microclimate and feed quality. In addition, veterinaries are crucial in the role of education and awareness of livestock farmers for progressive reduction of antimicrobial usage and maintenance of the health and welfare of animals (Di Martino et al. 2019). Another Italian study reported that administration of tetracyclines and polymyxins significantly decreased but sulfamides, bacitracin, and quinolones increased in 32 rabbit farms, between 2010 and 2015. In 2015, the antimicrobial classes more administrated were tetracyclines (28%), polymyxins (17%), pleuromutilins (14%), sulfonamides (13%), and bacitracin (12%). The remaining administrated classes included quinolones, diaminopyrimidines, aminoglycosides, macrolides, phenicols, and β-lactams (Agnoletti et al. 2018). In addition, the production of 1 kg of meat rabbit at slaughter age needs 72 animal daily doses (ADD), between 2010 and 2015. Although antibiotic use has been decreased during this period, tetracyclines and polymyxins were the antimicrobial classes more administrated (Agnoletti et al. 2018).

A Canadian study reported that 65% of rabbit farms routinely used antimicrobials when new animals were introduced or replaced stock on the farm, and 71% of farms reported the frequent use of antibiotic combinations, in which 58% used salinomycin sodium-virginiamycin combination. The intermittent use of trimethoprim/sulfamethoxazole was described by one farm (Kylie et al. 2017).

2.1 Route of Antimicrobial Administration

Antibiotics for rabbits can be administrated through various methods, including orally through their feed or drinking water (in pill form but crushed and mixed on feed/water). Additionally, antibiotics can be administered parenterally, which involves intravenous injections into a cephalic or lateral saphenous vein, and intramuscular and subcutaneous injections similar to those given to pets). Topical application is another method of administering antibiotics to rabbits (Hillyer 1994). Antimicrobials approved for rabbit use are limited (Hillyer 1994; Allen et al. 2005). Moreover, disease prevention in rabbit meat production has fewer vaccines available (Agnoletti et al. 2018). Although the antimicrobials are frequently used by oral route for infection treatment in rabbits, the orally administrated antimicrobials can easily cause intestinal microflora dysbiosis and, consequently, diarrhea (Table 1). For instance, β-lactams are rarely prescribed in rabbits because they are associated with diarrhea (Agnoletti et al. 2018). In the UK, there are very few antibiotics authorized for rabbit use like enrofloxacin, fusidic acid, and gentamicin, although other antibiotics not authorized can be used in very specific situations (Ramsey 2011; EFSA, ECDC 2016).

The second report of the Australian Pesticides and Veterinary Medicines Authority (APVMA) showed that 76% of antibiotics used for therapeutic purposes in food animals (without reported rabbits) were administrated in feed, 18% were administrated in water, 4% were injectable, and 2% were administrated by intramammary and other routes (APVMA 2014). The ANSES annual report showed specifically for French rabbits that these animals were treated mainly by the oral route. The administration by injections was lower between 2011 and 2018. Antimicrobial use as medicated premixes has decreased by 60% and 13% from 2011 and 2017, respectively, to 2018. Additionally, administration via oral powders and solutions showed a reduction (20%) from 2011, although an increase (15%) was observed in 2017 (ANSES 2020).

In France, fluoroquinolones that comprise danofloxacin, enrofloxacin, marbofloxacin, orbifloxacin, and pradofloxacin antibiotics are authorized for use in rabbits (oral and parenteral administration), representing 0.5% of antibiotic administration in 2018. Colistin, which represented 2% of antibiotic administration in 2018, is also used mainly in rabbits in oral forms (powders and solutions) (ANSES 2020). Agnoletti et al. (2018) reported that 99% of the antibiotics were administrated orally, 84% by medicated feed, and 15% by drinking water, and 0.7% were administrated parenterally. In 23 analyzed Canadian rabbit farms, it was reported that 74% of farms used at least one antibiotic routinely in feed and 4% in water for all rabbit age groups (Kylie et al. 2017).

3 Antimicrobial Resistance in Farm Rabbits

A variety of studies, as well as international associations like the European Centre for Disease Prevention and Control (ECDC), the European Food Safety Authority (EFSA), and the European Medicines Agency (EMA), have reported a significant association between antimicrobial usage and the emergence of antimicrobial resistance. These have also emphasized and highlighted this global problem, high rates of antimicrobial use, and antibiotic resistance present in humans, animals, and the environment. All demonstrated as the main aim a need to mitigate antibiotic usage in food-producing animals through multidisciplinary interventions (JIACRA 2017; Agnoletti et al. 2018).

According to Lengliz et al. (2022), the rearing conditions and the genetic background have demonstrated a high impact on the emergence and dissemination of antibiotic-resistant bacteria. These resistances can spread over the rabbit production chain consisting of input suppliers (genetic companies and livestock feed), rabbit farmers for meat producers, slaughterhouses/cutting plants, distributors (retailer and wholesaler), and finally consumers (Fig. 1) (Cullere and Dalle Zotte 2018). Since a majority of antimicrobials are administrated orally, it can result in the under-dosing of a high number of farmed animals, supporting a rapid antimicrobial resistance development (EFSA, ECDC 2016).

Studies regarding antibiotic resistance in rabbits are scarce (Agnoletti et al. 2018). Based on the French Resapath report, antibiotic-resistant *Pasteurella multocida* obtained from food-producing rabbits with respiratory infections was below 10%, during 2012–2017 (Table 2). In particular, *P. multocida* showed resistance to quinolone, tilmicosin, tetracycline, trimethoprim-sulfamethoxazole, gentamicin, and amoxicillin (ANSES 2020).

In the study of Agnoletti et al. (2018), 96% of *Escherichia coli* obtained from fecal samples showed resistance to at least three antimicrobial classes defined as multidrug resistant (MDR), including tetracycline (98%), trimethoprim (91%), sulfadiazine (97%), and enrofloxacin (67%), although resistance to paromomycin, chloramphenicol, and colistin was also detected. In addition, 62% of the isolated *Enterococcus faecalis* and 15% of *Enterococcus hirae* were also classified as MDR, in which the most frequent resistance was to tetracycline (100% and 98%), erythromycin (90% and 81%), gentamicin (40% and 15%), and chloramphenicol (39% and 8%) in Italian farmed rabbits, respectively, between 2014 and 2015. Although chloramphenicol is prohibited or rarely used in food-producing animals, *Escherichia coli* and enterococci demonstrated persistent resistance to this antibiotic without apparent selective pressure (Agnoletti et al. 2018).

Clostridium difficile, occasionally associated with diarrhea episodes in rabbits, was isolated from the intestinal tract (3%) of animals affected or not with enteric infections in Italian rabbit farms, mainly showing resistance to erythromycin (66%),

Table 2 Antibiotic-resistant bacteria reported in rabbit meat production

Country, year	Sampling	Bacteria	Main antibiotic resistance	References
France, 2012–2017	Sick, respiratory tract	*Pasteurella multocida*	QN-TIL-TET-SXT-GEN-AM	ANSES (2020)
Italy, 2014–2015	Healthy, feces	*Escherichia coli*	ENR-CHL-COL-PAR-TET-TMP-SDI	Agnoletti et al. (2018)
		Enterococcus faecalis	CIP-CHL-ERY-GEN-TET-Zn BAC	
		Enterococcus hirae	CHL-ERY-GEN-TET-Zn BAC	
Italy, 2007–2013	Healthy/sick, cecal content	*Clostridium difficile*	CLI-ERY-MTZ-RIP	Drigo et al. (2015)
Italy, 2012–2013	Skin, ear	*Staphylococcus aureus*	MET-TET	Agnoletti et al. (2014)
Canada, –	Fecal samples	*E. coli*	TET-CHL-S-SUL-KAN-AMP-GEN-SXT	Kylie et al. (2017)
		Salmonella enterica Kentucky	AMP-AMC-CEF-CET-CEX-TET-S	
Spain, –	Fecal samples	*Escherichia coli* *Enterobacter cloacae*	NAL-SSS-TET	Blanc et al. (2006)

AMP ampicillin, *AM* amoxicillin, *AMC* amoxicillin/clavulanic acid, *CEF* ceftiofur, *CET* cefoxitin, *CEX* ceftriaxone, *CIP* ciprofloxacin, *CLI* clindamycin, *CTX* cefotaxime, *COL* colistin, *CHL* chloramphenicol, *ENR* enrofloxacin, *ERY* erythromycin, *GEN* gentamicin, *KAN* kanamycin, *NAL* nalidixic acid, *MTZ* metronidazole, *PAR* paromomycin, *QN* quinolone, *RIF* rifampicin, *SDI* sulfadiazine, *SSS* sulfonamide, *TET* tetracycline, *TIL* tilmicosin, *TMP* trimethoprim, *SXT* trimethoprim-sulfamethoxazole, *S* streptomycin, *SUL* sulfisoxazole, *Zn BAC* zinc-bacitracin

clindamycin (37%), metronidazole (32%), and rifampicin (3%). Moreover, 18% of *C. difficile* were classified as MDR (Drigo et al. 2015). *Staphylococcus aureus* resistant to methicillin (48%) and tetracycline (48%) were also detected in several rabbit meat production in 2012–2013, while half (25%) of farmers maintained positive after 5 months of sampling. Additionally, samples collected from farm workers and their close relatives as well as surface and air were also positive for methicillin-resistant *S. aureus* (MRSA), carrying the *mec*A gene (Agnoletti et al. 2014).

In Canada, 56% of fecal samples of commercial meat rabbits demonstrated *Escherichia coli* isolates resistant to at least one analyzed antibiotic agent, in particular to tetracycline. *Salmonella enterica* Kentucky also isolated showed resistance to ampicillin, amoxicillin/clavulanic acid, ceftiofur, cefoxitin, ceftriaxone, tetracycline, and streptomycin (Kylie et al. 2017). In Spain, two *Escherichia coli* and one *Enterobacter cloacae* isolates from a rabbit farm were positive for extended-spectrum β-lactamases (ESBL) by identification of CTX-M-9 and CTX-M-14 and plasmidic class C β-lactamases by detection of CMY-2. All isolates showed resistance to nalidixic acid, sulfonamide, and tetracycline. The fact that no β-lactam antibiotics are recommended in rabbits, causing enteritis, can explain this low value of β-lactamase-harboring strains in this study (Blanc et al. 2006).

Rabbits can be found in a variety of ecosystems, such as meat production, research for animal experiments, companions, and environmental wild (Kylie et al. 2017). For close contact and the increased popularity of these animals, antibiotic resistance also was reported in pet rabbits. In Austria, the presence of MRSA isolate was detected in a pet rabbit with swelling on the head and neck, which was administrated with rifampicin for 2 weeks (Loncaric and Künzel 2013).

4 Impact/Limits of Antimicrobial Control in Rabbit Production

Antibiotic use began with the therapeutic application to treat bacterial infections; however, other advantages of antibiotics were discovered, and it began to be administrated on animals, in particular in livestock production, with the following purposes: metaphylactic, prophylactic, and growth promotion, maintaining the animal welfare, and the reproductive performance (Miranda et al. 2022). In general, antibiotics with therapeutic use required high dose-short term administrated on feed or water and injected into animals. In opposite, antibiotics used as AGPs required low dose-long term administrated via feed. Moreover, the AGP more used in rabbit feed was zinc-bacitracin (Falcão-e-Cunha et al. 2007).

Due to inappropriate use of antibiotics, as well as the presence of antibiotic residues on by-products, the emergence of antimicrobial resistance, and a lack of antibiotics to treat specific diseases in humans and animals, antibiotic control and a global effort have been verified in the world. In this context, a variety of antibiotics have been banned from animal production, in particular AGPs. For instance, all antimicrobials as AGPs have currently been prohibited in Sweden, Europe, and China since 1986, 2006, and 2020, respectively (Wegener 2006; Ma et al. 2021; Miranda et al. 2022).

Therefore, antimicrobial usage is beneficial and useful for the health and welfare of animals; in the opposite, the emergence and dissemination of antimicrobial resistance is a serious problem for public health. The antimicrobial resistance is much documented in bacteria isolated from main food-producing species like pigs, cattle, and poultry but not in others like rabbits and fish (Kylie et al. 2017). In the particular case of rabbits, it is not possible to exclude them as potential reservoirs or sources of pathogenic agents for other animals (livestock, pets and wild), as well as for the environment. This includes humans, especially when considering professional groups like veterinarians, breeders, farmers, and slaughterers who may have direct or indirect contact with them (Drigo et al. 2015).

Agnoletti et al. (2014) highlighted the presence of antibiotic resistance in food-producing animals and its impact on public health and the food chain, reporting that the production system itself can help disseminate antimicrobial resistance. Pure-bred animals with desired genetics and the same origin are reared by specialized companies and distributed to several farmers on a large scale to produce rabbit meat for human consumption. This route can allow the introduction of rabbits harboring antibiotic-resistant bacteria in a variety of farmers. Another risk factor for spread resistance is the movement of animals between farmers.

As enteritis is one of the main causes of disease in rabbits. On the one hand, this point can require more consumption of antibiotics in rabbit production for intestinal infections and consequently support the emergence of the new resistance (Kylie et al. 2017). Moreover, antibiotic usage showed an increase in the final costs of rabbit products (Abd El-Aziz et al. 2022). On the other hand, many antibiotics cause diarrhea, limiting their use in rabbits and contributing to the mitigation of resistance (Kylie et al. 2017). According to Rampin et al. (2008), the main pathological lesions observed in the slaughterhouse were in the digestive, tegumentary (subcutaneous abscess), and urinary systems. The slaughterhouse represents a crucial control point of all animal production chains to monitor rabbit diseases and antimicrobial resistances.

In spite of the development of rabbit meat production with modern rearing techniques, high hygienic-sanitary levels, rabbit origin control, improved biosecurity/welfare, and environmental improvements, the mortality rates have not significantly decreased over 25 years (EFSA 2005). For instance, enteric disorders have significantly increased in the last 15 years, becoming the most common cause of mortality in rabbits (EFSA 2005). However, it is important to alert to the need for alternatives to antibiotics frequently used in rabbits without side effects, before most bacteria show resistance to them (Abd El-Aziz et al. 2022).

5 Alternatives to Antimicrobials in Farmed Rabbits

As an alternative to using dietary antibiotics as therapeutic or growth promoters, several feed additives (Table 3) like probiotics, synbiotics, organic acids, and postbiotics have emerged and been described without causing side effects on the consumers, although these substances are scarce in the rabbit meat industry compared with others. Immune system stimulators, as well as herbal extracts and plants, have been reported as alternative products in food-producing animals. A majority of feed alternatives are used in human nutrition, causing changes in the intestinal microbiota and the immune system (Falcão-e-Cunha et al. 2007). In food-producing animals, their intestinal microbiota can easily undergo dysbiosis or dysregulation by many factors, such as medication, stress, and changes in diet. In general, these products improve performance and enhance growth, infection resistance, and overall animal health benefits (Ramos et al. 2020; Miranda et al. 2021).

Table 3 Potential health-beneficial effects of feed additives to farmed animals

Feed additives	Advantages	Disadvantages	References
Probiotics (lactic acid bacteria)	– Reduce the colonization of pathogenic resistance – Change of the intestinal microbiota – Reduce the inflammatory reactions – Stimulate the immune system – Produce antimicrobial substances – Improve the carcass traits – Increase the productivity and performance – Increase the feed conversion, digestibility, and absorption – Decrease the excretion of urea and ammonia – Metabolize xenobiotics	– Spread of antibiotic resistance – Best act in vitro – Dysbiosis – Risk of toxicity for the host	Guillot (2001), Delia et al. (2012), Miranda et al. (2021)
Prebiotics (carbohydrates or oligosaccharides)	– Stimulate the native microbial population of the gastrointestinal tract – Improve the carcass traits – Improve the economic efficiency – Induce genetic changes: upregulation of immunity and growth genes – Increase the productivity and performance – Improve digestibility and intestinal villi conditions – Improve spleen white pulp hyperplasia and humoral immune – Decrease the genotoxicity (DNA-damaging effects) of toxic substances – Increase the volatile fatty acids – Improve the gut morphology		Attia et al. (2015), Abd El-Aziz et al. (2022)

(continued)

Table 3 (continued)

Feed additives	Advantages	Disadvantages	References
Synbiotics (probiotic plus prebiotic substances)	– Reduce the cecal lactobacillus count – Improve the growth performance – Improve the nutrient digestibility		Oso et al. (2013), Ewuola et al. (2011)
Acidifiers or organic acids (fumaric acid and capric and caprylic acid)	– Increase the growth rate – Increase food efficiency – Reduce the mortality	– No influence on animal performance	Scapinello et al. (2001), Michelan et al. (2002), Skrivanová and Marounek (2006)
Postbiotics (exopolysaccharides, cellular wall fragments, and metabolites)	– Strengthen the gut – Inhibit pathogens – Regulate the host immunity system	– Toxic risk of specific substrates or metabolites	Zhong et al. (2022)
Others (bilberry pomace and purple loosestrife)	– Antioxidant effect – Modify the intestinal microflora – Increase the volatile fatty acids – Reduce the ammonia levels	– No influence on animal performance	Biagini et al. (2016), Dabbou et al. (2019)

5.1 Probiotics

Probiotics are classified as live microorganisms with beneficial effects on health when adequately consumed by humans or animals (Hamilton-Miller et al. 2003). The majority of the commercial probiotics includes microorganisms classified as Gram-positive bacteria and belongs to the *Bacillales* order, including *Bacillus*, and *Lactobacillales* order, including *Lactobacillus*, *Pediococcus*, *Leuconostoc*, *Enterococcus*, and *Streptococcus* genera. As for the fungal microorganisms, *Saccharomyces cerevisiae* that belongs to the *Saccharomycetales* order is the most applied (Falcão-e-Cunha et al. 2007; Miranda et al. 2021).

The lactic acid bacteria (LAB) including the *Lactobacillales* bacteria are the best studied and reported probiotics in food-producing animals (pigs, poultry, and cattle), agriculture, and the food industry. In animals, these bacteria work as feed supplements and demonstrate health benefits with antimicrobial action through the production of antimicrobial substances, such as bacteriocins, organic acids, and hydrogen peroxide.

Potential beneficial effects of probiotics allow the resistance of pathogenic colonization or epithelial invasion, production of vitamins, reduction in inflammatory reactions, and stimulation of the immune system and enzymes of the host. These act against other microorganisms, changing the intestinal microbiota and reducing its metabolic reactions that originated harmful substances. In addition, these probiotics can improve carcass traits, animal performance, and nutrient digestibility and its absorption and mitigate the excretion of urea and ammonia (Falcão-e-Cunha et al. 2007; Pinheiro et al. 2007; Delia et al. 2012; Miranda et al. 2021).

In the control of infectious diseases, probiotics use different mechanisms to mitigate pathogens like *Escherichia coli*, *Salmonella*, and *Clostridium*. Probiotics can promote the development of desired intestinal microorganisms, inhibit the adhesion or colonization of undesired bacteria by competition for epithelial receptors and nutrients and for passive aggregation between pathogens and probiotic bacteria, change environmental conditions by reducing the pH by acid production, and produce antimicrobial substances (Delia et al. 2012). In addition, probiotics can reduce the bioaccessibility of toxic xenobiotics, such as mycotoxins, pesticides, or heavy metals, by their metabolization or direct interaction as binding agents (Hernández-Mendoza et al. 2022).

The action and success of probiotics depend on several factors; however, some microorganisms only showed good results in laboratory conditions and in animals can cause an unbalance between the native microbial population and the added probiotic microorganisms (Guillot 2001). Moreover, antibiotic-resistant LAB have been isolated in animal production and by-products, increasing the potential risk of transmission of resistant genes among microorganisms and its spread to the food chain and the environmental ecosystem (Miranda et al. 2021).

In rabbits, there are few recent studies on the application of probiotics, and the results depend on farm conditions (Nicodemus et al. 2004). According to Falcão-e-Cunha et al. (2007), only two probiotics for rabbits, *Bacillus cereus* var. *toyoi* and *Saccharomyces cerevisiae*, are approved in the EU. The diet supplemented with *Bacillus cereus* var. *toyoi* was investigated in growing rabbits, showing improvement in growth performance, as well as a reduction of morbidity rates, independent of the concentration of dietary probiotics and presence or absence of farmed health problems (Trocino et al. 2005). Another study with the same dietary supplementation on lactating rabbits showed a decrease in parturition interval and an increase in numerical productivity by cage and year in weaned animals and feed efficiency (Nicodemus et al. 2004). Pinheiro et al. (2007) also reported positive effects on productivity (feed conversion, weight gain ratio, and mortality) of this probiotic in intensive rabbit meat production. Currently, Bhatt et al. (2017) reported *Lactobacillus acidophilus* probiotic supplementation reduced nitrogen excretion and improved the digestibility, nutrient use, feed conversion ratio, and weight gain, without effect on fatty acid profile and carcass quality. Moreover, these results with a direct impact on production allowed reducing the feed costs. In the study of Rotolo et al. (2014), the rabbit diet was supplemented with *Saccharomyces cerevisiae* and showed an increase in mold and yeast population in the cecum but no mortality rate changes among treatments.

Probiotics also showed a positive impact on better microbial control on rabbit meat, according to Koné et al. (2018) that reported that Canadian diets containing *Carnobacterium maltaromaticum* reduce undesirable coliforms like

Enterobacteriaceae and *Listeria monocytogenes* in stored meat, increasing the food security. In Egypt, the rabbit dietary supplementation with fenugreek seeds (*Trigonella foenum-graecum* L., one of the important natural phytogenic feed additives) and commercial probiotic (mix of LAB: *Bifidobacterium bifidum*, *Lactobacillus plantarum*, *Lactobacillus acidophilus*, *Bacillus subtilis*, and *Aspergillus niger* fermentation extract) combination improved serum hormones, nutrient digestibility, and growth performance without influence on carcass quality of growing rabbits (Abdel-Wareth et al. 2021).

5.2 Prebiotics

Another potential option instead of antibiotic use is prebiotics. It includes carbohydrates that are non-digestible by the animal enzymes like oligosaccharides and inulin-type fructans (Table 3). These can be obtained from natural sources, such as yeasts, milk, or plants, and artificially from acid or enzymatic reactions. There are different oligosaccharides such as fructo-, galacto-, transgalacto-, mannan-, isomalto- and xylo-oligosaccharides that demonstrated promising prebiotic purposes in rabbit dietary supplementation. The main purpose of these feed additives is to stimulate the immune system of the host and their gut microbiota (Falcão-e-Cunha et al. 2007; Abd El-Aziz et al. 2020, 2022). In comparison to probiotics, prebiotics have technological characteristics (no critical problem there are with the acid conditions of the stomach and the thermal processing of the feed) and safety benefits by stimulating the native microbial population of the gastrointestinal tract, while with probiotics, foreign microbial populations are introduced in the host (Falcão-e-Cunha et al. 2007; Abd El-Aziz et al. 2022).

In rabbits, oligosaccharide usage has also demonstrated health benefits, in particular their intestinal health and performance (Abd El-Aziz et al. 2022). Currently, Abd El-Aziz et al. (2022) showed that supplemented diets with the following oligosaccharides, such as mannan-oligosaccharides (MOS) and isomalto-oligosaccharides (IMO), had a positive effect on productivity and economic efficiency in rabbit production. In addition, IMO showed the most increase in duodenal villi length and area, and both IMO and MOS showed higher spleen white bulb length and area and a positive impact on carcass quality, economic efficiency, and upregulation of immunity and growth genes.

Hafner et al. (2019) observed that MOS obtained from the cell wall of *Saccharomyces cerevisiae* or with *Bacillus cereus* var. *toyoi* decrease the genotoxicity (DNA-damaging effects) of T-2 toxin when weaned rabbits were exposed. Attia et al. (2013) observed that the administration of MOS-supplemented diets to growing rabbits was more effective than inulin, in comparison to unsupplemented diets and zinc-bacitracin antibiotic. The continuous MOS administration showed positive effects from weaning to slaughter age of growing rabbits on blood profiles and performance, no changes in tissue morphology were detected, and intermittently administration allowed for reducing the costs of production. In opposite, zinc-bacitracin administration demonstrated negative effects on tissue morphology like the ileum, kidney, and liver (Attia et al. 2015). Additionally, MOS increased volatile fatty acid (VFA) levels in the cecum when growing rabbit diets were supplemented with these prebiotics (Guedes et al. 2009). Also, Bovera et al. (2010) concluded that the MOS application improves growth performance with higher growth rates and feed conversion ratio and lower mortality than antibiotic-medicated rabbits during the epizootic rabbit enteropathy outbreak. In concordance, the study of Abdel-Hamid and Farahat (2016) showed that MOS supplementation increases the immunity of the host and it had a positive impact on health indicators, such as live and carcass weights and blood parameters. Mourão et al. (2006) obtained better results for intestinal morphology, cecal VFA, and mortality in MOS diet, no differences were observed between MOS and antibiotic supplementation for performance.

In general, oligosaccharides as antibiotic alternatives can be used in rabbit diets. The MOS is the most studied in this animal species; however, contradictory results were obtained depending on several factors, such as farm or hygienic-sanitary conditions, animal age, concomitant infections, feed, and growth phase. Then, the best prebiotic selection should be performed considering the desirable needs for each situation and rabbit life phase (Falcão-e-Cunha et al. 2007).

5.3 Synbiotics, Organic Acids, and Postbiotics

A synbiotic, consisting of a mixture of probiotics and prebiotics, shows benefits to the host through the improvement of the survival and colonization of live microbial dietary supplements in the gastrointestinal tract (Table 3). The synbiotic term suggests that there is a synergistic effect of both probiotics and prebiotics with health-promoting effects, such as stimulating growth and improving the host welfare (Oso et al. 2013; Swanson et al. 2020). According to Swanson et al. (2020), the synbiotics can be classified as complementary symbiotic (probiotics plus prebiotic combination) and synergistic symbiotic that comprises a selectively used substrate that works together with the co-administrated live microorganism to increase the health benefits. However, studies on synbiotic use in rabbit nutrition are limited. In Nigeria, rabbit-fed diets containing probiotics (*Bacillus cereus* or *Predicoccus acidilactici*) or prebiotics (MOS or arabinoxylan oligosaccharides (AX)) or synbiotics (*Bacillus cereus* and fructooligosaccharides) demonstrated that both prebiotics improve growth and gut morphology; in particular, MOS showed the best weight gain, live weight, and VFA ratio and the lowest cecal *Lactobacillus* count, according to Oso et al. (2013) study that recommended the MOS supplementation. Ewuola et al. (2011) recommend symbiotic and prebiotic supplementation as natural alternative growth promoters in rabbit diets free of antibiotics since they observed enhancement in growth performance through improvement of nutrient digestibility.

Another alternative to the animal diet that promotes beneficial effects are acidifiers or organic acids. In several animals, these substances increase growth rate and food efficiency (Scapinello et al. 2001; Michelan et al. 2002). However, Brazilian rabbit diets containing only acidifiers (fumaric acid) or in combination with

MOS (prebiotics) did not influence the animal performance (Scapinello et al. 2001). In this line, fatty acids such as capric and caprylic acid with antimicrobial activity and produced in the mammary gland were studied in rabbit production in the Czech Republic. These medium-chain fatty acids (MCFA) were supplied as triacylglycerols showing a reduction in mortality but no influence on their total weight gain or performance (Skřivanová and Marounek 2006).

New natural alternatives have been investigated and tested as feed additives to improve rabbit performance and their health status as bilberry pomace and purple loosestrife, in Italian farms; however, more studies are needed. Bilberry pomace was considered a potential option of protein and fiber sources in the growing rabbit production. This supplementation showed an antioxidant effect and tends to modify the intestinal microflora (predominance of *Ruminococcus*, *Oscillospira*, *Clostridium*, and *Ruminococcaceae* species) without influence on growth performance and cecal conditions and, consequently, allowing the residue to use and valorize of agro-industrial products (Dabbou et al. 2019). The inclusion of purple loosestrife leaves (*Lythrum salicaria*) in rabbit diets indicated an increase in VFA rate, the concentration of acetic acid and white blood cells, and mitigation of ammonia levels (Kovitvadhi et al. 2016). Moreover, extract of plant bio-waste like acid soluble bio-organic (SBO) substances have been tested as feed additives, but no enhancement on health and animal productivity were influenced; it only contributed to the reduction of gas production and digestate ammonia in stored manure of rabbit production (Biagini et al. 2016).

Postbiotics, another great potential alternative to antibiotics and probiotics in food-producing animals, consist of a preparation of inactive microorganisms and/or their components to achieve health-beneficial effects for the host. This can comprise diverse probiotic components, such as exopolysaccharides found in LAB; cellular wall fragments including polysaccharides, teichoic acids, and lipoteichoic acids; and bacterial DNA and metabolites that included vitamins, indole, and fatty acid. This showed to be safer and more stable than probiotics, allowing the gut barrier to strengthen, pathogen inhibition, and host immunity regulation. More studies are needed to evaluate the interaction among diverse compounds, in particular the toxic risk of substrates or metabolites in rabbits (Zhong et al. 2022).

6 Conclusions

Antibiotics consist of chemotherapeutic substances used mainly for the clinical treatment of bacterial infections in humans, animals, and plants. In animals, these can be still used as APGs to increase health and performance in some countries. The associated problem with the use of antibiotics emerged with antibiotic-resistant strains. Overall, in rabbit production, there are limited data and a number of scientific researchers compared to other food-producing animals. This fact can be associated with their peculiar digestive physiology, in which few antibiotics are administrated orally since enteritis is one of the major health problems, or due to lack of welfare measures. A high percentage of antibiotic usage in rabbits raised for meat production is described to benefit the inadequate conditions on farms. Moreover, the scientific studies published are not easy to compare because the units used to be different among them. Exploring and screening the role of antibiotic resistance is beneficial to know the true situation relative to the spread or persistence of resistance and help the implementation of urgent national or international measures to mitigate the antibiotic usage levels in the rabbit industry.

Nutritional strategies have been developed in alternative to antibiotic use. With the aim to improve the animal performance, productivity, and meat quality of rabbits, phytogenics and/or natural feed additives only or in combination among them were tested. However, urgent solutions are needed to increase the confidence of consumers in meat rabbits, while meat rabbit consumption has demonstrated a decrease in the last years. Therefore, this review highlighted different alternatives to antibiotics that have been applied in the rabbit industry. Prebiotic and postbiotic supplementation in rabbit feeding was demonstrated as promising alternatives. However, in most cases, alternatives to antibiotics were studied in isolation, and the same additive often yielded contradictory results among different studies. Then, more research is needed, to investigate the action, the interactions with other substances, and the efficacy of commercial substances, offering safe food on rabbit products.

Acknowledgments This work was supported by the Associated Laboratory for Green Chemistry (LAQV), which is financed by national funds from FCT/MCTES (UIDB/50006/2020 and UIDP/50006/2020) and by the projects UIDB/00772/2020 and LA/P/0059/2020 funded by the Portuguese Foundation for Science and Technology (FCT).

Conflict of Interest The authors declare no conflict of interest.

References

Abd El-Aziz A, Aboghanima M, Kamel S, El-Deeb E, Mahrous UE-S (2020) Influence of two dietary prebiotic oligosaccharides supplementation on productive performance and carcass traits with special attention to their biochemical alterations in two rabbit breeds. Damanhour J Vet Sci 3:1–7. https://doi.org/10.21608/djvs.2019.19337.1005

Abd El-Aziz AH, El-Kasrawy NI, Abd El-Hack ME, Kamel SZ, Mahrous UE, El-Deeb EM, Atta MS, Amer MS, Naiel MAE, Khafaga AF, Metwally AE, Abo Ghanima MM (2022) Growth, immunity, relative gene expression, carcass traits and economic efficiency of two rabbit breeds fed prebiotic supplemented diets. Anim Biotechnol 33(3):417–428. https://doi.org/10.1080/10495398.2020.1800485

Abdel-Hamid TM, Farahat MH (2016) Effect of dietary mannan-oligosaccharides on some blood biochemical, haematological parameters and carcass traits in purebred New Zealand white and crossbred rabbits. Anim Prod Sci 56(12):2133–2139. https://doi.org/10.1071/AN15032

Abdel-Wareth AA, Elkhateeb FS, Ismail ZS, Ghazalah AA, Lohakare J (2021) Combined effects of fenugreek seeds and probiotics on growth performance, nutrient digestibility, carcass criteria, and serum hormones in growing rabbits. Livestock Sci 251:104616. https://doi.org/10.1016/j.livsci.2021.104616

Agnoletti F, Mazzolini E, Bacchin C, Bano L, Berto G, Rigoli R, Muffato G, Coato P, Tonon E, Drigo I (2014) First reporting of methicillin-resistant *Staphylococcus aureus* (MRSA) ST398 in an industrial rabbit holding and in farm-related people. Vet Microbiol 170(1–2):172–177. https://doi.org/10.1016/j.vetmic.2014.01.035

Agnoletti F, Brunetta R, Bano L, Drigo I, Mazzolini E (2018) Longitudinal study on antimicrobial consumption and resistance in rabbit farming. Int J Antimicrob Agents 51:197–205. https://doi.org/10.1016/j.ijantimicag.2017.10.007

Allen DG, Dowling P, Smith DA (2005) Handbook of veterinary drugs, 3rd edn. Lippincott Williams & Wilkins. isbn: 0-7817-4126-2

ANSES (2020) Resapath. French surveillance network for antimicrobial resistance in bacteria from diseased animals. 2018 Annual Report. French Agency for Food, Environmental and Occupational Health & Safety (ANSES). Available from: www.resapath.anses.fr. Accessed July 10 2022

APVMA (2014) Australian pesticides and veterinary medicines authority. Quantity of antimicrobial products sold for veterinary use in Australia, 2005–2010. Available from: https://apvma.gov.au/sites/default/files/images/antimicrobial_sales_report_march-2014.pdf. Accessed July 10 2022

Attia Y, Hamed R, El-Hamid AA, Shahba H, Bovera F (2013) Effect of inulin and mannan-oligosaccharides compared with zinc-bacitracin on growing performance, nutrient digestibility and hematological profiles of growing rabbits. Anim Prod Sci 55(1):80–86. https://doi.org/10.1071/AN13286

Attia Y, Hamed R, Abd El-Hamid A, Al-Harthi M, Shahba H, Bovera F (2015) Performance, blood profile, carcass and meat traits and tissue morphology in growing rabbits fed mannanoligosaccharides and zinc-bacitracin continuously or intermittently. Anim Sci Paper Rep 33:85–101

Bhatt RS, Agrawal AR, Sahoo A (2017) Effect of probiotic supplementation on growth performance, nutrient utilization and carcass characteristics of growing chinchilla rabbits. J Appl Animal Res 45(1):304–309. https://doi.org/10.1080/09712119.2016.1174126

Biagini D, Gasco L, Rosato R, Peiretti PG, Gai F, Lazzaroni C, Montoneri C, Ginepro M (2016) Compost-sourced substances (SBO) as feedstuff additives in rabbit production. Animal Feed Sci Tech 14:66–76. https://doi.org/10.1016/j.anifeedsci.2016.02.005

Blanc V, Mesa R, Saco M, Lavilla S, Prats G, Miró E, Navarro F, Cortés P, Llagostera M (2006) ESBL- and plasmidic class C beta-lactamase-producing *E. coli* strains isolated from poultry, pig and rabbit farms. Vet Microbiol 118(3–4):299–304. https://doi.org/10.1016/j.vetmic.2006.08.002

Bovera F, Nizza A, Marono S, Mallardo K, Piccolo G, Tudisco R, De Martino L, Nizza A (2010) Effect of mannan oligosaccharides on rabbit performance, digestibility and rectal bacterial anaerobic populations during an episode of epizootic rabbit enteropathy. World Rabbit Sci 18(1):9–16. https://doi.org/10.4995/wrs.2010.18.02

Cullere M, Dalle Zotte A (2018) Rabbit meat production and consumption: state of knowledge and future perspectives. Meat Sci 143:137–146. https://doi.org/10.1016/j.meatsci.2018.04.029

Dabbou S, Ferrocino I, Kovitvadhi A, Bergagna S, Dezzuto D, Schiavone A, Cocolin LS, Gai F, Santoro V, Gasco L (2019) Bilberry pomace in rabbit nutrition: effects on growth performance, apparent digestibility, caecal traits, bacterial community and antioxidant status. Animal 13(1):53–63. https://doi.org/10.1017/S175173111800099X

Delia E, Tafaj M, Männer K (2012) Efficiency of probiotics in farm animals. In: Rigobelo EC (ed) Probiotic in animals. IntechOpen, London. https://doi.org/10.5772/50055

Di Martino G, Crovato S, Pinto A, Dorotea T, Mascarello G, Brunetta R, Agnoletti F, Bonfanti L (2019) Farmers' attitudes towards antimicrobial use and awareness of antimicrobial resistance: a comparative study among Turkey and rabbit farmers. Ital J Anim Sci 18:194–201. https://doi.org/10.1080/1828051X.2018.1504236

Drigo I, Mazzolini E, Bacchin C, Tonon E, Puiatti C, Bano L, Spigaglia P, Barbanti F, Agnoletti F (2015) Molecular characterization and antimicrobial susceptibility of *Clostridium difficile* isolated from rabbits raised for meat production. Vet Microbiol 181(3–4):303–307. https://doi.org/10.1016/j.vetmic.2015.10.005

EFSA (2005) The impact of the current housing and husbandry systems on the health and welfare of farmed domestic rabbits. EFSA J 267:1–31. https://doi.org/10.2903/j.efsa.2005.267

EFSA, ECDC (2016) The European Union summary report on antimicrobial resistance in antimicrobial resistance in zoonotic and indicator bacteria from humans, animals and food in the European Union in 2014. EFSA J 14:4380. https://doi.org/10.2903/j.efsa.2016.4380

EFSA Panel on Animal Health and Welfare (AHAW), Saxmose Nielsen S, Alvarez J, Bicout DJ, Calistri P, Depner K, Drewe JA, Garin-Bastuji B, Gonzales Rojas JL, Gortázar Schmidt C, Michel V, Miranda Chueca MÁ, Roberts HC, Sihvonen LH, Spoolder H, Stahl K, Velarde Calvo A, Viltrop A, Buijs S, Edwards S, Candiani D, Mosbach-Schulz O, Van der Stede Y, Winckler C (2020) Health and welfare of rabbits farmed in different production systems. EFSA J 18(1):e05944. https://doi.org/10.2903/j.efsa.2020.5944

EMA (2020) European medicines agency, European surveillance of veterinary antimicrobial consumption, 2020. Sales of Veterinary Antimicrobial Agents in 31 European Countries in 2018. Tenth ESVAC Report (EMA/24309/2020)

European Parliament (EP) (2017) European Parliament Resolution of 14 March 2017 on Minimum Standards for the Protection of Farm Rabbits (2016/2077(INI)). Available online: https://www.europarl.europa.eu/doceo/document/TA-8-2017-0077_EN.pdf. Accessed July 22, 2022

Ewuola E, Amadi C, Imam T (2011) Performance evaluation and nutrient digestibility of rabbits fed dietary prebiotics, probiotics and synbiotics. Int J Appl Agric Apic Res 7:107–117

Falcão-e-Cunha L, Castro-Solla L, Maertens L, Marounek M, Pinheiro V, Freire J, Mourão J (2007) Alternatives to antibiotic growth promoters in rabbit feeding: a review. World Rabbit Sci 15: 127–140. https://doi.org/10.4995/wrs.2007.597

FAO (2020) FAOSTAT. The statistics division of the FAO. http://www.fao.org/faostat/en/#data

FDA (2019) US food drug administration, summary report on antimicrobials sold or distributed for use in food-producing animals, 2018. Available from: https://www.fda.gov/media/133411/download. Accessed July 10, 2022

Guedes C, Mourão JL, Silva S, Gomes MJ, Rodrigues M, Pinheiro V (2009) Effects of age and mannan-oligosaccharides supplementation on production of volatile fatty acids in the caecum of rabbits. Anim Feed Sci Technol 150(3–4):330–336. https://doi.org/10.1016/j.anifeedsci.2008.10.004

Guillot JF (2001) Consequences of probiotics release in the intestine of animal. In Brufeau J (Ed.) Feed manufacturing in the Mediterranean region improving safety: from feed to food. Zaragoza CIHEAM-IAMZ, 17–21

Hafner D, Tuboly T, Mézes M, Bloch-Bodnár Z, Balogh K, Vántus V, Bóta B, Szabó-Fodor J, Matics Z, Szabó A, Kovács M (2019) Effect of feeding Bacillus cereus var. toyoi and/or mannan oligosaccharide (MOS) on blood clinical chemistry, oxidative stress, immune response and genotoxicity in T-2 toxin exposed rabbits. Ital J Animal Sci 18:1239–1251. https://doi.org/10.1080/1828051X.2019.1641165

Hamilton-Miller JMT, Gibson GR, Bruck W (2003) Some insights into the derivation and early uses of the word "probiotic". Brit J Nutr 90:845. https://doi.org/10.1079/bjn2003954

Hernández-Mendoza A, González-Córdova AF, Martínez-Porchas M (2022) Influence of probiotics on the animal gut microbiota and their impact on the bioavailability of toxic agents: an opinion paper. Front Nutr 9:870162. https://doi.org/10.3389/fnut.2022.870162

HFSDG (2018) Health and food safety directorate-general. Overview report commercial rabbit farming in the European Union. isbn: 978-92-79-43540-9. Available online: http://ec.europa.eu/dgs/health_food-safety/index_en.htm. Accessed July 22 2022. https://doi.org/10.2772/62174

Hillyer EV (1994) Pet rabbits. Vet Clin N Am Small Anim Pract 24(1):25–65. https://doi.org/10.1016/S0195-5616(94)50002-0

JIACRA (2017) Joint interagency antimicrobial consumption and resistance analysis report. ECDC/EFSA/EMA second joint report on the integrated analysis of the consumption of antimicrobial agents and occurrence of antimicrobial resistance in bacteria from humans and food-producing animals. EFSA J 15(7):4872. https://doi.org/10.2903/j.efsa.2017.4872

Koné AP, Zea JMV, Gagné D, Cinq-Mars D, Guay F, Saucier L (2018) Application of *Carnobacterium maltaromaticum* as a feed additive for weaned rabbits to improve meat microbial quality and safety. Meat Sci 135:174–188. https://doi.org/10.1016/j.meatsci.2017.09.017

Kovitvadhi A, Gasco L, Ferrocino I, Rotolo L, Dabbou S, Malfatto VM, Gai F, Peiretti PG, Falzone M, Vignolini C, Cocolin LS, Zoccarato I (2016) Effect of purple loosestrife (*Lythrum salicaria*) diet supplementation in rabbit nutrition on performance, digestibility, health and meat quality. Animal 10(1):10–18. https://doi.org/10.1017/S1751731115001822

Kylie J, McEwen SA, Boerlin P, Reid-Smith RJ, Weese JS, Turner PV (2017) Prevalence of antimicrobial resistance in fecal *Escherichia coli* and *Salmonella enterica* in Canadian commercial meat, companion, laboratory, and shelter rabbits (*Oryctolagus cuniculus*) and its association with routine antimicrobial use in commercial meat rabbits. Prev Vet Med 147:53–57. https://doi.org/10.1016/j.prevetmed.2017.09.004

Lengliz S, Cheriet S, Raddaoui A, Klibi N, Ben Chehida N, Najar T, Abbassi MS (2022) Species distribution and genes encoding antimicrobial resistance in *Enterococcus* spp. isolates from rabbits residing in diverse ecosystems: a new reservoir of linezolid and vancomycin resistance. J Appl Microbiol 132(4):2760–2772. https://doi.org/10.1111/jam.15461

Li S, Zeng W, Li R, Hoffman LC, He Z, Sun Q, Li H (2018) Rabbit meat production and processing in China. Meat Sci 145:320–328. https://doi.org/10.1016/j.meatsci.2018.06.037

Loncaric I, Künzel F (2013) Sequence type 398 meticillin-resistant *Staphylococcus aureus* infection in a pet rabbit. Vet Dermatol 24(3):370–372. https://doi.org/10.1111/vde.12023

Ma F, Xu S, Tang Z, Li Z, Zhang L (2021) Use of antimicrobials in food animals and impact of transmission of antimicrobial resistance on humans. Biosaf Health 3(1):32–38. https://doi.org/10.1016/j.bsheal.2020.09.004

Michelan AC, Scapinello C, Natali MRM, Furlan AC, Sakaguti ES, Faria HG, Santolin MLR, Hernandes AB (2002) Utilização de probiotico, ácido orgânico e antibiótico em dietas para coelhos em crescimento: ensaio de digestibilidade, avaliação da morfometria intestinal e desempenho. Rev Bras Zootec 31:2227–2237. https://doi.org/10.1590/S1516-35982002000900011

Miranda C, Contente D, Igrejas G, Câmara SPA, Dapkevicius MLE, Poeta P (2021) Role of exposure to lactic acid bacteria from foods of animal origin in human health. Foods 10:2092. https://doi.org/10.3390/foods10092092

Miranda C, Igrejas G, Capita R, Alonso-Calleja C, Poeta P (2022) Worldwide colistin use and spread of resistant-*Enterobacteriaceae* in animal production. In: the global antimicrobial resistance epidemic – innovative approaches and cutting-edge solutions. Guillermo Téllez (ed). IntechOpen. Isbn: 978-1-80356-041-0. https://doi.org/10.5772/intechopen.102722

Mondin C, Trestini S, Trocino A, Di Martino G (2021) The economics of rabbit farming: a pilot study on the impact of different housing systems. Animals 11(11):3040 https://doi.org/10.3390/ani11113040

Mourão JL, Pinheiro V, Alves A, Guedes CM, Pinto L, Saavedra MJ, Spring P, Kocher A (2006) Effect of mannan oligosaccharides on the performance, intestinal morphology and cecal fermentation of fattening rabbits. Anim Feed Sci Technol 126:107–120. https://doi.org/10.1016/j.anifeedsci.2005.06.009

NethMap-MARAN (2020) Consumption of antimicrobial agents and antimicrobial resistance among medically important bacteria in The Netherlands. Monitoring of antimicrobial resistance and antibiotic usage in animals in the Netherlands in 2019. Available from: https://swab.nl/en/nethmap. Accessed July 16, 2022

Nicodemus N, Carabano R, Garcia J, De Blas JC (2004) Performance response of doe rabbits to Toyocerin® (*Bacillus cereus* var. *toyoi*) supplementation. World Rabbit Sci 12:109–118. https://doi.org/10.4995/wrs.2004.577

OIE (2017) Annual report on the use of antimicrobial agents intended for use in animals. Better Understanding of the Global Situation. Second report. Available from: https://www.oie.int/

fileadmin/home/eng/our_scientific _expertise/docs/pdf/amr/annual_report_amr_2.pdf. Accessed July 16 2022

Oso AO, Idowu OMO, Haastrup AS, Ajibade AJ, Olowonefa KO, Aluko AO, Ogunade IM, Osho SO, Bamgbose AM (2013) Growth performance, apparent nutrient digestibility, caecal fermentation, ileal morphology and caecal microflora of growing rabbits fed diets containing probiotics and prebiotics. Livestock Sci 157(1):184–190. https://doi.org/10.1016/j.livsci.2013.06.017

Pinheiro V, Mourao JL, Jimenez G (2007) Influence of Toyocerin® (*Bacillus cereus* var. *toyoi*) on breeding performance of primiparous rabbit does. World Rabbit Sci 15:179–187. https://doi.org/10.4995/wrs.2007.590

Ramos OY, Basualdo M, Libonatti C, Veja MF (2020) Current status and application of lactic acid bacteria in animal production systems with a focus on bacteria from honey bee colonies. J Appl Microbiol 128:1248–1260. https://doi.org/10.1111/jam.14469

Rampin F, Piccirillo A, Schiavon E, Poppi L, Grilli G (2008) Detection of pathological lesions in slaughtered rabbits. Ital J Anim Sci 7(1):105–111. https://doi.org/10.4081/ijas.2008.105

Ramsey (2011) Chapter 3. Therapeutics. BSAVA Small Animal Formulary, 7th edition

Rotolo L, Gai F, Peiretti P, Ortoffi M, Zoccarato I, Gasco L (2014) Live yeast (*Saccharomyces cerevisiae* var. *boulardii*) supplementation in fattening rabbit diet: effect on productive performance and meat quality. Livestock Sci. 162:178–184. https://doi.org/10.1016/j.livsci.2014.01.022

Scapinello C, Garcia FH, Furlan AC, Michelan AC (2001) Efeito da utilização de oligossacarídeo manose e acidificantes sobre o desempenho de coelhos em crescimento. Rev Bras Zootec 30: 1272–1277

SDa-report (2021) The Netherlands veterinary medicines authority. Usage of antibiotics in agricultural livestock in the Netherlands in 2020. https://www.autoriteitdiergeneesmiddelen.nl/en/news/37/available-now-sda-report-usage-of-antibiotics-in-agricultural-livestock-in-the-netherlands-in-2020. Accessed July 22, 2022

Skřivanová V, Marounek M (2006) A note on the effect of triacylglycerols of caprylic and capric fatty acid on performance mortality and digestibility of nutrients in young rabbits. Anim Feed Sci Technol 127(1-2):161–168. https://doi.org/10.1016/j.anifeedsci.2005.07.001

Swanson KS, Gibson GR, Hutkins R et al (2020) The International Scientific Association for Probiotics and Prebiotics (ISAPP) consensus statement on the definition and scope of synbiotics. Nat Rev Gastroenterol Hepatol 17:687–701. https://doi.org/10.1038/s41575-020-0344-2

Trocino A, Xiccato G, Carraro L, Jimenez G (2005) Effect of diet supplementation with Toyocerin® (*Bacillus cereus* var. *toyoi*) on performance and health of growing rabbits. World Rabbit Sci 13:17–28. https://doi.org/10.4995/wrs.2005.532

Trocino A, Cotozzolo E, Zomeño C, Petracci M, Xiccato G, Castellini C (2019) Rabbit production and science: the world and Italian scenarios from 1998 to 2018. Ital J Anim Sci 18(1): 1361–1371. https://doi.org/10.1080/1828051X.2019.1662739

Wegener HC (2006) Use of antimicrobial growth promoters in food animals: the risk outweigh the benefits. In: Barug D, de Jong J, Kies AK, Verstegen MWA (eds) Antimicrobial growth promoters. Wageningen Academic Publishers, Wageningen, pp 53–58

Zhong Y, Wang S, Di H, Deng Z, Liu J, Wang H (2022) Gut health benefit and application of postbiotics in animal production. J Anim Sci Biotechnol 13(1):38. https://doi.org/10.1186/s40104-022-00688-1

Respiratory Complex in Rabbits (Pasteurellosis)

João Simões, Sónia Saraiva,
Maria da Conceição Medeiros Castro Fontes, José Manuel Monteiro,
and Teresa Tavares

Abstract

Pasteurella multocida is one of the most common isolated bacteria from respiratory complex condition in rabbitries. Infection with this bacterium is characterized by a respiratory syndrome identified primarily by rhinitis (contagious coryza), by pneumonia, or by a rhinopneumonia complex. *P. multocida* is an opportunist and contagious microorganism mainly transmitted by direct contact and aerosol. Pasteurellosis is frequently endemic in rabbits in different countries and can cause considerable economic losses in rabbitries throughout the world due to high morbidity, poor performances, and mortality. Dissemination of metastatic and septicemic forms contributes to the establishment of

J. Simões (✉) · S. Saraiva · M. da Conceição Medeiros Castro Fontes
Veterinary and Animal Research Centre (CECAV), Associate Laboratory for Animal and Veterinary Sciences (AL4AnimalS), Department of Veterinary Sciences, School of Agricultural and Veterinary Sciences, University of Trás-os-Montes and Alto Douro (UTAD), Vila Real, Portugal

Department of Veterinary Sciences, School of Agricultural and Veterinary Sciences, University of Trás-os-Montes and Alto Douro (UTAD), Vila Real, Portugal
e-mail: jsimoes@utad.pt

J. M. Monteiro
José Azevedo Monteiro, Lda., Vila Boa do Bispo, Portugal

Faculty of Veterinary Medicine, Lusófona University, Lisbon, Portugal

CBIOS, Lusófona University, Lisbon, Portugal

T. Tavares
Veterinary and Animal Research Centre (CECAV), Associate Laboratory for Animal and Veterinary Sciences (AL4AnimalS), Department of Veterinary Sciences, School of Agricultural and Veterinary Sciences, University of Trás-os-Montes and Alto Douro (UTAD), Vila Real, Portugal

J. Simões, J. M. Monteiro (eds.), *Veterinary Care of Farm Rabbits*,
https://doi.org/10.1007/978-3-031-44542-2_19

infection and lesions in several body tissues and organs. Treatments with antibiotics are not fully effective and do not avoid relapses. The control of environmental factors such as air renewal and animals' thermal comfort are crucial as prevention measures. Genetic selection of more resistant rabbits is, also, a new approach to reduce the prevalence of pasteurellosis since. This chapter highlights the main issues related to the diagnosis, control, and prevention of pasteurellosis in rabbits.

Keywords

Pasteurella multocida · Respiratory distress · Rhinopneumonia · Septicemia

1 Introduction

Respiratory diseases in rabbits are divided into upper respiratory tract infections and lower respiratory tract infections. Upper respiratory tract infections are usually of mild course and chronic mucopurulent discharge such as rhinitis and conjunctivitis (Deeb and DiGiacomo 2000). Apart from chronic form, lower respiratory tract infections can cause an acute or subacute bronchopneumonia leading to high mortality rates (Wang et al. 2019). Both upper and lower respiratory tract can be simultaneously affected and is denominated rabbit rhinopneumonia whose terminology is often used as synonymous of rabbit pasteurellosis. Rabbit respiratory complex or rabbit respiratory syndrome is a more fair designation than pasteurellosis, since other microorganisms may be involved in this syndrome, like *Bordetella bronchiseptica* or *Staphylococcus aureus*, which are also very frequently isolated, often concomitantly with *Pasteurella multocida.*

Pasteurellosis is caused by *P. multocida*, and it is responsible for severe economic losses in rabbitries throughout the world (Jaglic et al. 2011; Wang et al. 2019). It is an opportunistic pathogen that normally resides as commensal in oral, nasopharyngeal, and upper respiratory tracts of rabbits, birds, and other species (Wilson and Ho 2013). Environmental variations in rabbit farms, namely, with high density populations, like sudden changes in temperature and humidity, can cause stress in animals, which may be responsible for the increased growth of bacteria in the upper respiratory tract (Taylor et al. 2010).

Infection with *P. multocida* is associated with important diseases in various species, including fowl cholera in poultry, hemorrhagic septicemia in cattle, and atrophic rhinitis in pigs (Stahel et al. 2009; Wilson and Ho 2013). Snuffle is typical in rabbit's infection, which is characterized by nasal discharge, sneezing, and conjunctivitis. *P. multocida* is the pathogen usually associated with the presence of these symptoms in rabbits (Langan et al. 2000). Nonetheless, even if the initial local of *P. multocida* development is the upper respiratory tract, metastatic and septicemic forms of dissemination originate infection of internal and tegument tissues. These disseminations can cause a multitude of non-respiratory lesions and clinical signs, like abscesses in several organs, metritis, mastitis, otitis, and subcutaneous abscesses (like the typical external ear abscesses). These forms of dissemination may have great impact in the disease transmission in rabbitries. Septicemic forms of the disease may lead to death.

With the current reduction in antimicrobial use in Europe, we have observed an increase in pneumonia incidence in rabbitries, most often associated with *P. multocida*, but also with *S. aureus*.

This chapter mainly aims to describe the most relevant aspects contributing to the diagnosis, control, and prevention of respiratory complex in rabbits, with special focus on pasteurellosis.

2 Etiology and Epidemiology

P. multocida is the main pathogenic agent of rabbits' respiratory complex, an economically major disease in rabbit meat industry (Qiu et al. 2022). It is a gram-negative, nonmotile coccobacillus of the family *Pasteurellaceae* (Kuhnert and Christensen 2008). Stahel et al. (2009) observed that *P. multocida* subsp. *multocida* was the most common organism in Swiss rabbits being present in 82% of all characterized isolates from nares, sinus, and other affected organs.

High incidence (100% of the isolates) of this bacterium isolated from clinical cases was also found in Italy (Massacci et al. 2018). In the study of Stahel et al. (2009), another bacterium of *Pasteurellaceae* family has been found in clinical material, namely, *P. multocida* subsp. *septica*, which represented a small number of the isolates (3%).

P. multocida strains are currently classified into 5 serogroups (capsular types A, B, D, E, and F) and further classified into 16 lipopolysaccharide (LPS) serotypes (1–16) or 8 LPS genotypes (L1–L8) (Harper et al. 2015). Colonization and disease development are influenced by factors related to both host and pathogen, which are classified by capsular type and serotype as reported by Asway et al. (2008). In their study, three strains of *P. multocida* were isolated from rabbits and classified as A:12, A:3, D:2, and B. Even though A:12 was the most commonly found serotype in rabbits, the greatest disease severity was associated with A:3 serotype (Asway et al. 2008). In Iberian Peninsula, García-Alvarez et al. (2015) observed a prevalence (n = 100 isolates) of 47.0%, 28.0%, and 25.0% of capsular types A, D, and F, respectively. In China, the frequency of serotype detection (n = 205 isolates) was 51.2% (A:L3), 24.9% (A:L6), 19.0% (D:L6), and 4.9% (D:L3) (L: lipopolysaccharide genotypes; Wang et al. 2019).

In addition, it is also known that *P. multocida* strains that cause fowl cholera in poultry typically belong to the serovars 1, 3, and 4 (Jaglic et al. 2011). In their study, Du et al. (2020) observed a large-scale mortality of rodents that was caused by serogroups A and F of *P. multocida*, highlighting the importance of greater surveillance in *P. multocida* in rabbits worldwide.

Nevertheless, other bacteria can be involved (Table 1). Other bacteria, such as *Escherichia coli*, *Chlamydophila*, and *Mycobacterium*, can play a role in lower respiratory tract infections (Hedley 2014). In skin abscesses observed as

Table 1 Respiratory pathogens in rabbits

Agent	Pathogenesis
Pasteurella multocida	Nasal discharge, sneezing, and conjunctivitis (snuffles). Lesions in the upper and lower respiratory tract (Patiño et al. 2022)
Bordetella bronchiseptica	Rhinitis mucoid or pneumonia. Disease in the upper respiratory tract. Mild symptoms such as sneezing, lacrimation, and ocular/nasal discharge (Espinosa et al. 2020)
Staphylococcus aureus	Dacryocystitis, pneumonia, and abscesses (Lennox 2012; Espinosa et al. 2020). Subcutaneous abscesses, lung and heart abscesses, conjunctivitis, rhinitis, and fibrinous pneumonia in adult animals (Wang et al. 2019)
Mycobacterium tuberculosis, *M. bovis*, and *M. avium*	Lesions in the lower respiratory tract. Pneumonia and septicemia (Lennox 2012)
Pseudomonas aeruginosa	Lesions in the upper respiratory tract. Eye and surrounding tissues (Hedley 2014)
Klebsiella pneumoniae	Pneumonia and septicemia (Sumitha and Sukumar 2014)
Moraxella catarrhalis	Lesions in the upper respiratory tract (Hedley 2014)
Chlamydia spp.	Lesions in the lower respiratory tract. Occasionally cause mild pneumonia in rabbits (Lennox 2012; Hedley 2014)
Mycoplasma pulmonis	Upper respiratory tract disease (Hedley 2014)

subcutaneous masses, *Pasteurella* spp. and *Staphylococcus* spp. have been isolated (Espinosa et al. 2020). Moreover, *S. aureus* infections are related with the presence of widespread subcutaneous abscesses, lung and heart abscesses, as well as conjunctivitis, rhinitis, and fibrinous pneumonia in adult animals (Wang et al. 2019).

High infection rates with *P. multocida* are often possible in rabbitries, as already referred by some authors, but stress factors are still the main trigger to the development of the disease. Environmental factors are the most impacting in respiratory disease, particularly the ones that influence rabbits' felt air temperature, like temperature, humidity, and air speed, as well as the air renewal. The most critical period is the night, when the external temperature falls.

Pneumonia is more frequent in young females showing depression, anorexia, respiratory distress with mucopurulent rhinitis, and blepharoconjunctivitis with a concurrent infection of *P. multocida* and *Bordetella* spp. (Espinosa et al. 2020). In some cases, *Staphylococcus* spp. are also involved.

Klebsiella pneumoniae infection in rabbits is rare but has been reported in an outbreak associated with respiratory form along with septicemia in a rabbit farm (Sumitha and Sukumar 2014).

The transmission of *P. multocida* can occur by aerosol, direct contact, or fomites (DiGiacomo et al. 1987). However, transmission is more frequent by contact between rabbits with acute infection. It's main route of entry is nasal exposure or via wounds, and infection can then spread to contiguous tissues or hematogenous

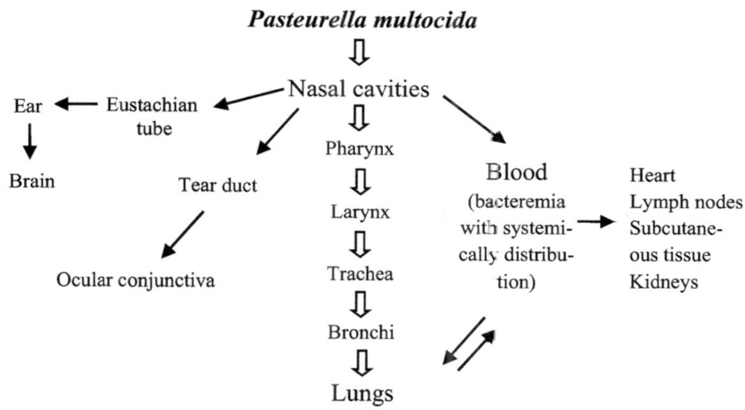

Fig. 1 Dissemination of *Pasteurella multocida* from nasal cavities to several organs

route (Deeb 2004). If the infecting bacteria enter the bloodstream, it can lead to fatal septicemia (Fig. 1).

The incubation period of the infection with *P. multocida* is difficult to define because many rabbits are subclinical carriers of infection; however, in experimental studies, rhinitis occurred 1 to 2 weeks after intranasal inoculation.

If genital infection is present, kits may become infected with *P. multocida* immediately after birth (Deeb 2004; Lennox 2012). The prevalence of the disease progressively increases with age, until about 5 months (Palócz et al. 2014). Infected rabbits may carry this microorganism for an indefinite time, without any clinical evidence of infection (D'Amico et al. 2022), though most adult rabbits are believed to be infected with *P. multocida* (Palócz et al. 2014). Morbidity is frequently high, and mortality can achieve 60% in accordance with Yang et al. (2022) findings.

Many rabbits are subclinical carriers of *P. multocida*, and the development of clinical disease is triggered by some factor of stress (Lennox 2012; Peng et al. 2019). In *S. aureus* or *Moraxella catarrhalis* infection, the disease arises if the mucosa is compromised, e.g., by high ammonia levels (Lennox 2012). In addition, *P. multocida* can cause a zoonotic infection in humans characterized by soft tissue infection as a result of bites or scratches (Martin et al. 2018).

3 Pathophysiology

According to LPS, *P. multocida* is able to cause lesions in the upper and lower respiratory tract. It binds to and is internalized by respiratory epithelial cells and also traverses the intercellular spaces to reach the blood vessels, where it binds to and is internalized by neutrophil and red blood cells. These cells may then move to the lungs where the LPS induces typical diffuse alveolar damage (Patiño et al. 2022), playing an important role in the pathogenicity.

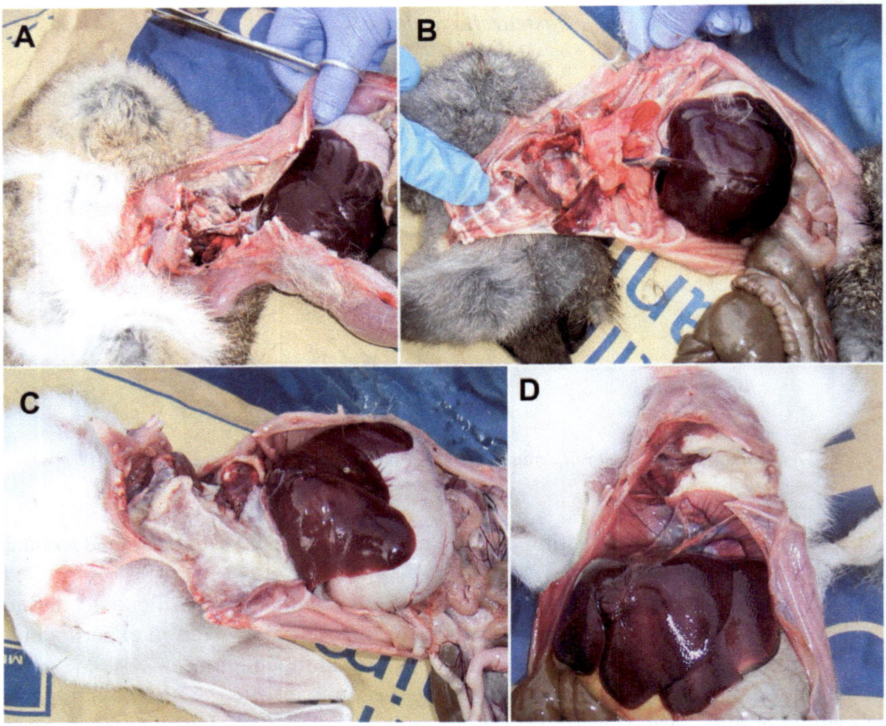

Fig. 2 Gross lesions of pneumonia (**a**, **b**) and pleuritis (**a**, **c**, **d**) at necropsy in rabbits with isolation of *Pasteurella multocida* from bacterial cultures

Due to its dissemination, *P. multocida* causes infection and inflammation in several tissues (Box 1). Pathological changes associated with pasteurellosis, involving the head and respiratory tract, are chronic rhinitis, suppurative otitis media and conjunctivitis, and cranioventrally located acute necrotizing, fibrinopurulent bronchopneumonia with possible fibrinous pleuritis and pericarditis (Lennox 2012) (Fig. 2).

Box 1 Dissemination of *P. multocida*

Metastatic forms	Septicemic forms
• Rhinitis	• Metritis/pyometra
• Sinusitis	• Mastitis
• Keratitis	• Hepatitis
• Otitis (otitis media and internal)	• Pericarditis/myocarditis
• Meningoencephalitis	• Orchitis/epididymitis
• Pneumonia	• Subcutaneous abscesses

(continued)

Box 1 (continued)

Metastatic forms	Septicemic forms
• Pericarditis	• Mucosal bleeding (acute form)
• Pleurisy	• Pneumonia
	• Pleurisy

P. multocida gains entry to the host primarily through the nares or wounds, and if the host does not resist infection, the bacteria colonize the nares and may cause production of nasal exudates (Wilson and Ho 2013). Once established, infection spreads to contiguous tissues, namely, paranasal sinuses, nasolacrimal duct and conjunctiva, Eustachian tube and middle ears, trachea, bronchi, and lungs (Wilson and Ho 2013). Hematogenous spread also accounts for infection that reaches the middle ears, lungs, and internal organs as the heart and kidneys (Deeb 2004). Pathological changes can occur in the lungs, trachea, and thymus, particularly, consolidation and abscesses in the lung (Yang et al. 2022). *P. multocida* can rapidly replicate in various tissues of rabbit and cause bacteremia after infection, and the expression of proinflammatory cytokines can be the main cause of respiratory inflammation and septicemia (Yang et al. 2022).

P. multocida serotype A:3 has been associated with deciliation of ciliated cells and hyperplasia of goblet cells of nasal mucosa, thickening of alveolar septa, swelling of capillary lining cells, and infiltration of inflammatory cells in the lungs (Al-Haddawi et al. 2000). Lesional and histological findings after *P. multocida* serovar A experimental inoculation were reported in Fig. 3.

In the past, serogroup A and D strains of *P. multocida* have been considered the only causative agents of rabbit pasteurellosis. However, new studies arose which confirmed that rabbits are highly sensitive to disease caused by infection with avian *P. multocida* serogroup F strains (Du et al. 2020), and, for this reason, avian hosts cannot be excluded as a source of rabbit infection with serogroup F (Fig. 4; Jaglic et al. 2008, 2011). García-Alvarez et al. (2015) observed in their molecular study that isolates of capsular type D and *hgbB* gene were associated (odds ratio = 5.91; 95% CI, 1.63–21.38) with pneumonia. Moreover, from the four virulence-associated genes (*hgbB*, *pfhA*, *tbpA*, and *toxA*) evaluated in their study, also *pfhA* was associated with isolates of capsular type F (García-Alvarez et al. 2015).

Rabbits with chronic and severe infections usually have high immunoglobulin G titers (Deeb et al. 1990). After inoculation, *P. multocida* can be isolated postmortem by culture from tracheal swab specimen of rabbits, also finding high IgG levels (Asway et al. 2008). Moreover, the secretory immune response (IgA) does not protect against nasal infection, although it may play a role in limiting spread (Hu et al. 2021).

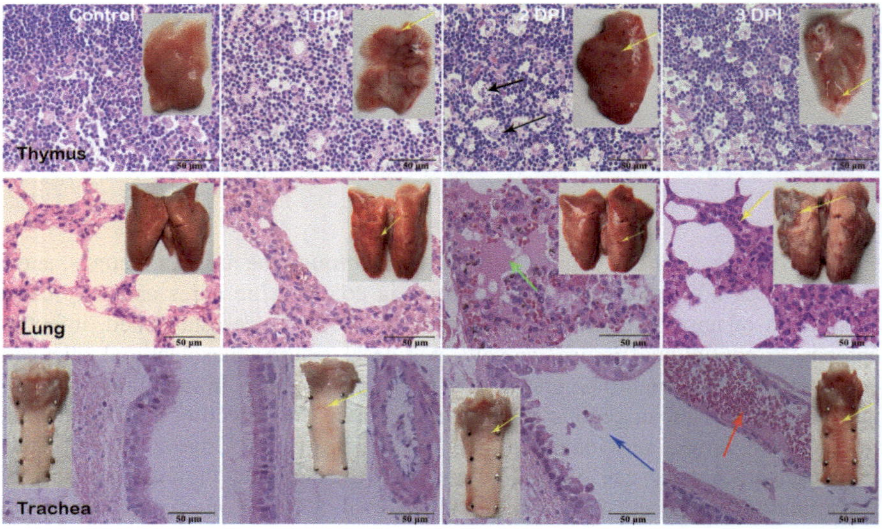

Fig. 3 Evolution of lesions of the thymus, lung, and trachea (at necropsy) and histologic changes from rabbits challenged with *P. multocida* serovar A (C51-17 strain). Modified from Yang et al. (2022). Legend: Macroscopic lesions (consolidation and abscesses) and hemorrhage (yellow arrows) can be observed in the lung and in the trachea and thymus, respectively. Histology: thymus, decreasing of lymphocyte number and cell necrosis (black arrow); lung, edema, hemorrhage (green arrow), and neutrophil infiltration (yellow arrow on the histological slide); trachea, necrotic, shed (blue arrow), and congestion (red arrow). DPI, days post-infection

4 Clinical Signs

The clinical signs in rabbits range from asymptomatic or mild upper respiratory inflammation to acute pneumonic and/or disseminated disease and death (Peng et al. 2019) (see Box 1). The main expression of the syndrome is an upper respiratory disease in the form of rhinitis and lower respiratory disease characterized by pneumonia (Wilson and Ho 2013).

Rhinitis ("snuffles"), pneumonia, septicemia, abscesses, and mastitis are some of the clinical manifestations caused by different strains of *P. multocida* (Fig. 5) (Coudert et al., 2006). Otitis, pyometra, orchitis, genital tract infections, meningitis, and conjunctivitis can also occur (Lennox 2012; Nassar et al. 2013).

It is also very frequent to find vestibular syndrome, with the so-called head tilt, as a consequence of otitis and vestibular abscesses.

Submandibular abscesses are also characteristics of pasteurellosis in rabbits.

After experimental-induced *P. multocida* infection, Yang et al. (2022) observed 1-day post-infection acute clinical signs such as depression, fever, anorexia, dyspnea, serous nasal discharge, as well as diarrhea. These symptoms peaked on the second day post-infection. Also, the mortality rate was 10% at the first day, >40% at

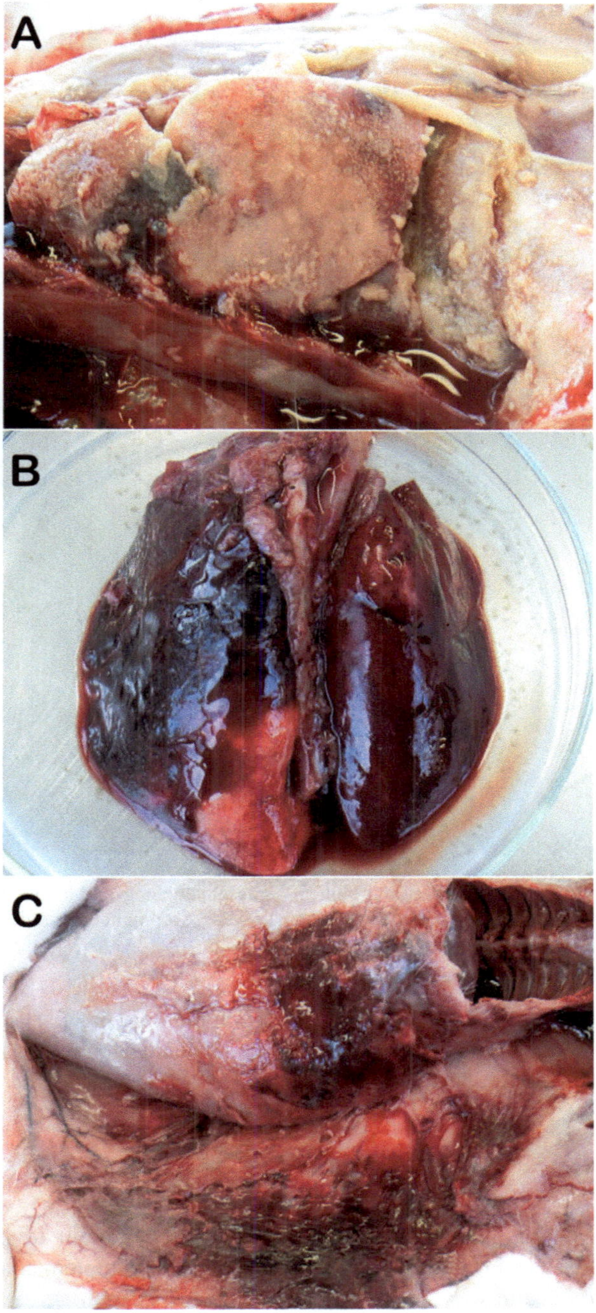

Fig. 4 Gross anatomy of lungs and pleura at necropsy from an intranasally challenged rabbit with *P. multocida* serovar F isolate J-4103 (CAPM 6431). Adapted from Jaglic et al. (2008) with permission from Elsevier. Legend: (**a**) A severe fibrinopurulent pleuropneumonia with extensive fibrinous adhesions on parietal pleura and lung surfaces and serohemorrhagic exudate in the thorax can be observed. Alveoli (and some bronchioli) were partly or completely filled with fibrin and mixed cellular infiltrate (erythrocytes, leukocytes, and desquamated epithelial cells). (**b**) Intranasally challenged rabbit, immunosuppressed with dexamethasone (2 mg/kg). Pulmonary hyperemia and hemorrhagic pneumonia can be observed. Moderate to very extensive penetration of erythrocytes into the alveolar lumen was the only relevant histological finding. (**c**) Subcutaneously challenged rabbit. Extensive hemorrhages, edematous and "dirty-colored" in subcutis

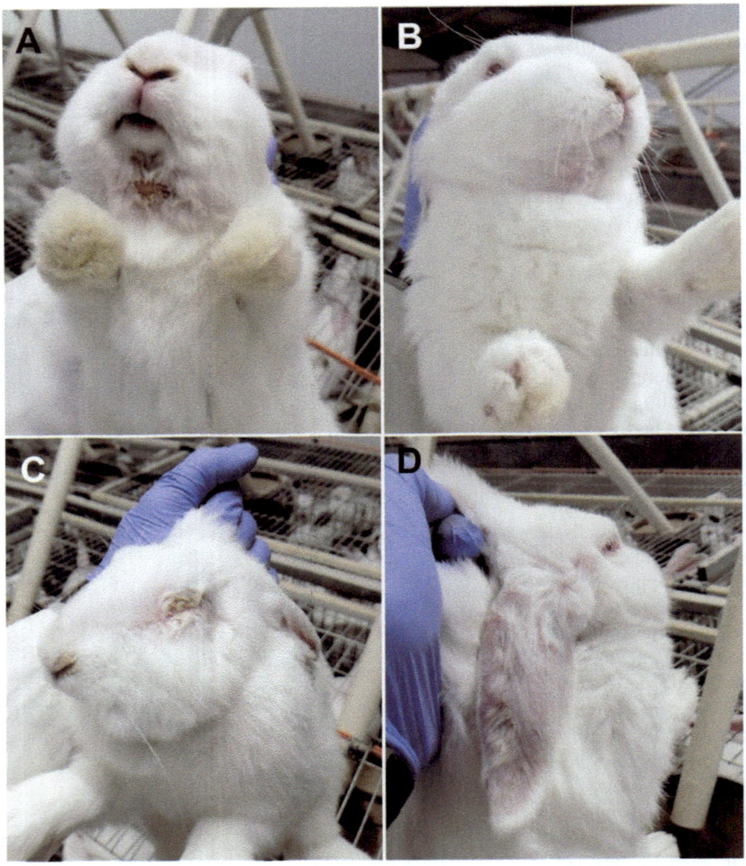

Fig. 5 Skin lesions and clinical signs of fattening rabbits affected by the respiratory complex. Legend: (**a**) and (**b**) submandibular abscess, (**c**) conjunctivitis, (**d**) ear abscess, (**b**) and (**c**) with nasal discharge

the second day, and reaching 60% at the third day post-infection. It is very important to value morbidity (not only mortality) and subclinical disease as the greatest causes of economic loss in this disease.

5 Diagnosis

Epidemiological studies are extremely important to highlight the occurrence of disease in rabbits and to support in the implementation of preventive control procedures.

Pasteurellosis should be suspected when an outbreak of respiratory problems occurs in the rabbitry and/or in presence of the non-respiratory clinical signs previously mentioned.

Culture and sensitivity results of nasal swabs and respiratory exudates, as well as infected tissues, are the best way to diagnose *P. multocida*. It is important that samples are collected from live or recently dead animals. Serology may be used for cases when infection is suspected in organs for which cultures are not attainable such as internal abscesses or when culture results are negative; however, positive serology only indicates exposure to *P. multocida*, not necessarily infection (Lennox 2012).

For phenotypical and genotypical analysis of postmortem isolates of rabbits' lesions, they can be subjected to biochemical tests and repetitive extragenic palindromic polymerase chain reaction (REP-PCR) (Stahel et al. 2009). REP-PCR typing reflects the sequencing results and, therefore, represents a suitable tool for genetic characterization of rabbit *Pasteurellaceae* isolates while also allowing epidemiological investigations (Stahel et al. 2009).

The respiratory form of myxomatosis is one of the differential diagnoses of pasteurellosis, as well as staphylococcosis.

6 Treatment, Control, and Prevention

Treatment of pasteurellosis can be difficult, because it may be disseminated throughout the body. Antibiotic therapy should be based on culture and sensitivity results, when possible. Most isolates of *P. multocida* tested on Mueller-Hinton agar with 5% sheep blood have been sensitive to amikacin (does not exist commercially and has little intestinal absorption), chloramphenicol (prohibited in animals for consumption), ciprofloxacin, doxycycline, enrofloxacin, gentamicin, tetracycline, and trimethoprim-sulfa. Enrofloxacin was commonly used as the best choice for treatment. However, in Europe, with the categorization of antibiotics (EMA/CVMP/CHMP/682198/2017; Pomba et al. 2018; EMA 2020), enrofloxacin and ciprofloxacin can only be used with the justification of an antibiogram, as these molecules are the only antimicrobials to which *P. multocida* is sensitive. Similar approach has been adopted by the FDA's Center for Veterinary Medicine in the USA (see Gens et al. 2022).

Antibiotic treatment should always be implemented as established in the summary of product characteristics. Administration should be discontinued and the patient re-evaluated if signs of enteritis (anorexia, soft or abnormal feces) are discovered. Antibiotherapy may lead to resolution of clinical signs, but elimination of the organism is difficult.

Appropriate antibiotic therapy, good nutrition, and supportive care should all be part of the therapeutic plan. Supportive care includes eliminating stress and providing high-quality nutrition.

In companion rabbits, it is feasible to ensure rehydration, clean the animal and environment, and provide oxygen therapy if needed. Hygiene conditions are also important to prevent the spread of the disease, and animals in treatment should be kept in quarantine in an area separate from healthy animals.

Alternative substances, e.g., pre- and probiotics and phytonutrients, to the use of antimicrobial have been evaluated. Positive effects on treatment of pasteurellosis were observed feeding rabbits with (1-3), (1-6) fungal β-glucans (Palócz et al. 2014), grape (*Vitis vinifera* L.) seed hydro-alcoholic extract (El-Sheikh et al. 2021), or allicin (active substance of garlic clove extracts) (Alam et al. 2018).

Vaccines are the most promising solution for controlling *Pasteurella* spp. outbreaks (Oslan et al. 2022). There has been considerable work in the development of a vaccine for *P. multocida* in rabbits. A recent study reports 70% and 90% successful protection from death with monovalent and bivalent (*P. multocida* and rabbit hemorrhagic disease virus) vaccines, respectively (El-Jakee et al. 2020).

Animal welfare is a very important issue in the control of respiratory complex, namely, in pasteurellosis, in rabbits. As it is an opportunistic bacterium, preventive medicine is even more important. Risk factors that increase immunosuppression should be controlled, such as sudden temperature variations, cold thermal sensation, high animal density, and excess of harmful gases (such as ammonia). Rabbit producers must be focused on thermal comfort, correct air renewal, and quarantine rooms.

It's crucial that the minimal ventilations are correctly adjusted and heating equipment needs to be on, to avoid drafts and abrupt temperature drops. It's important to make a constant monitorization and adjust the environmental control parameters daily, to face the external changes.

Nests are also an important matter. The nest needs to be kept warm and dry. For this purpose, wood shavings or other material placed in the nest should be abundant and dry. Ideally, this material should be mixed with mothers' hair and treated with drying powder.

The selection of rabbit breeds or lines for more resistance to infectious (Gunia et al. 2015) and non-specific (Gunia et al. 2018) diseases was approached in the last years. A 4–6% reduction of disease incidence per generation can be expected (Gunia et al. 2018). A moderate heritability for pasteurellosis resistance-related traits (extent of abscesses and bacteria growth from the tissue samples and the dead/alive status) without affecting commercial traits (kits born alive and weaning weight) was observed (Shrestha et al. 2020). This suggests the possibility to include pasteurellosis resistance traits on breeding programs keeping high productivity of the rabbitries.

7 Conclusions

Pasteurellosis is one of the most frequent diseases observed in rabbitries and causes great damage, not only in terms of mortality and morbidity but also with the reduction of productive factors, forcing the producer to a great renewal of the herd. It also causes many carcass rejections at the slaughterhouse, due to abscesses, pneumonia, and insufficient carcass yield.

In a subclinical stage of the disease, before being detected, pasteurellosis has already caused a lot of economic damage to the producer. It may even become a chronic disease and is very difficult to eradicate. Achieving the best control of the disease in the rabbitry must be our aim.

During the last 40 years, genetic improvement centers have focused on prolificacy, average daily weight gain, and productive results, neglecting the capacity to resist to opportunistic agents, like *P. multocida*, and leading to the rising of pasteurellosis in rabbitries. In recent years, genetic centers began to focus on making hybrids more resistant to bacteria such as *P. multocida*.

Acknowledgments The authors thank Springer for the use of the AJE Digital Editing. This work was supported by projects UIDP/CVT/00772/2020 (CECAV) and LA/P/0059/2020 (AL4AnimalS), funded by the Portuguese Foundation for Science and Technology (FCT).

References

Alam RTM, Fawzi EM, Alkhalf MI, Alansari WS, Aleya L, Abdel-Daim MM (2018) Anti-inflammatory, immunomodulatory, and antioxidant activities of allicin, norfloxacin, or their combination against *Pasteurella multocida* infection in male New Zealand rabbits. Oxidative Med Cell Longev 2018:1780956. https://doi.org/10.1155/2018/1780956

Al-Haddawi MH, Jasni S, Zamri-Saad M, Mutalib AR, Son R, Sheikh-Omar AR (2000) Ultrastructural observation of nasal and pulmonary intracellular *Pasteurella multocida* A:3 in rabbits. Vet Res Commun 24:153–167. https://doi.org/10.1023/a:1006452230452

Asway A, Ahmed A, Shalby H (2008) Bacteriological and serological studies on *Pasteurella multocida* infection in rabbits. Assiut Vet Med J 54:1–12. https://doi.org/10.21608/avmj.2008.175954

Coudert P, Rideaud P, Virag G, Cerrone A (2006) Pasteurellosis in rabbits. In: Maertens L, Coudert P (eds) Recent advances in rabbit science. ILVO, Melle, pp 147–162. ISBN: 978-92-898-0030-3

D'Amico F, Casalino G, Bozzo G, Camarda A, Lombardi R, Dimuccio MM, Circella E (2022) Spreading of *Pasteurella multocida* infection in a pet rabbit breeding and possible implications on healed bunnies. Vet Sci 9:301. https://doi.org/10.3390/vetsci9060301

Deeb BJ (2004) Respiratory disease and Pasteurellosis. In: Quesenberry KE, Carpenter JW (eds) Ferrets, rabbits, and rodents, 2nd edn. WB Saunders, St. Louis, Missouri, pp 172–182. isbn: 0-7216-9377-6. https://doi.org/10.1016/B0-72-169377-6/50019-8

Deeb BJ, DiGiacomo RF (2000) Respiratory diseases of rabbits. The veterinary clinics of North America. Exot Anim Pract 3:465–vii. https://doi.org/10.1016/s1094-9194(17)30082-8

Deeb BJ, DiGiacomo RF, Bernard BL, Silbernagel SM (1990) *Pasteurella multocida* and *Bordetella bronchiseptica* infections in rabbits. J Clin Microbiol 28:70–75. https://doi.org/10.1128/jcm.28.1.70-75.1990

DiGiacomo RF, Jones CD, Wathes CM (1987) Transmission of *Pasteurella multocida* in rabbits. Lab Anim Sci 37:621–623

Du C, Huang Y, Jiang J, Gao Z, Jia N, Yang S, Chen C, Ma S, Liang Y, Jiang B, Sang W, Wang P, Liang Y, Zhang Y, Zhou J, Chen S, Jongejan F, Song Z, Tong Y, Cao W (2020) A large-scale fatal outbreak of *Pasteurella multocida* among wild rodents in Southwestern China. J Biosaf Biosecurity 2:91–98. https://doi.org/10.1016/j.jobb.2020.07.001

El-Jakee JK, Moussa IM, Omran MS, Ahmed BM, Elgamal MA, Hemeg HA, Mubarak AS, Al-Maary KS, Kabli SA, Marouf SA, Haji Alhaaji J (2020) A novel bivalent pasteurellosis-RHD vaccine candidate adjuvanted with Montanide ISA70 protects rabbits from lethal challenge. Saudi J Biol Sci 27:996–1001. https://doi.org/10.1016/j.sjbs.2019.12.042

El-Sheikh SMA, Youssef FM, Mohamed HI, El-Saber Batiha G, Albrakati A, Galal AAA (2021) Efficacy of grape seed hydro-alcoholic extract in the treatment of experimentally *Pasteurella multocida* infected rabbits. Vet Med Sci 7:923–934. https://doi.org/10.1002/vms3.446

EMA (2020) Categorisation of antibiotics for use in animals for prudent and responsible use. Accessed from https://www.ema.europa.eu/en/documents/report/infographic-categorisation-antibiotics-use-animals-prudent-responsible-use_en.pdf (30-03-2023)

EMA/CVMP/CHMP/682198/2017 – Committee for Medicinal Products for Veterinary use (CVMP), Committee for Medicinal Products for Human Use (CHMP): Categorisation of antibiotics in the European Union. Accessed from https://www.ema.europa.eu/en/documents/other/answer-request-european-commission-updating-scientific-advice-impact-public-health-animal-health-use_en.pdf. 12 March 2023

Espinosa J, Ferreras MC, Benavides J, Cuesta N, Pérez C, García Iglesias MJ, García Marín JF, Pérez V (2020) Causes of mortality and disease in rabbits and hares: a retrospective study. Animals 10:158. https://doi.org/10.3390/ani10010158

García-Alvarez A, Chaves F, Fernández A, Sanz C, Borobia M, Cid D (2015) An ST11 clone of *Pasteurella multocida*, widely spread among farmed rabbits in the Iberian Peninsula, demonstrates respiratory niche association. Infect Genet Evol 34:81–87. https://doi.org/10.1016/j.meegid.2015.07.018

Gens KD, Singer RS, Dilworth TJ, Heil EL, Beaudoin AL (2022) Antimicrobials in animal agriculture in the United States: a multidisciplinary overview of regulation and utilization to foster collaboration: on behalf of the Society of Infectious Diseases Pharmacists. Open Forum Infect Dis 9:ofac542. https://doi.org/10.1093/ofid/ofac542

Gunia M, David I, Hurtaud J, Maupin M, Gilbert H, Garreau H (2015) Resistance to infectious diseases is a heritable trait in rabbits. J Anim Sci 93:5631–5638. https://doi.org/10.2527/jas.2015-9377

Gunia M, David I, Hurtaud J, Maupin M, Gilbert H, Garreau H (2018) Genetic parameters for resistance to non-specific diseases and production traits measured in challenging and selection environments; application to a rabbit case. Front Genet 9:467. https://doi.org/10.3389/fgene.2018.00467

Harper M, John M, Turni C, Edmunds M, St Michael F, Adler B, Blackall PJ, Cox AD, Boyce JD (2015) Development of a rapid multiplex PCR assay to genotype *Pasteurella multocida* strains by use of the lipopolysaccharide outer core biosynthesis locus. J Clin Microbiol 53:477–485. https://doi.org/10.1128/JCM.02824-14

Hedley J (2014) Respiratory disease. In: Meredith A, Lord B (eds) BSAVA manual of rabbit medicine. BSAVA, Gloucester, pp 160–167

Hu J, Li W, Qiao X, Li W, Xie K, Wang Y, Huang B, Zhao Q, Liu L, Fan X (2021) Characterization of microRNA profiles in *Pasteurella multocida*-infected rabbits and identification of miR-29-5p as a regulator of antibacterial immune response. Front Vet Sci 8:746638. https://doi.org/10.3389/fvets.2021.746638

Jaglic Z, Jeklova E, Leva L, Kummer V, Kucerova Z, Faldyna M, Maskova J, Nedbalcova K, Alexa P (2008) Experimental study of pathogenicity of *Pasteurella multocida* serogroup F in rabbits. Vet Microbiol 126:168–177. https://doi.org/10.1016/j.vetmic.2007.06.008

Jaglic Z, Jeklova E, Christensen H, Leva L, Register K, Kummer V, Kucerova Z, Faldyna M, Maskova J, Nedbalcova K (2011) Host response in rabbits to infection with *Pasteurella multocida* serogroup F strains originating from fowl cholera. Can J Vet Res 75:200–208

Kuhnert P, Christensen H (2008) *Pasteurellaceae*: biology, genomics and molecular aspects. Caister Academic Press, Norfolk. isbn: 978-1-913652-12-8

Langan GP, Lohmiller JJ, Swing SP, Wardrip CL (2000) Respiratory diseases of rodents and rabbits. The Veterinary Clinics of North America. Small Animal Pract 30:1309–1vii. https://doi.org/10.1016/s0195-5616(00)06009-5

Lennox AM (2012) Respiratory disease and pasteurellosis. In: Quesenberry KE, Carpenter JW (eds) Ferrets, rabbits, and rodents: clinical medicine and surgery, 3rd edn. WB Saunders, Philadelphia, pp 205–216. ISBN: 978-1-4160-6621-7

Martin TCS, Abdelmalek J, Yee B, Lavergne S, Ritter M (2018) *Pasteurella multocida* line infection: a case report and review of literature. BMC Infect Dis 18:420. https://doi.org/10.1186/s12879-018-3329-9

Massacci FR, Magistrali CF, Cucco L, Curcio L, Bano L, Mangili P, Scoccia E, Bisgaard M, Aalbæk B, Christensen H (2018) Characterization of *Pasteurella multocida* involved in rabbit infections. Vet Microbiol 213:66–72. https://doi.org/10.1016/j.vetmic.2017.11.023

Nassar SA, Mohamed AH, Soufy H, Nasr SM (2013) Protective effect of Egyptian propolis against rabbit pasteurellosis. Biomed Res Int 2013:163724. https://doi.org/10.1155/2013/163724

Oslan SNH, Tan JS, Yusoff AH, Sulaiman AZ, Awang MA, Lazim AM, Lim SJ, Oslan SN, Saad MZ, Ariff AB (2022) Pasteurellosis vaccine commercialization: physiochemical factors for optimum production. PRO 10:1248. https://doi.org/10.3390/pr10071248

Palócz O, Gál J, Clayton P, Dinya Z, Somogyi Z, Juhász C, Csikó G (2014) Alternative treatment of serious and mild *Pasteurella multocida* infection in New Zealand white rabbits. BMC Vet Res 10:276. https://doi.org/10.1186/s12917-014-0276-6

Patiño P, Gallego C, Martíneza N, Rey A, Iregui C (2022) Intranasal instillation of *Pasteurella multocida* lipopolysaccharide in rabbits causes interstitial lung damage. Res Vet Sci 152:115–126. https://doi.org/10.1016/j.rvsc.2022.07.026

Peng Z, Wang X, Zhou R, Chen H, Wilson BA, Wu B (2019) *Pasteurella multocida*: genotypes and genomics. Microbiol Mol Biol Rev 83:e00014–e00019. https://doi.org/10.1128/MMBR.00014-19

Pomba C, Catry B, Edo JT, Jukes H (2018) Licensing and approval of antimicrobial agents for use in animals. Microbiology spectrum 6. https://doi.org/10.1128/microbiolspec.ARBA-0016-2017

Qiu R, Wei H, Hu B, Chen M, Song Y, Xu W, Fan Z, Wang F (2022) Experimental pathogenicity and comparative genome analysis of high- and low-virulence strains of rabbit-origin *Pasteurella multocida*. Comp Immunol Microbiol Infect Dis 90-91:101889. https://doi.org/10.1016/j.cimid.2022.101889

Shrestha M, Garreau H, Balmisse E, Bed'hom B, David I, Guitton E, Helloin E, Lenoir G, Maupin M, Robert R, Lantier F, Gunia M (2020) Genetic parameters of resistance to pasteurellosis using novel response traits in rabbits. Genetics, Selection, Evolution: GSE 52:34. https://doi.org/10.1186/s12711-020-00552-8

Stahel AB, Hoop RK, Kuhnert P, Korczak BM (2009) Phenotypic and genetic characterization of *Pasteurella multocida* and related isolates from rabbits in Switzerland. J Vet Diagn Investig 21:793–802. https://doi.org/10.1177/104063870902100605

Sumitha P, Sukumar K (2014) An outbreak of *Klebsiella pneumoniae* infection in a rabbit farm. Int J Curr Microbiol App Sci 3:789–790

Taylor JD, Fulton RW, Lehenbauer TW (2010) The epidemiology of bovine respiratory disease: what is the evidence for predisposing factors? Can Vet J 51:1095–1102

Wang J, Sang L, Sun S, Chen Y, Chen D, Xie X (2019) Characterization of *Pasteurella multocida* isolated from dead rabbits with respiratory disease in Fujian, China. BMC Vet Res 15:438. https://doi.org/10.1186/s12917-019-2191-3

Wilson BA, Ho M (2013) *Pasteurella multocida*: From Zoonosis to Cellular Microbiology. Clin Microbiol Rev 26:631–655. https://doi.org/10.1128/CMR.00024-13

Yang W, Li M, Zhang C, Zhang X, Guo M, Wu Y (2022) Pathogenicity, colonization, and innate immune response to *Pasteurella multocida* in rabbits. BMC Vet Res 18:416. https://doi.org/10.1186/s12917-022-03517-9

Rabbit Enteric Syndrome

João Simões, José Manuel Monteiro, Teresa Tavares, and Maria da Conceição Fontes

Abstract

This chapter aims to approach the most relevant enteric disorders that affect rabbits and have great economic impact on rabbit farms as a syndrome. Enteric diseases are a major cause of losses in rabbits reared for meat production. Many digestive disorders of rabbits may be related to diet, environmental comfort, stress, or other management factors, and only a few are caused by enteric pathogens. There is a complex interrelationship between stress, diet, gut motility, and infectious agents involved in the etiopathogenesis of enteric syndrome. This syndrome is manifested by a disruption in normal fecal production. Digestible (fermentable) and indigestible (nonfermentable) fiber play a relevant role in digestive physiology. Additionally, the rabbit cecum is an ecosystem containing a microbiota that is essential to the health of the rabbit. Alterations in the balance of microorganisms within the cecum can result in the proliferation of pathogenic bacteria. The management of this syndrome is based on the control of the risk factors.

J. Simões (✉) · T. Tavares · M. da Conceição Fontes
Animal and Veterinary Research Centre (CECAV), Associate Laboratory for Animal and Veterinary Sciences (AL4AnimalS), Department of Veterinary Sciences, School of Agrarian and Veterinary Sciences, University of Trás-os-Montes and Alto Douro (UTAD), Vila Real, Portugal
e-mail: jsimoes@utad.pt

J. M. Monteiro
José Azevedo Monteiro, Lda., Vila Boa do Bispo, Portugal

Faculty of Veterinary Medicine, Lusófona University, Lisbon, Portugal

CBIOS, Lusófona University, Lisbon, Portugal

J. Simões, J. M. Monteiro (eds.), *Veterinary Care of Farm Rabbits*,
https://doi.org/10.1007/978-3-031-44542-2_20

Keywords

Enteropathy · Dysbiosis · Gastrointestinal tract · Gut microbiota · Nutrition

1 Introduction

In intensive or semi-intensive production systems, the high animal density in restricted spaces, even with controlled environmental conditions and homogeneous diets, facilitates the emergence of a set of biological and physicochemical factors, developing ecopathology that culminates in the onset of syndromes. Enteric syndrome (Anholt et al. 2015), as well as respiratory syndrome (see Chapter "Respiratory Complex in Rabbits (Pasteurellosis)"), involves a set of common pathophysiological mechanisms representing the most important multifactorial diseases due to their incidence and losses caused in rabbitries. In fact, digestive diseases are the main cause of morbidity and mortality in growing rabbits (Licois et al. 2006) and play a major role in the success or failure of raising rabbits. They can cause severe death loss as well as growth depression due to poor feed conversion (Gidenne et al. 2010).

The multifactorial concept of enteric syndrome (also called multifactorial enteropathy) was introduced by Whitney in 1976 to designate the occurrence of several diseases, such as typhlitis, typhlitis-diarrhea, hemorrhagic enteritis, and mucoid enteritis, now called epizootic rabbit enteropathy (García-Rubio et al. 2017).

A variety of events may disrupt the microbial balance within the gut and lead to clinical disease. These factors include exposure to pathogenic organisms, therapy with narrow-spectrum antibiotics, and dietary or drug-induced alterations. Although microorganisms may be involved in provoking specific nosological occurrences, the treatment, control, and prevention of these syndromes require a comprehensive approach toward risk factor management. Currently, there is a tendency to approach each disease individually and thus establish specific criteria for the control and prevention of each nosological entity (see next chapters of this book). Nevertheless, herd health management should incorporate different variables at the environmental and nutritional levels to mitigate stress and to be successful.

Enteric syndrome designates a set of changes in the balance of the intestinal microbiota, with abnormal overgrowth of other microorganisms present, which cause clinically expressed gastrointestinal disorders. They represent a set of conditions that cause the greatest economic losses in rabbit production.

In this chapter, the etiopathophysiology, clinical approach, and management of enteric alteration as a digestive syndrome are addressed.

2 Etiopathophysiology

In enteric syndrome, a variety of bacterial, parasitic, or viral agents can be identified, with emphasis on some *Escherichia coli* (enteropathogenic) serotypes and coccidia (Percy et al. 1993; García-Rubio et al. 2017). However, intestinal changes are closely linked to triggering factors of environmental origin, management, and food and linked to the animals' own rustic capacity (Box 1; Fig. 1).

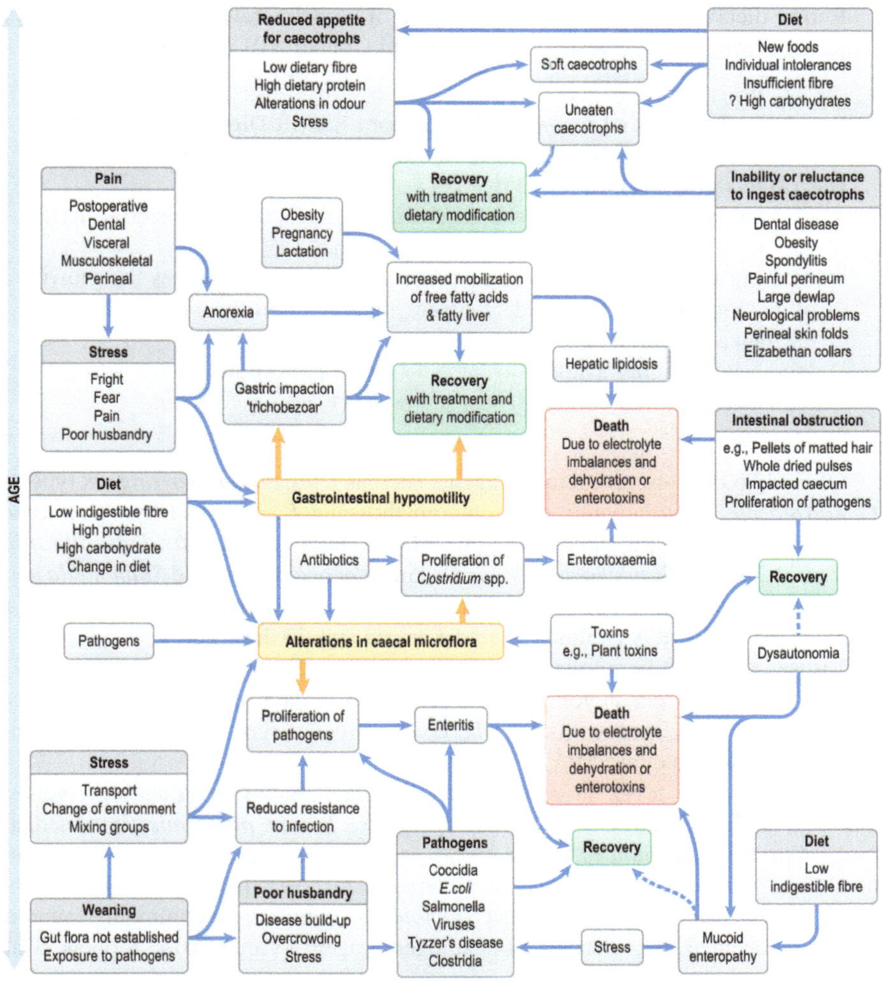

Fig. 1 Interrelationship between the predisposing factors and causes of digestive disease in (pet) rabbits. Adapted from Varga (2014) with permission from Elsevier

Box 1 Enteropathy Predisposing Factors

- Dietary imbalances.
- Chemical and microbiological changes in water.
- Stress situations and environmental variations.
- Intercurrent illnesses (e.g., staphylococci, respiratory syndrome).
- Immunosuppressive causes (e.g., myxomatosis and parasitism).
- Toxins.
- Incorrect antibiotherapy.

Rabbit diets containing a high percentage of carbohydrates and low fiber are a major risk factor for dysbiosis (DeCubellis and Graham 2013) (Box 2).

Box 2 Mechanisms of Adverse Effects from Rabbit Diets with a High Carbohydrate/Fiber Ratio on Gut Health (Oglesbee and Jenkins 2012; DeCubellis and Graham 2013)

- Forming dense masses that cause inadequate digestion and bacterial sterilization by the low pH gastric juice and enzymes
- Altering the cecal pH and fermentation environment promotes the growth of pathogenic species by decreasing hindgut motility and delaying the clearance of luminal bacteria and fermentation byproducts
- Increasing luminal availability of glucose as a substrate for opportunistic organisms (e.g., *E. coli* and *Clostridium* spp.)

Often, two or more proliferative agents such as *Clostridium perfringens* type A, *C. spiroforme*, *Pseudomonas* spp., enteropathogenic *E. coli* (EPEC), *Aeromonas* spp., *Enterococcus* spp. *Klebsiella* spp., *Mannheimia* spp., *Clostridium piliforme* (Tyzzer's disease), *Staphylococcus aureus*, *Streptococcus* spp., and *Salmonella* spp. (may cause abortions in the breeding stock) are present (Percy et al. 1993; García-Rubio et al. 2017). It is common to find *Eimeria* spp. in these cases.

Viruses may also be implicated, mainly rotavirus (Reynoso Utrera et al. 2019), causing yellow diarrhea in young rabbits from 7 to 21 days old, with high mortality, and are also associated with *E. coli*, coronavirus, parvovirus, adenovirus, and astrovirus (Martella et al. 2011).

At the parasitic level, in addition to coccidia, *Cryptosporidium* spp. (yellow diarrhea, high mortality), *Passalurus ambiguus*, *Taenia pisiformis* (intermediate host), *Cittotaenia ctenoides*, *Mosgovoyia pectinata*, *Taenia serialis* (intermediate host), and *Echinococcus granulosus* (intermediate host) may be involved in the rabbit enteric syndrome. The fungus *Cyniclomyces guttulatus* is commonly found in rabbit feces but is considered a nonpathogenic agent (Varga 2014; Sioutas et al. 2021).

A relative percentage of concurring enteropathogens were identified from 58 rabbits presenting clinical enteropathy in Mexico (García-Rubio et al. 2017). In

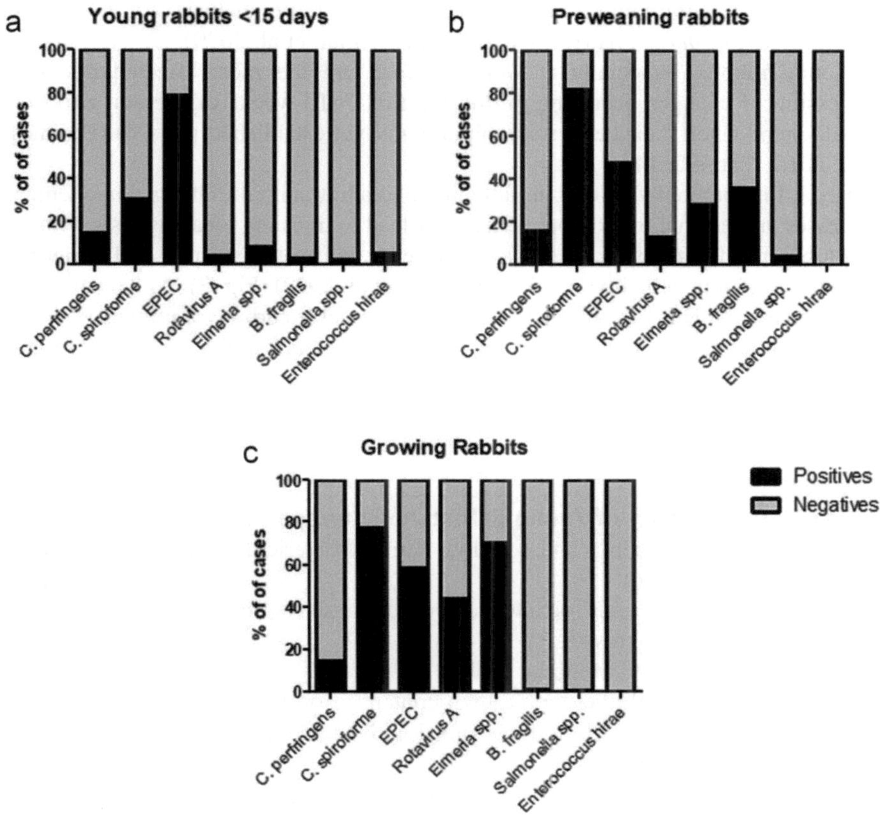

Fig. 2 Relative percentage of enteropathogens detected in rabbit clinical cases and according to age (young rabbits, $n = 95$, **a**; preweaning rabbits, $n = 117$, **b**; and growing rabbits, $n = 545$, **c**). Adapted from Solans et al. (2019). *EPEC* enteropathogenic *E. coli*

this study, *Eimeria magna* was the most frequent pathogen identified alone (43.1%) or in association with *E. perforans* + *Aeromonas* spp. (15.5%); however, in several other cases, bacteria (*S. aureus*, *Streptococcus* spp., *Klebsiella* spp., *Salmonella* spp.) or rotavirus was concomitantly observed with *Eimeria* species. In the Iberian Peninsula (2018–2019), a similar study was reported, but according to rabbit age: young rabbits were <15 days old, preweaning rabbits 15–35 days old, and growing rabbits >35 days old (Fig. 2) (Solans et al. 2019). *C. spiroforme* was the most frequently detected enteropathogen in preweaning and growing rabbits and was mainly associated (59%) with other agents. In contrast, EPEC was observed in 79% of young rabbits, and a relevant frequency was also found in preweaning (48%) and growing (58%) rabbits. *Eimeria* spp. still present in 30% and 70% of preweaning and growing rabbits, respectively, and mostly in association with *C. spiroforme*, EPEC, or rotavirus A (Solans et al. 2019). In the case of *Eimeria* spp., the most commonly found species in laboratory analysis are *E. magna*, *E. media*, and *E. perforans* (Reusch 2005; El-Ashram et al. 2020).

Young rabbits, especially at weaning time, are the most affected due to a succession of changes occurring during the first 3–10 weeks of life and exposure to pathogens when their cecal microbiota is not yet established (Box 3) (Harkness et al. 2010; Varga 2014).

Regarding pathophysiology, in fattening and adult animals, changes in chemical processes and normal microbiota action may lead to transient local acidosis with the development of pathogenic bacteria and a reduction in the number and variety of normal microbiota, which vary with age (Fang et al. 2020) (Fig. 3). As a result, there is an increase in pH in the presence of easily fermentable carbohydrates and potential production of bacterial toxins according to the species and strains present (Harkness et al. 2010). Colibacillosis is the most frequent process, and it depends on the serotypes present, giving rise to the so-called nest colibacillosis or postweaning colibacillosis, depending on the age of the affected young rabbits.

Box 3 Susceptibility of Young Rabbits to Enteropathies

- Gastric pH between 5 and 6.5 (low natural bactericidal capacity to allow cecal colonization).
- Decrease in fatty acids (octanoic and decanoic acids present in milk), which prevent bacterial growth.
- Neonates with live weight at birth < 60 g.
- Changing the milk diet with a high content of protein and fats chemically digestible to a diet rich in fibrous feed with bacterial digestion.
- Loss of passive immunity.
- Separation/movement of young rabbits at weaning.

3 Antibiotherapy and Cecal Dysbiosis

It is well known that antimicrobials modulate gut resident bacteria. For example, Abecia et al. (2007a) observed that chlortetracycline (400 ppm), contrary to tiamulin (100 ppm) and bacitracin (100 ppm), shifts the resident bacterial community toward a decrease in butyrate and an increase in propionate production rates. Nevertheless, tiamulin but not bacitracin at these therapeutic doses can significantly affect cecal microbiota in lactating does (Abecia et al. 2007b).

Iatrogenic situations such as the use of antibiotics, mainly narrow-spectrum (gram-positive) antibiotics (Box 4), may also induce dysbiosis, a condition in which there is an alteration in the natural gut flora (Harkness et al. 2010; Quesenberry and Carpenter 2012; DeCubellis and Graham 2013). Moreover, some antibiotics have the potential to cause enteritis in rabbit dysbiosis (Quesenberry et al. 2020) by selectively killing certain resident bacteria within the gut and allowing pathogenic species to proliferate (Ramirez et al. 2020).

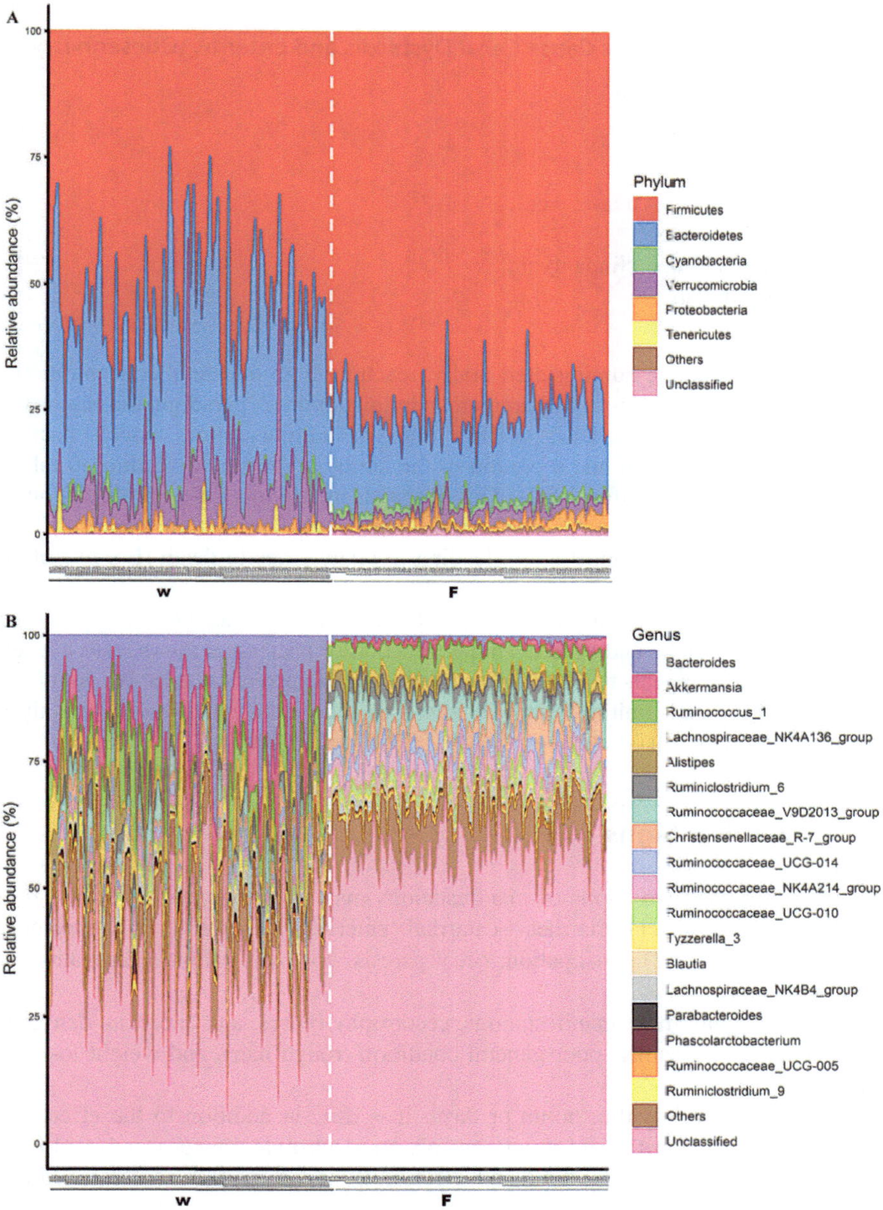

Fig. 3 The dynamic distributions of gut microbiota at the phylum (**a**) and genus (**b**) levels. Adapted from Fang et al. (2020). Legend: Gut microbiota (phyla and genera) from 105 healthy Ira rabbits (meat rabbits). "W" and "F" represent the same rabbit at the weaning (28 ± 2 days) and finishing (72 ± 2 days) stages, respectively

Box 4 Antibiotics that Cause Cecal Dysbiosis and Enteritis (Quesenberry et al. 2020)

- Procaine penicillin.
- Ampicillin.
- Amoxicillin.
- Amoxicillin-clavulanic acid.
- Cephalosporins.
- Lincomycin and clindamycin.
- Erythromycin.

Although some are administered orally, reaching high intestinal concentrations, dysbiosis can also be caused by parenteral administration. A paradigmatic case is the use of ampicillin, regardless of the route of administration, which has an enterohepatic cycle with bile secretion of the antibiotic in nature, leading to rabbit death from dysbiosis (Olfert 1981). This dysbiosis usually occurs by proliferation of clostridia or coliforms (e.g., *E. coli*), with diarrhea (or constipation) occurring 1–3 days after antibiotic administration, leading in many cases to death (Escoula et al. 1981; Morris 1995).

The excessive or inappropriate use of antibiotics is also related to the appearance of bacterial resistance and even eventual selection of bacteria at the digestive level. An example that appears to be associated with these occurrences (selective antibiotic pressure) is the self-limiting diarrhea caused by *Klebsiella pneumoniae* (mainly a respiratory tract agent) (Coletti et al. 2001; Stiefel et al. 2006).

4 Clinical Signs

In this syndrome, two forms of manifestations may occur, depending on the predominant process: (1) diarrhea—catarrhal enteritis, dysentery, or hyperacute typhlocolitis—and (2) constipation: cecal paresis, epizootic rabbit enteropathy, or enterotoxemia.

Accompanying these manifestations, cecotrophy ceases, and anorexia, dehydration, abdominal bloating, poor general condition (rough hair), and weight loss are observed.

Death can occur within hours or days. It is due, in addition to the effects of potential toxins (hepatic and renal), mainly to electrolyte changes and acid-base disorders (Box 5), due to fluid and electrolyte hypersecretion (enterotoxemia), increased mucosal permeability (enterotoxemia and coccidiosis), or intestinal malabsorption (virus) (Harkness et al. 2010).

Box 5 Blood Disorders: Electrolytic, Biochemical, and Acid-Base Balance

- Dehydration >5%.
- Acidosis.
- Decreased levels of glucose, protein, albumin, calcium, and alkaline phosphatase.
- Increased urea, phosphorus, cholesterol, uric acid, and aspartate transferase.

In pet rabbits, a similar syndrome called rabbit gastrointestinal syndrome (RGIS), which mainly causes gastrointestinal stasis, is well known. Many of them have concurrent noninfectious disease (e.g., intestinal obstruction, intestinal adhesions, pancreatitis, hepatic lipidosis). The main clinical signs of this syndrome are impactions or gas accumulation in the stomach, intestines, or cecum (Lichtenberger and Lennox 2010).

5 Diagnosis

In addition to the observation of clinical signs, dying animals should be sacrificed to perform immediate necropsy and thus reduce the onset of cadaveric phenomena and consequently their confusion with lesions arising from enteric diseases.

Fecal examinations (parasitism), bacterial smears and cultures, polymerase chain reaction (PCR), and real-time PCR (Agnoletti 2012) from intestinal and cecal contents, histopathology of the intestinal walls, and even specific serological analyses and serotyping of *E. coli* are essential in trying to determine the triggering processes of the outbreak.

6 Treatment, Control, and Prevention

Treatment and prevention depend on decreasing or even eliminating the effects of the predisposing factors and pathogens involved. Particular attention should be given to the diet, decreasing crude protein (as well as energy) and increasing crude fiber. A crucial aspect of prevention is food and nutrition management with adequate dietary balance between easily fermentable fiber and carbohydrates, as described in Chapter "Rabbit Nutrition and Feeding". In general, and whenever possible, dietary change should be gradually introduced over 4–5 days (recently delivered does and fattening rabbits) (Harkness et al. 2010).

Feed restriction for 2 weeks in weaned rabbits with high fiber is essential to decrease the incidence of enteric diseases in these groups. Only after this adaptation period should the energy density of the diet be increased to maximize the average daily gain (ADG) and feed conversion ratio (FCR). Several strategies to mitigate the adverse effect of feed intake restriction on rabbit performance and meat characteristics have been developed with relative success to ensure gut health without antibiotic use (Gidenne et al. 2012; Knudsen et al. 2014; Alabiso et al. 2017; Birolo et al. 2020).

Probiotics (e.g., *Lactobacillus acidophilus*, *Bacillus subtilis*, *Bifidobacterium bifidum*, *Saccharomyces cerevisiae*) are feed additives that interact with gastrointestinal microbiota (Mancini and Paci 2021) and have been studied to improve gut health and general health by decreasing rabbit morbidity and mortality and improving performance and meat quality. It was observed that *Lactobacillus casei RABX*1 induced a decrease in the *Escherichia-Shigella* percentage, increasing the proportion of luminal lactobacilli in suckling New Zealand White kits (Shen et al. 2020). An increased amount of intestinal and cecal yeast cells and *Lactobacillus* concomitantly with a decreased number of *Salmonella* and coliforms was observed in growing rabbits (40 days old) receiving nanoencapsulated 11×10^{12} CFU *Saccharomyces cerevisiae* yeast + 0.15 g *Moringa oleifera* leaf extract per kg of diet (Hashem et al. 2021). *Lactobacillus buchneri* at a 10^4 CFU/g dose can prevent diarrhea in weaning Rex rabbits (Zhou et al. 2018). According to Mancini and Paci (2021), gut microbiota modulation by appropriate probiotic supplementation can be a suitable tool to improve immunity and nutrient digestibility.

Weaker animals should be eliminated, and disinfection of surfaces and material must be performed carefully. Presently, medicated feed is also used as a metaphylactic, with the inclusion of only one antibiotic, always following the recommendations in the summary of product characteristics (SPC; see Regulation (EU) 2019/6 of the European Parliament and of the Council of 11 December 2018; EU Regulation 2019).

Although individual treatment is not generally introduced on farms, parenteral hydration with Ringer lactate solutions can be used in pet rabbits, which tends to correct the present acidosis. In this case, other therapeutic substances (antibacterial, mucosal astringents such as bismuth subsalicylate, or even analgesics in case of fever) are also advised.

Recently, Montoro-Dasi et al. (2022) observed that the exclusion of antibiotics (neomycin sulfate at 250 ppm and valnemulin hydrochloride at 35 ppm) from the diet of a rabbit's robust genetic line, between 28 and 61 days of age, did not affect cecal microbiota and performance. These results indicate that selecting rabbit lines for resilience can contribute to mitigating gut health problems as well as the use of antibiotics in weaned rabbits.

In the case of antibiotic-induced dysbiosis, it is important to stop using the antibiotic. After stopping the iatrogenic cause, it is important to reconstitute the microbiota of the cecum, acidify the water, and use probiotics. In some cases, an antibiotic appropriate for the bacterial disease that has developed must be used.

The rabbit producer must focus on animal welfare and thermal comfort and must control the thermal sensation of cold in rabbits.

7 Conclusions

Gut microbiota equilibrium is essential to maintain gut health in rabbits. Its disruption leads to dysbiosis and enteropathies. Several abiotic and biotic factors induce enteric syndrome, with quick development of one or more proliferative enteropathogens, such as parasites, bacteria, and/or viruses. The diagnostic approach of this multifactorial occurrence involves investigations of abiotic risk factors as well as laboratory detection of biologic agents. More than a specific treatment, which can be deceptive, the control of this syndrome is the best approach in rabbitries. Special focus on nutritional management (e.g., dietary balance, feed restriction, probiotic supplementation) and internal biosecurity measures (e.g., hygiene and sanitation) must be considered. The introduction of robust genetic lines in rabbits promises to control syndromes with low morbidity and mortality rates, mitigating or avoiding the use of antibiotics on farms.

Rabbit farms that prioritise animal welfare, environmental control, thermal comfort, and low levels of harmful gases (e.g., ammonia) mitigate thermal sensation of cold and sudden temperature variations inside the holding. Intestinal dysbiosis in rabbits is far less common in these rabbitries, and their productivity is significantly higher.

Acknowledgments The authors thank Springer for the use of the following AI tools in this chapter: Dimensions AutoSummarization 3.0 and AJE Digital Editing. This work was supported by projects UIDP/CVT/00772/2020 (CECAV) and LA/P/0059/2020 (AL4AnimalS), funded by the Portuguese Foundation for Science and Technology (FCT).

References

Abecia L, Fondevila M, Balcells J, Edwards JE, Newbold CJ, McEwan NR (2007a) Effect of antibiotics on the bacterial population of the rabbit caecum. FEMS Microbiol Lett 272:144–153. https://doi.org/10.1111/j.1574-6968.2007.00746.x

Abecia L, Fondevila M, Balcells J, Lobley GE, McEwan NR (2007b) The effect of medicated diets and level of feeding on caecal microbiota of lactating rabbit does. J Appl Microbiol 103:787–793. https://doi.org/10.1111/j.1365-2672.2007.03309.x

Agnoletti F (2012) Update on rabbit enteric diseases: despite improved diagnostic capacity, where does disease control and prevention stand? In: Proceedings 10th World Rabbit Congress; Sep 2012. Sharm El-Sheikh, Egypt

Alabiso M, Di Grigoli A, Mazza F, Maniaci G, Vitale F, Bonanno A (2017) A 3-week feed restriction after weaning as an alternative to a medicated diet: effects on growth, health, carcass and meat traits of rabbits of two genotypes. Animal 11:1608–1616. https://doi.org/10.1017/S175173111600269X

Anholt RM, Berezowski J, Robertson C, Stephen C (2015) Spatial-temporal clustering of companion animal enteric syndrome: detection and investigation through the use of electronic medical records from participating private practices. Epidemiol Infect 143:2547–2558. https://doi.org/10.1017/S0950268814003574

Birolo M, Trocino A, Zuffellato A, Xiccato G (2020) Effects of time-based feed restriction on morbidity, mortality, performance and meat quality of growing rabbits housed in collective systems. Animal 14:626–635. https://doi.org/10.1017/S1751731119002283

Coletti M, Passamonti F, Del Rossi E, Franciosini MP, Setta B (2001) Klebsiella pneumoniae infection in Italian rabbits. Veterinary Record 149:626–627. https://doi.org/10.1136/vr.149.20.626

DeCubellis J, Graham J (2013) Gastrointestinal disease in guinea pigs and rabbits. Vet Clin North Am Exot Anim Pract 16:421–435. https://doi.org/10.1016/j.cvex.2013.01.002

El-Ashram S, Aboelhadid SM, Abdel-Kafy E-SM, Hashem SA, Mahrous LN, Farghly EM, Kamel AA (2020) Investigation of pre- and post-weaning mortalities in rabbits bred in Egypt, with reference to parasitic and bacterial causes. Animals 10:537. https://doi.org/10.3390/ani10030537

Escoula L, Camguilhem R, Larrieu G, More J (1981) Sensibility of rabbits to treatment with ampicillin and gentamycin. Ann Rech Vet 12:11–17

EU Regulation (2019) Regulation (EU) 2019/6 of the European Parliament and of the Council of 11 December 2018 on veterinary medicinal products and repealing Directive 2001/82/EC. http://data.europa.eu/eli/reg/2019/6/oj

Fang S, Chen X, Pan J, Chen Q, Zhou L, Wang C, Xiao T, Gan QF (2020) Dynamic distribution of gut microbiota in meat rabbits at different growth stages and relationship with average daily gain (ADG). BMC Microbiol 20:116. https://doi.org/10.1186/s12866-020-01797-5

García-Rubio VG, Bautista-Gómez LG, Martínez-Castañeda JS, Romero-Núñez C (2017) Multi-causal etiology of the enteric syndrome in rabbits from Mexico. Revista Argentina de Microbiología 49:132–138. https://doi.org/10.1016/j.ram.2017.03.001

Gidenne T, García J, Lebas F, Licois D (2010) Nutrition and feeding strategy: interactions with pathology. In: Nutrition of the rabbit, 2nd edn. CAB International, Wallingford, pp 179–199

Gidenne T, Combes S, Fortun-Lamothe L (2012) Feed intake limitation strategies for the growing rabbit: effect on feeding behaviour, welfare, performance, digestive physiology and health: a review. Animal 6:1407–1419. https://doi.org/10.1017/S1751731112000389

Harkness JE, Turner PV, Vandewounde S, Wheler CL (2010) Harkness and Wagner's Biology and Medicine of Rabbits and Rodents, 5th edn. Wiley Blackwell, Hoboken, NJ

Hashem NM, Hosny NS, El-Desoky NI, Shehata MG (2021) Effect of Nanoencapsulated Alginate-Synbiotic on Gut Microflora balance, immunity, and growth performance of growing rabbits. Polymers 13:4191. https://doi.org/10.3390/polym13234191

Knudsen C, Combes S, Briens C, Coutelet G, Duperray J, Rebours G, Salaun J-M, Travel A, Weissman D, Gidenne T (2014) Increasing the digestible energy intake under a restriction strategy improves the feed conversion ratio of the growing rabbit without negatively impacting the health status. Livest Sci 169:96–105. https://doi.org/10.1016/j.livsci.2014.08.015

Lichtenberger M, Lennox A (2010) Updates and advanced therapies for gastrointestinal stasis in rabbits. Vet Clin North Am Exot Anim Pract 13:525–541. https://doi.org/10.1016/j.cvex.2010.05.008

Licois D, Coudert P, Marlier D (2006) Epizootic rabbit enteropathy. In: Recent advances in rabbit science. Institute of Agricultural and Fisheries Research (ILVO), Merelbeke, Belgium, pp 63–170

Mancini S, Paci G (2021) Probiotics in rabbit farming: growth performance, health status, and meat quality. Animals 11:3388. https://doi.org/10.3390/ani11123388

Martella V, Moschidou P, Pinto P, Catella C, Desario C, Larocca V, Circella E, Bànyai K, Lavazza A, Magistrali C, Decaro N, Buonavoglia C (2011) Astroviruses in Rabbits. Emerg Infect Dis 17:2287–2293. https://doi.org/10.3201/eid1712.110967

Montoro-Dasi L, Lorenzo-Rebenaque L, Ramon-Moragues A, Pérez-Gracia MT, de Toro M, Marin C, Villagra A (2022) Antibiotic removal does not affect cecal microbiota balance and productive parameters in LP robust rabbit line. Front Vet Sci 9:1038218. https://doi.org/10.3389/fvets.2022.1038218

Morris TH (1995) Antibiotic therapeutics in laboratory animals. Lab Anim 29:16–36. https://doi.org/10.1258/002367795780740393

Oglesbee BL, Jenkins JR (2012) Gastrointestinal diseases. In: Ferrets, Rabbits, and Rodents. Elsevier, St. Louis, pp 193–204. https://doi.org/10.1016/B978-1-4160-6621-7.00015-4

Olfert ED (1981) Ampicillin toxicity in rabbits. Can Vet J 22:217

Percy DH, Muckle CA, Hampson RJ, Brash ML (1993) The enteritis complex in domestic rabbits: a field study. Can Vet J 34:95–102

Quesenberry KE, Carpenter JW (eds) (2012) Ferrets, rabbits, and rodents: clinical medicine and surgery, 3rd edn. Elsevier/Saunders, St. Louis, Mo

Quesenberry K, Mans C, Orcutt C, Carpenter JW (eds) (2020) Ferrets, rabbits, and rodents: clinical medicine and surgery, 4th edn. Elsevier, Inc, Philadelphia

Ramirez J, Guarner F, Bustos Fernandez L, Maruy A, Sdepanian VL, Cohen H (2020) Antibiotics as major disruptors of gut microbiota. Front Cell Infect Microbiol 10:572912. https://doi.org/10.3389/fcimb.2020.572912

Reusch B (2005) Rabbit gastroenterology. The veterinary clinics of North America. Exotic Anim Pract 8:351–375. https://doi.org/10.1016/j.cvex.2005.01.007

Reynoso Utrera E, Bautista Gómez LG, Martínez Castañeda JS, López Aguado Almazán G, Romero Núñez C, García Rubio VG, Hernández García PA, Espinosa Ayala E (2019) Análisis de la presencia de Rotavirus en conejos del Estado de México. Revista Mexicana de Ciencias Pecuarias 10:511. https://doi.org/10.22319/rmcp.v10i2.4638

Shen XM, Cui HX, Xu XR (2020) Orally administered Lactobacillus casei exhibited several probiotic properties in artificially suckling rabbits. Asian-Australas J Anim Sci 33:1352–1359. https://doi.org/10.5713/ajas.18.0973

Sioutas G, Evangelou K, Vlachavas A, Papadopoulos E (2021) Deaths Due to Mixed Infections with *Passalurus ambiguus*, *Eimeria* spp. and *Cyniclomyces guttulatus* in an Industrial Rabbit Farm in Greece. Pathogens 10:756. https://doi.org/10.3390/pathogens10060756

Solans L, Arnal JL, Sanz C, Benito A, Chacón G, Alzuguren O, Fernández AB (2019) Rabbit Enteropathies on Commercial Farms in the Iberian Peninsula: Etiological Agents Identified in 2018–2019. Animals 9:1142. https://doi.org/10.3390/ani9121142

Stiefel U, Rao A, Pultz MJ, Jump RLP, Aron DC, Donskey CJ (2006) Suppression of gastric acid production by proton pump inhibitor treatment facilitates colonization of the large intestine by vancomycin-resistant *Enterococcus* spp. and *Klebsiella pneumoniae* in clindamycin-treated mice. Antimicrob Agents Chemother 50:3905–3907. https://doi.org/10.1128/AAC.00522-06

Varga M (2014) Chapter 8 - Digestive disorders. In: Varga M (ed) Textbook of rabbit medicine, 2nd edn. Butterworth-Heinemann, Oxford. isbn 978-0-7020-4979-8, pp 303–349. https://doi.org/10.1016/B978-0-7020-4979-8.00008-X

Zhou Y, Ni X, Wen B, Duan L, Sun H, Yang M, Zou F, Lin Y, Liu Q, Zeng Y, Fu X, Pan K, Jing B, Wang P, Zeng D (2018) Appropriate dose of *Lactobacillus buchneri* supplement improves intestinal microbiota and prevents diarrhoea in weaning Rex rabbits. Beneficial Microbes 9: 401–416. https://doi.org/10.3920/BM2017.0055

Epizootic Rabbit Enteropathy (ERE)

Teresa Tavares, José Manuel Monteiro, Maria da Conceição Fontes, and João Simões

Abstract

The present chapter mainly highlights the etiology, pathophysiology, and clinical approach of epizootic rabbit enteropathy (ERE). This syndrome mainly affects young rabbits around weaning and even adult animals reared on farms. More than two decades after its emergence (in France, 1996), the etiopathogenesis of this affection remains unclear. The hallmarks of ERE are cecal impaction and mucus secretion into the colon, without notable associated inflammation. The diagnosis is based on clinical signs and necropsy findings. Under a histological approach, mild to moderate atrophy and fusion of the villi associated with moderate transepithelial migration of heterophilic cells and mucosal infiltration of inflammatory cells can be observed. Therapeutic and even metaphylactic oral antibiotherapy can be performed to control the gut microbiome, but the implementation of feed restriction or feeding rabbits a high-fiber diet can be suitable to reduce morbidity and mortality. The control and prevention of ERE is a major key to managing gastrointestinal health and reducing morbidity and mortality rates.

T. Tavares (✉)
José Azevedo Monteiro, Lda., Vila Boa do Bispo, Portugal

J. M. Monteiro
José Azevedo Monteiro, Lda., Vila Boa do Bispo, Portugal

Faculty of Veterinary Medicine, Lusófona University, Lisbon, Portugal

CBIOS, Lusófona University, Lisbon, Portugal

M. da Conceição Fontes · J. Simões
Animal and Veterinary Research Centre (CECAV), Associate Laboratory for Animal and Veterinary Sciences (AL4AnimalS), Department of Veterinary Sciences, School of Agricultural and Veterinary Sciences, University of Trás-os-Montes and Alto Douro (UTAD), Vila Real, Portugal

J. Simões, J. M. Monteiro (eds.), *Veterinary Care of Farm Rabbits*,
https://doi.org/10.1007/978-3-031-44542-2_21

Keywords

Enteropathy · Dysbiosis · Cecal impaction · Abdominal distension · Mucus · Dietary restriction

1 Introduction

Epizootic rabbit enteropathy (ERE) mainly affects farmed rabbits. Although there are also reports of ERE in pet rabbits, these are considered to be rare (Haligur et al. 2009), as well as in wild rabbits (Harcourt-Brown 2002).

ERE is a specific noninflammatory enteric condition of rabbits (Licois et al. 2005; Pérez de Rozas et al. 2005) with unusual characteristics, which distinguishes this affection from the other usual rabbit intestinal diseases (Licois et al. 2005). However, many similarities exist with a disease first previously described (Flatt et al. 1974) as mucoid enteritis or mucoid enteropathy, which has decimated large numbers of colonies in many different countries (Licois et al. 2000). To date, any pathogen responsible for ERE, as well as mucoid enteritis, has been identified (Licois et al. 2005; Szalo et al. 2007; Huybens et al. 2013; Puón-Peláez et al. 2020).

Licois et al. (2006) consider that the mucus excretion and cecal impaction present in "mucoid enteritis" or "mucoid enteropathy" in rabbits are pathophysiological reactions common to several diseases (i.e., syndromes), and these denominations are misleading. There are even other researchers who do not distinguish between the two syndromes (Pérez de Rozas et al. 2005; Puón-Peláez et al. 2018). In China, Hu et al. (2018) reported different cecal microbiota profiles in rabbits experimentally inoculated with cecal content from dead rabbits affected by a mucoid enteropathy syndrome, suggesting a new syndrome similar to ERE.

This chapter mainly aims to describe the etiology, pathophysiology, and clinical approach of ERE.

2 Origin and Impact of ERE

ERE emerged first in late 1996 from France, with rabbits presenting abdominal bloat, watery diarrhea, and high morbidity (up to 100%) and mortality (30–80%) rates (Licois et al. 2005; Licois and Marlier 2008; Bäuerl et al. 2014). Nevertheless, genetic variability in resistance to ERE morbidity and mortality occurs (de Rochambeau et al. 2006; Garreau et al. 2006). Pérez et al. (2015) estimated that ERE is one of the most frequent enteropathies, representing approximately 31% of rabbitries. Its incidence is close to that of colibacillosis (33%) (Pérez et al. 2015). ERE often affects young rabbits, especially around the time of weaning (3–7 weeks of age), usually within the first 2 weeks after weaning, although adult animals can also be affected. This syndrome is characterized by the impaction of the cecum and by the presence of large amounts of mucus in the colon (Richardson 2000; Harkness et al. 2010; Oglesbee and Lord 2020).

Survivor growing rabbits show growth retardation and poor performance, and it is possible to observe a decrease in fertility (as much as 25%) and libido (15%) in bucks and does (Pérez de Rozas et al. 2005; Finzi et al. 2008; Puón-Peláez et al. 2018). The consequence is a decrease in the number of rabbits produced, which leads to a decrease in the economy (Licois et al. 2000; Puón-Peláez et al. 2018).

Other digestive diseases, such coccidiosis, colibacillosis, enterotoxemia, viral enteritis, and Tyzzer's disease, can occur concurrently (Licois et al. 2005), increasing the impact of ERE on performance and health.

Weaning ≥35 days old, young rabbit transfer at weaning, ≥ 0.14 m^3/kg (air volume/rabbit) fattening at weaning time, and mortality rate ≥ 10.5% are considered risk factors for ERE occurrence in rabbitries (Le Bouquin et al. 2009).

3 Etiopathophysiology

ERE etiopathogenesis remains unclear, and some assays involving inoculations of intestinal contents from diseased rabbits have even been performed in laboratory rabbits (Dewrée et al. 2007; Djukovic et al. 2018). Licois et al. (2005) demonstrated that the intestinal content was infectious as early as the second day after infection. Szalo et al. (2007) and Huybens et al. (2008) confirmed that ERE does not have a parasitic or viral etiology and suggested a bacterial cause, which could be a sporulated bacterium or a nonsporulated bacillus. Licois et al. (2005) and Puón-Peláez et al. (2020) have shown that several bacterial phyla are associated with this disease and suggested the necessity of further etiological investigations, particularly in bacteriology. Some researchers looking for a possible pathogenic agent have only revealed the presence of rotavirus and *Clostridium perfringens*, but their role remains questionable (Licois et al. 2005; Marlier et al. 2006; Puón-Peláez et al. 2018). In 2018, Djukovic et al. studied the changes in intestinal microbial communities before, during, and after the onset of ERE and observed an expansion of a single novel *Clostridium* species (*Clostridium cuniculi*) that was detected the day of ERE onset. *C. cuniculi* encodes several putative toxins and is phylogenetically related to the two well-characterized pathogens *C. botulinum* and *C. perfringens*. Although these researchers stated that further experimental validation is needed, they considered bacterial infection as a plausible cause of ERE rather than an initial dysbiosis to trigger the disease. What is clear is that ERE is characterized by high contagiousness, in agreement with Licois et al. (2005).

The specific role of diet, stress, pathogens, or cecal microbiota in the disease is unknown, although it seems that all these factors could be implicated (Carabaño et al. 2008) and as many times is suggested (Licois et al. 2000; Marlier et al. 2006) by ERE occurrence in field conditions.

ERE is also characterized by a marked imbalance in the cecal microbial community (Bäuerl et al. 2014; Puón-Peláez et al. 2020), which, in turn, changes the profile of the metabolites produced and absorbed in the cecum (Badiola et al. 2016). Alterations in the resident bacterial communities and volatile fatty acid (VFA) production in the intestine occur in association with ERE. According to Jin et al.

(2018), the overgrowth (*Bacteroides fragilis*, *C. perfringens*, *Enterobacter sakazakii*, and *Akkermansia muciniphila*) and inhibition (*Bifidobacterium* spp. and *Butyrivibrio fibrisolvens*) of some bacteria in the stomach, small intestine, and cecum decrease the amount of butyrate, resulting in an increase in ERE incidence. Significant ($p < 0.05$) abundance increases in the phylum *Saccharibacteria* or decreases in the phylum *Tenericutes* were also observed by Puón-Peláez et al. (2020).

Some researchers have proposed intestinal hypomotility or changes in the acidity of the cecum as etiological factors (Pérez de Rozas et al. 2005, among others). Perhaps a lack of immune defense against an overgrowth of different pathogens housed in the gut of animals may partially explain the pathophysiology of the syndrome (Bäuerl et al. 2014).

Abundant secretion of mucins by mucinase-producing bacteria, e.g., *Bacteroides thetaiotaomicron*, *Bacteroides fragilis*, *Clostridium perfringens*, and *Akkermansia muciniphila* (Badiola et al. 2016), occurs in the cecum and colon and is responsible for copious amounts of mucus. Bäuerl et al. (2014) identified two distinct profiles of this syndrome: one, inflammatory, consisting of very high expression of cytokines (TNF-α, IL-6, and IL-8) probably reacting to the presence of bacteria and moderate expression of mucins (MUC1, MUC4, and MUC13), and, the other, secretory, presenting very high expression of mucins and moderately high expression of cytokines. Both profiles may represent different stages of the syndrome.

4 Lesions and Clinical Signs

Lesions and clinical signs evolve quickly, appearing 1 day after inoculation, and the disease peaks 4 to 6 days later, usually as a subacute form (Licois et al. 2005). The first signs can be sudden anorexia ($> 50\%$; Djukovic et al. 2018), with rabbits first stopping drinking and later stopping eating (Dewrée et al. 2007), evolving to abdominal distension and depression (Licois et al. 2005). In the acute form, death occurs 1 to 3 days after the first symptoms (Bäuerl et al. 2014).

The clinical signs develop gradually and include subnormal body temperature or no fever (Jobert et al. 2001), depression, and hunched posture (Jin et al. 2018). It also exhibits the "water bottle" sound if picked up and shaken. A characteristic of the disease is tooth grinding (with rambling noise), presumably due to abdominal pain (Licois et al. 2005).

In the early stages, diarrhea might be present, but in the later stages, large amounts of mucus are excreted, either on its own or mixed with fecal material. As the disease progresses, fecal production can cease completely, and appetite is reduced to the point of complete anorexia (Harcourt-Brown 2014).

Rabbits between 3 and 7 weeks show a reduction in their daily feed intake of 50%, going from 110 g per day to 55 g a day for approximately 7 days (Pérez de Rozas et al. 2005). The growth can be strongly lowered from as early as the second day until the end of the second week, until the animals have recovered (Licois et al. 2005).

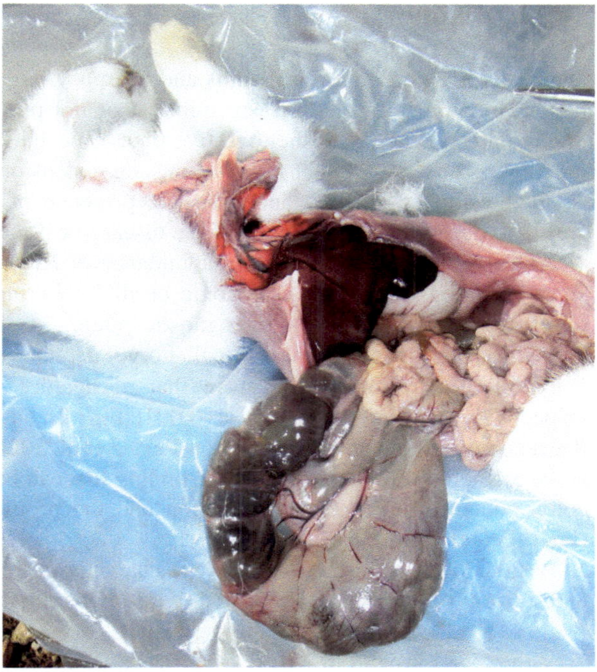

Fig. 1 Macroscopic lesions of rabbit digestive tracts with epizootic rabbit enteropathy. Legend: Cecal impaction and absence of perceptible inflammation in the small intestine and cecum

The most common cause of death from acute intestinal obstruction is hypovolemic shock, caused by a sudden loss of fluid into the lumen of the gastrointestinal tract proximal to the obstruction (Oglesbee and Lord 2020).

Even though the rabbits that had previously ERE made a full recovery in terms of no longer having diarrhea, they were still smaller than animals that never had diarrhea and had a different gut microbiota and metabolome produced by these organisms (Puón-Peláez et al. 2020).

In necropsy, at initial stages, the main features were the absence of obvious gross lesions of the mucosa of the different parts of the digestive tract, more precisely an acute inflammatory process, or congestion of the small intestine, cecum, or large intestine (Licois et al. 2005). In more advanced stages, inflammation, including intestinal necrosis, occurs. During the first 4–5 days, distension of the stomach, duodenum, or both by a very liquid and gaseous content was almost the only pathologic change. These main features, associated with the distension of the abdomen observed on live animals or before necropsy, should be taken into consideration for field diagnosis of ERE (Dewrée et al. 2007) (Fig. 1). The frequency of cecal impaction (due to dehydration) is very variable: generally, it is found in 20–30% of dead animals, but it can vary from 0 to 70% (Licois et al. 2000). The distal colon is generally empty (Licois et al. 2005).

5 Diagnosis

Diagnosis is based on clinical signs and necropsy findings. At postmortem examination, inflammatory changes are minimal or not present, and an impaction with dried contents and gas of the digestive tract, especially in the cecum and ileocecal junction, is observed. Other main lesions are distension of the stomach, sometimes small intestine, with fluid and gas. The presence of mucus in the colon is not constant, depending on the stage of the disease or if the animals are treated.

At microbiological analysis, an abundance of *Escherichia coli* and *Clostridium* spp., including *C. cuniculi* (Djukovic et al. 2018), among other bacteria, can be observed (Table 1) but without diagnostic value.

With a histological approach, smaller crypt depths were observed in the epithelial colon of affected animals than in that of healthy animals (173 vs. 308 μm, respectively; SEM: ±11 μm; $p < 0.001$; Fig. 2) but similar ($p > 0.05$) in the ileoceco-colic valve and cecum (Puón-Peláez et al. 2020). The atrophy of the crypts at the epithelial colon level is an ERE finding that can be attributed to the absence of colon content due to cecum impaction and the proliferation of *Clostridium* spp. (Puón-Peláez et al. 2020).

Lesions of the intestinal mucosa are characterized by mild to moderate atrophy and fusion of the villi, associated with moderate transepithelial migration of

Table 1 Percentage of bacteria observed in mixed fecal samples (stomach, small intestine, and cecum) at necropsy of 12 rabbits presenting epizootic rabbit syndrome from rabbitries. Adapted from Marlier et al. (2006) with permission from Elsevier

Bacterial strain	Percentage of positive examinations
Enteropathogenic *Escherichia coli*[a]	0
Escherichia coli (>10^4 FCU/gf)	66.7% (8/12)
Escherichia coli (<10^4 UFC/gf)	25.0% (3/12)
No *Escherichia coli*	8.3% (1/12)
Enterococcus casseliflavus and/or *E. durans*	25.0%
Staphylococcus spp.	25.0%
Streptococcus spp.	8.3%
Bacillus licheniformis	8.3%
Bacillus sphaericus	16.7% (2/12)
Clostridium perfringens	83.3% (10/12)
Clostridium paraputrificum	8.3%
Clostridium ramosum	8.3%
Clostridium fallax	8.3%
Clostridium histolyticum	16.7%
Clostridium baratii	16.7%
Capnocytophaga spp.	8.3%
Bacteroides spp.	25.0%

[a] Both biotyping and serotyping and PCR amplification for the eaeA gene were negative (no enteropathogenic *Escherichia coli* strains were observed). Bacteriological cultures were performed under aerobic and anaerobic conditions

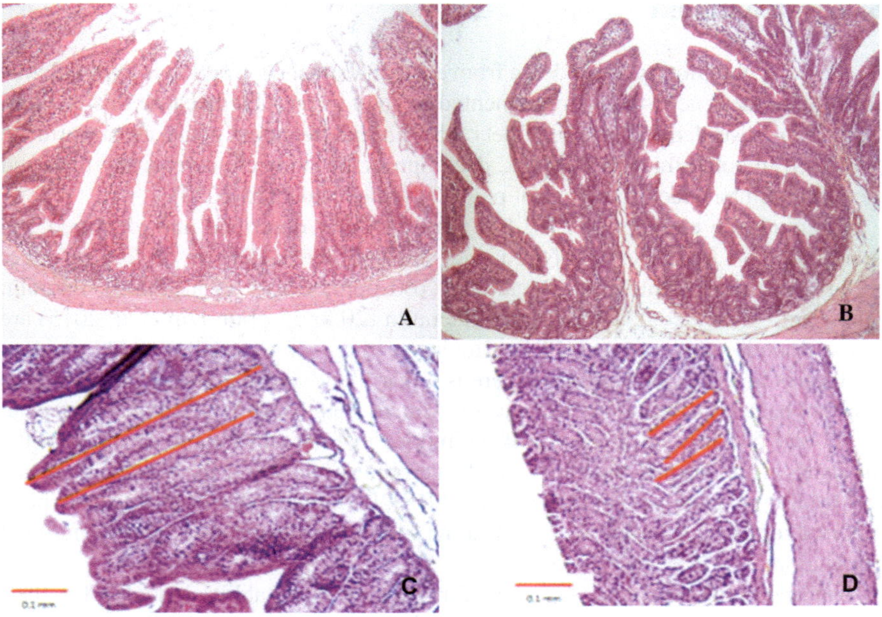

Fig. 2 Histological lesions of the jejunum and colon from rabbits presenting epizootic rabbit enteropathy. Legend: (**a** and **b**) Histological examination of the jejunum of 6-week-old rabbits slaughtered 10 days after experimental reproduction of epizootic rabbit enteropathy (ERE). (**a**) Healthy noninoculated rabbit; (**b**) disease-inoculated rabbit. Compared with the control, at the same magnification, the villi of the jejunum part of the small intestine appeared shortened with irregular shape and size. The superficial epithelium of the villi, crypts, and lamina propria are normal without any degenerative, hyperplastic, or inflammatory visible changes. The lesions are mild and devoid of etiological specificity (Adapted from Licois et al. 2005). (**c** and **d**) Histopathological comparison of rabbit digestive tracts without (**c**) and with (**d**) ERE. Histological section (5 μm) of villi of the colon for both cases. The red lines show the length of the crypts; hematoxylin and eosin staining. Adapted from Puón-Peláez et al. (2020)

heterophilic cells and mucosal infiltration of inflammatory cells, and are mostly observed in some inoculated rabbits sampled for histological examination. Nevertheless, the intestinal lesions appear nonconstant, with some inoculated rabbits being totally devoid of histological intestinal changes (Licois et al. 2005). Moreover, the lesions always remain moderate and clearly devoid of etiological specificity and, more particularly, without lesion kinetics, which usually characterizes the development of infectious pathogenic agents.

6 Treatment

As the most common cause of death from acute intestinal obstruction is hypovolemic shock, treatment needs to be implemented as soon as possible because this is a life-threatening disorder. In the case of pet rabbits, initial treatment goals are to warm the patient (if hypothermic), treat shock, decompress the stomach, correct any fluid and electrolyte imbalances, and control pain (Oglesbee and Lord 2020).

Although the etiology of ERE has not been fully described and identified, the most common treatment involves using oral antibiotics (Dip et al. 2015), either metaphylactic or therapeutic. Bacitracin (100 ppm; Boisot et al. 2004), tiamulin (100 ppm; Abecia et al. 2007), and valnemulin (20 or 35 ppm; Dip et al. 2015) are selected antibiotics to treat ERE. However, there is a trend toward avoiding antibiotics in rabbit diets because there is a fear that overuse could help to promote resistant bacterial strains (EFSA Panel on Animal Health and Welfare (AHAW) et al. 2021) and because the inclusion of antibiotics in food can lead to changes in the diversity of the normal gastrointestinal microbiota, causing a microbial imbalance and leading to dysbiosis (Djukovic et al. 2018). Different levels of effects have been reported for different antibiotics, with some showing a clear impact on the microbial population.

Antimicrobial treatment, not only for growing sick rabbits but also for their mothers, often has good results, namely, for the treatment of ERE, avoiding its appearance in the following groups of weaned rabbits.

In commercial units, affected rabbits are sometimes not treated with antibiotics because it is possible to control the problem only by implementing feed restriction or a high-fiber diet (Gomez-Conde et al. 2007; Martignon et al. 2010). This happens mostly when we know that ERE is present in a farm and we are able to prevent the problem but also when the problem is detected early.

7 Prevention

Nutritional management is one of the keys to controlling and preventing ERE. In rabbitries where the incidence of ERE is high, changing to a new batch of feed with a higher fiber content will usually help to control or eliminate the disease. Although there is some potential in the use of these high-fiber diets, Puón-Peláez et al. (2022) demonstrated that these diets are not able to produce the same digestive tract properties as those seen in rabbits that have never had the condition, which is true even after the rabbits have recovered from epizootic rabbit enteropathy.

The feed restriction strategies after weaning, for 1 to 4 weeks, consist of feeding rabbits 70% to 90% of their ad libitum daily intake (Gidenne et al. 2012) and seem to be one of the most efficient ways to control almost all gastrointestinal diseases, namely, ERE.

The effect of water pH on rabbits has been studied. In weaned rabbits (35 days old), pH = 4.3 did not reduce diarrhea or mortality rates but affected the cecal microbiome (*E. coli* decrease and tendency for *Bacteroides-Prevotella* or

Table 2 Main polysaccharides of interest in macroalgae (Lopez-Santamarina et al. 2020)

Polysaccharide	Chemical structure	Biological properties
Ulvan	L-Rhamnose, D-xylose, D-glucose, and D-glucuronic acid	Antioxidant, anticoagulant, anti-inflammatory, immune-stimulatory, antibacterial, antihyperlipidemic
Fucoidan	1,3-α-Fucopyranoside backbone with branching of α-1,2-fucopyranoside	Antioxidant, anticoagulant, anti-inflammatory, antiviral, immune-modulatory
Laminarin	1,3-β-D-Glucan with β-1,6-linkages	Antioxidant, anticoagulant, anti-inflammatory, antiviral, immune-modulatory, prebiotic
Alginate	1,4-β-D-Mannuronic acid and α-L-guluronic acid residues	Antioxidant, anti-inflammatory, antitumor

Lactobacillus increase proportion regarding the total bacteria), resulting in performance improvement (Zhu et al. 2014). In practice, pH water reduction (acetic acid 5%/L) is used for gut health control in enteropathy outbreaks, but some care is needed due to water palatability exchange, which can reduce intake.

Probiotics, e.g., *Bacillus subtilis* (Phuoc and Jamikorn 2017), *Aspergillus awamori* (El-Deep et al. 2021), and *Lactobacillus acidophilus* (D2/CSL) (Colombino et al. 2022), have been reported and used to mitigate the adverse effect of ERE (see review of Mancini and Paci 2021). Dietary prebiotics, such as marine macroalgae, which contain polysaccharides with antioxidant and anti-inflammatory activities (Table 2), can serve as substrates to modulate the gut microbiome, reducing rabbit morbidity and mortality by ERE (Lopez-Santamarina et al. 2020). Currently, these additives are used in the feed industry.

For the prevention of ERE, biosecurity measures and general good management practices are essential *all-in all-out* system, washing with detergent, disinfecting each time with a different substance, minimizing the adoptions among the does, adequate environmental conditions, and improving rabbits' welfare in general.

8 Conclusions

There is still a long way to go in the knowledge of this disease, and the suspicions instead of suspects of many authors about its bacterial origin need to be confirmed. Acute and mainly subacute forms of this ERE cause significant rabbit morbidity and mortality and economic losses on farms. Following antibiotic use restrictions in some worldwide regions, such as the EU, alternatives to control and prevent this syndrome have been developed. In practice, currently, the main way to control this disease is by nutritional management procedures such as the supply of a low-fiber diet, feed restriction, use of pro- and prebiotics, and improving biosecurity measures and animal welfare. In serious and persistent cases, when talking about commercial farms, it is necessary to consider the hypothesis of reducing the production rhythm or even culling all the animals, emptying the barns for cleaning, and halting production.

Acknowledgments The authors thank Juan Rosell for his suggestions and review of this chapter. The authors also thank Springer for the use of the following AI tools in this chapter: Dimensions Auto-Summarization 3.0 and AJE Digital Editing. This work was supported by projects UIDP/CVT/00772/2020 (CECAV) and LA/P/0059/2020 (AL4AnimalS), funded by the Portuguese Foundation for Science and Technology (FCT).

References

Abecia L, Fondevila M, Balcells J, Edwards JE, Newbold CJ, McEwan NR (2007) Effect of antibiotics on the bacterial population of the rabbit caecum. FEMS Microbiol Lett 272:144–153. https://doi.org/10.1111/j.1574-6968.2007.00746.x

Badiola I, Perez de Rozas A, Gonzalez J, Aloy N, García J, Carabaño R (2016) Recent advances in ERE in growing rabbits. Proceedings of the 11th World Rabbit Congress, Qingdao (China) (2016), pp 491–502

Bäuerl C, Collado MC, Zúñiga M, Blas E, Pérez Martínez G (2014) Changes in Cecal microbiota and mucosal gene expression revealed new aspects of epizootic rabbit enteropathy. PLoS One 9: e105707. https://doi.org/10.1371/journal.pone.0105707

Boisot P, Duperray J, Guyonvarch A, Richard A, Licois D, Coudert P (2004) Evaluation of the effectiveness of soluble bacitracin (Bacivet s®) in drinking water compared to bacitracin in the feed (Albac®), during an experimental reproduction of epizootic rabbit enteropathy syndrome. Proceedings of the 8th World Rabbit Congress, Puebla (Mexico) (2004), pp 475–462

Carabaño R, Badiola I, Chamorro S, García JA, García-Ruiz I, García-Rebollar P, Gómez-Conde MS, Gutiérrez I, Nicodemus N, Villamide MJ, de Blas JC (2008) Review. New trends in rabbit feeding: influence of nutrition on intestinal health. Span J Agric Res 6:15–25

Colombino E, Biasato I, Michetti A, Rubino MG, Franciosa I, Giribaldi M, Antoniazzi S, Bergagna S, Paliasso G, Ferrocino I, Cavallarin L, Gasco L, Capucchio MT (2022) Effects of dietary supplementation of lactobacillus acidophilus on blood parameters and gut health of rabbits. Animals 12:3543. https://doi.org/10.3390/ani12243543

de Rochambeau H, Licois D, Gidenne T, Verdelhan S, Coudert P, Elsen JM (2006) Genetic variability of the resistance for three types of enteropathy in the growing rabbit. Livest Sci 101:110–115. https://doi.org/10.1016/j.livprodsci.2005.10.019

Dewrée R, Meulemans L, Lassence C, Desmecht D, Ducatelle R, Mast J, Licois D, Vindevogel H, Marlier D (2007) Experimentally induced epizootic rabbit enteropathy: clinical, histopathological, ultrastructural, bacteriological and haematological findings. World Rabbit Sci 15:91–102. https://doi.org/10.4995/wrs.2007.602

Dip R, Nemet Z, Schiessl B, Klein U, Strehlau G (2015) Efficacy and tolerability of early administration of valnemulin hydrochloride premix on epizootic rabbit enteropathy. Vet J 204:309–314. https://doi.org/10.1016/j.tvjl.2014.12.036

Djukovic A, Garcia-Garcera M, Martínez-Paredes E, Isaac S, Artacho A, Martínez J, Ubeda C (2018) Gut colonization by a novel *clostridium* species is associated with the onset of epizootic rabbit enteropathy. Vet Res 49:123. https://doi.org/10.1186/s13567-018-0617-8

EFSA Panel on Animal Health and Welfare (AHAW), Nielsen SS, Bicout DJ, Calistri P, Canali E, Drewe JA, Garin-Bastuji B, Gonzales Rojas JL, Gortazar Schmidt C, Herskin M, Michel V, Miranda Chueca MA, Padalino B, Pasquali P, Roberts HC, Spoolder H, Stahl K, Velarde A, Viltrop A, Winckler C, Dewulf J, Guardabassi L, Hilbert F, Mader R, Baldinelli F, Alvarez J (2021) Assessment of animal diseases caused by bacteria resistant to antimicrobials: rabbits. EFSA J 19:e06999. https://doi.org/10.2903/j.efsa.2021.6999

El-Deep MH, Dawood MAO, Assar MH, Ahamad Paray B (2021) *Aspergillus awamori* positively impacts the growth performance, nutrient digestibility, antioxidative activity and immune responses of growing rabbits. Vet Med Sci 7:226–235. https://doi.org/10.1002/vms3.345

Finzi A, Macchioni P, Negretti P (2008) Rabbit health control by management. Proceedings of 9 world rabbit congress, Verona (Italy) (2008), pp 950–952

Flatt RE, Weisbroth SH, Kraus AL (1974) Metabolic, traumatic, mycotic and miscellaneous disease of rabbits. In: Weisbroth SH, Flatt RE, Kraus AL (eds) The biology of the laboratory rabbit, 1st edn. Academic Press, New York, pp 435–451, ISBN 0127421505. https://doi.org/10.1016/B978-0-12-742150-6.50022-0

Garreau H, Licois D, Rupp R, de Rochambeau H (2006) Genetic variability of the resistance to epizootic rabbit enteropathy (ERE): new results. Proceedings of the 8th world congress on genetics applied to livestock production, Belo Horizonte (Brazil) (2006), pp 15–28

Gidenne T, Combes S, Fortun-Lamothe L (2012) Feed intake limitation strategies for the growing rabbit: effect on feeding behaviour, welfare, performance, digestive physiology and health: a review. Animal 6:1407–1419. https://doi.org/10.1017/S1751731112000389

Gomez-Conde MS, Garcia J, Chamorro S, Eiras P, Rebollar PG, Perez de Rozas A, Badiola I, de Blas C, Carabaño R (2007) Neutral detergent-soluble fiber improves gut barrier function in 25 d old weaned rabbits. J Anim Sci 85:3313–3321

Haligur M, Ozmen O, Demir N (2009) Pathological and ultrastructural studies on mucoid enteropathy in New Zealand rabbits. J Exot Pet Med 18:224–228. https://doi.org/10.1053/j.jepm.2009.06.006

Harcourt-Brown F (2002) Digestive disorders. In: Harcourt-Brown F (ed) Textbook of rabbit medicine. Butterworth-Heinemann, Oxford, pp 249–292, ISBN 9780750640022. https://doi.org/10.1016/B978-075064002-2.50013-8

Harcourt-Brown F (2014) Digestive system disease. In: Meredith A, Lord B (eds) BSAVA manual of rabbit medicine. British Small Animal Veterinary Association, Gloucester, pp 168–190, ISBN 9781905319497. https://doi.org/10.22233/9781910443217.12

Harkness JE, Turner PV, VandeWounde S, Wheler CL (2010) Harkness and Wagner's biology and medicine of rabbits and rodents, 5th edn. Wiley Blackwell, Hoboken. ISBN 9780813815312

Hu B, Fan Z, Wei H, Song Y, Qiu R, Chen M, Xu W, Xue J, Wang F (2018) Detection of mucoid enteropathy syndrome disease in rabbit farms in East China. Res Vet Sci 119:259–261. https://doi.org/10.1016/j.rvsc.2018.07.011

Huybens N, Houeix J, Szalo M, Licois D, Mainil J, Marlier D (2008) Is epizootic rabbit enteropathy (ERE) a bacterial disease? Proceedings of the 9th World Rabbit Congress, Verona (Italy) (2008), pp 971–976

Huybens N, Houeix J, Licois D, Mainil J, Marlier D (2013) Pyrosequencing of epizootic rabbit enteropathy inocula and rabbit caecal samples. Vet J 196(1):109–110. https://doi.org/10.1016/j.tvjl.2012.08.014

Jin DX, Zou HW, Liu SQ, Wang LZ, Xue B, Wu D, Tian G, Cai J, Yan TH, Wang ZS, Peng QH (2018) The underlying microbial mechanism of epizootic rabbit enteropathy triggered by a low fiber diet. Sci Rep 8:12489. https://doi.org/10.1038/s41598-018-30178-2

Jobert JL, Eono F, Legall-Reculé G, Guittet M (2001) Experimental reproduction of an epizootic enterocolitis syndrome in healthy rabbits: a clinico-pathological and hematological study. World Rabbit Sci 9:24

Le Bouquin S, Jobert JL, Larour G, Balaine L, Eono F, Boucher S, Huneau A, Michel V (2009) Risk factors for an acute expression of epizootic rabbit enteropathy syndrome in rabbits after weaning in French kindling-to-finish farms. Livest Sci 125:283–290. https://doi.org/10.1016/j.livsci.2009.05.010

Licois D, Marlier D (2008) Pathologies infectieuses du lapin en élevage rationnel. INRAE Prod Anim 21:257–268. https://doi.org/10.20870/productions-animales.2008.21.3.3400

Licois D, Coudert P, Ceré N, Vautherot JF (2000) Epizootic enterocolitis of the rabbit: a review of current research. World Rabbit Sci 8(Suppl 1):187–194

Licois D, Wyers M, Coudert P (2005) Epizootic rabbit enteropathy: experimental transmission and clinical characterization. Vet Res 36:601–613. https://doi.org/10.1051/vetres:2005021

Licois D, Coudert P, Marlier D (2006) Epizootic rabbit enteropathy. In: Maertens L, Coudert P (eds) Recent advances in rabbit science. Institute of Agricultural and Fisheries Research (ILVO), Merelbeke, pp 63–170. ISBN 92-898-0030.EPS

Lopez-Santamarina A, Miranda JM, Mondragon ADC, Lamas A, Cardelle-Cobas A, Franco CM, Cepeda A (2020) Potential use of marine seaweeds as prebiotics: a review. Molecules 25:1004. https://doi.org/10.3390/molecules25041004

Mancini S, Paci G (2021) Probiotics in rabbit farming: growth performance, health status, and meat quality. Animals 11:3388. https://doi.org/10.3390/ani11123388

Marlier D, Dewrée R, Lassence C, Licois D, Mainil J, Coudert P, Meulemans L, Ducatelle R, Vindevogel H (2006) Infectious agents associated with epizootic rabbit enteropathy: isolation and attempts to reproduce the syndrome. Vet J 172:493–500. https://doi.org/10.1016/j.tvjl.2005.07.011

Martignon MH, Combes S, Gidenne T (2010) Digestive physiology and hindgut bacterial community of the young rabbit (*Oryctolagus cuniculus*): effects of age and short-term intake limitation. Comp Biochem Physiol Part A 156:156–162. https://doi.org/10.1016/j.cbpa.2010.01.017

Oglesbee BL, Lord B (2020) Gastrointestinal diseases of rabbits. In: Quesenberry KE, Orcutt CJ, Mans C, Carpenter JW (eds) Ferrets, rabbits, and rodents, 4th edn. W.B. Saunders, pp 174–187, ISBN: 978-0-323-48435-0. https://doi.org/10.1016/B978-0-323-48435-0.00014-9

Pérez de Rozas A, Carabaño R, García J, Rosell J, Diaz J, Barbe J, Pascual J, Badiola I (2005) Etiopatogenia de La EnteropatíaEpizoótica Del Conejo. Boletín de Cunicultura 139:167–174

Pérez C, Pérez A, Coscelli A, Azevedo AM, Prieto A, Díaz JM, Gullón J, Prieto C, Baña J, Sáiz A, Panadero R, Quiroga MI, Fernández G, Díaz-Baños P (2015) Identificación de infecciones digestivas en conejos: Orientaciones para optimizar la prevención y el control. Boletín de Cunicultura 176:27–32

Phuoc TL, Jamikorn U (2017) Effects of probiotic supplement (*Bacillus subtilis* and *lactobacillus acidophilus*) on feed efficiency, growth performance, and microbial population of weaning rabbits. Asian Australas J Anim Sci 30:198–205. https://doi.org/10.5713/ajas.15.0823

Puón-Peláez X-HD, McEwan NR, Olvera-Ramírez AM (2018) Epizootic Rabbit Enteropathy (ERE): a review of current knowledge. Eur Sci J 14:137. https://doi.org/10.19044/esj.2018.v14n36p137

Puón-Peláez X-HD, McEwan NR, Gómez-Soto JG, Álvarez-Martínez RC, Olvera-Ramírez AM (2020) Metataxonomic and histopathological study of rabbit epizootic enteropathy in Mexico. Animals 10:936. https://doi.org/10.3390/ani10060936

Puón-Peláez XD, McEwan NR, Álvarez-Martínez RC, Mariscal-Landín G, Nava-Morales GM, Mosqueda J, Olvera-Ramírez AM (2022) Effect of feeding insoluble fiber on the microbiota and metabolites of the caecum and feces of rabbits recovering from epizootic rabbit enteropathy relative to non-infected rabbits. Pathogens 11(5):571. https://doi.org/10.3390/pathogens11050571

Richardson VCG (ed) (2000) The digestive system. In: Rabbits: health, husbandry and diseases. Blackwell Science Ltd., Oxford, pp 81–107, ISBN 9780632052219. https://doi.org/10.1002/9780470693780.ch9

Szalo IM, Lassence C, Licois D, Coudert P, Poulipoulis A, Vindevogel H, Marlier D (2007) Fractionation of the reference inoculum of epizootic rabbit enteropathy in discontinuous sucrose gradient identifies aetiological agents in high density fractions. Vet J 173:652–657. https://doi.org/10.1016/j.tvjl.2005.12.013

Zhu KH, Xu XR, Sun DF, Tang JL, Zhang YK (2014) Effects of drinking water acidification by organic acidifier on growth performance, digestive enzyme activity and caecal bacteria in growing rabbits. Anim Feed Sci Technol 190:87–94. https://doi.org/10.1016/j.anifeedsci.2014.01.014

Escherichia coli in the Rabbit: Commensal or Pathogenic

Fábio Abade dos Santos, Teresa Tavares, Laurentina Pedroso, Ângela Dâmaso, and José Manuel Monteiro

Abstract

Colibacillosis is one of the most significant enteric diseases in rabbits, particularly in intensive production, and responsible for significant economic losses due to growing performance loss, mortality, poor growth performances, treatment, and veterinary costs. This book chapter provides an overview of the epidemiology, pathogenesis, and clinical signs of colibacillosis in rabbits. The disease primarily affects young rabbits, with two clinical courses observed depending on their age. The main risk factors associated with the disease are poor environmental control,

F. A. dos Santos (✉)
Faculty of Veterinary Medicine, Lusófona University, Lisbon, Portugal

CBIOS, Lusófona University, Lisbon, Portugal

National Agrarian and Veterinarian Research Institute (INIAV), Oeiras, Portugal
e-mail: fabio.santos@ulusofona.pt

T. Tavares
Veterinary and Animal Research Centre (CECAV), Associate Laboratory for Animal and Veterinary Sciences (AL4AnimalS), Department of Veterinary Sciences, School of Agricultural and Veterinary Sciences, University of Trás-os-Montes and Alto Douro (UTAD), Vila Real, Portugal

José Azevedo Monteiro, Lda., Vila Boa do Bispo, Portugal

L. Pedroso · Â. Dâmaso
Faculty of Veterinary Medicine, Lusófona University, Lisbon, Portugal

CBIOS, Lusófona University, Lisbon, Portugal

J. M. Monteiro
Faculty of Veterinary Medicine, Lusófona University, Lisbon, Portugal

CBIOS, Lusófona University, Lisbon, Portugal

José Azevedo Monteiro, Lda., Vila Boa do Bispo, Portugal

J. Simões, J. M. Monteiro (eds.), *Veterinary Care of Farm Rabbits*,
https://doi.org/10.1007/978-3-031-44542-2_22

hygiene (biosecurity), crowding and stress, poor nutrition (including inadequate or imbalanced diets), genetics, other concomitant diseases (e.g. intestinal or respiratory), and environmental factors, among others. The most significant serotypes in rabbits include O103, O15:H, O109:H2, O103:H2, O128, and O132. The main macroscopic lesions observed at necropsy consist of two main presentations: congested epithelium and oedema with fluid content. The attaching and effacing lesions on the small intestine are considered the hallmark of infection, which may be found with histopathological analysis. Clinical diagnosis can be achieved through information collection with the anamnesis, observation of clinical signs, and laboratorial analysis. This book chapter will serve as a useful resource for veterinarians and rabbit farmers in understanding the epidemiology, pathogenesis, and clinical signs of colibacillosis in rabbits.

Keywords

Colibacillosis · EPEC · Virulence factors · Rabbit production

1 Introduction

Colibacillosis is a common disease in rabbitries. There are pathogenic strains for kittens and fattening animals (Reusch 2005). Adult animals (above 3 months) are rarely affected by colibacillosis but may be the reservoir of the infection that perpetuate the infection in the rabbit farm. There are several non-pathogenic strains isolated from rabbit farms, which need a catalyst (an agent that opens the door) to give *Escherichia coli* the ability to cause morbidity and mortality. More than 50% of clinically relevant cases on rabbit production are related to digestive problems (Solans et al. 2019). Many of these predisposing factors are iatrogenic, stress, coccidiosis and other parasite infection, great temperature variations, sudden changes in diet formulation, and poor water quality, among others (Aboelhadid et al. 2021).

Whenever faecal samples from rabbits with diarrhea are sent to the laboratory, *E. coli* is present, often without detecting the *eae* gene. The fact that these *E. coli* strains have, usually, many antimicrobial resistances suggests that they are commensal bacteria of the cecal microbiota, and their multiplication and possible involvement in the disease were triggered by one or more risk factors. In these cases, the response to antimicrobials is often effective, and the animals recover quickly. Unfortunately, commercial laboratories do not typify *E. coli* strains by routine. The pathogenic strains of *E. coli* are less common. It is difficult to typify them in the laboratory, and the treatment is less effective or even ineffective. After controlling the risk factors, antimicrobial sensitivity should be evaluated by an antibiogram, and the producer must consider the possibility of a sanitary period according to all-in-all-out systems, which is already practised in the major part of industrial rabbitries. In practice, enteropathogenic *E. coli* (EPEC) in rabbits are the most relevant pathogenic strains affecting rabbitries (Freitas-Silva et al. 2018; Sakr et al. 2019).

EPEC was detected in 80% of the pathogenic enteric diseases of young rabbits with less than 15 days in Spain during 2018–2019 and was the cause of around 50% of enteric disease in preweaning and growing animals (Solans et al. 2019).

The emergence of mobile antibiotic resistance genes (e.g. colistin resistance gene mcr) is another challenge to animal health, particularly in the era of the reduction of antibiotics in food-producing animals. These genes can be disseminated in different countries and different environments (Freitas-Silva et al. 2018). Colistin is an example of an antibiotic that is no longer being used and is now considered a last-resort option for treating infections by the World Health Organization (WHO) for human medicine.

This chapter aims to highlight the diagnosis and prevention of pathologies associated with *E. coli* in industrial rabbits (namely, colibacillosis), not detailing specific microbiological aspects excessively when not relevant in clinical practice or animal production.

2 The Pathogen *Escherichia coli*: Structure, Reservoir, and Transmission

Escherichia, the type genus of the *Enterobacteriaceae* family, a potential zoonotic agent, predominant facultative anaerobe, are straight, gram-negative, rod-shaped bacilli (2.0–6.0 μm in length and 1.1–1.5 μm wide) with rounded ends (Bardiau et al. 2010; Quinn et al. 2003). *E. coli* is a natural inhabitant (commensal) of the large and lower small intestine of all warm-blooded mammals—"primary habitat"—and is excreted in feces and can survive in faecal particles, dust, and water for weeks or even months: "secondary habitat" (Markey et al. 2013; Moxley 2022). Although this agent can be an innocuous resident by colonization (Abade dos Santos et al. 2021) of the gastrointestinal tract, it also has the pathogenic capacity to cause significant diarrheal diseases (Croxen et al. 2013). The importance of reservoirs was demonstrated by the detection of EPEC in 33% of asymptomatic animals (Swennes et al. 2012). In infected does, subsequent litters may have passive immunity (Quesenberry and Carpenter 2012).

Rabbits can contract this pathogen, which is commonly transmitted through contaminated food and water (water- and food-borne pathogen), by ingesting it orally. When infected naturally, rabbits develop thrombotic microangiopathy, which is characterized by the presence of Shiga toxin and is thought to be responsible for the infrequent occurrence of acute renal failure (Panda et al. 2010). The severity of diarrhea in rabbits, which is caused by villus atrophy and malabsorption resulting from the infection, depends on the age of the rabbit and the particular serogroup of the pathogen involved and transmission to the environment and therefore to other animals (Quesenberry and Carpenter 2012).

EPEC strains can be found in rabbits, humans, bovines, dogs, cats, pigs, goats, and sheep among other non-domestic species (Krause et al. 2005). It is very associated with diarrhea, although in goats and sheep, EPEC still has not been associated with the same syndrome (Mainil 1999).

3 Pathogenesis

The disease due to *E. coli* in animals is called colibacillosis. This bacterium is currently classified according to its virulence factors, presenting different pathotypes (Nowland et al. 2015). Each pathotype of *E. coli* is associated with different general clinical syndromes: enteric, urinary, and sepsis/meningitis (Kaper et al. 2004).

Pathogenic strains of *E. coli* that cause gastrointestinal infections are categorized based on their virulence properties (Nataro and Kaper 1998). Officially, there are six enteric pathotypes recognized by the Centers for Disease Control and Prevention: enteropathogenic *E. coli* (EPEC), Shiga toxin (Stx)-producing *E. coli* (STEC, also known as enterohemorrhagic *E. coli* (EHEC) or verocytotoxin-producing *E. coli* (VTEC)), enterotoxigenic *E. coli* (ETEC), enteroaggregative *E. coli* (EAEC), enteroinvasive *E. coli* (EIEC), and diffusely adherent *E. coli* (DAEC) (CDC 2023). The EPEC and STEC pathotypes were already described as pathogenic in rabbits in addition to the necrotoxigenic *E. coli* (NTEC) (Cantey and Blake 1977; García et al. 2002; Clermont et al. 2011).

Of these pathotypes, EPEC and STEC are associated with natural diseases in rabbits (Cantey and Blake 1977; García et al. 2002). In addition, necrotoxigenic *E. coli* (NTEC) was also described as a cause of disease in rabbits (Blanco et al. 1996). Diarrheagenic *E. coli* is classified into six groups: enterotoxigenic *E. coli*, enteropathogenic *E. coli* (EPEC), enterohaemorrhagic *E. coli*, enteroinvasive *E. coli*, enteroaggregative *E. coli*, and diffusely adherent *E. coli* (Vilchez et al. 2009; Vendruscolo et al. 2017).

The capacity of *E. coli* to produce disease is associated with genes encoding virulence factors. The genes can be acquired by transduction, conjugation, or transformation, processes that lead to the development of pathogenic potential in the non-pathogenic strains (Moxley 2022).

EPEC is one of the main diarrheagenic *E. coli* pathotypes capable of developing colibacillosis in rabbits (REPEC), inducing a secretory diarrhea. This agent causes attaching and effacing (A/E) lesions with the classical definition of the characteristic intimate adherence and effacement of the intestinal microvilli. The chromosomal locus of enterocyte effacement (LEE) encodes the attaching and effacing virulence mechanisms (Elliott et al. 1998; Zhu et al. 2001). Apart from the LEE gene, rabbit EPEC isolates also contain colonization factors encoded by genes such as *afr2*, *ral*, *lifA*, and *paa*. However, the specific functions of these genes remain uncertain as some EPEC isolates lacking these rabbit-specific fimbrial adhesins have been identified. This suggests that there may be other potential adhesins present in these isolates that are yet to be discovered (Badagliacca et al. 2018).

The pathogenicity is directly associated with eae gene and adhesins/fimbriae (e.g. AF/R1, AF/R2, Ral) (Blanco et al. 1996; Fiederling et al. 1997; Boullier and Milon 2006). Typical strains carry the virulence factors intimin (eae gene) and bundle-forming pili (bfp gene), while atypical ones lack the last one (Pearson et al. 2016). Among the EPEC pathovar, the expression of O and H antigens defines the different serovar (Boullier and Milon 2006). For example, between 1990 and 2000, the epidemiological reports in France and Spain reported a predominance of O103: K-:H2 rhamnose-negative *E. coli* (Leroy et al. 1994; Blanco et al. 1996), and in Italy, the O103 and O2 were found in more than 50% of enteritis in the first years of this century (D'Incau et al. 2004).

The pathogenic process starts with the loss of adhesion mediated by adhesins, fitting the bacteria onto the epithelium. Then, several genes of bacterial virulence are activated—bacteria closely attached to the cell—also mediated by the intimin (an adhesin) (Jerse et al. 1991). At this moment, the enterocyte loses the microvilli and intimate adherence (Knutton et al. 1989; Finlay et al. 1992). Several virulence mechanisms and molecules are involved from here, not being explored here exhaustively because it is not the scope of this book. Diarrhea results from complex mechanisms: an increase in intestinal permeability and inflammation, loss of absorptive area, and active ion secretion (Boullier and Milon 2006).

The immune response to colonization by EPEC is not well studied. In another way, EPEC showed important immunomodulatory properties interfering with IL-2, IL-3, and IL-5 and also with IFN-γ (Klapproth et al. 1995). The main host immune response appears to be associated with mucosal immunity at the intestinal level—IgA (Boullier et al. 2003)—and is able to be acquired after vaccination, from the infection, or by ingestion of breast milk.

4 Epidemiology

A large part of the pathological condition present in rabbit farms is due to diseases of the digestive tract, varying from 30% to more than 50% of emergency veterinary interventions. Epizootic rabbit enteropathy (ERE) and enteritis-diarrhea were the most prevalent diseases in the last years in Spain (Rossel 2021a).

Colibacillosis affects mostly lactating young rabbits (does excrete EPEC in the first 10 days after birth) and weaned rabbits. So, two clinical courses can be seen in rabbits, depending on their age (Mayer 2022):

(a) **Rabbits 1–2 weeks old (colibacillosis strains of infants (lactating young rabbits))**—Severe yellowish diarrhea that results in high mortality with, habitually, entire litters succumbing. At this age, they may also present haemorrhagic lesions and hyper-acute mortality. It is possible to find a stomach full of milk. The great mortality rate happens among the biggest and healthiest rabbits.

(b) **Rabbits with 4–6 weeks old (fattening or adult colibacillosis—weaned rabbits)**—Rabbits with 3 weeks have acute diarrhea, haemorrhagic lesions in the cecum and small intestine, and acute mortality. Rabbits after weaning, with more than 5 weeks of life, present acute diarrhea and mortality peaks and do not present haemorrhagic lesions in the cecum or intestine. There is oedema and the mesenteric ganglion is enlarged.

The main risk factors associated with colibacilosis are poor environmental control and lack of drinking water treatment (excessive water treatment may also be harmful). The disease is a major enteric disease in Europe, responsible for significant economic losses (Boullier and Milon 2006), being less important in the United States (Lukefahr et al. 2022). The costs include growing performance loss, mortality, treatment, and veterinary costs. Despite the growing knowledge, the disease continues to be a huge problem and may increase with the reduction of the use of antimicrobials.

Swennes et al. (2013) observed that 4.3% (6/141) of the commercially-acquired laboratory rabbits were EPEC carriers. This finding that may let us infer that this microorganism is present at low levels in commercially acquired rabbits (Swennes et al. 2013), which may be one of the main sources of the infection.

Serotypes O103, O15:H, O109:H2, O103:H2, O128, and O132 are the most significant in rabbits (Mayer 2022). A large field study from Belgium and the Netherlands published in 1988 (Peeters et al. 1988) reported the serogroup 1+/O109 as mainly pathogenic to suckling rabbits, and the serogroups 3-/O15, 4+/O26, and 8+O103 as main pathogenic and 2+/O128 and 2+/O132 as less pathogenic serogroups to weaned rabbits. In general, in Europe outbreaks, O-103 strains are the most frequently detected (Agnoletti et al. 2006; Boullier and Milon 2006). The pathogenicity behaviour of the strains requires the evaluation of phenotypic trait (O-biotype) and genetic traits (e.g., EEAE, AF/R1, AF/R2) (Agnoletti 2012). Some EPEC with rabbit origin have zoonotic potential, e.g. EPEC O145:H2 (Swennes et al. 2013).

Strains with low pathogenicity are a major problem in rabbitries with low biosecurity measures and poor animal welfare conditions, while the highly pathogenic strains are a problem in any rabbitry and often whole litters can die (Peeters et al. 1988).

5 Clinical Signs and Lesions

Clinical diagnosis can be achieved by information collection with the anamnesis (age, history, productive results, mortality rates) and through observation of clinical signs. Young rabbits with 21–42 days of age affected by colibacillosis show fever, secretory diarrhea, and anorexia being the weaning stress a main factor. Death may occur in 5–14 days (Mayer 2022). When surviving, the animals show growth delay. The disease is associated with proliferation of REPEC in the distal ileum, cecum, and proximal colon (Boullier and Milon 2006).

When talking about REPEC infections, the main macroscopic lesions observed at necropsy consist with congested epithelium and oedema with fluid content (Fig. 1). Hepatic necrosis may occurs and rabbits die promptly impeding the treatment (Rosell 2000a, 2000b; Swennes et al. 2012; Lukefahr et al. 2022). These haemorrhagic lesions appear more characteristically in infant rabbits (lactating young rabbits), normally when rabbits start to eat, from 18 to 25 days of life (Fig. 2).

Fig. 1 Lactating rabbits (16 days old). Animals with good body condition, with stomach full of milk. The entire intestinal tract is fluid-filled with congestion

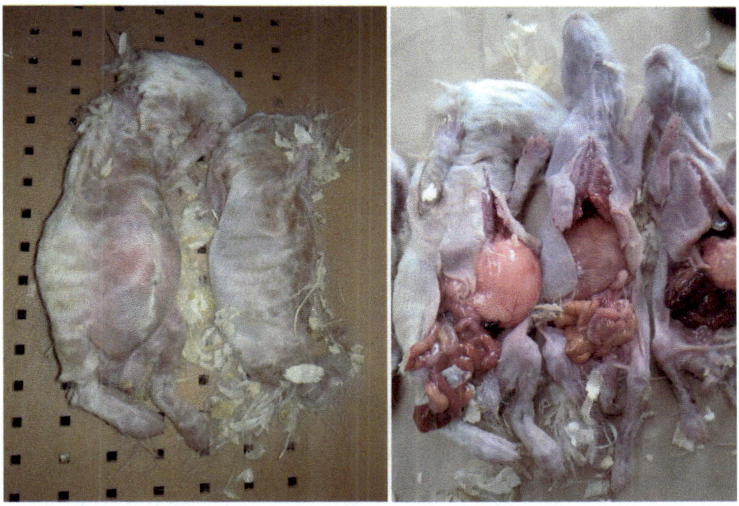

Fig. 2 Lactating rabbits, 4 days of age. Different lesional aspects of colibacillosis, with and without haemorrhagic/congestive lesions, in rabbits with yellowish diarrhea

After weaning, the most common injuries are very watery and brownish diarrhea, and it's possible to find many dirty rabbits in the affected cages. At necropsy, they usually don't present haemorrhagic lesions in the cecum, small intestine, or colon. Serosal oedema and enlargement of the mesenteric ganglion may exist (Fig. 3).

The attaching and effacing (A/E) lesions on the small intestine are considered the hallmark of infection (Cepeda-Molero et al. 2017), which may be found with histopathological analysis. Microscopically, *villi* injury is observed with

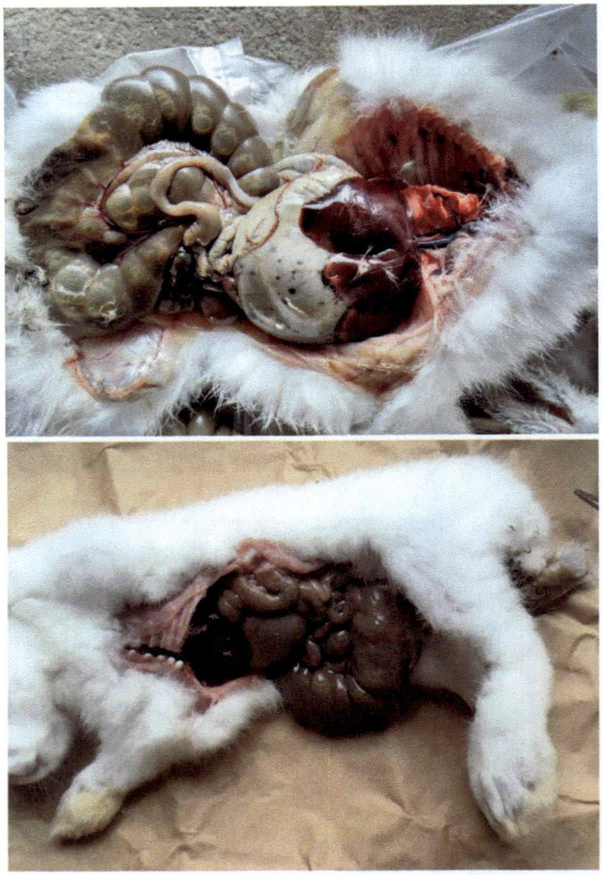

Fig. 3 Fattening rabbit (40 days old). (**a**) Fluid cecal content, without haemorrhagic or congestive lesions. (**b**) Fluid cecal content. Dirtiness of the legs and tail with diarrheic feces

haemorrhages on the cecal epithelium (Peeters et al. 1984a). The gross and microscopic pattern is associated with the serovar (Peeters et al. 1988). For example, O109:K-:H2 serovar leads to yellow diarrhea in suckling rabbits (3–12 days old) with high mortality within 20–48 hours (Peeters et al. 1984b) with adults being healthy carriers.

As a rule, in field clinical cases, *E. coli* is not typified at laboratory, and the most common colibacilloses are infections with non-pathogenic *E. coli* or which became pathogenic through the genetic modifying mechanisms that permit the virulence factors to appear.

6 Laboratory Diagnosis

Differential diagnosis with other intestinal diseases such as coccidiosis and clostridiosis is mandatory. Definitive diagnosis is important because an inadequate treatment may result in aggravation of dysbiosis, with even high mortality rates.

Sections of the ileum should be sent to a laboratory for histopathological examination. In the case of EPEC, it is expectable to detect the presence of A/E bacteria and the characteristic distortion of the *microvilli* and the effacement of the mucosal surface (Markey et al. 2013), but will not result in definitive identification of the pathogen. For that purpose, faecal isolates may be obtained from a more general-purpose selective medium for *E. coli* (e.g. MacConkey agar), and these can be tested by PCR.

Leukogram and specifically the heterophil to lymphocyte (H/L) ratio and phagocytic activity (using a nitroblue tetrazolium (NBT) heterophil slide adhesion test) can be used during clinical diagnosis although not specific (Panda et al. 2010). Biotype can be determined by the ability to ferment cellobiose, dulcitol, raffinose, rhamnose, sorbose, and sucrose (Peeters et al. 1988; Camguilhem and Milon 1989).

Ideally, the final diagnosis is carried out through microbiological culture (Fig. 4) and identification of the agent, the pathotype, and the serovar.

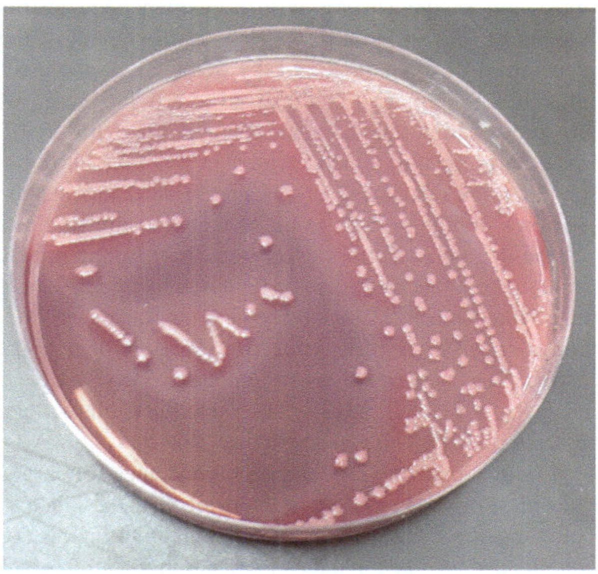

Fig. 4 Culture of *Escherichia coli* on MacConkey agar for diagnosis

7 Treatment and Prevention

The treatment of colibacillosis is almost virtually impossible, due to the difficult balance between pathogenic bacteria and the animal's normal microbial population, and without which, the animal cannot live. On the other hand, *E. coli* strains frequently acquire multiple resistances during the treatment (Boullier and Milon 2006)—conjugative plasmid-borne resistances (Blanco et al. 1994).

EPEC is most often resistant to ampicillin, streptomycin, tetracycline, and triple sulfa or its components (Nisa et al. 2013). Fluoroquinolones are generally the most effective (Pisoni et al. 2004).

The combination of enrofloxacin with colistin was quite effective in the past. Nowadays, only one antibiotic molecule can be used at a time, and with the categorization of antibiotics, both of the antibiotics mentioned can only be used after isolation of the agent with an antibiogram in which the only molecule to which that bacterium is sensitive is one of these antimicrobials. So, in practice, fluoroquinolones or colistin is not already being used. Neomycin or apramycin has shown good results in rabbitries in the treatment of this disease.

Prevention, essentially through biosecurity and animal welfare practices, is the fastest, cheapest, and simplest way to fight this and other diseases. Other factors should be taken into account such as housing, genetics, feeding, and general management practices. Digestive processes are often multifactorial. One of the most important measures is to implement feed restriction strategies for fattening rabbits or, once it's already being practised, adjust the feeding programme to the necessities of the animal and the treatment of the disease. It is about the adequate quantity of compound feed needed to provide the growing preconized by genetics.

Another important measure refers to the exposure of the doe, by placing her into the environment in which parturition will take place or by vaccinating the doe with preparations containing the antigenic determinants of *E. coli*, so that the female acquires immunity, as a result of the exposure to the antigenic determinants of the various virulence factors expressed on or by the infecting strains. The non-specific immune resistance can also be favoured by microbiota barriers—probiotics (Guarner and Schaafsma 1998)—namely with the most used *Bacillus cereus* (e.g., var. toyoi) and *Saccharomyces cerevisiae*. Some authors also reported positive effects of *Rumex crispus* L., *Potentilla anserina*, and *Polygonum aviculare* in the treatment (Kupczyński et al. 2019). Oils, acidifiers, clays, beta-galactomannan, and other prebiotics can be also used (Rossel 2021b).

The use of complete inactivated vaccines (often herd vaccines), live attenuated vaccines, or more recently subunit or virulence factor vaccines will be the future, allowing the disuse of antibiotics and minimizing costs and residues in meat. It should also be noted that the colonization of animals by bacteria, namely, pathogenic *E. coli* or other agents capable of suppressing the immune system, can lead to failures in vaccination against other diseases, including rabbit haemorrhagic disease (Eladl et al. 2019). Key points in control and treatment are reported in Box 1.

Box 1 Key Points in Control and Treatment

- Trying to solve the iatrogenic cause(s) or predisposing factor(s), when that's the case (e.g. nutritional, environmental).
- Trying to improve animals' environmental comfort, rising temperature, and reducing relative humidity in the shed.
- Ensure water permanent disinfection with appropriate products in appropriate concentrations.
- Separate sick animals, locating them at the end of the shed, in the cages near the extractors.
- Hand washing and disinfection during animal handling, especially after touching affected animals.
- In affected nests, it is necessary to remove dirty nest material and substitute for new material (dry and clean).
- Using dry powder (preferably with natural disinfectants) in nest material.
- Treating does with an appropriate antimicrobial molecule (antibiogram), to avoid future infection/affection of fattening groups (not fully effective; some animals remain silent carriers).
- It is possible to check these carriers, using rectal swabs (virulent strain infections).
- Testing new breeders when they arrive to the farm may also be a good practice while in the quarantine room.
- Limit adoptions (e.g. during the same module), during the actual and following groups.
- Limit group calibration.
- After emptying the fattening shed, try to reinforce cleaning and disinfection procedures (using foaming detergent and hot water to wash, had new disinfectant substances to the programme, fumigating).
- In cases of EPEC infections that affect suckling rabbits, vaccination of does with an autogenous vaccine may be vitally important.
- Evaluate carefully the presence of concomitant infections (e.g. coccidiosis), which may be the cause of the colibacillosis (when talking about nonvirulent strain infections).
- Delay productive rhythm may help animals to become more immunocompetent in relation to this disease and may be part of the strategy to eliminate it.
- As a last resource, in cases of very virulent strain infection, sanitary emptying of the farm may be the only choice.
- Choosing genetics with higher resistance to disease.
- Breeders should be provided by companies that guarantee exemption of these pathogens.

8 Conclusions

E. coli infections in rabbits are a huge problem, with potentially fatal consequences in addition to the economic and financial costs. Laboratory diagnosis is crucial for identifying the specific strain of *E. coli* causing the infection and guiding appropriate treatment. However, the typification of the strains is still little available to the veterinarian/breeder and is still mainly used in scientific research.

It is essential to manage rabbits' environment and diet to prevent the spread of infection. Treatment for *E. coli* infections in rabbits typically involves the administration of antibiotics, but it is important to note that not all strains of *E. coli* are susceptible to the same antibiotics. The major part of the occurrences needs an immediate treatment, based on a presumptive diagnosis achieved through clinical signs and lesions observed. Antimicrobial susceptibility testing may be also useful to ensure future appropriate treatment if the problem happens again in the next production cycle. Unhappily, there are a very limited number of molecules available for use in rabbits to treat colibacillosis, a fact that makes the choice easier but sometimes limits the success of the treatment. Supportive care, such as fluid therapy and nutritional support, may also be necessary to manage the symptoms and complications of the infection.

Prevention is critical to control the spread of *E. coli* in rabbit populations. Proper sanitation and hygiene practices should be followed in rabbit housing and handling, and appropriate measures should be taken to prevent contamination of food and water sources. Vaccination may also be an effective preventive measure in high-risk population.

In conclusion, *E. coli* infections in rabbits can have serious consequences, and prompt diagnosis and treatment are essential for successful management. Proper prevention measures, including good sanitation and hygiene practices and vaccination, are also critical to control the spread of infection and avoid occurrences in future groups.

References

Abade dos Santos FA, Portela SJ, Nogueira T, Carvalho CL, de Sousa R, Duarte MD (2021) Harmless or threatening? Interpreting the results of molecular diagnosis in the context of virus-host relationships. Front Microbiol 12:1–8. https://doi.org/10.3389/fmicb.2021.647730

Aboelhadid SM, Hashem S, Abdel-Kafy E-S, Mahrous LN, Farghly EM, Abdel-Baki A-AS, Al-Quraishy S, Kamel AA (2021) Prebiotic supplementation effect on Escherichia coli and Salmonella species associated with experimentally induced intestinal coccidiosis in rabbits. PeerJ 9:e10714. https://doi.org/10.7717/peerj.10714

Agnoletti F (2012) Update on rabbit enteric diseases: despite improved diagnostic capacity, where does disease control and prevention stand? In Proceedings 10th World Rabbit Congress, September 3–6, pp 1113–1127

Agnoletti F, Bano L, Cocchi M, Mazzolini E (2006) Aggiornamenti sulle enteriti batteriche del coniglio. A Riv Zoot Vet 34(1):29–38

Badagliacca P, Agnoletti F, Drigo I, Merildi V, Lopes F, Pompilii C, Mangone I, Scacchia M, Tonelli A, Masson L (2018) Virulence gene profiles of rabbit enteropathogenic Escherichia coli strains isolated in Northern Italy. Vet Ital 54(3):189–196. https://doi.org/10.12834/VetIt.859.4260.2

Bardiau M, Szalo M, Mainil JG (2010) Initial adherence of EPEC, EHEC and VTEC to host cells. Vet Res 41(5):57. https://doi.org/10.1051/vetres/2010029

Blanco JE, Blanco M, Blanco J, Rioja L, Ducha J (1994) Serotypes, toxins and antibiotic resistance of Escherichia coli strains isolated from diarrhoeic and healthy rabbits in Spain. Vet Microbiol 38(3):193–201. https://doi.org/10.1016/0378-1135(94)90001-9

Blanco JE, Blanco M, Blanco J, Mora A, Balaguer L, Mouriño M, Juarez A, Jansen WH (1996) O serogroups, biotypes, and eae genes in Escherichia coli strains isolated from diarrheic and healthy rabbits. J Clin Microbiol 34(12):3101–3107. https://doi.org/10.1128/jcm.34.12.3101-3107.1996

Boullier S, Milon A (2006) Rabbit colibacillosis. In: Maertens L, Coudert P (eds) Recent advances in rabbit sciences. ILVO, Merelbeke, pp 171–179

Boullier S, Nougayrède J-P, Marchès O, Tasca C, Boury M, Oswald E, De Rycke J, Milon A (2003) Genetically engineered enteropathogenic Escherichia coli strain elicits a specific immune response and protects against a virulent challenge. Microbes Infect 5(10):857–867. https://doi.org/10.1016/S1286-4579(03)00175-8

Camguilhem R, Milon A (1989) Biotypes and O serogroups of Escherichia coli involved in intestinal infections of weaned rabbits: clues to diagnosis of pathogenic strains. J Clin Microbiol 27(4):743–747. https://doi.org/10.1128/jcm.27.4.743-747.1989

Cantey JR, Blake RK (1977) Diarrhea due to Escherichia coli in the rabbit: a novel mechanism. J Infect Dis 135(3):454–462. https://doi.org/10.1093/infdis/135.3.454

CDC (2023) *E.coli* webpage. https://www.cdc.gov/ecoli/general/index.html

Cepeda-Molero M, Berger CN, Walsham ADS, Ellis SJ, Wemyss-Holden S, Schüller S, Frankel G, Fernández LÁ (2017) Attaching and effacing (A/E) lesion formation by enteropathogenic E. coli on human intestinal mucosa is dependent on non-LEE effectors. PLoS Pathog 13(10):e1006706. https://doi.org/10.1371/journal.ppat.1006706

Clermont O, Olier M, Hoede C, Diancourt L, Brisse S, Keroudean M, Glodt J, Picard B, Oswald E, Denamur E (2011) Animal and human pathogenic Escherichia coli strains share common genetic backgrounds. Infect Genet Evol 11(3):654–662. https://doi.org/10.1016/j.meegid.2011.02.005

Croxen MA, Law RJ, Scholz R, Keeney KM, Wlodarska M, Finlay BB (2013) Recent advances in understanding enteric pathogenic Escherichia coli. Clin Microbiol Rev 26(4):822–880. https://doi.org/10.1128/CMR.00022-13

D'Incau M, Pennelli D, Pacciarini M, Maccabiani G, Lavazza A, Tagliabue S (2004) Characterization of E. coli strains isolated from rabbits with enteritis in Lombardia and Emilia Romagna (North Italy) during the period 2000–2003. http://www.Dcam.Upv.Es/8wrc/ (Ed.), Proc. 8th World Rabbit Congress In: Mexico, pp 526–531

Eladl AH, Farag VM, El-Shafei RA, Elkenany RM, Elsayed MM, Mona MM, Ali HS, Saif MA (2019) Effect of colibacillosis on the immune response to a rabbit viral haemorrhagic disease vaccine. Vet Microbiol 238:108429. https://doi.org/10.1016/j.vetmic.2019.108429

Elliott SJ, Wainwright LA, McDaniel TK, Jarvis KG, Deng YK, Lai LC, McNamara BP, Donnenberg MS, Kaper JB (1998) The complete sequence of the locus of enterocyte effacement (LEE) from enteropathogenic Escherichia coli E2348/69. Mol Microbiol 28(1):1–4. https://doi.org/10.1046/j.1365-2958.1998.00783.x

Fiederling F, Boury M, Petit C, Milon A (1997) Adhesive factor/rabbit 2, a new fimbrial adhesin and a virulence factor from Escherichia coli O103, a serogroup enteropathogenic for rabbits. Infect Immun 65(2):847–851. https://doi.org/10.1128/iai.65.2.847-851.1997

Finlay BB, Rosenshine I, Donnenberg MS, Kaper JB (1992) Cytoskeletal composition of attaching and effacing lesions associated with enteropathogenic Escherichia coli adherence to HeLa cells. Infect Immun 60(6):2541–2543. https://doi.org/10.1128/iai.60.6.2541-2543.1992

Freitas-Silva J, Inácio ÂS, Mourão J, Antunes P, Mendes Â, de Carvalho AP, Vasconcelos V, Peixe L, da Costa PM (2018) Occurrence of mcr-1 in Escherichia coli from rabbits of intensive farming. Vet Microbiol 227:78–81. https://doi.org/10.1016/j.vetmic.2018.10.020

García A, Marini RP, Feng Y, Vitsky A, Knox KA, Taylor NS, Schauer DB, Fox JG (2002) A naturally occurring rabbit model of enterohemorrhagic Escherichia coli-induced disease. J Infect Dis 186(11):1682–1686. https://doi.org/10.1086/345371

Guarner F, Schaafsma GJ (1998) Probiotics. Int J Food Microbiol 39(3):237–238. https://doi.org/10.1016/s0168-1605(97)00136-0

Jerse AE, Gicquelais KG, Kaper JB (1991) Plasmid and chromosomal elements involved in the pathogenesis of attaching and effacing Escherichia coli. Infect Immun 59(11):3869–3875. https://doi.org/10.1128/iai.59.11.3869-3875.1991

Kaper JB, Nataro JP, Mobley HLT (2004) Pathogenic Escherichia coli. Nat Rev Microbiol 2(2): 123–140. https://doi.org/10.1038/nrmicro818

Klapproth JM, Donnenberg MS, Abraham JM, Mobley HL, James SP (1995) Products of enteropathogenic Escherichia coli inhibit lymphocyte activation and lymphokine production. Infect Immun 63(6):2248–2254. https://doi.org/10.1128/iai.63.6.2248-2254.1995

Knutton S, Baldwin T, Williams PH, McNeish AS (1989) Actin accumulation at sites of bacterial adhesion to tissue culture cells: basis of a new diagnostic test for enteropathogenic and enterohemorrhagic Escherichia coli. Infect Immun 57(4):1290–1298. https://doi.org/10.1128/iai.57.4.1290-1298.1989

Krause G, Zimmermann S, Beutin L (2005) Investigation of domestic animals and pets as a reservoir for intimin- (eae) gene positive Escherichia coli types. Vet Microbiol 106(1–2): 87–95. https://doi.org/10.1016/j.vetmic.2004.11.012

Kupczyński R, Szumny A, Bednarski M, Piasecki T, Śpitalniak-Bajerska K, Roman A (2019) Application of Pontentilla anserine, Polygonum aviculare and Rumex crispus mixture extracts in a rabbit model with experimentally induced E. coli infection. Animals 9(10):774. https://doi.org/10.3390/ani9100774

Leroy SM, Lesage MC, Chaslus-Dancla E, Lafont JP (1994) Presence of eaeA sequences in pathogenic and nonpathogenic Escherichia coli strains isolated from weaned rabbits. J Med Microbiol 40(2):90–94. https://doi.org/10.1099/00222615-40-2-90

Lukefahr SD, McNitt JI, Cheeke PR, Patton NM (2022) *Rabbit production* (10th Editi). CABI

Mainil J (1999) Shiga/verocytotoxins and Shiga/verotoxigenic Escherichia coli in animals. Vet Res 30(2–3):235–257

Markey B, Leonard F, Archambault M, Cullinane A, Maguire D (2013) Chapter 17: Enterobacteriaceae. In Clinical Veterinary Microbiology, 2nd Editio, pp 245–255

Mayer J (2022) Bacterial and mycotic diseases of rabbits. MSD Veterinary Manual. https://www.msdvetmanual.com/exotic-and-laboratory-animals/rabbits/bacterial-and-mycotic-diseases-of-rabbits

Moxley RA (2022) Enterobacteriaceae. In: Veterinary Microbiology. John Wiley & Sons, Ltd, pp 56–74. https://doi.org/10.1002/9781119650836.ch6

Nataro JP, Kaper JB (1998) Diarrheagenic Escherichia coli. Clin Microbiol Rev 11(1):142–201. https://doi.org/10.1128/CMR.11.1.142

Nisa S, Scanlon KM, Donnenberg MS (2013) Chapter 4: Enteropathogenic *Escherichia coli*. In: Donnenberg MS (ed.) *Escherichia coli*, 2nd edition, Academic Press, pp 75-119. ISBN 9780123970480. https://doi.org/10.1016/B978-0-12-397048-0.00004-8

Nowland MH, Brammer DW, Garcia A, Rush HG (2015) Biology and diseases of rabbits. Lab Anim Med 2015:411–461. https://doi.org/10.1016/B978-0-12-409527-4.00010-9

Panda A, Tatarov I, Melton-Celsa AR, Kolappaswamy K, Kriel EH, Petkov D, Coksaygan T, Livio S, McLeod CG, Nataro JP (2010) Escherichia coli O157:H7 infection in Dutch belted and New Zealand white rabbits. Comp Med 60(1):31–37

Pearson JS, Giogha C, Lung TWF, Hartland EL (2016) The genetics of enteropathogenic Escherichia coli virulence. Annu Rev Genet 50:493-513. https://doi.org/10.1146/annurev-genet-120215-035138

Peeters JE, Charlier GJ, Halen PH (1984a) Pathogenicity of attaching effacing enteropathogenic Escherichia coli isolated from diarrheic suckling and weanling rabbits for newborn rabbits. Infect Immun 46(3):690–696. https://doi.org/10.1128/iai.46.3.690-696.1984

Peeters JE, Geeroms R, Glorieux B (1984b) Experimental Escherichia coli enteropathy in weanling rabbits: clinical manifestations and pathological findings. J Comp Pathol 94(4):521–528. https://doi.org/10.1016/0021-9975(84)90056-2

Peeters JE, Geeroms R, Orskov F (1988) Biotype, serotype, and pathogenicity of attaching and effacing enteropathogenic Escherichia coli strains isolated from diarrheic commercial rabbits. Infect Immun 56(6):1442–1448. https://doi.org/10.1128/iai.56.6.1442-1448.1988

Pisoni AM, Piccirillo A, Gallazzi D, Agnoletti F, Grilli G (2004) Biotype and susceptibility to antimicrobial agents of rabbit. In Proceedings – 8th World Rabbit Congress – September 7–10, pp 608–613

Quesenberry KE, Carpenter JW (2012) Ferrets, rabbits and rodents. Clinical Medicine and Surgery, 3rd ed. https://doi.org/10.1016/C2009-0-60282-8

Quinn P, Markey B, Carter M, Donnelly W, Leonard F (2003) Veterinary microbiology and microbial disease. Canadian Vet J 44(12):986

Reusch B (2005) Rabbit gastroenterology. Vet Clin North Am Exot Anim Pract 8(2):351–375. https://doi.org/10.1016/j.cvex.2005.01.007

Rosell M (2000a) Enfermedades del conejo. Tomo II Ediciones Mundi Prensa, pp 168–169, 360–370

Rosell M (2000b) Enfermedades del conejo. Tomo I Ediciones Mundi-Prensa, pp 41–263

Rossel JM (2021a) Situation in rabbit farms [mainly in Spain]. In Proceedings 12th World Rabbit Congress, November 3–5

Rossel JM (2021b) The rabbit farming in the post-antibiotic era: a challenge that can be won? In Proceedings 12th World Rabbit Congress, November 3–5. http://world-rabbit-science.com/WRSA-Proceedings/Congress-2021-Nantes/Videos/Round-Table_1.mp4

Sakr M, Ebied S, Ibrahim M (2019) Detection of diarrheagenic Escherichia coli in rabbits. Alexandria Journal of Veterinary Sciences 62(1):137. https://doi.org/10.5455/ajvs.45817

Solans L, Arnal JL, Sanz C, Benito A, Chacón G, Alzuguren O, Fernández AB (2019) Rabbit enteropathies on commercial farms in the Iberian Peninsula: etiological agents identified in 2018-2019. Animals 9(12). https://doi.org/10.3390/ani9121142

Swennes AG, Buckley EM, Parry NMA, Madden CM, García A, Morgan PB, Astrofsky KM, Fox JG (2012) Enzootic enteropathogenic *Escherichia coli* infection in laboratory rabbits. J Clin Microbiol 50(7):2353–2358. https://doi.org/10.1128/JCM.00832-12

Swennes AG, Buckley EM, Madden CM, Byrd CP, Donocoff RS, Rodriguez L, Parry NM, Fox JG (2013) Enteropathogenic *Escherichia coli* prevalence in laboratory rabbits. Vet Microbiol 163(3–4):395–398. https://doi.org/10.1016/j.vetmic.2013.01.006

Vendruscolo JW, Waldrich TL, Saikawa GIA, Pelayo JS, Kobayashi RKT, Nakazato G, Rocha SPD (2017) Novel multiplex PCR for detection of diarrheagenic Escherichia coli strains isolated from stool and water samples. Genetics and Molecular Research: GMR 16(3). https://doi.org/10.4238/gmr16039760

Vilchez S, Reyes D, Paniagua M, Bucardo F, Möllby R, Weintraub A (2009) Prevalence of diarrhoeagenic Escherichia coli in children from León, Nicaragua. J Med Microbiol 58(Pt 5):630–637. https://doi.org/10.1099/jmm.0.007369-0

Zhu C, Agin TS, Elliott SJ, Johnson LA, Thate TE, Kaper JB, Boedeker EC (2001) Complete nucleotide sequence and analysis of the locus of enterocyte effacement from rabbit diarrheagenic Escherichia coli RDEC-1. Infect Immun 69(4):2107–2115. https://doi.org/10.1128/IAI.69.4.2107-2115.2001

Clostridium spp. in Rabbits: The Underappreciated Killer

Teresa Tavares, Ângela Dâmaso, Laurentina Pedroso, Maria da Conceição Fontes, Fábio Abade dos Santos, João Simões, and José Manuel Monteiro

Abstract

Digestive disorders are the most frequent cause of veterinarian visits to rabbit industrial farms. Despite multiple agent infections in rabbits, enterotoxemia is one of the most relevant. Enterotoxemia is the designation given to the infectious disease caused by some *Clostridium* species and the toxins they produce. *Clostridium spiroforme* is a highly prevalent pathogen responsible for rabbit enteric disease and is frequently isolated in pre-weaning and growing rabbits. Contrary to other *Clostridium* species, such as *C. perfringens* type A and type E, which are part of the normal rabbit microbiota, *C. spiroforme* is usually considered a

T. Tavares
Veterinary and Animal Research Centre (CECAV), Associate Laboratory for Animal and Veterinary Sciences (AL4AnimalS), Department of Veterinary Sciences, School of Agricultural and Veterinary Sciences, University of Trás-os-Montes and Alto Douro (UTAD), Vila Real, Portugal

José Azevedo Monteiro, Lda., Vila Boa do Bispo, Portugal

Â. Dâmaso · L. Pedroso · F. A. dos Santos
Faculty of Veterinary Medicine, Lusófona University, Lisbon, Portugal

CBIOS, Lusófona University, Lisbon, Portugal

M. da Conceição Fontes · J. Simões
Veterinary and Animal Research Centre (CECAV), Associate Laboratory for Animal and Veterinary Sciences (AL4AnimalS), Department of Veterinary Sciences, School of Agricultural and Veterinary Sciences, University of Trás-os-Montes and Alto Douro (UTAD), Vila Real, Portugal

J. M. Monteiro (✉)
José Azevedo Monteiro, Lda., Vila Boa do Bispo, Portugal

Faculty of Veterinary Medicine, Lusófona University, Lisbon, Portugal

CBIOS, Lusófona University, Lisbon, Portugal

J. Simões, J. M. Monteiro (eds.), *Veterinary Care of Farm Rabbits*,
https://doi.org/10.1007/978-3-031-44542-2_23

stranger bacterium. Predisposing factors that lead to rabbit stress are the triggers of disease development. Enterotoxemia treatment is often very difficult. In many cases, antimicrobial treatment is ineffective or even deteriorates the health status of the rabbit and complicates the evolution of the disease. The control of risk factors assumes crucial importance in preventing the disease. Feed restriction strategies, environmental comfort, and good biosecurity practices improve gut health and rabbit welfare. This chapter addresses the main features of enterotoxemia in rabbits and rabbitries and highlights the infection by *C. spiroforme*.

Keywords

Clostridium spiroforme · Enterotoxemia · Dysbiosis · Typhlitis · Enterotyphlitis · Risk factors

1 Introduction

The digestive microbiota plays an essential role in rabbit health and performance, not only because of the direct effects of its balance on gastrointestinal health but also because these microorganisms are very important nutrient producers and suppliers for the rabbit, producing amino acids, complex B vitamins, and some essential fatty acids in the cecum. Intestinal health is also crucial for the function of the immune response and protection against pathogens. The vital role of this fermentation chamber, the cecum and its microbiota, makes rabbit an animal with specific nutritional, environmental, and management-related needs that, when unsatisfied, may lead to cecal dysbiosis, often followed by clostridial proliferation and enterotoxemia. A large part of the enterotoxemic presentations observed in rabbits originates from cecal dysbiosis.

Clostridium is one of the most relevant genera normally identified in cecal microbiota (Velasco-Galilea et al. 2018). Under certain gut conditions, *Clostridium* spp. quickly grow, causing enteric diseases in rabbits (Vántus et al. 2014). Nevertheless, its importance is often underestimated. This may be a lethal agent, and enterotoxemia is sometimes difficult to explain to the farmer, as it is also difficult to find the predisposing factor or factors that may be implicated in the beginning of the disease. It is often difficult to find the best way to classify these bacteria's involvement in the disease, either as opportunistic agents or as pathogens, when dealing with such problems. In the opinion of the authors, clostridiosis is often a conditioned disease, which means it is triggered by predisposing factors, such as environmental stress, incorrect nutritional practices, or concomitant parasitic or infectious diseases, among others.

There have been some advances in the knowledge of clostridia in recent years, namely, about new toxins and other virulence factors, which have helped to understand the full extent of these microorganisms' role in the course of enteric diseases in rabbits. However, there is still a long way to go in this field of study.

The present chapter aims to describe the main clinical features to approach clostridiosis in rabbits and rabbitries, with special emphasis on *Clostridium spiroforme*.

2 The Pathogens *Clostridium* spp.: General Features

The majority of bacteria from the genus *Clostridium* are obligate anaerobic and Gram-positive rods that form heat-resistant endospores (Dürre 2014; Cruz-Morales et al. 2019). The relationship within *Clostridium* species can be studied based on 16S rRNA sequences (Prescott et al. 2016). Nevertheless, new phylogenetic classification (Fig. 1) has been suggested (Cruz-Morales et al. 2019). In fact, *C. cocleatum*, *C. ramosum*, *C. spiroforme*, and *C. saccharogumia* were proposed to be classified as *Thomasclavelia* gen. nov. (Lawson et al. 2023). The reclassification is not yet official, so it will not be adopted in this chapter.

Most of these microorganisms are commensal or soil bacteria and do not cause disease. In this chapter, we focus on rabbit pathogenic clostridia, which are summarized in Table 1. *C. spiroforme*, *C. perfringens* type A, and *C. perfringens* type E are the most isolated from enteric samples of rabbits presenting enterotoxemia clinical signs (Prescott et al. 2016).

Enterotoxemia is defined as the disease caused by the absorption of a clostridial toxin from the intestine. Other species, such as *C. chauvoei*, *C. difficile*, and *C. piliforme* (Tyzzer's disease), are also described as possible pathogenic agents for these animals, but according to the authors' experience, they are not very often found in the clinical practice of farm rabbits, at least on the Iberian Peninsula.

C. spiroforme is, phylogenetically, the most distinct among the species that affect rabbits (see Fig. 1). This bacillus has various degrees of coiling, and its iota-like toxin, named *C. spiroforme* toxin (CST), was first identified in the cecal content of rabbits with spontaneous diarrhea in 1983 by Borriello and Carman (1983).

This bacterium causes characteristic enterotoxemia signs and enterotyphlitis, which cannot be considered pathognomonic, but it is the most characteristic clinical presentation, especially in animals around the weaning period (Songer and Uzal 2016; Solans et al. 2019; Oglesbee and Lord 2020).

In 2018, a single novel *Clostridium* species (*C. cuniculi*) was isolated and suggested by those authors to be associated with the onset of ERE (epizootic rabbit enteropathy), as described in Chapter "Epizootic Rabbit Enteropathy (ERE)". This *Clostridium* strain encodes several toxins and was shown to be phylogenetically related to *C. botulinum* and *C. perfringens* (Djukovic et al. 2018).

3 Epidemiology of Rabbit Enterotoxemia

All pathogenic clostridia can undergo a cellular morphogenesis process known as sporulation, which leads to the production of resistant spores that substitute metabolically active vegetative bacteria in the process of transmission (Shen et al. 2019).

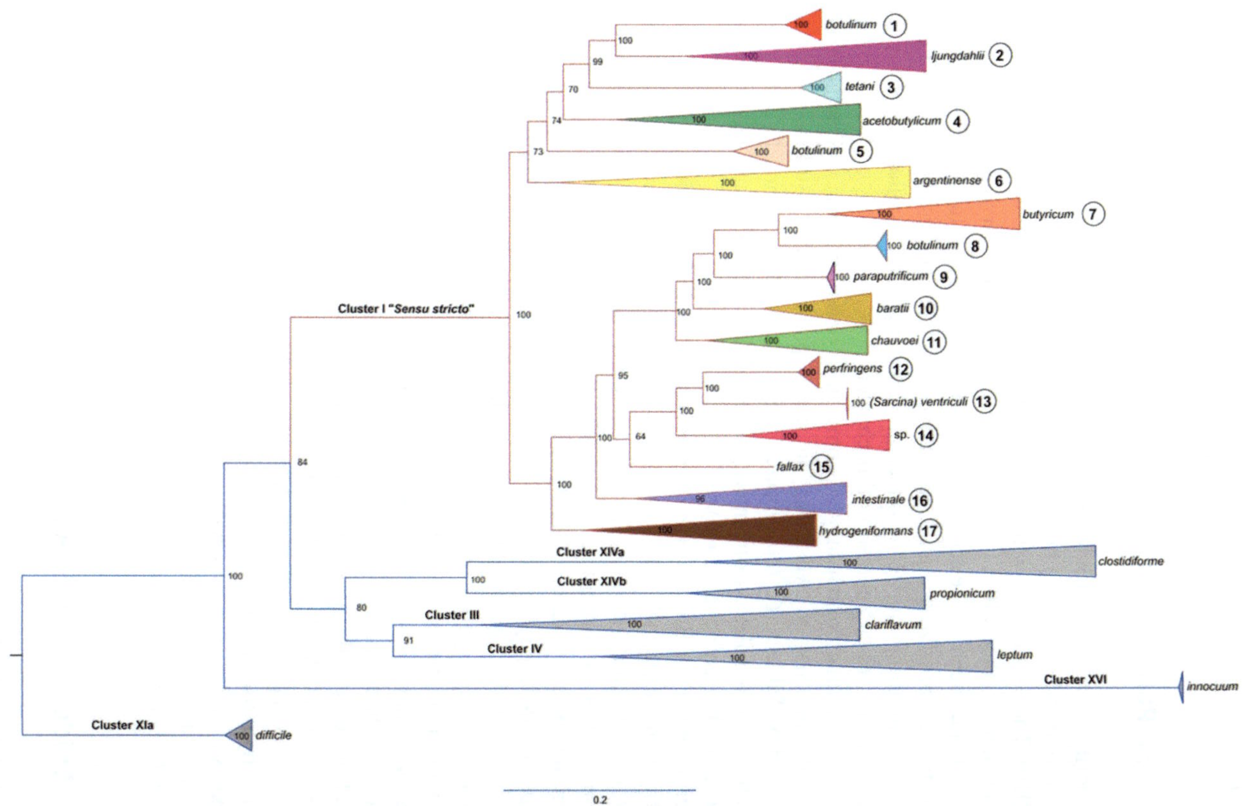

Fig. 1 Phylogenetic reconstruction of *Clostridium* species. The phylogeny was constructed using 27 markers. Branch support is shown at each node (Cruz-Morales et al. 2019). An important species, *C. spiroforme*, on rabbits was not included in this phylogenetic analysis (see the text)

Table 1 The most relevant clostridial pathogens of the rabbit and its main characteristics (brief description) (adapted from Prescott et al. (2016))

Species	Types	Major toxin(s)	Main disease(s)	Main source(s)	Main hosts affected
C. perfringens type A	A	CPA (alpha-toxin)	Gas gangrene (malignant edema) in several mammalian species; yellow lamb disease in sheep	Soil, large intestine, manured soil, widespread in nature; specific toxin types are likely also specific host-associated, found in small and large bowel	All mammalian and avian species; the full range of *C. perfringens*-associated enteric diseases and their toxin-mediated basis is still being characterized
		CPE (enterotoxin)	Food poisoning in humans; possible gastroenteritis in several mammalian species		Human food poisoning; possible gastroenteritis, including antibiotic- associated diarrhea in humans and possible other mammalian species
		NetB (necrotic enteritis B-like toxin)	Necrotic enteritis in poultry		Avian, especially chickens
		NetF (necrotic enteritis F-like toxin)	Necrotizing and hemorrhagic enteritis in dogs and foals		Foal necrotizing enteritis; dog hemorrhagic gastroenteritis
C. perfringens type E	E	ITX (iota toxin)	Bovine hemorrhagic gastroenteritis and possible enterotoxemia in rabbits		Possibly cattle and rabbits
C. spiroforme	–	CST, binary toxin actin-ADP-ribosylating toxin family (as *C. perfringens* ITX)	Typhlocolitis of rabbits	Humans' feces, ceca of chickens and rabbits	Rabbits

When the release of the formed spore occurs from the mother cell, it may also result in the release of toxins into the gut lumen, where it can cause pathological effects, namely, diarrhea, facilitating the excretion of spores into the environment and transmission. Therefore, infection usually occurs via the fecal-oral route with bacteria present in the environment (e.g., soil) (Songer and Uzal 2016).

C. spiroforme is the most prevalent pathogen causing enterotoxemia in farmed rabbits (Barthold et al. 2016; Drigo et al. 2008; Harkness et al. 2010; Marlier et al. 2003). According to Solans et al. (2019), *C. spiroforme* was one of the most frequently identified clostridia in pre-weaning and growing rabbits with enteric disease. It is commonly isolated concomitantly with other gut pathogens (Solans et al. 2019) (Fig. 2). These co-infections pose diagnostic challenges in defining the biotic cause. It would be important to search for clostridia toxins, which are the main trigger for enteric disease, to conclude about the causative agent implicated in the process.

The presence of *C. spiroforme* and its toxin (CST) is a consistent finding in the cecal content of diarrheic animals but is not commonly observed in healthy rabbits (Songer and Uzal 2016), despite some authors stating that the presence of this clostridia may occur in healthy rabbits (Benato et al. 2014). Some researchers consider that *C. spiroforme* is non-pathogenic at low levels (Borriello and Carman 1983). However, this microorganism has been reported to elicit fatal enteritis in rabbits that develop cecal dysbiosis (Benato et al. 2014; Jenkins 2004) and may be responsible for high morbidity and mortality rates in rabbit farms.

The role of *C. perfringens* in rabbit enteritis is still under debate. The involvement of alpha toxin (CPA) produced by *C. perfringens* in intestinal disease of mammals is poorly documented, but there is evidence that this toxin is essential for the production of gas, namely, in gangrene, in several animal species (Uzal et al. 2010). The findings of Garcia et al. (2014) support *C. perfringens* type A enterotoxin (CPE) effects on the rabbit colon as a potential contributor to CPE-mediated gastrointestinal disease. *C. perfringens* beta-toxin (CPB) was documented in a rabbit ileal loop model to cause bloody necrotic enteritis in *C. perfringens* type C infections (Vidal et al. 2008).

C. perfringens type B and C infections primarily involve the small intestine, with CPB acting as a very potent and fast-acting toxin (Vidal et al. 2008). The same authors concluded that 1 microgram of purified CPB was sufficient to cause a significant accumulation of hemorrhagic luminal fluid in duodenal, jejunal, or ileal loops. The role of iota toxin (ITX) in diseased animals is poorly understood (Navarro and Mcclane 2018), although it is usually assumed that the pathogenesis of intestinal diseases produced by *C. perfringens* type E is mediated by this toxin (Uzal et al. 2010).

C. perfringens type E has been considered one of the agents of enterotoxemia. However, when enterotoxemia diagnosis was based on toxin detection, some previously reported outbreaks of enterotoxemia associated with *C. perfringens* type E may have been due to *C. spiroforme* because there is 100% cross-reactivity between *C. perfringens* type E and *C. spiroforme* toxins, with the diagnostic methods currently available in rabbits (Songer and Uzal 2016).

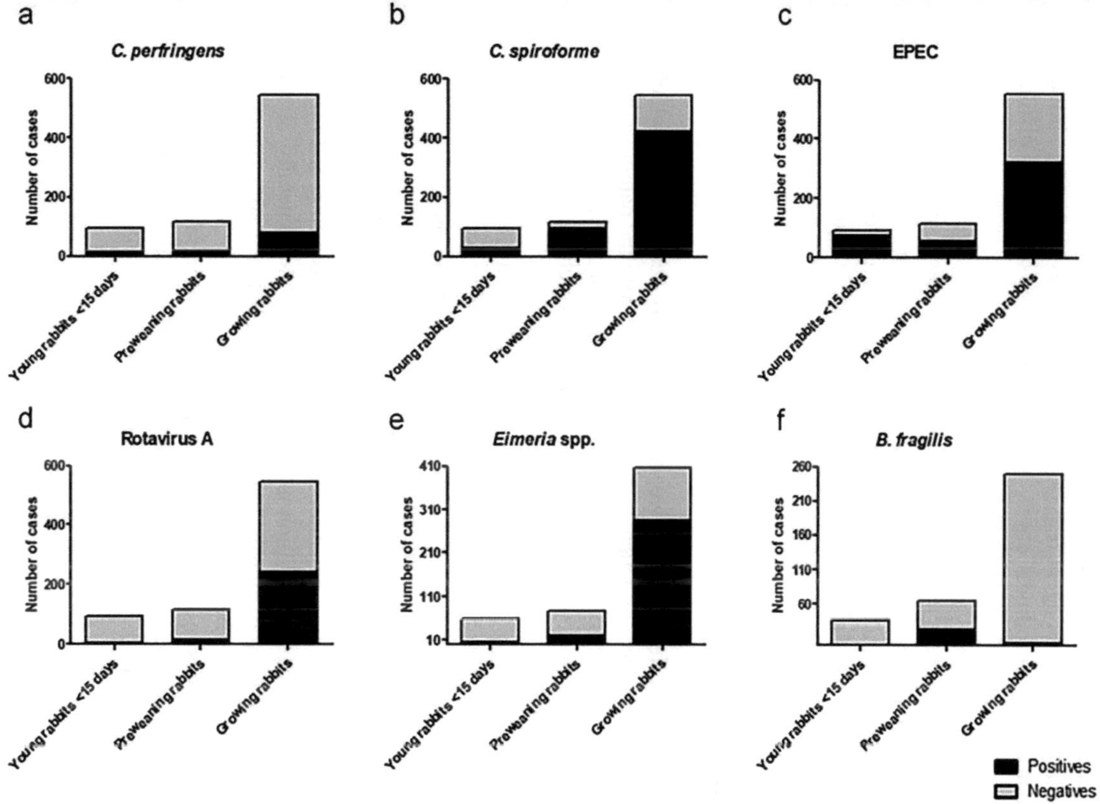

Fig. 2 Rate of detection (qPCR) of individual pathogens distributed by age in rabbits presenting clinical enteric disease. Adapted from Solans et al. (2019). Legend: The distributions of positive and negative cases within age groups are represented using black for positives and gray for negatives. (**a**) *C. perfringens* analysis where the total number of cases for each age group was young rabbits (<15 days old), 95; pre-weaning rabbits, 117; and growing rabbits, 545. (**b**) *C. spiroforme* analysis where the total number of cases for each age group was young rabbits (<15 days old), 95; pre-weaning rabbits, 117; and growing rabbits, 545. (**c**) EPEC analysis where the total number of cases for each age group was young rabbits (<15 days old), 95; pre-weaning rabbits, 117; and growing rabbits, 545. (**d**) *C. perfringens* analysis where the total number of cases for each age group was young rabbits (<15 days old), 95; pre-weaning rabbits, 117; and growing rabbits, 545. (**e**) *Eimeria* spp. analysis where the total number of cases for each age group was young rabbits (<15 days old), 59; pre-weaning rabbits, 77; and growing rabbits, 406. (**f**) Enterotoxigenic *B. fragilis* analysis where the total number of cases for each age group was young rabbits (<15 days old), 35; pre-weaning rabbits, 64; and growing rabbits, 205

Both diseases are clinically characterized by diarrhea and characteristic necropsy changes, including hemorrhages of the cecal serosa and mucosa, sometimes with the involvement of the distal ileum and proximal colon, and the presence of watery mucoid content in these organs. Histologically, there is acute necrosis of the superficial epithelium with relative sparing of the crypt epithelium, submucosal hemorrhage, and transmural hemorrhage (Uzal et al. 2010). From the authors' experience, distinct clinical appearances may be found between different clinical cases, while isolating the same agent (e.g., *C. spiroforme*), or even finding similar clinical appearance between clinical cases, when there is the isolation of distinct agents in each (e.g., *C. spiroforme* and *C. perfringens*).

In general, the scientific community highlights the role of *C. spiroforme* in rabbit enteric disease, much more than the effects of *C. perfringens* types A and E and even more than B and C types. Despite this position, it would be necessary to study enteric diseases of rabbits more accurately and intensively, namely, enterotoxemia, to better comprehend their origin and development and to achieve more adequate treatments and forms of control and prevention, such as vaccines.

4 Pathogenesis and Clinical Signs

C. spiroforme frequently affects recently weaned rabbits (3–6 weeks of age), which have the highest mortality rate, in which case simple contact with the pathogen may lead to enterotoxemia (Oglesbee and Lord 2020). Disease in neonates and weanlings may be associated with high gastric pH, between 5 and 6.5, which allows clostridial proliferation (Reusch 2005), and an underdeveloped population of normal gastrointestinal microbiota, allowing *C. spiroforme* to proliferate. Nevertheless, this disease may affect rabbits of all ages. Adult rabbits, in general, develop the disease in the presence of predisposing factors that work as triggers to dysbiosis and precede the colonization of the pathogen (Oglesbee and Lord 2020).

In a large part of the cases described, enteritis caused by clostridia is considered conditioned pathology due, for example, to inadequate diet or sudden changes in diet, stress factors (namely, environmental), specific drug administration, and ingestion of toxins or toxic substances (e.g., ionophores), among others. Generally, all the causes that may lead to perturbation in bowel microbiota balance are suitable to permit the overgrowth of *Clostridium* spp.

Diet-related disease and stress-related disease (Box 1) result in immunosuppression and decreased gastrointestinal motility, and the control of these risk factors should represent a large part of a preventive approach (Reusch 2005).

Box 1 Risk Factors for Enterotoxemia

- High temperature ranges with sudden cold feelings (great increase in feed consumption)
- Inadequate diet (nutrient balance, e.g., high starch proportion)
- Sudden changes in diet
- Antimicrobial or parasiticide treatments
- Ingestion of toxins or toxic substances (e.g., mycotoxins)
- Parasitosis (e.g., coccidiosis)
- Weaning
- Accumulation of feces above the cages
- Contact with feces
- Poor hygiene and biosecurity practices

Control feeding with the implementation of feed restriction strategies is one of the most efficacious methods used to prevent intestinal diseases in rabbits, namely, enterotoxemia. If only the necessary and adequate feed portion is administered to each rabbit, it is possible to avoid excessive ingestion and digestive problems, guaranteeing good health. Despite this, it is necessary to always ensure good environmental comfort.

The compound feed formula needs to be adjusted for the rabbit and its stage of life. Feed with high protein or high carbohydrate levels, namely, starch, is potentially dangerous for rabbits since it may trigger *Clostridium* spp. multiplication (Fernandez-Miyakawa and Redondo 2016).

Poor biosecurity practices, in general, may also have a significant negative impact on this disease (Kylie et al. 2017). The accumulation of feces above the cages (when daily or frequent removal during production is not possible), for example, represents a risk factor since the microbial concentration inside the shed is higher. It also leads to high humidity and acts as a medium for insect proliferation. Flies represent themselves a stress factor and act as vectors of this and other diseases.

The use of some antibiotics in rabbits, such as beta-lactams (such as penicillin, amoxicillin, and ampicillin) or some macrolides (such as lincomycin), is completely forbidden, even in parenthetical ways, in some worldwide regions. Their oral administration may be the cause of enterotoxemia. Treatments with a tetracycline used alone (e.g., for the treatment of respiratory complex) must be very carefully decided through a strict risk-benefit assessment of its use, since it may potentiate clostridial disease.

Large-spectrum antimicrobials, as well as different molecule associations, are tendentially safer and better tolerated than narrow-spectrum molecules used alone. The best example is the association of penicillin and streptomycin, which has been used subcutaneously in rabbits over the years, mainly for the treatment of *Staphylococcus* infections (e.g., mastitis, abscesses), with no significant risk. Otherwise, when used alone, and even more if it is used by oral administration, penicillin may be harmful (Licois 1996).

However, from the antimicrobial resistance point of view, antimicrobial associations must be avoided, as well as large-spectrum molecules. From the authors' point of view, this dichotomy, as well as the small number of molecules available and registered for use in rabbits, makes therapeutics in this species a very difficult task. Every treatment must be indispensable and chosen carefully.

Intestinal coccidiosis may also have not only a great impact as a trigger for dysbiosis and enterotoxemia but also a direct impact on the immune system. Nevertheless, despite *Eimeria* spp. oocysts are often found in diarrheic domestic rabbits concomitantly with *C. spiroforme* (Harkness et al. 2010; Barthold et al. 2016), these results should have a critical interpretation. Different *Eimeria* species have different pathogenicity. In our opinion, evaluating the real impact of coccidiosis on intestinal disease is useful for identifying *Eimeria* species and counting the number of oocysts *per* gram of feces (OPG). In addition, the increase in oocyst number may be a cause but may also be a consequence of bacterial enteritis. Simply, when pathogenic clostridia overcome rabbits' innate and acquired immune defenses, they can grow, multiply, and produce toxins that cause cell and tissue damage (Rood 2007).

CST (*C. spiroforme* toxin) and iota toxin (*C. perfringens* type E) are binary toxins that consist of two independent proteins, one being the binding component, which mediates the internalization of the intracellularly active component into the cell, and an enzyme component that initiates apoptosis once inside the cell cytoplasm. Nevertheless, the specific role of iota toxins in this disease remains unclear (Navarro and Mcclane 2018).

Enterotoxemia is normally an acute or hyperacute process. The animals start to present anorexia, followed by watery diarrhea, with fecal contamination of the perineum and legs. Sudden death can occur (Varga 2014), and therefore, it is also possible to find dead animals with no external evidence of diarrhea. Depression, signs of dehydration, and hypothermia are also found. As the disease progresses, rabbits become hypothermic and moribund and die within 1 or 2 days (Oglesbee and Lord 2020).

Changes in fecal size are one of the main clinical signs of gastrointestinal diseases in rabbits (Reusch 2005; Benato et al. 2014), namely, in clostridiosis, but are not easy to detect. It is also possible to find cecotrophs under the cages that were not ingested by the sick rabbit. Chronic cases are occasionally seen with intermittent diarrhea and weight loss (Reusch 2005). Nursing does can be responsible for "milk enterotoxemia" in kits. This is believed to be caused by clostridial endotoxin produced in the does' cecum and passed to the bunnies in the milk (Oglesbee and Lord 2020).

This disease is often characterized by increased fluidity of the cecal content, concomitant fermentation, and more or less gas production. In cases of sudden death, within a few hours, dead animals present good body condition. Mortality rates can reach very high values. From our experience, we may tell that the other animals in the group remain in good body condition, sometimes with no signs of morbidity or even the presence of diarrhea, and, in the next day, more rabbits are going to be found dead.

When assessing a rabbit farm with a problem of enterotoxemia, animals usually look active and in good condition, and it is common to wrongly foresee a drop in mortality in the following days. However, mortality frequently remains high or even increases. Moreover, different clostridia and even within the same species (such as *C. spiroforme*) can present various clinical presentations (lesions and clinical signs), affecting different ages, which may have different predisposing factors. More study in this field would be necessary to systematize the approach to this disease.

5 Diagnosis

Many cases of enterotoxemia on farm rabbits are clinically diagnosed through symptoms and postmortem findings. In general, necropsy is one of the most important tools for the veterinarian of rabbit production and is of great diagnostic value for this disease, which must have laboratory confirmation. Postmortem evaluation of these rabbits revealed the characteristic gross signs of clostridial enterotoxemia, including a swollen and edematous cecum with petechiae and ecchymoses on the serosal surface (Songer and Uzal 2016). Lesions can also comprise the appendix and proximal colon. Various amounts of gas throughout the intestinal tract may be present in the ileus, cecum, and colon. Hemorrhaged pseudomembranes or mucus may be present in the mucosa of the cecum and proximal colon (Oglesbee and Lord 2020). We cannot say it is pathognomonic, but the image of typhlitis with hemorrhagic ecchymoses such as brush strokes on the cecal wall, especially in rabbits right after weaning (Fig. 3), caused by *C. spiroforme* is very characteristic of this infection, despite this microorganism being present in animals with distinct enteritis aspects.

There is evidence that CPA toxin is important for gas production (Uzal et al. 2010); therefore, we supposed that a distended cecum filled with gas, which appears very frequently in enterotoxemia cases, may be an indicator of abnormal clostridia multiplication. Figure 4 shows a distended cecum filled with gas in rabbits with enterotoxemia.

For the isolation of *C. spiroforme* or other clostridia, cecal content samples may be collected from diarrheic rabbits and cultured. *C. spiroforme* is a fastidious microorganism, requiring enriched, pre-reduced culture media, strictly anaerobic conditions, and prolonged incubation time to grow (Agnoletti et al. 2009; Oglesbee 2011; Paul and Friend 2019).

The distinctive helically coiled shape of *C. spiroforme* allows its identification on Gram-stained smears from feces. Since this organism is not considered to be part of the normal microbiota of the rabbit, it is possible to achieve a presumptive diagnosis based on its presence in fecal samples, together with clinical signs and gross lesions found in necropsy (Agnoletti et al. 2009; Songer and Uzal 2016; Varga 2014; Paul and Friend 2019).

PCR (polymerase chain reaction) for *C. spiroforme* may not be widely available (Paul and Friend 2019). However, if available, it should be favored, as it is a rapid,

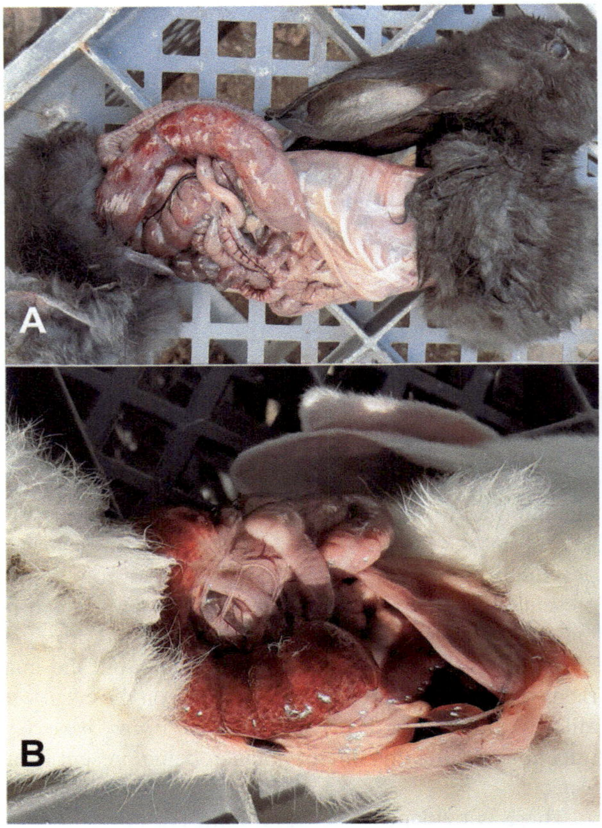

Fig. 3 Typhlitis in farmed rabbits with characteristic hemorrhagic lesions on the serosa with the aspect of brush strokes caused by *C. spiroforme* in 5-week-old animals immediately after weaning. Distal (**a**) and proximal (**b**) views. These animals may present slightly dirty perineum or even no evidence of diarrhea

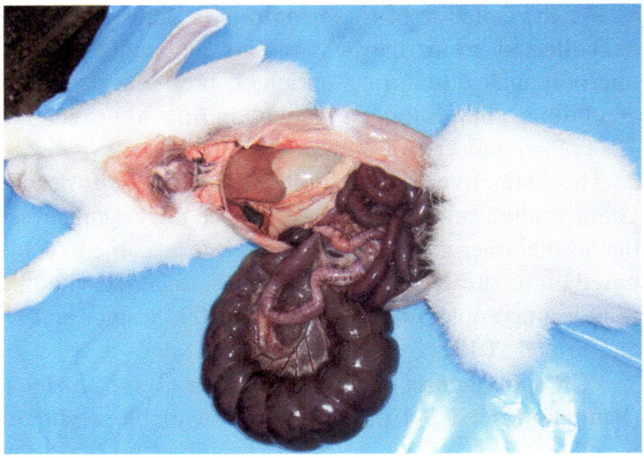

Fig. 4 Enterotyphlitis in farmed rabbits with enterotoxemia with great gas accumulation in the small intestine and cecum, severe wall congestion, and inflated stomach filled with fluids and gas. This and other 9-week-old animals from the same group presented with sudden death and no signs of diarrhea

reliable tool for the diagnosis of *C. spiroforme*-related enterotoxemia (Drigo et al. 2008).

Despite the detection of clostridia being useful, it has not true diagnostic value since *Clostridium* species (namely, *C. perfringens*) are often part of the normal gastrointestinal microbiota. Even in the case of *C. spiroforme* (which does not make up part of the physiological microbiota), it is possible to be present in the animal without causing disease. Great diagnostic value is given by the detection of toxin genes, which can be performed by PCR and other methods. Again, precise identification of the toxin may be hindered by the lack of both biomolecular techniques and rapid commercial kits (Paul and Friend 2019).

In the past, cases of enterotoxemia in rabbits were attributed to *C. perfringens* type E based on the detection of ITX in the intestine of affected animals. Nevertheless, it has been hypothesized that those cases may have been caused by an iota-like toxin produced by *C. spiroforme* (CST) that cross-reacts with ITX by currently available toxin detection methods (Navarro and Mcclane 2018).

ELISA (enzyme-linked immunoassay) and latex agglutination assay are commercially available for the detection of some clostridia toxins. However, there are doubts about their sensitivities and specificities (Reusch 2005).

Samples of the cecal wall may also be sent to the laboratory for histopathology. Acute epithelial necrotizing typhlitis is shown, with marked mucosal edema (Paul and Friend 2019).

6 Treatment, Control, Prophylactic Measures, and Prognosis

If suspected risk factors are identified as the potential trigger for a clostridial problem, the primary intervention should involve their removal. As already mentioned, some can be incorrect management procedures – nutrition, feeding, stress, cold sensation (wide temperature amplitudes, high-speed air flow), or others. Concomitantly, it is necessary to stop or at least slow mortality and restructure the intestinal microbiota.

Poor environmental control during cold seasons and the great ranges between minimum and maximum temperatures during the same day, with direct involvement in feed consumption, are also important stressful conditions. If feed restriction strategies are implemented on the farm, the impact of thermic amplitudes on feed consumption will be lower, as well as the possibility of enteric disease. High ammonia levels lead to pulmonary injury and to overall injury of the immune system.

Years of experience in this field made us tell that feed restriction strategies must be implemented in the context of prevention and that they may also be used as a complement for treatment. Feed restriction for growing rabbits after weaning is the best way to prevent digestive disease during this period of their lives. They ingest only what is necessary to gain the weight that is preconized by the genetics, which is monitored by the farmer. Rabbits are weighed, ideally, twice a week, and the portion of feed is adjusted in conformity. This procedure avoids feed overconsumption, which could lead to dysbiosis and enterotoxemia.

The apport of an extra fiber source in rabbits' diet, such as good microbial quality hay, may be a good option, but this may be very carefully decided because of the sanitary risk.

Our experience also allows us to say that individual treatment is not useful in these cases due to the rapid onset and dissemination of the disease and its fatal outcome and because in acute or hyperacute cases, there are no signs present in live animals that suddenly die.

Therefore, the aim of antimicrobial treatment is metaphylactic, attempting to treat some sick rabbits and stop the dissemination of the disease. Antimicrobial therapy is often necessary. Molecules able to control the Gram-positive population must be selected, taking special care with the toxicity of the chosen molecule. Rabbits with enterotoxemia have general organic effects, namely, the liver and kidneys, that are damaged and under great pressure because of the effect of bacterial toxins. Therefore, ideally, the clinician must choose an antimicrobial molecule with poor intestinal absorption and low toxicity. From this point of view, the oral administration of apramycin or bacitracin, for example, may be a possible choice, according to our experience. In the past, the combination of doxycycline with apramycin in water has shown good results in the field. Very good results may be obtained when treatment with metronidazole begins early in the onset of the disease (Paul and Friend 2019). Unfortunately, this molecule is not available for rabbits for meat production, but it may be a solution for pet rabbits.

Antibiotic treatment may be as useful as harmful, so the choice must be made very carefully.

Data on clostridial antimicrobial susceptibility are insufficient, perhaps because of its difficult culture and identification (Agnoletti et al. 2009), and insufficient are also studies about this subject. The same authors verified that the *C. spiroforme* field strain is, in vitro, resistant or had low susceptibility to sulfadimethoxine, spiramycin, norfloxacin, lincomycin, and tiamulin, with the strains they used in their experiment.

The veterinarian must keep in line with modern production systems, focusing on prevention rather than on the treatment of clostridial infections and working with the farmers to mitigate risk factors, also using antimicrobials prudently. Regular review of biosecurity practices in each farm (e.g., adoptions after parturition) is a crucial tool for improvement. The infectious agent together with poor biosecurity practices may lead to disease transmission within and between farms (Kylie et al. 2017). In severe or persistent cases, it may even be necessary to reduce the production rhythm, delaying insemination and weaning for 1 or 2 weeks. In this way, it is possible to avoid part of the natural immunodepression around weaning and its influence on clostridial disease.

In pet rabbits, aggressive fluid and supportive therapy are essential. Cholestyramine (2 g in 20 mL water, by gavage, every 24 hours) to bind clostridial toxins, loperamide hydrochloride (0.1 mg/kg, orally, every 8 hours for the first 3 days and then every 24 hours, for the fourth and fifth days after the onset of diarrhea), and a high-fiber diet may improve recovery (Reusch 2005).

Regarding prevention, in addition to feed restriction programs, probiotics, prebiotics, and vaccines must be the tools used in the fight against clostridial enterotoxemia.

Water or feed acidification, with organic acids that can reach the cecum (e.g., butyric acid) and control its pH, may also bring some benefits (Zhu et al. 2014).

Preliminary evidence from experiments performed in rabbitries suggests that probiotics may bring health benefits to rabbits, such as boosting the immune system or enhancing productivity and weight gain (Benato et al. 2014). There is an incomplete understanding of the mechanisms of probiotic activity (Reusch 2005). Suggested mechanisms for the reduction of pathogens include competition for nutrients, competition for enterocyte adhesion sites, and production of inhibitory substances (Reusch 2005).

Bacteria such as *Bacteroides* are associated with exerting an inhibitory effect on potentially pathogenic bacteria, including *Clostridium* and coliforms (Reusch 2005), which makes probiotics a good practice to implement to prevent enteric problems such as enterotoxemia. *Enterococcus faecalis* and *Enterococcus faecium*, which are part of the predominant enterococcal species in the rabbit gastrointestinal tract, appear to be promising solutions as probiotics for this species, but further work is needed in this field.

These products are already routinely administered concomitantly with conventional treatments. In pet rabbits, probiotics are used to alleviate the clinical signs of gastrointestinal disease and to reduce the potential adverse effects of antibiotic treatment (Benato et al. 2014).

In farm rabbits, their use is currently widely implemented, often included in feed, to prevent enteritis events and enhance the immune system, as well as the use of prebiotics. Furthermore, it is important to note that many of the available products on the market are not specifically formulated to meet the unique needs of rabbits but rather are generic products designed for multiple animal species or products formulated for other species and adapted to rabbits.

Reusch (2005) suggested the transfaunation of fresh cecotrophs from healthy rabbits as a source of microbiota to help reestablish cecal homeostasis in sick animals, but further studies should be carried out.

Maternal immunity lasts less than 1 month in young rabbits, which makes vaccination a useful practice in the prevention and control of enterotoxemia. In the future, the use of clostridia toxoid vaccines in rabbits may become very important in the prophylaxis of the disease, as happens in other species such as ruminants and swine.

A vaccine against *C. perfringens* toxins that could induce effective humoral and cellular immune responses would be the best prophylactic solution for the prevention of enterotoxemia. Oral vaccination would be able to achieve both types of immune responses, providing effective host protection (Vela Ramirez et al. 2017). In addition, oral vaccination (through drinking water) would avoid many hours of labor necessary for individual vaccination of fattening rabbits. The results presented by Bai et al. (2020) were promising, as well as the results obtained by other authors

(Peng et al. 2020; Saadh et al. 2022). However, this is an issue that requires more interest and investment from the pharmaceutical industry.

In rabbitries that present recurrent cases of enterotoxemia, special consideration must be made regarding animal welfare and farm economics. Assessing and mitigating risk factors is the key. Sometimes, to mitigate recurrent cases, it is necessary to reduce the production rhythm.

Other factors that limit the control of diseases such as clostridiosis include the intensive nature of rabbit farming, as well as the lack of solutions regarding vaccination for common diseases.

7 Conclusions

Enterotoxemia is one of the major important causes of gastrointestinal disease in rabbits, with a great impact on productive performance and an enormous economic impact. Its nature of conditioned pathology allows us to try to control some of the possible causes (environmental, nutritional, stressful, therapeutic) and create prophylactic protocols. Therapeutics should be very carefully chosen to treat this or any other disease of the rabbit, since many drugs (especially oral antibiotherapy) contribute to dysbiosis and enterotoxemia.

Feed restriction strategies are currently the most efficient way to control and even treat gastrointestinal problems in rabbits.

Rabbitries must focus on animal welfare and thermal comfort. The thermal sensation of cold potentiates enterotoxemia. It is urgent to have a vaccination solution against the toxins of these pathogenic agents, namely, *C. spiroforme*.

Acknowledgments The authors thank Springer for the use of the AJE Digital Editing. This work was supported by projects UIDP/CVT/00772/2020 (CECAV) and LA/P/0059/2020 (AL4AnimalS), funded by the Portuguese Foundation for Science and Technology (FCT).

References

Agnoletti F, Ferro T, Guolo A, Marcon B, Cocchi M, Drigo I, Mazzolini E, Bano L (2009) A survey of *Clostridium spiroforme* antimicrobial susceptibility in rabbit breeding. Vet Microbiol 136: 188–191. https://doi.org/10.1016/j.vetmic.2008.10.020

Bai J, Qiao X, Ma Y, Han M, Jia S, Huang X, Han B, Wang L, Li Y, Xu Y (2020) Protection Efficacy of Oral Bait Probiotic Vaccine Constitutively Expressing Tetravalent Toxoids against Clostridium perfringens Exotoxins in Livestock (Rabbits). Vaccines 8(1):17. https://doi.org/10.3390/vaccines8010017

Barthold SW, Griffey SM, Percy DH (2016) Rabbit. In: Pathology of laboratory rodents and rabbits, 4th edn. John Wiley & Sons, Ltd, Ames, Iowa, pp 253–324. isbn: 978-1-118-92404-4. https://doi.org/10.1002/9781118924051.ch06

Benato L, Hastie P, O'Shaughnessy P, Murray J-A, Meredith A (2014) Effects of probiotic *Enterococcus faecium* and *Saccharomyces cerevisiae* on the faecal microflora of pet rabbits. J Small Anim Pract 55:442–446. https://doi.org/10.1111/jsap.12242

Borriello SP, Carman RJ (1983) Association of Iota-like Toxin and *Clostridium spiroforme* with both spontaneous and antibiotic-associated diarrhea and colitis in rabbits. J Clin Microbiol 17: 414–418. https://doi.org/10.1128/jcm.17.3.414-418.1983

Cruz-Morales P, Orellana CA, Moutafis G, Moonen G, Rincon G, Nielsen LK, Marcellin E (2019) Revisiting the evolution and taxonomy of clostridia, a phylogenomic update. Genome Biol Evol 11:2035–2044. https://doi.org/10.1093/gbe/evz096

Djukovic A, Garcera MG, Paredes EM, Isaac S, Artacho A, Martínez J, Ubeda C (2018) Gut colonization by a novel *Clostridium* species is associated with the onset of epizootic rabbit enteropathy. Vet Res 49:123. https://doi.org/10.1186/s13567-018-0617-8

Drigo I, Bacchin C, Cocchi M, Bano L, Agnoletti F (2008) Development of PCR protocols for specific identification of *Clostridium spiroforme* and detection of sas and sbs genes. Vet Microbiol 131:414–418. https://doi.org/10.1016/j.vetmic.2008.04.013

Dürre P (2014) Physiology and sporulation in *clostridium*. Microbiol Spectrum 2:TBS-0010-2012. https://doi.org/10.1128/microbiolspec.TBS-0010-2012

Fernandez-Miyakawa ME, Redondo LM (2016) Role of *Clostridium perfringens* alpha, beta, epsilon and iota toxins in enterotoxemia of monogastrics and ruminants. In: Gopalakrishnakone P, Stiles B, Alape-Girón A, Dubreuil JD, Mandal M (eds) Microbial toxins. Springer, Dordrecht, pp 1–26. https://doi.org/10.1007/978-94-007-6725-6_16-1

Garcia JP, Li J, Shrestha A, Freedman JC, Beingesser J, McClane BA, Uzal FA (2014) Clostridium perfringens Type A Enterotoxin Damages the Rabbit Colon. Infect Immun. 82(6):2211–2218. https://doi.org/10.1128/IAI.01659-14

Harkness JE, Turner P, Vandewoude S, Wheler CL (2010) Chapter 4: clinical signs and differential diagnoses. Harkness and Wagner's biology and medicine of rabbits and rodents, 5th edn. Wiley & Sons Ltd, Hoboken, pp 195–247. isbn: 978-0-8138-1531-2

Jenkins JR (2004) Chapter 16 - gastrointestinal diseases. In: Quesenberry KE, Carpenter JW (eds) Ferrets, rabbits, and rodents, 2nd edn. W.B. Saunders, pp 161–171. isbn: 9780721693774. https://doi.org/10.1016/B0-72-169377-6/50018-6

Kylie J, Brash M, Whiteman A, Tapscott B, Slavic D, Weese JS, Turner PV (2017) Biosecurity practices and causes of enteritis on Ontario meat rabbit farms. Can Vet J 58:571–578

Lawson PA, Perez LS, Sankaranarayanan K (2023) Reclassification of Clostridium cocleatum Clostridium ramosum Clostridium spiroforme and Clostridium saccharogumia as Thomasclavelia cocleata gen. nov. comb. nov. Thomasclavelia ramosa comb. nov. gen. nov. Thomasclavelia spiroformis comb. nov. and Thomasclavelia saccharogumia comb. nov. Int J Syst Evol Microbiol. 73(1). https://doi.org/10.1099/ijsem.0.00569

Licois D (1996) Risques Associes a l'utilisation des antbiotiques chez le lapin: une mini revue. World Rabbit Sci 4(2):63–68

Marlier D, Dewrée R, Delleur V, Licois D, Lassence C, Poulipoulis A, Vindevogel H (2003) Description des principales étiologies des maladies digestives chez le lapin européen (*Oryctolagus cuniculus*). Annales de Medecine Veterinaire 147:385–392

Navarro MA, Mcclane BA (2018) Mechanisms of action and cell death associated with *Clostridium perfringens* toxins. Toxins 10:212. https://doi.org/10.3390/toxins10050212

Oglesbee BL (2011) Diarrhea, acute. In: Oglesbee BL (ed) Blackwell's five-minute veterinary consult: small mammal, 2nd edn. Wiley-Blackwell, Hoboken, pp 388–390. ISBN: 978-0-470-96101-8

Oglesbee BL, Lord B (2020) Gastrointestinal diseases of rabbits. In: Quesenberry KE, Orcutt CJ, Mans C, Carpenter JW (eds) Ferrets, rabbits, and rodents, 4th edn. W.B. Saunders, pp 174–187. isbn: 9780323484350. https://doi.org/10.1016/B978-0-323-48435-0.00014-9

Paul GC, Friend DG (2019) Clostridial Enterotoxemia and coccidiosis in weanling cottontail rabbits (*Sylvilagus audubonii, Sylvilagus floridanus, Sylvilagus nuttallii*) from Colorado, USA. J Wildl Dis 55:189–195. https://doi.org/10.7589/2018-02-026

Peng X, Peng G, Li X, Feng L, Dong L, Jiang Y (2020) Immunization of rabbits with recombinant Clostridium perfringens alpha toxins CPA-C and CTB-CPA-C in a bicistronic design expression system confers strong protection against challenge. Protein Expr Purif 167:105550. https://doi.org/10.1016/j.pep.2019.105550

Prescott JF, MacInnes JI, Wu AKK (2016) Taxonomic relationships among the clostridia. In: Uzal FA, Songer JG, Prescott JF, Popoff MR (eds) Clostridial diseases of animals. John Wiley & Sons, Inc, Ames, Iowa, pp 1–5. isbn: 9781118728406. https://doi.org/10.1002/9781118728291.ch1

Reusch B (2005) Rabbit gastroenterology. The veterinary clinics of North America. Exot Anim Prac 8:351–375. https://doi.org/10.1016/j.cvex.2005.01.007

Rood JI (2007) In: Dworkin M, Falkow S, Rosenberg E, Schleifer KH, Stackebrandt E (eds) *Clostridium perfringens* and histotoxic disease. The Prokaryotes: A Handbook on the Biology of Bacteria, 3rd edn. Springer, New York, pp 753–770. isbn: 9780387254999

Saadh MJ, Lafi FF, Dahadha AA, Albannan MS (2022) Immunogenicity of a newly developed vaccine against *Clostridium perfringens* alpha-toxin in rabbits and cattle. Vet World 15:1617–1623. https://doi.org/10.14202/vetworld.2022.1617-1623

Shen A, Edwards AN, Sarker MR, Paredes-Sabja D (2019) Sporulation and germination in clostridial pathogens. Microbiol Spectr 7(6). https://doi.org/10.1128/microbiolspec.GPP3-0017-2018

Solans L, Arnal JL, Sanz C, Benito A, Chacón G, Alzuguren O, Fernández AB (2019) Rabbit enteropathies on commercial farms in the Iberian Peninsula: etiological agents identified in 2018–2019. Animals 9(12):1142. https://doi.org/10.3390/ani9121142

Songer JG, Uzal FA (2016) In: Uzal FA, Songer JG, Prescott JF, Popoff MR (eds) Diseases produced by *Clostridium spiroforme*. Clostridial diseases of animals. John Wiley & Sons, Inc, Ames, Iowa, pp 221–227. isbn: 9781118728406. https://doi.org/10.1002/9781118728291.ch18

Uzal FA, Vidal JE, McClane BA, Gurjar AA (2010) *Clostridium perfringens* toxins involved in mammalian veterinary diseases. Open Toxinol J 2:24–42

Vántus VB, Kovács M, Zsolnai A (2014) The rabbit caecal microbiota: development, composition and its role in the prevention of digestive diseases – a review on recent literature in the light of molecular genetic methods. Acta Agraria Kaposváriensis 18:55–65

Varga M (2014) Chapter 14 - Infectious diseases of domestic rabbits. In: Varga M (ed) Textbook of Rabbit Medicine, 2nd edn. Butterworth-Heinemann, pp 435–471. ISBN: 978070204979. https://doi.org/10.1016/B978-0-7020-4979-8.00014-5

Vela Ramirez JE, Sharpe LA, Peppas NA (2017) Current state and challenges in developing oral vaccines. Adv Drug Deliv Rev 114:116–131. https://doi.org/10.1016/j.addr.2017.04.008

Velasco-Galilea M, Piles M, Viñas M, Rafel O, González-Rodríguez O, Guivernau M, Sánchez JP (2018) Rabbit microbiota changes throughout the intestinal tract. Front Microbiol 9:2144. https://doi.org/10.3389/fmicb.2018.02144

Vidal JE, McClane BA, Saputo J, Parker J, Uzal FA (2008) Effects of *Clostridium perfringens* Beta-toxin on the rabbit small intestine and colon. Infect Immun 76:4396–4404. https://doi.org/10.1128/IAI.00547-08

Zhu KH, Xu XR, Sun DF, Tang JL, Zhang YK (2014) Effects of drinking water acidification by organic acidifier on growth performance, digestive enzyme activity and caecal bacteria in growing rabbits. Anim Feed Sci Technol 190:87–94. https://doi.org/10.1016/j.anifeedsci.2014.01.014

Staphylococcosis in Rabbits and Rabbitries

Teresa Tavares, João Simões, José Manuel Monteiro,
and Juan Manuel Corpa

Abstract

Staphylococcal infections represent a major health and production problem in livestock and have a potential negative impact on public health. This chapter mainly focuses on the role of *Staphylococcus aureus* and its virulence in rabbits. *S. aureus* is a ubiquitous bacterium found in the environment and tegument of animals. The infection is mainly caused by high-virulent strains causing outbreaks and endemicity in rabbits of all ages. Direct and indirect contact with the bacterium allows colonization of the tegument and different rabbit organs in

T. Tavares
Veterinary and Animal Research Centre (CECAV), Associate Laboratory for Animal and Veterinary Sciences (AL4AnimalS), Department of Veterinary Sciences, School of Agricultural and Veterinary Sciences, University of Trás-os-Montes and Alto Douro (UTAD), Vila Real, Portugal

José Azevedo Monteiro, Lda., Vila Boa do Bispo, Portugal

J. Simões (✉)
Veterinary and Animal Research Centre (CECAV), Associate Laboratory for Animal and Veterinary Sciences (AL4AnimalS), Department of Veterinary Sciences, School of Agricultural and Veterinary Sciences, University of Trás-os-Montes and Alto Douro (UTAD), Vila Real, Portugal
e-mail: jsimoes@utad.pt

J. M. Monteiro
José Azevedo Monteiro, Lda., Vila Boa do Bispo, Portugal

Faculty of Veterinary Medicine, Lusófona University, Lisbon, Portugal

CBIOS, Lusófona University, Lisbon, Portugal

J. M. Corpa
Biomedical Research Institute (PASAPTA-Pathology group), Facultad de Veterinaria, Universidad Cardenal Herrera-CEU, CEU Universities, C/Tirant lo Blanc 7, 46115 Alfara del Patriarca, Valencia, Spain

J. Simões, J. M. Monteiro (eds.), *Veterinary Care of Farm Rabbits*,
https://doi.org/10.1007/978-3-031-44542-2_24

an ascending manner or septicemia of an initial focus. Suppurative dermatitis and subcutaneous abscesses, multisystemic abscessation, pododermatitis, purulent mastitis and metritis, and pneumonia are the prevalent clinical manifestations of the disease. The diagnosis is based on lesions and clinical signs followed by *S. aureus* identification and virulence typing. The success of treatment is mostly restricted to low-virulent *S. aureus* infections, as they usually affect only a few animals in rabbitries. The control of staphylococcosis is also based on sanitation measures and culling of affected animals, prioritizing those with suppurative lesions. Nevertheless, total depopulation and sanitary break measures are needed, such as incorporating non-carrier does and bucks to prevent high-virulent strains in affected rabbitries.

Keywords

Staphylococcus aureus · Subcutaneous abscesses · Pododermatitis · MRSA · Virulence factors · Enterotoxins

1 Introduction

Staphylococci are Gram-positive bacteria involved in different types of infections, such as cutaneous staphylococcosis with skin and subcutaneous lesions (e.g., suppurative dermatitis, furuncles, abscesses), and abscesses in several internal organs or tissues (e.g., heart, lungs, brain, kidneys), frequently causing multisystemic abscessation, mastitis, pneumonia, and metritis. The clinical signs of staphylococcal infection can be observed in more than 60% of rabbitries (Rosell and de la Fuente 2009).

Staphylococci, especially high-virulent strains of *Staphylococcus aureus* (see Német et al. 2020), represent a significant challenge in rabbit production, as they can adapt well to the rabbit environment and can be extremely difficult to control, especially under favorable environmental conditions and in animals with integumentary and immunosuppressive alterations. It may even pose a public health problem due to methicillin-resistant *Staphylococcus* species and toxins. Out of 324 samples, *S. aureus* was isolated in most of the suppurative lesions (69.2%), pododermatitis (95%), mastitis (78.6%), subcutaneous abscesses (78.5%), pneumonia (30.8%), and pyometra (15.9%), which are the most important causes of culling in rabbit does (Segura et al. 2007). More recently, using multilocus sequence typing, Wang et al. (2019) identified two sequence types (ST398 and ST121) responsible for 8.37% to 52.9% of the mortality rate in nine rabbitries (Fujian Province in southeastern China), confirming the relevance of *S. aureus* infections. In this study, *S. aureus* was isolated from 83.9% (78/93) of the mastitis, 83.3% (110/132) of the pododermatitis, and 10.0% (93/466) of the lung samples of dead rabbits.

The transmission route occurs through environmental (injured) tegument contact and ascending vias (uterus, udder, respiratory system, and auricular pavilions). The second route is via septicemia from an initial focus. However, the presence of specific *S. aureus* virulence factors and bacterium–host interactions are the main determinants for infection occurrence (Attili et al. 2020a).

This chapter mainly aims to describe the role of staphylococci in rabbit health according to different pathologies as well as their control.

2 Staphylococcosis

2.1 Etiopathophysiology

Caused by *S. aureus*, low-, classic high-, and atypical high-virulent strains can be identified. Severe outbreaks in rabbitries, as well as chronic staphylococcal infections from the multilocus sequence typing (MLST) ST121 lineage (Wang et al. 2019; Guerrero et al. 2015), are caused by high-virulence strains. Low-virulence strains usually occur in a small number of animals with little impact on rabbitries (Vancraeynest et al. 2004).

It is known that the combination of more than 30 virulence factors is a major determinant to improve the ability of *S. aureus* to cause disease (Haveri et al. 2008). Virulence factors, such as toxic shock syndrome toxin-1 (33.2% of the isolates), staphylococcal enterotoxin-C (17.0%), DNase (84.0%), α-hemolysin (69.9%), β-hemolysin (36.8%), δ-hemolysin (100%), protein A (78.3%), and clumping factor (54.7%), were detected from *S. aureus* isolates ($n = 106$) of rabbits suffering from acute clinical form (combined clinical signs of mastitis, pododermatitis, and subcutaneous abscesses), subacute clinical form (either mastitis, subcutaneous abscesses, or pododermatitis), or subclinical form (no clinical signs) (Tirpude and Batra 2012). The thermonuclease, Panton–Valentine leukocidin, and fibronectin-binding protein (fnbpB) genes have also been identified (Wang et al. 2019). Tirpude and Batra (2012) observed that the production of TSST-1, SC-C, and protein A was associated ($p < 0.01$) with the acute clinical form, which is related to clinical syndromes contributing to inflammatory reactions through strong mitogenic activity in T cells.

Selva Martínez et al. (2015) reported proportions of 84.2% (48/57) and 28.8% (17/59) of nasal carriers in rabbits with compatible gross lesions caused by *S. aureus* (mastitis, pododermatitis, abscess, and vulvovaginitis) and apparently healthy rabbits, respectively. Due to genotype similarity between nasal and lesion samples, these researches suggested that nasal carriers can play a relevant role in the pathogenesis of staphylococcal infection. Moreover, in a randomly selected sample of 400 rabbits, Attili et al. (2020b) observed a *S. aureus* prevalence of 29.2% in the nose, 31.5% in the ear, 11.4% in the axilla, and 19.9% in each of the groin and perineum regions. The risk of lesions due to *S. aureus* was progressively higher (odds ratio varying between 4.5 and 14.3) according to the increase in the number of sites (one to five) colonized by *S. aureus* identified by samplings (reference = 0 sites; Attili et al. 2020b). These findings reinforce the important role of nasal carriers in staphylococcal infections and suggest that more than one location should be sampled to improve diagnosis accuracy. Therefore, in addition to lesions, nose secretions should be sampled at the time of diagnosis to evaluate the presence of *S. aureus* in rabbitries.

The cutaneous form is mainly transmitted by direct contact of purulent secretions (e.g., skin abrasions), even indirect contacts (vectors, fomites), or previous septicemia. It is normally dependent on the occurrence of risk factors (Box 1).

Box 1 Staphylococci's Predisposing Factors

- Poor ventilation and excess ammonia, high relative humidity, and humid places
- Use of irritating disinfectants, poor nest preparation, and stress from different sources (noise, strange animals)
- Deficiencies in vitamins (vitamin A and B complex) and trace elements (zinc, copper, manganese, and cobalt)
- Traumatic skin and limb injuries, as well as breakouts caused by scabies, ringworm, scratches by other rabbits, plantar necrobacillosis, etc.
- Introduction of infected animals and poor elimination of non-curable positives

The umbilical stump is a relevant entry portal for kits. Healthy rabbits can be asymptomatic carriers. Ferreira et al. (2014) isolated *S. aureus* in 25.9% of subcutaneous abscesses ($n = 27$), being *Pasteurella* spp. the most isolated genera (59.3%) at the slaughterhouse, showing the importance of this bacterium from suppurative lesions.

2.2 Lesions and Clinical Signs

Two distinct forms of evolution can occur, the cutaneous form and the septicemic form.

In the cutaneous form (suppurative dermatitis), pockets of pus appear that evolve into boils on the abdomen, limbs, neck, and back of kits (pinpoint lesions developing into firm caseous pustules between 3 and 5 days old) (Corpa et al. 2010), which often leads to high mortality rates. Exudative dermatitis is the most typical lesion found in suckling kits. In adults, abscesses of variable size appear in similar locations (Fig. 1). At the histological level, dermatitis shows dystrophic alterations in the epidermis with infiltration of leukocytes in the dermal connective tissue. Additionally, multifocal necrotic areas can be detected in the lungs and heart following septicemia (Corpa et al. 2010).

The septicemic form occurs by spreading from a focal point (e.g., skin, lung). It causes otitis media and interna (torticollis), purulent mastitis (characteristic), purulent metritis, sporadic visceral abscesses (lung, pleura, kidney, and liver), abortions, and necrotic enteritis. Inflammation of the ocular conjunctiva and eyelids is also common.

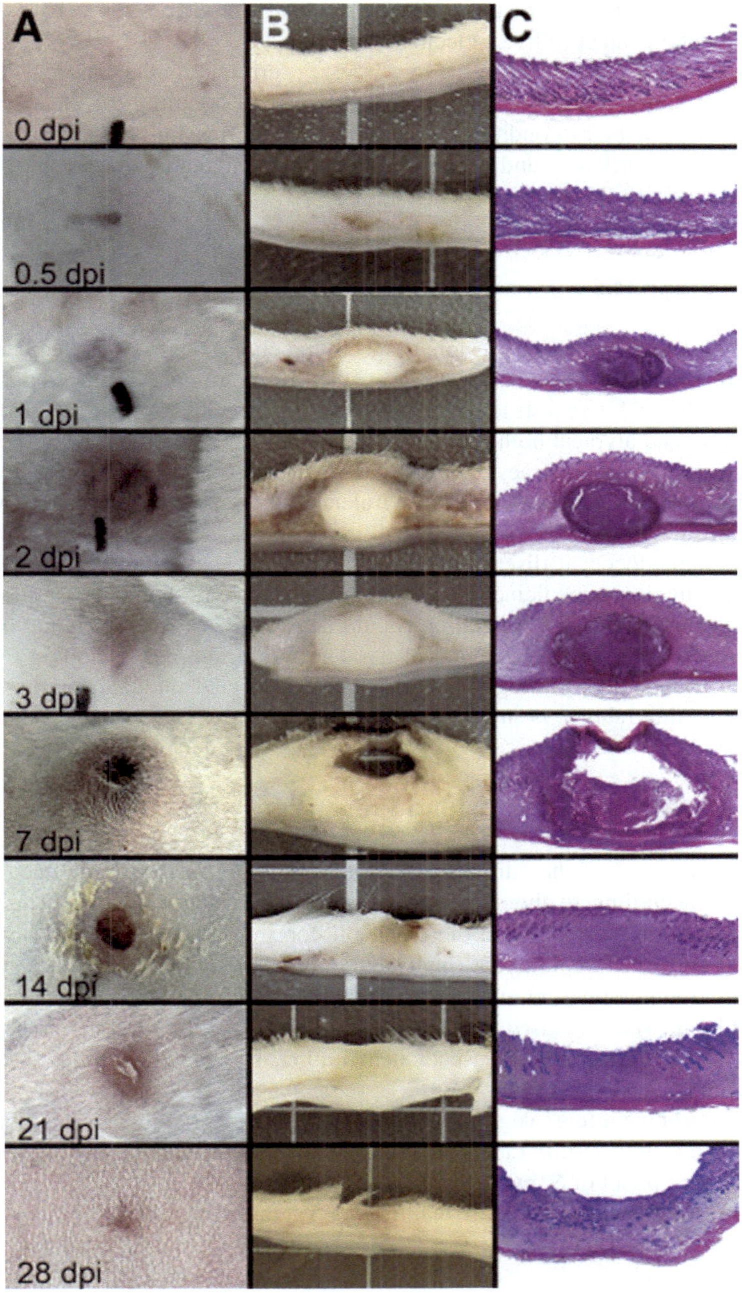

Fig. 1 Abscess evolution after intradermal inoculation of 300 colony-forming units (CFUs) of *S. aureus* (Jwt strain). Legend: (**a**) External aspect. (**b**) Sagittal cut of formalin-fixed tissue. (**c**) Microscopically low-magnification view stained with hematoxylin–eosin. Dpi, days post-infection. Adapted from Muñoz-Silvestre et al. (2020) with permission from Elsevier. For more figures and details, see Muñoz-Silvestre et al. (2020)

S. aureus infection of the lung, mammary gland, and uterus can occur by direct environmental contact or by septicemia.

Chronic mastitis is characterized by abscesses surrounding the teats and cellulitis causing purulent secretions and affecting the general condition of the does (e.g., lethargy, loss of body condition). Acute gangrenous mastitis is characterized by edematous, hemorrhagic, and necrotic changes in the mammary tissue with a rapid and fatal outcome (Viana et al. 2011).

2.3 Diagnosis

Clinical diagnosis is based on lesional aspects and clinical signs. It is complemented by microbiological tests and antibiograms, at least from samples of infected tissue or fluid, as well as DNAse tests to confirm the bacterial species (Corpa et al. 2010). The low-, high-, and atypical high-virulent strains can be identified by a multiplex PCR assay as reported by Vancraeynest et al. (2007), which is able to differentiate at least 26 typical high-virulence and 50 low-virulence rabbit *S. aureus* strains as well as atypical virulent strains.

Once staphylococcal disease has been diagnosed at the individual and group levels (populational medicine), herd health management should be applied. The health status of the rabbit population should be evaluated for concomitant diseases and predisposing factors. Screening of nasal secretion samples, including from the ear (see "Etiopathophysiology" section) from suspected and apparently healthy young and adult rabbits, is strongly recommended.

2.4 Treatment, Control, and Prevention

Treatment consists of the administration of antimicrobials for a prolonged period ($\geq$7 days) according to the susceptibility profile of the antibiogram. The use of antimicrobials in the feed or in the injectable form depends on the severity of the disease. Penicillin–streptomycin (injectable only), spiramycin, sulfamides–trimethoprim, fluoroquinolones, or even the classical tetracycline (e.g., 800 mg tetracycline HCl per kg feed) can be used to treat this condition. Subcutaneous abscesses can be drained and cleaned, and in case of dehydration, subcutaneous fluids should be administered (pets).

The environment (e.g., cages, materials, floor, and other surfaces) should also be cleaned and disinfected. In rabbitries, affected and suspected animals may be culled to reduce the spread of *S. aureus*.

These therapeutic measures are mainly aimed at infections caused by low-virulent *S. aureus* strains and are ineffective in the presence of high-virulent strains. Although morbidity and mortality can be reduced, the presence of staphylococcosis is still a major problem in rabbitries (Corpa et al. 2010). Infections caused by high-virulent strains require a sanitary break with complete depopulation and complete disinfection of installations, equipment, and materials.

Control and prevention are essentially based on hygiene, prophylaxis, and elimination of affected animals (Box 2). One of the most valuable measures to prevent staphylococcosis is to prevent the introduction of infected replacement animals colonized by high-virulent *S. aureus* strains, but animals are usually carriers, so preventive measures must target environmental and animal factors.

Several experimental vaccines against *S. aureus* have been progressively tested in rabbits, including as model for other animal species (Reinoso et al. 2002) and humans (Malachowa et al. 2022). Nevertheless, up to date, no commercial vaccines reaching protective immunological levels suitable to adequately control staphylococcosis in rabbitries were achieved. Autogenous vaccines are used as an alternative to commercial vaccines when staphylococcosis represents a significant problem in farms (Meulemans et al. 2011).

Box 2 Staphylococcal Control and Prevention Measures

- Rigorous culling of does (breeding females), affected by abscesses, metritis, mastitis, foot disease, and respiratory symptoms, as well as kits.
- Application of disinfectants and drying powders and pH changes in the nests.
- Removal of affected animals and litters (or complete sanitation).
- Pay particular attention to replacement animals introduced into the rabbitry at the time of first calving.
- Possible use of polyvalent vaccines (*Pasteurella*, *Staphylococcus*) or autogenous vaccines in replacement breeders with primary vaccination at 2.5 months and booster vaccination after 2 weeks.
- Suspend adoption techniques and keep the nests clean and dry.
- Avoid or reduce injuries and concomitant diseases.
- Improve welfare measures in general.

3 Pododermatitis: A Specific Occurrence

3.1 Etiopathophysiology

Pododermatitis, also known as "sore hocks," is caused by *S. aureus*, although other microorganisms such as *Fusobacterium necrophorum* and *Corynebacterium* spp. may also be present.

Triggering by a continuous or acute traumatic occurrence is necessary to develop this condition (Box 3), except in the presence of highly contagious agents (e.g., *F. necrophorum*).

It is most commonly observed in adult animals (does and bucks) and in the hind limbs due to their higher weight. Medium-weight breeds, such as New Zealand and Californian breeds, and animals with abundant hair on the underside of the limbs to protect the skin are less likely to the develop this condition.

Box 3 Predisposing Factors for Pododermatitis

- Poorly meshed cages (wire floors).
- Animal weight (large breeds, males, pregnant females).
- Humid environment with excess ammonia (e.g., urine in the bedding).
- Other bacteria (responsible for the severity of lesions).

Normally, it is limited to the skin and connective tissues, causing parakeratosis and the appearance of crusts. Lack of hygiene on the cage floor can lead to secondary infection, which can invade the metatarsals, making the process purulent.

3.2 Lesions and Clinical Signs

The presence of ulcerated and hemorrhagic areas and crusts in the plantar and palmar areas of the foot are observed. Multiple interdigital abscesses may be observed.

Difficulty with limb support and weight loss are observed in affected animals. Lactating does stop breastfeeding the kits.

3.3 Diagnosis

Clinical diagnosis involves inspection of lesions and lameness. A swab should be taken from the lesions for bacteriology and typing.

3.4 Treatment, Control, and Prevention

In severely affected animals, i.e., when the abscesses become suppurative or the forelimbs are also affected, the infection becomes difficult to treat, and the animals must be culled to avoid environmental spread of the bacteria in the rabbitry.

Slightly and moderately affected animals can be treated with topical iodine or antimicrobials (if approved). Persistent traumatic aggressions must be removed for meaningful recovery.

Parenteral antibiotherapy (e.g., penicillin 40,000 IU/kg/SID SC for at least 10 days) is recommended in cases of multiple limb involvement.

Cages should be washed and disinfected frequently. The affected animals should be provided with a mat to prevent continuous bruising of the injured area.

Rubber mats are usually used to modify the floor of the maternity cage (Fig. 2).

For cages, it is important to choose metal mesh with thick, welded, and galvanized wires, with a mesh width between 13 and 15 mm in diameter.

Fig. 2 Maternity cage with a rubber mat to prevent pododermatitis

4 Public Health Concerns of *Staphylococcus aureus* Infections

S. aureus is a commensal bacterium commonly found on the skin, nasopharynx, and mucous membranes of healthy animals and humans (Guerrero et al. 2015). Approximately 30% of the human population carries *S. aureus* on the nasal mucosa (Krismer et al. 2017).

Most species of *Staphylococcus* spp. are agents of opportunistic infections, but *S. aureus* is recognized as a primary pathogen, being the most clinically relevant species (Fisher et al. 2018). Although this pathogen is considered a commensal bacterium, it can cause several diseases ranging from mild skin and soft tissue infections, osteomyelitis, endocarditis, or food poisoning to severe and potentially fatal diseases (Crespo-Piazuelo and Lawlor 2021; Phiri et al. 2022).

S. aureus pathogenicity relies on several virulence factors, including extracellular proteins such as enzymes and toxins. Currently, *S. aureus* infections are a major public health problem worldwide due to increasing antimicrobial resistance (Pal et al. 2020).

4.1 Methicillin-Resistant *Staphylococcus aureus*

Antimicrobial resistance results from the general overuse of antibiotics (including on farms), and the spread of these bacteria is facilitated by poor infection prevention and control measures. This contributes to the spread of resistant bacteria from animals to humans, either by direct contact or via the food chain through contact with or ingestion of animal food products (Lakhundi and Zhang 2018).

Methicillin-resistant *S. aureus* (MRSA) is considered a serious human pathogen due to its high morbidity and mortality and its resistance to most beta-lactams and many other antibiotics (Lakhundi and Zhang 2018; Pal et al. 2020; Smith et al. 2021). MRSA strains can be divided into three categories: healthcare-associated MRSA (HA-MRSA), community-associated MRSA (CA-MRSA), and livestock-associated MRSA (LA-MRSA) (Pal et al. 2020; Crespo-Piazuelo and Lawlor 2021; Smith et al. 2021).

LA-MRSA strains have been detected in farm animals, including rabbits (Agnoletti et al. 2014; Attili et al. 2020a; Bencardino et al. 2021). These strains can be transmitted between different animal species and to humans through prolonged close contact with colonized animals, such as veterinarians, slaughterhouse personnel, and farm workers (Lakhundi and Zhang 2018; Pal et al. 2020; Crespo-Piazuelo and Lawlor 2021; Smith et al. 2021). Human-to-human and environment-to-human transmission must also be considered, as well as through contaminated meat and meat products (Lakhundi and Zhang 2018).

The detection and geographical spread of LA-MRSA is increasing, with implications for human and veterinary medicine, highlighting the importance of these strains as a One Health issue (Kinross et al. 2017). The evidence to date suggests that farm animals may become a permanent reservoir of human MRSA infections.

4.2 Staphylococcal Enterotoxins

S. aureus is recognized as an important foodborne pathogen. Staphylococcal food poisoning is caused by the ingestion of food contaminated with one or more preformed, heat-stable staphylococcal enterotoxins (SEs) (Pal et al. 2020; Mahros et al. 2021; Phiri et al. 2022). There are several SEs and enterotoxin-like (SEl) toxins, including the classical (SEA-SEE), that are responsible for approximately 95% of foodborne intoxication outbreaks, and the newer (SEG-SEIY) (Fisher et al. 2018; Mahros et al. 2021).

SEs are resistant to proteolytic enzymes, heat, and low pH, which allows them to maintain their functionality in the gastrointestinal tract after ingestion. The mechanism by which enterotoxins cause foodborne intoxication is not fully understood, but some authors suggest that SEs act directly on the intestinal epithelium and the vagus nerve, stimulating the emetic center. The amount of SEs required to cause staphylococcal food poisoning in humans ranges from 20 ng to 1 μg (Kadariya et al. 2014; Mahros et al. 2021).

Nausea, vomiting, abdominal cramps, and diarrhea are the most common symptoms associated with staphylococcal food intoxication, which is usually self-limiting and resolves in 24 to 48 hours. Elderly individuals, infants, and immunocompromised persons may develop severe illness (Pal et al. 2020; Mahros et al. 2021).

Various food matrices have been associated with staphylococcal food poisoning, including meat and meat products, poultry and egg products, unpasteurized milk and milk products, salads, cream-filled bakery products, and sandwich fillings (Pal et al. 2020; Bencardino et al. 2021).

During the slaughter of production animals, microbial contamination of carcasses can occur. In the case of MRSA-positive animals, the carcass and later the meat may be contaminated by this bacterium. Many authors have reported the detection of MRSA in different foods, including meat and meat products (de Boer et al. 2009; Lakhundi and Zhang 2018), which represents an imminent risk to human health and a major global public health concern and challenge.

5 Conclusions

Staphylococcosis is a common disease in rabbits with high health, welfare, and economic implications in farm rabbits. This complex infection depends on bacterium and host–bacterium interactions. Staphylococcal infections present in a variety of clinical forms, including cutaneous and septicemic forms, involving various internal and external organs.

The management of this problem in rabbitries is difficult, especially when high-virulent strains of *S. aureus* are widespread in animals and the environment, constituting a source of persistent infections.

S. aureus also plays a relevant role in food poisoning in the human population, mainly through heat-stable preformed SEs. Special care is required to control LA-MRSA strains in rabbitries.

Further research on animal resilience and herd health management is required to achieve adequate levels of control and prevention.

Acknowledgments The authors thank Springer for the use of the AJE Digital Editing. This work was supported by projects UIDP/CVT/00772/2020 (CECAV) and LA/P/0059/2020 (AL4AnimalS), funded by the Portuguese Foundation for Science and Technology (FCT).

References

Agnoletti F, Mazzolini E, Bacchin C, Bano L, Berto G, Rigoli R, Muffato G, Coato P, Tonon E, Drigo I (2014) First reporting of methicillin-resistant *Staphylococcus aureus* (MRSA) ST398 in an industrial rabbit holding and in farm-related people. Vet Microbiol 170:172–177. https://doi.org/10.1016/j.vetmic.2014.01.035

Attili A-R, Bellato A, Robino P, Galosi L, Papeschi C, Rossi G, Fileni E, Linardi M, Cuteri V, Chiesa F, Nebbia P (2020a) Analysis of the antibiotic resistance profiles in methicillin-sensitive *S. aureus* pathotypes isolated on a commercial rabbit farm in Italy. Antibiotics 9:673. https://doi.org/10.3390/antibiotics9100673

Attili A-R, Nebbia P, Bellato A, Galosi L, Papeschi C, Rossi G, Linardi M, Fileni E, Cuteri V, Chiesa F, Robino P (2020b) The effect of age and sampling site on the outcome of *Staphylococcus aureus* infection in a rabbit (Oryctolagus cuniculus) farm in Italy. Animals 10:774. https://doi.org/10.3390/ani10050774

Bencardino D, Amagliani G, Brandi G (2021) Carriage of *Staphylococcus aureus* among food handlers: an ongoing challenge in public health. Food Control 130:108362. https://doi.org/10.1016/j.foodcont.2021.108362

Corpa JM, Hermans K, Haesebrouck F (2010) Main pathologies associated with Staphylococcus aureus infections in rabbits: a review. World Rabbit Sci 17:115–125. https://doi.org/10.4995/wrs.2009.651

Crespo-Piazuelo D, Lawlor PG (2021) Livestock-associated methicillin-resistant *Staphylococcus aureus* (LA-MRSA) prevalence in humans in close contact with animals and measures to reduce on-farm colonisation. Ir Vet J 74:21. https://doi.org/10.1186/s13620-021-00200-7

de Boer E, Zwartkruis-Nahuis JTM, Wit B, Huijsdens XW, de Neeling AJ, Bosch T, van Oosterom RAA, Vila A, Heuvelink AE (2009) Prevalence of methicillin-resistant *Staphylococcus aureus* in meat. Int J Food Microbiol 134:52–56. https://doi.org/10.1016/j.ijfoodmicro.2008.12.007

Ferreira A, Monteiro JM, Vieira-Pinto M (2014) The importance of subcutaneous abscess infection by *Pasteurella* spp. and *Staphylococcus aureus* as a cause of meat condemnation in slaughtered commercial rabbits. World Rabbit Sci 22:311. https://doi.org/10.4995/wrs.2014.2238

Fisher EL, Otto M, Cheung GYC (2018) Basis of virulence in enterotoxin-mediated staphylococcal food poisoning. Front Microbiol 9:436. https://doi.org/10.3389/fmicb.2018.00436

Guerrero I, Ferrian S, Penadés M, García-Quirós A, Pascual JJ, Selva L, Viana D, Corpa JM (2015) Host responses associated with chronic staphylococcal mastitis in rabbits. Vet J 204:338–344. https://doi.org/10.1016/j.tvjl.2015.03.020

Haveri M, Hovinen M, Roslöf A, Pyörälä S (2008) Molecular types and genetic profiles of *Staphylococcus aureus* strains isolated from bovine intramammary infections and extramammary sites. J Clin Microbiol 46:3728–3735. https://doi.org/10.1128/JCM.00769-08

Kadariya J, Smith TC, Thapaliya D (2014) Staphylococcus aureus and staphylococcal food-borne disease: an ongoing challenge in public health. Biomed Res Int 2014:1–9. https://doi.org/10.1155/2014/827965

Kinross P, Petersen A, Skov R, Van Hauwermeiren E, Pantosti A, Laurent F, Voss A, Kluytmans J, Struelens MJ, Heuer O, Monnet DL, the European Human LA-MRSA Study Group (2017) Livestock-associated methicillin-resistant *Staphylococcus aureus* (MRSA) among human MRSA isolates, European Union/European economic area countries, 2013. Eur Secur 22(44): 16-00696. https://doi.org/10.2807/1560-7917.ES.2017.22.44.16-00696

Krismer B, Weidenmaier C, Zipperer A, Peschel A (2017) The commensal lifestyle of Staphylococcus aureus and its interactions with the nasal microbiota. Nat Rev Microbiol 15:675–687. https://doi.org/10.1038/nrmicro.2017.104

Lakhundi S, Zhang K (2018) Methicillin-resistant *Staphylococcus aureus*: molecular characterization, evolution, and epidemiology. Clin Microbiol Rev 31:e00020–e00018. https://doi.org/10.1128/CMR.00020-18

Mahros MA, Abd-Elghany SM, Sallam KI (2021) Multidrug-, methicillin-, and vancomycin-resistant *Staphylococcus aureus* isolated from ready-to-eat meat sandwiches: an ongoing food and public health concern. Int J Food Microbiol 346:109165. https://doi.org/10.1016/j.ijfoodmicro.2021.109165

Malachowa N, McGuinness W, Kobayashi SD, Porter AR, Shaia C, Lovaglio J, Smith B, Rungelrath V, Saturday G, Scott DP, Falugi F, Missiakas D, Schneewind O, DeLeo FR (2022) Toward optimization of a rabbit model of *Staphylococcus aureus* (USA300) skin and soft tissue infection. Microbiol Spectr 10:e0271621. https://doi.org/10.1128/spectrum.02716-21

Meulemans G, Haesebrouck F, Lipinska U, Duchateau L, Hermans K (2011) Efficacy of an autogenous vaccine against highly virulent "Staphylococcus aureus" infection in rabbits. World Rabbit Sci 19:1–9. https://doi.org/10.4995/wrs.2011.812

Muñoz-Silvestre A, Penadés M, Selva L, Pérez-Fuentes S, Moreno-Grua E, García-Quirós A, Pascual JJ, Arnau-Bonachera A, Barragán A, Corpa JM, Viana D (2020) Pathogenesis of intradermal staphylococcal infections. Am J Pathol 190:1188–1210. https://doi.org/10.1016/j.ajpath.2020.01.019

Német Z, Albert E, Dán Á, Balka G, Szenes Á, Sipos R, Bódizs S, Biksi I (2020) Genomic analysis of *Staphylococcus aureus* strains originating from Hungarian rabbit farms reinforce the clonal origin of various virulence types. Animals 10:1128. https://doi.org/10.3390/ani10071128

Pal M, Kerorsa GB, Marami LM, Kandi V (2020) Epidemiology, pathogenicity, animal infections, antibiotic resistance, public health significance, and economic impact of *Staphylococcus aureus*: a comprehensive review. Am J Public Health Res 8:14–21

Phiri BSJ, Hang'ombe BM, Mulenga E, Mubanga M, Maurischat S, Wichmann-Schauer H, Schaarschmidt S, Fetsch A (2022) Prevalence and diversity of *Staphylococcus aureus* in the Zambian dairy value chain: A public health concern. Int J Food Microbiol 375:109737. https://doi.org/10.1016/j.ijfoodmicro.2022.109737

Reinoso E, Magnano G, Giraudo J, Calzolari A, Bogni C (2002) Bovine and rabbit models for the study of a *Staphylococcus aureus* avirulent mutant strain, RC122. Can J Vet 66:285–288

Rosell JM, de la Fuente LF (2009) Culling and mortality in breeding rabbits. Prev Vet Med 88:120–127. https://doi.org/10.1016/j.prevetmed.2008.08.003

Segura P, Martinez J, Peris B, Selva L, Viana D, Penades JR, Corpa JM (2007) Staphylococcal infections in rabbit does on two industrial farms. Vet Rec 160(25):869-872. https://doi.org/10.1136/vr.160.25.869

Selva Martínez L, Viana D, Corpa Arenas JM (2015) *Staphylococcus aureus* nasal carriage could be a risk for development of clinical infections in rabbits. World Rabbit Sci 23:181. https://doi.org/10.4995/wrs.2015.3960

Smith RP, Sharma M, Gilson D, Anjum M, Teale CJ (2021) Livestock-associated methicillin-resistant *Staphylococcus aureus* in slaughtered pigs in England. Epidemiol Infect 149:e236. https://doi.org/10.1017/S0950268821002260

Tirpude RJ, Batra HV (2012) Characteristics of *Staphylococcus aureus* isolated from acute, sub-acute and sub-clinical staphylococcosis in rabbits. World Rabbit Sci 20:215–221. https://doi.org/10.4995/wrs.2012.1203

Vancraeynest D, Hermans K, Martel A, Vaneechoutte M, Devriese LA, Haesebrouck F (2004) Antimicrobial resistance and resistance genes in Staphylococcus aureus strains from rabbits. Vet Microbiol 101:245–251. https://doi.org/10.1016/j.vetmic.2004.03.021

Vancraeynest D, Haesebrouck F, Hermans K (2007) Multiplex PCR assay for the detection of high virulence rabbit Staphylococcus aureus strains. Vet Microbiol 121:368–372. https://doi.org/10.1016/j.vetmic.2006.12.011

Viana D, Selva L, Callanan JJ, Guerrero I, Ferrian S, Corpa JM (2011) Strains of *Staphylococcus aureus* and pathology associated with chronic suppurative mastitis in rabbits. Vet J 190:403–407. https://doi.org/10.1016/j.tvjl.2010.11.022

Wang J, Sang L, Sun S, Chen Y, Chen D, Xie X (2019) Characterisation of *Staphylococcus aureus* isolated from rabbits in Fujian. China Epidemiol Infect 147:e256. https://doi.org/10.1017/S0950268819001468

Rabbit Viral Diseases: Myxomatosis and Rabbit Hemorrhagic Disease

Margarida D. Duarte, Teresa Fagulha, Margarida Henriques, Silvia Barros, Fernanda Ramos, Ana Duarte, Tiago Luís, and Fábio A. Abade dos Santos

Abstract

Infectious diseases remain the major cause of morbidity and mortality in rabbits, causing significant losses in the industry, traditional rabbit husbandry and wildlife. In the industry, the two major viral threats, myxomatosis and rabbit hemorrhagic disease, can be controlled mainly by vaccination and biosecurity measures. However, the relentless emergence of new viral variants poses an enduring risk to the prevention and control of infectious diseases. This chapter highlights the relevant aspects of both diseases in terms of aetiology, epidemiology, lesions and clinical signs, diagnosis and control. Although the main focus is on domestic rabbits of the genus *Oryctolagus*, infections in other rabbit and hare species are also discussed.

M. D. Duarte (✉) · F. A. Abade dos Santos
Instituto Nacional de Investigação Agrária e Veterinária (INIAV, I.P.), Oeiras, Portugal

Centre for Interdisciplinary Research in Animal Health (CIISA), Faculdade de Medicina Veterinária, Universidade de Lisboa, Lisbon, Portugal

Associate Laboratory for Animal and Veterinary Sciences (AL4AnimalS), Lisbon, Portugal

Faculdade de Medicina Veterinária, Universidade Lusófona de Humanidades e Tecnologia, Lisbon, Portugal
e-mail: margarida.duarte@iniav.pt

T. Fagulha · M. Henriques · S. Barros · F. Ramos · T. Luís
Instituto Nacional de Investigação Agrária e Veterinária (INIAV, I.P.), Oeiras, Portugal

A. Duarte
Instituto Nacional de Investigação Agrária e Veterinária (INIAV, I.P.), Oeiras, Portugal

Centre for Interdisciplinary Research in Animal Health (CIISA), Faculdade de Medicina Veterinária, Universidade de Lisboa, Lisbon, Portugal

Associate Laboratory for Animal and Veterinary Sciences (AL4AnimalS), Lisbon, Portugal

J. Simões, J. M. Monteiro (eds.), *Veterinary Care of Farm Rabbits*,
https://doi.org/10.1007/978-3-031-44542-2_25

Keywords

European rabbit · *Oryctolagus cuniculus* · Hare · Myxoma virus · Fomites · RHDVGI.1 · RHDV2 · GI.2

1 Introduction

Since their introduction in Europe in the early 1950s to mid-1980s, myxoma virus (MYXV) and rabbit hemorrhagic disease (RHD) virus (RHDV/RHDV2) have had a significant impact on the rabbit industry, particularly before effective commercial vaccines became available and vaccination became a regular practice. Control of myxomatosis and RHD is complicated by the role of blood-sucking/biting insects in the transmission of MYXV and RHDV/RHDV2 between domestic and wild rabbits and vice versa. The circulation of MYXV and RHDV/RHDV2 in wild rabbit populations provides a constant source of infection and represents an ongoing risk of disease transmission to the industry. Furthermore, the impact of myxomatosis and RHD on the health, size and dynamics of wild rabbit populations, particularly in Iberia, where these diseases are endemic, has had dramatic consequences on the equilibrium of the Mediterranean ecosystems.

Here, we review the two main viral diseases of domestic rabbits, myxomatosis and RHD. Because of the epidemiological importance that the wild rabbit has in the maintenance and evolution of the aetiological agents and in their dynamic transmission to the domestic rabbit, data on wild populations are also presented.

2 Myxomatosis

2.1 Etiology

Myxomatosis is caused by a large, double-stranded DNA (approximately 160,000 bp long) virus that replicates only in the cytoplasm of the host cells. It belongs to the genus *Leporipoxvirus* of the family *Poxviridae* (Fenner 1990). The brick-shaped virus has a diameter of about 250 nm.

Infection of the European rabbit (*Oryctolagus cuniculus*) has been associated with systemic disease with high mortality rates of up to 100% (Fenner and Ratcliffe 1965), although the severity of the disease is highly dependent on the strain of the virus and the species of rabbit (Sobey 1969; Sobey et al. 1970). In 1950, around five decades after its first detection, MYXV was deliberately introduced into Australia as a biological weapon to control the non-native populations of European rabbits, a major animal pest that was threatening Australia's native fauna and flora (Kerr and Best 1998), with initial rabbit mortality of approaching 100%. In subsequent years, however, less virulent strains and more resistant hosts were selected, resulting in a reduction in the mortality, a classic example of host-pathogen co-evolution following a species jump to a new host (the European rabbit). In 1952, the virus was also introduced into France from where it spread throughout continental Europe (Kerr et al. 2015). Myxomatosis has been endemic in Iberia since 1956 (Fenner and Fantini 1999).

2.2 The Disease Origin and Its Natural Hosts

Myxomatosis is an often fatal disease of the European rabbit (*O. cuniculus*) caused by the myxoma virus (MYXV), which affects both domestic and wild rabbits. However, in its natural hosts, the *Sylvilagus brasiliensis* (brush rabbits of South and Central America), the *S. bachmani* (Californian brush rabbits of the western coast of North America) and the jackrabbits (*Lepus californicus*), the disease is limited to the formation of benign fibromas from which the animals usually recover completely.

MYXV was first described in Uruguay in 1898 in European laboratory rabbits used for serum production, which had contracted the disease from infected brush rabbits (*S. brasiliensis*) probably via *Aedes* mosquitoes, which promote mechanical transport of the virus into the laboratory (de Aragão 1942; Fenner and Ratcliffe 1965; Sanarelli 1898).

2.3 Susceptible Species

Animal experiments showed that all other mammals were considered refractory to MYXV (Sanarelli 1898). The few reports of infection in hare species (mainly in *L. europaeus* and sporadically in *L. timidus* and *L. granatensis*) were rarely associated with disease (Fenner and Ratcliffe 1965; Wibbelt and Frölich 2005; Barlow et al. 2014). However, in 2018, this scenario changed drastically with the emergence of a recombinant MYXV (ha-MYXV) in Spain (Dalton et al. 2019), which was alsoi detected in Portugal a few months later (Carvalho et al. 2020). This new virus was able to infect European hares (*L. granatensis*) causing mortality rates of 54% (García-Bocanegra et al. 2020). The impact on the wild populations was so severe that hunters estimated a population reduction of more than 70% in the population in the following year (Abade dos Santos, personal communication). Among other genetic differences, compared to classical strains of MYXV, namely, in the Lausanne strain, ha-MYXV has a 2.8 Kb insertion that disrupts the M009L gene (Dalton et al. 2019; Pinto et al. 2019). This additional genetic material includes several genes with homology to cervidpoxviruses. The recombinant ha-MYXV was subsequently detected in domestic (Abade dos Santos et al. 2020a) and wild rabbits (Abade dos Santos et al. 2020b), demonstrating its ability to infect the European rabbit. Classic MYXV strains have also been found in the Iberian hare (*L. granatensis*), in co-infection with ha-MYXV strains (Abade dos Santos et al. 2022b). A retrospective study using Iberian hare serum samples from the 1990s showed the presence of MYXV-DNA and antibodies in some animals indicating that the hare populations were already in contact with MYXV or with a MYXV-like virus at that time (Abade dos Santos et al. 2022a, b).

2.4 Clinical Signs and Post-mortem Macroscopic and Microscopic Lesions

Two forms of the disease are observed, namely, the classical nodular form, characterized by the development of mucinous skin lesions, called as myxomas or myxoeodemas (Fig. 1), and a less commonly observed amyxomatous form of myxomatosis, characterized by the absence of cutaneous nodules and the presence of respiratory disease (Marlier et al. 2000b).

Clinically, typical myxomatosis is characterized by nodules located where the fur is shorter, allowing insects, mainly arthropod vectors, such as mosquitoes, fleas and biting flies, to bite. These nodules are usually seen at the site of infection (ears, head, body and legs) and may become congested or necrotic (DiGiacomo and John Maré 1994) accompanied by edema of the eyelids, ears, lips, nose, anus and genitals (vulva and scrotum) and blepharoconjunctivitis (Fig. 1). Early signs include prostration, anorexia and a fever up to 42 °C. In acute forms of the disease, rabbits may die within 48 hours of the onset of clinical signs. Occasionally, domestic rabbits survive for several weeks, and the skin nodules become fibrotic. Pneumonia associated with secondary bacterial infection is quite common (Patton and Holmes 1977) and may account for a severe cough.

The amyxomatous form of myxomatosis is clinically milder and generally not fatal (Marlier et al. 2000a, b) and is transmitted by direct contact (Joubert et al. 1972). Respiratory signs, such as serous or purulent rhinitis, are usually present, but the cutaneous lesions are smaller and less frequent. Oedema of the perineum and eyelids and purulent blepharoconjunctivitis may also be present.

On necropsy, in the acute forms, the disease is fulminant, with few characteristic gross lesions. In rabbits that survive longer, subcutaneous oedema and nodular skin tumours are seen. The spleen is occasionally enlarged.

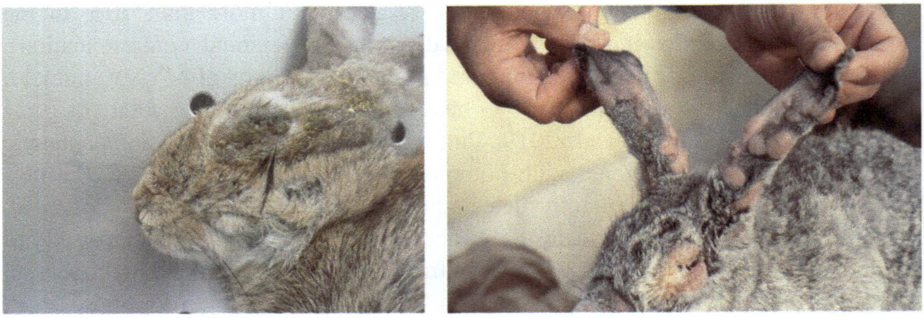

Fig. 1 Rabbits with nodular form of myxomatosis. Left—severe blepharoconjunctivitis in a wild rabbit (Photopgraph by Fábio A. Santos). Right—blepharoconjunctivitis and a large number of myxomas in the external ears (photograph kindly provided by the Pathology Laboratory of INIAV, I.P.)

Histological lesions in the skin consist of proliferation of stellate (myxoma) mesenchymal cells surrounded by an abundant mucinous matrix. In advanced lesions, endothelial cell proliferation; large, eosinophilic intracytoplasmic inclusion bodies, in different cell types; and epidermal cell hyperplasia or degeneration may be observed (Fenner and Marshall 1957; Fenner and Ratcliffe 1965). The spleen is almost always devoid of lymphocytes.

2.5 Laboratory Diagnosis

Laboratory diagnosis is important to establish a conclusive diagnosis. This may include anatomical and histopathological examination, direct detection of viral proteins by antigen ELISA of tissue extracts or detection of specific antibodies, usually carried out by commercial ELISAs. Virus isolation can also be performed in rabbit epithelial kidney cells (RK13 cells). Electron microscopy can also be used to confirm the presence of viral particles with the characteristic brike shape core in infected tissues. Inoculation of rabbits with a suspension of suspected infected tissue is no longer recommended for animal welfare reasons.

Direct detection of the viral genome by molecular methods (mainly real-time PCR (qPCR)) is recommended because of its high specificity and sensitivity, providing results in a few hours (Duarte et al. 2014). As a quantitative method, qPCR can also provide information on the viral load in the samples.

Recently, a quadruplex qPCR has been developed to differenciate between classical (MYXV) and recombinant (ha-MYXV) strains (Abade dos Santos et al. 2021a).

2.6 Preventive Measures

Since MYXV is spread by insects, mosquito nets that protect rabbits from exposure to arthropods are an important measure to control the disease in domestic rabbits. Isolation of the infected animals is necessary to prevent the spread of disease, as myxoma virus is transmitted by direct contact between infected and uninfected animals, in addition to the bite of blood-sucking insects. Prevention of the disease is achieved through biosecurity measures and vaccination with attenuated strains, which generally induces immunity that lasts up to 6 months (EMA 2019). Industrial vaccination is strongly recommended in endemic areas.

As commercial vaccines are designed for parenteral administration, they are not suitable for use in wild rabbits. The impact of myxomatosis in wild populations is difficult to quantify due to predation of the affected animals, which become blind from oedema of the eyelids and blepharoconjunctivitis.

As there is no effective treatment, euthanasia and burial or burning of infected rabbits is recommended for the industry. However, treatment is possible in pet rabbits consisting of supportive care, although the prognosis is variable. Myxoma virus does not infect humans.

3 Rabbit Hemorrhagic Disease

3.1 Etiology

RHD is caused by RHDV and RHDVa, also designated by GI.1, hereafter referred to as RHDV, or by RHDV2 (or GI.2), both single-stranded, 7.4-kb long positive-sense RNA viruses (Meyers et al. 1991; Ohlinger et al. 1990; Parra and Prieto 1990; Le Gall-Reculé et al. 2011a). Their genomic RNA consists of two open reading frames (ORFs): ORF1, which encodes a polyprotein that is cleaved into several non-structural proteins (viral proteins necessary for viral replication but that are not found in the viral particle) and the major structural capsid protein, VP60 (60 KDa, which is highly immunogenic), and ORF2, which encodes only the minor structural protein called VP10 or VP2 (Ohlinger et al. 1990; Meyers et al. 1991; Wirblich et al. 1996). The virus capsid has a diameter of ~40 nm, is icosahedral in symmetry and comprises 90 dimers of a single capsid subunit, the VP60 protein. The absence of an outer envelope makes RHDV/RHDV2 extremely resistant, particularly when protected by organic material, such as decomposing rabbit carcasses, remaining infectious for up to 20 days (Rocchi and Dagleish 2018), facilitating new infections and disease persistence.

RHDV (GI.1) and RHDV2 (GI.2) are related but genetically and antigenically distinct and are considered to be different viruses. Both belong to the genus *Lagovirus* of the family *Caliciviridae* (International Committee on Taxonomy of Viruses (ICTV)) (Le Gall-Reculé et al. 2011a; Vinjé et al. 2019). The genus *Lagovirus* also includes several non- or moderately pathogenic rabbit caliciviruses (Capucci et al. 1996; Forrester et al. 2007; Bergin et al. 2009; Jahnke et al. 2010; Le Gall-Reculé et al. 2011b), which are genetically related but more distant from RHDV (Strive et al. 2009; Abrantes et al. 2012), and the European brown hare syndrome virus (EBHSV).

Non-pathogenic rabbit caliciviruses (RCV) comprise two distinct genotypes, namely, GI.3 (European rabbit calicivirus, RCV-E1, including strain 06-11) and GI.4 (Australian RCV-A1 and European RCV-E2 strains).

No cross-protection (Le Gall-Reculé et al. 2011b), partial cross-protection (Tanja Strive et al. 2013) or full-cross protection (Capucci et al. 1996) against RHDV infection has been demonstrated by previous exposure to non-pathogenic lagoviruses.

EBHS affects wild and farmed hares (*L. timidus* and *L. europaeus*) in many European countries. Clinical signs are similar to RHD in rabbits, and mortality is high (Gavier-Widén 1994; Wirblich et al. 1994). Experimental infection of domestic rabbits with EBHSV neither caused clinical disease nor protected the animals from subsequent challenge with RHDV (Nauwynck et al. 1993).

3.2 The Disease Origin and Its Natural Hosts

RHD is a highly infectious disease of domestic and wild rabbits characterized by fulminant necrotizing hepatitis, disseminated intravascular coagulation and internal hemorrhages in multiple organs. RHD is of socio-economic importance in rabbit-producing countries affecting international trade in animals and animal products. It also impacts wildlife and biodiversity.

RHD was first reported in 1984 in the People's Republic of China, in Angora rabbits imported from Germany (Liu et al. 1984). In 1986, the disease was recorded in Italy (Cancellotti and Renzi 1991) and soon spread through the domestic and wild rabbit populations of continental Europe, becoming endemic in several countries, such as Spain (Arguello-Villares et al. 1988) and Portugal (Anonymous 1989). In the following years, the disease was reported in the Azorean islands of Faial (1988), São Jorge (1989) and Santa Maria (1990) (de Carvalho and de Almeida 1991; de Carvalho et al. 1993, 1994) and throughout Europe (Abrantes et al. 2012). It has also been reported in the Middle East, in North Africa and in the Western Hemisphere (namely, in Mexico City in 1988, although being successfully eradicated in 1992 (Gregg et al. 1991) and in Cuba in 1993, 1997, 2000–2001 and 2004–2005 (Toledo et al. 1995; Farnós et al. 2007)) as well as in the USA in 2000 (Campagnolo et al. 2003; McIntosh et al. 2007).

The disease has had a severe impact on the rabbit industry in Europe and in North Africa (Morisse et al. 1991) and on the wild rabbit populations, particularly on the Iberian Peninsula (Arguello-Villares et al. 1988; Villafuerte et al. 1995; Delibes-Mateos et al. 2007).

In Australia and New Zealand, where rabbits are pests, the virus was deliberately introduced for rabbit biocontrol in 1995 and 1997, respectively (Motha and Clark 1998; Mutze et al. 1998; O'Keefe et al. 1998).

RHDV2 was first reported in France, in 2010 (Le Gall-Reculé et al. 2011a, 2013), and quickly replaced the circulating RHDV strains in most European countries, namely, in Spain (Dalton et al. 2012, 2014; Puggioni et al. 2013), mainland Portugal (Abrantes et al. 2013; Lopes et al. 2014a) and islands (Duarte et al. 2015a; Abade dos Santos et al. 2017), Italy (Le Gall-Reculé et al. 2013), Belgium (Hermans et al. 2016), the Netherlands (WOAH 2021), England and Wales (Westcott et al. 2014), Scotland (Baily et al. 2014), the Republic of Ireland (Rocchi et al. 2016), Poland (Fitzner and Niedbalski 2018) and Sweden (Neimanis et al. 2018), among many other countries across Europe.

RHDV2 has been reported outside Europe in Canada (WOAH 2016) and Australia (Hall et al. 2015), the USA (Asin et al. 2021) and South Africa (WOAH notification, 18 November 2022).

Replacement of RHDV by RHDV2 has been observed in both wild and in domestic populations, with the new virus becoming the dominant pathogenic calicivirus in domestic and wild rabbits worldwide (Calvete et al. 2014; Duarte et al. 2018, 2021), demonstrating its notorious ability to overcome, at least in part, the immunity conferred by previously circulating RHDV variants.

The origin of RHDV/RHDV2 is still unclear. However, the genetic relationships between RHDV/RHDV2 and other members of the *Lagovirus* genus (namely, the rabbit caliciviruses (RCV), which include several non-pathogenic or low-pathogenic rabbit lagoviruses that were recognized long before the emergence of pathogenic strains) provide some information on the possible emergence of pathogenicity in this group of viruses.

Molecular clock-based studies have proposed different dates for the emergence of RHDV, with some suggesting that RHDV emerged just before (Eden et al. 2015) or decades before (Moss et al. 2002; Kerr et al. 2009; Kinnear and Linde 2010) its first detection in China, in 1984.

In the latter scenario, the most recent common ancestor (MRCA) for a group of RHDV and rabbit calicivirus sequences has been estimated to have emerged 550 years ago, while the MRCA for the RHDV isolates that have spread around the world after 1984 was estimated to have emerged 150 years ago (Kerr et al. 2009). As multiple lineages of RHDV were circulating before the 1984 Chinese outbreak, it has been suggested that the evolution of RHDV virulence began in the early 20th century, long before the emergence of rabbit haemorrhagic disease in 1984.

3.3 Susceptible Species

RHD is not a zoonotic disease, and, with regard to RHDV, no species other than the European rabbit (*O. cuniculus*) have been reported to be fatally susceptible to infection. Cottontail rabbits (*S. floridanus*), jackrabbits (*L. californicus*) and volcano rabbits (*Romerolagus diazi*) are not susceptible to RHDV. RHDV was detected in two specimens of Iberian hare (*L. granatensis*) collected in Portugal in 1990 (Lopes et al. 2014b). Other hare species from Europe, namely, the European brown hare (*L. europaeus*), the mountain hare (*L. timidus*), the Italian hare (*L. corsicanus*) and the Sardinian Cape hare (*L. capensis*), do not appear to be affected by RHDV/RHDVa classical strains. Furthermore, RHDV replication has never been reported in rabbit predators, such as the red fox, although seroconversion can occur (Simón et al. 1994; Leighton et al. 1995; Frölich et al. 1998), indicating contact with the virus. Experimental inoculation of RHDV-infected tissues into various vertebrate species other than rabbits failed to induce disease, and no viral replication was detected in the tissues. However, RHDV has recently been detected in an Alpine musk deer (*Moschus sifanicus*) in China, which showed haemorrhages in the internal organs. The full sequence of the RHDV strain showed 98.7% identity with a Chinese rabbit strain HB (Bao et al. 2020). RHDV-RNA was also detected in natural populations of two sympatric mouse species (*Mus spretus* and *Apodemus sylvaticus*) captured alive in central Spain in the summer of 2005 (Merchán et al. 2011).

RHDV2 can infect and cause a RHD-like disease in some hare species, namely, the Sardinian Cape hare (Puggioni et al. 2013), the Italian hare (Camarda et al. 2014), the European brown hare (Velarde et al. 2017) and the mountain hare (Buehler et al. 2020; Kennedy et al. 2021).

Infectious RHDV2 has also been detected in small mammals, namely, in the white-toothed shrew (order Eulipotyphla) and the Mediterranean pine vole (order Rodentia), demonstrating the development of disease in rabbits inoculated with tissues from these small mammals (Calvete et al. 2019).

Evidence that RHDV2 can infect the Eurasian badger (*Meles meles*) (order Carnivora) has also been provided more recently (Abade dos Santos et al. 2021b), alhough its role as a potential reservoir is not yet fully understood.

3.4 Clinical Signs and Post-mortem Macroscopic and Microscopic Lesions

RHDV causes high mortality in both domestic and wild adult animals, although the clinical ourse of the disease may be peracute, acute, subacute or chronic. The incubation period ranges from 1 to 5 days, despite longer periods of 9 days have been described.

It varies depending on the type of causative agent, and death may occur 12–36 hours after the onset of fever (>40 °C) (WOAH 2021). During this time, the virus is already highly contagious.

Rabbits may show no clinical signs if the disease is fulminant (peracute form) (Fig. 2). If present (in acute and subacute forms), clinical signs may be non-specific such as fever, anorexia, apathy, dullness and prostration, and cyanosis of mucous membranes. Nervous signs, such as convulsions, seizures, ataxia, paralysis, opisthotonos and paddling, and respiratory signs such as dyspnoea, and frothy and bloody nasal discharge may be observed. Moaning and cries may be heard.

During outbreaks in the industry, about 5–10% of the animals in the case of RHDV, and a higher percentage when the infection is caused by RHDV2, may

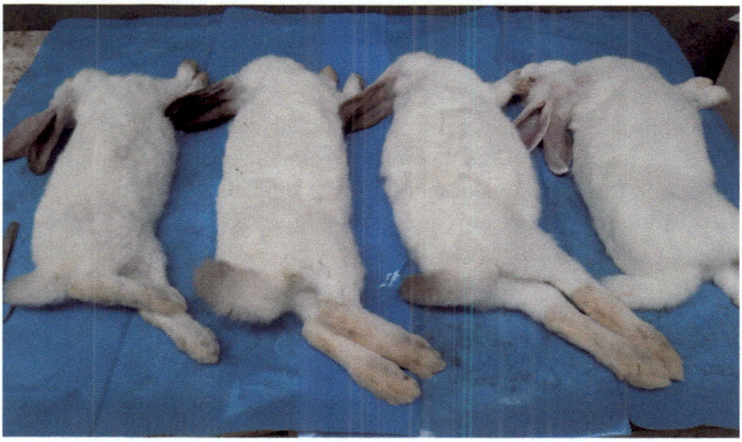

Fig. 2 Rabbits with RHDV in good body condition and without any apparent symptoms (photograph kindly provided by José Monteiro)

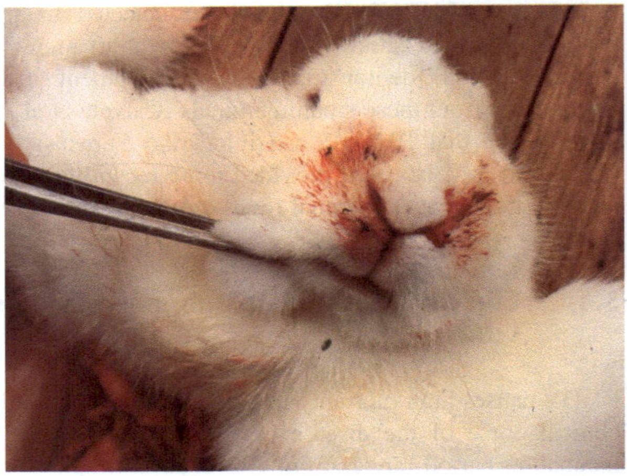

Fig. 3 Rabbit with RHD showing epistaxis (photograph kindly provided by the Pathology Laboratory of INIAV, I.P.)

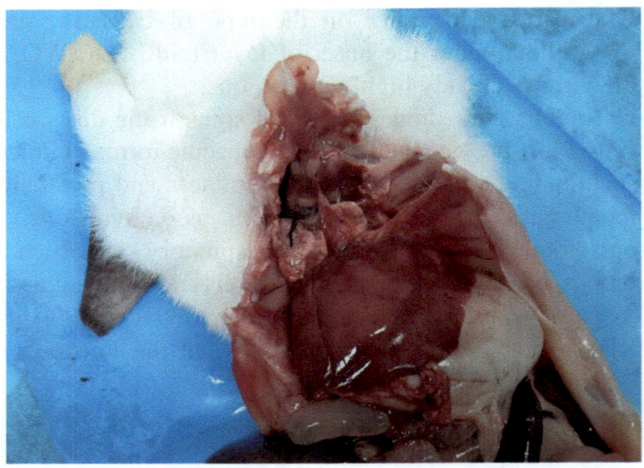

Fig. 4 Rabbit with RHD showing liver discolouration (photograph kindly provided by José Monteiro)

present with chronic or subclinical forms of the disease, characterized by lethargy, loss of weight and generalized jaundice. Animals often die within 1-2 weeks, probably due to liver dysfunction. Surviving animals have very high levels of antibodies. Mortality varies from 5 to 70% (WOAH 2021).

In addition to the broader host range of RHDV2, there are differences in the clinical presentation depending on the age of the animals. While RHDV2 can infect and cause disease in rabbits less than 9 days old, RHDV only causes subclinical infections in rabbits less than 6–8 weeks old (WOAH 2021).

At necropsy, gross pathological lesions are variable. Blood-stained nasal discharge may be present (Fig. 3). Liver necrosis, characterized by a yellowish-brown colour, brittle and degenerated appearance and a marked lobular pattern, and hepatomegaly are observed (Figs. 4 and 5). Splenomegaly is common. Hemorrhages on the organs may be observed. The tracheal mucosa is usually hyperaemic, with

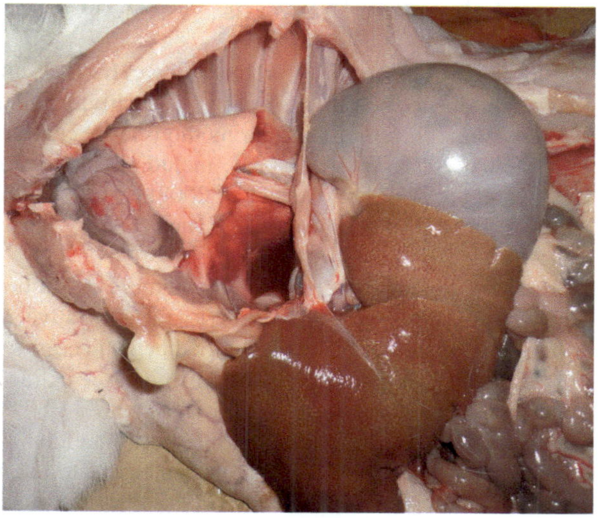

Fig. 5 Rabbit with RHD with accentuated hepatic lobular pattern (photograph kindly provided by José Monteiro)

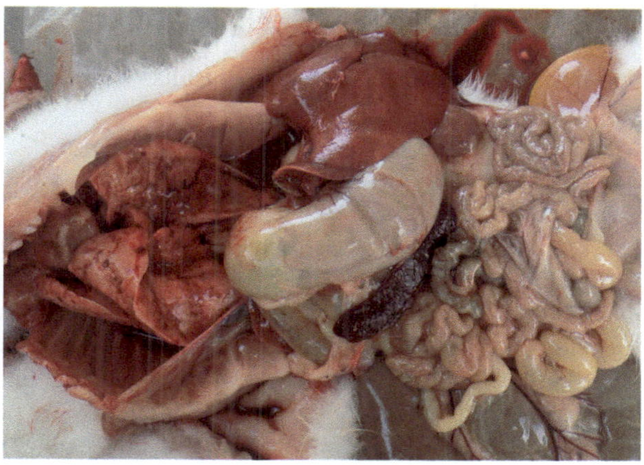

Fig. 6 Rabbit with RHD showing congestion and edema of the lungs with diffuse petechiae (photograph kindly provided by José Monteiro)

abundant fluid, and the lungs are edematous and congested (Fig. 6). The extension of the hemorrhages, which may also involve the pericardium and intestine, and the presence of clotted blood in the blood vessels are due to disseminated intravascular coagulopathy and account for sudden death (Marcato et al. 1991). In subacute and chronic forms of the disease, conjunctival icterus and jaundice of the ears and skin are observed.

Histopathologically, the disease is characterized by acute hepatocellular necrosis often accompanied by signs of disseminated coagulopathy in other organs.

3.5 Laboratory Diagnosis

Confirmation of the disease can be achieved by molecular methods, which are very convenient due to their high specificity, sensitivity and automation, proiding results in a few hours. Several molecular methods are recommended in the WOAH Terrestrial Manual 2021 (WOAH 2021), including conventional RT-PCR (Guittre et al. 1995; Gould et al. 1997; Yang et al. 2008; Le Gall-Reculé et al. 2013) and real-time RT-PCR for RHDV (Gall et al. 2007) and RHDV2 detection (Duarte et al. 2015a, b; Dalton et al. 2018).

The hemagglutination test is less commonly used because not all isolates induce hemagglutination, which can lead to false-negative results, and because of the difficulty in obtaining human red blood cells, necessary for the assay.

Antigen ELISA can be performed using tissue extracts as a positive control. Viral isolation is not an option for diagnosis as neither RHDV nor RHDV2 grows in cell culture. There are also some commercial field tests based on immunochromatography (Fig. 7), which may can allow the rapid diagnosis of an outbreak, although the results must always be confirmed by laboratory diagnosis.

Histopathology of the liver, spleen and other organs can also be used as a means of laboratory diagnosis.

As animals often die before seroconversion, diagnosis based on antibody detection may also lead to false-negative results.

3.6 Disease Control and Prevention

Disease control in rabbitries relies on the application of biosecurity measures and vaccination of the animals. Due to the exceptional resistance of the virus, and the fact that it can be passively transmitted from humans to animals via fomites (contaminated hands, footwear and clothing, car wheels, etc.), additional care must be taken when workers support different production units.

It has been repeatedly suggested that humans were pointed as the possible origin of natural RHD outbreaks in Portugal (Carvalho et al. 2017), Spain (Villafuerte et al. 1995) and the UK (Chasey 1994). In endemic areas, hunters need to be aware of the recommended biosecurity measures, as there is a high risk of carrying the virus from infected areas to industry or domestic rabbitries. In the wild, spread may occur through the natural movement of wild rabbit populations or through translocations of populations as a result of management activities.

The spread of the disease has been also attributed to rabbit exhibitions, which bring together rabbits from different locations (Loliger et al. 1989) and to the trade of infected live rabbits (Schlüter et al. 1990; Gregg et al. 1991) or contaminated products, namely, as fur (Morisse et al. 1991; Carvalho et al. 2017). In addition to mechanical vectors (Crosby and McLennan 1996; Barratt et al. 1998), the virus can be spread by predators and scavengers and carcasses (often carried long distances by birds of prey), and rodents, hares and badgers may also play a role in the dissemination of the disease.

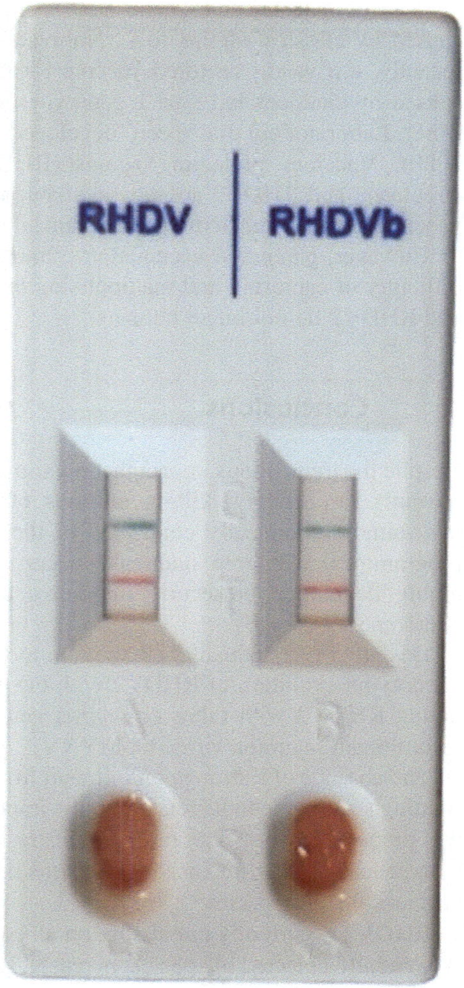

Fig. 7 Pen side test (immunochromatographic test) for rapid RHD diagnosis. This example shows a positive result for RHDV2 (green bands confirm test validity; red bands in both sides indicate the presence of RHDV2) (photograph by Fábio A. Santos)

RHD viruses are extremely resistant but can be inactivated by 1% sodium hydroxide, 1–2% formalin, sodium hypochlorite (10% household bleach) and chloramine.

In terms of prevention, effictive commercial anactivated vaccines against RHDV were introduced in the early 1990s and provided good coverage until 2010 because all circulating strains at that time were of the same genotype. Similarly, after the emergence and spread of RHDV2, RHDV2 inactivated vaccines were commercialized by many laboratories (e.g., Eravac (Hipra Laboratories), Filavac VHD K C + V (Filavie Laboratories), Novarvilap (Ovejero Laboratories), Cunipravac RHDV2 (Hipra Laboratories)).

Recombinant vaccines have also been developed for myxomatosis and RHD caused by classical strains (e.g., Nobivac Myxo-RHD, MSD Laboratories). More recently, a trivalent vectored vaccine for the control of myxomatosis and disease caused by classical and type 2 genotypes of RHDV (Nobivac Myxo-RHD PLUS, MSD Laboratories) has been developed and commercialized (Reemers et al. 2020). Vaccines protecting against RHDV (for countries where RHDV still circulates) and RHDV2 are now available in several countries. Rabbits are usually vaccinated at 10 weeks of age, and annual boosters are recommended.

However, parenteral vaccination cannot be used in wild populations due to the difficulty of capturing and manipulating these extremely sensitive animals. RHDV and RHDV2 do not infect humans.

4 Conclusions

Despite the tremendous impact that myxomatosis and RHD had from 1984 and 2010 onwards respectively, killing millions of meat rabbits, these viral diseases are now almost completely controlled in the industry through intensive vaccination programmes, with rare outbreaks occurring in vaccinated animals. In addition, highly developed biosecurity measures allow some vaccination failures to be overcome.

Despite these valuable prevention tools, we have witnessed the continuous phenotypic evolution of RHDV2 with the emergence of recombinant viruses combining RHDV2 with other *Lagovirus* genotypes, and the recent appearance of a recombinant myxoma virus (ha-MYXV).

The absence of a strong commitment from the pharmaceutical sector to the rabbit farming sector, due to its relatively low importance, raises additional concerns about the preparedness of the rabbit industry for the potential threat of new viral emergencies, taking into account the time needed to develop and bring to the market vaccines for this species.

RHDV2 is currently circulating on all continents except Antarctica and has been sequentially detected in species increasingly distant in evolutionary terms from the rabbits, such as the hares, rodents and Eurasian badgers among others, providing an ideal scenario for the emergence of new virus strains with differnt phenotypes and consequently with unpredictable pathogenic effects.

The rabbit industry must protect itself by mantaining close links with the laboratory sector, monitoring the serological response of the animals to vaccination, testing promptly all suspicians and implementing strict biosecurity measures to prevent the introduction and spread of the disease.

References

Abade dos Santos FA, Carvalho C, Nuno O, Correia JJ, Henriques M, Peleteiro MC, Fevereiro M, Duarte MD (2017) Detection of rabbit haemorrhagic disease virus 2 during the wild rabbit

(*Oryctolagus cuniculus*) eradication from the Berlengas archipelago, Portugal. BMC Vet Res 13:336. https://doi.org/10.1186/s12917-017-1257-3

Abade dos Santos FA, Carvalho C, Monteiro M, Carvalho P, Mendonça P, Peleteiro MC, Duarte MD (2020a) Recombinant myxoma virus infection associated with high mortality in rabbit farming (*Oryctolagus cuniculus*). Transbound Emerg Dis 68:2616–2621. https://doi.org/10.1111/tbed.13899

Abade dos Santos FA, Carvalho C, Pinto A, Rai R, Monteiro M, Carvalho P, Mendonça P, Peleteiro MC, Parra F, Duarte MD (2020b) Detection of recombinant hare myxoma virus in wild rabbits (*Oryctolagus cuniculus algirus*). Viruses 12:1–12. https://doi.org/10.3390/v12101127

Abade dos Santos FA, Carvalho CL, Parra F, Dalton KP, Peleteiro MC (2021a) A quadruplex qPCR for detection and differentiation of classic and natural recombinant myxoma virus strains of leporids. Int J Mol Sci 22:12052. https://doi.org/10.3390/ijms222112052

Abade dos Santos FA, Pinto A, Burgoyne T, Dalton KP, Carvalho CL, Ramilo DW, Carneiro C, Carvalho T, Peleteiro MC, Parra F, Duarte MD (2021b) Spillover events of rabbit hemorrhagic disease virus 2 (recombinant GI.4P-GI.2) from Lagomorpha to Eurasian badger. Transbound Emerg Dis 69:1030–1045. https://doi.org/10.1111/tbed.14059

Abade dos Santos FA, Santos N, Carvalho CL, Martinez-Haro M, Gortázar C, García-Bocanegra I, Capucci L, Duarte M, Alves PC (2022a) Retrospective serological and molecular survey of myxoma or antigenically related virus in the Iberian hare, *Lepus granatensis*. Transbound Emerg Dis 69:3637–3650. https://doi.org/10.1111/tbed.14734

Abade dos Santos FA, Dalton KP, Casero M, Álvarez ÁL, Parra F, Carvalho CL, Duarte MD (2022b) Co-infection by classic MYXV and ha-MYXV in Iberian hare (*Lepus granatensis*) and European wild rabbit (*Oryctolagus cuniculus algirus*). Transbound Emerg Dis 69:1684–1690. https://doi.org/10.1111/tbed.14540

Abrantes J, van der Loo W, Le Pendu J, Esteves PJ (2012) Rabbit haemorrhagic disease (RHD) and rabbit haemorrhagic disease virus (RHDV): a review. Vet Res 43:12. https://doi.org/10.1186/1297-9716-43-12

Abrantes J, Lopes AM, Dalton KP, Melo P, Correia JJ, Ramada M, Alves PC, Parra F, Esteves PJ (2013) New variant of rabbit hemorrhagic disease virus, Portugal, 2012-2013. Emerg Infect Dis 19:1900–1902. https://doi.org/10.3201/eid1911.130908

Anonymous (1989) Doença hemorrágica a vírus do Coelho em Portugal. Revista Portuguesa De Ciências Veterinárias 84:57–58

Arguello-Villares JL, Llanos Pellitero A, Perez Ordoyo Garcia LM (1988) Enfermedad vírica hemorrágica del conejo en Espana. Med Vet 5:645–650

Asin J, Rejmanek D, Clifford DL, Mikolon AB, Henderson EE, Nyaoke AC, Macías-rioseco M, Streitenberger N, Beingesser J, Woods LW, Lavazza A, Capucci L, Crossley B, Uzal FA (2021) Early circulation of rabbit haemorrhagic disease virus type 2 in domestic and wild lagomorphs in southern California, USA (2020–2021). Transbound Emerg Dis 69:e394–e405. https://doi.org/10.1111/tbed.14315

Baily JL, Dagleish MP, Graham M, Maley M, Rocchi MS (2014) RHDV variant 2 presence detected in Scotland. Vet Rec 174:411. https://doi.org/10.1136/vr.g2781

Bao S, An K, Liu C, Xing X, Fu X, Xue H, Wen F, He X, Wang J (2020) Rabbit hemorrhagic disease virus isolated from diseased alpine musk deer (Moschussifanicus). Viruses 12:897. https://doi.org/10.3390/v12080897

Barlow A, Lawrence K, Everest D, Dastjerdi A, Finnegan C, Steinbach F (2014) Confirmation of myxomatosis in a European brown hare in Great Britain. Vet Rec 175:75–76. https://doi.org/10.1136/vr.g4621

Barratt BIP, Ferguson CM, Heath ACG, Evans AA, Logan RAS (1998) Can insects transmit rabbit haemorrhagic disease virus? Proceedings of the Fifty First New Zealand Plant Protection Conference, Hamilton, New Zealand, pp 245–250

Bergin IL, Wise AG, Bolin SR, Mullaney TP, Kiupel M, Maes RK (2009) Novel calicivirus identified in rabbits, Michigan, USA. Emerg Infect Dis 15:1955–1962. https://doi.org/10.3201/eid1512.090839

Buehler M, Jesse ST, Kueck H, Lange B, Koenig P, Jo WK, Osterhaus A, Beineke A (2020) Lagovirus europeus GI.2 (rabbit hemorrhagic disease virus 2) infection in captive mountain hares (Lepus timidus) in Germany. BMC Vet Res 16:166. https://doi.org/10.1186/s12917-020-02386-4

Calvete C, Sarto P, Calvo AJ, Monroy F, Calvo JH (2014) Could the new rabbit haemorrhagic disease virus variant (RHDVb) be fully replacing classical RHD strains in the Iberian Peninsula ? World Rabbit Sci 22:91–91. https://doi.org/10.4995/wrs.2014.1715

Calvete C, Mendoza M, Sarto MP, de Bagüés MPJ, Luján L, Molín J, Calvo AJ, Monroy F, Calvo JH (2019) Detection of rabbit hemorrhagic disease virus GI.2/RHDV2/b in the Mediterranean pine vole (Microtus duodecimcostatus) and white-toothed shrew (Crocidura russula). J Wildl Dis 55:467–472. https://doi.org/10.7589/2018-05-124

Camarda A, Pugliese N, Cavadini P, Circella E, Capucci L, Caroli A, Legretto M, Mallia E, Lavazza A (2014) Detection of the new emerging rabbit haemorrhagic disease type 2 virus (RHDV2) in Sicily from rabbit (*Oryctolagus cuniculus*) and Italian hare (*Lepus corsicanus*). Res Vet Sci 97:642–645. https://doi.org/10.1016/j.rvsc.2014.10.008

Campagnolo ER, Ernst MJ, Berninger ML, Gregg DA, Shumaker TJ, Boghossian AM (2003) Outbreak of rabbit hemorrhagic disease in domestic lagomorphs. J Am Vet Med Assoc 223: 1151–1155. https://doi.org/10.2460/javma.2003.223.1151

Cancellotti F, Renzi M (1991) Epidemiology and current situation of viral haemorrhagic disease of rabbits and the European brown hare syndrome in Italy. Rev Sci Tech 10:409–422

Capucci L, Fusi P, Lavazza A, Pacciarini ML, Rossi C (1996) Detection and preliminary characterization of a new rabbit calicivirus related to rabbit hemorrhagic disease virus but nonpathogenic. J Virol. 70:8614–8623

de Carvalho GF, Fonseca A, Cruz A, Célio P, Mantua P, Simões C, Silva S, Arruda G (1994) Estudo preliminar de alguns parâmetros de uma população de coelho selvagem (Oryctolagus cuniculus) da ilha do Faial-Açores, vol 22. Relatórios e Comunicaçãoes Do Departamento de Biologia, Universidade Dos Açores, Ponta Delgada, Portugal, pp 49–60

Carvalho CL, Silva S, Gouveia P, Costa M, Duarte EL, Henriques AM, Barros SS, Luís T, Ramos F, Fagulha T, Fevereiro M, Duarte MD (2017) Emergence of rabbit haemorrhagic disease virus 2 in the archipelago of Madeira, Portugal (2016–2017). Virus Genes 53:922–926. https://doi.org/10.1007/s11262-017-1483-6

Carvalho CL, Abade Dos Santos FA, Monteiro M, Carvalho P, Mendonça P, Duarte MD (2020) First cases of myxomatosis in Iberian hares (*Lepus granatensis*) in Portugal. Vet Rec Case Rep 8:1–5. https://doi.org/10.1136/vetreccr-2019-001044

Chasey D (1994) Possible origin of rabbit haemorrhagic disease in the United Kingdom. Vet Rec 135:469–499

Crosby TK, McLennan J (1996) Potential vectors of rabbit Calicivirus disease (RCD) in New Zealand: a review. MAF Policy, MAF, Wellington, New Zealand, p 29

Dalton KP, Nicieza I, Balseiro A, Muguerza MA, Rosell JM, Casais R, Álvarez ÁL, Parra F (2012) Variant rabbit hemorrhagic disease virus in young rabbits, Spain. Emerg Infect Dis 18:2009–2012. https://doi.org/10.3201/eid1812.120341

Dalton KP, Nicieza I, Abrantes J, Esteves PJ, Parra F (2014) Spread of new variant RHDV in domestic rabbits on the Iberian Peninsula. Vet Microbiol 169:67–73. https://doi.org/10.1016/j.vetmic.2013.12.015

Dalton KP, Arnal JL, Benito AA, Chacón G, Martín Alonso JM, Parra F (2018) Conventional and real time RT-PCR assays for the detection and differentiation of variant rabbit hemorrhagic disease virus (RHDVb) and its recombinants. J Virol Methods 251:118–122. https://doi.org/10.1016/j.jviromet.2017.10.009

Dalton KP, Martín JM, Nicieza I, Podadera A, Llano D, Casais R, Gimenez S, Badiola I, Agüero M, Duran M, Buitrago D, Romero LJ, García E, Parra F (2019) Myxoma virus jumps species to the Iberian hare. Transbound Emerg Dis 66:2218–2226. https://doi.org/10.1111/tbed.13296

de Aragão HB (1942) Sensibilidade do coelho do mato ao virus do mixoma; transmissio pelo “Aedcs scapularis” e pelo “Stagnmya”. Brasil Médico 56:207

de Carvalho GF, de Almeida LM (1991) Contribuição para o estudo de uma população de coelhos selvagens (*Oryctolagus cuniculus* L.) na ilha de Santa Maria e o impacto do RVHD na população local, vol 19. Relatórios e Comunicações do Departamento de Biologia, Universidade Dos Açores, Ponta Delgada, Portugal, pp 61–67

de Carvalho GF, Ferrand N, Fonseca A, Branco M, Azevedo M, Mendes R, Batista P, Mântua P (1993) Estudo de uma população de coelhos selvagens (*Oryctolagus cuniculus* L.) na ilha de São Jorge, Açores, vol 21. Relatórios e Comunicações do Departamento de Biologia, Universidade Dos Açores, Ponta Delgada, Portugal, pp 8–20

Delibes-Mateos M, Redpath SM, Angulo E, Ferreras P, Villafuerte R (2007) Rabbits as a keystone species in southern Europe. Biol Conserv 137:149–156. https://doi.org/10.1016/j.biocon.2007.01.024

DiGiacomo RF, John Maré C (1994) Viral diseases. In: Manning PJ, Ringler DH, Newcomer CE (eds) The biology of the laboratory rabbit, 2nd edn. Academic Press, New York. ISBN: 9780124692350, pp 171–204. https://doi.org/10.1016/B978-0-12-469235-0.50015-4

Duarte MD, Barros SC, Henriques AM, Fagulha MT, Ramos F, Luís T, Fevereiro M (2014) Development and validation of a real time PCR for the detection of myxoma virus based on the diploid gene M000.5L/R. J Virol Methods 196:219–224. https://doi.org/10.1016/j.jviromet.2013.11.014

Duarte MD, Carvalho CL, Barros SC, Henriques AM, Ramos F, Fagulha T, Luís T, Duarte EL, Fevereiro M (2015a) A real time Taqman RT-PCR for the detection of rabbit hemorrhagic disease virus 2 (RHDV2). J Virol Methods 219:90–95. https://doi.org/10.1016/j.jviromet.2015.03.017

Duarte M, Henriques M, Barros S, Fagulha T, Ramos F, Luis T, Fevereiro M, Benevides S, Flor L, Barros S, Bernardo S (2015b) Detection of RHDV variant 2 in the Azores. Vet Rec 176:130

Duarte MD, Carvalho CL, Abade dos Santos FA, Gomes J, Alves PC, Esteves PJ, Abrantes J, Lopes AM, Serronha A, Santos N, Santos PT, Vaz Y, Carvalho J, Pinto FC, Amaro J, Cunha MV (2018) +Coelho: Avaliação Ecossanitária das Populações Naturais de Coelho Bravo Visando o Controlo da Doença Hemorrágica Viral. Grupo de Trabalho +Coelho. Fundo Florestal Permmanente. Projecto 2017014300001

Duarte MD, Carvalho CL, Abade dos Santos FA, Gomes J, Alves PC, Esteves PJ, Abrantes J, Lopes AM, Serronha A, Santos N, Santos PT, Vaz Y, Carvalho J, Pinto FC, Amaro J (2021) +Coelho 2: Desenvolvimento e implementação de medidas práticas impulsionadoras da recuperação dos leporídeos silvestres em Portugal. +Coelho 2 GT. https://www.iniav.pt/images/Noticias/Coelho2_Final-Report-web.pdf. Accessed 05 April 2023

Eden J-S, Read AJ, Duckworth JA, Strive T, Holmes EC (2015) Resolving the origin of rabbit hemorrhagic disease virus: insights from an investigation of the viral stocks released in Australia. J Virol 89:12217–12220. https://doi.org/10.1128/JVI.01937-15

EMA (2019) CVMP assessment report for Nobivac Myxo-RHD plus (EMEA/V/C/004989/0000). https://www.ema.europa.eu/en/documents/assessment-report/nobivac-myxo-rhd-plus-epar-public-assessment-report_en.pdf. Accessed 05 April 2023

Farnós O, Rodríguez D, Valdés O, Chiong M, Parra F, Toledo JR, Fernández E, Lleonart R, Suárez M (2007) Molecular and antigenic characterization of rabbit hemorrhagic disease virus isolated in Cuba indicates a distinct antigenic subtype. Arch Virol 152:1215–1221. https://doi.org/10.1007/s00705-006-0926-5

Fenner F (1990) Wallace P. Rowe lecture. Poxviruses of laboratory animals. Lab Anim Sci 40:469–480

Fenner F, Fantini B (1999) History of myxomatosis – an experiment in evolution. Biological Control of Vertebrate Pests (CABI), CABI Publishing, Oxford. ISBN: 9780851993232

Fenner BYF, Marshall ID (1957) A comparison of the virulence for European rabbits (*Oryctolagus cuniculus*) of strains of myxoma virus recovered in the field in Australia, Europe and America. J Hyg 55:149–191

Fenner F, Ratcliffe FN (1965) Myxomatosis. Cambridge University Press, London. ISBN: 9780521049917

Fitzner A, Niedbalski W (2018) Detection of rabbit haemorrhagic disease virus 2 (GI.2) in Poland. Pol J Vet Sci 21:451–458. https://doi.org/10.24425/122618

Forrester NL, Trout RC, Gould EA (2007) Benign circulation of rabbit haemorrhagic disease virus on Lambay Island, Eire. Virology 358:18–22. https://doi.org/10.1016/j.virol.2006.09.011

Frölich K, Klima F, Dedek J (1998) Antibodies against rabbit hemorrhagic disease virus in free-ranging red foxes from Germany. J Wildl Dis 34:436–442. https://doi.org/10.7589/0090-3558-34.3.436

Gall A, Hoffmann B, Teifke JP, Lange B, Schirrmeier H (2007) Persistence of viral RNA in rabbits which overcome an experimental RHDV infection detected by a highly sensitive multiplex real-time RT-PCR. Vet Microbiol 120:17–32. https://doi.org/10.1016/j.vetmic.2006.10.006

Le Gall-Reculé G, Zwingelstein F, Boucher S, Le Normand B, Plassiart G, Portejoie Y, Decors A, Bertagnoli S, Guerin J-L, Marchandeau S (2011a) Detection of a new variant of rabbit haemorrhagic disease virus in France. Vet Rec 168:137–138. https://doi.org/10.1136/vr.d697

Le Gall-Reculé G, Zwingelstein F, Fages MP, Bertagnoli S, Gelfi J, Aubineau J, Roobrouck A, Botti G, Lavazza A, Marchandeau S (2011b) Characterisation of a non-pathogenic and non-protective infectious rabbit lagovirus related to RHDV. Virology 410:395–402. https://doi.org/10.1016/j.virol.2010.12.001

García-Bocanegra I, Camacho-Sillero L, Caballero-Gómez J, Aguero M, Gómez-Guillamón F, Ruiz-Casas J, Díaz-Cao J, García E, Ruano M, Haza R (2020) Monitoring of emerging myxoma virus epidemics in Iberian hares. Transbound Emerg Dis 68:1275–1282. https://doi.org/10.1111/tbed.13781

Gavier-Widén D (1994) Morphologic and immunohistochemical characterization of the hepatic lesions associated with European brown hare syndrome. Vet Pathol 31:327–334. https://doi.org/10.1177/030098589403100305

Gould AR, Kattenbelt JA, Lenghaus C, Morrissy C, Chamberlain T, Collins BJ, Westbury HA (1997) The complete nucleotide sequence of rabbit haemorrhagic disease virus (Czech strain V351): use of the polymerase chain reaction to detect replication in Australian vertebrates and analysis of viral population sequence variation. Virus Res 47:7–17

Gregg D, House C, Meyer R, Berninger M (1991) Viral haemorrhagic disease of rabbits in Mexico: epidemiology and viral characterization. Rev Sci Tech 10:435–451

Guittre C, Baginski I, Le Gall G, Prave M, Trepo C, Cova L (1995) Detection of rabbit haemorrhagic disease virus isolates and sequence comparison of the N-terminus of the capsid protein gene by the polymerase chain reaction. Res Vet Sci 58:128–132. https://doi.org/10.1016/0034-5288(95)90065-9

Hall RN, Mahar JE, Haboury S, Stevens V, Holmes EC, Strive T (2015) Emerging rabbit hemorrhagic disease virus 2 (RHDVb), Australia. Emerg Infect Dis 21:2276–2278. https://doi.org/10.3201/eid2112.151210

Hermans K, Moeremans I, Verlinden M, Garmyn A (2016) Also Belgian rabbits are affected by the rabbit hemorrhagic disease type 2 virus. Vlaams Diergeneeskundig Tijdschrift 85:309–314. https://doi.org/10.21825/vdt.v85i5.16323

Jahnke M, Holmes EC, Kerr PJ, Wright JD, Strive T (2010) Evolution and phylogeography of the nonpathogenic calicivirus RCV-A1 in wild rabbits in Australia. J Virol 84:12397–12404. https://doi.org/10.1128/JVI.00777-10

Joubert L, Leftheriosis E, Mouchet J (1972) La Myxomatose (in French). In: Goret P (ed) Maladies Animales a virus Lepine, vol 643. L'Expansion Scientifique Française, Paris, France. ISBN: 2-7D46-D7D2-8

Kennedy A, Britton L, Byrne AW, Byrne C, Casey M, Flynn O, Lozano JM, Marnell F, McElroy M, Reid N, Wilson M, FitzGerald W (2021) First detected case of rabbit haemorrhagic disease virus 2 (RHDV2) in the Irish hare (*Lepus timidus hibernicus*). Ir Vet J 74:25. https://doi.org/10.1186/s13620-021-00205-2

Kerr PJ, Best SM (1998) Myxoma virus in rabbits. Revue scientifique et technique (International Office of Epizootics) 17:256–268. https://doi.org/10.20506/rst.17.1.1081

Kerr PJ, Kitchen A, Holmes EC (2009) Origin and phylodynamics of rabbit hemorrhagic disease virus. J Virol 83:12129–12138. https://doi.org/10.1128/jvi.01523-09

Kerr PJ, Liu J, Cattadori I, Ghedin E, Read AF, Holmes EC (2015) Myxoma virus and the leporipoxviruses: An evolutionary paradigm. Viruses 7:1020–1061. https://doi.org/10.3390/v7031020

Kinnear M, Linde CC (2010) Capsid gene divergence in rabbit hemorrhagic disease virus. J Gen Virol 91:174–181. https://doi.org/10.1099/vir.0.014076-0

Le Gall-Reculé G, Lavazza A, Marchandeau S, Bertagnoli S, Zwingelstein F, Cavadini P, Martinelli N, Lombardi G, Guérin JL, Lemaitre E, Decors A, Boucher S, Le Normand B, Capucci L (2013) Emergence of a new lagovirus related to rabbit haemorrhagic disease virus. Vet Res 44:1–13. https://doi.org/10.1186/1297-9716-44-81

Leighton FA, Artois M, Capucci L, Gavier-Widén D, Morisse JP (1995) Antibody response to rabbit viral haemorrhagic disease virus in red foxes (*Vulpes vulpes*) consuming livers of infected rabbits (*Oryctolagus cuniculus*). J Wildl Dis 31:541–544

Liu S, Xue H, Pu B, Qian N (1984) A new viral disease in rabbit. Anim Husb Vet Med (Xumu yu Shouyi) 16:253–255

Loliger HC, Matthes S, Liess B (1989) Uber das Auftreten einer infektiosen hamorrhagischen Erkrankung bei Hauskaninchen in tier Bundesrepublik Deutschland - Tierarztliche. Umschau 44:22–25

Lopes AM, Correia J, Abrantes J, Melo P, Ramada M, Magalhães MJ, Alves PC, Esteves PJ (2014a) Is the new variant RHDV replacing genogroup 1 in Portuguese wild rabbit populations? Viruses 7:27–36. https://doi.org/10.3390/v7010027

Lopes AM, Marques S, Silva E, Magalhães MJ, Pinheiro A, Alves PC, Le Pendu J, Esteves PJ, Thompson G, Abrantes J (2014b) Detection of RHDV strains in the Iberian hare (Lepus granatensis): earliest evidence of rabbit lagovirus cross-species infection. Vet Res 45:1–7. https://doi.org/10.1186/s13567-014-0094-7

Marcato P, Benazzi C, Vecchi G, Galeotti M, Della Salda L, Sarli G, Lucidi P (1991) Clinical and pathological features of viral haemorrhagic disease of rabbits and the European brown hare syndrome. Revue scientifique et technique (International Office of Epizootics) 10:371–392

Marlier D, Mainil J, Linde A, Vindevogel H (2000a) Infectious agents associated with rabbit pneumonia: isolation of amyxomatous myxoma virus strains. Vet J 159:171–178. https://doi.org/10.1053/tvjl.1999.0413

Marlier D, Mainil J, Sulon J, Beckers JF, Linden A, Vindevogel H (2000b) Study of the virulence of five strains of amyxomatous myxoma virus in crossbred New Zealand White/Californian conventional rabbits, with evidence of long-term testicular infection in recovered animals. J Comp Pathol 122:101–113. https://doi.org/10.1053/jcpa.1999.0345

McIntosh MT, Behan SC, Mohamed FM, Lu Z, Moran KE, Burrage TG, Neilan JG, Ward GB, Botti G, Capucci L, Metwally SA (2007) A pandemic strain of calicivirus threatens rabbit industries in the Americas. Virol J 4:96. https://doi.org/10.1186/1743-422X-4-96

Merchán T, Rocha G, Alda F, Silva E, Thompson G, de Trucios SH, Pagés A (2011) Detection of rabbit haemorrhagic disease virus (RHDV) in nonspecific vertebrate hosts sympatric to the European wild rabbit (*Oryctolagus cuniculus*). Infect Genet Evol 11:1469–1474. https://doi.org/10.1016/j.meegid.2011.05.001

Meyers G, Wirblich C, Thiel HJ (1991) Rabbit hemorrhagic disease virus-molecular cloning and nucleotide sequencing of a calicivirus genome. Virology 184:664–676. https://doi.org/10.1016/0042-6822(91)90436-F

Morisse J, Le Gall G, Boilletot E (1991) Hepatitis of viral origin in Leporidae: introduction and aetiological hypotheses. Revue scientifique et technique (International Office of Epizootics) 10: 283–295

Moss SR, Turner SL, Trout RC, White PJ, Hudson PJ, Desai A, Armesto M, Forrester NL, Gould EA (2002) Molecular epidemiology of rabbit haemorrhagic disease virus. J Gen Virol 83:2461–2467. https://doi.org/10.1099/0022-1317-83-10-2461

Motha MX, Clark RG (1998) Confirmation of rabbit haemorrhagic disease in wild New Zealand rabbits using the ELISA. N Z Vet J 46:83–84. https://doi.org/10.1080/00480169.1998.36065

Mutze G, Cooke B, Alexander P (1998) The initial impact of rabbit hemorrhagic disease on European rabbit populations in South Australia. J Wildl Dis 34:221–227. https://doi.org/10.7589/0090-3558-34.2.221

Nauwynck H, Callebaut P, Peeters J, Ducatelle R, Uyttebroek E (1993) Susceptibility of hares and rabbits to a Belgian isolate of European brown hare syndrome virus. J Wildl Dis 29:203–208. https://doi.org/10.7589/0090-3558-29.2.203

Neimanis AS, Ahola H, Zohari S, Larsson Pettersson U, Bröjer C, Capucci L, Gavier-Widén D (2018) Arrival of rabbit haemorrhagic disease virus 2 to northern Europe: emergence and outbreaks in wild and domestic rabbits (*Oryctolagus cuniculus*) in Sweden. Transbound Emerg Dis 65:213–220. https://doi.org/10.1111/tbed.12650

O'Keefe JS, Tempero J, Atkinson PH, Pacciarini L, Fallacara F, Horner GW, Motha J (1998) Typing of rabbit haemorrhagic disease virus from New Zealand wild rabbits. N Z Vet J 46:42–43. https://doi.org/10.1080/00480169.1998.36053

Ohlinger VF, Haas B, Meyers G, Weiland F, Thiel H-J (1990) Identification and characterization of the virus causing rabbit hemorrhagic disease. J Virol 64:3331–3336

Parra F, Prieto M (1990) Purification and characterization of a calicivirus as the causative agent of a lethal hemorrhagic disease in rabbits. J Virol 64:4013–4015

Patton NM, Holmes HT (1977) Myxomatosis in domestic rabbits in Oregon. J Am Vet Med Assoc 171:560–562

Pinto A, De Matos AL, Abrantes M, Kraberger S, Risalde MA, Gort C, Mcfadden G, Varsani A, Esteves PJ (2019) Genetic characterization of a recombinant myxoma virus in the Iberian hare (*Lepus granatensis*). Viruses 11:1–16. https://doi.org/10.3390/v11060530

Puggioni G, Cavadini P, Maestrale C, Scivoli R, Botti G, Ligios C, Le Gall-Reculé G, Lavazza A, Capucci L (2013) The new French 2010 rabbit hemorrhagic disease virus causes an RHD-like disease in the Sardinian Cape hare (*Lepus capensis mediterraneus*). Vet Res 44:1–7. https://doi.org/10.1186/1297-9716-44-96

Reemers S, Peeters L, van Schijndel J, Bruton B, Sutton D, van der Waart L, van de Zande S (2020) Novel trivalent vectored vaccine for control of Myxomatosis and disease caused by classical and a new genotype of rabbit haemorrhagic disease virus. Vaccine 8:441. https://doi.org/10.3390/vaccines8030441

Rocchi MS, Dagleish MP (2018) Diagnosis and prevention of rabbit viral haemorrhagic disease 2. In Pract 40:11–16. https://doi.org/10.1136/inp.k54

Rocchi M, Maley M, Dagleish M, Vick C, Ryan D, Lee A, Jahns H (2016) RHDV-2 on the Isle of Man and in the Republic of Ireland. Vet Rec 179:389. https://doi.org/10.1136/vr.i5530

Sanarelli G (1898) Das myxomatogene virus. Beitrag Zum Stadium Der Krankheit-Serreger Ausserhalbdes Sichtbaren. Zentralblatt für Bakteriologie, Parasitenkunde, Infektionskrankheiten und Hygiene. Zweite Naturwissenschaftliche Abteilung: Allgemeine, Landwirtschaftliche und Technische Mikrobiologie 23:865–873

Schlüter H, Schirrmeier H, Bohme R (1990) Stand der Diagnostik und Bekämpfung der Hämorrhagischen Septikäemie der Kaninchen in der DDR [in German]. Monatsh Veterinarmed 45:286–289

Simón MC, Muguruza R, Alonso JL, Muzquiz JL, Girones O, Haffar A (1994) Rechererche du virus de la maladie hémorrhagique virale du lapin (RHD) chez le renard et rôle des canides domesiques dans la transmission de la maladie. Recueil de Medecine Veterinaire 170:841–845

Sobey WR (1969) Selection for resistance to myxomatosis in domestic rabbits (*Oryctolagus cuniculus*). J Hyg 67:743–754. https://doi.org/10.1017/s0022172400042194

Sobey WR, Conolly D, Haycockp, Edmonds JW (1970) Myxomatosis. The effect of age upon survival of wild and domestic rabbits (*Oryctolagus cuniculus*) with a degree of genetic resistance and unselected domestic rabbits infected with myxoma virus. J Hyg 68:137–149. https://doi.org/10.1017/s002217240002859x

Strive T, Wright JD, Robinson AJ (2009) Identification and partial characterisation of a new lagovirus in Australian wild rabbits. Virology 384:97–105. https://doi.org/10.1016/j.virol.2008.11.004

Strive T, Elsworth P, Liu J, Wright JD, Kovaliski J, Capucci L (2013) The non-pathogenic Australian rabbit calicivirus RCV-A1 provides temporal and partial cross protection to lethal rabbit haemorrhagic disease virus infection which is not dependent on antibody titres. Vet Res 44:1–11. https://doi.org/10.1186/1297-9716-44-51

Toledo M, Seoane G, Serrano E, Encinosa A, Delgado C, González S (1995) Estudio epizootiológico y control de la enfermedad hemorrágica viral del conejo en la República de Cuba. Revista Cubana de Ciencias Veterinarias 24:5–9

Velarde R, Cavadini P, Neimanis A, Cabezón O, Chiari M, Gaffuri A, Lavín S, Grilli G, Gavier-Widén D, Lavazza A, Capucci L (2017) Spillover events of infection of brown hares (*Lepus europaeus*) with rabbit haemorrhagic disease type 2 virus (RHDV2) caused sporadic cases of an European brown hare syndrome-like disease in Italy and Spain. Transbound Emerg Dis 64: 1750–1761. https://doi.org/10.1111/tbed.12562

Villafuerte R, Calvete C, Blanco JC, Lucientes J (1995) Incidence of viral hemorrhagic disease in wild rabbit populations in Spain. Mammalia 59:651–660. https://doi.org/10.1515/mamm.1995.59.4.651

Vinjé J, Estes MK, Esteves P, Green KY, Katayama K, Knowles NJ, L'Homme Y, Martella V, Vennema H, White PA, Consortium IR (2019) ICTV virus taxonomy profile: caliciviridae. J Gen Virol 100:1469–1470. https://doi.org/10.1099/jgv.0.001332

Westcott DG, Frossard J-P, Everest D, Dastjerdi A, Duff JP, Choudhury B (2014) Incursion of RHDV2-like variant in Great Britain. Vet Rec 174:333. https://doi.org/10.1136/vr.g2345

Wibbelt G, Frölich K (2005) Infectious diseases in European Brown hare (*Lepus europaeus*). Wildl Biol Pract 1:86–93. https://doi.org/10.2461/wbp.2005.1.11

Wirblich C, Meyers G, Ohlinger VF, Capucci L, Eskens U, Haas B, Thiel HJ (1994) European brown hare syndrome virus: relationship to rabbit hemorrhagic disease virus and other caliciviruses. J Virol 68:5164–5173. https://doi.org/10.1128/2Fjvi.68.8.5164-5173.1994

Wirblich C, Thiel HJ, Meyers G (1996) Genetic map of the calicivirus rabbit hemorrhagic disease virus as deduced from in vitro translation studies. J Virol 70:7974–7983. https://doi.org/10.1006/viro.2000.0579

WOAH (2016) Rabbit haemorrhagic disease, Canada–immediate notification report. World Organisation of Animal Health, Paris, France. (Information received on 25/08/2016 from Dr Martine Dubuc, OIE Delegate for Canada, Chief Food Safety Officer Vice-President Science Branch, Health Ministry, Canadian Food Inspection Agency, Ottawa, Canada)

WOAH (2021) Rabbit haemorrhagic disease. Manual of diagnostic tests and vaccines for terrestrial animals 2022, Chapter 3.7.2. https://www.woah.org/fileadmin/Home/eng/Health_standards/tahm/3.07.02_RHD.pdf. Accessed 05 April 2023

Yang L, Wang F, Hu B, Xue J, Hu Y, Zhou B, Wang D, Xu W (2008) Development of an RT-PCR for rabbit hemorrhagic disease virus (RHDV) and the epidemiology of RHDV in three eastern provinces of China. J Virol Methods 151:24–29. https://doi.org/10.1016/j.jviromet.2008.04.003

Rabbit Viral Diseases (II): Other Viruses

Margarida Duarte, Teresa Fagulha, Margarida Henriques, Silvia Barros, Fernanda Ramos, Ana Duarte, Tiago Luís, and Fábio A. Abade dos Santos

Abstract

This chapter focuses on viral diseases of rabbits which, compared with myxomatosis and rabbit haemorrhagic disease have a lower pathogenic impact, lower associated morbidity and mortality or are rarely observed However, they may cause severe disease in association with other viral or nonviral agents Rabbitpox, rabbit (Shope) fibroma virus, papillomavirus, rotavirus, coronavirus, parvovirus, herpesvirus, and rabies, are reported under the headings. The main aspects of etiology, epidemiology, pathogenesis and clinical signs, laboratory diagnosis, control, and prevention are highlighted.

M. Duarte (✉) · A. Duarte
Instituto Nacional de Investigação Agrária e Veterinária (INIAV, I.P.), Oeiras, Portugal

Centre for Interdisciplinary Research in Animal Health (CIISA), Faculdade de Medicina Veterinária, Universidade de Lisboa, Lisbon, Portugal

Associate Laboratory for Animal and Veterinary Sciences (AL4AnimalS), Lisbon, Portugal
e-mail: margarida.duarte@iniav.pt

T. Fagulha · M. Henriques · S. Barros · F. Ramos · T. Luís
Instituto Nacional de Investigação Agrária e Veterinária (INIAV, I.P.), Oeiras, Portugal

F. A. Abade dos Santos
Instituto Nacional de Investigação Agrária e Veterinária (INIAV, I.P.), Oeiras, Portugal

Centre for Interdisciplinary Research in Animal Health (CIISA), Faculdade de Medicina Veterinária, Universidade de Lisboa, Lisbon, Portugal

Associate Laboratory for Animal and Veterinary Sciences (AL4AnimalS), Lisbon, Portugal

Faculdade de Medicina Veterinária, Universidade Lusófona de Humanidades e Tecnologia, Lisbon, Portugal

J. Simões, J. M. Monteiro (eds.), *Veterinary Care of Farm Rabbits*,
https://doi.org/10.1007/978-3-031-44542-2_26

Keywords

European rabbit · *Oryctolagus cuniculus* · Rabbit fibroma virus · Rabbitpox · Papillomavirus · Rotavirus · Coronavirus · Parvovirus · Herpesvirus and rabies

1 Introduction

Viral diseases of the European domestic rabbit (*Oryctolagus cuniculus*) have very different clinical presentations and clinical significance, ranging from those that severely impact the health and welfare of rabbits, associated with mortality rates approaching 100% and registered almost worldwide, to those of lesser relevance, reported only sporadically.

The 1st group is represented by rabbit hemorrhagic disease (RHD) and myxomatosis (i.e., the most pathogenic strains), both of which are included in the list of the 117 notifiable diseases of terrestrial animals by the World Organization for Animal Health (WOAH) for their importance (see Sect 3.7.1. for Myxomatosis (2021) and Sect. 3.7.2. for RHD (2021)). Other viruses, such as rabbitpox, rabbit (Shope) fibroma virus, papillomavirus, rotavirus, coronavirus, parvovirus, herpesvirus, and rabies, among others, are far less commonly described in rabbits. However, some of these, such as rotavirus, coronavirus, and parvovirus, can have a strong impact when associated with other pathogens.

In the present chapter, we review eight less common viral diseases with tropisms for different tissues, e.g., skin, gastrointestinal tract, and nervous system.

2 Rabbit (Shope) Fibroma Virus

Etiology

Rabbit fibroma, also known as Shope fibroma, is a benign tumor caused by a poxvirus closely related to myxoma virus (Fenner 1953), known as rabbit fibroma virus (RFV). RFV is a dsDNA virus with up to 260 bp which, like myxoma virus, belongs to the *Leporipoxvirus* genus, *Poxviridae* family. The induced subcutaneous formations are flat and mobile and are usually found on the legs, feet, ears, and muzzle and around the eyes' thinly fur-haired areas. These tumors may measure several centimeters in diameter and disappear spontaneously within a few months.

The Disease Origin and Its Natural Hosts

Rabbit fibroma virus was first in 1932 (Shope 1932) in an eastern cottontail rabbit (*Sylvilagus floridanus*), its natural host (Dalmat 1959; Pulley and Shively 1973; McLeod and Langlinais 1981).

Susceptible Species

European rabbit (*O. cuniculus*) is also susceptible to infection (Fenner and Ratcliffe 1965; DiGiacomo and John Maré 1994). In domestic rabbits, the infection with RFV has been associated with outbreaks resulting in high morbidity and mortality in newborn animals due to systemic disease (Joiner et al. 1971).

In areas where RFV is endemic in wild rabbits, fibromas may be found in domestic rabbits where poor husbandry practices allow contact with biting arthropod vectors such as mosquitoes and fleas, which mechanically transmit the virus to healthy animals (Kilham and Dalmat 1955; Dalmat 1959).

Clinical Signs and *Postmortem* Macroscopic and Microscopic Lesions
The tumor begins as a slight thickening of the subcutaneous tissue, usually on the legs, feet, and ears and develops into a demarcated soft swelling. These tumors may persist for several months before regressing without sequelae.

Histopathological examination of sections of the tumour reveals. Microscopic examination of the skin lesions shows proliferation of mesenchymal cells which become stellate or ovoid, multifocal areas of necrosis, and mononuclear and polymorphonuclear infiltrates (Raflo et al. 1973; McLeod and Langlinais 1981).

Laboratory Diagnosis
Diagnosis of RFV is based on the clinical presentation, gross lesion characteristics, and histopathological data from tumor biopsies. Differentiation of rabbit fibroma virus from myxoma virus may be achieved by polymerase chain reaction (PCR) directed to the M0005L/R gene which does not detect RFV (Duarte et al. 2014).

Disease Control and Prevention
Control of the disease can be achieved by limiting exposure to arthropod vectors through the use of nets. As tumors spontaneously regress, treatment is generally not necessary (Jenkins 2001).

3 Rabbitpox

Etiology
Rabbitpox virus (RPV) infection is an acute, systemic, generalized, and contagious rare disease reported only in laboratory rabbits (*Oryctolagus* spp.), caused by a dsDNA virus (~200 bp) from the genus *Orthopoxvirus*, family *Poxviridae* (rabbitpox virus), with 95% genetic similarity to vaccinia virus (VV) (Li et al. 2005). VV confers cross-protection against RPV infection demonstrating the close relationship between these two viruses.

The Disease Origin and Its Natural Hosts
The original discovery of RPV in rabbit colonies was thought to have originated from a strain of VV selected by an animal passage from a strain of VV used at the time (Bedson and Duckworth 1963). RPV was first isolated at the Rockefeller Institute in New York, USA, between 1930 and 1933, and then from a rabbit colony outbreak in a laboratory in Utrecht, the Netherlands, in 1941 (Greene 1935b; Jenkins

2001; Esposito and Fenner 2011). RPV has never been recognized in wild rabbits (*Sylvilagus* spp., *O. cuniculus algirus*, or others).

Susceptible Species

Rabbitpox emerged as a consequence of inoculation of laboratory rabbits (*O. cuniculus*), with VV. Differences in susceptibility between rabbit breed have been previously reported (Greene 1935a).

Clinical Signs and *Postmortem* Macroscopic and Microscopic Lesions

The disease is usually lethal, and death may occur before clinical signs develop. However, affected rabbits may develop clinical signs including an erythematous rash progressing to papules and crusty nodules on the skin and mucous membranes (resembling smallpox in humans), fever, lymphadenitis, nasal discharge, blepharitis, keratitis, conjunctivitis, and orchitis. Rabbits may also show respiratory signs rather than skin lesions. Mortality rates are usually high.

At necropsy, in addition to the alterations referred above, subcutaneous edema and edema of the oral cavity, scrotum, and vulva may be present (Bedson and Duckworth 1963; McLeod and Langlinais 1981; Brabb and Di Giacomo 2012).

Laboratory Diagnosis

Virological diagnosis can be based on the observation of characteristic skin lesions, isolation of the virus in rabbit kidney cells (RK13 cells), and on fluorescent antibody test (FAT) of viral antigens in infected tissues which can be detected by VV antibodies.

Histopathological analysis of the lesions shows lymphoid necrosis and mononuclear cell infiltration.

Serological diagnosis can be made using methods developed for the detection of vaccinia virus antibodies.

Rabbitpox may be confused with myxomatosis which can be excluded by molecular methods specific for the detection of myxoma virus (Duarte et al. 2014; Abade dos Santos et al. 2021a).

Disease Control and Prevention

RPV is transmitted by ingestion or inhalation of infected tissue. Unlike myxoma virus or fibroma virus, arthropods have not been shown to play a role in the epidemiology of rabbitpox infection. Treatment is only supportive.

4 *Sylvilagus floridanus* Papillomavirus *1* (Shope Papillomavirus)

Etiology

First named Shope papillomavirus (SPV), it has been renamed and is officially known as *S. floridanus* papillomavirus 1 (SfPV1). It is an oncogenic, non-enveloped, circular double-stranded DNA virus (Giri et al. 1985) of the

Papillomaviridae family (Van Regenmortel et al. 2002), which is transmitted by biting arthropods (Harvey 1990).

The Disease Origin and Its Natural Hosts

Rabbit (Shope) papillomavirus was the first tumorigenic virus to be identified in mammals and was associated with the occurence of wartlike tumors in the cottontail rabbit (*S. floridanus*) (Shope and Hurst 1933).

Susceptible Species

Although the cottontail rabbit is the natural host of SfPV1, this virus is transmissible to the European rabbit (*O. cuniculus*) (Kidd and Rous 1938; Shope 1935).

Clinical Signs and *Postmortem* Macroscopic and Microscopic Lesions

Clinical signs in *Sylvilagus* species include skin lesions, located predominantly on the neck, shoulders, and abdomen, which initially appear as rough, red, wart-like protuberances and later develope into keratinized, often pigmented and large horn-shaped growths which may disappear within a few months or become neoplastic being replaced by squamous cell carcinomas (Shope and Hurst 1933). In the domestic rabbit, spontaneous infection was characterized by horny protuberances that were located most commonly on hairless areas such as the ears and eyelids, suggesting that transmission from the cottontail rabbit to the domestic rabbit may be primarily mosquito-borne (Hagen 1966).

Laboratory Diagnosis

S. floridanus papillomavirus 1 is diagnosed by histopathological examination of typical gross lesions characterized by hyperkeratotic tissues in domestic rabbits. In contrast to other papillomaviruses, inclusion bodies are rarely observed.

In the differential diagnosis, Shope papillomas must be distinguished from spontaneous, nonviral papillomas. In contrast, pappilomas in the natural host (*Sylvilagus* species), usually contain infectious viral particles.

Disease Control and Prevention

In endemic areas, the natural infection and spread of the disease can be controlled through the control of the arthropod vectors, mainly rabbit ticks, mosquitoes, and reduviid bugs (Larson et al. 1936; Dalmat 1959), by mosquito nets, as the main mode of transmission is by arthropod vectors, although the virus can also be transmitted by direct contact.

Treatment involves the surgical removal of the papilloma (Jenkins 2001).

5 Rotavirus Infection

Etiology

Rotavirus infections are characterized by mild to severe diarrhea and are caused by a double-stranded segmented RNA (dsRNA) virus, belonging to the family *Reoviridae*, subfamily *Sedoreovirinae*, genus *Rotavirus*. There are seven groups (A to G) of rotaviruses (Theil and Saif 1985; Saif 1990; Bridger 1994). Rotaviruses (RV) are highly contagious, mild viruses that are spread by the fecal-oral route (Baker 1998; Martella et al. 2005; Schoondermark-van de Ven et al. 2013). Although the infection rate with rabbit rotavirus (RRV) is high, most infections are mild or undetected (subclinical forms of infection). However, coinfections with other pathogens, either bacteria such as *Clostridium* spp. or *Escherichia coli*, viruses, or parasites, can cause severe enteritis with the excretion of high viral loads in the feces which can lead to new infections. Mixed infections result in a much more severe clinical picture.

The Disease Origin and Its Natural Hosts

Rabbit rotavirus was first isolated from diarrheic rabbits in England, in 1976 (Bryden et al. 1976), and has since been isolated in many different countries.

Although the incidence of rotaviral infection in rabbits is not known, serological data from the United States demonstrated seroprevalences in wild leporidae (29% in *S. floridanus*, 52% in *L. americanus*) and in domestic rabbits (98% in New Zealand white rabbits (*O. cuniculus*)) (Petric et al. 1978). Seroprevalence was extremely high (90%) in all age groups (i.e., neonates, weanlings, young adults and breeding adults), in contrast to preweanling animals where the percentage of positives was much lower (25 to 69%) (DiGiacomo and Thouless 1984, 1986). Rotavirus infection appears to be age-dependent, being more common in rabbits between 35 and 50 days of age, when maternal antibodies decline. At this age, the disease is characterized by high morbidity and clinical signs such as dehydration, diarrhea, depression, and anorexia.

Susceptible Species

Rabbit rotaviral infections are caused by Group A rotaviruses, which are also common in humans and several other animal species (Castrucci et al. 1985; Thouless et al. 1986; Tanaka et al. 1988).

Clinical Signs and *Postmortem* Macroscopic and Microscopic Lesions

The most common clinical signs are diarrhea and dehydration (Schoeb et al. 1986; Thouless et al. 1988, 1996). However, rotavirus infections may be subclinical, with virus shedding in feces without any signs of disease (Bryden et al. 1976; DiGiacomo and Thouless 1986; Conner et al. 1988; Hambraeus et al. 1989). Experimental infections with rotavirus isolated from the feces of animals with severe diarrhea resulted in only mild clinical signs (Thouless et al. 1988, 1996; Hambraeus et al. 1989; Ciarlet et al. 1998), suggesting that other factors (e.g., other pathogens) are involved in the pathogenesis of the disease. This is supported by a study showing more severe signs and increased mortality when rabbits were co-infected with *E. coli* (Thouless et al. 1996). At necropsy, the small intestine and cecum may be distended and filled with fluid (Petric et al. 1978; Thouless et al. 1988). Histopathological findings concerned the jejunum and the ileum where shortened, fused villi with reduced epithelium and an increased crypt depth may be observed (Schoeb et al. 1986; Thouless et al. 1988).

Laboratory Diagnosis

Clinical diagnosis can be made on the basis of clinical signs in the appropriate age group.

Laboratory diagnosis can be made by molecular methods, namely, RT-PCR (Gentsch et al. 1992; Elschner et al. 2002; Iturriza Gómara et al. 2002) or Luminex x-TAG assay (Wu et al. 2018), virus isolation from the feces of animals with diarrhea, and antigen ELISA, such as those used in diagnosis in cattle (Castrucci et al. 1984) or in humans (Fushuku and Fukuda 2006), given the similarity between genotypes.

Differential diagnosis should include coccidiosis (*Eimeria* spp.), colibacillosis (*E. coli*), Tyzzer's disease (*Clostridium piliforme*), clostridial enteritis (*Clostridium perfringens*), toxemia, and dysbiosis (Martinic et al. 1995).

Disease Control and Prevention

While treatment of an individual rabbit relies on fluid therapy, administered orally or subcutaneously, breeding in a rabbit farm can be interrupted for 4 to 6 weeks to stop the influx of naïve animals and to allow infected animals to seroconvert and stop shedding of virus in feces (DiGiacomo and John Maré 1994). Rabbits that recover become resistant to subsequent infections (Castrucci et al. 1984; Hambraeus et al. 1989).

6 Rabies

Etiology

Although natural rabies infections in rabbits are rarely reported, rabbits are highly susceptible to rabies, having been used for many years in the development of the first rabies vaccine by Louis Pasteur in the 1880s and in diagnostic tests.

Therefore, rabies should not be excluded in animals with a history of rapid progression of neurologic disease (also associated with herpes simplex infection) and with a history of outdoor housing or possible exposure to wildlife in areas where the disease is endemic.

Rabies virus (RV) is a bullet-shaped virus with a non-segmented, negative-stranded RNA of approximately 12 Kb, belonging to the order *Mononegavirales*, *Rhabdoviridae* family, *Lyssavirus* genus which also includes Lagos bat virus, Mokola virus, Duvenhage virus, European bat virus 1 and 2, and Australian bat virus.

The Disease Origin and Its Natural Hosts

The history of rabies can be traced back more than 5000 years ago (Fu 1997). The earliest record of a disease affecting humans describing signs and symptoms compatible with rabies dates back to the eighteenth and nineteenth centuries BC, and is found in the *Eshnunna cuneiform law tablets* from ancient Mesopotamia, Middle East (Fales 2009).

Susceptible Species

Though all species of mammals are susceptible to rabies, only a few species play a role as virus reservoirs, namely, insectivorous bats, raccoons, skunks, and foxes.

Clinical Signs and Postmortem Macroscopic and Microscopic Lesions

In rabbits, the initial signs of infection may be nonspecific such as anorexia, prostration and lethargy, fever, restlessness, and weight loss. As the disease progresses, neurological signs become evident, including teeth grinding, head tremors, blindness, poor coordination of the hind limbs, and ascending paralysis, with loss of deep tendon reflexes and sphincter tone. Rabbits usually die within a few days of infection (Karp et al. 1999).

Laboratory Diagnosis

Although clinical signs are good indicators of rabies in small animals, confirmation by accurate laboratory diagnostic tests is mandatory for a definitive diagnosis. Laboratory diagnosis can only be performed in state health departments with BSL3 facilities.

Rabies is diagnosed postmortem by direct immunofluorescent antibody testing (FAT) of brain tissue, molecular methods such as RT-PCR or real-time RT-PCR, or virus isolation in susceptible cell lines, i.e., mouse neuroblastoma cells (*Neuro 2A*) and baby hamster kidney cells (*BHK-21*) (WOAH, Manual of Diagnostic Tests and Vaccines for Terrestrial Animals 2022, Chapter Rabies (infection with rabies virus and other lyssaviruses).

Disease Control and Prevention

There are no licensed rabies vaccines for rabbits in Europe or the United States. In areas where rabies is enzootic, such as in the United States, rabbits that are kept outdoors should be protected from potential exposure to wildlife, such as raccoons, skunks, bats, and foxes (Krebs et al. 2001).

7 Lapine Parvovirus (LPV)

Etiology

Parvovirus infection in rabbits has been shown to cause mild clinical disease (Matsunaga and Chino 1981; Metcalf et al. 1989). It is caused by Lapine parvovirus (LPV), which belongs to the *Parvoviridae* family, a group of small nonenveloped, icosahedral viruses, approximately 18 to 26 nm in diameter, with genomes consisting of a single molecule of linear, single-stranded DNA, between 4.5 kb and 5.5 kb.

The Disease Origin and Its Natural Hosts

LPV was first isolated from rabbits in Japan in 1977, with 46.7% of the animals being found seropositive for LPV (Matsunaga et al. 1977). A few years later, 75% of 46 laboratory rabbits (*O. cuniculus*) in the United States were found seropositive for LPV (Metcalf et al. 1989). Despite the high percentage of seropositive animals, the clinical signs are mild.

Susceptible Species

Only rabbits are susceptible to LPV.

Clinical Signs and *Postmortem* Macroscopic and Microscopic Lesions

Clinical signs in experimentally inoculated rabbits included listlessness and inappetence or anorexia for a few days and diarrhea. Young rabbits, 4 to 10 weeks old, may show the most pronounced signs. Intestinal congestion and inflammation may be observed at necropsy.

Histologically, mild to moderate catarrhal enteritis with hyperemia, exfoliation of enterocytes, and luminal fluid accumulation may be observed in the small intestine (Matsunaga and Chino 1981).

Laboratory Diagnosis

Diagnosis of parvoviral disease in rabbits can prove to be difficult, particularly when associated with other pathogens. Laboratory diagnosis can be achieved by virus isolation in rabbit cell lines or by observation of viral particles by electron microscopy of feces. To our knowledge, no PCR has been described for molecular diagnosis.

Disease Control and Prevention

As infection is spread by fecal-oral contamination, proximity between animals favors virus transmission. Treatment of mildly symptomatic parvoviral infections is supportive and includes electrolyte replacement and probiotic and antibiotic administration.

8 Coronavirus

Etiology

Rabbit coronavirus (RbCV) and rabbit enteric coronavirus (RECV) belong to the genus *Coronavirus*, family *Coronaviridae*. Like other coronaviruses, RbCV and RECV are enveloped positive-sense RNA viruses, characterized by the presence of club-like spikes on their surface.

Antibodies against RbCV cross-reacted by radioimmunoassay with feline coronavirus virus (FCoV), canine coronavirus (CCoV), and transmissible gastroenteritis virus (TGEV) (Small and Woods 1987). Also, the lack of cross-reactivity of RECV with mouse hepatitis virus (type II coronavirus) and its reactivity with canine coronavirus (CCoV) also suggest that RECV is a type I coronavirus (Deeb et al. 1993).

The Disease Origin and Its Natural Hosts

Rabbit coronavirus (RbCV) was first recognized in the early 1960s in rabbits used to propagate the Nichols strain of *Treponema pallidum* that was contaminated with coronavirus (Small and Woods 1987). Mortality rates in the inoculated animals rose steadily to 75% in subsequent years, leading to the discovery of a new coronavirus with cardiac tropism by electron microscopy examination. It was not until 1980 that rabbit coronaviruses were associated with diarrhea, when viral particles suggestive of coronaviruses were also observed by electronmicroscopy on the feces of diarrheic rabbits (Lapierre et al. 1980).

Susceptible Species

No information is currently available.

Clinical Signs and *Postmortem* Macroscopic and Microscopic Lesions

Coronavirus infections in rabbits can have two different clinical presentations, namely, enteritis (caused by RECV) or a pleural effusion disease associated with cardiomyopathy (caused by RbCV) (Osterhaus et al. 1982). RECV has been detected in the feces of clinically healthy adult rabbits, indicating subclinical infection.

On the other hand, reported clinical signs of experimental RECV infection comprise anorexia, watery diarrhea, abdominal distension, and even sudden death (Lapierre et al. 1980; Descôteaux and Lussier 1990). As with rotavirus infection, coinfection with other organisms may exacerbate the clinical signs caused by rabbit coronaviruses.

Rabbits between 3 and 10 weeks of age appear to be most affected by RECV (Lapierre et al. 1980).

Cardiomyopathy has been reported in experimental infections, although it is not clear whether this type of lesion also occurs in natural infections (Alexander et al. 1992). Gross lesions in experimentally infected animals include congested small intestine and fluid cecal contents.

Histopathology of the small intestine shows necrosis of the villous epithelial cells, small intestinal villous blunting, crypt hypertrophy, complete M-cell necrosis, and necrosis of villous epithelial cells overlying the gut-associated lymphoid tissue (Krogstad et al. 2005).

Laboratory Diagnosis

Diagnosis can be made by observation of viral particles by EM in the feces of rabbits that have diarrhea or in characteristic lesions collected at necropsy. RECV can be detected by commercial available test for canine coronavirus (Castrucci et al. 1984). Serology is not recommended for diagnostic purposes since animals with diarrhea are unlikely to have mounted an antibody response to RECV. Seroprevalence studies in the United States and southwestern Canada have shown that 3% to 33% of the rabbits tested had antibodies to RECV. Rabbit farms with a history of diarrhea were more likely to have antibodies to rabbit enteric coronavirus (RECV) (Descôteaux and Lussier 1990).

Control and Prevention

There is no vaccine available for rabbit coronavirus. Managment of individual animals includes supportive therapy and isolation from healthy rabbits to prevent virus spread. In rabbitries, management strategies such as dietary changes, delayed weaning, and antibiotic treatment have failed to prevent morbidity and mortality (Eaton 1984).

9 Rabbit and Hare Herpesviruses

Etiology

Five herpesviruses have been described so far in rabbits and hares (known as leporids), namely, Leporid herpesvirus 1 (LeHV-1) to Leporid herpesvirus 5 (LeHV-5).

Leporid herpesviruses belong to the *Herpesviridae* family, which comprises subfamilies *Alphaherpesvirinae*, *Betaherpesvirinae*, and *Gammaherpesvirinae*, corresponding to different biological and phylogenetic properties, including the pattern of disease and lytic cycle induced (Pellett and Roizman 2013). While members of the *Alphaherpesvirinae* subfamily cause rapid lysis in cell culture, members of the *Betaherpesvirinae* subfamily grow slowly in cell culture and induce the formation of multinucleated giant cells (syncytia). *Gammaherpesvirinae* are further divided into four genera, *Macavirus*, *Percavirus*, *Lymphocryptovirus*, and *Rhadinovirus*, and infect lymphoid tissues, leading to lymphoproliferative diseases (Cesarman 2011) and oncogenesis.

LeHV-1, LeHV-2, LeHV-3, and LeHV-5 belong to the *Gammaherpesvirinae* subfamily (Abade dos Santos et al. 2020; Roizman and Pellett 2001), with LeHV-2 and LeHV-3 being the most commonly described herpesviruses in rabbits (Jin et al. 2008; Pellett and Roizman 2013) and LeHV-5 being the only herpesvirus detected in hares (Abade dos Santos et al. 2020). In rabbits, LeHV-4, an alphaherpesvirus, is the most pathogenic herpesvirus causing fatal respiratory infections (Sunohara-Neilson 2013). The virus particle consists of a central core containing a linear 124–295-kb dsDNA, surrounded by an icosahedral capsid with approximately 125 nm in diameter, which forms the nucleocapsid. The nucleocapsid is surrounded by the tegument – an amorphous layer – and externally by an envelope containing several surface glycoproteins (spikes). The size of the mature virion ranges from 120 to 260 nm (Roizman and Pellett 2001). The virion acquires the envelope by budding from the nuclear membrane, which is modified by the inclusion of several unique viral glycoproteins.

The Disease Origin, Natural Hosts, and Susceptible Species

LeHV-1 has been found in *S. floridanus* (Schmidt et al. 1992), and there are no reports of disease in domestic rabbits.

LeHV-2 (previously known as *Herpesvirus cuniculi*) was first recognized in 1924 and causes asymptomatic infections in domestic rabbits (*O. cuniculus*) (Zygraich and Berge 1972; Matsunaga and Yamazaki 1976).

LHV-3 (previously known as *Herpesvirus sylvilagus*) is endemic in cottontail rabbits (*S. floridanus*) where it causes lymphoproliferative disease (tumor-like lesions in the lymph nodes, kidney, spleen, and liver) (Hinze 1971a, b; Hesselton et al. 1988; Swan et al. 1991). However, it does not cause disease in domestic rabbits (*O. cuniculus*) (Cohrs and Rouhandeh 1987; Hesselton et al. 1988).

LeHV-4 was first identified in domestic rabbits in the early 1990s in Canada (Swan et al. 1991; Onderka et al. 1992;) following two outbreaks of acute, fatal herpesvirus infection. Later, in mid-2006, it was associated with outbreaks in a commercial pet farm and in a meat rabbit farm (*O. cuniculus*) in Alaska (Jin et al. 2008).

It has been suggested that LHV-4 may be present in a wildlife reservoir found in northwestern North America, where it causes asymptomatic infections that sporadically come into contact with domestic rabbits and cause disease (Jin et al. 2008).

The LeHV-5 was described in Iberian hares in 2020, and is the only herpesvirus described so far in hares to date (Abade dos Santos et al. 2020). When associated with ha-MYXV infection, which induces immunosuppression (Jeklova et al. 2008; Kerr and Donnelly 2013), LeHV-5 increases morbidity and mortality rates by causing more aggressive disease.

Sporadically, human herpesvirus 1 (HHV-1, also known as herpes simplex virus) infections in domestic rabbits result in severe encephalitis (de Matos et al. 2014).

Clinical Signs and *Postmortem* Macroscopic and Microscopic Lesions

LeHV-1 does not affect domestic rabbits. However, in *S. floridanus*, macroscopic lesions involved the lungs (numerous multifocal, circumscribed, tan foci), spleen (splenomegaly with a mottled reddish tan color), and brain (red to tan friable tract) (Hesselton et al. 1988).

Microscopically, severe bronchiolitis was characterized by epithelial cells containing eosinophilic intranuclear inclusion bodies and pneumonia. Encephalomalacia of the cerebral cortex showed tissue disruption and intranuclear inclusion bodies in astrocytes and neurons (Hesselton et al. 1988).

LeHV-2 is apathogenic.

LeHV-3 causes lymphoproliferative disease in the cottontail rabbit (*S. floridanus*) (Hinze 1971a, b; Hesselton et al. 1988; Swan et al. 1991) but does not cause disease in the domestic rabbit (*O. cuniculus*) (Hinze 1971a; Kramp et al. 1985; Cohrs and Rouhandeh 1987; Hesselton et al. 1988).

Clinical signs of LeHV-4 infection in domestic rabbits appear 3 to 8 days postinfection and include anorexia, oculonasal discharge, and dyspnea.

Gross lesions include severe necro-hemorrhagic bronchopneumonia and marked pulmonary edema and multifocal splenic necrosis progressing to submassive necrosis. Eosinophilic herpesviral intranuclear inclusion bodies were seen in the nasal mucosa and skin (Cowdry type A inclusions).

Clinical expression of LeHV-5 infection was only observed in Iberian hares with myxomatosis probably due to the associated immunosuppression. LeHV-5 DNA was detected in tissues of healthy hares, suggesting latent asymptomatic infection in these wild populations. Gross lesions attributed to LeHV-5 included necrosis of the genitalia, more evident in males affecting the glans of the penis, but also observed in the vulva in females, and herpetic-like skin vesicles on the lips and nose. Other lesions included marked enlargement of the penis and thickening of the eyelids and lips due to the accumulation of mucopurulent exudate or to multinodular ulceration (Abade dos Santos et al. 2020).

Typical microscopic lesions of herpesvirus included (1) slightly eosinophilic inclusion bodies in the nuclei (Cowdry type A inclusions) of stromal cells, together with the proliferation of pleomorphic spindle cells; (2) replacement of the penile epithelium by a thick band of necrotic cells, heterophils, and red blood cells; (3) multifocal detachment of the eyelids, lips, and prepuce epidermis; and (4) coalescent intraepidermal and subepidermal vesicopustules filled with fibrin and necrotic cell debris.

Laboratory Diagnosis

Molecular diagnosis can be achieved by PCR amplification of a region within the DNA polymerase gene, using primers flanking variable regions (Devanter et al. 1996) between different herpesviruses. Sequencing and analysis of the PCR products allow discrimination between herpesviruses and determines which virus is being detected.

Since herpesviruses replicate in the nucleus of the cells, histopathological examination is useful to observe intranuclear bodies, which are highly suggestive of herpesvirus infection but does not provide information on which herpesvirus is present.

A specific PCR has been developed for the detection of LeHV-5 (Abade dos Santos et al. 2021a, b).

Disease Control and Prevention

Biosecurity measures in the industry are recommended. There are no preventive measures to control leporid herpesviruses.

10 Conclusions

As shown in the previous chapter (Chap. 25), RHD and myxomatosis are, by far, the biggest problems for the rabbit sector because to their epidemic and devastating effects.

However, due to the intensification of animal production and the search for higher growth rates and profitability on rabbit farms, other viruses with less obvious effects, but which may reduce the growth rate of the animal by affecting its health, should also be investigated and taken into account. Examples of these viruses are rotavirus, coronavirus, and herpesvirus, which are often known for their insidious effects and about which very little is known in rabbit farming. It is therefore imperative that laboratory studies of the rabbit virome are carried out in different geographical locations and where animals are exposed to different habitat conditions, to assess the prevalence, role, and impact of these lesser known viruses on the physiology and health status of the animal.

References

Abade dos Santos FA, Monteiro M, Pinto A, Carvalho CL, Peleteiro MC, Carvalho P, Mendonça P, Carvalho T, Duarte MD (2020) First description of a herpesvirus infection in genus Lepus. PLoS One 15:e0231795. https://doi.org/10.1371/journal.pone.0231795

Abade dos Santos FA, Carvalho CL, Parra F, Dalton KP, Peleteiro MC (2021a) A Quadruplex qPCR for detection and differentiation of classic and natural recombinant myxoma virus strains of Leporids. Int J Mol Sci 22:12052. https://doi.org/10.3390/ijms222112052

Abade dos Santos FA, Carvalho CL, Peleteiro MC, Parra F, Duarte MD (2021b) A versatile qPCR for diagnosis of Leporid Gammaherpesvirus 5 using Evagreen® or Taqman® technologies. Viruses 13. https://doi.org/10.3390/v13040715

Alexander LK, Small JD, Edwards S, Baric RS (1992) An experimental model for dilated cardiomyopathy after rabbit coronavirus infection. J Infect Dis 166:978–985. https://doi.org/10.1093/infdis/166.5.978

Baker DG (1998) Natural pathogens of laboratory mice, rats, and rabbits and their effects on research. Clin Microbiol Rev 11:231–266. https://doi.org/10.1128/CMR.11.2.231

Bedson HS, Duckworth MJ (1963) Rabbit pox: an experimental study of the pathways of infection in rabbits. J Pathol Bacteriol 85:1–20

Brabb T, Di Giacomo RF (2012) Chapter 14 - Viral diseases. In: Suckow MA, Stevens KA, Wilson RP (eds) The laboratory rabbit, guinea pig, hamster, and other rodents. Academic Press, New York, pp 365–413. https://doi.org/10.1016/B978-0-12-380920-9.00014-6

Bridger JC (1994) A definition of bovine rotavirus virulence. J Gen Virol 75:2807–2812. https://doi.org/10.1099/0022-1317-75-10-2807

Bryden AS, Thouless ME, Flewett TH (1976) Rotavirus and rabbits. Vet Rec 99:323. https://doi.org/10.1136/vr.99.16.323-a

Castrucci G, Frigeri F, Ferrari M, Cilli V, Aldrovandi V, Caleffi F, Gatti R (1984) Comparative study of rotavirus strains of bovine and rabbit origin. Comp Immunol Microbiol Infect Dis 7: 171–178. https://doi.org/10.1016/0147-9571(84)90023-7

Castrucci G, Ferrari M, Frigeri F, Cilli V, Perucca L, Donelli G (1985) Isolation and characterization of cytopathic strains of rotavirus from rabbits. Brief report. Arch Virol 83:99–104. https://doi.org/10.1007/BF01310967

Cesarman E (2011) Gammaherpesvirus and lymphoproliferative disorders in immunocompromised patients. Cancer Lett 305:163–174. https://doi.org/10.1016/j.canlet.2011.03.003

Ciarlet M, Gilger MA, Barone C, McArthur M, Estes MK, Conner ME (1998) Rotavirus disease, but not infection and development of intestinal histopathological lesions, is age restricted in rabbits. Virology 251:343–360. https://doi.org/10.1006/viro.1998.9406

Cohrs R, Rouhandeh H (1987) Characterization of the genome of herpesvirus sylvilagus. Intervirology 28:181–184. https://doi.org/10.1159/000150014

Conner ME, Estes MK, Graham DY (1988) Rabbit model of rotavirus infection. J Virol 62:1625–1633. https://doi.org/10.1128/JVI.62.5.1625-1633.1988

Dalmat HT (1959) Arthropod transmission of rabbit fibromatosis (Shope). J Hyg 57:1–30. https://doi.org/10.1017/s0022172400019860

de Matos R, Russell D, Van Alstine W, Miller A (2014) Spontaneous fatal human herpesvirus 1 encephalitis in two domestic rabbits (*Oryctolagus cuniculus*). J Vet Diagn Invest. 26:689–694. https://doi.org/10.1177/1040638714545111

Deeb BJ, DiGiacomo RF, Evermann JF, Thouless ME (1993) Prevalence of coronavirus antibodies in rabbits. Lab Anim Sci 43:431–433

Descôteaux JP, Lussier G (1990) Experimental infection of young rabbits with a rabbit enteric coronavirus. Can J Vet Res 54:473–476

Devanter DRVAN, Warrener P, Bennett L, Schultz ER, Coulter S, Garber RL, Rose TM, Black J (1996) Detection and analysis of diverse Herpesviral species by consensus primer PCR. J Clin Microbiol 34:1666–1671. https://doi.org/10.1128/JCM.34.7.1666-1671.1996

DiGiacomo RF, John Maré C (1994) Viral diseases. In: Manning PJ, Ringler DH, Newcomer CE (eds) The biology of the laboratory rabbit. Academic Press, pp 171–204. ISBN 9780124692350. https://doi.org/10.1016/B978-0-12-469235-0.50015-4

DiGiacomo RF, Thouless ME (1984) Age-related antibodies to rotavirus in New Zealand rabbits. J Clin Microbiol 19:710–711. https://doi.org/10.1128/jcm.19.5.710-711.1984

DiGiacomo RF, Thouless ME (1986) Epidemiology of naturally occurring rotavirus infection in rabbits. Lab Anim Sci 36:153–156

Duarte MD, Barros SC, Henriques AM, Fagulha MT, Ramos F, Luís T, Fevereiro M (2014) Development and validation of a real time PCR for the detection of myxoma virus based on the diploid gene M000.5L/R. J Virol Methods 196:219–224. https://doi.org/10.1016/j.jviromet.2013.11.014

Eaton P (1984) Preliminary observations on enteritis associated with a coronavirus-like agent in rabbits. Lab Anim 18:71–74. https://doi.org/10.1258/002367784780864938

Elschner M, Prudlo J, Hotzel H, Otto P, Sachse K (2002) Nested reverse transcriptase-polymerase chain reaction for the detection of group A rotaviruses. Journal of Veterinary Medicine. J Vet Med B Infect Dis Vet Public Health 49:77–81. https://doi.org/10.1046/j.1439-0450.2002.00510.x

Esposito J, Fenner F (2011) Poxviruses. In: Knipe DE, Howley PM (eds) Fields virology. Lippincott Williams & Wilkins, Philadelphia, PA, pp 2885–2922. ISBN 9780781760607

Fales FM (2009) Chapter 2: Mesopotamia. In: Aminoff MJ, Boller F, Swaab DF (eds) Handbook of clinical neurology. Elsevier, pp 15–27. ISBN 9780444520098. https://doi.org/10.1016/S0072-9752(08)02102-7

Fenner F (1953) Classification of Myxoma and fibroma viruses. Nature 171:562–563. https://doi.org/10.1038/171562a0

Fenner F, Ratcliffe FN (1965) Myxomatosis. Cambridge University Press, London. ISBN 978-0521049917

Fu ZF (1997) Rabies and rabies research: past, present and future. Vaccine 15:S20–S24. https://doi.org/10.1016/s0264-410x(96)00312-x

Fushuku S, Fukuda K (2006) Examination of the applicability of a commercial human rotavirus antigen detection kit for use in laboratory rabbits. Exp Anim 55:71–74. https://doi.org/10.1538/expanim.55.71

Gentsch JR, Glass RI, Woods P, Gouvea V, Gorziglia M, Flores J, Das BK, Bhan MK (1992) Identification of group a rotavirus gene 4 types by polymerase chain reaction. J Clin Microbiol 30:1365–1373. https://doi.org/10.1128/jcm.30.6.1365-1373.1992

Giri I, Danos O, Yaniv M (1985) Genomic structure of the cottontail rabbit (Shope) papillomavirus. Proc Natl Acad Sci U S A 82:1580–1584. https://doi.org/10.1073/pnas.82.6.1580

Greene HS (1935a) Rabbit pox: IV. Susceptibility as a function of constitutional factors. J Exp Med 62:305–329. https://doi.org/10.1084/jem.62.3.305

Greene HSN (1935b) Rabbit pox: III. Report of an epidemic with especial reference to epidemiological factors. J Exp Med 61:807–831. https://doi.org/10.1084/jem.61.6.807

Hagen KW (1966) Spontaneous papillomatosis in domestic rabbits 1. Bull Wildl Dis Assoc 2:108–110. https://doi.org/10.7589/0090-3558-2.4.108

Hambraeus BA, Hambraeus LE, Wadell G (1989) Animal model of rotavirus infection in rabbits--protection obtained without shedding of viral antigen. Arch Virol 107:237–251. https://doi.org/10.1007/BF01317920

Harvey RG (1990) Demodex cuniculi in dwarf rabbits (*Oryctolagus cuniculus*). J Small Anim Pract 31:204–207. https://doi.org/10.1111/j.1748-5827.1990.tb00774.x

Hesselton RM, Yang WC, Medveczky P, Sullivan JL (1988) Pathogenesis of herpesvirus sylvilagus infection in cottontail rabbits. Am J Pathol 133:639–647

Hinze HC (1971a) New member of the herpesvirus group isolated from wild cottontail rabbits. Infect Immun 3:350–354. https://doi.org/10.1128/iai.3.2.350-354.1971

Hinze HC (1971b) Induction of lymphoid hyperplasia and lymphoma-like disease in rabbits by herpesvirus sylvilagus. Int J Cancer 8:514–522. https://doi.org/10.1002/ijc.2910080320

Iturriza Gómara M, Wong C, Blome S, Desselberger U, Gray J (2002) Molecular characterization of VP6 genes of human rotavirus isolates: correlation of genogroups with subgroups and evidence of independent segregation. J Virol 76:6596–6601. https://doi.org/10.1128/jvi.76.13.6596-6601.2002

Jeklova E, Leva L, Matiasovic J, Kovarcik K, Kudlackova H, Nevorankova Z, Psikal I, Faldyna M (2008) Characterisation of immunosuppression in rabbits after infection with myxoma virus. Vet Microbiol 129:117–130. https://doi.org/10.1016/j.vetmic.2007.11.039

Jenkins JR (2001) Skin disorders of the rabbit. The veterinary clinics of North America. Exotic Anim Pract 4:543–563. https://doi.org/10.1016/s1094-9194(17)30042-7

Jin L, Löhr CV, Vanarsdall AL, Baker RJ, Moerdyk-Schauwecker M, Levine C, Gerlach RF, Cohen SA, Alvarado DE, Rohrmann GF (2008) Characterization of a novel alphaherpesvirus associated with fatal infections of domestic rabbits. Virology 378:13–20. https://doi.org/10.1016/j.virol.2008.05.003

Joiner GN, Jardine JH, Gleiser CA (1971) An epizootic of shope fibromatosis in a commercial rabbitry. J Am Vet Med Assoc 159:1583–1587

Karp BE, Ball NE, Scott CR, Walcoff JB (1999) Rabies in two privately owned domestic rabbits. J Am Vet Med Assoc 215(1806):1824–1827

Kerr PJ, Donnelly TM (2013) Viral infections of rabbits. The veterinary clinics of North America. Exotic Anim Pract 16:437–468. https://doi.org/10.1016/j.cvex.2013.02.002

Kidd JG, Rous P (1938) The carcinogenic effect of a papilloma virus on the tarred skin of rabbits: II. Major factors determining the phenomenon: the manifold effects of tarring. J Exp Med 68: 529–562. https://doi.org/10.1084/jem.68.4.529

Kilham L, Dalmat HT (1955) Host-virus-mosquito relations of Shope fibromas in cottontail rabbits. Am J Hyg 61:45–54. https://doi.org/10.1093/oxfordjournals.aje.a119737

Kramp WJ, Medveczky P, Mulder C, Hinze HC, Sullivan JL (1985) Herpesvirus Sylvilagus infects both B and T lymphocytes. In Vivo 56:60–65

Krebs JW, Mondul AM, Rupprecht CE, Childs JE (2001) Rabies surveillance in the United States during 2000. J Am Vet Med Assoc 219:1687–1699. https://doi.org/10.2460/javma.2001.219.1687

Krogstad AP, Simpson JE, Korte SW (2005) Viral diseases of the rabbit. The veterinary clinics of North America. Exotic Anim Pract 8:123–138. https://doi.org/10.1016/j.cvex.2004.09.002

Lapierre J, Marsolais G, Pilon P, Descôteaux JP (1980) Preliminary report on the observation of a coronavirus in the intestine of the laboratory rabbit. Can J Microbiol 26:1204–1208. https://doi.org/10.1139/m80-201

Larson CL, Schillinger JE, Green RC (1936) Transmission of rabbit papillomatosis by the rabbit tick, *Haemaphysalis leporis palustris*. Proc Soc Exp Biol Med 33:536–538. https://doi.org/10.3181/00379727-33-8441P

Li G, Chen N, Roper RL, Feng Z, Hunter A, Danila M, Lefkowitz EJ, Buller RML, Upton C (2005) Complete coding sequences of the rabbitpox virus genome. J Gen Virol 86:2969–2977. https://doi.org/10.1099/vir.0.81331-0

Martella V, Ciarlet M, Lavazza A, Camarda A, Lorusso E, Terio V, Ricci D, Cariola F, Gentile M, Cavalli A, Camero M, Decaro N, Buonavoglia C (2005) Lapine rotaviruses of the genotype P [22] are widespread in Italian rabbitries. Vet Microbiol 111:117–124. https://doi.org/10.1016/j.vetmic.2005.10.005

Martinic G, Harkness-Author JE, Wagner-Author JE (1995) The biology and medicine of rabbits and rodents, 4th edn ISBN 0-683-03919-9
Matsunaga Y, Chino F (1981) Experimental infection of young rabbits with rabbit parvovirus. Arch Virol 68:257–264. https://doi.org/10.1007/BF01314579
Matsunaga Y, Yamazaki S (1976) Studies on herpesvirus cuniculi. I. Virus isolation and antibody response in rabbits with experimental infection (author's transl). Uirusu 26:11–19
Matsunaga Y, Matsuno S, Mukoyama J (1977) Isolation and characterization of a parvovirus of rabbits. Infect Immun 18:495–500. https://doi.org/10.1128/iai.18.2.495-500.1977
McLeod CGJ, Langlinais PC (1981) Pox virus keratitis in a rabbit. Vet Pathol 18:834–836. https://doi.org/10.1177/030098588101800615
Metcalf JB, Lederman M, Stout ER, Bates RC (1989) Natural parvovirus infection in laboratory rabbits. Am J Vet Res 50:1048–1051
Onderka DK, Papp-Vid G, Perry AW (1992) Fatal herpesvirus infection in commercial rabbits. Can Vet J 33:539–543
Osterhaus ADME, Teppema JS, van Steenis G (1982) Coronavirus-like particles in laboratory rabbits with different syndromes in The Netherlands [Oryctolagus]. Lab Anim Sci 32:663–665
Pellett PE, Roizman B (2013) Herpesviridae. In: Knipe DM, Howley PM (eds) Fields virology, 6th edn. Wolters Kluwer Health/Lippincott Williams & Wilkins, Philadelphia, pp 1802–1822. ISBN 9781451105636
Petric M, Middleton PJ, Grant C, Tam JS, Hewitt CM (1978) Lapine rotavirus: preliminary studies on epizoology and transmission. Can J Comp Med 42:143–147
Pulley LT, Shively JN (1973) Naturally occurring infectious fibroma in the domestic rabbit. Vet Pathol 10:509–519. https://doi.org/10.1177/030098587301000604
Raflo CP, Olsen RG, Pakes SP, Webster WS (1973) Characterization of a fibroma virus isolated from naturally-occurring skin tumors in domestic rabbits. Lab Anim Sci 23:525–532
Roizman B, Pellett P (2001) The family Herpesviridae: a brief introduction. In: Knipe DM, Howley P (eds) Fields virology, 4th edn. Lippincott Williams & Wilkins, Philadelphia, pp 2381–2397
Saif LJ (1990) A review of evidence implicating bovine coronavirus in the etiology of winter dysentery in cows: an enigma resolved? Cornell Vet 80:303–311
Schmidt SP, Bates GN, Lewandoski PJ (1992) Probable herpesvirus infection in an eastern cottontail (*Sylvilagus floridanus*). J Wildl Dis 28:618–622. https://doi.org/10.7589/0090-3558-28.4.618
Schoeb TR, Casebolt DB, Walker VE, Potgieter LN, Thouless ME, DiGiacomo RF (1986) Rotavirus-associated diarrhea in a commercial rabbitry. Lab Anim Sci 36:149–152
Schoondermark-van de Ven E, Van Ranst M, de Bruin W, van den Hurk P, Zeller M, Matthijnssens J, Heylen E (2013) Rabbit colony infected with a bovine-like G6P[11] rotavirus strain. Vet Microbiol 166:154–164. https://doi.org/10.1016/j.vetmic.2013.05.028
Shope RE (1932) A transmissible tumor-like condition in rabbits. J Exp Med 56:793–802. https://doi.org/10.1084/jem.56.6.793
Shope RE (1935) Serial transmission of virus of infectious papillomatosis in domestic rabbits. Proc Soc Exp Biol Med 32:830–832. https://doi.org/10.3181/00379727-32-7875
Shope RE, Hurst EW (1933) Infectious papillomatosis of rabbits : with a note on the histopathology. J Exp Med 58:607–624. https://doi.org/10.1084/jem.58.5.607
Small JD, Woods RD (1987) Relatedness of rabbit coronavirus to other coronaviruses. Adv Exp Med Biol 218:521–527. https://doi.org/10.1007/978-1-4684-1280-2_68
Sunohara-Neilson JR (2013) Investigation of Leporid herpesvirus 4, an emerging pathogen of rabbits: infection and prevalence studies. The University of Guelph
Swan C, Perry A, Papp-Vid G (1991) Alberta. Herpesvirus-like viral infection in a rabbit. Can Vet J 32:627–628
Tanaka TN, Conner ME, Graham DY, Estes MK (1988) Molecular characterization of three rabbit rotavirus strains. Arch Virol 98:253–265. https://doi.org/10.1007/BF01322173
Theil KW, Saif LJ (1985) In vitro detection of porcine rotavirus-like virus (group B rotavirus) and its antibody. J Clin Microbiol 21:844–846. https://doi.org/10.1128/jcm.21.5.844-846.1985

Thouless ME, DiGiacomo RF, Neuman DS (1986) Isolation of two lapine rotaviruses: characterization of their subgroup, serotype and RNA electropherotypes. Arch Virol 89:161–170. https://doi.org/10.1007/BF01309886

Thouless ME, DiGiacomo RF, Deeb BJ, Howard H (1988) Pathogenicity of rotavirus in rabbits. J Clin Microbiol 26:943–947. https://doi.org/10.1128/jcm.26.5.943-947.1988

Thouless ME, DiGiacomo RF, Deeb BJ (1996) The effect of combined rotavirus and *Escherichia coli* infections in rabbits. Lab Anim Sci 46:381–385

Van Regenmortel MHV, Fauquet C, Bishop D, Calisher C, Carsten E, Estes M, Lemon S, Maniloff J, Mayo M, McGeoch D, Pringle C, Wickner R (2002) Virus taxonomy. Seventh report of the International Committee for the Taxonomy of viruses. Academic Press, New-York. ISBN 978-0123702005

Wu M, Zhu Y, Cong F, Rao D, Yuan W, Wang J, Huang B, Lian Y, Zhang Y, Huang R, Guo P (2018) Rapid detection of three rabbit pathogens by use of the Luminex x-TAG assay. BMC Vet Res 14:127. https://doi.org/10.1186/s12917-018-1438-8

Zygraich N, Berge E (1972) Experimental infection of rabbits and monkeys with Herpesvirus cuniculi. Res Vet Sci 13:241–244. https://doi.org/10.1016/S0034-5288(18)34035-9

Dermatophytosis in Rabbits

Ana Cláudia Coelho, Daniel Morais, José Manuel Monteiro, and João Simões

Abstract

Dermatophytosis or ringworm in rabbits is a zoonotic disease. Dermatophytes are keratinophilic fungi. *Trichophyton mentagrophytes* is the main microbiological agent that causes infection in rabbit farms, affecting a wide variety of breeds and causing morbidity that diminishes the performance of animals. The development of dermatophytosis in rabbits is strongly related to risk factors such as animal density and immunosuppression conditions. Rabbits may present with clinical signs such as focal alopecia, erythema, and crusting. Risk factor management and disinfection of the environment to decrease fungal loads are important components in control and prevention. As such, the sanitary measures carried out by the rabbit farmer are of extreme importance for the control of the disease. This chapter highlights the main features of dermatophytosis presence in rabbitries.

A. C. Coelho (✉) · J. Simões
CECAV – Veterinary and Animal Research Centre, University of Trás-os-Montes and Alto Douro, Vila Real, Portugal

Associate Laboratory for Animal and Veterinary Sciences (AL4AnimalS), Lisbon, Portugal
e-mail: accoelho@utad.pt

D. Morais
Nanta, SA, Alverca do Ribatejo, Portugal

J. M. Monteiro
José Azevedo Monteiro, Lda., Vila Boa do Bispo, Portugal

Faculty of Veterinary Medicine, CBIOS, Lusófona University, Lisbon, Portugal

J. Simões, J. M. Monteiro (eds.), *Veterinary Care of Farm Rabbits*,
https://doi.org/10.1007/978-3-031-44542-2_27

Keywords

Ringworm; *Trichophyton* · *Microsporum* · Skin · Antimycotic agent · Immunosuppression · Animal density

1 Introduction

Dermatophytosis or ringworm is a superficial cutaneous disease mainly caused by fungi of the genera *Trichophyton* and *Microsporum* (Chang et al. 2022). The disease is widespread throughout the world, and a high prevalence of dermatophytes in rabbits is described worldwide (Cafarchia et al. 2010; Chang et al. 2022). The number of rabbit farms affected by dermatophytosis is very high (Rosell 2000; Cafarchia et al. 2010). It occurs, essentially, on farms with high animal density and low hygiene measures (Espinosa et al. 2020).

Fungi cause infections of the superficial and keratinized layers of the skin and its annexes (i.e., hairs and nails) (OIE 2005). It can be defined as a fungal infection of the superficial layers of the skin caused by keratinophilic fungi with tropism for the keratinized tissues of the skin and the stratum corneum in humans and animals (White et al. 2014). The disease depends on the geographical location and management conditions of the animals. It is therefore a common zoonosis and a highly contagious infectious disease (Carpouron et al. 2022). In rabbit farms, it rarely causes death but can cause high morbidity and large economic losses associated with malnutrition, stunted growth, treatments, and a decrease in performance (Dey et al. 2016). There are other costs that are not normally accounted for, such as costs associated with infection in farmers, fear and concern of being infected during work activities, and fear of losing their job or being stigmatized, mainly when skin lesions are apparent (Benedict et al. 2019). At the slaughterhouse level, animals presenting lesions compatible with dermatophytosis are subject to total rejection (Rosell 2000).

This chapter mainly highlights the features related to dermatophytosis diagnosis in rabbits and its management in rabbitries.

2 Dermatophyte Phylogeny and Nomenclature

Dermatophytes are a specialized group of filamentous fungi capable of infecting and degrading keratinized tissues, including the skin, hair, nails, and feathers, and use the products resulting from this metabolism as a source of nutrients (Mercer and Stewart 2019; Gnat et al. 2019). Taxonomically, dermatophytes belong to the order *Onygenales* and family *Arthrodermataceae* and until recently were divided into three genera: *Trichophyton*, *Microsporum*, and *Epidermophyton* (Segal and Elad 2021).

Fungal nomenclature is constantly evolving and is based on phylogenetic studies (Naranjo-Ortiz and Gabaldón 2019). With the advent of genomics and phylogenomics, the taxonomy and nomenclature of dermatophytes have been revised. Based on current knowledge, there are nine genera of dermatophytes: *Epidermophyton*, *Microsporum*, *Arthroderma*, *Lophophyton*, *Nannizzia*, *Paraphyton*, *Trichophyton*, *Guarromyces*, and *Ctenomyces* (de Hoog et al. 2017; Gnat et al. 2019; Porras-López et al. 2021). The two genera of fungi that are the main cause of dermatophytoses in animals are *Microsporum* spp. and *Trichophyton* spp. (Paryuni et al. 2020). The genus *Microsporum* includes three species: *Microsporum audouinii*, *M. canis*, and *M. ferrugineum*. The genus *Arthroderma* consists of 21 species. *Trichophyton terrestre* and *T. ajelloi* are now included in this genus. The remaining geophilic and zoophilic species and the other species previously classified as *Microsporum* have been transferred to the genera *Lophophyton* (*L.*), *Nannizzia* (*N.*), and *Paraphyton* (*P.*). *Epidermophyton* is an exclusively anthropophilic genus and includes one species (*E. floccosum*). The genus *Lophophyton* has a single species – *Lophophyton gallinae* (synonym *M. gallinae*) – as do the genera *Guarromyces* and *Ctenomyces*. In contrast, the genus *Nannizzia* includes nine different species. The most frequent are *N. gypsea*, *N. persicolor*, *N. fulva*, and *N. nana*. Previously, these species belonged to the genus *Microsporum*. *Trichophyton* includes 16 species. The *T. mentagrophytes* complex currently includes anthropophilic and zoophilic species (de Hoog et al. 2017; Ciesielska and Stączek 2020; Metin and Heitman 2020). In humans and animals, more than 50 different species of fungi have been described on the skin. Dermatophyte species in animals are isolated as asexual forms, called anamorphs, identified as belonging to the genus *Microsporum* or *Trichophyton* based on the macroscopic and microscopic characteristics of these organisms in culture. The sexual states (teleomorphs) of some species of dermatophytes allowed them to be defined in the genus *Arthroderma*, phylum *Ascomycota* (Metin and Heitman 2017, 2020).

The dermatophyte species with the greatest importance in dermatophytosis in rabbit farms are *T. mentagrophytes*, *M. canis*, and *N. gypsea* (Cafarchia et al. 2012; Dey et al. 2016), but most infections are caused by *T. mentagrophytes* (Coelho et al. 2008a, b; Dey et al. 2016). Other dermatophytes were sporadically identified in rabbit farms as *N. gypsea*, *M. canis*, *M. Audouinii*, and *T. verrucosum*. As a rule, infected animals from the same rabbit farm have similar etiologies (Dey et al. 2016; Chupia et al. 2019).

3 Epidemiology

The incidence of dermatophytosis in rabbit farms is high (Cafarchia et al. 2012). Infection is seasonal in nature and depends on the climate and the general conditions of the farms. Transmission occurs by direct or indirect contact with infected animals and by contact with fomites with fur and skin scales of infected rabbits (Rosell 2000; Samanta 2015). The lesions caused by dermatophytes present a high quantity of infectious material (Dalis et al. 2019), which can contaminate facilities and all surfaces in contact with the animal for long periods of time. Vectors, handlers, domestic animals, rodents, and birds are also identified as sources of contagion (Guo et al. 2020).

Dermatophytosis most often enters a farm when breeding animals are introduced. Sick animals are the most important source of infection to other animals through contact between infective spores and the skin (Weitzman and Summerbell 1995; Donnelly et al. 2000).

The infection can have a prolonged course, with chronic and progressive phases, but in general is self-limiting with spontaneous healing (Moriello 2014).

T. mentagrophytes is the main cause of infection in rabbit farms, affecting a wide variety of breeds. *N. gypsea* is normally found in animalized soils or soils with high concentrations of organic matter. The soil is therefore considered a reservoir of *N. gypsea*, where under optimum conditions of temperature and humidity, the multiplication of the fungus is favored (Prochnau et al. 2004; Iorio et al. 2007; Coelho et al. 2008b).

Farms with infected animals may be potential sources of infection to other farms, since dermatophytoses are extremely contagious and easily transmissible to humans (Van Rooij et al. 2006).

Spores (arthrospores) are present in the environment and in contaminated materials that infect tissues and cause the destruction of the hairs, resulting in a typical dermatophytosis lesion. The infection is a consequence of the reaction of the host to the metabolic products of the fungus and the invasion of the tissue by the living organism. The spores can survive in the environment for long periods of time (months or years) when they are found inside capillary segments or epidermal scales (Zagnoli et al. 2004).

Rabbits without clinical signs, escaping clinical detection, function as an environmental contaminant, complicating the treatment and eradication of the disease. In many cases, females present lesions in the mammary region, and males present lesions in the pelvic zone near the testicles. The occurrence of dermatophytosis in a litter or at fattening is a sign that the mother is a carrier, passing it directly from one generation to the next. Lesions on the mother are usually found on the abdomen, near the teats, and the kits become infected when they contact the teats during suckling (Rosell 2000; Boucher and Nouaille 2002).

In rabbits, there are several predisposing factors, such as age, housing conditions, immune status, and genetics (breed), among others (Espinosa et al. 2020; Segal and Elad 2021). Dermatophytosis mainly affects kits between 30 and 60 days old. Housing conditions during the fattening phase are more precarious, the animal density is higher, there is more fighting, and there is more humidity and heat (Rosell 2000; Boucher and Nouaille 2002; Rauterberg et al. 2019).

The immune status determines the susceptibility of the animals. Young, old, or immunosuppressed animals (disease, gestation, etc.) are more vulnerable (Suckow et al. 2002). Young animals are more susceptible, which may be associated with their immune status and the fact that they have not been previously exposed, without having acquired immunity (Rosser 1995). Rabbits are immunocompromised during few weeks after weaning. During this period, growing rabbits are more likely to show skin lesions, sometimes exuberant, which decrease toward the end of fattening, when their immune system gets more competent. Genetics is closely related to immune status. Thus, animals with a certain genetic breed are more susceptible to dermatophytosis than others (Morisse and Maurice 1997; Morisse et al. 1999).

Other physiological or clinical processes, such as lactation, affect the prevalence of dermatophytosis, since lactating rabbits often present lesions near the mammary region due to physiological alopecia and scratching caused by hair removal (Boucher and Nouaille 2002).

Dermatophytosis in rabbits is normally self-limiting. Reducing stress and improving housing and environmental conditions are important considerations in the regression of lesions (Van Rooij et al. 2006; Chang et al. 2022).

4 Pathogenesis

Dermatophytoses are infections of the superficial keratinized tissues, epidermis, hair, and nails caused by keratinolytic and keratinophilic fungi. Hair is usually the most affected part, resulting in premature alopecia (Hainer 2003; Martinez-Rossi et al. 2017).

When infection occurs, viable spores or conidia penetrate the skin and, through chemical stimuli, germinate and produce arthrosporic hyphae, which proliferate through the stratum corneum and invade the keratin of the hair. Arthrospores adhere to keratinocytes. This process can be triggered relatively quickly (2 h) (Zurita and Hay 1987). The spores can be viable in the environment for between 6 and 12 months but can be longer (Fig. 1).

Two situations may occur: not producing detectable lesions (asymptomatic carriers) or producing clinical lesions. Infection does not always occur. Dermatophytes may be eliminated mechanically or simply not establish any resistance due to an inability to compete with the resident microbiota (Connole 1967; Moriello 1996). The skin represents the first barrier against dermatophytes. The mechanisms involved in generating immunologic reactions in the skin are complex, with epidermal Langerhans cells, other dendritic cells, lymphocytes, microvascular endothelial cells, and keratinocytes (Wagner and Sohnle 1995; Rouzaud et al. 2015).

Through enzymatic and mechanical activity, dermatophytes penetrate along the hair, destroying the keratin. The hair becomes increasingly weakened and fractures (Mercer and Stewart 2019). Arthrospore production occurs inside the hair by both endothrix and ectothrix infections. Only in ectothrix invasion is there production of arthrospore masses around the hair. T-cell mediated immunity is associated with resistance to infection. However, there was no relationship between the circulating antibody titer and the degree of resistance to dermatophyte infection. The environment, local factors, skin and hair integrity, and the antifungal activity of sebum and sweat are important barriers to fungal invasion and development (Moriello 1996).

The maintenance of infection and the survival of dermatophytes largely depend on their ability to avoid the inflammatory response. Throughout the process, the fungi adapt to the adverse conditions provided by each host by releasing minute quantities of toxins and allergens (Simpanya 2000). *T. mentagrophytes* infections are

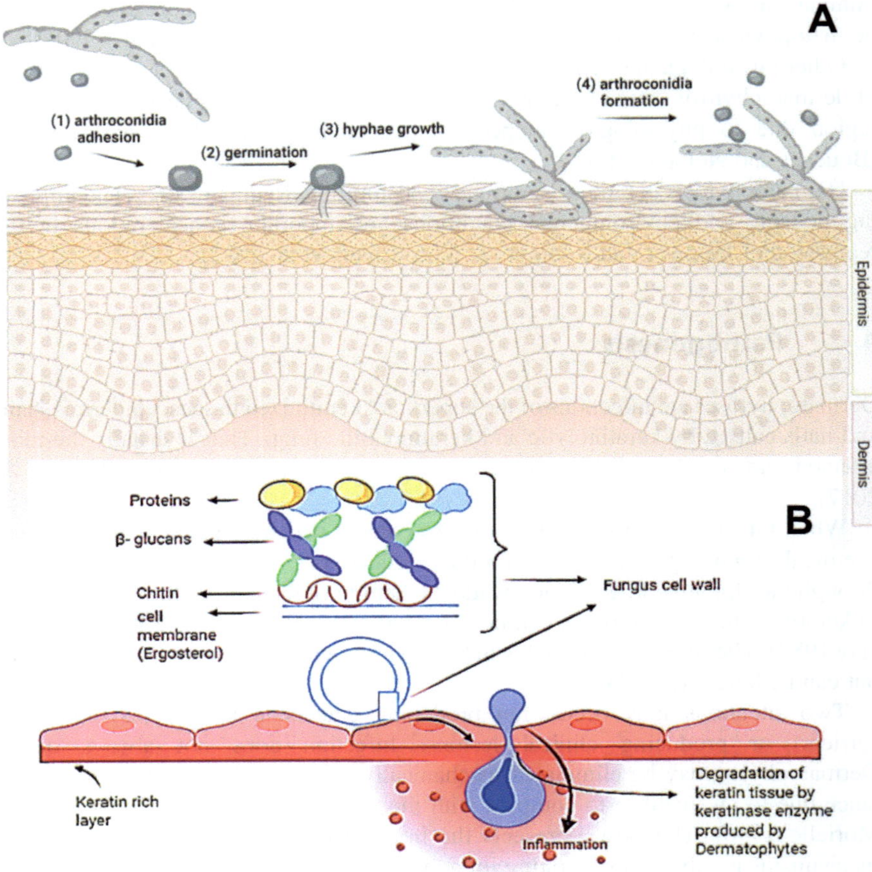

Fig. 1 Illustrated mechanisms of skin adhesion and pathogeny of hyphae. (**a**) Adhesion of arthroconidia (infectious fragments of hypha) to the skin occurs between 2 and 6 h after contact (1). Hyphal germination starts on the top layer of the epidermis (2), forming germ tubes toward the epidermis (3). The infection cycle is complete within 7 days (4). Adapted from Moskaluk and VandeWoude (2022). (**b**) After skin invasion, dermatophytes produce keratinase enzyme to degrade keratin for fungal growth, causing local inflammation (potential mechanism). Adapted from Kaur et al. (2021)

the most common in rabbit farms, and the inflammatory reaction is more pronounced, causing more intense lesions (White et al. 2002).

Fig. 2 Dermatophytosis skin lesions in early weaned rabbits. Diffuse scaling on the head and ears (large photo) and focal alopecia (small photo). Weaning stress is a probable cause of immunosuppression driving a higher dermatophytosis incidence in these rabbit groups than in late-fattening rabbits or breeding females

5 Lesions and Clinical Features

Infected animals normally present diffuse scaling mainly on the head, ears, and extremities of the limbs, sometimes involving extensive areas of the body (Fig. 2). This is most frequently observed in large litters, suckling rabbits, or recently weaned rabbits (Rosell 2000).

Lesions are nonpruritic and may be circular or alopecic, focal or multifocal, and solitary or generalized, and their location varies greatly according to the intensity of infection. They may be present on the head (ears, nape, and muzzle), back, abdomen, pelvic area, tail, and limbs (Moriello 1996).

The affected area becomes alopecic, and lesions can be circular (Rosser 1995). Infections may occur with variable frequency and intensity, never forgetting asymptomatic animals (Patro et al. 2019). Dermatophytoses caused by *T. mentagrophytes* are usually inflammatory and often give rise to kerion-like lesions, which may cover a large part of the body surface (Van Rooij et al. 2006). Dermatophytosis does not cause mortality, but the economic losses are very high due to growth delays and an increased conversion index (Rosell 2000).

Sometimes, there are lesions on the extremities that make it difficult to remove the skin at the slaughterhouse, even leading the slaughterhouse to reject its supplier (the rabbit producer). In farm, when the producer removes the nests, a considerable amount of dirt can remain into the cage, in contact with the weaned rabbits. This occurrence can lead to exuberant lesions on the paws (extremities), with no visible lesions on the other parts of the body.

6 Diagnostic Approaches to Dermatophytosis

6.1 Clinical-Epidemiological Diagnosis

Dermatophytosis is a very contagious disease that affects a large number of animal species. In a rabbit farm, carrier females are a primordial risk factor for the dissemination of the infection in the farm, affecting animals of all sexes and all ages (Pinard and Chermette 1987; Rosell 2000).

The differential diagnoses of dermatophytosis in rabbits include sarcoptic mange and hypertrichosis due to excess sulfur (Rosell 2000). Dermatophytoses in industrial production are a problem for all rabbit farmers due to the difficulty of treatment and control. Clinical diagnosis can be easily performed by visualizing the clinical signs. Clinical diagnosis is a challenge when the number of infected animals is small and the infection is at an early stage. The clinical appearance and location of the lesions are the first step toward a suspicion of dermatophytosis (Pinard and Chermette 1987; Rosser 1995).

The correct sample collection increases diagnostic accuracy. Samples must be carried out on the periphery of well-defined skin lesions, as the center of the lesions may contain unfeasible material (Weitzman and Summerbell 1995). Laboratory diagnosis is based on direct observation and culture of hairs and skin scaling. The appearance and abundance of lesions make bacterial complications possible, making this type of diagnosis difficult. It is recommended to collect samples from the head and ear area, as these are the potentially most affected sites (Medleau and Ristic 1992; Sparkes et al. 1993).

The on-farm sampling technique is extremely important. Hair samples can be collected either by plucking or using the "Mackenzie" brush technique and are never cut with scissors or shaved (Mackenzie 1963; Moskaluk and VandeWoude 2022). Disinfection of the site with alcohol at 70 °C is important to reduce the number of contaminants, and collection should be performed in several places where there are suspicious lesions. If possible, broken hairs, intrafollicular portions, and skin peels should also be collected (Moriello et al. 2017).

Collection is performed with forceps, scalpel blades, or tape, and the sample should be sent to the laboratory dry in an envelope and placed in a sterile, closed container (Moriello et al. 2017).

6.2 Wood's Lamp

Hairs, infected by some species of dermatophytes, emit green fluorescence when exposed to Wood's lamp (UV light with wavelength ranges between 320 and 400 nm) due to a metabolite of tryptophan (Asawanonda and Taylor 1999; Klatte et al. 2015).

As not all species of dermatophytes fluoresce, a negative Wood's lamp examination cannot rule out dermatophytosis as a diagnosis (Klatte et al. 2015).

6.3 Direct Microscopic Examination

Direct observation of hairs and shedding by microscopy allows diagnosis in a few minutes or hours and is inexpensive. This type of examination is used as a presumptive diagnosis, and preparations are made with various clarifying preparations, depending on the material and the fungal species involved (Bhattarai et al. 2021).

In the analysis with potassium hydroxide (KHO), at 10% to 40%, the collected hairs are placed on slides or coverslips and visualized under an optical microscope in 10× and 40× up to 400× objectives. The function of KHO is to dissolve tissue structures and lighten pigments without affecting fungal structures. Microscopic examination can be enhanced by the use of lactophenol-cotton blue (Coelho et al. 2008b; Park et al. 2020).

Visualization of the arthrosporic hyphae invading the keratin of the stratum corneum or hair. The base of the hair is surrounded by arthrospores, but this situation also varies according to species. In the case of *Microsporum*, the spores surround the base of the hair in an irregular mosaic, but in the case of *Trichophyton*, the spores are arranged in clusters. Many false negatives occur in this type of examination, even when performed by experienced mycologists or dermatologists (Moriello et al. 2017).

6.4 Fungal Culture

Fungal cultures are the basis for the definitive identification of the causative agent. Two types of culture media can be used: nonselective and selective media. Nonselective media allow the growth of virtually all fungal species. Examples are Sabouraud agar (SDA) and potato dextrose agar (Black 2020).

Dermatophyte test medium® contains nutrients, gentamicin, chlortetracycline, and cycloheximide, which inhibit the growth of bacteria and other saprophytic fungi that may be present in contaminated samples. The composition of this medium facilitates visualization because in the case of dermatophyte growth, the medium changes color (yellow to red) due to the production of alkaline substrates (Kaufmann et al. 2016). The change in color of the medium cannot be regarded as unequivocal proof of the presence of dermatophytes but only as an indicator (Weitzman and Summerbell 1995).

Microscopic examination is the basis of diagnosis for most fungi. Visualization of the reproductive structures and spores, as well as their respective morphological differences, is the starting point for the identification process. A fragment of the primary culture, or subculture, is taken and placed on the slide and coverslip or an adhesive tape, with an appropriate mounting fluid such as lactophenol with cotton blue. Dermatophytes, being hyaline fungi, do not possess pigments and should therefore be examined with lactophenol-cotton blue, which stains the structures blue (Upadhyay et al. 2019).

6.5 Histopathological Diagnosis

Histopathological diagnosis is possible to demonstrate the fungal elements present in the tissues. Three types of staining can be used: hematoxylin-eosin (HES), periodic acid-Schiff (PAS), and Grocott (GMS) (Reza Kermanshahi and Rhatigan 2010).

There are several histological alterations compatible with dermatophytosis, namely, the presence of neutrophils, massive orthokeratosis, perifolliculitis, folliculitis, furunculosis, perivascular dermatitis, and vesicular/pustular intraepidermal dermatitis (Curtis 2001).

Septate and arthroconid hyphae are observed in the stratum corneum and hair follicles within or around the hairs. This type of examination makes it possible to give a result when positive, almost immediately, differentiating the type of dermatophyte involved (Chabasse and Contet-Audonneau 2003).

6.6 Molecular Diagnosis

The polymerase chain reaction (PCR) technique for the detection of dermatophyte DNA can be useful; however, a positive result does not necessarily mean an active infection. In veterinary medicine, this technique is not yet optimized. This is because nonviable dermatophytes resulting from successful treatment will be detected by PCR. A negative PCR result in a treated cat is compatible with a cure (Petrucelli et al. 2020; Faramarzi et al. 2021).

The vast majority of genotyping studies of dermatophyte species strains have been carried out to characterize the degree of genetic variation within a population of isolates, providing an initial characterization of the diversity within a given species (Salehi et al. 2021).

Among these techniques, the ones that have been most used to study dermatophytes are those that involve the amplification of unidentified regions of the genome and consist of obtaining a pattern of bands to create a unique "fingerprinting" for each strain. The RAPD ("random amplified polymorphic DNA") and the AFLP ("amplified fragment length polymorphism") can be highlighted. Fingerprinting methods can be used to detect genomic variations in sequences even when no previous information is available. However, they have many limitations due to the use of poorly specific markers. For example, RAPD and AFLP have allowed the identification of dermatophyte species but do not allow for strain differentiation (Mohammadi et al. 2015).

Sequencing of ITS (internal transcribed spacer) rDNA and TEF-1α gene sequencing analyses provided accurate identification of *T. rubrum*, *T. interdigitale*, and *T. tonsurans* at the genus and species levels (Salehi et al. 2021).

7 Treatment

Two compounds, enilconazole and sulfur, are mainly used as therapies. Enilconazole is the drug of choice, but sulfur is the most widely used due to its quality/price ratio. It is a good antifungal, not as specific and effective as enilconazole, and has the disadvantage of being keratinolytic (Aneke et al. 2018). When used excessively or incorrectly, it burns the hair and ears of the animals, leaving them with a lumpy appearance. The abuse of sulfur may cause the death of the animals, mainly lactating kits. Oral use should be done with great precautions, and it is recommended to lightly spray the nest and feed in each cage (Van Cutsen et al. 1985). In practice, good results are achieved in controlling the disease with the application of sulfur in the nests before parturition. In the cases when does do not enter the nests before giving birth, the sulfur must be removed to avoid burns in the newly born rabbits.

In Europe, enilconazole is no longer available for use in rabbits.

Griseofulvin is another antifungal for the treatment of dermatophytosis in breeding and future breeding rabbits. In most farms, systemic treatment is not used due to the toxicity of the molecules available, which may fatally compromise the animals, mainly due to its repercussions at the renal level. This substance is teratogenic and should not be administered to pregnant females. Its use is associated with diets with doses varying from 200 to 250 ppm for 10 days or 100 to 125 ppm for 20 days (Barnes and Boothroyd 1961; Scarff 2002; Mohamad et al. 2020). Ketoconazole can also be used in therapy, but it has the same disadvantages as griseofulvin, which, like all systemic antifungals, may cause infertility (Bereda 2022).

Farms working with breeding animals should have a specific prevention program. In the hatchery, the use of sulfur in the nests is fundamental, and treatment with enilconazole is sometimes necessary. The animals, material, and ship should be sprayed weekly at least once (Chang et al. 2022).

The application of a treatment depends on the prevalence and the interest of the owner. Topical treatment with diluted enilconazole, three times per week, during the 10 days following weaning is the most recommended. In females, it is advisable to apply the treatment three times in 10 days in the mammary region with the aid of a sponge (Van Cutsen et al. 1985).

Treating breeding rabbits with 0.3 mL of ivermectin 1%, 9 days after giving birth, and applying sulfur with a sponge in breeders' belly 15 days after artificial insemination are practices that also increased the success in dermatophytosis control.

Dermatophytosis is self-limiting in a relatively short period of time, but fattening rabbits are slaughtered at approximately 70 days of age, which means that there is no time for the pathology to regress (Boucher and Nouaille 2002).

Dermatophytoses are often difficult to eradicate, and it may take months for remission (Greene and Watson 1998). Success of therapy does not guarantee elimination of the infection, and asymptomatic carriers may remain. Only the absence of lesions associated with negative culture tests constitutes a remission criterion.

8 Control and Prophylaxis

Control must take into account various aspects, mainly during pre-weaning, when the females remain with their kits in fattening, which is a critical phase of greatest infection. Straw or hay should be removed at this stage, or if infected animals are present, as they may contain contaminated material (Torres-Rodriguez et al. 1992; Cafarchia et al. 2010).

Fungal elements of resistance present in the coat and eliminated in skin shedding remain viable as sources of infection and therefore need to be eliminated. An important prevention program for dermatophytosis includes the use of dry heat (Fig. 3) followed by the application of disinfectants and antiseptics (Hammer et al. 2011; Akhoundi et al. 2022). The aim is the destruction of hair and spores found in cages with the aid of blowtorches or other means of heat, and areas are also frequently affected, such as the roofs of farms, windows, and air extractors. Decontamination of the environment is an important element in the control of these infections (Peters et al. 2022).

Biosecurity is extremely important regarding dermatophytosis control and prophylaxis. Disinfection is central to prophylaxis. Disinfectants such as formaldehyde, glutaraldehyde, or chlorinated and antiseptics such as bleach and ionophores should be applied frequently. Formaldehyde is often used in cases of voiding, and bleach is a good antifungal agent, but it must be used with caution because it is very aggressive to the skin (Moriello et al. 2004; Moriello 2015).

During the sanitary process, it is important to use high pressure washers, with high temperature water, and apply descaling foaming detergent to inactivate spores. After washing and disinfecting, using disinfectants suitable for combating dermatophytosis, fumigation with antifungal products must be carried out. After that, it's also essential to dry the shed and warming it conveniently before receiving the does.

Breeder certification by the genetic multiplication centers, about being free from dermatophytosis, should be mandatory.

Immunoprophylaxis is available in some European countries, and vaccines (e.g., autogenous vaccines) against dermatophytes have already been used quite successfully. Vaccination appears to be a very useful means of infection control in farms that

Fig. 3 Flame thrower for elimination of hair from the mesh of cages

are frequently exposed to this problem, mainly breeding farms (Al-Janabi and Al-Khikani 2021).

9 Zoonotic Potential of Dermatophytoses in Productive Rabbits

Dermatophytosis is a zoonosis that should be addressed according to the One Health paradigm (Corrêa-Junior et al. 2022). The disease has a significant impact on the quality of life of human patients and can affect psychological, economic, and social aspects. The disease causes low mortality with high morbidity. It is estimated to affect 25% of the human population (Segal and Elad 2021). It is a disease that is associated with anxiety, depression, and low self-esteem, mainly due to discomfort in relation to itching and esthetic issues (Patro et al. 2019; Patel et al. 2020). Most affected people are asymptomatic, heal spontaneously, and do not require medical assistance (Moskaluk and VandeWoude 2022).

Dermatophytosis is the main zoonosis in rabbit farms, where 90% of the keepers working in affected farms have also been affected. The control of this pathology is only possible if there is collaboration between all those involved in this field (Boucher and Nouaille 2002).

Clinical manifestations in human medicine include circular, erythematous, scaly lesions on the skin, or in the case of fungal infections (tinea unguium or onychomycosis), there may be discoloration, thickening, scaling, and separation of the nail bed. There are other associated factors, such as socioeconomic status, occupation, climate change, and predisposing factors (such as immunosuppression/disability and diabetes) (Gnat et al. 2019).

Other underlying conditions associated with dermatophytosis with a complex clinical picture include solid organ transplantation, HIV infection, use of immunosuppressive drugs such as systemic corticosteroids, and immunodeficiencies. In the case of rabbit farms affected with dermatophytosis, and since most of them are contaminated with *T. mentagrophytes*, the presence of infection by this species in farmers and veterinarians should also be suspected (Tizzani et al. 2007).

10 Conclusions

Rabbit ringworm is a zoonotic disease mainly caused by two main fungal genera, *Trichophyton* and *Microsporum*, which are ubiquitous. Even dermatophytosis only affects cutaneous structures, and it is related to poor rabbit performance in rabbitries, resulting in significant economic losses. High animal densities as well as immunosuppressive conditions are major risk factors for developing the condition.

The clinical-epidemiological and laboratory diagnosis of dermatophytosis is easy to perform. Topical enilconazole to all animals and sulfur application to structures, or oral administration of griseofulvin to breeding rabbits, are effective pharmacological substances to treat affected animals and control the disease. Nevertheless, their use is strongly restricted in some worldwide regions, such as the EU.

Decontamination of the environment through disinfection and management of risk factors are keys to controlling and preventing dermatophytosis. A continuous effort has been made to obtain effective vaccines for rabbit immunization.

Acknowledgments The authors thank Springer for the use of the AJE Digital Editing service. This work was supported by the projects UIDP/CVT/00772/2020 (CECAV) and LA/P/0059/2020 (AL4AnimalS), funded by the Portuguese Foundation for Science and Technology (FCT).

References

Akhoundi M, Nasrallah J, Marteau A, Chebbah D, Izri A, Brun S (2022) Effect of household laundering, heat drying, and freezing on the survival of dermatophyte conidia. J Fungi (Basel) 8(5):546. https://doi.org/10.3390/jof8050546

Al-Janabi AAHS, Al-Khikani FHO (2021) Prophylaxis and therapeutic ability of inactivated dermatophytic vaccine against dermatophytosis in the rabbits as an animal model. Turk J Pharm Sci 18(3):326–331. https://doi.org/10.4274/tjps.galenos.2020.81226

Aneke CI, Otranto D, Cafarchia C (2018) Therapy and antifungal susceptibility profile of *Microsporum canis*. J Fungi (Basel) 4(3):107. https://doi.org/10.3390/jof4030107

Asawanonda P, Taylor CR (1999) Wood's light in dermatology. Int J Dermatol 38(11):801–807. https://doi.org/10.1046/j.1365-4362.1999.00794.x

Barnes MJ, Boothroyd B (1961) The metabolism of Griseofulvin in mammals. Biochem J 78:41–43

Benedict K, Jackson BR, Chiller T, Beer KD (2019) Estimation of direct healthcare costs of fungal diseases in the United States. Clin Infect Dis 68(11):1791–1797. https://doi.org/10.1093/cid/ciy776

Bereda G (2022) Antimycolytic agents: fungistatic and fungicide. Ann Dermatol Res 6:001–009. https://doi.org/10.29328/journal.adr.1001019

Bhattarai D, Banday AZ, Sadanand R, Arora K, Kaur G, Sharma S, Rawat A (2021) Hair microscopy: an easy adjunct to diagnosis of systemic diseases in children. Appl Microsc 51(1):18. https://doi.org/10.1186/s42649-021-00067-6

Black WD (2020) A comparison of several media types and basic techniques used to assess outdoor airborne fungi in Melbourne, Australia. PLoS One 15(12):e0238901

Boucher S, Nouaille L (2002) Maladies des Lapins, 2nd edn. France Agricole, Paris, France

Cafarchia C, Camarda A, Coccioli C, Figueredo LA, Circella E, Danesi P, Capelli G, Otranto D (2010) Epidemiology and risk factors for dermatophytoses in rabbit farms. Med Mycol 48(7): 975–980. https://doi.org/10.3109/13693781003652620

Cafarchia C, Weigl S, Figueredo LA, Otranto D (2012) Molecular identification and phylogenesis of dermatophytes isolated from rabbit farms and rabbit farm workers. Vet Microbiol 154(3–4): 395–402. https://doi.org/10.1016/j.vetmic.2011.07.021

Carpouron JE, de Hoog S, Gentekaki E, Hyde KD (2022) Emerging animal-associated fungal diseases. J Fungi (Basel) 8(6):611. https://doi.org/10.3390/jof8060611

Chabasse D, Contet-Audonneau N (2003) Examen direct et place D'histology en mycology. Rev Fr Lab 357:49–54

Chang CC, Wechtaisong W, Chen SY, Cheng MC, Chung CS, Lin LS, Lien YY, Tsai YL (2022) Prevalence and risk factors of zoonotic dermatophyte infection in pet rabbits in northern Taiwan. J Fungi (Basel, Switzerland) 8(6):627

Chupia V, Jirasarunyanon P, Sriwises S, Na Lampang K, Buranapim N (2019) Type of dermatophytes on rabbit skin in rabbit Café in Chiang Mai. Vet Integr Sci 17(1):75–85

Ciesielska A, Stączek P (2020) A new molecular marker for species-specific identification of *Microsporum canis*. Braz J Microbiol 51:1505–1508

Coelho AC, Alegria N, Rodrigues J (2008a) Isolamento de Dermatófitos em Animais Domésticos em Vila Real, Portugal. Arquivo Brasileiro de Medicina Veterinária e Zootecnia 60:1017–1020

Coelho AC, Pinto ML, Coelho AM, Fontes MC, Mourão JL, Pinheiro V, Rodrigues J (2008b) Laboratory limits on dermatophyte diagnosis in rabbits with clinical lesions. In: Pathology and Hygiene, 9th World Rabbit Congress, pp 925–29

Connole MD (1967) Review of animal mycosis in Australia. Vet Bull 37(3):133–164

Corrêa-Junior D, Almeida Paes R, Frases S (2022) One health and dermatophytes, what is the link between them? Biomed J Sci Tech Res 44(1):2022

Curtis CF (2001) Diagnostic techniques and sample collection. Clin Tech Small Anim Pract 16(4): 199–206. https://doi.org/10.1053/svms.2001.26998

Dalis JS, Kazeem HM, Kwaga JKP, Kwanashie CN (2019) Prevalence and distribution of dermatophytosis lesions on cattle in plateau state, Nigeria. Vet World 12(9):1484–1490. https://doi.org/10.14202/vetworld.2019.1484-1490

de Hoog GS, Dukik K, Monod M, Packeu A, Stubbe D, Hendrickx M, Kupsch C, Stielow JB, Freeke J, Göker M, Rezaei-Matehkolaei A, Mirhendi H, Gräser Y (2017) Toward a novel multilocus phylogenetic taxonomy for the dermatophytes. Mycopathology 182(1–2):5–31. https://doi.org/10.1007/s11046-016-0073-9

Dey JC, Rahman MK, Rumi MA, Dutta A, Sayeed A, Halder BC, Mannan A, Hassan MM, Chittagong SH (2016) Prevalence of dermatophytosis in rabbits at Saqtvh, Chittagong, Bangladesh. J Dairy Vet Anim Res 3(6):201–205. https://doi.org/10.3390/jof8060627

Donnelly TM, Rush EM, Lackneg PA (2000) Ringworm in small exotic pets. Semin Avian Exotic Pet Med 9(2):82–93. https://doi.org/10.1053/AX.2000.5045

Espinosa R, Tago D, Treich N (2020) Infectious diseases and meat production. Environ Resour Econ 76:1019–1044. https://doi.org/10.1007/s10640-020-00484-3

Faramarzi S, Motamedi M, Rezaei-Matehkolaei A, Aboutalebian S, Ansari S, Didehdar M, Bahadoran M, Mirhendi H (2021) A simple multiplex polymerase chain reaction assay for rapid identification of the common pathogenic dermatophytes: *Trichophyton interdigitale*, *Trichophyton rubrum*, and *Epidermophyton floccosum*. Curr Med Mycol 7(2):1–7. https://doi.org/10.18502/cmm.7.2.7030

Gnat S, Nowakiewicz A, Zięba P (2019) Taxonomy of dermatophytes the classification systems may change but the identification problems remain the same. Adv Microbiol 1:49–58. https://doi.org/10.21307/PM-2019.58.1.049

Greene CE, Watson ADJ (1998) Antifungal chemotherapy. In: Greene CE (ed) Infectious diseases of the dog and cat. W.B. Saunders, Philadelphia, PA, pp 357–362

Guo Y, Ge S, Luo H, Rehman A, Li Y, He S (2020) Occurrence of *Trichophyton verrucosum* in cattle in the Ningxia Hui autonomous region, China. BMC Vet Res 16(1):187. https://doi.org/10.1186/s12917-020-02403-6

Hainer BL (2003) Dermatophyte infections. Am Fam Physician 67(1):101–108

Hammer TR, Mucha H, Hoefer D (2011) Infection risk by dermatophytes during storage and after domestic laundry and their temperature-dependent inactivation. Mycopathologia 171:43–49

Iorio R, Cafarchia C, Capelli G, Fasciocco D, Otranto D, Giangasperos A (2007) Dermatophytoses in cats and humans in Central Italy: epidemiological aspects. Mycoses 50(6):491–495

Kaufmann R, Blum SE, Elad D, Zur G (2016) Comparison between point-of-care dermatophyte test medium and mycology laboratory culture for diagnosis of Dermatophytosis in dogs and cats. Vet Dermatol 27(4):284-e68. https://doi.org/10.1111/vde.12322

Kaur N, Bains A, Kaushik R, Dhull SB, Melinda F, Chawla P (2021) A review on antifungal efficiency of plant extracts entrenched polysaccharide-based nanohydrogels. Nutrients 13:2055. https://doi.org/10.3390/nu13062055

Klatte JL, van der Beek N, Kemperman PM (2015) 100 years of wood's lamp revised. J Eur Acad Dermatol Venereol 29(5):842–847

Mackenzie DWR (1963) "Hairbrush diagnosis" in detection and eradication of non-fluorescent scalp ringworm. BMJ 2:363–365

Martinez-Rossi NM, Peres NT, Rossi A (2017) Pathogenesis of dermatophytosis: sensing the host tissue. Mycopathologia 182(1–2):215–227

Medleau L, Ristic Z (1992) Diagnosing dermatophytosis in dogs and cats. J Vet Med 87:1086–1091. https://doi.org/10.1111/vde.12440

Mercer DK, Stewart CS (2019) Keratin hydrolysis by dermatophytes. Med Mycol 57(1):13–22

Metin B, Heitman J (2017) Sexual reproduction in dermatophytes. Mycopathologia 182(1–2): 45–55

Metin B, Heitman J (2020) She loves me, she loves me not: on the dualistic asexual/sexual nature of dermatophyte fungi. Mycopathologia 185(1):87–101

Mohamad A, Hassan A, Amer M, AbdelAziz ES (2020) The effects of Oral ketoconazole and Griseofulvin on the fertility of male rabbits. Mansoura Vet Med J 21:32–38. https://doi.org/10.35943/mvmj.2020.21.2.0203

Mohammadi R, Abastabar M, Mirhendi H, Badali H, Shadzi S, Chadeganipour M, Pourfathi P, Jalalizand N, Haghani I (2015) Use of restriction fragment length polymorphism to rapidly identify dermatophyte species related to dermatophytosis. Jundishapur J Microbiol 8(6):e17296. https://doi.org/10.5812/jjm.8(5)2015.17296

Moriello KA (1996) Dermatophytosis part 1: aetiology, pathogenesis, and clinical aspects in small animals. Department of Medical Sciences, School of Veterinary Medicine, University of Wisconsin-Madison, Madison, pp 33–41

Moriello K (2014) Feline Dermatophytosis: aspects pertinent to disease management in single and multiple cat situations. J Feline Med Surg 16(5):419–431

Moriello KA (2015) Kennel disinfectants for *Microsporum canis* and *Trichophyton* sp. Vet Med Int 2015:853937. https://doi.org/10.1155/2015/853937

Moriello KA, Deboer DJ, Volk LM, Sparkes A, Robinson A (2004) Development of an in vitro, isolated, infected spore testing model for disinfectant testing of *Microsporum canis* isolates. Vet Dermatol 15(3):175–180

Moriello KA, Coyner K, Paterson S, Mignon B (2017) Diagnosis and treatment of dermatophytosis in dogs and cats: clinical consensus guidelines of the World Association for Veterinary Dermatology. Vet Dermatol 28(3):266-e68

Morisse JP, Maurice R (1997) Influence of stocking density or group size on behavior of fatting rabbits kept under intensive conditions. Appl Anim Behav Sci 54:71–80

Morisse JP, Boilletot E, Martrenchar A (1999) Preference testing in intensively kept meat prodution rabbits for straw on wire grid floor. Appl Anim Behav Sci 64:71–80

Moskaluk AE, VandeWoude S (2022) Current topics in dermatophyte classification and clinical diagnosis. Pathogens 11(9):957. https://doi.org/10.3390/pathogens11090957

Naranjo-Ortiz MA, Gabaldón T (2019) Fungal evolution: diversity, taxonomy and phylogeny of the fungi. Biol Rev 94(6):2101–2137. https://doi.org/10.1111/brv.12550

OIE, Institute for International Cooperation in Animal Biologics (2005) Dermatophytosis, ringworm, tinea. Dematomycosis 2005:1–6

Park SH, Seol JE, Kim DH, Kim H (2020) Analysis of microscopic examination of pulled out hair in Telogen effluvium patients. Ann Dermatol 32(2):141–145. https://doi.org/10.5021/ad.2020.32.2.141

Paryuni AD, Indarjulianto S, Widyarini S (2020) Dermatophytosis in companion animals: a review. Vet World 13(6):1174–1181. https://doi.org/10.14202/vetworld.2020.1174-1181

Patel NH, Padhiyar JK, Patel AP, Chhebber AS, Patel BR, Patel TD (2020) Psychosocial and financial impact of disease among patients of Dermatophytosis, a questionnaire-based observational study. Indian Dermatol Online J 11(3):373–377. https://doi.org/10.4103/idoj.IDOJ_331_19

Patro N, Panda M, Jena AK (2019) The menace of superficial dermatophytosis on the quality of life of patients attending referral Hospital in Eastern India: a cross-sectional observational study. Indian Dermatol Online J 10(3):262–266. https://doi.org/10.4103/idoj.IDOJ_342_18

Peters A, Schmid MN, Parneix P, Lebowitz D, de Kraker M, Sauser J, Zingg W, Pittet D (2022) Impact of environmental hygiene interventions on healthcare-associated infections and patient colonization: a systematic review. Antimicrob Resist Infect Control 11:38

Petrucelli MF, Abreu MH, Cantelli BAM, Segura GG, Nishimura FG, Bitencourt TA, Marins M, Fachin AL (2020) Epidemiology and diagnostic perspectives of dermatophytoses. J Fungi (Basel) 6(4):310. https://doi.org/10.3390/jof6040310

Pinard M, Chermette R (1987) Diagnostic et prophylaxie des tignes des carnivores domestique, Étude critique à partir d'une enquête à l'Ecole nationale vétérinaire d'Alfort, France. Recueil de Medicine Veterinaire 1987:1107–1116

Porras-López C, Martínez-Herrera E, Frías-De-León MG, Moreno-Coutiño G, Reyes-Montes M, Arenas R, González-Cespón JL, Rodríguez-Cerdeira C (2021) Dermatophytosis caused by *Nannizzia nana*. J Med Mycol 31(2):101047. https://doi.org/10.1016/j.mycmed.2020.101047

Prochnau A, Almeida HL Jr, Souza PR, Vetoratto G, Duquia R, Defferrari R (2004) Scutular tinea of the scrotum: report of two cases. Mycoses 48:162–164

Rauterberg SL, Bill J, Kimm S, Kemper N, Fels M (2019) Effect of a new housing system on skin lesions, performance and soiling of fattening rabbits: a German case study. Animals 9(9):650

Reza Kermanshahi T, Rhatigan R (2010) Comparison between PAS and GMS stains for the diagnosis of onychomycosis. J Cutan Pathol 37(10):1041–1044. https://doi.org/10.1111/j.1600-0560.2009.01468.x

Rosell JM (2000) Enfermedades del conejo. Ediciones Mundi-Prensa, Barcelona, Spain

Rosser EJ (1995) Infectious crusting dermatoses. Vet Clin N Am Equine Pract 11(1):53–59

Rouzaud C, Hay R, Chosidow O, Dupin N, Puel A, Lortholary O, Lanternier F (2015) Severe Dermatophytosis and acquired or innate immunodeficiency: a review. J Fungi (Basel) 2(1):4

Salehi Z, Shams-Ghahfarokhi M, Razzaghi-Abyaneh M (2021) Molecular epidemiology, genetic diversity, and antifungal susceptibility of major pathogenic dermatophytes isolated from human dermatophytosis. Front Microbiol 12:643509. https://doi.org/10.3389/fmicb.2021.643509

Samanta I (2015) Veterinary mycology. Springer, Berlin. ISBN: 978-81-322-2280-4

Scarff DH (2002) Dermatoses. In: Meredith A, Flecknell P (eds) BSAVA manual of rabbit medicine and surgery, vol 9, 4th edn. BSAVA, London, p 76

Segal E, Elad D (2021) Human and zoonotic dermatophytoses: epidemiological aspects. Front Microbiol 12:713532. https://doi.org/10.3389/fmicb.2021.713532

Simpanya MF (2000) Dermatophytes: their taxonomy, ecology and pathogenicity. Revista Iberoamericana de Micologia 2000:1–12

Sparkes AH, Gruffydd-Jones TJ, Shaw SE (1993) Epidemiological and diagnostic features of canine and feline dermatophytosis in the United Kingdom from 1956 to 1991. Vet Rec 133: 57–61

Suckow MA, Brammer DW, Rush HG, Chrisp CE (2002) Biology and diseases of rabbits. Lab Anim Med 2022:329–364

Tizzani P, Gallo G, Peano A, Molinar Min A, Martínez-Carrasco Pleite C, Meneguz PG (2007) Dermatophytosis caused by Microsporum canis in eastern cottontail (Sylvilagus floridanus). Eur J Wildl Res 53:238–240

Torres-Rodriguez JM, Dronda MA, Rosell J, Madrenys N (1992) Incidence of dermatophytes in rabbits farms in Catalonia, Spain, and its repercussion on human health. Eur J Epidemiol 8:326–329

Upadhyay V, Kumar A, Singh AK, Pandey J (2019) Epidemiological characterization of dermatophytes at a tertiary care hospital in eastern Uttar Pradesh, India. Curr Med Mycol 5(1):1–6

Van Cutsen J, Van Gerven F, Geerts H (1985) Treatment with enilconazole spray of dermatophytosis in rabbit farms. Mykosen 28:400–407

Van Rooij P, Detandt M, Nolard N (2006) *Tricophyton mentagrophytes* of rabbit origin causing family incidence of kerion: an environmental study. Mycoses 50(2):150

Wagner DK, Sohnle PG (1995) Cutaneous defenses against dermatophytes and yeasts. Clin Microbiol Rev 8(3):317–335

Weitzman I, Summerbell RC (1995) The dermatophytes. Clin Microbiol Rev 8(2):240–260
White SD, Bourdeau PJ, Meredith A (2002) Dermatologic problems of rabbits. Semin Avian Exotic Pet Med 11(3):141–151
White TC, Findley K, Dawson TL Jr, Scheynius A, Boekhout T, Cuomo CA, Xu J, Saunders CW (2014) Fungi on the skin: dermatophytes and *Malassezia*. Cold Spring Harb Perspect Med 4(8): a019802
Zagnoli A, Chevalier B, Sassolas B (2004) Dermatophyties et dermatophytes. EMC-Pédiatrie 2:96–116
Zurita J, Hay R (1987) Adherence of dermatophyte microconidia and arthroconidia to human keratinocytes in vitro. J Investig Dermatol 89:529

Encephalitozoon cuniculi: An Emergent Pathogen

Ana Cláudia Coelho, Ana Patrícia Lopes, Daniel Morais, Filipa Teixeira Rodrigues, José Manuel Monteiro, and Luís Cardoso

Abstract

Encephalitozoon cuniculi is a pathogen with a worldwide distribution, belonging to the phylum *Microsporidia* (kingdom *Fungi*), which infects rabbits (*Oryctolagus cuniculus*) and several other animals, including humans. *Encephalitozoon cuniculi* was the first microsporidium to be described in mammals. The global population of rabbits is exposed to this pathogen, and, although most infections are subclinical, *E. cuniculi* can cause severe illness, with neurological, ocular, and renal involvement. In humans, especially in immunocompromised individuals, *E. cuniculi* has been implicated as an opportunistic agent. Spore proliferation and migration can be reduced by treatment, and preventive antimicrosporidial medications may be considered. Different control measures are necessary for decreasing the spread of *E. cuniculi*, not only to other rabbits but also to humans. This chapter provides a review of the literature regarding the pathogen *E. cuniculi*, infection epidemiology, pathogenesis, lesions and clinical signs of disease, and diagnostic and control methods including treatment and prevention of encephalitozoonosis.

A. C. Coelho · A. P. Lopes · F. T. Rodrigues · L. Cardoso (✉)
Department of Veterinary Sciences, School of Agrarian and Veterinary Sciences (ECAV), and Animal and Veterinary Research Centre (CECAV), University of Trás-os-Montes e Alto Douro (UTAD), Vila Real, Portugal

Associate Laboratory for Animal and Veterinary Sciences (AL4AnimalS), Lisbon, Portugal
e-mail: lcardoso@utad.pt

D. Morais
Nanta, SA, Alverca do Ribatejo, Portugal

J. M. Monteiro
José Azevedo Monteiro, Lda., Vila Boa do Bispo, Portugal

Faculty of Veterinary Medicine, CBIOS, Lusófona University, Lisbon, Portugal

J. Simões, J. M. Monteiro (eds.), *Veterinary Care of Farm Rabbits*,
https://doi.org/10.1007/978-3-031-44542-2_28

Keywords

Encephalitozoonosis · Microsporidia · Rabbits · Zoonosis

1 Introduction

Encephalitozoon cuniculi is a spore-forming unicellular pathogen belonging to the phylum *Microsporidia* (kingdom *Fungi*) with a worldwide distribution, which infects rabbits (*Oryctolagus cuniculus*) and several other mammalians, including humans, and also some bird species (Mathis et al. 2005; Hinney et al. 2016; Han and Weiss 2017; Künzel and Fisher 2018). *E. cuniculi* was described as an infectious agent in rabbits in 1922 (Wright and Craighead 1922), the first microsporidium to be recognized in mammals. Current seroprevalence studies demonstrate that the global population of captive rabbits is considerably exposed to this pathogen (Latney et al. 2014).

Although most infections are subclinical, *E. cuniculi* can cause severe illness in rabbits, with neurological (Fukui et al. 2013; Shin et al. 2015; Künzel and Fisher 2018), ocular (Giordano et al. 2005; Sandmeyer et al. 2011), and renal (Harcourt-Brown 2007; Csokai et al. 2009a; Maestrini et al. 2017) involvement, and also in immunocompromised people, in which it has been implicated as an opportunistic agent (Wasson and Peper 2000). The disease caused by *E. cuniculi*, encephalitozoonosis, may affect different populations, including wild, domestic, farm, industrial, pet, and laboratory rabbits, and lead to chronic infections that persist for years (Saviotti et al. 2000; Fukui et al. 2013; Mathis et al. 2005; Sadeghi-Dehkordi et al. 2019), with possible excretion of spores (Kimura et al. 2013; Maestrini et al. 2017; Askari et al. 2015).

Diagnosis is challenging because a considerable number of infected rabbits have a chronic subclinical infection (Künzel et al. 2008). Spore proliferation and migration can be reduced by treatment, and preventive antimicrosporidial medications can be considered. Other control measures are necessary for decreasing the spread of *E. cuniculi*, not only to other rabbits but also to humans, especially immunocompromised individuals (Fukui et al. 2013; Latney et al. 2014; Künzel and Fisher 2018).

This chapter provides a review of the literature, by primarily searching MEDLINE® bibliographic database through PubMed (and also by checking reference lists of retrieved articles), and may serve as a guide regarding the pathogen *E. cuniculi*, infection epidemiology, pathogenesis, lesions and clinical signs of disease, diagnostic methods and procedures, and control measures including treatment and prevention of encephalitozoonosis in rabbits.

2 The Pathogen *Encephalitozoon cuniculi*

2.1 General Features and Cell Structure

E. cuniculi, from the phylum *Microspora*, is a unicellular, intracellular obligate eukaryotic spore-forming microorganism (Morsy et al. 2020). It is very much related to fungi (Cali et al. 2011), due to the presence of fungal components, such as tubulins, chitin, and trehalose, and the analyses of multiple gene sequences (Didier and Weiss 2006).

This microorganism was first identified, in 1922, in the United States in a colony of research rabbits (Wright and Craighead 1922). Moreover, it was the first microsporidia to be cultivated, which has contributed to a better understanding of the pathogenesis of microsporidiosis (Anane and Attouchi 2010). Furthermore, this parasite is considered a good model system for studies on eukaryotic genome evolution (Peyretaillade et al. 2011).

2.2 Genotypes

Based on the rRNA internal transcribed spacers (ITS), four *E. cuniculi* genotypes have been identified: EcI, EcII, EcIII, and EcIV. Genotype EcI has been cultivated in vitro and in laboratory rabbits ("rabbit strain"). Genotype EcII originated from rats ("murine strain") but has also been identified in other rodents. Genotype EcIII, whose natural host is dogs ("canine strain"), has also been identified in monkeys. Genotype EcIV, whose main hosts are humans ("human strain"), has also been reported in dogs and cats (Talabani et al. 2010; Nell et al. 2015; Hinney et al. 2016). Genotype EcIV has nonspecific diarrhea as its main clinical sign (Hinney et al. 2016). Although the names refer to the genotypes' host preference, there is no strict specificity (Hinney et al. 2016). For example, the "rabbit strain," genotype I, has been already isolated in horses (Wagnerová et al. 2012) and gorillas and chimpanzees (Sak et al. 2011b).

The four genotypes were only possible to detect in the early 1990s when morphological and biochemical analyses allowed the identification and discrimination of the various types by analysis of the ITS. The four genotypes have no morphological differences and can only be differentiated through their DNA, by the number of repeats of the 5′-GTTT-3′ chair in the transcription organism (ITS), and all can infect humans apart from genotype EcIV, where type I has been reported in Switzerland and type III in the United States (Hinney et al. 2016).

2.3 Reservoir and Life Cycle

Microsporidia are known as the most efficient and developed intracellular agents for infection of invertebrates and vertebrates, due to the fact that in immunocompetent

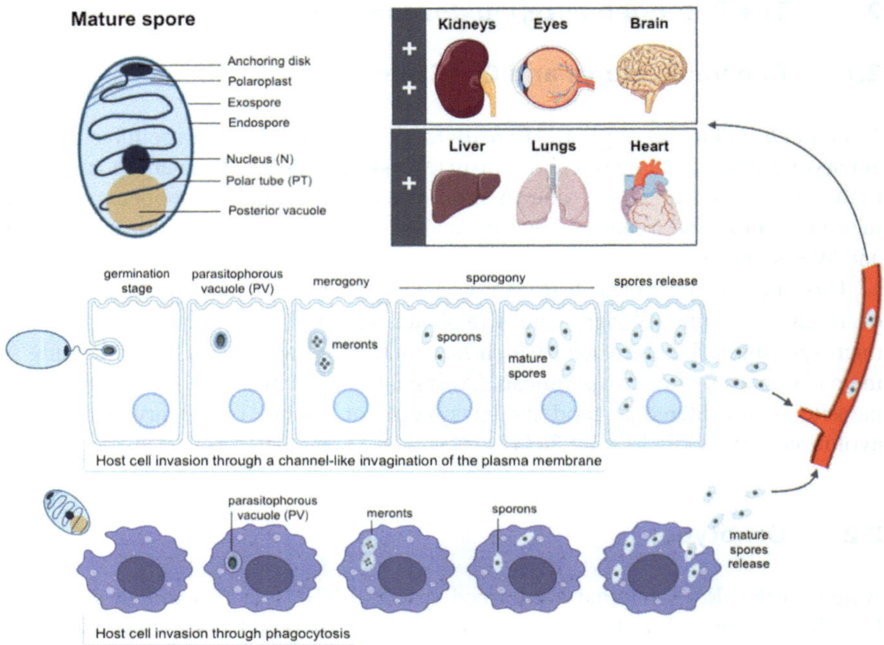

Fig. 1 Life cycle of the *Encephalitozoon cuniculi*. Adapted from Magalhães et al. (2022) with permission from Springer Nature

individuals, the interaction between host and pathogen causes mild clinical disease (Dalboni et al. 2021).

This pathogen can infect different mammalian hosts around the world (Mäkitaipale et al. 2022). *E. cuniculi* has zoonotic potential, because it can infect also humans, especially immunocompromised adults and children (Mathis et al. 2005).

The spores, which are environmentally resistant small oval- or pyriform-shaped structures, with an average size varying between 1 and 4 μm length in mammals, are responsible for transmitting *E. cuniculi* (Han et al. 2020a; Morsy et al. 2020). They contain various components that link it with microsporidians such as the polar filament or tubule, the anchoring disk, the lamellar polaroplast, and the tubular polaroplast (Han et al. 2020a).

E. cuniculi has a direct life cycle that takes 3–5 weeks to be completed (Richardson 2000). It is divided in three different stages (Fig. 1):

- Infective phase: the spore is present in the host cell and it is fully developed (Cali et al. 2011); after an increase in the osmotic pressure, the polar tube suffers an extrusion that allows the penetration into the hosts' cell membrane and the release of the infective sporoplasm. The specialization of the polar tube represents one of the most advanced infection mechanisms among eukaryotes (Peyretaillade et al. 2011).

- Proliferative phase: transformation, inside the newly formed parasitophorous vacuole, of the sporoplasm into a new structure (i.e., the meront), by a process of asexual reproduction called merogony (Fasshauer et al. 2005; Bohne et al. 2011).
- Sporogonic phase: formation of mature spores by acquisition of the polar tube and all the other components, like the spore wall (Doboşi et al. 2022). The accumulation of spores inside the host's cell will eventually cause its rupture and the release of spores to close cells and through vascular system to other places (Wasson and Peper 2000).

The transmission can be either horizontal (ingestion or inhalation of spores from infected rabbit urine or feces) or vertical (intrauterine infection), and the *E. cuniculi* spores can survive outside the host for up to 6 weeks at 22 °C (Künzel and Fisher 2018) and under humid conditions for several months (Li et al. 2003). Horizontal transmission by excretion of urine is the main route of infection, and, once ingested or inhaled, spores invade enterocytes and then enter the bloodstream or the lymphatic system, so that they can reach high flow organs like the lungs, liver, and kidney and, in a later phase of the disease, the brain (Maddox et al. 2000). In vertical transmission, the transplacental transmission of *E. cuniculi* occurs during the first trimester period of gestation when embryological development of lens capsule is active, which causes ocular encephalitozoonosis and cataract (Özkan et al. 2019).

3 Epidemiology of Infection in Rabbits

E. cuniculi, like other Microsporidia, survive only by living in other cells and can produce environmentally resistant spores, which infect host cells (Jeklova et al. 2020). *E. cuniculi* has the rabbit as its natural host and infects both domestic and wild rabbits. In rabbits, horizontal transmission by ingestion or inhalation of spores occurs most frequently (Didier et al. 2000), but ocular infections have also been documented (Jeklova et al. 2010b, 2019; Özkan et al. 2019). Vertical transmission via transplacental route also needs to be considered (Hunt et al. 1972; Weiss 2001; Baneux and Pognan 2003).

After 35 days of infection (depending on the age of the infected rabbit), spores are excreted intermittently in the urine of infected rabbits for up to 3 months if not longer (Cox et al. 1979; Jeklova et al. 2010a, b). Even in an immunocompetent host, *E. cuniculi* persists despite an active immune response. However, latent infection remains subclinical as long as this microsporidium multiplication and the host immune response are balanced (Schmidt and Shadduck 1983). If immune status is compromised, microsporidia can proliferate rapidly, and clinical disease may occur, which contributes to the infection of other animals (Shadduck and Orenstein 1993; Jeklova et al. 2010b).

3.1 Prevalence

This parasite affects rabbits worldwide, with high seroprevalence levels reported in Europe. It is recognized as a common pathogen and an important cause of disease among pet rabbits (Latney et al. 2014; Morsy et al. 2020). Rabbits (*O. cuniculus*) are considered the natural host of *E. cuniculi*, but a wide range of species can be affected, including rodents and carnivores, birds, and primates (Jeklova et al. 2020), which makes it a pathogen with low host specificity (Deplazes et al. 2000).

In rabbits, its natural host, the infection by this pathogen is mostly subclinical. Nevertheless, it is a common cause of economic losses in rabbit farms, mostly on those whose hygiene and biosecurity are not well performed. *E. cuniculi* is also considered as a zoonotic opportunistic pathogen for immunocompromised individuals (Deplazes et al. 1996; Mathis et al. 2005; Didier and Weiss 2006; Jeklova et al. 2020).

Although human infections have been described to be caused by the same strain isolated in rabbits, a direct zoonotic connection has yet to be found, and it has been postulated human infections are mainly of environmental origin via contaminated water sources or other infected humans or animals. The contact between pet owners and susceptible pet species could, therefore, increase the risk of exposure in humans (Keeble 2011).

The prevalence in rabbits is variable but can reach very high values. In pet rabbits, prevalence can range between 0% and 93.8% and in commercially reared rabbits may vary between 7.6% and 75.4%, depending on the population and the moment when have been studied (Magalhães et al. 2022).

In Europe seroprevalence estimates ranged from 47.2% in Portugal in subclinical pet rabbits (Patrício 2014) to 59.6% in Italy (Lavazza et al. 2016) and to 52% in England (Keeble and Shaw 2006). In the rest of the world, seroprevalence levels are highly variable, ranging from 7.7% in New Zealand (Özkan et al. 2011) to 81.7% in Brazil (Baldotto et al. 2015).

High seropositivity in apparently healthy rabbits is a public health problem as, apart from going unnoticed by the owner, it represents a potential source of infection because the animal may be excreting microsporidia spores without showing clinical signs. The agent has also been detected in wild animals such as hares. Seroprevalence density values in wild animals are significantly less likely because the comparison with the rabbit population is smaller compared with that of domestic rabbits (Bártová et al. 2015).

3.2 Hosts

Although each genotype has a preferred host, it has been proven that this microsporidium is not stenoxenous and can infect various types of animals. Cases of encephalitozoonosis are described in reptiles such as bearded dragons (Richter et al. 2013), guinea pigs (Percy and Barthold 2007), chinchillas, horses (Goodwin et al. 2006; Hollyer et al. 2014), and cats (Benz et al. 2011), among others.

This feature is important from the One Health approach as encephalitozoonosis is increasingly spreading and infecting both domestic and wild animals, adapting to different species managing to survive in erratic hosts for long periods of time (Hinney et al. 2016).

3.3 Effects on Production

These microorganisms are responsible for direct and indirect losses in rabbit production that may be of great economic impact. The clinical features, morbidity, and mortality have varied among farms and between sex and age groups.

Acute illness and death, premature slaughter, decreased growth rate, weight loss, and late maturity of slaughter stock are the most frequent findings in growing rabbits. In pregnant rabbits, a marked decrease in body weight can be observed as well as repeated abortions. Morsy et al. (2020) also reported morbidity in adult and young rabbits of 76.7% and 81.5%, respectively. The highest mortality was recorded in offspring (12.3%), followed by dams (5.6%), and the lowest recorded mortality was in males (0.04%).

Additional nonspecific clinical signs such as lethargy, anorexia, and/or weight loss or a combination of these signs were observed in 5.2% of the rabbits, while 0.26% of the pregnant female rabbits had repeated abortions (Morsy et al. 2020).

4 Disease

4.1 Pathogenesis

E. cuniculi is considered an opportunistic zoonotic agent in immunocompromised humans (Deplazes et al. 1996; Mathis et al. 2005). In rabbits, horizontal transmission by ingestion or inhalation of spores occurs more frequently (Didier et al. 2000; Mathis et al. 2005), but intrauterine (Hunt et al. 1972; Baneux and Pognan 2003) and ocular infections have also been documented (Jeklova et al. 2010b; Özkan et al. 2019; Jeklová et al. 2019).

The success of *E. cuniculi* relates to some adaptations that include the development of a polar tube, intracytoplasmic vacuole formation, and immunologic-based strategies to evade destruction in the host (Latney et al. 2014).

4.1.1 Infection, Dissemination, and Invasion

After ingestion, the spores invade the enterocytes, where parasites multiply by merogony and by sporogony (Didier et al. 2004). Spores can also be internalized by intercellular phagocytes, which are located in the intestinal epithelium (Maddox et al. 2000). Infective free spores or proliferative forms are disseminated by infected macrophages or through the lymphatic system and bloodstream, reaching various organs including the brain, kidney, liver, lungs, and myocardium (Cox et al. 1979; Marcato and Rosmini 1986; Didier et al. 2004; Künzel and Joachim 2010; Maddox et al. 2000). Those organs with high blood flow such as the kidneys, liver, and lungs are the first places for *E. cuniculi* replication in rabbits (Leipig et al. 2013; Csokai et al. 2009a); however, the final predilection sites are the kidneys and the brain (Cox et al. 1979).

Dissemination of *E. cuniculi*, especially to nervous tissue, is more rapid than previously reported (Jeklova et al. 2020). As early as 2 weeks after infection, *E. cuniculi* may be detected in various nervous tissue, including the cerebrum, medulla oblongata, and leptomeninges, where the microsporidia are most frequently detected, but also cerebellum and lumbar medulla (Jeklova et al. 2020). At the same time, the kidneys, liver, and lungs may also be infected (Özkan and Alcigir 2018a; Jeklova et al. 2020), apart from the intestine, mesenteric lymph nodes, spleen, and heart (Jeklova et al. 2020).

Phagocytosis of the spore by host cells has been demonstrated, in in vitro infection of rodent tissue models, but this invasion mechanism is not the primary infectious route (Wasson and Peper 2000; Orlik et al. 2010).

Invasion of cells is mainly by a process described as germination, through the extrusion of the polar filament (Didier et al. 2004) followed by transference of infectious sporoplasm (genetic machinery and DNA), through this polar filament, directly into the host cells (Wasson and Peper 2000; Didier et al. 2004; Orlik et al. 2010).

Changes in pH and osmotic pressure cause the ejection of the polar filament and sporoplasm from the spore (Didier et al. 2000). After the sporoplasm injection, and within a host-derived parasitophorous vacuole (PV) in the host cell cytoplasm, the sporoplasm develops into its proliferative form, the meront, and divides through merogony (Didier et al. 2000). The meronts differentiate into sporonts and subsequently into sporoblasts (Didier et al. 2000). Sporoblasts then mature into infective spores after the development of several organelles including the polar tube, the disc, the anterior and posterior vacuoles, and the plasma membrane (Didier et al. 2000). After successive divisions, PV and host cell disruption occur and the infective spores are released into the tissue, causing inflammation, and continue the dissemination in the host, with the subsequent shed of spores in the urine (Didier et al. 2000). As early as 35 days after infection, spores are intermittently shed in the urine of infected rabbits for up to 3 months or more (Cox et al. 1979; Mathis et al. 2005; Jeklova et al. 2010b).

Multiple PVs may be formed in a single host cell, and the PVs can undergo merging (fusion) or splitting (fission) during development (Lee and Heitman 2017). It is hypothesized that genetically distinct isolates can infect the same host cell, interact with each other, and possibly undergo sexual development (genetic exchange) (Lee and Heitman 2017).

4.1.2 Vertical Transmission

Baneux and Pognan (2003) reported, at 28 days of gestation, the simultaneous infection of the pregnant mother, the placentas, and the unborn fetuses. Nevertheless, some authors, including Baneux and Pognan (2003), also emphasized that intrauterine transmission of *E. cuniculi* may not be systematic, playing a minor role (Scharmann et al. 1986; Baneux and Pognan 2003).

Intraocular infection in young rabbits is believed to occur by this intrauterine route (Wolfer et al. 1993; Stiles et al. 1997; Felchle and Sigler 2002; Giordano et al. 2005). During embryologic development, the lens capsule is thin and strongly vascularized, allowing *E. cuniculi* to invade the eye after intrauterine transmission (Ashton et al. 1976; Scharmann et al. 1986). Unilateral phacoclastic uveitis results when spores induce rupture of the lens and lenticular contents are released, causing a granulomatous uveitis (Wolfer et al. 1993; Stiles et al. 1997; Felchle and Sigler 2002; Giordano et al. 2005).

4.1.3 Immunity

Despite the clinical significance worldwide of encephalitozoonosis in rabbits (Mathis et al. 2005; Csokai et al. 2009b), studies of the immune response to this infection are based mainly on a murine model and only rarely in the natural rabbit host (Omalu et al. 2007; Al-Sadi and Al-Mahmood 2014; Latney et al. 2014).

E. cuniculi induces both cellular and humoral immune responses (Latney et al. 2014). According to Jeklova et al. (2020), oral infection in immunocompetent rabbits leads to subclinical infection with the activation of humoral and cell-mediated immune responses.

All immune competent animals exposed to *E. cuniculi* develop antibodies that may persist for the life of the host; however, humoral-mediated responses alone (from B-cells) do not prevent reinfection. Although the humoral immune response alone does not appear to be protective, the antibodies contribute to host resistance, and persistence of antibodies in serum reflects latent or chronic infection; consequently, a rabbit can be infected with *E. cuniculi* more than once in its lifetime (Bywater and Kellett 1979; Schmidt and Shadduck 1983, 1984; Didier et al. 2000; Khan et al. 2001; Sak et al. 2004, 2006; Jordan et al. 2006; Künzel and Joachim 2010; Latney et al. 2014; Maestrini et al. 2017; Künzel and Fisher 2018).

In a study by Jeklova et al. (2020), specific IgM and IgG antibody responses were detected in the sera of experimentally orally infected rabbits 2 weeks after infection, and both antibody isotypes persisted up to 8 weeks (Jeklova et al. 2020). The presence of IgM specific antibodies is indicative of active infection. The presence of only IgG specific antibodies represents a chronic or latent infection (Jeklova et al. 2010a). In a former report by Kunstýr et al. (1986a), IgM indicating acute infection was the first antibody detected in all inoculated rabbits, around 20–30 days after infection, and declined within 8 to10 days. There was no consistent IgG pattern noted since some rabbits had a plateau in IgG levels, others had a linear elevation in IgG levels, and one group had no detectable IgG levels at the end of 68 days, and therefore the authors concluded that these different responses may be influenced by *E. cuniculi* infection load and individual variation in immune response (Kunstýr et al. 1986a).

As a successful intracellular pathogen, several studies have shown that the host response to disease is primarily cell-mediated (Bywater and Kellett 1979; Cox et al. 1979; Didier et al. 2000). Cell-mediated immune response is vital for protection against *E. cuniculi* infection and prevention of lethal encephalitozoonosis, with CD8+ and CD4+ T lymphocyte subpopulations playing an important role (Khan et al. 2001; Didier et al. 2000; Braunfuchsová et al. 2002). Latney et al. (2014) demonstrated that CD8 + T-cells alone have been successful at preventing lethal *E. cuniculi* disease in infected immunodeficient mice, which is in line with previous studies reporting CD8+ lymphocytes (and possibly γδ-T lymphocytes which regulate them) as being essential for the host's survival, while CD4+ cells do not seem to play a significant role (Khan and Moretto 1999; Khan et al. 1999).

The cytokine interferon gamma (IFN-γ) is essential for the survival of intraperitoneally or orally infected mice, promoting CD8+ T-cell immunity (Salát et al. 2004). Elevated serum IFN-γ levels have recently been detected in naturally infected rabbits (Rodríguez-Tovar et al. 2019; Jeklova et al. 2020) and also, in the past, in murine studies (Khan et al. 1999).

In addition to preventing tissue damage, IFN-γ also has an important role in the survival and persistence of *E. cuniculi* in the host, in a balanced relationship between the host and the microsporidia (Jeklova et al. 2020).

In rabbits, the chronic presentation is the most frequent, and a positive antibody titer is indicative of infection but not of acute disease (Csokai et al. 2009a). According to several studies, conducted in immunocompetent rabbits, the infection was subclinical and no neurological, renal, or ocular signs were observed (Kunstýr et al. 1986b; Jeklova et al. 2010b, 2020). However, if immune competence is compromised, a latent infection can be reactivated, with rapid proliferation of *E. cuniculi* leading to clinical manifestations, mainly neurological, renal, and ocular forms (Shadduck and Orenstein 1993; Jeklova et al. 2010b; Maestrini et al. 2017; Künzel and Fisher 2018).

4.2 Lesions

Infection by *E. cuniculi* is common in rabbits causing lesions mainly in the central nervous tissue, including the brain, kidneys, liver, and lungs (Baneux and Pognan 2003; Csokai et al. 2009a; Jeklova et al. 2020). However, lesions are most frequently recognized in the brain and kidneys (Cox and Gallichio 1978; Rodríguez-Tovar et al. 2017). The eye is also affected in a large percentage of infected rabbits (Felchle and Sigler 2002; Harcourt-Brown and Holloway 2003, Özkan and Alcigir 2018b; Morsy et al. 2020). The heart and spleen may also be involved, though generally to a more limited extent (Cox et al. 1979; Jeklova et al. 2020). In intrauterine infection, lesions occur more frequently in the placenta, followed by the lungs in the fetus; the brain and kidneys are less frequently affected. The degree of the lesions varies from mild, moderate, and severe microscopic alterations (Baneux and Pognan 2003).

According to several authors, the organs involved and the extent of the lesions vary according to the time elapsed after the infection. In general, the first tissue alterations occur in the kidneys, liver, and lung, around 31 days after infection, while the brain is not affected at the beginning, but, after about 3 months, the most significant alterations are present in the kidneys and the brain (Cox et al. 1979;

Jeklova et al. 2020). At that point, lesions decrease in the lungs, liver, and heart, and *E. cuniculi* is not usually identified in these organs and were of minor severity (Cox et al. 1979).

These observations are in line with those obtained by Jeklova et al. (2020), in which *E. cuniculi* was detected in the kidneys, liver, lungs, heart and spleen, intestine, and mesenteric lymph nodes and in one rabbit in the brain and other nervous tissues, as soon as the second week after infection. The same authors also reported that in the course of infection, the pathogen persisted in the aforementioned organs, with an increase in nervous tissue, and at 98 days after infection, *E. cuniculi* was frequently observed in the brain and kidneys while disappearing from other organs (Jeklova et al. 2020).

In the brain, *E. cuniculi* seems to prevail for a longer period of time, compared to the presence of the microsporidia in the kidneys (Flatt and Jackson 1970; Csokai et al. 2009a). However, Csokai et al. (2009a) hypothesize that, in naturally infected rabbits, the kidney is infected first than the brain, since histological lesions in the kidney are usually more severe than those located in the brain and are, therefore, presumed to exist for a longer period of time and because, in some cases, inflammatory lesions were found in the kidney but not in the brain. According to Rodríguez-Tovar et al. (2016), kidney lesions can be observed within 4 weeks and brain lesions within 8 weeks after infection.

In the majority of rabbits, microscopic lesions in the brain and kidney consist of focal nonsuppurative granulomatous meningoencephalitis and chronic granulomatous and focal to segmental interstitial nephritis, with variable degrees of fibrosis, characterized by macroscopically visible scars where the tissue is retracted (Flatt and Jackson 1970; Cox and Gallichio 1978; Csokai et al. 2009a; Rodríguez-Tovar et al. 2017; Morsy et al. 2020). Usually, granulomas are commonly located within the brain and kidneys of serologically positive rabbits (Cox and Gallichio 1977); however, in the study by Csokai et al. (2009a), they were rarely identified in the kidneys.

In latent or chronic infections, severe lesions are rarely observed and may even be absent, except in cases of chronic interstitial nephritis (Jeklova et al. 2020).

4.2.1 Central Nervous System (CNS)

In rabbits, including in pet rabbits, the most frequently affected nervous tissues are, in descending order, the cerebrum (cortex and medulla), medulla oblongata (Csokai et al. 2009a; Rodríguez-Tovar et al. 2016; Jeklova et al. 2020), and leptomeninges (Csokai et al. 2009a; Jeklova et al. 2020). Although vestibular disease is described as being common in encephalitozoonosis, Csokai et al. (2009a) and Rodríguez-Tovar et al. (2016) reported the cerebellum and the vestibular cores as less commonly affected. However, it is important to emphasize that the distribution of lesions associated with *E. cuniculi* infection is not always directly correlated with the severity of neurological signs (Künzel et al. 2008; Csokai et al. 2009a; Gruber et al. 2009). According to Rodríguez-Tovar et al. (2016), clinical manifestations were directly associated to the severe inflammatory reaction observed in the cerebral tissue.

Extensive proliferation of *E. cuniculi* in the brain may lead to the rupture of the cysts, with the release of infective spores into the adjacent nervous tissue (Csokai et al. 2009a; Rodríguez-Tovar et al. 2016). Damage in the nervous tissue includes focal or multifocal granulomatous lesions, mainly in the brain, and mild to moderate tissue and perivascular lymphocyte infiltrates (Baneux and Pognan 2003; Jeklova et al. 2020). These findings were also reported by Csokai et al. (2009a) who verified cerebral lesions, generally characterized by perivascular inflammatory infiltrates and by formation of granulomas. According to Harcourt-Brown (2004) and Morsy et al. (2020), granulomas contain mainly glia cell lymphocytes, plasma cells with central necrosis.

Leipig et al. (2013) reported focal brain lesions limited, in the vast majority of cases, to the gray matter and to a lesser extent in the white matter. In the cerebral cortex, the inner layers were often affected as opposed to the middle or outer layers. Rabbits with granulomatous lesions may also display lymphoplasmacytic meningoencephalitis most frequently affecting, in descending order, the neopallium, the hippocampus, the basal nuclei, and the thalamus (Leipig et al. 2013).

4.2.2 Kidneys

Kidney lesions are characterized in various degrees from mild and moderate to severe and may include tubular degeneration and interstitial infiltrates of mononuclear cells with lymphocytes (Jeklova et al. 2020) and nephropathy (Baneux and Pognan 2003).

Lesions may be classified into acute, chronic, or both. Usually, damage to the kidneys shows a chronic course, due to fibrosis, with typical kidney lesions, manifesting as chronic interstitial nephritis and macroscopic depressed scar tissue formation on the outer surface which are more evident when the capsule is detached, giving the kidney a pitted appearance (Csokai et al. 2009a; Rodríguez-Tovar et al. 2016).

Chronic kidney lesions occurred primarily in the cortex and are characterized by interstitial mononuclear cell infiltration, interstitial fibrosis (Leipig et al. 2013; Rodríguez-Tovar et al. 2016; Morsy et al. 2020) which varied from small subcapsular foci to large areas reaching into the medulla, and diffuse interstitial fibrosis in the cortex-medulla junction in areas adjacent to chronic inflammation. In the kidneys with predominantly chronic lesions, no spores were detectable by Leipig et al. (2013).

A low percentage of the animals demonstrates acute interstitial nephritis and multifocal granulomas (Csokai et al. 2009a). Acute lesions, most frequently found of mild severity in the medulla in the form of round foci, were characterized by focal tubular necrosis with inflammatory infiltrates consisting of heterophils, macrophages, lymphocytes, and plasma cells, dilated tubules frequently filled with detritus, and regeneration of the tubular epithelial cells (Leipig et al. 2013; Morsy et al. 2020).

Within PV, spores multiply in the tubular epithelium with the subsequent rupture of cells (Csokai et al. 2009a; Leipig et al. 2013; Rodríguez-Tovar et al. 2016) and the observation of ruptured cysts and free spores within the lumen of renal tubules

(Rodríguez-Tovar et al. 2016). Spores were observed in the urine of rabbits with severe renal tissue damage, suggesting a recurrent excretion of spores in urine (Rodríguez-Tovar et al. 2016), which is contrary to what has been reported by other authors, who indicate an absence of spores in the urine of rabbits with severe renal fibrosis scar tissue (Harcourt-Brown 2004; Furuya et al. 2009). As the infection progresses, the detection of spores in the tissues is less frequent (Flatt and Jackson 1970; Csokai et al. 2009a).

4.2.3 Eyes

In addition to the central nervous system and the kidneys, the eye is also considered a predilection site (Cox and Gallichio 1978; Harcourt-Brown and Holloway 2003, Özkan and Alcigir 2018b). Ocular lesions caused by *E. cuniculi* colonization of the lens are very common in rabbits with encephalitozoonosis (Felchle and Sigler 2002; Wolfer et al. 1993; Stiles et al. 1997; Giordano et al. 2005; Morsy et al. 2020).

Phacoclastic uveitis, due to intrauterine infection, is characterized by the infiltration of the eye lens by various inflammatory cells, including granulocytes, macrophages, and giant cells, leading to a disruption of the fibers and of the lens capsule (Harcourt-Brown and Holloway 2003; Giordano et al. 2005; Csokai et al. 2009a; Leipig et al. 2013; Özkan and Alcigir 2018b). Inflammation of the anterior lens capsule can lead to spontaneous rupture, and phacoclastic uveitis develops due to perilenticular fibroplasia along with rupture of the lens capsule (Wolfer et al. 1993; Felchle and Sigler 2002; Özkan and Alcigir 2018a).

Other repercussions are granulomatous lens uveitis, as well as unilateral or bilateral cataracts (Harcourt-Brown and Holloway 2003; Csokai et al. 2009a; Özkan and Alcigir 2018b; Morsy et al. 2020). The iris and ciliary body are infiltrated by plasma cells and lymphocytes (Giordano et al. 2005). Spores may also be present (Leipig et al. 2013) and be observed in the retina (Morsy et al. 2020). Other eye lesions include focal lens fiber necrosis (Csokai et al. 2009a) and hypopyon (Harcourt-Brown and Holloway 2003; Özkan and Alcigir 2018b).

4.2.4 Other Organs

Lesions in the lungs from infected rabbits may include mild to severe alveolar inflammatory infiltrations of lymphocytes, plasma cells, and monocytes and alveolar destruction (Jeklova et al. 2020) and nonsuppurative inflammatory infiltrations in the lung that may be associated with a granulomatous inflammation (Csokai et al. 2009a; Leipig et al. 2013).

Liver lesions can range from mild to moderate focal granulomatous inflammation in the bile ducts (Baneux and Pognan 2003; Csokai et al. 2009a; Leipig et al. 2013) to mild to moderate periportal and perivascular infiltrations by lymphocytes and plasma cells (periportal lymphoplasmacytic infiltrates) and mild vacuolar degeneration (Leipig et al. 2013; Jeklova et al. 2020; Morsy et al. 2020) and among more moderate to severe liver lesions, including nonsuppurative periportal interstitial infiltration, fibrosis, and necrosis of the periportal area (Rodríguez-Tovar et al. 2016).

Focal granulomatous infiltrates and lymphoplasmacytic myocarditis were observed in the heart (Csokai et al. 2009a; Leipig et al. 2013). In the spleen, hyperemia of the red pulp and lymphocyte infiltration were described (Jeklova et al. 2020).

4.3 Clinical Signs

In many rabbits, infection usually has a chronic or latent course (Valencakova et al. 2008; Rich 2010), and only a low percentage of infected animals develop clinical manifestations (Igarashi et al. 2008; Valencakova et al. 2008; Maestrini et al. 2017). In the vast majority of cases, the development of clinical signs occurs if the hosts are unable to control the infection (Csokai et al. 2009a). The severity of clinical signs depends on the ability of the host immune system and the virulence of the agent (Jeklova et al. 2020). In many cases, the onset of clinical signs is sudden and often follows a stressful situation on the rabbit's part (Harcourt-Brown and Holloway 2003; Künzel and Joachim 2010). As mentioned earlier, it is important to note that the severity of clinical signs may not be associated with the severity of organ damage (Csokai et al. 2009a).

In rabbits with encephalitozoonosis (most studies conducted on pet rabbits), the main forms of presentation include neurological manifestations, ocular lesions like phacoclastic uveitis, and renal failure. All these three manifestations may occur separately or in association, with neurological signs being the predominant clinical manifestations of *E. cuniculi* infection (Harcourt-Brown and Holloway 2003; Percy and Barthold 2007; Künzel et al. 2008; Jeklova et al. 2010b; Maestrini et al. 2017; Mancinelli and Lennox 2017; Künzel and Fisher 2018; Morsy et al. 2020), and cardiac and respiratory signs have also been reported (Percy and Barthold 2007) in addition to clinical disease in immunosuppressed people (Deplazes et al. 1996).

In rabbit farms, especially in industrial housing systems, encephalitozoonosis can cause considerable financial loss, due to mortality, reduced carcass weight, and increased number of rejected animals at the slaughterhouse (Saviotti et al. 2000). In laboratory rabbits, the infection is a frequent problem and can affect the health status of the animals and interfere with experimental activity (Künzel and Joachim 2010).

4.3.1 Central Nervous System

Neurological findings are common manifestations of the infection and can be observed in rabbits of any age (Gruber et al. 2009; Künzel and Joachim 2010; Zietek et al. 2014; Özkan and Alcigir 2018a). Due to brain lesions, infected rabbits may develop signs of central vestibular disease and paresis (Kunstýr et al. 1986a, b; Genz and Carpenter 1997; Thomas 2000; Harcourt-Brown and Holloway 2003; Jass et al. 2008; Künzel et al. 2008; Rodríguez-Tovar et al. 2016; Künzel and Joachim 2010; Morsy et al. 2020); however, the severity of tissue inflammation in affected rabbits does not reflect neurologic signs (Latney et al. 2014).

According to Csokai et al. (2009a), because central vestibular disease (brainstem and cerebellar location) is in the vast majority of cases clinically more apparent than brain disease, its higher number of reports may be a reflection of this situation. These same authors indicate that brain disease is most often diagnosed when the rabbit presents with seizures or cortical blindness. Additionally, brain disease may also be limited to only slight changes in mental status, which may be more difficult to identify by neurological examination (Csokai et al. 2009a).

The most frequently reported central vestibular signs of encephalitozoonosis are ataxia, head tilt (torticollis), circling, and nystagmus (Thomas 2000; Harcourt-Brown and Holloway 2003; Harcourt-Brown 2004; Jass et al. 2008; Künzel et al. 2008; Valencakova et al. 2008; Csokai et al. 2009a; Rich 2010; Rodríguez-Tovar et al. 2016; Mancinelli and Lennox 2017; Künzel and Fisher 2018; Özkan and Alcigir 2018a; Morsy et al. 2020). According to observations by Künzel et al. (2008), the degree of head tilt is a major factor in assessing prognosis in rabbits exhibiting vestibular disease. Only a slight head tilt may be seen at the beginning, without any other signs (Özkan and Alcigir 2018a). Over time, the degree of the head tilt may considerably increase, until the rabbit cannot maintain its balance and is unable to stand up (Özkan and Alcigir 2018a). Nevertheless, even in rabbits with severe head tilt, the general condition and food intake may be normal (Gruber et al. 2009; Künzel and Joachim 2010; Zietek et al. 2014; Özkan and Alcigir 2018a).

Other common manifestations include rotational movements around the body length axis (longitudinal rolling), with the animal often unable to stand up straight (Harcourt-Brown and Holloway 2003; Thomas 2000; Harcourt-Brown 2004; Jordan et al. 2006; Jass et al. 2008; Künzel et al. 2008; Rich 2010; Rodríguez-Tovar et al. 2016; Özkan and Alcigir 2018a; Morsy et al. 2020), seizures, head tremors, swaying or nodding at rest, and behavior changes (hyperactivity, aggressiveness, jumping against cage-fences, disorientation, and absences) (Harcourt-Brown and Holloway 2003; Jordan et al. 2006; Künzel et al. 2008; Csokai et al. 2009a; Rich 2010; Latney et al. 2014; Özkan and Alcigir 2018a).

The observation of paresis or paralysis in *E. cuniculi* infected rabbits has also been described in a large number of cases (Harcourt-Brown and Holloway 2003; Harcourt-Brown 2004; Jordan et al. 2006; Künzel et al. 2008; Valencakova et al. 2008; Csokai et al. 2009a; Rich 2010; Morsy et al. 2020). Occasionally, focal spinal radiculitis is observed which might be responsible for atypical neurological symptoms (Nast et al. 1996). Jass et al. (2008) reported a case of tetraparalysis associated with multifocal disease denoted by an abnormal mental status, upper motor neuron paresis of the thoracic limbs, and lower motor neuron paresis of the pelvic limbs (Fig. 2).

4.3.2 Kidneys

In contrast with vestibular disease, with usually no impact on the general animal condition, rabbits with signs of renal failure are frequently apathic and hiporexic (Harcourt-Brown and Holloway 2003; Künzel et al. 2008; Künzel and Joachim 2010). In serious cases, especially during chronic renal failure, encephalitozoonosis can lead to fatal outcome in rabbits (Künzel and Joachim 2010) with overall

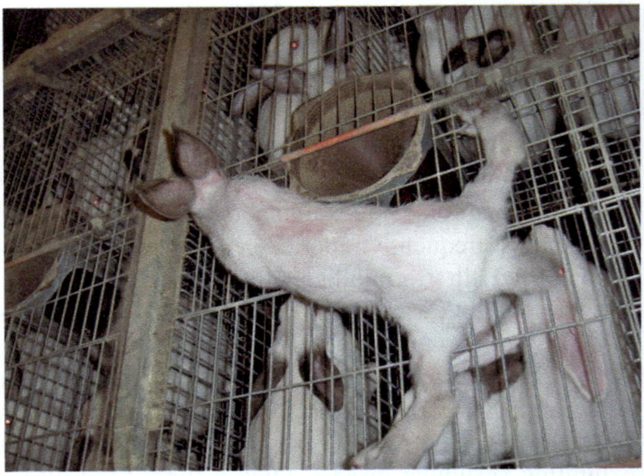

Fig. 2 Rabbit with encephalitozoonosis and paresis of the pelvic limbs (original from Dr. José Manuel Monteiro)

mortality rates around 50%, despite etiological treatment (Künzel et al. 2008). In the study by Künzel et al. (2008), almost all rabbits that demonstrated renal failure had to be euthanized, while more than 50% of the rabbits with neurological signs survived.

Chronic renal failure, due to *E. cuniculi* infection, although quite common in infected rabbits, tends to be subclinical or leads to less specific signs such as lethargy, inappetence, anorexia, weight loss, and dehydration (Harcourt-Brown and Holloway 2003; Harcourt-Brown 2004; Künzel et al. 2008; Valencakova et al. 2008; Künzel and Joachim 2010; Keeble 2011). These renal disorders are usually detected through blood examination and the evidence of azotaemia (elevated blood creatinine and urea) (Harcourt-Brown and Holloway 2003; Künzel et al. 2008; Csokai et al. 2009a). Chronic renal failure in rabbits can also cause anemia or even osteodystrophy resulting in pathologic fractures of the long bones (Harcourt-Brown and Holloway 2003).

Despite frequent reports of polyuria, polydipsia, pollakiuria, and urinary incontinence (Harcourt-Brown and Holloway 2003; Künzel and Joachim 2010; Morsy et al. 2020), the presence of these manifestations may not be systematic (Künzel et al. 2008).

4.3.3 Eyes

Signs of phacoclastic uveitis due to encephalitozoonosis have been widely reported and may include cataract, of various degrees, uveitis, and granuloma (intraocular masses). These ophthalmological signs may be present separately, in combinations, or concomitantly (Wolfer et al. 1993; Stiles et al. 1997; Felchle and Sigler 2002; Baneux and Pognan 2003; Harcourt-Brown and Holloway 2003; Harcourt-Brown 2004; Giordano et al. 2005; Valencakova et al. 2008; Csokai et al. 2009a; Jeklova et al. 2010b; Künzel and Joachim 2010; Sanchez et al. 2018; Özkan et al. 2019; Morsy et al. 2020). The majority of rabbits suffering from phacoclastic uveitis are younger than those with renal failure (Grahn et al. 1991; Wolfer et al. 1993; Giordano et al. 2005).

Ocular signs that commonly occur such as unilateral and bilateral phacoclastic uveitis are only occasionally documented (Ashton et al. 1976; Felchle and Sigler 2002; Harcourt-Brown and Holloway 2003; Giordano et al. 2005; Künzel et al. 2008; Sanchez et al. 2018; Morsy et al. 2020), mostly in young rabbits (Ashton et al. 1976; Giordano et al. 2005).

Usually, physical examination of rabbits with phacoclastic uveitis reveals no other abnormalities, and concerned rabbits do not have considerable loss of visual function or influence of their quality of life (Felchle and Sigler 2002; Harcourt-Brown and Holloway 2003; Künzel et al. 2008).

4.3.4 Other Clinical Signs

Additional nonspecific clinical signs include lethargy, anorexia, and/or weight loss or a combination of these signs and, in some pregnant females, also repeated abortions (Morsy et al. 2020). Facial paralysis and dry eye have also been reported in rabbits infected with *E. cuniculi* (Künzel et al. 2008; Lewis 2010). These same authors suggest testing for *E. cuniculi* infection in every case of dental disease in rabbits, e.g., incisor malocclusion or molar malocclusion, particularly the unexpected cases in older rabbits that have never previously had dental disease and that have had no changes in their diet or husbandry.

5 Diagnosis

The methods established for the detection of encephalitozoonosis vary from microscopy, molecular methods, and serology, and tissue samples and fluids such as serum, urine, and feces may be used. Microscopy combined with the use of specific dyes allows the detection of the phylum *Microsporidia*, but does not allow knowing to which species they belong. Using the electron microscope, it is possible to know the phylum. The detection of antibodies by serological techniques (immunofluorescence antibody test [IFAT], enzyme-linked immunosorbent assay [ELISA], Western blotting [WB], and direct agglutination test [DAT]) also allows the detection of the infecting microsporidia (Hinney et al. 2016).

The diagnostic methods may simply involve the observation of the animal and its clinical signs in conjunction with other complementary methods. The examinations available to the veterinarian may be carried out *antemortem* or *postmortem*, and it is the set of examinations of various natures (collection of clinical signs, serology, detection of spores, polymerase chain reaction [PCR]) which allows to ascertain that the presumptive diagnosis may be encephalitozoonosis (Zietek et al. 2014).

5.1 *Antemortem/Postmortem* Findings (Necropsy)

The infection is routinely diagnosed with *antemortem* clinical findings, whose onset, in many cases, is sudden and often follows a stressful situation (Harcourt-Brown and Holloway 2003; Künzel and Fisher 2018; Jeklova et al. 2020).

There are several factors involved in the diagnosis of this microsporidium, starting with the interval of spore excretion by the host, which is short (about 3 months, in which excretion is intermittent and decreasing in intensity) and can lead to difficulties in perceiving whether the animal has a chronic, latent, or active infection. Another factor is having or not clinical signs and relating them with the histopathological findings and/or antibody titration, because depending on the host immune response, clinical signs may appear only at a late phase or be fulminant (Csokai et al. 2009a; Vergneau-Grosset and Larrat 2015).

The presence of neurologic (head tilt/torticollis, trembling, convulsions, nystagmus or motor incoordination), ocular (phacoclastic uveitis), or even respiratory signs (that may occur in the beginning of the infection) and poor productive performances may trigger the suspect of infection with *E. cuniculi* (Doboşi et al. 2022).

Clinical diagnosis is difficult in living rabbits because *E. cuniculi* infection has a subclinical course (Özkan et al. 2019).

The life cycle of *E. cuniculi* and pathogenesis of associated infection allow *E. cuniculi* to escape a serological titer correlation to clinical disease severity, evade humoral-mediated immunity in its rabbit host, remain impervious to therapeutic attempts to reduce sporont-mediated inflammation, and escape therapeutic attempts to eliminate germination in chronically infected patients (Latney et al. 2014). The diagnosis is more difficult in pet rabbits than in commercially reared rabbits, because in the latter there is the possibility of getting useful samples in dead animals and because of the collection of samples from multiple animals. Frequently, the infection is suspected if there are in the group poor reproductive and growth performances (long time), most common in farms with deficient hygiene and biosecurity measures, e.g., farms where the does are kept always in the same place and there is no possibility of environmental cleaning and disinfection (Doboşi et al. 2022).

Necropsy is the first diagnostic method used by the veterinarian for exploring the origin of the problem. Lesions in the kidney and brain are usually found about 4 and 8 weeks, respectively, after initial seroconversion (Csokai et al. 2009a, b). When lesions appear, spores can be found in the urine (Özkan et al. 2019).

In latent infections, gross lesions are rare or even absent except in cases of chronic interstitial nephritis (Jeklova et al. 2020). It is possible to find lesions such as fibrosis, subcapsular pitting, and adherence of the capsule to the underlying parenchyma (Csokai et al. 2009a, b), indicative of chronic nephritis, which are expressed by multiple depressed and irregular dark red foci on the surface of the renal cortical tissue (Fig. 3).

Fig. 3 Multiple depressed and irregular dark red foci on the surface of the renal cortical tissue, indicative of chronic nephritis, compatible with *Encephalitozoon cuniculi* infection (original from Dr. José Manuel Monteiro)

5.2 Histopathology

Microsporidia stained with hematoxylin-eosin may go undetected even by experienced pathologists. Gram, Brown-Brenn, or Brown-Hopps stains appear to be reliable in identifying microsporidia in sections of paraffin-embedded tissues. Staining of microsporidium spores varies with Gram stains; however, they stain dark blue or red on a yellowish-brown background. Fluorescent stains are easy and quick to perform and their sensitivity is very high (Franzen and Müller 1999).

E. cuniculi infection is associated with focal, granulomatous lesions mainly in the brain and kidneys (Jeklova et al. 2020). In the kidney, histopathology may reveal inflammatory lesions, including nonsuppurative to granulomatous meningoencephalomyelitis and chronic interstitial nephritis on hematoxylin and eosin stain of target organs. Lymphoplasmacytic to granulomatous infiltrates and accompanying fibrosis may be seen in other target organs, such as the liver, lung, and heart. Occasionally spores can be identified but are best highlighted by a Gram stain (Latney et al. 2014).

It is important to remember the histologic severity, and distribution of lesions associated with *E. cuniculi* infection is not directly correlated with the severity of neurological clinical signs or the neuroanatomic localization of antemortem neurological disease (Künzel et al. 2008; Csokai et al. 2009a, b; Latney et al. 2014).

Immunohistochemistry and nested PCR may help obtain a more definitive diagnosis. Transmission electron microscopy (TEM) can be used to confirm encephalitozoonosis by detecting the polar filament within spores and demonstrating ultrastructural features. This technique may also contribute to the taxonomic organization of the microsporidia (Franzen et al. 2005).

5.3 Serological Diagnosis

Over the years, several tests have been performed to make a more concrete diagnosis of encephalitozoonosis. The association between acute phase proteins with encephalitozoonosis and C-reactive protein was recognized as having statistical significance, with titers being about ten times higher in animals suspected of encephalitozoonosis compared to the group of nonsuspect animals. Serum haptoglobin amyloid protein was also studied, but did not show statistically significant values (Cray et al. 2013).

Serological tests prove to be quite important in the detection of an *E. cuniculi* infection, as the production of an immune response by the host is produced up to 30 days postinfection and can be detected up to 8 weeks before spores are detected in the urine or lesions in the brain or kidney (Paré and Paul-Murphy 2004; Harcourt-Brown 2007). Thus, serological tests are the most sensitive method to detect encephalitozoonosis at an early stage of infection (Csokai et al. 2009b). Studies have shown that rabbits develop circulating antibodies 2–3 weeks after oral, nasal, or intratracheal infection, with a peak near 9 weeks (Jordan et al. 2006).

The clinician should always bear in mind that animals before 8 weeks of age rarely seroconvert due to the maternal immunity they may still have received vertically. The immune response is not developed by the animal until around 10 weeks of age. The most commonly used serological tests for the detection of anti-*E. cuniculi* antibodies are IFAT, carbon immunoassay (CIA), and ELISA. Peak seroconversion is between 6 and 9 weeks postinfection, and IgM decreases up to 35 days postinfection. IgM is the first antibody to be produced when there is an active infection, appearing at 17 days postinfection. It disappears at around 35 days postinfection. IgG, which is produced by the animal's own cellular immunity, becomes the dominant antibody from 17 days postinfection, and it continues to be produced even when it is already a chronic process, with fluctuations (Csokai et al. 2009a, b; Tee et al. 2011). However, the detection of antibodies does not tell whether the animal is infected or has only had contact with the agent. It is the ratio between IgG and IgM that tells whether the animal has an active and recent infection (Vergneau-Grosset and Larrat 2015).

Studies show that there is a strong correlation between serological response and histopathological lesions, although there are cases of seronegative rabbits presenting with interstitial nephritis (Csokai et al. 2009b). It should be noted that in rabbits with acute infection, there is no correlation between antibody titer and the amount of spores excreted or the severity of lesions found at necropsy (Harcourt-Brown 2013). Another important point will be to remember that the seroconversion of an animal is only made around 30 days, i.e., if a serological test is done before these days, it will be negative, not because the animal is not infected but because it has not yet had time to develop an immune response (Vergneau-Grosset and Larrat 2015).

Another obstacle that encephalitozoonosis presents is that it spreads mainly through subclinically infected animals, which makes the work difficult for the veterinarian because, as such, there are no pathognomonic clinical signs of this pathology (Künzel and Fisher 2018).

Serologic assays, such as ELISA, are the most useful tools for antemortem diagnosis of the infection in different populations (Cray et al. 2013; Özkan et al. 2019). Antibody detection is useful to confirm exposure to the parasite, but antibody serostatus is also reported not to be correlated with clinical signs of a true infection (Rodríguez-Tovar et al. 2016), which ranges from a total lack of clinical signs to sudden death in infected rabbits (Özkan et al. 2019).

There is typically an increase in IgM titer from day 0 to day 35 postexposure and an increase in IgG titers only at 2 and 3 weeks after exposure. They only excrete *E. cuniculi* by 6 weeks postinfection (Morsy et al. 2020), when histopathological changes are visible in the kidney. Histopathological changes in the brain are generally seen even much later, usually more than 8 weeks after antibodies are detectable. Antibodies are also passively transmitted from an infected dam to its offspring, which can show antibodies until approximately 4 weeks of age. After maternal antibodies wane, they become susceptible to natural infection. If infected at this point, after an initial seronegative period, seroconversion occurs at 8–10 weeks (Harcourt-Brown and Holloway 2003).

However, the measurement of serum IgG to *E. cuniculi* cannot distinguish between active, early, reactivated, or chronic infection (Boot et al. 2000; Csokai et al. 2009a). IgG are therefore merely considered indication of exposure or infection status. Rabbits may show a considerable individual variation in their immune response, with some animals showing persistently high antibody levels for years – even in the absence of clinical signs – and others becoming seronegative soon after initial infection (Keeble 2011). Differing responses may also be influenced by *E. cuniculi* exposure load (Latney et al. 2014).

Simultaneous screening of IgM, which is more indicative of active infection (IgM titers will be elevated from day 0 to 35 postexposure and can persist for up to 18 weeks in rabbits), in combination with IgG, gives a better indication of the infective status of the affected rabbit (Jeklova et al. 2010a).

Positive serology is strong evidence of infection, but conventional titers are not necessarily correlated with the degree of infection, nor are they predictive or indicative of clinical signs of disease. Second, seroconversion does not result in a protective response or immunity for the patient (Latney et al. 2014). In fact, all healthy animals exposed to *E. cuniculi* develop antibodies that may persist for the life of the host; however, they do not offer protection against reinfection (Didier et al. 2000). If the humoral immune response alone is not protective, the cell-mediated immunity is instead considered essential to eliminate the parasite. In human, rabbit, nonhuman primate, and rodent studies of microsporidial infections, humoral-mediated responses to *E. cuniculi* have not been shown to provide immunity. *E. cuniculi* is an intracellular parasite, and T-cell-mediated cytolytic processes that aim to eliminate infected cells from the host appear to be most important in conferring resistance Therefore, the humoral response of the production of antibodies (from B-cells) alone is not sufficient to prevent reinfection; it only indicates the rabbit has a persistent *E. cuniculi* infection (Valencakova and Halanova 2012).

A rabbit can be infected with *E. cuniculi* more than once in its lifetime. This point is very important to remember when trying to determine whether an animal has been exposed to *E. cuniculi* by assessing a patient's serological reactivity against *E. cuniculi*. In a rabbit farm, it is necessary to choose a representative sample (in number and distribution in farm area), to get the better results for a good "picture" of the immune status and probable stage of infection of our farm.

5.4 Polymerase Chain Reaction (PCR)

PCR-based methods that typically utilize primers for amplification of microsporidial rDNA have been commonly utilized to enhance diagnostic sensitivity and specificity (Morsy et al. 2020).

PCR and real-time PCR can be used for detection of *Encephalitozoon* DNA in a rabbit's urine and cerebrospinal fluid (Zietek et al. 2014). However, Cox et al. (1979) demonstrated that urine spore excretion progressively declines and ceases in acutely infected rabbits at 98 days postinfection. Urine excretion and elimination rates will need to be studied in chronically infected rabbits to ascertain the true value of PCR urine spore detection. Spores are shed in the urine of infected rabbits from 3 to 5 weeks post-seroconversion and only sporadically, so their detection appears to not always be possible and not necessarily correlated with the severity of the disease (Csokai et al. 2009a).

PCR methods were found to be more sensitive and specific when performed on samples from eyes with phacoclastic uveitis, possibly due to the higher spore concentration in lens material (Csokai et al. 2009a). PCR spore detection in cerebrospinal fluid samples has demonstrated variable sensitivity. PCR appears to be an unreliable diagnostic test for in vivo detection of *E. cuniculi* when performed on samples from body fluids and tissues, other than in the ocular contents of rabbits with phacoclastic uveitis (Csokai et al. 2009a; Künzel and Joachim 2010). When interpreting PCR results, it is important to note that PCR spore detection will rely heavily on stage of infection (acute, latent, or reinfection) and on dose-dependent spore burden within the patient. PCR spore detection in cerebrospinal fluid or urine samples in seropositive animals could therefore result in false negative results (Künzel et al. 2008).

5.5 Creatinine and Blood Urea Nitrogen (BUN)

As the first and most significant alterations in rabbits are in the kidneys, serum creatinine and BUN levels were found to have a statistically significant relationship with the serological status of rabbits (Özkan et al. 2019), and this may be useful for the evaluation of general health status and renal function of the seropositive rabbits, for clinical interpretation of the serology, and also in predicting the disease in clinically healthy rabbits (Rich 2010; Abu-Akkada and Oda 2016). Therefore, high levels of urea and especially high creatinine levels could be associated with *E. cuniculi* infection in rabbits, which can be very useful in companion rabbits. In these animals, it is frequent to note also polyuria-polydipsia syndrome (Boucher and Nouaille 1996).

5.6 Differential Diagnosis

Based on the variety of clinical signs possible in the manifestation of this disease, the differential diagnoses are presented by categories:

- Infectious and noninfectious causes may lead to neurological signs: bacterial, viral, protozoal, and parasitic agents. Pasteurelosis (infection of the internal ear) and otoacariasis (auricular mange) are the most common causes, namely, of head tilt (Boucher and Nouaille 1996). Other bacterial infections were also reported and include *Staphylococcus aureus*, *Escherichia coli*, *Pseudomonas aeruginosa*, and *Listeria monocytogenes* (Latney et al. 2014). Clinical differential diagnoses for peripheral vestibular disease include also traumatic rupture to the tympanic bulla, aminoglycoside mediated ototoxicity, and idiopathic vestibular disease.
- Noninfectious causes include congenital, metabolic, neoplastic, inflammatory, toxic, and traumatic causes and vascular disease. In pet rabbits, the most common include lymphosarcoma, lead toxicity, hydrocephalus, and cerebral infarcts (Latney et al. 2014).

In the presence of poor performances in rabbit production, the veterinarian also needs to look for other factors that may influence the poor weights, like general management, parturition assistance and nest care, as well as nutritional management (water and feed) and general hygiene and biosecurity protocols.

E. cuniculi is rarely a cause of clinically appreciated primary renal dysfunction. However, chronic azotaemia can be associated with weight loss and emaciation despite a normal dietary intake. If this affection is associated with a positive serology (IgG, IgM) or even with a positive PCR, it makes *E. cuniculi* infection a likely diagnosis.

6 Control

Control is understood as the reduction of disease morbidity and mortality and is a general term encompassing all measures aimed at interfering with the unrestrained occurrence of disease. In this section, prevention is synonymous with primary prevention, which modifies determinants to prevent new cases of disease and therefore to reduce incidence, whereas treatment is related to the so-called secondary (and also tertiary) prevention, which detects and treats disease promptly to shorten disease duration or to prolong life (Thrusfield and Christley 2018).

6.1 Prevention

E. cuniculi spores are environmentally resistant but can be destroyed using several disinfection protocols (Latney et al. 2014; Magalhães et al. 2022). Spores are susceptible to 0.1% bleach at a contact time of 10 minutes, and ethanol (70%) effectively destroys spores with a contact time of 30 seconds. Sodium hydroxide (1%), formaldehyde (0.3%), and hydrogen peroxide (1%) effectively destroy spores with a contact time of 30 minutes. An effective measure to reduce the exposure of rabbits to *E. cuniculi* spores is the regular cleaning and disinfection of housing facilities with bleach and ethanol (Jordan et al. 2006).

Antibodies to *E. cuniculi* can be detected 3–4 weeks before spores appear in the urine (Deplazes et al. 2000), and the separation of seronegative from seropositive rabbits, especially before introduction into new herds, will help to prevent transmission of the pathogen (Okewole 2008; Fukui et al. 2013). For the early detection of infected individuals, regular serological screening is recommended (Magalhães et al. 2022). An "all-out and all-in" approach may be implemented at rabbit production systems if an outbreak is detected (Fukui et al. 2013).

Fenbendazole administered orally (20 mg/kg bw) from 7 days before until 2 or 21 days after experimental infection of rabbits with *E. cuniculi* spores prevented seropositivity after 120 days and agent isolation from brain tissue. In addition, *E. cuniculi* was not isolated from seropositive, naturally infected rabbits given fenbendazole for 4 weeks, but the same did not happen with a majority of non-medicated animals (Suter et al. 2001). Based on these findings, current treatment practices include the off-label use of fenbendazole for *E. cuniculi* prevention (Latney et al. 2014).

More recently, fenbendazole (at the same dose as above) was administered as a preventative for 7 consecutive days before experimental infection or as a prophylactic treatment for 4 weeks after infection, in both situations, to immunosuppressed rabbits. Spores of *E. cuniculi* were detected in the urine of all infected rabbits from the 28th day postinfection (PI) until the end of the study (8 weeks PI). Fenbendazole as a preventative was effective to some extent against infection and disease, but when given as a prophylactic treatment, no significantly different effects were seen compared to a control group (Abu-Akkada and Oda 2016).

Considering that environmental changes can be a triggering factor for the development of clinical signs in infected rabbits, an additional preventative measure is to keep rabbits in a stable and stress-free condition (Künzel and Fisher 2018). Still in terms of prevention for rabbits and also other animals, there are no vaccines available for immunization against *E. cuniculi*, and studies are needed for this purpose (Magalhães et al. 2022).

E. cuniculi is an important opportunistic zoonotic pathogen, with higher prevalence of infection in immunocompromised people (Hocevar et al. 2014). The agent can spread to different organs, originating varied clinical signs, and syndromes, a circumstance which may have a fatal outcome (Fournier et al. 2000; Abu-Akkada et al. 2015; Kicia et al. 2019). Infection may also be present in a small percentage in immunocompetent individuals but is mostly subclinical (Didier 2005; Didier and Weiss 2006; Sak et al. 2011a).

In addition to immunocompromised people, animal handlers and slaughterhouse workers may also be at increased risk of infection (Halanova et al. 2003; Özkan et al. 2011; Sak et al. 2011b; Carhan et al. 2015). Due to the increased risk of infection and disease for owners with some degree of immunosuppression, a serological screening for *E. cuniculi* would be recommended for their pet rabbits (Dipineto et al. 2008).

Taking into consideration the risk for the immunocompromised individuals and also for those who professionally contact with rabbits and their waste, biosafety training and practices must be adopted (Fukui et al. 2013). Awareness must be raised under a One Health perspective in order to reduce the risk also to the general population. Reducing the spread of *E. cuniculi* among rabbits and also to immunocompromised humans is of utmost veterinary and public health importance (Latney et al. 2014; Doboşi et al. 2022).

6.2 Treatment and Prognosis

Clinical signs associated with *E. cuniculi* infection may or may not be associated with the presence of spores, but they are definitively associated with inflammatory response and cell damage (Latney et al. 2014). Therefore, treatment of encephalitozoonosis is aimed at reducing spore-induced inflammation, inhibiting spore formation and migration, managing concurrent disease, and controlling severe clinical signs. Response to treatment varies greatly, depending on whether the infection is acute or chronic (Keeble 2011; Latney et al. 2014), but also depends on the affected organs, since uveitis and encephalitis may be considered globally more curable than chronic renal failure (Künzel et al. 2008; Desoubeaux et al. 2017a, b).

Latent infections can occur, and some clinical cases improve spontaneously without treatment, as the rabbit's immune response, especially that mediated by cytotoxic CD8+ T-cells, plays an important role against this microsporidium (Khan et al. 2001; Latney et al. 2014; Künzel and Fisher 2018; Han et al. 2020b).

Current recommended therapeutic options, which directly inhibit microsporidian infections, are quite limited and include a class of drugs called benzimidazoles (Desoubeaux et al. 2017a, b; Wei et al. 2022). These are microtubule inhibitors and prevent extrusion of the polar filament, thus stopping the host cell from becoming infected by the pathogen (Keeble 2011).

Fenbendazole is currently the drug of choice in rabbits with encephalitozoonosis, allowing the elimination, or at least reduction, of spores present in the affected tissues and thus controlling the infection and reducing clinical signs, at a dose of 20 mg/kg orally, once a day, over a period of 28 days (Suter et al. 2001; Künzel and Joachim 2010; Shin et al. 2015).

The potential use of albendazole and oxibendazole is also occasionally reported (Giordano et al. 2005; Desoubeaux et al. 2017a, b; Fisher and Graham 2018).

Benzimidazole toxicosis in rabbits has been reported, with likely poor prognosis, providing evidence that benzimidazoles should be used cautiously in rabbits, even at recommended doses and treatment intervals, and not without warning of potential risks with this class of drugs. Special care should be taken, mainly with albendazole and oxibendazole, and when doses, higher than the standard reference range, are eventually given (Graham et al. 2014). Bone marrow suppression, characterized by anemia, leukopenia, or thrombocytopenia noted in rabbits during benzimidazole therapy, may need immediate cessation of the drug, with monitoring complete blood counts and aggressive supportive therapy (Keeble 2011; Graham et al. 2014; Fisher and Graham 2018).

In rabbits presented with CNS signs early on in the disease, associated with granulomatous encephalitis, corticosteroids have been indicated by some authors, once a diagnosis has been reached (Keeble 2011), to help reduce the inflammatory reaction, but their use is controversial, because of their potential immunosuppressive effects (Künzel and Joachim 2010; Latney et al. 2014). A single shot of dexamethasone (2 mg/kg) was used with benefit in positive rabbits with neurologic signs (head tilt and nystagmus) in association with fenbendazole (Shin et al. 2015). However, in a study by Sieg et al. (2012), treatment with dexamethasone showed no effect on neurological score on survival, providing no evidence of efficacy in the treatment of suspected encephalitozoonosis with neurological signs.

The use of systemic nonsteroidal anti-inflammatory drugs, instead of corticosteroids, is questionable to manage inflammation. And nonsteroidal anti-inflammatory drug therapy should also be used cautiously in animals that have concurrent renal disease, as the kidney is a target organ for *E. cuniculi* (Latney et al. 2014; Künzel and Fisher 2018).

In rabbits showing seizures or severe vestibular disease that is causing flailing/rolling, sedatives such as diazepam or midazolam are recommended (Künzel and Joachim 2010).

As mentioned above, response to treatment and prognosis depends on the affected organs. A study by Künzel et al. (2008) reported that rabbits exhibiting neurological signs had a 54.2% overall recovery when undergoing anti-*E. cuniculi* treatment with additional antibiotic and corticosteroid therapy, while 87.5% of the rabbits suffering from renal failure died or had to be euthanized. Chronic cases usually present with severe CNS signs, and treatment may not be successful (Suter et al. 2001; Keeble 2011).

Environmental modifications should be made to provide severely affected animals with protection. In order to avoid injury, padded cages should be provided for rabbits with severe vestibular dysfunction (Keeble 2011). Soft padding and nonmetal cage barriers should be employed to prevent accidental ocular and limb trauma. These rabbits should be placed in a quiet and stress-free environment and able to reach the offered food (Künzel and Joachim 2010; Latney et al. 2014; Künzel and Fisher 2018).

Physiotherapy can have an important impact on the improvement of clinical disease, in rabbits with vestibular dysfunction, in the form of a gradually increased exercise, like providing assistance to maintain an upright position, minimizing grade of head tilt, and helping the animal making straightforward movements (Künzel and Fisher 2018).

Antibiotic therapy can be initially indicated, before confirmation results, for rabbits showing vestibular signs, because *Pasteurella multocida* infection can be an important differential diagnosis, and also for secondary bacterial infections. Either enrofloxacin or oxytetracycline was used to treat rabbits showing vestibular signs (Harcourt-Brown and Holloway 2003; Keeble 2011).

E. cuniculi may cause cataract development and/or spontaneous lens capsule rupture, resulting in a phacoclastic uveitis, the second most common clinical manifestation. Chronic uveitis may also lead to *phthisis bulbi* and secondary glaucoma (Giordano et al. 2005; Sandmeyer et al. 2011).

Conservative systemic and local treatment of lens rupture, with anti-inflammatory and antibiotic therapy alone, has little if any effect on the severe inflammation; thus, surgical removal of the lens by phacoemulsification can treat the uveitis and restore vision and is considered the treatment of choice for affected rabbits (Felchle and Sigler 2002; Csokai et al. 2009a; Latney et al. 2014). Unfortunately, rabbits often present late in the course of disease, and, therefore, in this cases, enucleation of the affected eye is recommended (Künzel and Joachim 2010; Keeble 2011). Most rabbits with *E. cuniculi* infection presenting ocular manifestations rarely have concurrent neurological signs or renal disease, and chances of survival are good (Sandmeyer et al. 2011).

7 Concluding Remarks

- *E. cuniculi* has a worldwide distribution, infecting rabbits and other mammalians, including humans.
- Domestic, farm, industrial, pet, and laboratory rabbits are considerably exposed to this pathogen.
- *E. cuniculi* can cause severe illness in rabbits, with neurological, ocular, and renal involvement.
- In immunocompromised adult and children, *E. cuniculi* has been implicated as an opportunistic agent.
- In rabbits, horizontal transmission by ingestion or inhalation of spores occurs most frequently.
- *E. cuniculi* is a common cause of economic losses in rabbit farms, mostly on those whose hygiene and biosecurity are not well performed.
- In many rabbits, infection usually has a chronic or latent course, and only a low percentage of infected animals develop clinical manifestations.

- Chronic presentation is the most frequent in rabbits, and a positive antibody titer is indicative of infection but not of acute disease.
- Diagnostic methods for encephalitozoonosis include microscopy, molecular techniques, and serology, and samples such as serum, urine, and feces may be used.
- Diagnosis is challenging because a considerable number of infected rabbits have a chronic subclinical infection.
- Spore proliferation and migration can be reduced by treatment, and preventive antimicrosporidial medications can be considered.
- *E. cuniculi* spores are environmentally resistant but can be destroyed using several disinfection protocols.
- Fenbendazole is the drug of choice for rabbits with encephalitozoonosis, allowing the elimination, or at least reduction, of spores present in the affected tissues.
- Other control measures are necessary for decreasing the spread of *E. cuniculi*, not only to other rabbits but also to humans, especially immunocompromised individuals.
- From a One Health perspective, encephalitozoonosis is increasingly spreading.

References

Abu-Akkada SS, Oda SS (2016) Prevention and treatment of *Encephalitozoon cuniculi* infection in immunosuppressed rabbits with fenbendazole. Iran J Vet Res 17(2):98–105

Abu-Akkada SS, El Kerdany ED, Mady RF, Diab RG, Khedr GA, Ashmawy KI, Lotfy WM (2015) *Encephalitozoon cuniculi* infection among immunocompromised and immunocompetent humans in Egypt. Iran J Parasitol 10(4):561–570

Al-Sadi HI, Al-Mahmood SS (2014) Pathology of experimental *Encephalitozoon cuniculi* infection in immunocompetent and immunosuppressed mice in Iraq. Pathol Res Int 2014:857036. https://doi.org/10.1155/2014/857036

Anane S, Attouchi H (2010) Microsporidiosis: epidemiology, clinical data and therapy. Gastroenterol Clin Biol 34:450–464

Ashton N, Cook C, Clegg F (1976) Encephalitozoonosis (nosematosis) causing bilateral cataract in a rabbit. Br J Ophthalmol 60(9):618–631. https://doi.org/10.1136/bjo.60.9.618

Askari Z, Mirjalali H, Mohebali M, Zarei Z, Shojaei S, Rezaeian T, Rezaeian M (2015) Molecular detection and identification of zoonotic microsporidia spore in fecal samples of some animals with close-contact to human. Iran J Parasitol 10(3):381–388

Baldotto SB, Cray C, Giannico AT, Reifur L, Montiani-Ferreira F (2015) Seroprevalence of *Encephalitozoon cuniculi* infection in pet rabbits in Brazil. J Exotic Pet Med 24(4):435–440. https://doi.org/10.1053/j.jepm.2015.08.010

Baneux PJR, Pognan F (2003) In utero transmission of *Encephalitozoon cuniculi* strain type I in rabbits. Lab Anim 37(2):132–138. https://doi.org/10.1258/00236770360563778

Bártová E, Marková J, Sedlák K (2015) Prevalence of antibodies to *Encephalitozoon cuniculi* in European hares (*Lepus europaeus*). Ann Agric Environ Med 22(4):674–676. https://doi.org/10.5604/12321966.1185773

Benz P, Maass G, Csokai J, Fuchs-Baumgartinger A, Schwendenwein I, Tichy A, Nell B (2011) Detection of *Encephalitozoon cuniculi* in the feline cataractous lens. Vet Ophthalmol 14(Suppl. 1):37–47. https://doi.org/10.1111/j.1463-5224.2011.00873.x

Bohne W, Böttcher K, Gross U (2011) The parasitophorous vacuole of *Encephalitozoon cuniculi*: biogenesis and characteristics of the host cell-pathogen interface. Int J Med Microbiol 301(5): 395–399. https://doi.org/10.1016/j.ijmm.2011.04.006

Boot R, Hansen AK, Hansen CK, Nozari N, Thuis HC (2000) Comparison of assays for antibodies to *Encephalitozoon cuniculi* in rabbits. Lab Anim 34(3):281–289. https://doi.org/10.1258/002367700780384726

Boucher S, Nouaille L (1996) Maladies des lapines. France Agricole

Braunfuchsová P, Salát J, Kopecký J (2002) Comparison of the significance of CD4+ and CD8+ T lymphocytes in the protection of mice against *Encephalitozoon cuniculi* infection. J Parasitol 88(4):797–799. https://doi.org/10.1645/0022-3395(2002)088[0797:COTSOC]2.0.CO;2

Bywater JE, Kellett BS (1979) Humoral immune response to natural infection with *Encephalitozoon cuniculi* in rabbits. Lab Anim 13(4):293–297. https://doi.org/10.1258/002367779780943305

Cali A, Neafie RC, Takvorian PM (2011) Microsporidiosis. In: Meyers WM, Firpo A, Wear DJ (eds) Topics on the pathology of protozoan and invasive arthropod diseases, 3rd edn. Springer, Washington, DC, pp 61–76

Carhan A, Özkan O, Ozkaya E (2015) The first identification of *Encephalitozoon cuniculi* infection in an animal care worker in Turkey. Iran J Parasitol 10(2):280–285

Cox JC, Gallichio HA (1977) An evaluation of indirect immunofluorescence in the serological diagnosis of *Nosema cuniculi* infection. Res Vet Sci 22(1):50–52

Cox JC, Gallichio HA (1978) Serological and histological studies on adult rabbits with recent, naturally acquired encephalitozoonosis. Res Vet Sci 24(2):260–261

Cox JC, Hamilton RC, Attwood HD (1979) An investigation of the route and progression of *Encephalitozoon cuniculi* infection in adult rabbits. J Protozool 26(2):260–265. https://doi.org/10.1111/j.1550-7408.1979.tb02776.x

Cray C, Rodriguez M, Fernandez Y (2013) Acute phase protein levels in rabbits with suspected *Encephalitozoon cuniculi* infection. J Exotic Pet Med 22(3):280–286. https://doi.org/10.1053/j.jepm.2013.08.008

Csokai J, Gruber A, Künzel F, Tichy A, Joachim A (2009a) Encephalitozoonosis in pet rabbits (*Oryctolagus cuniculus*): pathohistological findings in animals with latent infection versus clinical manifestation. Parasitol Res 104(3):629–635. https://doi.org/10.1007/s00436-008-1239-2

Csokai J, Joachim A, Gruber A, Tichy A, Pakozdy A, Künzel F (2009b) Diagnostic markers for encephalitozoonosis in pet rabbits. Vet Parasitol 163(1–2):18–26. https://doi.org/10.1016/j.vetpar.2009.03.057

Dalboni LC, Alvares Saraiva AM, Konno FTC, Perez EC, Codeceira JF, Spadacci-Morena DD, Lallo MA (2021) *Encephalitozoon cuniculi* takes advantage of efferocytosis to evade the immune response. PLoS One 16(3):e0247658. https://doi.org/10.1371/journal.pone.0247658

Deplazes P, Mathis A, Baumgartner R, Tanner I, Weber R (1996) Immunologic and molecular characteristics of *Encephalitozoon*-like microsporidia isolated from humans and rabbits indicate that *Encephalitozoon cuniculi* is a zoonotic parasite. Clin Infect Dis 22(3):557–559. https://doi.org/10.1093/clinids/22.3.557

Deplazes P, Mathis A, Weber R (2000) Epidemiology and zoonotic aspects of microsporidia of mammals and birds. Contrib Microbiol 6:236–260. https://doi.org/10.1159/000060363

Desoubeaux G, Pantin A, Peschke R, Joachim A, Cray C (2017a) Application of Western blot analysis for the diagnosis of *Encephalitozoon cuniculi* infection in rabbits: example of a quantitative approach. Parasitol Res 116(2):743–750. https://doi.org/10.1007/s00436-016-5343-4

Desoubeaux G, Piqueras MC, Pantin A, Bhattacharya SK, Peschke R, Joachim A, Cray C (2017b) Application of mass spectrometry to elucidate the pathophysiology of *Encephalitozoon cuniculi* infection in rabbits. PLoS One 12(7):e0177961. https://doi.org/10.1371/journal.pone.0177961

Didier ES (2005) Microsporidiosis: an emerging and opportunistic infection in humans and animals. Acta Trop 94(1):61–76. https://doi.org/10.1016/j.actatropica.2005.01.010

Didier ES, Weiss LM (2006) Microsporidiosis: current status. Curr Opin Infect Dis 19(5):485–492. https://doi.org/10.1097/01.qco.0000244055.46382.23

Didier ES, Didier PJ, Snowden KF, Shadduck JA (2000) Microsporidiosis in mammals. Microbes Infect 2(6):709–720. https://doi.org/10.1016/s1286-4579(00)00354-3

Didier ES, Stovall ME, Green LC, Brindley PJ, Sestak K, Didier PJ (2004) Epidemiology of microsporidiosis: sources and modes of transmission. Vet Parasitol 126(1–2):145–166. https://doi.org/10.1016/j.vetpar.2004.09.006

Dipineto L, Rinaldi L, Santaniello A, Sensale M, Cuomo A, Calabria M, Menna LF, Fioretti A (2008) Serological survey for antibodies to *Encephalitozoon cuniculi* in pet rabbits in Italy. Zoonoses Public Health 55(3):173–175. https://doi.org/10.1111/j.1863-2378.2007.01097.x

Doboşi A-A, Bel L-V, Paştiu AI, Pusta DL (2022) A review of *Encephalitozoon cuniculi* in domestic rabbits (*Oryctolagus cuniculus*) – biology, clinical signs, diagnostic techniques, treatment, and prevention. Pathogens 11(12):1486. https://doi.org/10.3390/pathogens11121486

Fasshauer V, Gross U, Bohne W (2005) The parasitophorous vacuole membrane of *Encephalitozoon cuniculi* lacks host cell membrane proteins immediately after invasion. Eukaryot Cell 4(1):221–224. https://doi.org/10.1128/EC.4.1.221-224.2005

Felchle LM, Sigler RL (2002) Phacoemulsification for the management of *Encephalitozoonon cuniculi*-induced phacoclastic uveitis in a rabbit. Vet Ophthalmol 5(3):211–215. https://doi.org/10.1046/j.1463-5224.2002.00240.x

Fisher P, Graham J (2018) Rabbits. In: Carpenter JW (ed) Exotic animal formulary, 5th edn. Elsevier, pp 494–531

Flatt RE, Jackson SJ (1970) Renal nosematosis in young rabbits. Pathol Vet 7(6):492–497. https://doi.org/10.1177/030098587000700603

Fournier S, Liguory O, Sarfati C, David-Ouaknine F, Derouin F, Decazes JM, Molina JM (2000) Disseminated infection due to *Encephalitozoon cuniculi* in a patient with AIDS: case report and review. HIV Med 1(3):155–161. https://doi.org/10.1046/j.1468-1293.2000.00022.x

Franzen C, Müller A (1999) Molecular techniques for detection species differentiation and phylogenetic analysis of microsporidia. Clin Microbiol Rev 12(2):243–285. https://doi.org/10.1128/CMR.12.2.243

Franzen C, Müller A, Hartmann P, Salzberger B (2005) Cell invasion and intracellular fate of *Encephalitozoon cuniculi* (microsporidia). Parasitology 130(Pt. 3):285–292. https://doi.org/10.1017/s003118200400633x

Fukui D, Bando G, Furuya K, Yamaguchi M, Nakaoka Y, Kosuge M, Murata K (2013) Surveillance for an outbreak of *Encephalitozoon cuniculi* infection in rabbits housed at a zoo and biosecurity countermeasures. J Vet Med Sci 75(1):55–61. https://doi.org/10.1292/jvms.12-0231

Furuya K, Asakura T, Igarashi M, Morita T (2009) Microsporidian *Encephalitozoon cuniculi* antibodies in rabbit urine samples. Vet Rec 165(3):85–86. https://doi.org/10.1136/vetrec.165.3.85

Genz EJ, Carpenter JW (1997) Rabbits – neurologic and musculoskeletal disease. In: Hillyer EV, Queensberry KE (eds) Ferrets, rabbits and rodents: clinical medicine and surgery. Saunders, pp 221–225

Giordano C, Weigt A, Vercelli A, Rondena M, Grilli G, Giudice C (2005) Immunohistochemical identification of *Encephalitozoon cuniculi* in phacoclastic uveitis in four rabbits. Vet Ophthalmol 8(4):271–275. https://doi.org/10.1111/j.1463-5224.2005.00394.x

Goodwin D, Gennari SM, Howe DK, Dubey JP, Zajac AM, Lindsay DS (2006) Prevalence of antibodies to *Encephalitozoon cuniculi* in horses from Brazil. Vet Parasitol 142(3–4):380–382. https://doi.org/10.1016/j.vetpar.2006.07.006

Graham JE, Garner MM, Reavill DR (2014) Benzimidazole toxicosis in rabbits: 13 cases (2003–2011). J Exotic Pet Med 23(2):188–195. https://doi.org/10.1053/j.jepm.2014.02.012

Grahn B, Wolfer J, Keller C (1991) Diagnostic ophthalmology. Can Vet J 32(6):372–373

Gruber A, Pakozdy A, Weissenböck H, Csokai J, Künzel F (2009) A retrospective study of neurological disease in 118 rabbits. J Comp Pathol 140(1):31–37. https://doi.org/10.1016/j.jcpa.2008.09.009

Halanova M, Cislakova L, Valencakova A, Balent P, Adam J, Travnicek M (2003) Serological screening of occurrence of antibodies to *Encephalitozoon cuniculi* in humans and animals in eastern Slovakia. Ann Agric Environ Med 10(1):117–120

Han B, Weiss LM (2017) Microsporidia: obligate intracellular pathogens within the fungal kingdom. Microbiol Spectr 5(2). https://doi.org/10.1128/microbiolspec.FUNK-0018-2016

Han B, Takvorian PM, Weiss LM (2020a) Invasion of host cells by microsporidia. Front Microbiol 11:172. https://doi.org/10.3389/fmicb.2020.00172

Han Y, Gao H, Xu J, Luo J, Han B, Bao J, Pan G, Li T, Zhou Z (2020b) Innate and adaptive immune responses against microsporidia infection in mammals. Front Microbiol 11:1468. https://doi.org/10.3389/fmicb.2020.01468

Harcourt-Brown FM (2004) *Encephalitozoon cuniculi* infection in rabbits. Semin Avian Exotic Pet Med 13(2):86–93

Harcourt-Brown F (2007) Radiographic signs of renal disease in rabbits. Vet Rec 160(23):787–794. https://doi.org/10.1136/vr.160.23.787

Harcourt-Brown FM (2013) Diagnosis of renal disease in rabbits. Vet Clin North Am Exotic Anim Pract 16(1):145–174. https://doi.org/10.1016/j.cvex.2012.10.004

Harcourt-Brown FM, Holloway HK (2003) *Encephalitozoon cuniculi* in pet rabbits. Vet Rec 152(14):427–431. https://doi.org/10.1136/vr.152.14.427

Hinney B, Sak B, Joachim A, Kváč M (2016) More than a rabbit's tale – *Encephalitozoon* spp. in wild mammals and birds. Int J Parasitol Parasit Wildl 5(1):76–87. https://doi.org/10.1016/j.ijppaw.2016.01.001

Hocevar SN, Paddock CD, Spak CW, Rosenblatt R, Diaz-Luna H, Castillo I, Luna S, Friedman GC, Antony S, Stoddard RA, Tiller RV, Peterson T, Blau DM, Sriram RR, da Silva A, de Almeida M, Benedict T, Goldsmith CS, Zaki SR, Visvesvara GS, Kuehnert MJ, for the Microsporidia Transplant Transmission Investigation Team (2014) Microsporidiosis acquired through solid organ transplantation: a public health investigation. Ann Intern Med 160(4): 213–220. https://doi.org/10.7326/M13-2226

Hollyer JA, McGuinness E, Bowers LC, Didier ES, Giudice C, Perl DP, Fogarty U (2014) *Encephalitozoon cuniculi*-associated equine encephalitis: a case report. J Equine Vet 34(11–12):1348–1351. https://doi.org/10.1016/j.jevs.2014.09.002

Hunt RD, King NW, Foster HL (1972) Encephalitozoonosis: evidence for vertical transmission. J Infect Dis 126(2):212–214. https://doi.org/10.1093/infdis/126.2.212

Igarashi M, Oohashi E, Dautu G, Ueno A, Karya T, Furuya K (2008) High seroprevalance of *Encephalitozoon cuniculi* in pet rabbits in Japan. J Vet Med Sci 70(12):1301–1304. https://doi.org/10.1292/jvms.70.1301

Jass A, Matiasek K, Henke J, Küchenhoff H, Hartmann K, Fischer A (2008) Analysis of cerebrospinal fluid in healthy rabbits and rabbits with clinically suspected encephalitozoonosis. Vet Rec 162(19):618–622. https://doi.org/10.1136/vr.162.19.618

Jeklova E, Jekl V, Kovarcik K, Hauptman K, Koudela B, Neumayerova H, Knotek Z, Faldyna M (2010a) Usefulness of detection of specific IgM and IgG antibodies for diagnosis of clinical encephalitozoonosis in pet rabbits. Vet Parasitol 170(1–2):143–148. https://doi.org/10.1016/j.vetpar.2010.01.029

Jeklova E, Leva L, Kovarcik K, Matiasovic J, Kummer V, Maskova J, Skoric M, Faldyna M (2010b) Experimental oral and ocular *Encephalitozoon cuniculi* infection in rabbits. Parasitology 137(12):1749–1757. https://doi.org/10.1017/S0031182010000648

Jeklová E, Levá L, Kummer V, Jekl V, Faldyna M (2019) Immunohistochemical detection of *Encephalitozoon cuniculi* in ocular structures of immunocompetent rabbits. Animals (Basel) 9(11):988. https://doi.org/10.3390/ani9110988

Jeklova E, Leva L, Matiasovic J, Ondrackova P, Kummer V, Faldyna M (2020) Characterization of humoral and cell-mediated immunity in rabbits orally infected with *Encephalitozoon cuniculi*. Vet Res 51(1):79. https://doi.org/10.1186/s13567-020-00806-9

Jordan CN, Zajac AM, Lindsay DS (2006) *Encephalitozoon cuniculi* infection in rabbits. Compend Contin Educ Pract Vet 28(2):108–116

Keeble E (2011) Encephalitozoonosis in rabbits – what we do and don't know. In Pract 33(9): 426–435. https://doi.org/10.1136/inp.d6077

Keeble EJ, Shaw DJ (2006) Seroprevalence of antibodies to *Encephalitozoon cuniculi* in domestic rabbits in the United Kingdom. Vet Rec 158(16):539–544. https://doi.org/10.1136/vr.158.16.539

Khan IA, Moretto M (1999) Role of gamma interferon in cellular immune response against murine *Encephalitozoon cuniculi* infection. Infect Immunol 67(4):1887–1893. https://doi.org/10.1128/IAI.67.4.1887-1893.1999

Khan IA, Schwartzman JD, Kasper LH, Moretto M (1999) CD8 + CTLs are essential for protective immunity against *Encephalitozoon cuniculi* infection. J Immunol 162(10):6086–6091

Khan IA, Moretto M, Weiss LM (2001) Immune response to *Encephalitozoon cuniculi* infection. Microbes Infect 3(5):401–405. https://doi.org/10.1016/s1286-4579(01)01397-1

Kicia M, Szydłowicz M, Cebulski K, Jakuszko K, Piesiak P, Kowal A, Sak B, Krajewska M, Hendrich AB, Kváč M, Kopacz Ż (2019) Symptomatic respiratory *Encephalitozoon cuniculi* infection in renal transplant recipients. Int J Infect Dis 79:21–25. https://doi.org/10.1016/j.ijid.2018.10.016

Kimura M, Aoki M, Ichikawa-Seki M, Matsuo K, Yagita K, Itagaki T (2013) Detection and genotype of *Encephalitozoon cuniculi* DNA from urine and feces of pet rabbits in Japan. J Vet Med Sci 75(8):1017–1020. https://doi.org/10.1292/jvms.12-0567

Kunstýr I, Naumman S, Kaup F-J (1986a) Torticollis bei Kaninchen. Ätiologie, Pathologie, Diagnose und Therapie. Die Berliner und Münchener Tierärztliche Wochenschrift 99(1):14–19

Kunstýr I, Lev L, Naumannm S (1986b) Humoral antibody response of rabbits to experimental infection with *Encephalitozoon cuniculi*. Vet Parasitol 21(4):223–232. https://doi.org/10.1016/0304-4017(86)90048-8

Künzel F, Fisher PG (2018) Clinical signs, diagnosis, and treatment of *Encephalitozoon cuniculi* infection in rabbits. Vet Clin North Am Exot Anim Pract 21(1):69–82. https://doi.org/10.1016/j.cvex.2017.08.002

Künzel F, Joachim A (2010) Encephalitozoonosis in rabbits. Parasitol Res 106(2):299–309. https://doi.org/10.1007/s00436-009-1679-3

Künzel F, Gruber A, Tichy A, Edelhofer R, Nell B, Hassan J, Leschnik M, Thalhammer JG, Joachim A (2008) Clinical symptoms and diagnosis of encephalitozoonosis in pet rabbits. Vet Parasitol 151(2–4):115–124. https://doi.org/10.1016/j.vetpar.2007.11.005

Latney LV, Bradley CW, Wyre NR (2014) *Encephalitozoon cuniculi* in pet rabbits: diagnosis and optimal management. Vet Med (Auckl) 5:169–180. https://doi.org/10.2147/VMRR.S49842

Lavazza A, Chiari M, Nassuato C, Giardiello D, Tittarelli C, Grilli G (2016) Serological investigation on *Encephalitozoon cuniculi* in pet rabbits in north-Central Italy. J Exotic Pet Med 25(1): 52–59. https://doi.org/10.1053/j.jepm.2015.12.004

Lee SC, Heitman J (2017) Dynamics of parasitophorous vacuoles formed by the microsporidian pathogen *Encephalitozoon cuniculi*. Fungal Genet Biol 107:20–23. https://doi.org/10.1016/j.fgb.2017.07.006

Leipig M, Matiasek K, Rinder H, Janik D, Emrich E, Baiker K, Hermanns W (2013) Value of histopathology immunohistochemistry, and real-time polymerase chain reaction in the confirmatory diagnosis of *Encephalitozoon cuniculi* infection in rabbits. J Vet Diagn Investig 25(1):16–26. https://doi.org/10.1177/1040638712466394

Lewis W (2010) Exposure to *Encephalitozoon cuniculi* in rabbits with dental disease. Vet Rec 166(23):730. https://doi.org/10.1136/vr.c2549

Li X, Palmer R, Trout JM, Fayer R (2003) Infectivity of microsporidia spores stored in water at environmental temperatures. J Parasitol 89(1):185–188. https://doi.org/10.1645/0022-3395(2003)089[0185:IOMSSI]2.0.CO;2

Maddox JV, Brooks WM, Solter LF (2000) Bioassays of microsporidia. In: Navon A, K.R.S. (eds) Bioassays of entomopathogenic microbes and nematodes. CABI International, pp 197–228

Maestrini G, Ricci E, Cantile C, Mannella R, Mancianti F, Paci G, D'Ascenzi C, Perrucci S (2017) *Encephalitozoon cuniculi* in rabbits: serological screening and histopathological findings. Comp Immunol Microbiol Infect Dis 50:54–57. https://doi.org/10.1016/j.cimid.2016.11.012

Magalhães TR, Pinto FF, Queiroga FL (2022) A multidisciplinary review about *Encephalitozoon cuniculi* in a one health perspective. Parasitol Res 121(9):2463–2479. https://doi.org/10.1007/s00436-022-07562-z

Mäkitaipale J, Järvenpää E, Bruce A, Sankari S, Virtala AM, Näreaho A (2022) Seroprevalence of *Encephalitozoon cuniculi* and *toxoplasma gondii* antibodies and risk-factor assessment for *Encephalitozoon cuniculi* seroprevalence in Finnish pet rabbits (*Oryctolagus cuniculus*). Acta Vet Scand 64(2):2. https://doi.org/10.1186/s13028-022-00622-5

Mancinelli E, Lennox AM (2017) Management of otitis in rabbits. J Exotic Pet Med 26(1):63–73. https://doi.org/10.1053/j.jepm.2016.10.009

Marcato PS, Rosmini R (1986) Sistema nervoso. In: Marcato PS, Rosmini R (eds) Patologia del coniglio e della lepre. Società Editrice Esculapio

Mathis A, Weber R, Deplazes P (2005) Zoonotic potential of the microsporidia. Clin Microbiol Rev 18(3):423–445. https://doi.org/10.1128/CMR.18.3.423-445.2005

Morsy EA, Salem HM, Khattab MS, Hamza DA, Abuowarda MM (2020) *Encephalitozoon cuniculi* infection in farmed rabbits in Egypt. Acta Vet Scand 62(1):11. https://doi.org/10.1186/s13028-020-0509-6

Nast R, Middleton DM, Wheler CL (1996) Generalized encephalitozoonosis in a Jersey wooly rabbit. Can Vet J 37:303–305

Nell B, Csokai J, Fuchs-Baumgartinger A, Maaß G (2015) *Encephalitozoon cuniculi* causes focal anterior cataract and uveitis in dogs. Tierarztl Prax Ausg K Kleintiere Heimtiere 43(5):337–344. https://doi.org/10.15654/tpk-141053

Okewole EA (2008) Seroprevalence of antibodies to *Encephalitozoon cuniculi* in domestic rabbits in Nigeria. Onderstepoort J Vet Res 75(1):33–38. https://doi.org/10.4102/ojvr.v75i1.85

Omalu ICJ, Duhlinska DD, Anyanwu GI, Pam VA, Inyama PU (2007) Immune responsiveness associated with experimental *Encephalitozoon intestinalis* infection in immunocompetent rats. Indian J Med Microbiol 25(3):209–213. https://doi.org/10.4103/0255-0857.34760

Orlik J, Böttcher K, Groß U, Bohne W (2010) Germination of phagocytosed *E. cuniculi* spores does not significantly contribute to parasitophorous vacuole formation in J774 cells. Parasitol Res 106(3):753–755. https://doi.org/10.1007/s00436-010-1736-y

Özkan Ö, Alcigir ME (2018a) Encephalitozoonosis infection in a traditional rabbit farm with neurological manifestations. Vet Parasitol 262:26–29. https://doi.org/10.1016/j.vetpar.2018.09.009

Özkan Ö, Alcigir ME (2018b) Subacute stage of *Encephalitozoon cuniculi* infection in eye lesions of rabbit in Turkey. Iran J Parasitol 13(2):301–309

Özkan O, Özkan AT, Zafer K (2011) Encephalitozoonosis in New Zealand rabbits and potential transmission risk. Vet Parasitol 179(1–3):234–237. https://doi.org/10.1016/j.vetpar.2011.02.007

Özkan O, Karagoz A, Kocak N (2019) First molecular evidence of ocular transmission of Encephalitozoonoosis during the intrauterine period in rabbits. Parasitol Int 71:1–4. https://doi.org/10.1016/j.parint.2019.03.006

Paré JA, Paul-Murphy J (2004) Disorders of the reproductive and urinary systems. In: Queensberry KE, Carpenter JW (eds) Ferrets, rabbits, and rodents: clinical medicine and surgery, 2nd edn. Saunders, pp 183–193

Patrício RFG (2014) Seroprevalência de *Encephalitozoon cuniculi* em coelhos assintomáticos da região de Lisboa. Master's dissertation

Percy DH, Barthold SW (2007) Pathology of laboratory rodents and rabbits, 3rd edn. Blackwell

Peyretaillade E, El Alaoui H, Diogon M, Polonais V, Parisot N, Biron DG, Peyret P, Delbac F (2011) Extreme reduction and compaction of microsporidian genomes. Res Microbiol 162(6): 598–606. https://doi.org/10.1016/j.resmic.2011.03.004

Rich G (2010) Clinical update on testing modalities for *Encephalitozoon cuniculi* in clinically sick rabbits. J Exotic Pet Med 19(3):226–230. https://doi.org/10.1053/j.jepm.2010.07.007

Richardson VCG (2000) Rabbits: health, husbandry, and diseases. Wiley-Blackwell

Richter B, Csokai J, Graner I, Eisenberg T, Pantchev N, Eskens HU, Nedorost N (2013) Encephalitozoonosis in two inland bearded dragons (*Pogona vitticeps*). J Comp Pathol 148(2–3):278–282. https://doi.org/10.1016/j.jcpa.2012.05.009

Rodríguez-Tovar LE, Nevárez-Garza AM, Trejo-Chávez A, Hernández-Martínez CA, Hernández-Vidal G, Zarate-Ramos JJ, Castillo-Velázquez U (2016) *Encephalitozoon cuniculi*: grading the histological lesions in brain, kidney, and liver during primo infection outbreak in rabbits. J Pathol 2016:5768428. https://doi.org/10.1155/2016/5768428

Rodríguez-Tovar LE, Villarreal-Marroquín A, Nevárez-Garza AM, Castillo-Velázquez U, Rodríguez-Ramírez HG, Navarro-Soto MC, Zárate-Ramos JJ, Hernández-Vidal G, Trejo-Chávez A (2017) Histochemical study of *Encephalitozoon cuniculi* spores in the kidneys of naturally infected New Zealand rabbits. J Vet Diagn Investig 29(3):269–277. https://doi.org/10.1177/1040638716668559

Rodríguez-Tovar LE, Castillo-Velázquez U, Arce-Mendoza AY, Nevárez- Garza AM, Zarate-Ramos JJ, Hernández-Vidal G, Rodríguez-Ramírez HG, Trejo-Chávez A (2019) Interferon γ and interleukin 10 responses in immunocompetent and immunosuppressed New Zealand white rabbits naturally infected with *Encephalitozoon cunicul*i. Dev Comp Immunol 62:82–88. https://doi.org/10.1016/j.dci.2016.05.003

Sadeghi-Dehkordi Z, Norouzi E, Rezaeian H, Nourian A, Noaman V, Sazmand A (2019) First insight into *Encephalitozoon cuniculi* infection in laboratory and pet rabbits in Iran. Comp Immunol Microbiol Infect Dis 65:37–40. https://doi.org/10.1016/j.cimid.2019.04.005

Sak B, Saková K, Ditrich O (2004) Effects of a novel anti-exospore monoclonal antibody on microsporidial development in vitro. Parasitol Res 92(1):74–80. https://doi.org/10.1007/s00436-003-0988-1

Sak B, Salát J, Horká H, Saková K, Ditrich O (2006) Antibodies enhance the protective effect of CD4+ T lymphocytes in SCID mice perorally infected with *Encephalitozoon cuniculi*. Parasite Immunol 28(3):95–99. https://doi.org/10.1111/j.1365-3024.2005.00813.x

Sak B, Brady D, Pelikánová M, Květoňová D, Rost M, Kostka M, Tolarová V, Hůzová Z, Kváč M (2011a) Unapparent microsporidial infection among immunocompetent humans in The Czech Republic. J Clin Microbiol 49(3):1064–1070. https://doi.org/10.1128/JCM.01147-10

Sak B, Kvác M, Petrzelková K, Kvetonová D, Pomajbíková K, Mulama M, Kiyang J, Modrý D (2011b) Diversity of microsporidia (fungi: microsporidia) among captive great apes in European zoos and African sanctuaries: evidence for zoonotic transmission? Folia Parasitol (Praha) 58(2): 81–86. https://doi.org/10.14411/fp.2011.008

Salát J, Sak B, Le T, Kopecký J (2004) Susceptibility of IFN-gamma or IL-12 knock-out and SCID mice to infection with two microsporidian species, *Encephalitozoon cuniculi* and *E intestinalis*. Folia Parasitol (Praha) 51(4):275–282

Sanchez RF, Everson R, Hedley J, Dawson C, Lam R, Priestnall SL, Garcia de Carellan A, de Miguel C, Seymour C (2018) Rabbits with naturally occurring cataracts referred for phacoemulsification and intraocular lens implantation: a preliminary study of 12 cases. Vet Ophthalmol 21(4):399–412. https://doi.org/10.1111/vop.12525

Sandmeyer LS, Bauer BS, Grahn BH (2011) Diagnostic ophthalmology. Can Vet J 52(3):327–328

Saviotti M, Tamba M, Gallazzi D, Lavazza A (2000) Further data on the diffusion of *Encephalitozoon cuniculi* in Italian rabbitries. World Rabbit Sci 8(Suppl. 1):355–362

Scharmann W, Reblin L, Griem W (1986) Untersuchungen über die Infektion von Kaninchen durch *Encephalitozoon cuniculi*. Die Berliner und Münchener Tierärztliche Wochenschrift 99(1): 20–24

Schmidt EC, Shadduck JA (1983) Murine encephalitozoonosis model for studying the host-parasite relationship of a chronic infection. Infect Immunol 40(3):936–942. https://doi.org/10.1128/iai.40.3.936-942.1983

Schmidt EC, Shadduck JA (1984) Mechanisms of resistance to the intracellular protozoan *Encephalitozoon cuniculi* in mice. J Immunol 133(5):2712–2719

Shadduck JA, Orenstein JM (1993) Comparative pathology of microsporidiosis. Arch Pathol Lab Med 117(12):1215–1219

Shin J, Kim S, Kim S, Song K (2015) Head tilt associated with encephalitozoonosis in four pet rabbits. J Vet Clin 32(2):212–214. https://doi.org/10.17555/jvc.2015.04.32.2.212

Sieg J, Hein J, Jass A, Sauter-Louis C, Hartmann K, Fischer A (2012) Clinical evaluation of therapeutic success in rabbits with suspected encephalitozoonosis. Vet Parasitol 187(1–2): 328–332. https://doi.org/10.1016/j.vetpar.2011.12.014

Stiles J, Didier E, Ritchie B, Greenacre C, Willis M, Martin C (1997) *Encephalitozoon cuniculi* in the lens of a rabbit with phacoclastic uveitis: confirmation and treatment. Vet Comp Ophthalmol 7(4):233–238

Suter C, Müller-Doblies UU, Hatt JM, Deplazes P (2001) Prevention and treatment of *Encephalitozoon cuniculi* infection in rabbits with fenbendazole. Vet Rec 148(15):478–480. https://doi.org/10.1136/vr.148.15.478

Talabani H, Sarfati C, Pillebout E, van Gool T, Derouin F, Menotti J (2010) Disseminated infection with a new genovar of *Encephalitozoon cuniculi* in a renal transplant recipient. J Clin Microbiol 48(7):2651–2653. https://doi.org/10.1128/jcm.02539-09

Tee KY, Kao JP, Chiu HY, Chang MH, Wang JH, Tung KC, Cheng FP, Wu JT (2011) Serological survey for antibodies to *Encephalitozoon cuniculi* in rabbits in Taiwan. Vet Parasitol 183(1–2): 68–71. https://doi.org/10.1016/j.vetpar.2011.06.011

Thomas WB (2000) Vestibular dysfunction. Vet Clin N Am Small Anim Pract 30(1):227–249. https://doi.org/10.1016/s0195-5616(00)50011-4

Thrusfield M, Christley R (2018) Veterinary epidemiology, 4th edn. Wiley-Blackwell

Valencakova A, Halanova M (2012) Immune response to *Encephalitozoon* infection review. Comp Immunol Microbiol Infect Dis 35(1):1–7. https://doi.org/10.1016/j.cimid.2011.11.004

Valencakova A, Balent P, Petrovova E, Novotny F, Luptakova L (2008) Encephalitozoonosis in household pet Nederland Dwarf rabbits (*Oryctolagus cuniculus*). Vet Parasitol 153(3–4): 265–269. https://doi.org/10.1016/j.vetpar.2008.01.047

Vergneau-Grosset C, Larrat S (2015) Microsporidiosis in vertebrate companion exotic animals. J Fungi (Basel) 2(1):3. https://doi.org/10.3390/jof2010003

Wagnerová P, Sak B, Květoňová D, Buňatová Z, Civišová H, Maršálek M, Kváč M (2012) *Enterocytozoon bieneusi* and *Encephalitozoon cuniculi* in horses kept under different management systems in The Czech Republic. Vet Parasitol 190(3–4):573–577. https://doi.org/10.1016/j.vetpar.2012.07.013

Wasson K, Peper RL (2000) Mammalian microsporidiosis. Vet Pathol 37(2):113–128. https://doi.org/10.1354/vp.37-2-113

Wei J, Fei Z, Pan G, Weiss LM, Zhou Z (2022) Current therapy and therapeutic targets for microsporidiosis. Front Microbiol 13:835390. https://doi.org/10.3389/fmicb.2022.835390

Weiss LM (2001) Microsporidia: emerging pathogenic protists. Acta Trop 78(2):89–102. https://doi.org/10.1016/s0001-706x(00)00178-9

Wolfer J, Grahn B, Wilcock B, Percy D (1993) Phacoclastic uveitis in the rabbit. Prog Vet Comp Ophthalmol 3:92–97

Wright JH, Craighead EM (1922) Infectious motor paralysis in young rabbits. J Exp Med 36(1): 135–140. https://doi.org/10.1084/jem.36.1.135

Zietek J, Adaszek Ł, Dziegiel B, Kalinowski M, Bartnicki M, Kalinowska A, Jarosz Ł, Winiarczyk S (2014) Diagnosis of the *Encephalitozoon cuniculi* infections in pet rabbits with neurological symptoms. Pol J Vet Sci 17(2):361–363. https://doi.org/10.2478/pjvs-2014-0050

Rabbit Coccidiosis

Ana Patrícia Lopes and João Simões

Abstract

This chapter aims to highlight the main features of coccidiosis in rabbits. The *Eimeria* genus belongs to the Apicomplexa phylum causing intestinal and hepatic coccidiosis in rabbits. The complete life cycle of *Eimeria* is monoxenous leading to the development of successive protozoa stage inside the epithelial cells of the intestine or bile ducts of the host, until their rupture. Subclinical and clinical coccidiosis mainly affect fattening rabbits from approximately 6 weeks old. Additionally, rabbits are also reservoir hosts for this protozoan when their immunity level mitigates the intensity of the lesions and clinical signs. Due to their biologic complexity, a bidirectional cause-effect with other diseases can occur. The treatment of coccidiosis is likely, even when some limitations of drug effectiveness are observed during gametogony. Nevertheless, due to *Eimeria* ubiquity in rabbitries, only prevention is suitable to keep rabbit performance, health, and well-being at acceptable levels. Currently, coccidiosis control is anticoccidial drug-based, even when other sanitary measures are ensured. Immunization-based measures are under investigation to avoid anticoccidial use and resistance.

Keywords

Anticoccidials · *Eimeria* · Hepatic coccidiosis · Intestinal coccidiosis · Oocysts

A. P. Lopes (✉) · J. Simões
Animal and Veterinary Research Centre (CECAV), Associate Laboratory for Animal and Veterinary Sciences (AL4AnimalS), Department of Veterinary Sciences, School of Agrarian and Veterinary Sciences, University of Trás-os-Montes e Alto Douro (UTAD), Vila Real, Portugal
e-mail: aplopes@utad.pt

J. Simões, J. M. Monteiro (eds.), *Veterinary Care of Farm Rabbits*,
https://doi.org/10.1007/978-3-031-44542-2_29

1 Introduction

Coccidiosis is caused by obligate intracellular parasites belonging to the genus *Eimeria* (Schneider, 1875). Species of *Eimeria* are ubiquitous being one of the most important parasites that affect the health of animals and one of the most economically relevant problems in the rabbit management. It can be a serious concern in rabbit breeding on large commercial/industrial farms but also on small traditional farms, experimental facilities, and pet rabbits (De Vos 1970; Cowie-Whitney 1977; Peeters et al. 1987; Pakandl 2009; Oncel et al. 2011; Al-Quraishy et al. 2012; Hrženjak et al. 2021).

This protozoan is characterized by different degrees of virulence and pathogenicity and affinity for individual parts of the rabbit intestine. Due to these issues, the relative biologic complexity of their life cycle assumes special relevance (Taylor et al. 2016).

In rabbits, there are two distinct forms of the disease, intestinal coccidiosis, which can be caused by several species of the genus *Eimeria*, and hepatic coccidiosis caused only by *Eimeria stiedae* (Pakandl 2009). This chapter describes the main features related with both forms.

2 The Pathogens

The most prevalent species of the genus *Eimeria* identified with clinical significance in rabbits include *E. coecicola*, *E. exigua*, *E. flavescens*, *E. intestinalis*, *E. irresidua*, *E. magna*, *E. media*, *E. perforans*, *E. piriformis*, and *E. vejdovskyi*, parasites of the intestine, whereas *E. stiedae* exclusively infects the liver and the bile ducts (Coudert et al. 1995; Eckert et al. 1995; Bhat et al. 1996; Taylor et al. 2007; Li and Ooi 2009; Pakandl 2009). Additionally, other species have also been described to infect rabbits including *E. elongate*, *E. matsubayashi*, *E. nagpurensis*, and *E. neoleporis* (Baker 2007; Taylor et al. 2007; Li and Ooi 2009). More recently, Cui et al. (2017) reported the identification of a new species, *E. kongi* n. sp., in a domestic rabbit from China.

Within the mentioned species, *E. intestinalis* and *E. flavescens* are considered to be the most virulent, while the virulence of *E. stiedae* depends strictly on the intensity of infection. Other species are considered to have low virulence or to be non-virulent (Pakandl 2009).

Intestinal and hepatic coccidiosis is transmitted by the fecal-oral route, through the ingestion of sporulated oocysts in contaminated water or food (Al-Quraishy et al. 2012). Wild rabbits can be a potential source of infection for pet rabbits kept outdoors or that are fed grass and vegetation contaminated with oocysts excreted by wild rabbits (Harcourt-Brown 2014; Bochyńska et al. 2022).

After ingestion of sporulated oocysts by the rabbit, the wall is digested, and the released sporozoites either enter the enterocytes (species that parasitize the intestine) or, in the case of *E. stiedae*, penetrate the duodenal mucosa to migrate to the mesenteric lymph nodes and then undergo hepatic migration via mononuclear cells within lymphatics (Fig. 1) (Barthold et al. 2016; Taylor et al. 2016).

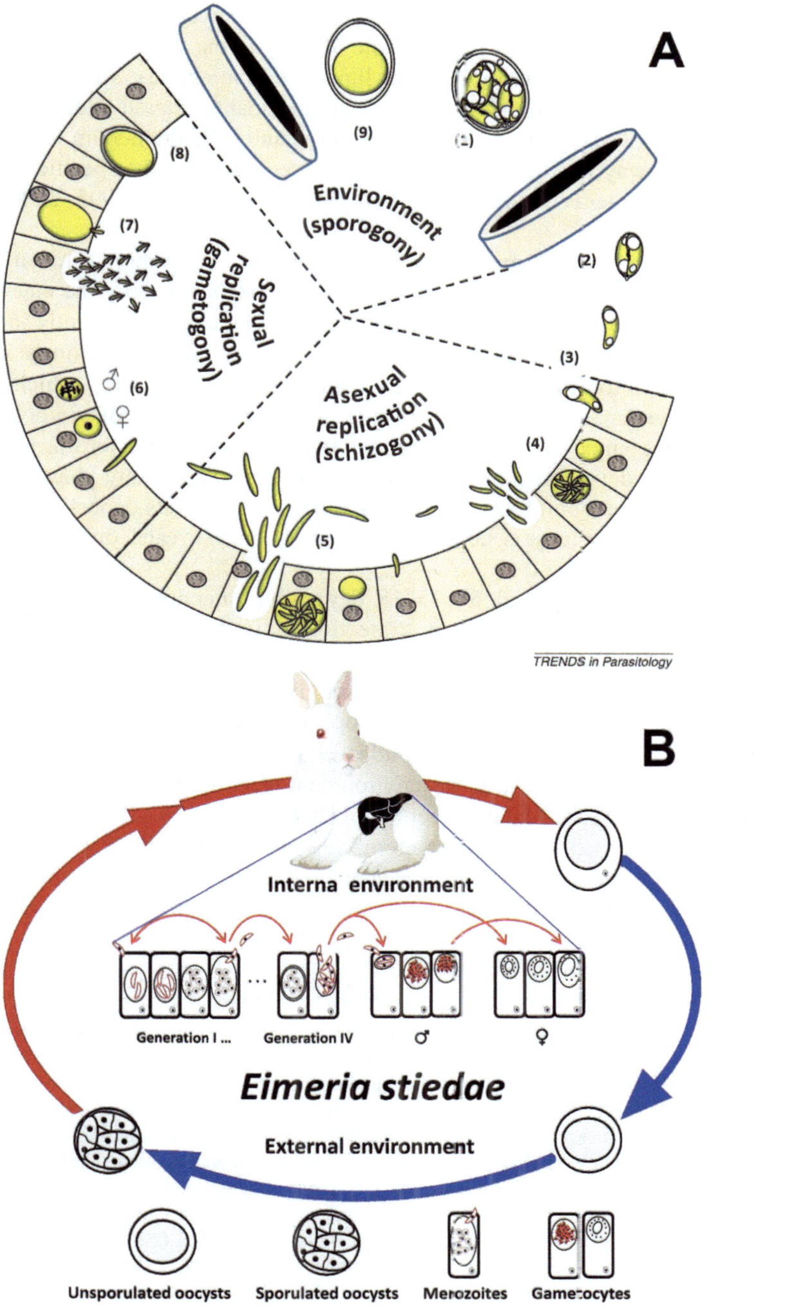

Fig. 1 Life cycle of *Eimeria*. (**a**) Major developmental stages of intestinal coccidiosis. (1) Ingestion of sporulated oocyst. (2) Excystation is activated by enzymatic reaction in the gut (stomach and proximal intestine) releasing the sporozoite. (3) The sporozoite penetrates into the enterocyte and

There are some peculiarities in rabbit coccidia life cycle: the most outstanding is the migration of sporozoites from the site of entry to the target site (Pakandl 2009). However, migration is apparently not limited to sporozoites, since merozoites of *E. flavescens* were also found to undergo an intestinal migration (Norton et al. 1979; Pakandl et al. 2003). The small intestine, especially the duodenum, seems to be a common gateway for rabbit coccidia to enter, regardless of the specific site of their further development (Drouet-Viard et al. 1994; Pakandl et al. 1995, 2006; Pakandl 2009).

Inside the cells, the asexual schizogony phase occurs forming schizonts that, by cell division, produce numerous merozoites. Merozoites destroy the enterocyte, or the epithelial cells of bile ducts, in case of *E. stiedae*, and subsequently infect other host cells. The gametogony begins when merozoites, resulting from schizogony, differentiate into macrogametocytes and microgametocytes, which develop macrogametes and microgametes, respectively. After rupture of the host cell, the microgametes enter the cells containing macrogametes, and fertilization occurs, leading to the development of unsporulated oocysts (sporogony phase), which are excreted with the rabbit feces. The sporulation of the oocysts occurs in the environment, and four sporocysts develop inside each oocyst, each one containing two sporozoites.

The cycle is completed by 15–18 days after infection (Taylor et al. 2016). Prepatent periods of 7 to 8 days postinfection (dpi) are described *for E. intestinalis* (5 days for *E. media* and *E. perforans*; Pakandl 2009; El-Shahawi et al. 2012) reaching maximum excretion of oocysts on 10 dpi, with approximately 5 million oocysts, and rapidly decrease from 11 dpi onward. Oocysts become infective at temperatures of 25–30 °C, in a humid, well-oxygenated environment. In adverse situations, they can persist in the environment (e.g., rabbit cage) for up to a year (Vadlejch et al. 2010; Abdel-Haleem et al. 2017).

The exception to the intestinal pattern life cycle includes the species *E. coecicola*. In fact, the life cycle of this species is characterized by the extraintestinal migration, probably through the lymphatic system, of the majority of the sporozoites to

---

Fig. 1 (continued) evolves to a trophozoite stage, developing a first merozoites' generation by schizogony (asexual multiple fission) (4), and ruptures the cell. (5) Merozoites invade new enterocytes for one, two, three, or more generations. (6) The final generation of merozoites (species-specific) differentiates into macrogametocytes and microgametocytes (gametogony). (7) Mature microgametes leave the enterocytes penetrating in neighboring cells, to fertilize mature macrogametes (zygotes). (8) Each zygote developed a resistant wall evolving to a new unsporulated (noninfectious) oocyst, which ruptures the enterocyte and is excreted through feces (9). The sporulation (sequential meiotic and mitotic nuclear divisions) occurs under specific temperature, oxygen, and moisture conditions. Adapted from Blake and Tomley (2014) with permission from Elsevier. (**b**) *Eimeria stiedae*. Sporozoites penetrate into epithelial cells and reach the liver and epithelial bile duct, where schizogony starts. Spindle-shaped merozoites replicate (new generations) in new epithelial cells until male and female gametocytes are differentiated. Free-liver gametogony occurs producing new unsporulated oocysts. The length of the cycle is approximately 7 days. Adapted from Xie et al. (2021)

mesenteric lymph nodes and spleen before invading different parts of the intestine and finally developing intracellularly in the cecum and appendix, the final target sites of *E. coecicola* (Bhat et al. 1996; Pakandl et al. 2006; Pakandl 2009). This extraintestinal migration appears to be crucial for the cycle to be completed (Pakandl et al. 1993; Renaux et al. 2001).

3 Epidemiology

Intestinal coccidiosis mainly affects animals between 6 weeks and 5 months of age and is generally attributed to management factors such as transport, weaning, noise, and other stressful situations, causing immunosuppression. Also, the severity of hepatic coccidiosis depends on the number of oocysts ingested and the age of the infected animal. Younger animals are more susceptible than older animals due to some degree of immunity from the latter's natural infection (Coudert et al. 1995; Taylor et al. 2016).

Several studies on the prevalence of *Eimeria* spp. infection in domestic rabbits have been reported. Studies conducted in Poland and Ukraine, using coprological analysis, revealed the presence of coccidian oocysts in all the sampled rabbit backyard farms (Basiaga et al. 2020). In Poland, four coccidian species were identified by means of sporulation and subsequent morphometry, namely, *E. media* (48.5%), *E. stiedae* (27.3%), *E. coecicola* (12.1%), and *E. flavescens* (12.1%) (Basiaga et al. 2020). In Ukraine, nine species were found to infect rabbits including *E. media* (25.1%), *E. coecicola* (14.2%), *E. magna* (13.1%), *E. piriformis* (11.2%), *E. stiedae* (10.3%), *E. irresidua* (9.6%), *E. exigua* (8.1%), *E. intestinalis* (6.3%), and *E. perforans* (2.1%) (Basiaga et al. 2020). In Iberian Peninsula, *E. magna*, *E. media*, and *E. perforans* are one of the most frequently detected in both domestic and wild rabbits (Silva et al. 2015).

An increase in prevalence of hepatic coccidiosis in the spring and summer was reported in wild rabbits (*Oryctolagus cuniculus*), probably as a result of a higher proportion of juvenile susceptible individuals (Stephens 1952). More recently, in the United Kingdom, a prevalence of 24% of *E. stiedae* was reported by Bochyńska et al. (2022) who also described a higher prevalence in juvenile (43%) wild rabbits than adults (17%). Additionally, females presented a higher prevalence (34%) than males (17%) (Bochyńska et al. 2022). In France, the prevalence varied between 4 and 21%, depending on the region (Grès et al. 2003).

In non-European countries, 11.5% of farmed rabbits in Kenya were infected with hepatic coccidiosis (Okumu et al. 2014). In Indonesia, prevalence of hepatic coccidiosis was 31% among farmed rabbits (Hamid et al. 2019) and 25.8% in samples from rabbits trapped in Western Australia (Hobbs and Twigg 1998). In China, the overall prevalence of infections varied between 41.9% and 87.4% (Jing et al. 2012; Chen et al. 2015). In Egyptian domestic rabbit farms, an overall prevalence, reaching 70%, was reported with the identification of *E. intestinalis* and *E. coecicola* as the most prevalent (El-Shahawi et al. 2012). The prevalence of intestinal coccidiosis in Taiwan was 41.9%, and ten species of *Eimeria* were identified, the most common being *E. perforans* (35.2%), *E. media* (31.3%), *E. magna* (28.8%), *E. irresidua* (19.4%), and *E. intestinalis* (1.8%) (Li et al. 2010). In Saudi Arabia, the level of infected home farm rabbits was 75% (Abdel-Baki and Al-Quraishy 2013). The most common species found were *E. coecicola* (70%), *E. magna* (60%), *E. perforans* (60%), and *E. media* (55%). *E. stiedae* (5%) and *E. exigua* (5%) were less prevalent (Abdel-Baki and Al-Quraishy 2013).

4 The Disease

4.1 Pathogenesis and Virulence

Although the severity of infection is dependent on the number of infective oocysts and their virulence, poor hygienic and sanitary situations of the farms and stressful situations implying a decrease in the immunity of the animals are considered relevant predisposing factors (Norton et al. 1979; Gregory and Catchpole 1986; Coudert et al. 1993; Licois et al. 1995). Oocyst infections are usually self-limiting. Rabbits can develop some degree of immunity against reinfection to the same *Eimeria* species (Coudert et al. 1995; Taylor et al. 2016).

According to Pakandl and Hlásková (2007), rabbits under 20 days of age are generally not susceptible to *Eimeria* infection, but there is an increase in susceptibility from 19 to 22 days of age, since in addition to milk, the animals generally start consuming vegetable feed, which leads to changes in the intestinal environment. Rose (1973), reviewed by Pakandl (2009), suggests that failure of sporozoite excystation and paraaminobenzoic acid deficiency in breast milk contribute to innate resistance to coccidia in very young mammals.

Concomitant infections due to abnormal multiplication of *Clostridium* spp. are also possible in cases of coccidiosis, similarly to what happen in other animal species. Enterotoxaemia, caused by *Clostridium* spp. and their toxins, is a consequence, or sometimes the cause, of *Eimeria* rising (Paul and Friend 2019; Espinosa et al. 2020).

According to Coudert et al. (1995), *Eimeria* species can be assigned to five distinct groups, taking into consideration their pathogenicity/virulence (Table 1): non-virulent (*E. coecicola*), slightly virulent (*E. perforans*, *E. exigua*, and *E. vejdovskyi*), mildly virulent or non-virulent (*E. media*, *E. magna*, *E. piriformis*, and *E. irresidua*), highly virulent (*E. intestinalis* and *E. flavescens*), and virulence depending on the infective dose (*E. stiedae*).

The pathogenicity/virulence of a specific species may be related to the localization of the coccidia. *Eimeria* spp. parasitize distinct parts of the intestine and at different depths of the mucosa (Table 1). Their developmental sites overlap in some cases, but despite this, it appears that each *Eimeria* species mostly inhabits different "niches" (Pakandl 2009).

The intestinal species considered to be among the most virulent are *E. intestinalis* and *E. flavescens*, which infect the crypts of posterior portion of the small intestine or caecum, respectively (Coudert et al. 1995; Pakandl et al. 2008; Vadlejch et al. 2010).

Table 1 Virulence and localization of the main rabbit *Eimeria* species. Reviewed by Coudert et al. (1995) and Pakandl (2009)

Species	Virulence	Localization
E. coecicola	Non-virulent	Appendix, *sacculus rotundus*, Peyer's patches (epithelium of domes and mushrooms)
E. exigua	Slightly virulent	Duodenum-ileum; successively moves from proximal to distal parts of the small intestine (tops of the villi)
E. flavescens	Highly virulent	Small intestine and caecum (crypts and superficial epithelium)
E. intestinalis	Highly virulent	Lower jejunum and ileum [crypts and villi (wall)]
E. irresidua	Mildly virulent or non-virulent	Jejunum and ileum (lamina propria and villous epithelium)
E. magna	Mildly virulent or non-virulent	Jejunum and ileum, in a lesser extent duodenum [intestinal villi (walls and tops)]
E. media	Mildly virulent or non-virulent	Duodenum-jejunum, low concentration of the parasite in the ileum [intestinal villi (walls and tops)]
E. perforans	Slightly virulent	Maximal parasite concentration in the duodenum, but also in the jejunum and ileum (villi and crypts)
E. piriformis	Mildly virulent or non-virulent	Colon (crypts)
E. vejdovskyi	Slightly virulent	Ileum (crypts and villi)
E. stiedae	Virulence depending on the infective dose	Liver (epithelium of biliary ducts)

The damage to the intestinal epithelium is considered most severe when the stem cells, located in the crypts, are infected and destroyed during life cycle stages, mainly during the final stages of schizogony and gametogony (Norton et al. 1979; Gregory and Catchpole 1986; Coudert et al. 1995; Pakandl et al. 2003). Coccidial enteritis is often associated with increased numbers of *Escherichia coli* and rotaviruses in the host gut, which may be an important interaction in animals raised outdoors (Peeters et al. 1984).

Data from Al-Quraishy et al. (2012) study indicate that intestinal infections with *E. coecicola* cause inflammatory and specific immune reactions in the liver, although the liver is not a target site of *E. coecicola*. According to these authors, the liver immune reactions promote the host immunity toward the intestine stages of the parasite, contributing to the resolution of *E. coecicola* intestinal infection.

Hepatic coccidiosis is also a major concern. *E. stiedae* may be a very pathogenic species of domestic rabbits that parasitizes the epithelial cells of the bile duct, causing severe liver damage and great economic losses (Hung et al. 1984). According to Çam et al. (2008), *E. stiedae* causes lipid peroxidation resulting from the destruction of the bile duct and subsequently the hepatic parenchyma. The peroxidation of cell membrane causes impairment in its permeability and triggers a series of reactions that may result in cell death (Comporti 1993).

Data from different studies showed a significant increase in the percentages of apoptotic cells (Abdel-Haleem et al. 2017). Green (2000) postulated that the parasite invasion and replication might cause considerable stress to the host cell and that this stress, in turn, may trigger apoptosis. The intracellular stages of the coccidian parasites are known to extensively modify their host cells and to induce host cell apoptosis in order to maximize their probability of survival (Lüder et al. 2001, 2009). This may be due to the fact that schizogony of species of *Eimeria* requires a stable intracellular environment to obtain essential nutrients and evade host immune attack (Abdel-Haleem et al. 2017).

4.2 Lesions

Histopathological findings are sometimes not correlated with pathogenicity/virulence criteria including mortality, weight gain, and clinical signs (Pakandl 2009). An example is *E. coecicola* which leads to severe lesions in the appendix but in relation to the general health of the animal is classified as being non-virulent (Vítovec and Pakandl 1989).

Studies on the pathomorphological changes caused by *E. intestinalis* revealed a severe atrophy of the intestinal villi between 7 and 10 dpi, whereby by the 9 dpi most of the villi and the entire mucosa were severely swollen, the epithelial cells were mostly detached, and the microvillus became irregular, thickened, and atrophied. However, as soon as parasite development was almost finished, the mucosa regained an outwardly normal morphological pattern (Peeters et al. 1984).

Similar results were observed by Gregory and Catchpole (1986) who characterized pathological changes in heavy infections with *E. flavescens* as invasion of the cecal epithelium with parasitic phases, loss of the superficial epithelium, a normal process in cell turnover, and inability of the infected crypts to replace the lost superficial epithelium, resulting in a generalized denudation of the cecal mucosa and severe inflammation. Nevertheless, in the final stages, reparative changes were observed. According to Gregory and Catchpole (1986), the death of the animals was apparently due to high dehydration and tissue invasion by bacteria.

Infections with *E. stiedae* may lead to severe lesions in the affected rabbits. A characteristic lesion of *E. stiedae* infection is hepatomegaly, ectasia of the bile ducts resulting from extensive biliary hyperplasia, with large numbers of intraepithelial gametocytes and oocysts, as well as fibrosis and mononuclear cell infiltrations (Barriga and Arnoni 1979; Singla et al. 2000; Abu-Akkada et al. 2010; Bradley et al. 2021; Bochyńska et al. 2022). Lesions consist of papillary hyperplasia, characteristic papillary luminal projections, necrosis, and desquamation of the biliary epithelium, probably due to liberated toxins or mechanical irritation induced by the parasite (Sastry 2000).

The livers may appear pale in color and studded with whitish nodules 0.2–0.5 cm in diameter (Fig. 2a), even in wild rabbits (Bochyńska et al. 2022). Upon sectioning, the nodules exude a thick, creamy, yellowish-white liquid. The gallbladder is distended with yellowish-white creamy fluid (Barriga and Arnoni 1979; Abu-Akkada et al. 2010). The microscopic examination of the gallbladder contents may lead to identification of *E. stiedae* oocysts (Fig. 2B).

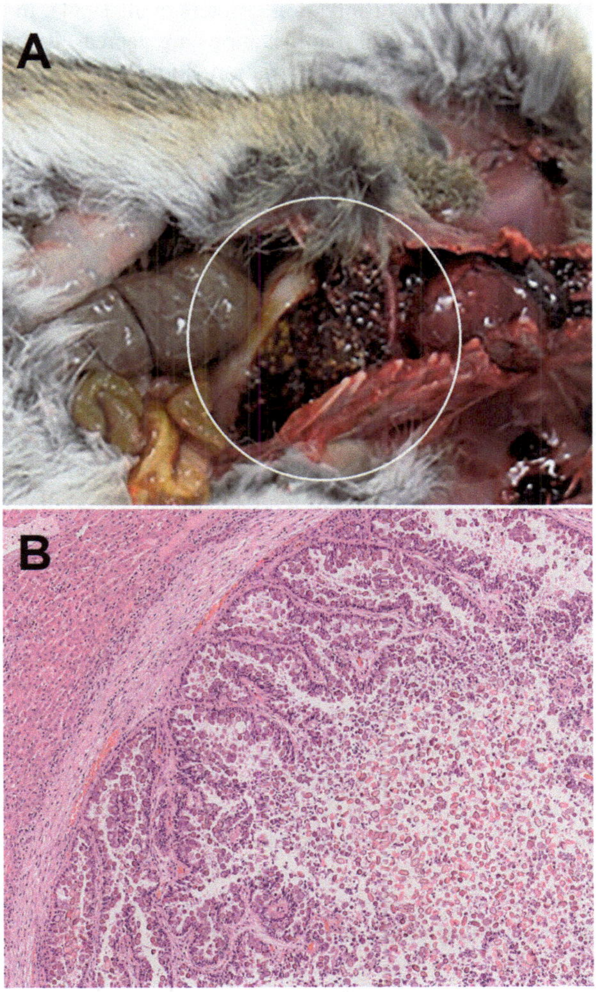

Fig. 2 Liver and gallbladder infection by *Eimeria stiedae*. (**A**) Macroscopic view of liver surface with pale, miliary areas corresponding to dilated bile ducts. (**b**) Microscopic view of parasitic cholangitis due to *Eimeria stiedae*. Hypertrophy of the bile ducts, with papillary proliferation and presence of parasitic forms (H&E, 200×). Photographs gently provided by Isabel Pires (University of Trás-os Montes e Alto Douro - UTAD)

4.3 Clinical Signs

Coccidiosis in rabbits has great economic impact and often leads to huge losses in industrial rabbit farms. Even subclinical forms of intestinal and hepatic coccidiosis can lead to a decrease in animal body weight as a result of digestion disturbances, intestinal and hepatic dysfunction, malabsorption of nutrients, dehydration, diarrhea, cachexia, secondary bacterial and viral infections, high mortality of rabbit offspring, and increased costs in the fattening period and treatment of sick animals (Bhat et al. 1996; Pakandl 2009; Al-Quraishy et al. 2012; Metwaly et al. 2013).

The severity of intestinal coccidiosis depends mainly on the infecting dose, the *Eimeria* species that infect the animal, as well as the immune status and age of the rabbits at the time of infection (Peeters et al. 1984). In the intestinalcoccidiosis, the most common clinical manifestations are decreased appetite (hyporexia), leading to reduced fecal excretion, dehydration, body weight loss; poor feed conversion, due to disturbance of food absorbance and digestion; lethargy and profuse diarrhea (Peeters et al. 1984; Vadlejch et al. 2010; Metwaly et al. 2013; Abdel-Haleem et al. 2017). Metwaly et al. (2013) attributed the significant reduction in rabbits' body weight during the course of intestinal coccidiosis to deterioration in the food conversion and factors including depletion of growth hormones, distributed metabolic status, and electrolyte homeostasis.

In the acute clinical form, affected rabbits, usually of 4 to 5 weeks of age, develop diarrhea, often with mucoid bloody feces due to intestinal lesions, dilated abdomen due to accumulation of gas and intestinal fluids, and emaciation leading to the rapid death of a large number of animals. Mortality is high in rabbits between 2 and 4 months of age (Peeters et al. 1984; Jithendran and Bhat 1996; Pakandl 2009; Abdel-Haleem et al. 2017), especially if the disease has been caused by one of the more virulent species (Vadlejch et al. 2010). Ion metabolism is affected and fecal losses of potassium lead to hypokalemia in blood plasma (Peeters et al. 1984).

Hepatic coccidiosis is potentially fatal and occurs mostly in young, weanling rabbits, resulting in poor weight gain. In the hepatic form, the main clinical sign is delayed growth, with consequent repercussions on the productive indices. Besides progressive weight loss, hepatomegaly and eventual death due to secondary infections can be observed. However, mortality is low. Healthy rabbits may be asymptomatic carriers (Turner et al. 2018). Also, in pet rabbits, *E. stiedae* has been documented to cause clinical disease, and in severe cases rabbits may develop diarrhea, anorexia, and lethargy (Hrženjak et al. 2021).

When there is a large load of oocysts, an increase in concomitant enteric disease, such as *E. coli*, enterotoxemia, or epizootic rabbit enteropathy, as well as a reduction in carcass yield at slaughterhouse, can be expected.

5 Diagnosis and Prognosis

Due to the stated differences in the individual species pathogenicity/virulence, an appropriate diagnosis of coccidiosis in rabbits is crucial and should cover not only the results of coprological methods but also the accurate species identification. The diagnosis can be made by examination of gastrointestinal tissues or microscopic fecal examination for the detection of *Eimeria* spp. stages, including oocysts, in case of intestinal and hepatic coccidiosis. The microscopic examination of the gallbladder contents may lead to identification of *E. stiedae* oocysts (Fig. 2b) (Hobbs and Twigg 1998; Grès et al. 2003).

Identification of *Eimeria* species is based on the observation of the morphology and the measurements, length, and breadth of the sporulated oocysts (morphometric parameters). However, there are some species that cannot be readily identified; for example, it is not possible to distinguish *E. perforans* from *E. media* and *E. flavescens* from *E. irresidua* only with the form. Then, the differentiation of these species can be difficult and time-consuming if the life cycle (site of development in the intestine, prepatent period) or other criteria (clinical signs) must be studied (Cere et al. 1996). New methods arise for direct and rapid identification of these parasites. The random amplified polymorphic DNA (RAPD) method allows the identification of coccidian parasites, demonstrating the efficiency of this technique in differentiating coccidia species (MacPherson and Gajadhar 1993; Cere et al. 1995). However, this method cannot be used when the species are mixed or in very small quantities in feces. In this case, amplification of specific DNA sequences by PCR and hybridization of these fragments with oligonucleotide probes provide a highly sensitive and specific tool for the detection of coccidia directly from feces (Cere et al. 1996).

At necropsy, in the intestinal form, inflammation and edema of the intestine with thickening wall, hemorrhage, and occasional ulceration of the mucosa are observed. In the hepatic form, the damaged parenchyma is replaced by fibrous tissue, and the animals present small yellowish-whitish nodules scattered throughout the hepatic parenchyma and hepatomegaly (Bochyńska et al. 2022).

E. stiedae induce adverse effects on the liver function. Regarding biochemistry parameters, serum aspartate transaminase (AST), alanine transaminase (ALT) and gamma-glutamyl transferase (GGT), globulin, and total protein levels increase (Joyner et al. 1983; San Martín-Núñez et al. 1988; Athanasiou et al. 2023), while albumin levels decrease (Gomez-Bautista et al. 1986). According to Çam et al. (2008), ultrasonography could be useful for the diagnosis of hepatic coccidiosis.

In rabbits, a macroscopically white-spotted liver can be indicative of one of several disease processes that cause similar gross hepatic changes that also occur in hepatic coccidiosis, due to *E. stiedae* (Hrženjak et al. 2021). Other parasitic causes of white-spotted livers in lagomorphs include *Cysticercus pisiformis* (Andrews and Davidson 1980) and *Calodium hepaticum* (Barlow and Mullineaux 2018). Bacterial agents that can cause hepatic abscesses include *Yersinia pseudotuberculosis* (Turner et al. 2018) and *Francisella tularensis* (Kim et al. 2010). However, the organism that is most typically associated with white-spotted livers in rabbits is *E. stiedae* (Turner et al. 2018).

The prognosis depends on the species of *Eimeria* involved in the infection, as well as the immune status of the animal and its age at the time of infection. According to Pakandl (2009), the most reliable criterion of the health status of the animals, which also allows estimation of the intensity of infection in growing rabbits, is weight gain.

6 Treatment and Prevention

Coccidiosis is controlled primarily through prophylaxis (Pakandl 2009). The disease affects mostly young rabbits right after weaning (5- to 6-week-old animals), so prevention should be started very early, between 20 and 30 days of age (Drouet-Viard et al. 1997).

Table 2 Anticoccidial drugs used in rabbit coccidiosis. Adapted from the review by Pakandl (2009)

Drugs	
Effective	Less effective or noneffective
Decoquinate	Amprolium
Diclazuril	Amprolium/ethopabate
Formaldehyde-sulphathiazol[a]	Buquinolate
Metichlorpindol/methylbenzoquate (Lerbek®)	Clopidol
Monensin	Furazolidone
Narasin	Lasalocid
Robenidine[b]	Methylbenzoquate
Salinomycin	Metichlorpindol
Sulphadimethoxine[a]	Monensin[b]
Sulphadimidine/diaveridine[a]	Nicarbazine
Sulphaquinoxaline[a]	Nitrofurazone/furazolidone
Sulphaquinoxaline[a]/pyrimethamine[b]	Robenidine[b] (hepatic coccidiosis)
Toltrazuril	Sulphaquinoxaline[a]/pyrimethamine[b]

[a]Sulfonamides
[b]Contradictory results have been reported regarding the efficacy of these drugs for coccidiosis

The control of coccidiosis depends on a number of abiotic and biotic factors. Housing in cages with effective feces disposal systems, as is usually the case in industrial rabbit farming, decreases the possibility of contamination by infective oocysts. The stocking density per cage, as well as the environmental and hygienic conditions of the pens, should be controlled. The hygiene of feeders and drinkers should be regular and the diet balanced. Good hygiene conditions, including the removal of oocysts from the premises before they finish sporulation, could in theory prevent coccidiosis; however, in practice it is impossible to remove all oocysts. However, a significant reduction in their number in the environment can reduce infective doses and, consequently, clinical signs of disease. Furthermore, daily infection with small doses is the best way to achieve a significant degree of immunity. However, rabbit producers cannot rely on good hygiene alone, and prevention, using coccidiostats, is currently used (Pakandl 2009).

Several drugs, including salinomycin, robenidine, and diclazuril, are used in rabbits; nonetheless, countries' legislations must be observed (Pakandl 2009). In the European Union, the use of coccidiostats, for the prevention and treatment of animals, is regulated by competent authorities (Regulation (EC) No. 1831/2003). Modern coccidiostats must be highly effective, be easy to administer, have a positive therapeutic index, and have a broad spectrum of action and a short period of use.

Anticoccidials are usually mixed in feeding pellets, but if application of some drugs, such as sulfonamides or toltrazuril, in drinking water is possible, it may be advantageous for small rabbitries (Pakandl 2009).

A list of anticoccidial drugs is presented in Table 2. Sulfonamides are generally used primarily for treatment of coccidiosis outbreak in some worldwide regions (Pakandl 2009).

Several studies highlight the beneficial effects of coccidiostats used in the rabbit production (Balicka-Ramisz et al. 2014). Economic performance, including weight gain and feed intake, improves considerable as well as parasitological parameters with a decrease in oocysts per gram of feces (OPG) and identification of species (Vereecken et al. 2012). However, in order to avoid and mitigate the emergence of drug resistance by coccidia, the implementation of strategies including programs of rotation is important (Vereecken et al. 2012).

Toltrazuril, used for the treatment of coccidiosis in rabbits, is effective on both schizont and gamont stages of *Eimeria*, including *E. stiedae*. It may be used at a dose of 25 ppm in drinking water for 2 days and repeated after 5 days if necessary (Peeters and Geeroms 1986; Lindansay and Blagburn 1995; Singla et al. 2000; Balicka-Ramisz et al. 2014).

Oral administration of toltrazuril or amprolium, or their use in association, ameliorates the adverse effects of intestinal coccidiosis in rabbits by preventing the onset of clinical signs, significantly reducing OPG and effectively controlling disease-related mortality. Additionally, there has been an improvement in feed intake, weight gain, and feed conversion ratio (El-Ghoneimy and El-Shahawy 2017).

In rabbitries, transmission mainly takes place between the third and fourth week of life. Anticoccidials are mainly effective in the immature stages of *Eimeria* spp. So, the best time to start treatment is when the rabbits begin to drink a considerable amount of water, what happens from the 26th day of life on.

Even with the rotation of coccidiostats, a regular surveillance of *Eimeria* infection should be kept, with counting and identification of *Eimeria* species. As soon as the number of oocysts begins to increase above a certain value (depending on the species of *Eimeria*), rotation of the coccidiostat used in the feed must be done. In some severe cases, combination with an anticoccidial drug in water may be necessary.

Arafa and Wanas (1996) reported that ivermectin decreases the severity of coccidiosis and reduces values of OPG in rabbits with intestinal coccidiosis, being effective for all developmental stages of *Eimeria*, but especially for the asexual stage. Still, contradictory results were observed when ivermectin was used for the treatment of hepatic coccidiosis. According to Çam et al. (2008), ivermectin at dose of 1 mg/kg body weight, administrated subcutaneously, might not be effective for the treatment of hepatic coccidiosis, in contrast to intestinal disease, when used alone, and especially after late schizogony and gametogony. However, toltrazuril was effective for the treatment of hepatic coccidiosis when used at a dose of 50 ppm in the drinking water (Çam et al. 2008).

For a long time, conventional control of rabbit coccidiosis was based mainly on continuous administration of anticoccidial drugs. However, its use has several disadvantages besides being expensive, such as possible toxicity to the hosts, negative impact on the environment due to the use of feces with residues as fertilizer, and the absorption of the drugs into the tissues of the host, including meat, having a negative impact on this product (Pakandl 2009). Additionally, the routinely continuous use and misuse of these drugs resulted in the emergence of drug resistant *Eimeria* strains (Peeters et al. 1988; Coudert 1989; Mortensen et al. 2003). In order to overcome this inconvenience, other methods to prevent coccidiosis include unconventional strategies such as the use of herbal supplements throughout the rearing period as alternatives or adjuncts to current antiparasitic drugs (Anthony et al. 2005; Kowalska et al. 2012; Nosal et al. 2014).

Garlic (*Allium sativum* L.) is reported to be a parasiticide, amebicide, acarifuge, vermifuge, and larvicide (Duke 2002) and has immunomodulatory activity (Clement et al. 2010). Several studies have demonstrated the efficacy of garlic against coccidiosis in rabbits (Yassin 2005; Toulah and Al-Rawi 2007; Worku et al. 2009). As an antioxidant, garlic inhibits lipid peroxidation, leading to a hepatoprotective effect (Khanum et al. 1998; Banerjee et al. 2003), improving liver fibrosis and oxidative damage induced by biliary obstruction (Toulah and Al-Rawi 2007).

Oral administration of crude garlic at a daily dose of 0.5 g/kg body weight for 5 consecutive days before challenge with *E. stiedae* ameliorated the adverse impacts of hepatic coccidiosis on rabbits when used as a prophylactic, revealing a valuable effect on reducing liver damage and the number of OPG, but was less effective as a treatment, since garlic was unable to reduce the OPG when administered after the beginning of oocyst shedding (Abu-Akkada et al. 2010).

One report on the use of Herba Cox®, a commercial herbal formulation containing extracts from *Bombax malabaricum*, *Aegle marmelos*, *Anethum foeniculum*, *Resina salvia*, *Ferula asafetida*, and *Papaver somniferum*, indicates that the product is safe and highly effective for the treatment of hepatic coccidiosis when given at a dose of 1 mL per liter of drinking water for 7 days before infection and continued for 28 days postinfection. The product improved growth performance and liver function, enhanced immunity, and exerted protective effects against oxidative stress in hepatic coccidiosis, reducing the clinical, parasitological, and production-related effects of hepatic coccidiosis when given continuously before and during infection (Eladl et al. 2020).

The use of herbal oils is also documented. According to the study by Abu El Ezz et al. (2020), both *Moringa oleifera* and *Thymus vulgaris* oils had an anticoccidial effect in rabbits experimentally infected with *E. stiedae*. Nonetheless, the highest protective activity was observed in rabbits treated with *Thymus vulgaris* oil. These authors suggest that *Thymus vulgaris* oil may be useful as an alternative product for the control of rabbit coccidiosis.

The administration of prebiotics as a preventive measure is also reported as diminishing adverse effects caused by *Eimeria* spp. through decreasing fecal oocyst counts, retaining body weight, and reducing the number of parasitic stages in the intestinal tissues (El-Ashram et al. 2019).

The development of effective methods, to be an alternative to continuous medication with anticoccidial drugs, for rabbit coccidiosis control, is a challenge for future research work (Pakandl 2009). Vaccination with live attenuated lines of coccidia, with significantly reduced pathogenicity (precocious lines), seems to be an alternative, since most coccidia species are highly immunogenic (Coudert 1994; Licois et al. 1994).

Vaccination trials performed with precocious line of *E. magna* or *E. media* give satisfactory results in terms of acquired protection leading to the absence of mortality and pathology, a decrease in oocyst output after the challenge inoculation, and a correct growth, although it did not give full protection (Drouet-Viard et al. 1997; Akpo et al. 2012; Bachene et al. 2018). More recently, potential recombinant vaccine candidates, such as membrane antigen 1 and immune mapped protein 1, against *E. stiedae*, are promising to develop cellular and humoral immunity (Xiao et al. 2022).

Because *Eimeria* species multiplies in intestinal cells, changes in the gut microbiota can occur; however, these changes are still largely unknown. An interesting study by Li et al. (2021) intended to evaluate the influence of *Eimeria* infection on this subject. For that purpose, SPF rabbits were infected with oocysts of a highly pathogenic species *E. intestinalis* wild type (WT) and a precocious line (EIP8). Also, vaccination with low doses of EIP8 oocysts was also performed in newly weaned rabbits, and a diclazuril control group was also included. The main results were as follows: (a) in EIP8-infected rabbits, the intestinal flora that changed postinfection accounted for less than 5.23% of the entire flora. In contrast, there was a change of 27.18% in the WT group on 14 dpi, and more than 20% in diclazuril control group on the 7 or 10 dpi; (b) vaccination had no significant impact on rabbit microbiota of newly weaned rabbits and resulted in sufficient protection against WT infection, since the body weight gain was similar in vaccinated and untreated animals. Additionally, a more than 80% reduction of oocyst excretion compared to unimmunized and challenged animals was observed; (c) treatment with diclazuril disrupted the balance of the microbiota, which may lead to increased susceptibility to some diseases and promote the dissemination of resistance genes. These same authors suggested that *E. intestinalis* precocious line could benefit rabbit mucosal immunity development, inducing a new fecal microbiota biodiversity in rabbits, and would be the most suitable component of live vaccine in the strategy for the control of rabbit coccidiosis.

The biggest problem related to the use of anticoccidial vaccines for rabbits is their cost for commercial laboratories and their lack of interest for this animal species, since rabbits are considered a minor specie in animal production.

7 Final Remarks and Conclusions

Coccidiosis is one of the most important parasitic diseases affecting the health of rabbits and one of the most economically important problems in the rabbit management. Subclinical form of coccidiosis can originate significant economic losses mainly due to low performances and cause susceptibility to other diseases.

Both intestinal and hepatic coccidiosis are transmitted by the fecal-oral route. The severity of hepatic coccidiosis depends on the number of oocysts ingested and the age of the infected animal. Moreover, the virulence of a specific species may be related to the localization of the coccidia. Nonetheless, *E. intestinalis* and *E. flavescens* are considered highly virulent, and *E. stiedae* virulence depends mainly on the infective dose.

An appropriate diagnosis of coccidiosis in rabbits is crucial. Weight gain is the most reliable criterion for prognosis. When coccidia acquire resistance to the coccidiostat, some bacteria, such as *E. coli* and *Clostridium* spp., can grow more likely, which are able to gain pathogenic capacity. The consequences are the increasing of diarrhea problems and mortality, mainly in fattening rabbits.

Coccidiosis is controlled primarily through prophylaxis. It is important to mitigate the emergence of drug resistance by *Eimeria* strains. Other methods to prevent rabbit coccidiosis, including unconventional strategies, should be investigated.

Acknowledgments The authors thank Springer for the use of the AJE Digital Editing. This work was supported by the projects UIDP/CVT/00772/2020 (CECAV) and LA/P/0059/2020 (AL4AnimalS), funded by the Portuguese Foundation for Science and Technology (FCT).

References

Abdel-Baki AA, Al-Quraishy S (2013) Prevalence of Coccidia (*Eimeria* spp.) infection in domestic rabbits, *Oryctolagus cuniculus* in Riyadh, Saudi Arabia. Pak J Zool 45(5):1329–1333

Abdel-Haleem HM, Aboelhadid SM, Sakran T, El-Shahawy G, El-Fayoumi H, Al-Quraishy S, Abdel-Baki AAS (2017) Gene expression, oxidative stress and apoptotic changes in rabbit ileum experimentally infected with *Eimeria intestinalis*. *Folia Parasitol* 64:012. https://doi.org/10.14411/fp.2017.012

Abu El Ezz NM, Aboelsoued D, Hassan SE, Abdel-Megeed KN, El-Metenawy TM (2020) Therapeutic effect of *Moringa oleifera* and *Thymus vulgaris* oils against hepatic coccidiosis in experimentally infected rabbits. *Trop Biomed* 37(4):1018–1028. https://doi.org/10.47665/tb.37.4.1018

Abu-Akkada SS, Oda SS, Ashmawy KI (2010) Garlic and hepatic coccidiosis: prophylaxis or treatment? Trop Anim Health Prod 42(7):1337–1343. https://doi.org/10.1007/s11250-010-9590-6

Akpo Y, Kpodékona MT, Djago Y, Licois D, Youssao IAK (2012) Vaccination of rabbits against coccidiosis using precocious lines of *Eimeria magna* and *Eimeria media* in Benin. Vet Parasitol 184(1):73–76. https://doi.org/10.1016/j.vetpar.2011.08.012

Al-Quraishy S, Metwaly MS, Dkhil MA, Abdel-Baki AA, Wunderlich F (2012) Liver response of rabbits to *Eimeria coecicola* infections. Parasitol Res 110(2):901–911. https://doi.org/10.1007/s00436-011-2574-2

Andrews CL, Davidson WR (1980) Endoparasites of selected populations of cottontail rabbits (*Sylvilagus floridanus*) in the southeastern United States. J Wildl Dis 16(3):395–401. https://doi.org/10.7589/0090-3558-16.3.395

Anthony JP, Fyfe L, Smith H (2005) Plant active components - a resource for antiparasitic agents? Trends Parasitol 21(10):462–468. https://doi.org/10.1016/j.pt.2005.08.004

Arafa MA, Wanas MQ (1996) The efficacy of ivermectin in treating rabbits experimentally infected with *Eimeria* as indicated parasitologically and histologically. J Egypt Soc Parasitol 26(3):373–380

Athanasiou LV, Tsokana CN, Doukas D, Kantere MC, Katsoulos PD, Papakonstantinou GI, Katsogiannou EG, Dedousi A (2023) Hepatic coccidiosis in wild rabbits in Greece: parasite detection on liver imprints and the associated biochemical profile. Vet Sci 10:248. https://doi.org/10.3390/vetsci10040248

Bachene MS, Temim S, Ainbaziz H, Bachene A, Suo X (2018) A vaccination trial with a precocious line of *Eimeria magna* in Algerian local rabbits *Oryctolagus cuniculus*. Vet Parasitol 261:73–76. https://doi.org/10.1016/j.vetpar.2018.08.013

Baker DG (2007) Flynn's parasitology of laboratory animals, 2nd edn. Blackwell, Ames, p 840

Balicka-Ramisz A, Wróbel M, Adadyńska K (2014) Epidemiology and economic benefits of treating rabbit's coccidiosis in small farms from West Pomerania province, Poland. Ann Parasitol 60(4):247–251

Banerjee SK, Mukherjee PK, Maulik SK (2003) Garlic as an antioxidant: the good, the bad and the ugly. Phytother Res 17(2):97–106. https://doi.org/10.1016/j.vetpar.2018.08.01310.1002/ptr.1281

Barlow A, Mullineaux L (2018) Zoonotic disease in a European brown hare in England. Vet Rec 183(5):163. https://doi.org/10.1136/vr.k3316

Barriga OO, Arnoni JU (1979) *E. stiedae* weight, oocyst output and hepatic function of rabbits with graded infection. Exp Parasitol 48(3):407–414. https://doi.org/10.1016/0014-4894(79)90125-5

Barthold SW, Griffey SM, Percy DH (2016) Pathology of laboratory rodents and rabbits, 4th edn. Wiley Blackwell, pp 297–300. https://doi.org/10.1002/9781118924051

Basiaga M, Levytska V, Kowal J, Nosal P, Wyrobisz-Papiewska A (2020) Coccidiosis – a problem in backyard rabbitries. Ann Parasitol 66(1):97–99. https://doi.org/10.17420/ap6601.242

Bhat TK, Jithendran KP, Kurade NP (1996) Rabbit coccidiosis and its control: a review. World Rabbit Sci 4(1):37–41. https://doi.org/10.4995/wrs.1996.269

Blake DP, Tomley FM (2014) Securing poultry production from the ever-present Eimeria challenge. Trends Parasitol 30(1):12–19. https://doi.org/10.1016/j.pt.2013.10.003

Bochyńska D, Lloyd S, Restif O, Hughes K (2022) *Eimeria stiedae* causes most of the white-spotted liver lesions in wild European rabbits in Cambridgeshire, United Kingdom. J Vet Diagn Investig 34(2):199–205. https://doi.org/10.1177/10406387211066923

Bradley AE, Wancket LM, Rinke M, Gruebbel MM, Saladino BH, Schafer K, Katsuta O, Garcia B, Chanut F, Hughes K, Nelson K, Himmel L, McInnes E, Schucker A, Uchidsa K (2021) International harmonization of nomenclature and diagnostic criteria (INHAND): nonproliferative and proliferative lesions of the rabbit. J Toxicol Pathol 34(3 Suppl):183S–292S. https://doi.org/10.1293/tox.34.183S

Çam Y, Alasever A, Eraslan G, Kibar M, Atalay O, Beyaz L, Inci A, Liman BC (2008) *Eimeria stiedae*: experimental infection in rabbits and the effect of treatment with toltrazuril and ivermectin. Exp Parasitol 119(1):164–172. https://doi.org/10.1016/j.exppara.2008.01.005

Cere N, Licois D, Humbert JF (1995) Study of the inter and intra-specific variation of *Eimeria* spp. from the rabbit using the random amplified polymorphic DNA. Parasitol Res 81(4):324–328. https://doi.org/10.1007/BF00931539

Cere N, Humbert JF, Licois D, Corvione M, Afanassieff M, Chanteloup N (1996) A new approach for the identification and the diagnosis of *Eimeria media* parasite of the rabbit. Exp Parasitol 82(2):132–138. https://doi.org/10.1006/expr.1996.0017

Chen JM, Zhang S, Wang X, Lin R, Chen LH, Chen ZL, Wen FX (2015) Epidemiological investigation and species identification of *Eimeria* of a rex rabbit farm in Fujian province. China Anim Husbandry Vet Med 42(8):2204–2209. https://doi.org/10.16431/j.cnki.1671-7236.2015.08.040

Clement F, Pramod SN, Venkatesh YP (2010) Identity of the immunomodulatory proteins from garlic (*Allium sativum*) with the major garlic lectins or agglutinins. Int Immunopharmacol 10(3):316–324. https://doi.org/10.1016/j.intimp.2009.12.002

Comporti M (1993) Lipid peroxidation, biopathological significance. Mol Asp Med 14(3):199–207. https://doi.org/10.1016/0098-2997(93)90006-y

Coudert P (1989) Some peculiarities of rabbit coccidiosis. In: Proceedings of the 5th international coccidiosis conference on Coccidia and intestinal coccidiomorphs, Tours, France, October 17–20, pp 481–488

Coudert P (1994) Potentials of precocious *Emeria* strains for vaccination of rabbits against coccidiosis. Cahiers Options Mediterraneennes 8:501–504

Coudert P, Lic D, Provôt F, Drouet-Viard F (1993) *Eimeria* sp. from rabbit (*Oryctolagus cuniculus*): pathogenicity and immunogenicity of *Eimeria intestinalis*. Parasitol Res 79(3):186–190. https://doi.org/10.1007/BF00931890

Coudert P, Lic D, Drouet-Viard F (1995) *Eimeria* species and strains of the rabbits. In: Eckert J, Braun R, Shirley MW, Coudert P (eds) Guidelines on techniques in coccidiosis research. European Commission, Directorate-General XII, Science, Research and Development Environment Research Programme, pp 52–73

Cowie-Whitney J (1977) Diseases of the commercial rabbit. Vet Rec 101(15):299–330. https://doi.org/10.1136/vr.101.15.299

Cui P, Liu H, Fang S, Gu X, Wang P, Liu C, Tao G, Liu X, Su X (2017) A new species of *Eimeria* (Apicomplexa: Eimeriidae) from Californian rabbits in Hebei Province, China. Parasitol Int 66(5):677–680. https://doi.org/10.1016/j.parint.2017.06.009

De Vos AJ (1970) Coccidiosis of rabbits at Ondersterpoort. J S Afr Vet Assoc 41(3):189–194

Drouet-Viard FD, Provôt F, Coudert P (1994) The invasion of rabbit intestinal tract by *Eimeria intestinalis* sporozoites. Parasitol Res 80(8):706–707. https://doi.org/10.1007/BF00932959

Drouet-Viard F, Coudert P, Licois D, Boivin M (1997) Acquired protection of the rabbit (*Oryctolagus cuniculus*) against coccidiosis using a precocious line of *Eimeria magna*: effect of vaccine dose and age at vaccination. Vet Parasitol 69(3–4):197–201. https://doi.org/10.1016/s0304-4017(96)01133-8

Duke JA (2002) Handbook of medicinal herbs, 2nd edn. CRC Press, Boca Raton, FL, p 896

Eckert J, Taylor M, Catchpole J, Licois D, Coudert P, Bucklar H (1995) Morphological characteristics of oocyts. In: Eckert J, Braun R, Shirley MW, Coudert P (eds) Guidelines on techniques in coccidiosis research. European Commission, Directorate-General XII, Science, Research and Development Environment Research Programme, pp 103–119

Eladl AH, Mahgoub HA, El-Shafei RA, Al-Kappany YM (2020) Comparative effects of Herba Cox®, a commercial herbal extract, on rabbits (*Oryctolagus cuniculus*) experimentally infected with *Eimeria stiedae*. Comp Immunol Microbiol Infect Dis 68:101378. https://doi.org/10.1016/j.cimid.2019.101378

El-Ashram SA, Aboelhadid SM, Abdel-Kafy EM, Hashem SA, Mahrous LN, Farghly EM, Moawad UK, Kamel AA (2019) Prophylactic and therapeutic efficacy of prebiotic supplementation against intestinal coccidiosis in rabbits. Animals 9(11):965. https://doi.org/10.3390/ani9110965

El-Ghoneimy A, El-Shahawy I (2017) Evaluation of amprolium and toltrazuril efficacy in controlling natural intestinal rabbit coccidiosis. Iran J Vet Res 18(3):164–169

El-Shahawi GA, El-Fayomi HM, Abdel-Haleem HM (2012) Coccidiosis of domestic rabbit (*Oryctolagus cuniculus*) in Egypt: light microscopic study. Parasitol Res 110:251–258. https://doi.org/10.1007/s00436-011-2479-0

Espinosa J, Ferreras MC, Benavides J, Cuesta N, Pérez C, García Iglesias MJ, García Marín JF, Pérez V (2020) Causes of mortality and disease in rabbits and hares: a retrospective study. Animals 10(1):158. https://doi.org/10.3390/ani10010158

Gomez-Bautista M, Garcia MV, Rojo-Vazquez FA (1986) The levels of total protein fractions in the serum of rabbits infected with *Eimeria stiedae*. Ann Parasitol Hum Comp 61(4):393–400. https://doi.org/10.1051/parasite/1986614393

Green DR (2000) Apoptotic pathways: paper wraps stone blunts scissors. Cell 102(1):1–4. https://doi.org/10.1016/s0092-8674(00)00003-9

Gregory MW, Catchpole J (1986) Coccidiosis in rabbits: the pathology of *Eimeria flavescens* infection. Int J Parasitol 16(2):131–145. https://doi.org/10.1016/0020-7519(86)90098-6

Grès V, Voza T, Chabaud A, Landau I (2003) Coccidiosis of the wild rabbit (*Oryctolagus cuniculus*) in France. Parasite 10(1):51–57. https://doi.org/10.1051/parasite/2003101p51

Hamid PH, Prastowo S, Kristianingrum YP (2019) Intestinal and hepatic coccidiosis among rabbits in Yogyakarta, Indonesia. Vet World 12(8):1256–1260. https://doi.org/10.14202/vetworld.2019.1256-1260

Harcourt-Brown F (2014) Digestive system disease. In: Meredith A, Lord B (eds) BSAVA manual of rabbit medicine. British Small Animal Veterinary Association, pp 168–190

Hobbs RP, Twigg LE (1998) Coccidia (*Eimeria* spp.) of wild rabbits in southwestern Australia. Aust Vet J 76(3):209–210. https://doi.org/10.1111/j.1751-0813.1998.tb10131.x

Hrženjak NM, Zadravec M, Švara T, Račnik J (2021) Hepatic coccidiosis in two pet rabbits. J Exotic Pet Med 36:53–56. https://doi.org/10.1053/j.jepm.2020.12.009

Hung LX, Chen FQ, Lin YG (1984) Studies on the coccidia of rabbit, with a note on the endogenous stages and histopathology of *Eimeria stiedae*. Wuyi Sci J 4:55–64

Jing F, Yin G, Liu X, Suo XY, Qin Y (2012) Large-scale survey of the prevalence of *Eimeria* infections in domestic rabbits in China. Parasitol Res 110(4):1495–1500. https://doi.org/10.1007/s00436-011-2653-4

Jithendran KP, Bhat KP (1996) Subclinical coccidiosis in angora rabbits—a field survey in Himachal Pradesh (India). World Rabbit Sci 4(1):29–32. https://doi.org/10.4995/wrs.1996.267

Joyner LP, Catchpole J, Berrett S (1983) *Eimeria stiedae* in rabbits: the demonstration of responses to chemotherapy. Res Vet Sci 34(1):64–67

Khanum F, Anilakumar KR, Sudarshanakrishna KR, Viswanathan KR (1998) Effects of feeding fresh garlic and garlic oil on detoxifying enzymes and micronuclei formation in rats treated with azoxymethane. Int J Vitam Nutr Res 68(3):208–213

Kim DY, Reilly TJ, Schommer SK, Spagnoli ST (2010) Rabbit tularemia and hepatic coccidiosis in wild rabbit. Emerg Infect Dis 16(12):2016–2017. https://doi.org/10.3201/eid1612.101013

Kowalska D, Bielański P, Nosal P, Kowal J (2012) Natural alternatives to coccidiostats in rabbit nutrition. Ann Anim Sci 12(4):561–574. https://doi.org/10.2478/v10220-012-0047-x

Li M, Ooi H (2009) Fecal occult blood manifestation of intestinal *Eimeria* spp. infection in rabbit. Vet Parasitol 161(3–4):327–329. https://doi.org/10.1016/j.vetpar.2009.01.009

Li MH, Huang HI, Ooi HK (2010) Prevalence, infectivity and oocyst sporulation time of rabbit coccidia in Taiwan. Trop Biomed 27(3):424–429

Li C, Liu X, Tao G, Wang Y, Wu S, Suo J, Lv Y, Yu F, Suo X (2021) A new landscape of rabbit gut microbiota shaped by the infection of precocious parasites of *Eimeria intestinalis*. Vet Parasitol 300:109579. https://doi.org/10.1016/j.vetpar.2021.109579

Licois D, Coudert P, Drouet-Viard F, Boivin M (1994) *Eimeria media*: selection and characterization of a precocious line. Parasitol Res 80(1):48–52. https://doi.org/10.1007/BF00932623

Licois D, Coudert P, Drouet-Viard F, Boivin M (1995) *Eimeria magna*: immunogenicity and selection and characterization of a precocious line. Vet Parasitol 60(1–2):27–35. https://doi.org/10.1016/0304-4017(94)00768-8

Lindansay DS, Blagburn BL (1995) Antiprotozoan drugs. In: Adams HR (ed) Veterinary pharmacology and therapeutics, 7th edn. Iowa State University Pres, Ames, pp 971–983

Lüder CK, Walter W, Beuerle B, Maeurer MJ, Gross U (2001) *Toxoplasma gondii* down-regulates MHC class II gene expression and antigen presentation by murine macrophages via interference with nuclear translocation of STAT1alpha. Eur J Immunol 31(5):1475–1484. https://doi.org/10.1002/1521-4141(200105)31:5<1475::AID-IMMU1475>3.0.CO;2-C

Lüder CK, Stanway RR, Chaussepied M, Langsley G, Heussler VT (2009) Intracellular survival of apicomplexan parasites and host cell modification. Int J Parasitol 39(2):163–173. https://doi.org/10.1016/j.ijpara.2008.09.013

MacPherson JM, Gajadhar AA (1993) Differentiation of seven *Eimeria* species by random amplified polymorphic DNA. Vet Parasitol 45(3–4):257–266. https://doi.org/10.1016/0304-4017(93)90080-7

Metwaly MS, Dkhil MA, Gewik MM, AlGhamdy AO, Al-Quraishy S (2013) Induced metabolic disturbance and growth depression in rabbits infected with *Eimeria coecicola*. Parasitol Res 112(9):3109–3114. https://doi.org/10.1007/s00436-013-3485-1

Mortensen LL, Williamson LH, Terrill TH, Kircher R, Larsen M, Kaplan RM (2003) Evaluation of prevalence and clinical implications of anthelmintic resistance in gastrointestinal nematodes of goats. J Am Vet Med Assoc 223(4):495–500. https://doi.org/10.2460/javma.2003.223.495

Norton CC, Catchpole J, Joyner LP (1979) Redescriptions of *Eimeria irresidua* Kessel & Jankiewicz, 1931 and *E. flavescens* Marotel & Guilhon, 1941 from the domestic rabbit. Parasitology 79(2):231–248. https://doi.org/10.1017/s0031182000053312

Nosal P, Kowalska D, Bielański P, Kowal J, Kornaś S (2014) Herbal formulations as feed additives in the course of rabbit subclinical coccidiosis. Ann Parasitol 60(1):65–69

Okumu PO, Gathumbi PK, Karanja DN, Mande JD, Wanyoike MM, Gachuiri CK, Kiarie N, Mwanza RN, Borter DK (2014) Prevalence, pathology and risk factors for coccidiosis in domestic rabbits (*Oryctolagus cuniculus*) in selected regions in Kenya. Vet Q 34(4):205–210. https://doi.org/10.1080/01652176.2014.978044

Oncel T, Gulegen E, Senlik B, Bakirci S (2011) Intestinal coccidiosis in angora rabbits (*Oryctolagus cuniculus*) caused by *Eimeria intestinalis*, *Eimeria perforans* and *Eimeria coecicola*. Yüzüncü Yıl Üniversitesi Veteriner Fakultesi Dergisi 22(1):27–29

Pakandl M (2009) Coccidia of rabbit: a review. *Folia Parasitol* 56(3):153–166. https://doi.org/10.14411/fp.2009.019

Pakandl M, Hlásková L (2007) The reproduction of *Eimeria flavescens* and *Eimeria intestinalis* in suckling rabbits. Parasitol Res 101(5):1435–1437. https://doi.org/10.1007/s00436-007-0646-0

Pakandl M, Coudret P, Licois D (1993) Migration of sporozoites and merogony of *Eimeria coecicola* in gut-associated lymphoid tissue. Parasitol Res 79(7):593–598. https://doi.org/10.1007/BF00932244

Pakandl M, Drouet-Viard F, Coudert P (1995) How do sporozoites of rabbit *Eimeria* species reach their target cells? C R Acad Sci III 318(12):1213–1217

Pakandl M, Černík F, Coudert P (2003) The rabbit coccidium *Eimeria flavescens* Marotel and Guilhon, 1941: an electron microscopic study of its life cycle. Parasitol Res 91(4):304–311. https://doi.org/10.1007/s00436-003-0946-y

Pakandl M, Sewald B, Drouet-Viard F (2006) Invasion of the intestinal tract by sporozoites of *Eimeria coecicola* and *Eimeria intestinalis* in naive and immune rabbits. Parasitol Res 98(4):310–316. https://doi.org/10.1007/s00436-005-0071-1

Pakandl M, Hlásková L, Poplštein M, Chromá V, Vodička T, Salát J, Mucksová J (2008) Dependence of the immune response to coccidiosis on the age of rabbit suckling. Parasitol Res 103:1265–1127. https://doi.org/10.1007/s00436-008-1123-0

Paul GC, Friend DG (2019) Clostridial enterotoxemia and coccidiosis in weanling cottontail rabbits (*Sylvilagus audubonii, Sylvilagus floridanus, Sylvilagus nuttallii*) from Colorado, USA. J Wildl Dis 55(1):189–195. https://doi.org/10.7589/2018-02-026

Peeters JE, Geeroms R (1986) Efficacy of toltrazuril against intestinal and hepatic coccidiosis in rabbits. Vet Parasitol 22(1–2):21–35. https://doi.org/10.1016/0304-4017(86)90004-x

Peeters BE, Charlier G, Antoine O, Mammericlex M (1984) Clinical and pathological changes after *Eimeria intestinalis* infection in rabbits. Zentralbl Veterinarmed B 31(1):9–24. https://doi.org/10.1111/j.1439-0450.1984.tb01275.x

Peeters JE, Geeroms R, Norton C (1987) *Eimeria magna* resistance against robenidine in the rabbit. Vet Rec 121(23):545–546

Peeters JE, Geeroms R, Halen P (1988) Evolution of coccidial infection in commercial and domestic rabbits between 1982 and 1986. Vet Parasitol 29(4):327–331. https://doi.org/10.1016/0304-4017(88)90149-5

Regulation (EC) No 1831/2003 of the European Parliament and of the Council of 22 September 2003 on additives for use in animal nutrition. Official Journal of the European Union 268:29–43

Renaux S, Viard FD, Chanteloup NK, Vern Y, Kerboeuf D, Pakandl M, Coudret P (2001) Tissues and cells involved in the invasion of rabbit intestinal tract by sporozoites of *Eimeria coecicola*. Parasitol Res 87(2):98–106. https://doi.org/10.1007/s004360000308

Rose ME (1973) Immunity. In: Hammond DM, Long PL (eds) The Coccidia. University Press, Baltimore, pp. 295–241

San Martín-Núñez BV, Ordóñez-Escudero D, Alunda JM (1988) Preventive treatment of rabbit coccidiosis with α-difluoromethylornithine. Vet Parasitol 30(1):1–10. https://doi.org/10.1016/0304-4017(88)90137-9

Sastry GA (2000) Veterinary clinical pathology, 1st edn. CBS Publishers & Distributors, New Delhi

Silva SM, Ferreira C, Paupério J, Silva RM, Alves PC, Lemos A (2015) Coccidiosis in European rabbit (*Oryctolagus cuniculus algirus*) populations in the Iberian Peninsula. Acta Parasitol 60(2):350–355. https://doi.org/10.1515/ap-2015-0049

Singla LD, Juyal PD, Sandhu BS (2000) Pathology and therapy in naturally *Eimeria stiedae*-infected rabbits. J Protozool Res 10(4):185–191

Stephens MN (1952) Seasonal observations on the wild rabbit (*Oryctolagus cuniculus cuniculus L.*) in West Wales. Proc Zool Soc London 122:417–434. https://doi.org/10.1111/j.1096-3642.1952.tb00319.x

Taylor MA, Coop RL, Wall RL (2007) Veterinary parasitology, 3rd edn. Blackwell Publishing, p 901

Taylor MA, Coop RL, Wall RL (2016) Veterinary parasitology, 4th edn. Wiley Blackwell Publishing Company, p 1006

Toulah FH, Al-Rawi MM (2007) Efficacy of garlic extract on hepatic coccidiosis in infected rabbits (*Oryctolagus cuniculus*): histological and biochemical studies. J Egypt Soc Parasitol 37(3):957–986

Turner PV, Brash ML, Smith DA (2018) Rabbits. In: Turner PV, Brash ML, Smith DA (eds) Pathology of small mammal pets. Wiley, pp 1–88. https://doi.org/10.1002/9781118969601.ch1

Vadlejch J, Knízková I, Makovcová K, Kunc P, Jankovská I, Janda K, Borkovcová M, Langrová I (2010) Thermal profile of rabbits infected with *Eimeria intestinalis*. Vet Parasitol 171(3–4):343–345. https://doi.org/10.1016/j.vetpar.2010.03.037

Vereecken D, Lavazza A, De Gussem K, Chiari M, Tittarelli C, Zuffellato A, Maertens L (2012) Activity of diclazuril against coccidiosis in growing rabbits: experimental and field experiences. World Rabbit Sci 20(4):223–230. https://doi.org/10.4995/wrs.2012.1232

Vítovec J, Pakandl M (1989) The pathogenicity of rabbit coccidium *Eimeria coecicola* Cheissin, 1947. Folia Parasitol 36(4):289–293

Worku M, Franco R, Baldwin K (2009) Efficacy of garlic as anthelmintic in adult boar goats. Arch Biol Sci 61(1):135–140. https://doi.org/10.2298/ABS0901135W

Xiao J, Zheng R, Bai X, Pu J, Chen H, Gu X, Xie Y, He R, Xu J, Jing B, Peng X, Yang G (2022) Preliminary evaluation of the protective effects of recombinant AMA1 and IMP1 against *Eimeria stiedae* infection in rabbits. Parasit Vectors 15(1):400. https://doi.org/10.1186/s13071-022-05492-4

Xie Y, Xiao J, Zhou X, Gu X, He R, Xu J, Jing B, Peng X, Yang G (2021) Global transcriptome landscape of the rabbit protozoan parasite *Eimeria stiedae*. Parasit Vectors 14(1):308. https://doi.org/10.1186/s13071-021-04811-5

Yassin MM (2005) Prophylactic efficacy of crushed garlic lobes, black seed or olive oils on cholinesterase activity in central nervous system parts and serum of lead intoxicated rabbits. Turk J Biol 29(3):173–180

Major Endoparasitoses in Rabbits: oxyuriosis and Metacestodosis

João Simões and Ana Patrícia Lopes

Abstract

Helminthosis involves both Nematoda and Cestoda parasites which infect rabbits. Pinworm infection is the most prevalent nematode found in rabbitries, even under environmental and nutritional controlled conditions. Monoxenous *Passalurus ambiguus* (Rudolphi, 1819) infect the cecum and colon of young and adult rabbits and can persist for a long time in rabbitries. Its nutritional depletive action can cause significant losses to poor performances, if not controlled by regular screenings and deworming. Also, metacestodoses represent biohazards in domestic and wild rabbits, mainly in small-scale or open populations. The larval form *Cysticercus pisiformis* is the most prevalent metacestode, but both *Coenurus serialis* and *Echinococcus polymorphus* are characterized by their zoonotic potential. Usually, these diseases are inapparent, and specific clinical signs are related to the tissue where the cysts are implanted. Surveillance at slaughterhouse level is required to monitoring these parasitoses. This chapter describes the main features of rabbit's oxyuriosis, cisticercosis, coenurosis, and hydatidosis.

Keywords

Antiparasitic drugs · Cisticercosis · Coenurosis · Hydatidosis · Metacestodes · Nematodes · Oxyuriosis · Pinworm infection · Poor performance · Weight loss

J. Simões (✉) · A. P. Lopes
Animal and Veterinary Research Centre (CECAV), Associate Laboratory for Animal and Veterinary Sciences (AL4AnimalS), Department of Veterinary Sciences, School of Agrarian and Veterinary Sciences, University of Trás-os-Montes e Alto Douro (UTAD), Vila Real, Portugal
e-mail: jsimoes@utad.pt

J. Simões, J. M. Monteiro (eds.), *Veterinary Care of Farm Rabbits*,
https://doi.org/10.1007/978-3-031-44542-2_30

1 Introduction

Endoparasites can cause significant losses in rabbits, mainly due to their chronic form related with poor performances. Other than protozoa (*Eimeria* spp.; see Chapter "Rabbit Coccidiosis"), helminths are major gut parasites. *Passalurus ambiguus* (Rudolphi, 1819), *Cysticercus pisiformis* (Bloch, 1780), and *Nematodirus leporis* (Ransom, 1907) have been observed in rabbit farms with a prevalence varying between 20.0% (breeding rabbits) and 27% (fattening rabbits) (Bogach et al. 2022). Similar to farm rabbits, high prevalence of helminths (26–54%) also occurs in pet (Hajipour and Zavarshani 2020) and wild (Fernández-Álvarez et al. 2014) rabbits. Other nematodes species, such as *Trichuris leporis* and *Graphidium strigosum*, are rarely observed in farms (Kornaś et al. 2015; Raue et al. 2017). Mixed infections are very common. As an example, Szkucik et al. (2014) observed mixed infections in 80% of rabbits at slaughter, most of them involving *Eimeria* and originating from small-scale farms.

Both oxyuriosis (i.e., *P. ambiguous* infection) and metacestodosis (e.g., *C. pisiformis*) remain the most prevalent helminthoses with high negative impact in rabbitries (Domínguez-Roldan et al. 2018; Sioutas et al. 2021). Usually, their presence is inapparent causing a subclinical chronic form. They can remain in farms (breeding rabbits) for long time if not watched.

This chapter highlights the main features of pinworm and metacestode infections.

2 Rabbits' Oxyuriosis (Pinworm Infection)

2.1 Introduction

Rabbits' pinworm (as a member of infraorder Oxyuridomorpha; Liu et al. 2016) infection is caused by *Passalurus ambiguus* (Rudolphi, 1819) (previously denominated *Oxyuris ambigua*; Hofing and Kraus 1994). It was first discovered in *Oryctolagus cuniculus* and *Lepus europaeus* Hall, 1916 (Hussein et al. 2022). *P. ambiguus* is found in the cecum and colon, in both young and adult rabbits and hares.

P. ambiguus is the nematode most prevalent in rabbits, independently of the production system or rearing purpose. It is frequent in intensive or semi-intensive production systems, where a high animal density exists, despite controlled environmental conditions and homogeneous diets. Nevertheless, it is also present in wild and companion rabbits, presenting a worldwide distribution (Szkucik et al. 2014).

2.2 The Pathogen *Passalurus ambiguus*

The adult size of *P. ambiguus* was well defined by Khorolskyi et al. (2021). In adult females, these researchers observed a length of parasite body from 9.0 to 11.3 mm and from 11.6 to 13.4 mm for short- and long-tailed specimens, respectively, which

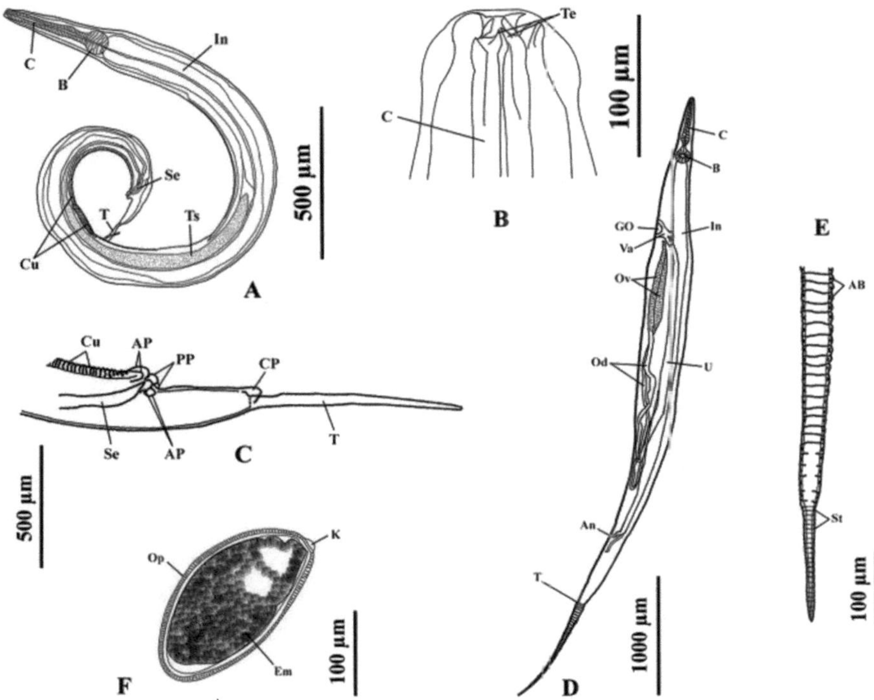

Fig. 1 Illustrated morphology of *Passalurus ambiguus*. Adapted from Hussein et al. (2022) with permission from Springer Nature. (**a**) Male *Passalurus ambiguus*. (**b**) The anterior portion of the worm. (**c**) The posterior portion of the male. (**d**) Female *Passalurus ambiguus*. (**e**) The tail of the female. (**f**) The egg. Scale bars: (**a**, **c**) 500 µm; (**b**, **d**, **e**, **f**) 100 µm. An, anus; AB, annular bands; AP, anterior papillae; B, bulb; C, corpus; Cu, cuticle; CP, caudal papilla; Em, embryo; GO, genital opening; In, intestine; K, knob; Od, oviducts; Op, operculum; Ov, ovaries; PP, posterior papillae; Se, spicule; St, striations; T, tail; Te, teeth; Ts, testis; U, uterus; Va, vagina

agree with previous studies about the *P. ambiguus* morphology. The maximum width of body was 602.2–680.2 µm and 718.0–822.2 µm for short- and long-tailed specimens, respectively, but this value can decrease until 464 µm. The body length of the male is about half of the female, between 3.6 and 6.1 mm and between 189.3–334.5 µm, respectively. The main structures of female and male *P. ambiguus* can be observed in Fig. 1 (illustration) as well as in Fig. 2 (microphotographs).

The life cycle is direct and infects specifically lagomorphs. Domestic and wild rabbits, as well as hares, can be infected by (infective) eggs of *P. ambiguous* (Schoeb et al. 2007).

The mature females migrate from the cecum and colon to the rectum of the rabbit and around the anus by the action of peristalsis. By uterus contraction, these females expel the unembryonated sticky eggs, together or not with hard or soft feces, which adhere to the host hair (perineum). The length and width of eggs in the uterus vary

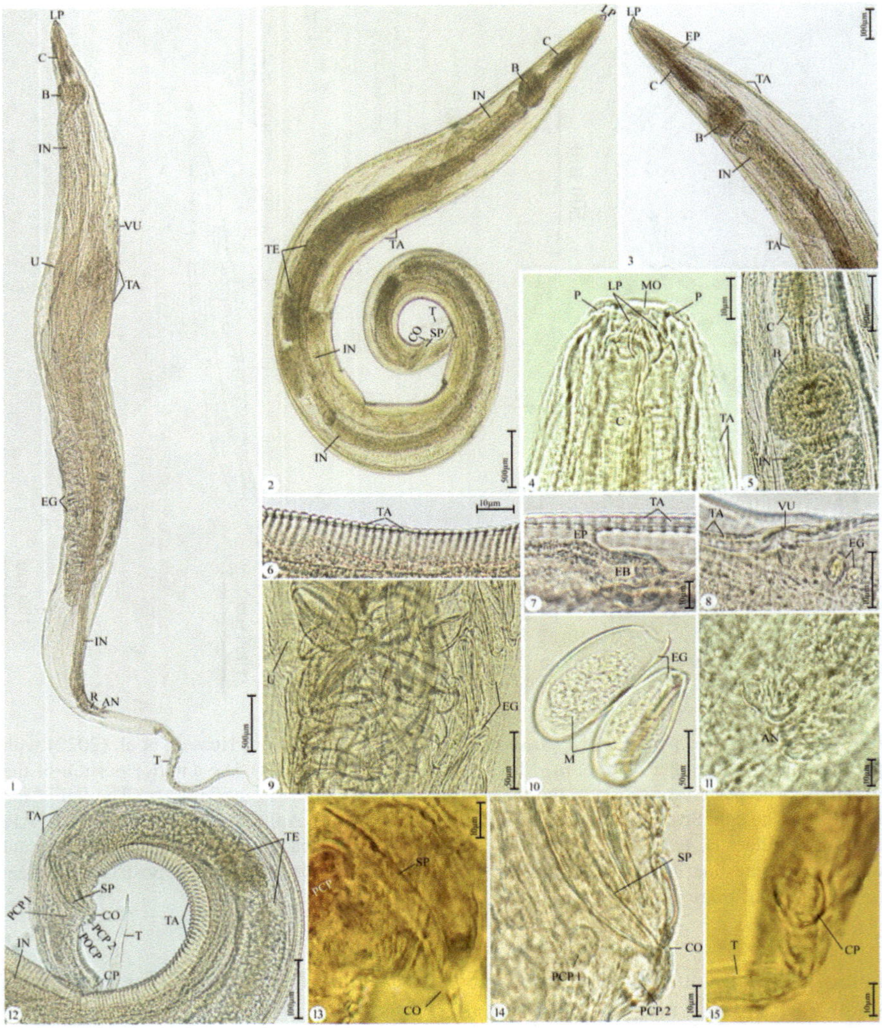

Fig. 2 The structure (photomicrographs) of female and male *Passalurus ambiguus*. Adapted from Abdel-Gaber et al. (2019) with permission from Springer Nature. (1) A whole mount of the female worm characterized by the anterior extremity with a triangular mouth opening surrounded with simple lips (LP), followed by the esophagus divided into the corpus (C) and bulb (B), leading to the large intestine (IN), and followed by the rectum (R) opening to the outside by an anal opening (AN). Note: the presence of the uterus (U) was filled with numerous eggs (EG), opened by the vulval opening (VU); and the body was covered by the transverse annulated cuticle (TA) and ended with a long tail (T). (2) A whole mount of the male worm was characterized by the anterior extremity with the mouth opening surrounded with simple lips (LP) followed by the esophagus divided into the corpus (C) and bulb (B), leading to the large intestine (IN), and followed by the rectum (R) opening to the outside by the cloacal opening (CO). Note the presence of a pair of testes (TE), a single spicule (SP), and the body covered by the transverse annulated cuticle (TA), ending with a long tail (T). (3) The anterior extremity of the recovered worm showing the mouth opening surrounded with simple lips (LP), followed by the esophagus divided into the corpus (C) and bulb

between 90.5 and 122.7 μm and between 41.2 and 47.4 μm, respectively (Khorolskyi et al. 2021).

During the exogenous cycle, the eggs of *P. ambiguus* develop into four stages: (1) zygote, (2) cleavage and formation of blastomeres, (3) formation of larvae, and (4) motile larvae (Fig. 3). The embryogenesis is completed within 4 and 9 days in dependence of environmental temperature (20 to 35 °C). At 35 °C, the optimal temperature, 72.7% of eggs can reach the motile larvae stage within 4 days (Yevstafieva et al. 2022). According to Rödel and Starkloff (2014), the eggs can be (immediately) re-ingested during grooming or coprophagy allowing permanent autoinfections. This can be a more important driver to infect rabbits than the weather conditions (rain and grass humidity) for egg's infective development of endoparasitic intestinal nematodes (Rödel and Starkloff 2014).

2.3 Epidemiology

P. ambiguus is more frequently detected in coprological examinations than other nematodes, such as *Trichostrongylus*, *Strongyloides*, *Obeliscoides cuniculi*, *Graphidium strigosum*, and *Trichuris leporis*, as well as other oxyurids, namely, *P. nonannulatus*, *Dermatoxys veligera*, and *D. hispaniensis* in wild rabbits and *D. vlakhaasi in* hares (see Schoeb et al. 2007). Its prevalence varies between 3% and more than 70%, depending on environmental conditions mainly related with poor sanitation (Table 1).

A seasonal peak is usually observed in summer season (Nosal et al. 2006; Elshahawy and Elgoniemy 2018) but also can occur in spring (Legendre et al. 2021).

The transmission is done by fecal-oral route (Hofing and Kraus 1994). The infective larva only leaves the eggs after ingestion, in the small intestine. The incubation period is 18 days (Varga 2014) and the prepatent period varies between 55 and 60 days (Mehlhorn 2016). More than 1000 adult parasites can be observed in a rabbit (Hofing and Kraus 1994).

Fig. 2 (continued) (B), leading to the large intestine (IN), the excretory bladder opening to the outside by the excretory pore (EP), and the body covered with transverse annulated cuticle (TA). (4–10) High magnification images of different body parts showing the following: (4) the anterior extremity with the mouth opening (MO) surrounded with simple lips (LP), cephalic papillae (P), and the buccal cavity followed by the corpus region of the esophagus (C). (5) Two portions of the esophagus were the corpus (C) and bulb (B), followed by the intestine (IN). (6) Transverse annulations (TA) of the cuticle. (7) The excretory bladder (EB) opened to the outside by the excretory pore (EP). (8) The vulval opening (VU) with eggs (EG) in the vagina. (9) The uterus (U) filled with numerous eggs (EG). (10) Eggs (EG). (11–15) The posterior ends of the recovered worms showing the following: (11) the anal opening (AN) of a female worm. (12) The posterior of a male ending with the cloacal opening (CO), a single spicule (SP), and two pairs of peri-cloacal papillae (PCP 1 and PCP 2), a pair of post-cloacal papillae (POCP), and a pair of caudal papillae (CP) before the tail region (T). (13, 14) The spicule (SP) and peri-cloacal papillae (PCP 1 and PCP 2) surrounding the cloacal opening (CO). (15) Caudal papillae (CP) before the tail region (T)

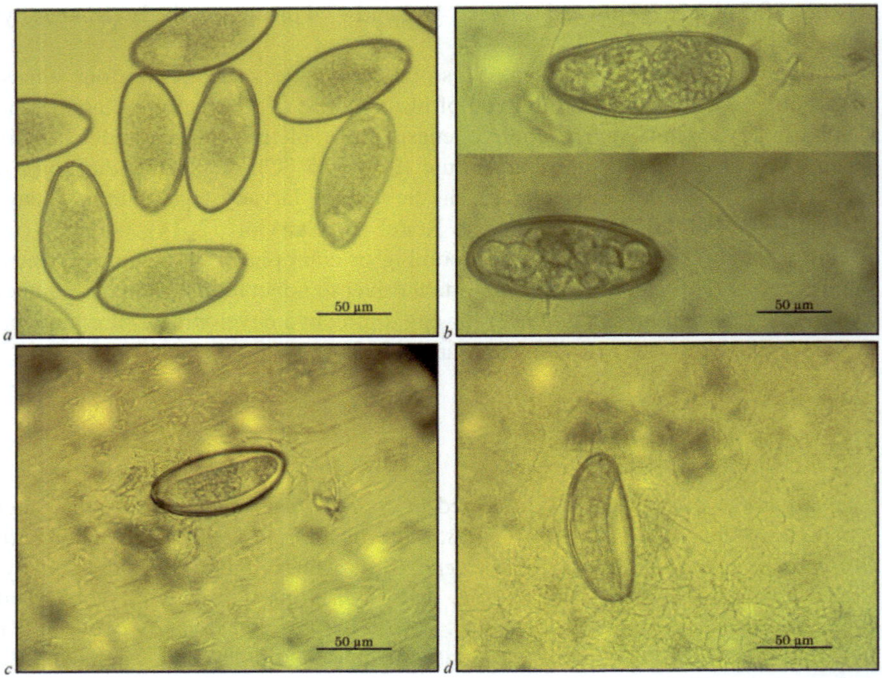

Fig. 3 Morphological characteristics of the development of *Passalurus ambiguus* eggs. Adapted from Yevstafieva et al. (2022). These eggs were obtained from the gonads of female helminths and maintained in experimental culture: (**a**) zygote, (**b**) cleavage and formation of blastomeres, (**c**) formation of immotile larva, (**d**) formation of motile larva

In Italy, it was observed the presence of an excretion circadian rhythm of *P. ambiguus* eggs which is more intense during evening [18:00–24:00 h; mean eggs per gram (EPG) = 55.4] and night (24:00–6:00 h; mean EPG = 53.4) than when samples were collected outside of these periods ($p < 0.05$; Rinaldi et al. 2007). This daily pattern implies that, whenever possible, the fecal sample should be collected at this time.

This oxyuriosis is characterized by a continuous reinfection (Mykhailiutenko et al. 2019) and will be persistent during rabbits' life if not treated.

2.4 Pathogeny, Lesions, and Clinical Signs

P. ambiguous penetrates into the Lieberkühn glands causing a mechanical irritation of the epithelial cells, but they are essentially coprophage (Mykhailiutenko et al. 2019).

The pathogenesis and symptoms are dependent on the infection intensity. They can be observed inside crypt cells, which can induce hyperplasia of the lymphoid

Table 1 Prevalence of *Passalurus ambiguus* in some countries

Origin	Country	Prevalence	Observations	References
Finnish pet rabbits	Finland	3.0%	Mean EPG = 65 (25–175)	Mäkitaipale et al. (2017)
New Zealand white rabbit farms	Poland	22.0%		Kornaś et al. (2015)
Small farms	Egypt	5.7%	Belgicie breed more susceptible	Elshahawy and Elgoniemy (2018)
Organic rabbit farms	France	>70.0%		Legendre et al. (2021)
Small farms		5.8%	Samples collected at slaughterhouse; mean EPG = 16	Szkucik et al. (2014)
Farms with poor sanitation	Iran	54.0%		Hajipour and Zavarshani (2020)
Small farms	Serbia	17.0%	Mixed infections (*Eimeria* spp., *Trichostrongylus* spp.)	Ilić et al. (2018)
Farms and wild (hares)	Italy	13.0%	*Lepus europaeus*	Sergi et al. (2018)
Large and small private households	Ukraine	35.0%	Mean EPG = 156–163	Prus et al. (2022)

EPG eggs per gram of feces

tissues and the follicular epithelium cells (Hussein et al. 2022). Villi and crypt cell destruction and lymphocytes and monocytes in the submucosa can be observed, as well as agranular dystrophy of smooth muscle cells of the muscle layer (Varga 2014; Mykhailiutenko et al. 2019; Hussein et al. 2022). Histological features caused by *P. ambiguus* on the cecal appendix are reported in Fig. 4.

Catarrhal or catarrhal-hemorrhagic inflammation can occur due to their mechanical action in intensive infections (Mykhailiutenko et al. 2019), causing diarrhea. Hyperemia, edema, and punctate hemorrhages can be observed in the cecum and rectum (Terentyeva et al. 2021). Small white nodules can be observed through the cecal serosa (Hussein et al. 2022), at necropsy.

Itches and scratches can be observed due to the migration of *P. ambiguus* female to the anal zone (Legendre et al. 2021). Diarrhea and weight loss are also common clinical signs (Mehlhorn 2016; Sioutas et al. 2021).

Due to continuous nutritional depletion, the rabbits present low body score. At necropsy, the mesenteric lymph nodes appear enlarged, hyperemic, and swollen (Terentyeva et al. 2021).

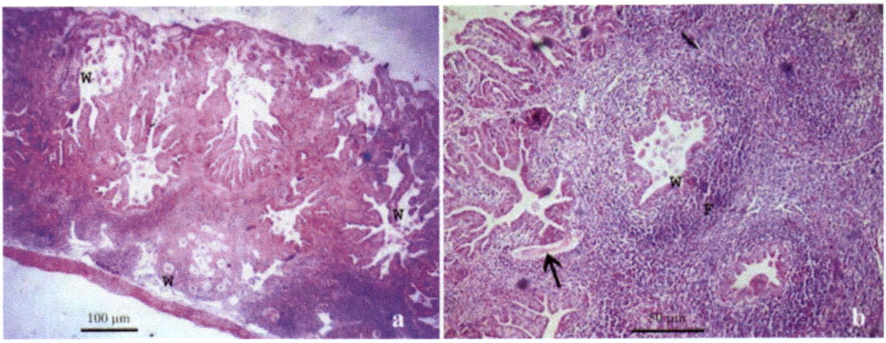

Fig. 4 Histological features of cecal appendix from *P. ambiguus* infected rabbits. Adapted from Hussein et al. (2022) with permission from Springer Nature. (**a**) Visualization of nematode worm (W) in the appendiceal layers (×40). (**b**) Cross section of *P. ambiguus* (W) inside the germinal layer of the lymphoid follicle (F) and appendiceal lumen (×100). Arrow: transverse section of worm attached to hyperplastic epithelium cells. Hematoxylin and eosin staining

2.5 Diagnosis

During necropsy, the females of *P. ambiguus* can be seen adhered to the feces inside the cecum and colon of infected rabbits. Contrarily to females, it is little probable to recover males of *P. ambiguus*, since they die after mating (Abdel-Gaber et al. 2019).

The qualitative and quantitative FLOTAC technique is preferable to McMaster technique due to the evaluation of a large amount of feces (10 mL), improving sensibility, when compared with the classic McMaster (Rinaldi et al. 2007). Also, FLOTAC technique can be used in practice, to evaluate different helminths and protozoa of rabbits. A high *K* agreement (0.821; $p < 0.001$) was observed between FLOTAC technique and the cellophane tape test (Rinaldi et al. 2007). Fecal flotation method (ZnSO4 solutio) is also suitable to establish a diagnosis (Varga 2014).

2.6 Prognosis

Usually, *P. ambiguus* is not pathogenic and the host can be infected for life. Nevertheless, due to nutrient depletion, including in the cecotrophs, and also some hematophagy, pinworm infection can induce thinness and cachexia affecting performance in animals. Also, physical obstruction can be observed due to great numbers of individuals in the cecum. After treatment, the prognosis is good. Nonetheless, rabbit death can occur (Sioutas et al. 2021).

2.7 Treatment and Prevention

The individual treatment is based on fenbendazole administration (10–20 mg/kg per os, repeated in 14 days; Reusch 2005; Sioutas et al. 2021). Due to the direct life cycle and quick reinfection, multiple doses are recommended, e.g., 50 mg/kg/daily for 5 days in feed (Taylor et al. 2015; laboratory rabbits). Thiabendazole (50 mg/kg per os) and albendazole (7.5–20 mg/kg per os), two doses 14 days apart, are also effective for treatment (Oglesbee and Jenkins 2012). Nevertheless, an administration at 20 mg/day per os for 10 days provoked some degree of hepatic and renal damages, as well as alterations on hematological and biochemical profiles (Elalfy et al. 2021).

The deworming of kits is recommended, when they start to leave the nest and ingest solid feed, i.e., due to environment contamination with infective eggs.

Sioutas et al. (2021) recommend regular screenings of *P. ambiguus* and other gastrointestinal parasites in industrial rabbit farms. The regular administration of fenbendazole or albendazole should be used after these screenings, and the evaluation of their efficacy is mandatory to prevent nematode resistance to drugs.

Biosecurity measures, such as regular sanitation (e.g., regular disinfection of environment) and quarantine (e.g., segregation) of newly acquired rabbits for breeding, are required.

3 Metacestodoses

3.1 Introduction

Diseases called metacestodoses are brought on by Taeniidae (species of *Taenia* and *Echinococcus*) larvae in the intermediate host. Domestic and wild lagomorphs are infected by three major larval cestodes: *Cysticercus pisiformis* (cysticercosis), *Coenurus serialis* (rabbit coenurose), and *Echinococcus polymorphus* (hydatidosis) (Taylor et al. 2015). The last two parasites also can affect humans (zoonotic disease). Usually undetectable, the infection is found at necropsy or as a slaughterhouse finding. Nevertheless, clinical signs, mainly due to cyst compression, might happen based on the tissue or organ involved as well as the number and size of vesicles.

The definite hosts are dogs and other carnivorous predators, who consume infected rabbit carcasses and develop *Taenia* or *Echinococcus* adults in their gastrointestinal tract producing eggs inside their proglottids.

When these adult parasites pass proglottids through the feces, the cycle in the definitive host is over.

C. pisiformis is the metacestode of *Taenia pisiformis*. The definitive hosts of this parasite are mainly dogs; *C. serialis*, the metacestode of *Taenia serialis*, has dogs and foxes as definite hosts; and *Echinococcus polymorphus* is the larval stage of *Echinococcus granulosus* whose definite hosts are dogs and other Canidae (e.g., dingo, jackal, hyena). Overall, these heteroxenous parasites have a similar life cycle.

3.2 The Pathogens

3.2.1 *Cysticercus pisiformis*

Mainly dogs and, more rarely, cats are the major definitive hosts of *T. pisiformis*; however, other Canidea and Felidae can also act as hosts (Remesar et al. 2021). The intermediate hosts are lagomorphs, and it is not considered a zoonosis.

Infective eggs are ingested by rabbits and quickly liberate the oncospheres in duodenum. The oncospheres cross the intestine wall and primarily reach hepatic parenchyma through the blood vessels. Meanwhile, the larvae stage starts. In days 3 to 6 postinfection (PI), small immature vesicles can be observed in the liver surface and migrate to adjacent abdominal structures, where they remain attached. The larvae are fully developed 5 to 6 weeks PI (Box 1; Radwan et al. 2014). *C. pisiformis* larvae develop usually attached to the mesenterium, greater omentum, liver capsule, and abdominal cavity (Fig. 5) but can also be observed in the heart and lungs, causing cysticercosis (Hallal-Calleros et al. 2016; Wang et al. 2021a).

Box 1 Developmental Stages of *C. pisiformis*: Egg to *Cysticercus* (Radwan et al. 2014)

- 15 min postinfection (PI): egg hatch the oncosphere in the duodenum.
- 60 min PI: oncosphere (ovoid shape) with three pairs of hooks.

(continued)

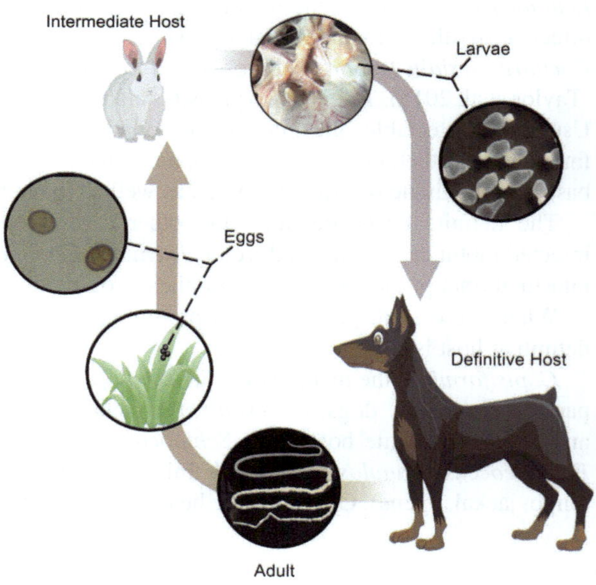

Fig. 5 The life cycle of *Taenia pisiformis*. The proglottids rupture and release the eggs into the environment. Eggs are ingested by the rabbits. The larva crosses the intestine of the intermediate host and migrates to the liver, and subsequently to the mesentery and omentum, where they develop into fluid-filled bladderlike cysts. The definitive host ingests *Cysticercus*, which develop and reproduce in the intestine. Adapted from Pu et al. (2022) with permission from Elsevier

Box 1 (continued)

- 2 h PI: oncosphere (oval shaped) with three pairs of marginal hooks; they attach to the duodenal mucosa and submucosa, starting to penetrate into the intestinal wall; metamorphose to cysticercus type starts.
- 24 h PI: reached the liver through the blood vessel and the hepatic portal vein.
- Day 3–6 PI: immature vesicle surges at liver surface. Early scolex development, an invaginated thickening of the wall at the anterior end. Hooks and suckers remain undifferentiated.
- 1 week PI: rostellar hook development (surge of minute conical bodies to form hooks and invagination of the canal). Scolex is projected into the internal cavity. Bladder develops. Suckers are not well differentiated.
- 2 weeks PI: complete development of the hooks and invaginated canal; suckers and rostellum are developed.
- 5–6 weeks PI: fully developed *Cysticercus*.

The shape of matured cysts varies from spherical to ellipsoid, fluid-filled, with a mean size of 5–7 × 5–12 mm with a single invaginated scolex (Loos-Frank 2000), and they prompt to infect the definitive host, when ingested.

3.2.2 *Coenurus serialis*

The definitive hosts of *T. serialis* are dogs and foxes (Fig. 6), but other canids can be susceptible. The intermediate hosts are nonspecific. Rabbits, hares, rodents, horses, cattle, sheep, goats, and primates, including humans, can be infected by *C. serialis*. Usually, the cysts are located in the subcutis and intermuscular connective tissue, but other sites, such as the central nervous system (Yamazawa et al. 2020) or eyes (O'Reilly et al. 2002), can be involved.

Mature cysts of *C. serialis* are fluid-filled, measuring up to 5 cm in diameter and presenting numerous scolices (Schoeb et al. 2007).

3.2.3 *Echinococcus polymorphus*

According to strains and genotype, *E. granulosus* was classified in different species, implicating differences within hosts and life cycle (Thompson 2008), but this issue remains unclear (Mousa et al. 2020) (Fig. 7).

The definite hosts of *E. granulosus* are domestic and wild canids mainly dogs, as well as predatory carnivores (i.e., wolf, lion, dingo, jackal, hyena) (Harcourt-Brown 2002; Eckert and Deplazes 2004).

Herbivores are intermediated hosts, and according to the genotype strain (G1 to G9), they can prevail in specific species (Eckert and Deplazes 2004). Hidatidosis is a zoonosis.

Similar to *Cysticercus*, oncospheres hatch from *E. polymorphus* eggs in the small intestine, attaching to the mucosa and surpassing the wall. In this case, they enter the blood and lymphatic vessels, reaching several body organs (e.g., liver, lungs,

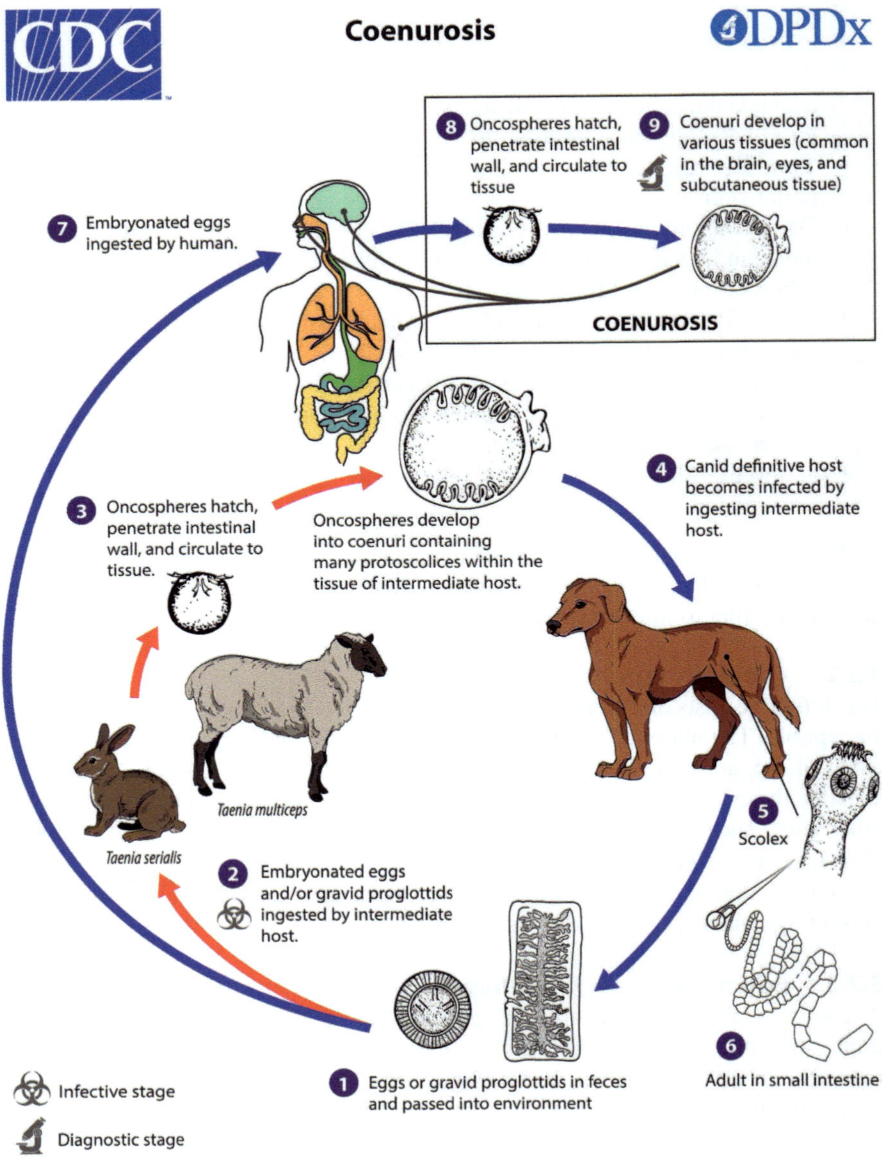

Fig. 6 The life cycle of *Taenia serialis*. Adapted from CDC (2019) with permission from Division of Parasitic Diseases and Malaria (CDC)

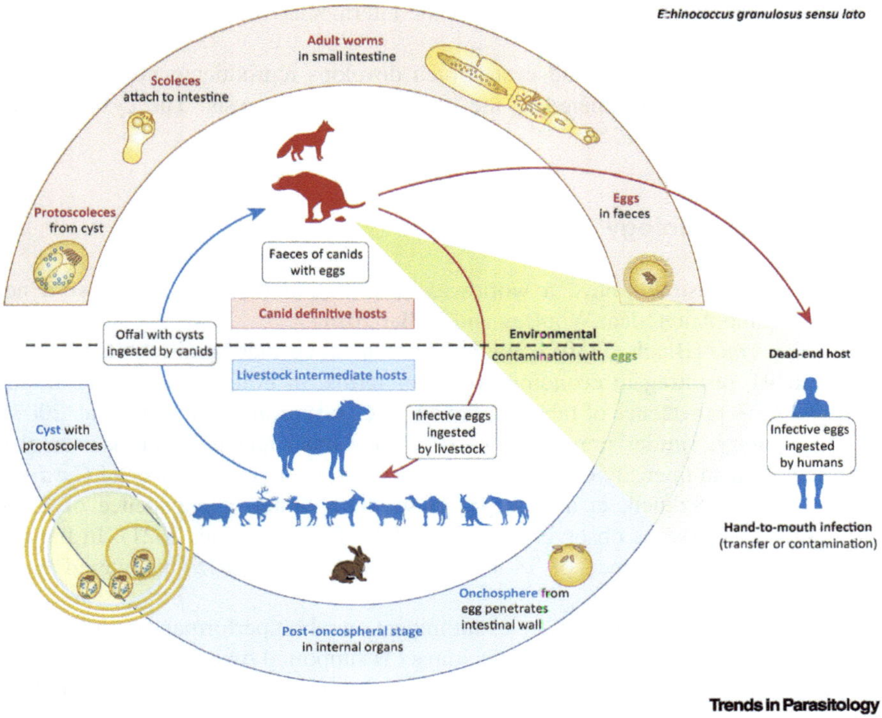

Fig. 7 The life cycle of *Echinococcus granulosus*. Modified from Casulli et al. (2019) with permission from Elsevier

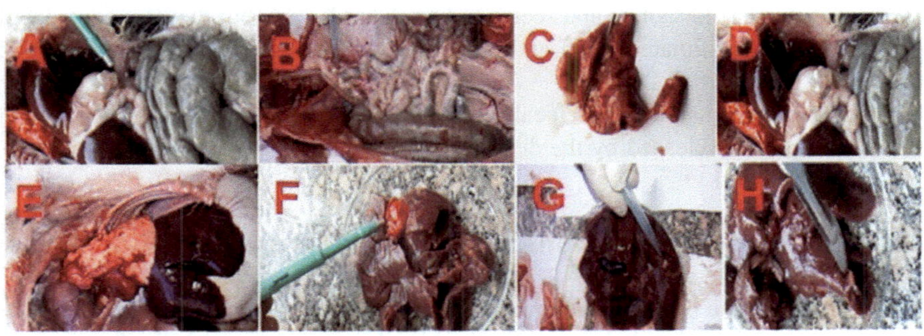

Fig. 8 The postmortem inspection after 5 months of the experimentally infected New Zealand white rabbits (male) with active oncospheres or viable protoscoleces of *Echinococcus granulosus*. (**a**) Hydatid cyst in the omentum. (**b**) Hydatid cyst in the omentum and liver. (**c**) Hydatid cyst in the liver. (**d**) Hydatid cyst in the lung. (**e**) Hydatid cyst in the lung and liver. (**f**) Hydatid cyst in the lung. (**g**, **h**) Hydatid cyst in the liver. Adapted from Kandil et al. (2020) with permission from Elsevier

kidneys, brain, spleen, bone marrow) of the intermediate host (Kandil et al. 2020) (Fig. 8).

Each larva forms a hydatid cyst, which develops reproducing new larva, i.e., daughter cysts, which adhered to the inner wall of the cysts. The cyst ruptures disseminating new larva.

3.3 Epidemiology

All the three cestodes have a worldwide distribution (*E. granulosus* were not detected in the Antarctica; Woolsey and Miller 2021).

C. pisiformis (Bloch, 1780) is the most prevalent cestodosis in rabbits (Szkucik et al. 2014), resulting in economic losses of farms. In Polish large farms, it was reported a 5% prevalence of rabbits infected by *C. pisiformis* (Sołtysiak et al. 2007). In this country, similar prevalence (4.7%) from small and large farms was also reported, with an intensity of invasion index (oocysts per gram/number of *Cysticercus*) of 18, by Szkucik et al. (2014). In southern Spain, a prevalence of 2.8% (95% CI: 2.2–3.4) was observed in wild rabbits (Remesar et al. 2021). In Italy, it was reported a prevalence of 14.8% (2013) and 3.3% (2015) in hunted hares (Stancampiano et al. 2019).

It has, currently in rabbitries, relevant impact on rabbit performances and welfare (Domínguez-Roldan et al. 2018). This impact is supported by the findings of Arias-Hernández et al. (2020), who evidenced the role of these vesicles on metabolism alteration using normal and obese rabbits infected or not by this larva. Domínguez-Roldan et al. (2018) reported an interesting association of *C. pisiformis* with wild rabbit's females, probably due to hormonal influence.

The prevalence of *C. serialis* and *E. polymorphus* in rabbits is not fully evaluated in domestic and wild rabbits and seems to be relatively rare, but they are considered potentially underestimated, such as in other hosts (Jenkins and Macpherson 2003; Schneider-Crease et al. 2017).

The route of transmission is fecal-oral through contaminated feed or water with gravid proglottids released by the definite host into the environments.

3.4 Pathogenesis, Lesions, and Clinical Signs

After ingestion, the oncospheres are discharged in the small intestine, penetrate the duodenal and jejunal epithelium, enter the blood vessel, and reach the liver parenchyma through the hepatic portal vein. At this point, the oncospheres migrate toward the liver capsule, originating hepatic granulomas (Rosa et al. 2022) and traumatic hepatitis (Volodymyrovych and Oleksiyovych 2022), and to the greater omentum and mesentery (Heath 1971; Owiny 2001; Yang et al. 2013). At necropsy, Rosa et al. (2022) observed that 52.6%, 31.6%, and 15.8% of the metacestodes were recovered in the peritoneum, pelvic region, and mesentery, respectively, after experimental infection of 40 rabbits.

Rabbits infected with *C. pisiformis* have a weak immunologic resistance, increasing the susceptibility to other diseases, have poor performance, and can appear emaciated and in some cases may die (Rajasekariah et al. 1985; Yang et al. 2013).

It was observed that rabbits have an increase of serum cortisol levels, probably due to high stress response, contributing to immunosuppression (Arias-Hernández et al. 2020). Nonetheless, exosome-like vesicles, produced by *C. pisiformis*, seem to modulate the host immune response (Wang et al. 2020; Wang et al. 2021b).

C. pisiformis induce behavior alterations. Rabbits spend less time grooming and drinking and more time lying stretched. Also, hematological and hepatic function profiles are modified during larva migration. The circulating leukocytes, lymphocytes, and basophils cells increase, and alanine aminotransferase, alkaline phosphatase, and total bilirubin decrease, at least temporarily (Betancourt-Alonso et al. 2011).

C. pisiformis affects negatively the prolificacy to half litter size (Hallal-Calleros et al. 2016). Moreover, it was observed a reduction of 33% and 19% of number and size of embryos, respectively, at day 8 postmating (Rosa et al. 2022).

Newborns of infected rabbit does are heavier than noninfected does but show a lowest body weight at weaning (Hallal-Calleros et al. 2016). These occurrences can have impact on the productivity of the rabbitries, according to the number of infected rabbits does.

A strong negative spearman correlation ($\rho = -0.8$; $p = 0.04$) was observed between the number of hepatic granulomas and the number of implanted embryos suggesting that the granulomas which exchange the expression of hormones and cytokines could affect embryo implantation (Rosa et al. 2022).

Usually, clinical signs are not perceptible, unless the number and size of the cyst are high or are located in specific organs. In this last case, the clinical signs are mainly due to the compression of the tissues. Painless nodules (e.g., *C. serialis*) can be detected in live animals (Zhang et al. 2018).

3.5 Diagnosis

The clinical diagnosis of metacestodosis is based on the location and morphology of the cysts as well their contents. *C. pisiformis* only contains a protoscolex, which can serve as differential diagnosis with *C. serialis* (multiple protoscoleces) (Nourani et al. 2012).

Differential diagnosis of *C. serialis* can include any subcutaneous mass such as abscesses, cellulitis, cysts, neoplasia, foreign bodies, granulomas, and traumatic hematomas (Templeton 1971).

Enzyme-linked immunosorbent assay (ELISA) and indirect fluorescent antibody test (IFAT), based on crude antigens, have been used to detect circulating antibodies to *C. pisiformis* (Craig 1984; Wang et al. 2009). Antibody circulating level starts to increase on week 3 and reaches the peak on week 10 PI (Wang et al. 2009). Meanwhile, new ELISA tests based on recombinant fatty acid-binding proteins (FABP) were developed as potential serodiagnostic tests. ELISA using a new recombinant FABP, rTpFABP, showed a sensitivity of 98.2% and specificity of 92.1% when compared with results of necropsy ($n = 169$) (Yang et al. 2013). Also, a Dot ELISA assay, to detect *C. pisiformis*, using a recombinant fusion protein (rTp1; from Pp1 gene) achieved a sensitivity from 92.9 to 97.6% and a specificity from 95.2 to 98.4% (Chen et al. 2014).

Usually, in low and moderate infections, *C. pisiformis* is detected macroscopically at necropsy or by surgical/laparoscopic procedures and at slaughterhouse.

PCR sequence analysis can also be performed (Graham-Brown et al. 2018; Samorek-Pieróg et al. 2021).

E. polymorphus ELISA, measuring IgG antibodies (Mousa et al. 2017), and PCR (Kandil et al. 2020) methods are available.

3.6 Prognosis

Usually, rabbits can live without evidence of parasitism.

3.7 Treatment and Prevention

Praziquantel, pyrantel, and febantel can be used to cause the larvae death and involution of the cysts, but its effectiveness is not consistent.

The surgical extraction of one or more cysts is performed in companion animals.

Regarding *E. polymorphus*, a protection of 83%, according to the decrease in cyst number, linked to a high IgG immune response, was obtained when rabbits were challenged with crude germinal layer antigen from infected camels (Kandil et al. 2020). Little increase in protection, 86–90%, was reported by Mousa et al. (2017) when rabbits were challenged with crude hydatid cyst fluid, protoscoleces, or partially purified hydatid cyst fluid, suggesting that a self-vaccine can be obtained in critical zones.

The prevention of metacestodosis in rabbits is related to the break of parasite life cycle, mainly avoiding contact of rabbits with gravid proglottids excreted by definitive hosts. The prevention effectiveness should be watched in rabbits reared as open populations, e.g., outdoor production systems.

4 Final Remarks and Conclusions

Rabbits' pinworm infection is caused by a specific host parasite, *P. ambiguus*, presenting a direct life cycle, allowing self-reinfections which perpetuate the disease as a chronic form. Even when low pathogenicity has been demonstrated, mainly in low to moderate parasite load, the impact on rabbit's performance can be relevant in rabbitries.

A regular screening, using flotation methods or similar techniques from feces samples of young and adult rabbits, is recommended before and after deworming to treat, control, and prevent the disease and mitigate its adverse effects on the farms.

Metacestodosis can have more impact in outdoor than indoor production systems due to the heteroxenous dynamics of their life cycles. Overall, they are still inapparent in farm rabbits. Nonetheless, some of them are important zoonoses and should be watched at slaughterhouse level.

Acknowledgments The authors thank Springer for the use of the AJE Digital Editing. This work was supported by the projects UIDP/CVT/00772/2020 (CECAV) and LA/P/0059/2020 (AL4AnimalS), funded by the Portuguese Foundation for Science and Technology (FCT).

References

Abdel-Gaber R, Ataya F, Fouad D, Daoud M, Alzuhairy S (2019) Prevalence, morphological and molecular phylogenetic analyses of the rabbit pinworm, *Passalurus ambiguus* Rudolphi 1819, in the domestic rabbits *Oryctolagus cuniculus*. Acta Parasitol 64:316–330. https://doi.org/10.2478/s11686-019-00047-7

Arias-Hernández D, García-Jiménez S, Domínguez-Roldan R, Murcia-Mejía C, Báez-Saldaña A, Hallal-Calleros C, Flores-Pérez I (2020) Effects of *Taenia Pisiformis* infection and obesity on clinical parameters, organometry and fat distribution in male rabbits. Pathogens 9:861. https://doi.org/10.3390/pathogens9110861

Betancourt-Alonso MA, Orihuela A, Aguirre V, Vázquez R, Flores-Pérez FI (2011) Changes in behavioral and physiological parameters associated with Taenia pisiformis infection in rabbits (*Oryctaolagus cuniculus*) that may improve early detection of sicks rabbits. World Rabbit Sci 18:21–30. https://doi.org/10.4995/wrs.2011.801

Bogach M, Paliy A, Horobei O, Perotska L, Kushnir V, Bohach DM (2022) Endoparasites of rabbits (*Oryctolagus cuniculus* domesticus) in southern Ukraine. Biosyst Divers 30:173–178. https://doi.org/10.15421/012218

Casulli A, Siles-Lucas M, Tamarozzi F (2019) *Echinococcus granulosus* sensu lato. Trends Parasitol 35:663–664. https://doi.org/10.1016/j.pt.2019.05.006

CDC (2019) Coenurosis. CDC's Division of Parasitic Diseases and Malaria (DPDM). Centers for Disease Control and Prevention. Retrieved from https://www.cdc.gov/dpdx/coenurosis/index.html (23-02-2023)

Chen L, Yang D, Gu X, Peng X, Yang G (2014) Evaluation of a novel dot-ELISA assay utilizing a recombinant protein for the effective diagnosis of *Taenia pisiformis* larval infections. Vet Parasitol 204:214–220. https://doi.org/10.1016/j.vetpar.2014.05.021

Craig PS (1984) Circulating antigens, antibodies and immune complexes in experimental *Taenia pisiformis* infections of rabbits. Parasitology 89:121–131. https://doi.org/10.1017/s0031182000001177

Domínguez-Roldan R, Pérez-Martínez M, Rosetti MF, Arias-Hernández D, Bernal-Fernández G, Flores-Pérez FI, Hallal-Calleros C (2018) High frequency of *Taenia pisiformis* metacestodes and high sex-associated susceptibility to cysticercosis in naturally infected wild rabbits. Parasitol Res 117:2201–2206. https://doi.org/10.1007/s00436-018-5907-6

Eckert J, Deplazes P (2004) Biological, epidemiological, and clinical aspects of echinococcosis, a zoonosis of increasing concern. Clin Microbiol Rev 17:107–135. https://doi.org/10.1128/CMR.17.1.107-135.2004

Elalfy E, Hassan A, Amer M (2021) Adverse effect of Nitroxynil and Albendazole on rabbits. Mansoura Vet Med J 22:147–156. https://doi.org/10.21608/mvmj.2021.215026

Elshahawy I, Elgoniemy A (2018) An epidemiological study on endoparasites of domestic rabbits (Oryctolagus cuniculus) in Egypt with special reference to their health impact. Sains Malaysiana 47:9–18. https://doi.org/10.17576/jsm-2018-4701-02

Fernández-Álvarez Á, Feliu C, Miquel J, Torres J, Foronda P (2014) Helminth fauna of wild rabbit *Oryctolagus cuniculus* in the Canary Islands, Spain. Helminthologia 50:155–160. https://doi.org/10.2478/s11687-013-0125-3

Graham-Brown J, Gilmore P, Harcourt-Brown F, Eastham H, Williams D (2018) Lethal cysticercosis in a pet rabbit. Vet Rec Case Rep 6:6e000634. https://doi.org/10.1136/vetreccr-2018-000634

Hajipour N, Zavarshani M (2020) Ectoparasites and Endoparasites of New Zealand white rabbits from north west of Iran. Iran J Parasitol 15:266–271

Hallal-Calleros C, Morales-Montor J, Orihuela-Trujillo A, Togno-Peirce C, Murcia-Mejía C, Bielli A, Hoffman KL, Flores-Pérez FI (2016) *Taenia pisiformis* cysticercosis induces decreased prolificacy and increased progesterone levels in rabbits. Vet Parasitol 229:50–53. https://doi.org/10.1016/j.vetpar.2016.09.015

Harcourt-Brown F (2002) Infectious diseases of domestic rabbits. In: Textbook of rabbit medicine. Springer, pp 361–385. https://doi.org/10.1016/B978-075064002-2.50019-9

Heath DD (1971) The migration of oncospheres of *Taenia pisiformis*, *T. serialis* and *Echinococcus granulosus* within the intermediate host. Int J Parasitol 1:145–152. https://doi.org/10.1016/0020-7519(71)90008-7

Hofing GL, Kraus AL (1994) In: Manning PJ, Ringler DH, Newcomer CE (eds) Arthropod and helminth parasites. Biology of the laboratory rabbit, 2nd edn. Academic Press, Orlando, FL, pp 335–353

Hussein NM, Rabie SAH, Abuelwafa WA, ElDin MMM (2022) Morphometry, molecular identification and histopathology of *Passalurus ambiguus* Rudolphi, 1819 in domestic rabbits (*Oryctolagus cuniculus*) in Qena, upper Egypt. J Parasit Dis 46:511–525. https://doi.org/10.1007/s12639-022-01477-3

Ilić T, Stepanović P, Nenadović K, Dimitrijević S (2018) Improving agricultural production of domestic rabbits in Serbia by follow-up study of their parasitic infections. Iran J Vet Res 19: 290–297

Jenkins DJ, Macpherson CN (2003) Transmission ecology of *Echinococcus* in wild-life in Australia and Africa. Parasitology 127(Suppl):S63–S72. https://doi.org/10.1017/s0031182003003871

Kandil OM, Abdelrahman KA, Mahmoud MS, Mahdy OA, Ata EB, Aloufi AS, Al-Olayan E (2020) Cystic echinococcosis: development of an intermediate host rabbit model for using in vaccination studies. Exp Parasitol 208:107800. https://doi.org/10.1016/j.exppara.2019.107800

Khorolskyi A, Yevstafieva V, Kravchenko S, Pishchalenko M, Vakulenko Y, Gutyj B (2021) Specifics of the morpho-logical identification of the pathogen of passaluariasis of rabbits. Regul Mech Biosyst 12:702–709. https://doi.org/10.15421/022197

Kornaś S, Kowal J, Wierzbowska I, Basiaga M, Nosal P, Niedbała P (2015) The Alice - "follow the white rabbit" - parasites of farm rabbits based on coproscopy. Ann Parasitol 61:257–261. https://doi.org/10.17420/ap6104.16

Legendre H, Goby J, Le Stum J, Hoste H, Cabaret J, Gidenne T (2021) Organic rabbit farming: should we be afraid of gastrointestinal parasites? In: 12th World Rabbit Congress. 12th World Rabbit Congress, November 2021, Nantes, Communication, p 24

Liu GH, Li S, Zou FC, Wang CR, Zhu XQ (2016) The complete mitochondrial genome of rabbit pinworm *Passalurus ambiguus*: genome characterization and phylogenetic analysis. Parasitol Res 115:423–429. https://doi.org/10.1007/s00436-015-4778-3

Loos-Frank B (2000) An up-date of Verster's (1969) 'Taxonomic revision of the genus Taenia Linnaeus' (Cestoda) in table format. Syst Parasitol 45:155–183. https://doi.org/10.1023/a:1006219625792

Mäkitaipale J, Karvinen I, Virtala AK, Näreaho A (2017) Prevalence of intestinal parasites and risk factor analysis for Eimeria infections in Finnish pet rabbits. Vet Parasitol Reg Stud Rep 9:34–40. https://doi.org/10.1016/j.vprsr.2017.04.005

Mehlhorn H (2016) In: Mehlhorn H (ed) Encyclopedia of parasitology, 4th edn. Springer, Berlin, Germany

Mousa W, Mahdy O, Abdel-Wahab A, El-Gameel S (2017) Evaluation of humeral immunity against three hydatid cyst antigens of camels using secondary cyst development in rabbit model. J Egypt Soc Parasitol 47:623–631. https://doi.org/10.21608/jesp.2017.77719

Mousa WM, Abdel-Wahab AM, El-Gameel Sohila M, Mahdy OA (2020) Genetic characterization of hydatid cysts of different intermediate hosts. Helminthologia 57:185–195. https://doi.org/10.2478/helm-2020-0031

Mykhailiutenko SM, Kruchynenko OV, Klymenko O, Serdioucov JK, Dmytrenko N, Tkachenko VV (2019) Pathomorphological changes in the large intestine of rabbits parasitised by *Passalurus ambiguus* (Nematoda, *Oxyuridae*). Regul Mech Biosyst 10:69–74. https://doi.org/10.15421/021911

Nosal P, Petryszak A, Nowosad B, Sobolewska M (2006) Pasozyty przewodu pokarmowego królików w badaniach koproskopowych [Gastrointestinal parasites of rabbits in coproscopic investigations]. Wiad Parazytol 52:327–330

Nourani H, Khosravi M, Pirali Kheirabadi KH (2012) Muscular coenurosis in a wild rabbit. Iran Vet J 8(3):73–76

O'Reilly A, McCowan C, Hardman C, Stanley R (2002) Taenia serialis causing exophthalmos in a pet rabbit. Vet Ophthalmol 5:227–230. https://doi.org/10.1046/j.1463-5224.2002.00230.x

Oglesbee BL, Jenkins JR (2012) Gastrointestinal diseases. In: Quesenberry KE, Carpenter JW (eds) Ferrets, rabbits, and rodents, 3rd edn. W.B. Saunders, pp 193–204. https://doi.org/10.1016/B978-1-4160-6621-7.00015-4

Owiny JR (2001) Cysticercosis in laboratory rabbits. Contemp Top Lab Anim Sci 40(2):45–48

Prus M, Duda Yu, Koreyba L, Borisevich B, Lisova V (2022) Seasonal and age dynamics of passalurosis invasion of rabbits and pathological and histological changes in this nematodosis. Scientific Horizons 25(11):9–19. https://doi.org/10.48077/scihor.25(11).2022.9-19

Pu L, Liu Z, Guo A, Wu B, Liu GH, Zhang S, Guo X, Li X, Jian J, Cai X, Wang S (2022) A chromosome-level genome assembly for the rabbit tapeworm *Taenia pisiformis*. Gene 834: 146650. https://doi.org/10.1016/j.gene.2022.146650

Radwan N, El Sefy MN, Noor El Din SA, Abou Shafeey HE, Sharafa SE, Khalil AI (2014) *Cysticercus pisiformis*: ultrastructural transformation of the tegument during development from oncosphere to cysticercus. Parasitol United J 7:13–26. https://doi.org/10.4103/1687-7942.139686

Rajasekariah GR, Rickard MD, O'Donnell IJ (1985) *Taenia pisiformis*: protective immunization of rabbits with solubilized oncospheral antigens. Exp Parasitol 59:321–327. https://doi.org/10.1016/0014-4894(85)90087-6

Raue K, Heuer L, Böhm C, Wolken S, Epe C, Strube C (2017) 10-year parasitological examination results (2003 to 2012) of faecal samples from horses, ruminants, pigs, dogs, cats, rabbits and hedgehogs. Parasitol Res 116:3315–3330. https://doi.org/10.1007/s00436-017-5646-0

Remesar S, Castro-Scholten S, Jiménez-Martín D, Camacho-Sillero L, Morrondo P, Rouco C, Gómez-Guillamón F, Cano-Terriza D, García-Bocanegra I (2021) Spatiotemporal monitoring of *Cysticercus pisiformis* in European wild rabbit (*Oryctolagus cuniculus*) in Mediterranean ecosystems in southern Spain. Prev Vet Med 197:105508. https://doi.org/10.1016/j.prevetmed.2021.105508

Reusch B (2005) Rabbit gastroenterology. Vet Clin North Am Exotic Anim Pract 8:351–375. https://doi.org/10.1016/j.cvex.2005.01.007

Rinaldi L, Russo T, Schioppi M, Pennacchio S, Cringoli G (2007) *Passalurus ambiguus*: new insights into copromicroscopic diagnosis and circadian rhythm of egg excretion. Parasitol Res 101:557–561. https://doi.org/10.1007/s00436-007-0513-z

Rödel HG, Starkloff A (2014) Social environment and weather during early life influence gastro-intestinal parasite loads in a group-living mammal. Oecologia 176:389–398. https://doi.org/10.1007/s00442-014-3017-4

Rosa DR, David AH, Emmanuel DG, Edda S, Virginio AF, Ivan FP, Claudia HC (2022) Decreased embryo implantation in rabbits infected with *Taenia pisiformis*. Parasitol Res 121:3689–3692. https://doi.org/10.1007/s00436-022-07694-2

Samorek-Pieróg M, Karamon J, Brzana A, Bilska-Zając E, Zdybel J, Cencek T (2021) Molecular confirmation of massive *Taenia pisiformis* cysticercosis in one rabbit in Poland. Pathogens 10: 1029. https://doi.org/10.3390/pathogens10081029

Schneider-Crease I, Griffin RH, Gomery MA, Dorny P, Noh JC, Handali S, Chastain HM, Wilkins PP, Nunn CL, Snyder-Mackler N, Beehner JC, Bergman TJ (2017) Identifying wildlife reservoirs of neglected taeniid tapeworms: non-invasive diagnosis of endemic Taenia serialis infection in a wild primate population. PLoS Negl Trop Dis 11:e0005709. https://doi.org/10.1371/journal.pntd.0005709

Schoeb TR, Cartner SC, Bake RA, Gerrity LW (2007) Parasites of rabbits. In: Baker DG (ed) Flynn's parasites of laboratory animals, Chapter 15. Blackwell, Oxford, UK, pp 451–499. https://doi.org/10.1002/9780470344552.ch15

Sergi V, Romeo G, Serafini M, Torretta E, Macchioni F (2018) Endoparasites of the European hare (*Lepus Europaeus*) (Pallas, 1778) in Central Italy. Helminthologia 55:127–133. https://doi.org/10.2478/helm-2018-001

Sioutas G, Evangelou K, Vlachavas A, Papadopoulos E (2021) Deaths due to mixed infections with *Passalurus ambiguus, Eimeria spp. and Cyniclomyces guttulatus* in an industrial rabbit farm in Greece. Pathogens 10(6):756. https://doi.org/10.3390/pathogens10060756

Sołtysiak Z, Bednarski M, Piekarska J (2007) *Cysticercosis pisiformis* in rabbit livers. Med Weter 63:1255–1257

Stancampiano L, Ravagnan S, Capelli G, Militerno G (2019) Cysticercosis by *Taenia pisiformis* in Brown hare (*Lepus europaeus*) in northern Italy: epidemiologic and pathologic features. Int J Parasitol Parasit Wildl 9:139–143. https://doi.org/10.1016/j.ijppaw.2019.04.004

Szkucik K, Pyz-Łukasik R, Szczepaniak KO, Paszkiewicz W (2014) Occurrence of gastrointestinal parasites in slaughter rabbits. Parasitol Res 113:59–64. https://doi.org/10.1007/s00436-013-3625-7

Taylor MA, Coop RL, Wall RL (2015) Parasites of laboratory animals, Chapter 15. In: Taylor MA, Coop RL, Wall RL (eds) Veterinary parasitology, 4th edn. Wiley, West Sussex, UK, pp 816–892. ISBN 9780470671627. https://doi.org/10.1002/9781119073680.ch15

Templeton AC (1971) Anatomical and geographical location of human coenurus infection. Trop Geogr Med 23:105–108

Terentyeva ZK, Taiguzin RS, Matveev OA, Shakhbiyev KK, Kryazhev AL (2021) Diagnostic studies for common invasive and infectious pathologies of rabbits in the Orenburg region. E3S Web Conf 282:03020. https://doi.org/10.1051/e3sconf/202128203020

Thompson RCA (2008) The taxonomy, phylogeny and transmission of Echinococcus. Exp Parasitol 119:439–446. https://doi.org/10.1016/j.exppara.2008.04.016

Varga M (2014) Infectious diseases of domestic rabbits. In: Textbook of rabbit medicine. Springer, pp 435–471. https://doi.org/10.1016/2FB978-0-7020-4979-8.00014-5

Volodymyrovych BM, Oleksiyovych HO (2022) Morphological parameters of rabbit blood during migration of *Cysticercus pisiformis*. Scientific Collection «InterConf+» 20:299–305. https://doi.org/10.51582/interconf.19-20.04.2022.030

Wang X, Luo X, Sun X, Yu S, Cai X (2009) Kinetics of antibodies in rabbits experimentally infected with *Cysticercus pisiformis*. Vet Sci China 39:283–286

Wang LQ, Liu TL, Liang PH, Zhang SH, Li TS, Li YP, Liu GX, Mao L, Luo XN (2020) Characterization of exosome-like vesicles derived from Taenia pisiformis cysticercus and their immunoregulatory role on macrophages. Parasit Vectors 13:318. https://doi.org/10.1186/s13071-020-04186-z

Wang ZX, Chen X, Han L, Liu HD, Guo JH, Zhao Y, Sun XL (2021a) Generation and application of a monoclonal antibody against the 18-kDa oncosphere antigen of *Taenia pisiformis*. Exp Parasitol 224:108096. https://doi.org/10.1016/j.exppara.2021.108096

Wang L, Liu T, Chen G, Li Y, Zhang S, Mao L, Liang P, Fasihi Harandi M, Li T, Luo X (2021b) Exosomal microRNA let-7-5p from *Taenia pisiformis* Cysticercus prompted macrophage to M2 polarization through inhibiting the expression of C/EBP δ. Microorganisms 9:1403. https://doi.org/10.3390/microorganisms9071403

Woolsey ID, Miller AL (2021) *Echinococcus granulosus* sensu lato and *Echinococcus multilocularis*: a review. Res Vet Sci 135:517–522. https://doi.org/10.1016/j.rvsc.2020.11.010

Yamazawa E, Ohno M, Satomi K, Yoshida A, Miyakita Y, Takahashi M, Satomi N, Asanome T, Maeshima A, Shiotsuka M, Iwata S, Yamasaki H, Morishima Y, Sugiyama H, Narita Y (2020) First case of human neurocoenurosis caused by Taenia serialis: a case report. Int J Infect Dis 92: 171–174. https://doi.org/10.1016/j.ijid.2020.01.004

Yang D, Chen L, Xie Y, Wu X, Nong X, Peng X, Lai W, Gu X, Wang S, Peng X, Yang G (2013) Expression and immunolocalisation of TpFABP as a candidate antigen for the serodiagnosis of rabbit *Taenia pisiformis* cysticercosis. Parasite (Paris, France) 20:53. https://doi.org/10.1051/parasite/2013053

Yevstafieva V, Khorolskyi A, Kravchenko S, Melnychuk V, Nikifo-rova O, Reshetylo O (2022) Features of the exogenic develop-ment of *Passalurus ambiguus* (Nematoda, Oxyuroidae) at different temperature regimes. Biosyst Divers 30:74–79. https://doi.org/10.15421/012207

Zhang XY, Jian YN, Ma LQ, Li XP, Karanis P (2018) A case of Coenurosis in a wild rabbit (*Lepus sinensis*) caused by *Taenia serialis* Metacestode in Qinghai Tibetan plateau area, China. Korean J Parasitol 56(2):195–198. https://doi.org/10.3347/kjp.2018.56.2.195

Ectoparasitoses in Rabbits

Filipa Teixeira Rodrigues, João Simões,
and Ana Patricia Antunes Lopes

Abstract

This chapter aims to highlight the main features of ectoparasites with impact in rabbits. Major acarioses are described. The Astigmata (Sarcoptiformes) are a large group of relatively similar mites. The order includes the families Sarcoptidae and Psoroptidae, which are of major veterinary importance as they contain the most common mite species causing mange and scab, in domestic animals including rabbits. *Psoroptes cuniculi* is a cosmopolitan parasite of rabbits that causes infestation primarily in the ears. *Psoroptes* mites are highly contagious being of importance with respect to general animal health and hygiene and also economic concerns. *Sarcoptes scabiei* var. *cuniculi* and *Notoedres cati* var. *cuniculi* are highly contagious burrowing mites that can be found in farms and in pet rabbit. They both represent an important cause of skin disease in rabbits. *Cheyletiella parasitovorax* and *Leporacarus gibbus* (formerly known as *Listrophorus gibbus*) are nonburrowing fur mites that affect both wild and domestic rabbits worldwide. Besides infesting domestic and wild rabbits, these mites have zoonotic potential. Being contagious skin parasites, all of these mites are generally transmitted by direct contact with infested animals, but indirect contact is also possible, e.g., with fomites or the contaminated environment. An overview about pediculosis, pulicosis, and ixodidosis is also presented in this chapter.

F. T. Rodrigues (✉) · J. Simões · A. P. A. Lopes
Animal and Veterinary Research Centre (CECAV), Associate Laboratory for Animal and Veterinary Sciences (AL4AnimalS), Department of Veterinary Sciences, School of Agrarian and Veterinary Sciences, University of Trás-os-Montes e Alto Douro (UTAD), Vila Real, Portugal
e-mail: filipar@utad.pt

J. Simões, J. M. Monteiro (eds.), *Veterinary Care of Farm Rabbits*,
https://doi.org/10.1007/978-3-031-44542-2_31

Keywords

Psoroptes · Otoacariosis · *Sarcoptes* · *Notoedres* · *Cheyletiella* · *Leporacarus* · Mange · Rabbit · Scabies · Pediculosis · Lice · Pulicosis · Flea · Ixodidosis · Tick

1 Introduction

Ectoparasitosis, especially that caused by Sarcoptiformes, have relevant impact in rabbit health. They usually colonize the skin or superficial layers of the epidermis. Clinical signs vary from erythema, scaling, and hair loss to crusting and pruritus causing excoriations and even auto-mutilation (Eshar 2019). They mainly are observed in wild rabbits (open populations), but their impact in domestic rabbits is sometimes undervalued. Economic losses in rabbitries are mainly due to weight loss, low fertility, fur damage, and even death of rabbits (Abd El-Ghany 2022). Additionally, some of these pathogens have zoonotic potential (Moroni et al. 2022).

Despite the importance of acari due to their relative frequency in rabbitries, some other less frequent arthropods such as fleas, lice, and ticks can also serve as vectors of other diseases.

This chapter aims to describe the most relevant ectoparasites and their clinical management in rabbits.

2 Psoroptic Mange

2.1 Introduction

Psoroptes cuniculi is a cosmopolitan parasite of rabbits that causes infestation primarily in the ears. This ear mange mite is of importance with respect to general animal health and hygiene and also economic concerns (Ulutas et al. 2005; Pan et al. 2006; Calderón-Arguedas et al. 2011).

2.2 The Pathogen

Mites of the genus *Psoroptes* are nonburrowing Acarina, up to 0.75 mm in length and oval in shape, responsible for ear and body mange of herbivores (Fehr and Koestlinger 2013). At least five variants within this genus have been recognized, based on the infested host, the infestation site, and differences in morphological features (Bates 1999; Pegler et al. 2005).

The parasite of rabbits *P. cuniculi*, synonym with *P. ovis* var. *cuniculi* (Sanders et al. 2000) and also considered conspecific with *P. ovis* (Zahler et al. 2000; Pegler et al. 2005), primarily lives on the inner surface of the pinna and is responsible for ear mange or otoacariasis (Romero Núñez et al. 2020).

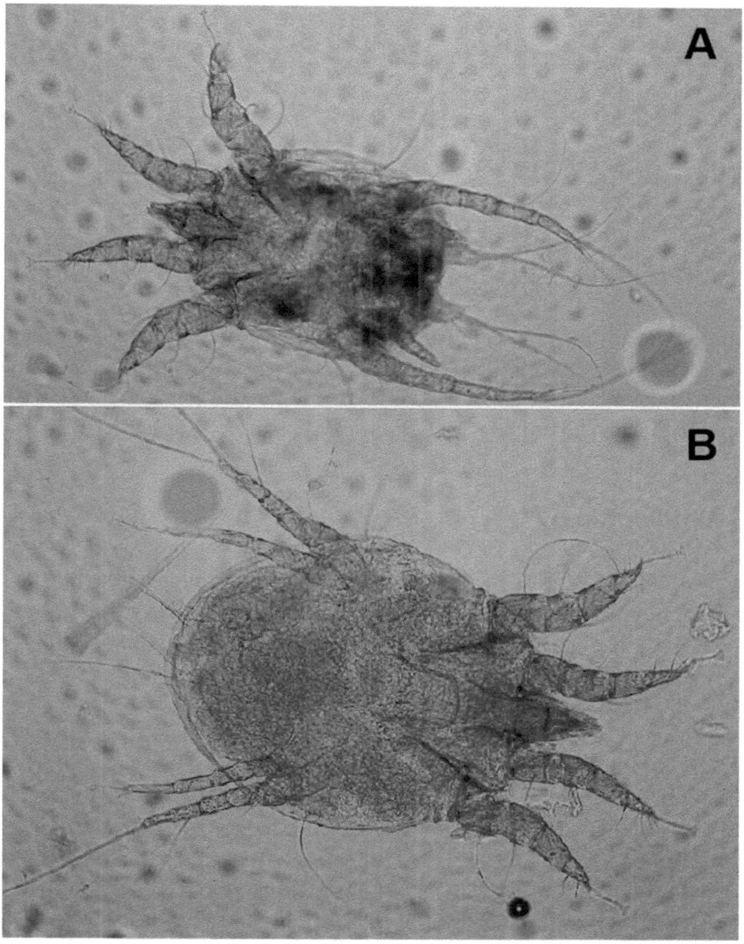

Fig. 1 *Psoroptes ovis* (magnification 100×): (**a**) male and (**b**) female. Laboratory of Parasitology, UTAD

All the eight legs are projected beyond the body margin, and most important recognition features are the pointed mouthparts and the three jointed pretarsi (pedicels) with funnel-shaped suckers (pulvilli) (Fig. 1). These mites have recognizable dimorphism between adult females and males (Taylor et al. 2016; Varga and Paterson 2020).

The 21-day life cycle of *P. cuniculi* (egg, larva, two nymphal stages, and adults) completely occurs on the host (Fig. 2), but mites can survive off the host at normal environmental temperatures (Sanders et al. 2000; Meredith 2006; Pritt et al. 2012). All parasite stages feed superficially on lipid emulsion of skin lymph, cells, and exudates (Rafferty and Gray 1987; Amer et al. 2015), which are produced on the

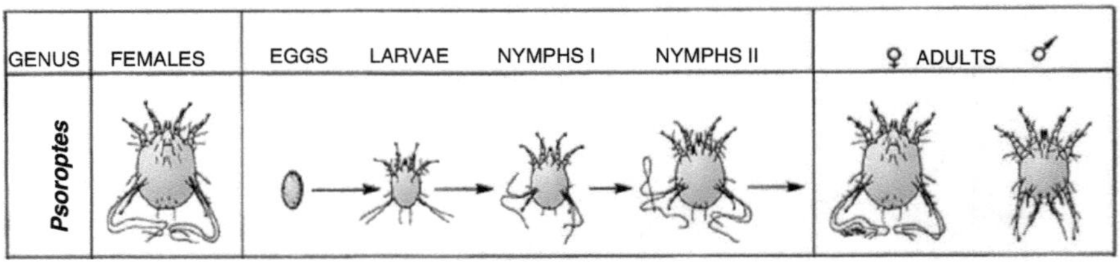

Fig. 2 Life cycle (larvae stages and adults) of *Psoroptes* spp. Modified from Mehlhorn (2016) with permission from Springer

host skin as a result of a hypersensitivity reaction to the presence of antigenic mite fecal material (Pegler et al. 2005). *Psoroptes* can survive and maintain their infectivity after 15 days off the host, while mite eggs were able to hatch for up to 7 days in the same conditions (Ebani and Mancianti 2021).

2.3 Epidemiology

Psoroptes mites are highly contagious and spread by direct contact with infested rabbits and fomites (especially bedding material). Healthy rabbits may harbor low-grade subclinical infestation for long periods and present with only mild pruritus (Harcourt-Brown 2002; Varga and Paterson 2020).

Psoroptic otocariosis in rabbits, also called psoroptic ear canker, was first described in Paris in 1858 by Delafond (Sweatman 1958) and has since then been recorded worldwide (Bates 1999). The prevalence of otocariosis in rabbits was reported with values of 12% in Serbia (Ilić et al. 2018) and 16.4% in Kenya (Okumu 2014). In China, the reported prevalence of ear mange mite in rabbits ranged from 7.9% to 15.7%, causing reduction of weight gain or even death of rabbits (Pan et al. 2006). Cases of psoroptic ear mange have also been recorded in rabbits in India (Sivajothi et al. 2014), Italy (Fichi et al. 2007a), Turkey (Kurtdede et al. 2007), and Korea (Kyung-Yeon and Oh-Deog 2010).

Within the genus *Psoroptes*, mites can be also identified in other hosts, including goats, horses, and sheep (Osman et al. 2006; Amorim et al. 2015; Makepeace et al. 2022), causing irritation and head shaking (Taylor et al. 2016). They are morphologically and behaviorally similar, and cross-infestation of these mites to rabbits is possible (Pritt et al. 2012).

Predisposing factors responsible for ear infestations of parasitic etiology in rabbits may be high temperatures and increased relative humidity but also poor hygiene, feeding methods, and way of handling (Morton et al. 2005; Swarnakar et al. 2014). Global warming climate conditions, with increasing number of days with high temperatures and abundant rain, can create tropical conditions favoring the appearance of psoroptic mange in rabbits (Ulutas et al. 2005; Ilić et al. 2018).

Psoroptic mange can potentially be an anthropozoonosis. In India, a case of infestation in humans was recorded in the staff that engaged on a rabbit farm, with direct contact with exposed sick animals (Swarnakar et al. 2014).

2.4 The Disease

Mites feed by piercing the skin and browsing on loose epidermal debris. Lesions in part may be caused by a hypersensitivity reaction to mite feces and saliva (Varga and Paterson 2020).

A grayish-white to brown exudate builds up in the external auditory canal and may progress for long periods before being noticed. This exudate consists of desquamated epithelial cells, serum, inflammatory cells, mite feces, and adult mites (Schoeb et al. 2007). The underlying tissue may be red and inflamed. As disease progresses, the exudate becomes thicker and gets more extensive, often filling the ear canal and extending onto the pinnae. Ears will often droop under the weight of the debris (Varga and Paterson 2020; Fehr and Koestlinger 2013). Alternatively, infested animals can be clinically normal until placed under stress (Bowman 2009; Pritt et al. 2012). The skin surface is moist and red and may be odoriferous, and the crustaceous lesions can be very painful when touched (Bowman 2009; Varga and Paterson 2020).

The allergic response to the mites may play an important pathological role, causing intense pruritus; thus, affected rabbits may shake or scratch at their ears, leading to excoriation and self-mutilation (Schoeb et al. 2007; Pritt et al. 2012) as well as to secondary bacterial infections (Ulutas et al. 2005). Although less frequent, medium otitis with head tilt, or aural hematoma, may also be present in psoroptic mange in rabbits (Fisher et al. 2007; Fehr and Koestlinger 2013).

In severe cases of otocariosis or in older, sick, and untreated animals, the parasites can colonize and infest other parts of the body, through grooming, including the head, neck, legs, ventral abdomen, and perianal region, causing erythematous and exudative pruritic skin lesions (Acar et al. 2007; Romero Núñez et al. 2020). Heavy infestations with *P. cuniculi* can also cause considerable weight loss, less favorable feed conversion rates, vestibular dysfunction, and meningitis, which is frequently fatal when complicated by the secondary bacterial pathogens (Ulutas et al. 2005; Varga and Paterson 2020).

2.5 Diagnosis

Observing the exudate/crusts covering the inner surface of the ear, even from afar, can be considered pathognomonic (Pritt et al. 2012). Mites can be seen with the naked eye, otoscope, or microscope. Microscopic examination of samples of exudate or the yellow-gray scales from affected ears, mounted in mineral oil, 10% potassium hydroxide, or lactophenol, reveals mites, mite feces, or mite eggs (White et al. 2002; Varga and Paterson 2020).

2.6 Treatment and Prevention

Psoroptic mange in rabbits can be treated with a wide range of parasiticides, some of them off-label, such as ivermectin (200–400 μg/kg, SC) once or repeated in 15–18 days (Bowman et al. 1992; Curtis et al. 1990; McKellar et al. 1992; Pandey 1989; Lu et al. 2018), 1% topically (pour-on) in a single or two doses (Elhawary et al. 2017), or one drop of concentrated ivermectin solutions inside the ear canal (Pritt et al. 2012); selamectin (6–18 mg/kg, topically) one or two doses 28 days apart (McTier et al. 2003; Fisher et al. 2007; Kurtdede et al. 2007); moxidectin (200–300 μg/kg, oral) two doses 10 days apart (Wagner and Wendlberger 2000; moxidectin with imidocloprid (1 mg/kg and 10 mg/kg, respectively, topically) every 4 weeks for three treatments (Hansen et al. 2005); eprinomectin (200–300 μg/kg, SC) as a single dose (Pan et al. 2006) or 0.5–1 mg/kg topically of 0.5% pour-on solution once or twice at 14-day interval (Ulutas et al. 2005; Wen et al. 2010); doramectin (200–400 μg/kg, SC) as a single dose (Elhawary et al. 2017; Santos et al. 2017); permethrin (2 mg) of 2% permethrin spray applied to the ear canal on a single occasion (Melo et al. 2008); fluralaner (25 mg/kg, oral) single dose (Sheinberg et al. 2017); and afoxolaner with milbemycin (2.5 mg/kg and 0.5 mg/kg, respectively, oral), single dose (Romero Núñez et al. 2020).

Recently, topical treatment of a dinotefuran (4.95%), pyriproxifen (0.44%), and permethrin (36%) spot-on combination product applied on the ears and back, or only on the back, was effective in rabbits naturally infested with *Psoroptes* (Borges et al. 2022).

Although these drugs have been leading to relatively good treatment outcomes, concerns have been raised due to public safety effects and environmental hazards, increased drug resistance of the target species, or risk of toxicity to the treated animals (Halley et al. 1993; Currie et al. 2004; Shan et al. 2013). Thus, integrated mite management using biological agents/plants has become an important research interest, particularly in the acaricidal properties (in vitro and in vivo) of essential oils. The acaricidal mechanism is still unclear and should be studied further, but potential compounds for effective clinical control of psoroptic mange have been proposed, including essential oil extracted from *Eugenia caryophyllata* (Fichi et al. 2007a), *Cinnamomum zeylanicum* (Fichi et al. 2007b), *Eupatorium adenophorum* (Nong et al. 2013a; Hu et al. 2014), and *Rhododendron nivale Hook* (Guo et al. 2017) and also methanol extract of *Adonis coerulea Maxim* (Shang et al. 2013, 2017) and oregano oil and its major component, carvacrol, thymol, and p-cymene (Shang et al. 2016).

Ear crusts should not be removed or softened initially, because this is very painful and can cause bleeding or result in necrosis of the thin and delicate structures of the pinna (Pritt et al. 2012). Analgesics, like meloxicam, can be administered as needed for pain and irritation. As the mite population decreases and irritation is resolved, any residual crusts can be carefully cleaned from the ear (White et al. 2002; Varga and Paterson 2020).

Affected rabbits should be placed in separate cages, to avoid any skin or fur contact with other animals, with adequate heat and ventilation conditions. All in-contact animals should be treated and the environment cleaned thoroughly to remove all organic material, including loose hair and crusts that will harbor mites. The grooming equipment and the equipment of the handlers can also be treated with an appropriate acaricidal spray discarded, to prevent a possible reinfestation from these sources, eventually (Ulutas et al. 2005; Varga and Paterson 2020).

Studies demonstrated that significant alteration of oxidant/antioxidant balance is a factor in the pathogenesis of *P. cuniculi*. Recovery can be enhanced by combining parasiticide treatment with vitamins A, D3, and E and biotin supplementation (Kanbur et al. 2008; Singh et al. 2012).

The application of restricted hygienic measures surrounding the animals is crucial to prevent infestation, including quarantine and prophylactic treatment of animals whose history is unknown (Morton et al. 2005; Pritt et al. 2012; Ilić et al. 2018). Humans also must take all precautions during handling rabbits to avoid zoonotic transmission (Abd El-Ghany 2022).

2.7 Final Remarks

The cosmopolitan parasite of rabbits *P. cuniculi* (syn. *P. ovis* var. *cuniculi*) causes infestation primarily in the ears and is of importance with respect to general animal health and hygiene and also economic concerns.

Nonburrowing mites of the genus *Psoroptes* feed superficially on lipid emulsion of skin lymph, cells, and exudates and complete their 21-day life cycle on the host. These oval mites have eight legs that projected beyond the body margin, pointed mouthparts, and three jointed pretarsi with funnel-shaped pulvilli.

Psoroptes mites are highly contagious and spread by direct contact with infested rabbits and fomites (especially bedding material). Psoroptic otocariosis in rabbits, also called psoroptic ear canker, is frequently associated with high temperatures and increased relative humidity but also poor hygiene, feeding methods, and way of handling.

In clinical disease, a grayish-white to brown exudate builds up in the external auditory canal, consisting of desquamated epithelial cells, serum, inflammatory cells, mite feces, and adult mites. As disease progresses, the exudate becomes thicker and gets more extensive, often filling the ear canal and extending onto the pinnae, being very painful when touched.

Microscopic examination of samples of exudate or the yellow-gray scales from affected ears can identify the psoroptic adult mites, mite feces, or mite eggs.

Psoroptic mange in rabbits can be treated with a wide range of parasiticides, such as ivermectin, moxidectin, eprinomectin, doramectin, permethrin, and afoxolaner. Biological agents/plants with acaricidal properties, i.e., essential oils, have been proposed for effective clinical control of psoroptic mange. Ear crusts should not be initially removed, and analgesics can be administered as needed for pain and irritation. All in-contact animals should be treated and the environment cleaned thoroughly to remove all organic material. Quarantine and prophylactic treatment of animals whose history is unknown is advisable.

3 Sarcoptic and Notoedric Mange

3.1 Introduction

Sarcoptes scabiei var. *cuniculi* and *Notoedres cati* var. *cuniculi* are highly contagious burrowing mites that can be found in farms and, less commonly, in pet rabbit. They both represent an important cause of skin disease in rabbits (Anderson et al. 2015; Panigrahi et al. 2016; Varga and Paterson 2020).

S. scabiei var. *cuniculi* is one of the most important veterinary ectoparasites in rabbits and results in considerable weight loss, productivity, and fur quality (Seddiek et al. 2013). Overall, *S. scabiei* is a nonspecific acari affecting numerous animal species and a well-known zoonotic ectoparasite (Moroni et al. 2022).

N. cati var. *cuniculi* can cause varied clinical signs like anorexia, patchy alopecia, intense pruritus, erythema, and dry scabs, on the ears, nose, face, forelimbs, and hindlimbs (Sharma et al. 2021).

3.2 The Pathogens

The Astigmata (Sarcoptiformes) are a large group of relatively similar mites. The order includes the families Sarcoptidae and Psoroptidae, which are of major veterinary importance as they contain the most common mite species causing mange and scab, in domestic animals such as dogs, cats, horses, ruminants, pigs, and rabbits and also humans. The two genera of veterinary importance, *Sarcoptes* and *Notoedres*, belong to the family Sarcoptidae (Miller et al. 2013; Taylor et al. 2016).

These astigmatic mites have circular bodies flattened ventrally and a cuticle covered with fine striations. The coxae are sunk into the body, creating a "short-leg" appearance. The legs have a clawlike empodium with the pulvillus borne on a stalklike pretarsus. Paired claws on the tarsus are absent (Taylor et al. 2016).

S. scabiei mites are characteristic globose mites with a flat ventral surface (Fig. 3). The dorsal idiosoma bears stout lateral and dorsal setae, cuticular spines, and coarse, transversely ridged, cuticular striations. Legs III and IV of both sexes do not extend beyond the lateral-posterior margin of the idiosoma, while legs I and II do extend beyond it (Arlian and Morgan 2017).

The mouthparts (capitulum) have a rounded appearance, with short, stout chelicerae and pedipalps. These mites have no eyes or stigmata. The dorsal surface of the body of *S. scabiei* is covered with transverse ridges but also bears a central patch of triangular scales. The dorsal setae are strong and spinelike. The anus is terminal and only slightly dorsal. Males are much smaller than female mites. There are a number of hosts-adapted varieties of *S. scabiei* that differ subtly in their morphology (Taylor et al. 2016; Arlian and Morgan 2017).

Notoedres closely resembles *Sarcoptes* with a circular outline and short legs with long unjointed pedicels, but it is distinguished by its concentric "thumbprint" striations and absence of spines. The dorsal scales are rounded and arranged

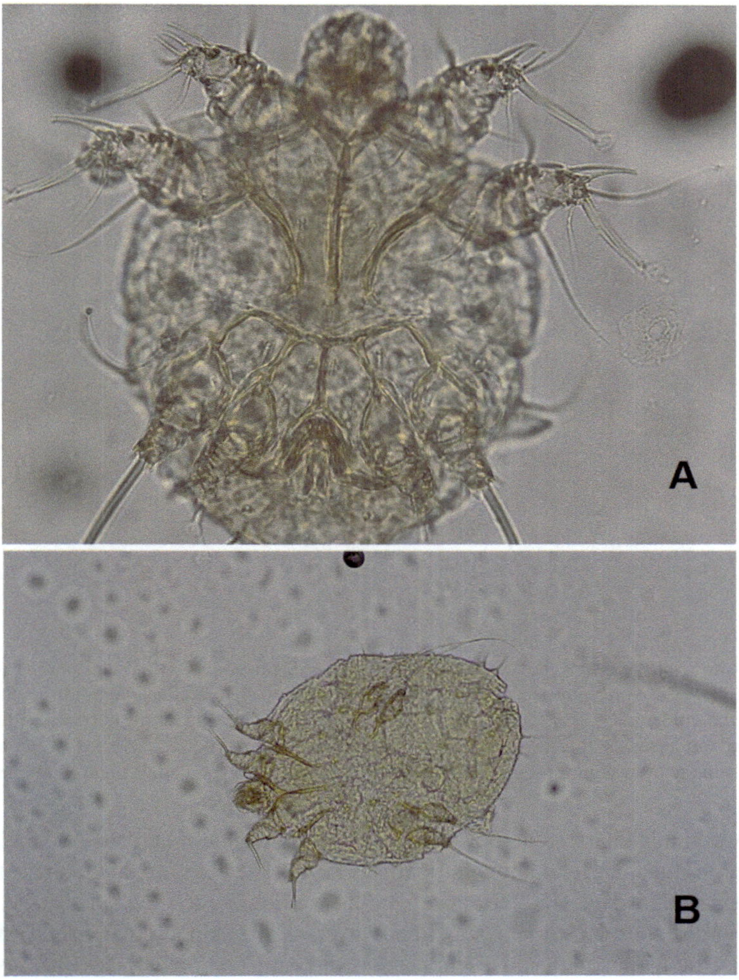

Fig. 3 *Sarcoptes scabiei*: (**a**) male (magnification 400×) and (**b**) female (magnification 100×). Laboratory of Parasitology, UTAD

transversely. Species of this genus are also smaller than *S. scabiei*. The anal opening is distinctly dorsal and not posterior (d'Ovidio and Santoro 2021).

These burrowing mites inhabit the epidermis of sparsely haired areas of the body, during their entire life cycle, where they form tunnels within the stratum corneum. The developmental stages consist of egg, larva, nymphs, and adult. The complete life cycle, from egg to adult, usually takes 17–21 days but may be as short as 14 days (Taylor et al. 2016; d'Ovidio and Santoro 2021).

After mating on the skin surface, the fertilized female mites create a winding burrow or tunnel in the upper layers of the epidermis, feeding on the damage skin

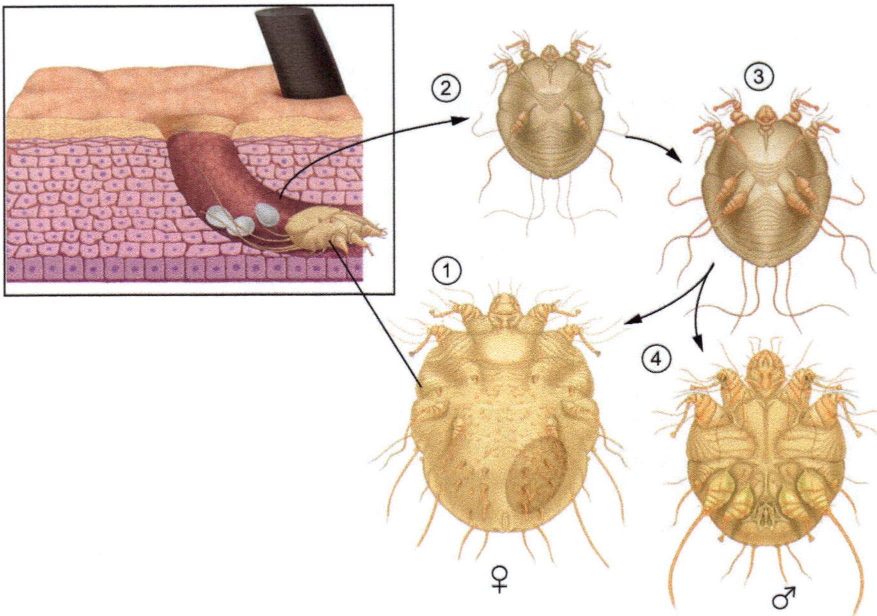

Fig. 4 Life cycle of *Sarcoptes scabiei.* Modified from Saari et al. (2019) with permission from Elsevier. 1 – adult *Sarcoptes scabiei* mites dig burrows into the keratin layer and the superficial layer of the epidermis. Fertilized females expel eggs; 2 – the eggs hatch in burrows and produce six-legged larval stages; this larva exit the tunnel and develop to an active eight-legged protonymph stage. 3 – the next molting produces a tritonymph; 4 – and finally, an adult male or female

and tissue fluids that result from their excavations, where it starts to lay their eggs, over a reproductive life of about 2 months (Fig. 4). The eggs are laid in these tunnels and hatch in 3–5 days, and the six-legged larvae migrate to the skin surface. These larvae, in turn, burrow into the superficial layers of the skin to create small "molting pockets," in which molts to nymph and adult are completed. The life cycle of *Notoedres* is similar to that of *Sarcoptes*, except that the females in the dermis are usually found in aggregations, and development from egg to adult takes 6–10 days (Taylor et al. 2016; Arlian and Morgan 2017).

Nowadays, it is accepted that each genetic variant of *S. scabiei* is associated with a single animal species or a limited number of phylogenetically related species, determining their high host specificity (Arlian and Morgan 2017; d'Ovidio and Santoro 2021). *Sarcoptes scabiei* var. *cuniculi* affects domestic (Voyvoda et al. 2005; Arul Prakash et al. 2017; Choe et al. 2020) and wild leporids (Millán et al. 2012; Cardells et al. 2021).

More than 40 species of *Notoedres* have been described, most of which infest bats or rodents, with a few species being reported from insectivores, marsupials, procyonids, and felids (Foley et al. 2016; Grunert et al. 2022). *Notoedres cati*, one of the most virulent, is found worldwide and most commonly infests felids, rarely dogs, foxes, and other canids (Mullen and O'Connor 2019). *Notoedres cati* var. *cuniculi* affects domestic rabbits (*Oryctolagus cuniculus*). A recent report described clinical cases of notoedric mange, by *N. centrifera*, in two wild species of rabbits from Florida, USA (Grunert et al. 2022).

3.3 Epidemiology

Mite infestation is highly contagious and an important problem in domestic pets and laboratory animals, like canines, felines, and lagomorphs, including rabbits, especially when reared in close proximity to moist and damp places. The morbidity and mortality in rabbit farms depends upon severity of the mite infestation (Sharma et al. 2021).

Being contagious skin parasites, sarcoptic and notoedric mites are generally transmitted by direct contact with infested animals, but indirect contact is also possible, e.g., with the contaminated environment (Arther 2009; Panigahi and Gupta 2013; Arlian and Morgan 2017). The ability of scabies mites to survive and to remain infective while off the host is key factor in the infestation of hosts with mites in the environment (Arlian and Morgan 2017). New hosts are infested by close contact, presumably from transferal of larvae, which are present more superficially than the other stages (Foley et al. 2016).

Sarcoptic mites seek the source of stimuli originating from the host when they are off the host but in close proximity to it. This behavior may facilitate finding a host if they are dislodged from it and contaminate the host environment. Thus, direct contact with an infested host may not be required for humans and other mammals to become infested with *S. scabiei* (Arlian and Morgan 2017).

For wild and domestic mammals, shared or common bedding places, stalls in barns and corrals, or close proximity in moist and damp place could serve as a source of infecting scabies mites (Arlian and Morgan 2017; Sharma et al. 2021).

Disease is more common at higher host density (Foley et al. 2016). Mange triggered by *S. scabiei* var. *cuniculi* can be a common occurrence among farm and pet rabbits and lead to considerable production losses (decreased productivity and weight loss) and occasionally to the death in conditions of exacerbated infestation (Radi 2004; Arul Prakash et al. 2017). Therefore, bearing that in mind, it is important to develop a rabbit breed with improved host resistance to *Sarcoptes*, which will help strengthen the rabbit industry (Wei et al. 2019). Some reports have suggested that it is more commonly found in some parts of the world, such as Africa and Asia (Miller et al. 2013; Elshahawy et al. 2016).

One study from India showed an overall 9.3% prevalence of sarcoptic mange infestation in rabbits (Soundararajan and Iyue 2005), while Arul Prakash et al. (2017) reported 23.6% infestation in a rabbit farm. Kumar et al. (2022) reported prevalence (9.23%) and crusting severity (low grade of 69.3% and medium grade of 30.8%) of scabies by *S. scabiei* on a rabbit farm in Kuala Lumpur, Malaysia.

The prevalence of *S. scabiei* in pet rabbits in pet stores in Israel was investigated and found in 20 of 245 rabbits (8.2%) (Eshar 2010). An epidemiological survey on mange mite of rabbits in the southern region of Egypt showed an overall prevalence, in 200 cases of dermatologic disease, of 22.5% of *S. scabiei* var. *cuniculi* (Elshahawy et al. 2016).

Both species, *S. scabiei* and *N. cati*, can also infest rabbits concurrently (Panigahi and Gupta 2013) and also with *Psoroptes cuniculi* mite (Elshahawy et al. 2016; Panigrahi et al. 2016).

Mange cases are not only a concern in rabbits for meat production but also for free-ranging wild populations (Millan 2009), with deleterious consequences for rabbit population viability (Navarro-Gonzalez et al. 2010; Abd El-Ghany 2022; Grunert et al. 2022).

Humans rarely become infested with *Notoedres* mites (Mullen and O'Connor 2019; Sharma et al. 2021). The risk that humans may be exposed to sarcoptic mange from newly adopted pet rabbits must be borne in mind, and owners should be advised of this possibility (Eshar 2010). Live mites in bedding, furniture, toys, and clothing can be a source of infestation. *Sarcoptes scabiei* var. *hominis* have been recovered from laundry bins in a nursing home (Ljunggren 2005; Arlian and Morgan 2017).

3.4 The Disease

The feeding behavior of mite larvae and nymphs may cause irritation, hypersensitivity reaction, and inflammation, resulting in different dermatological clinical signs (Scott et al. 2001). This usually occurs 3 weeks or more after the initial infestation, with the reaction time becoming much shorter (e.g., a few days) after subsequent exposures (Mullen and O'Connor 2019).

Sarcoptic mange lesions most commonly initiate on the face, neck, and ears of rabbits but can also spread to the limbs, trunk, and genitalia (Mullen and O'Connor 2019; Kumar et al. 2022). Clinical examination reveals alopecia, erythema, scales, and crusts (Eshar 2010; d'Ovidio and Santoro 2021), resulting in thickening of the skin, with possible disfigurement of face and ear (Arul Prakash et al. 2017). Intense pruritus causes alopecia and dermal abrasions, commonly resulting in serous encrustations and secondary bacterial dermatitis (Scott et al. 2001).

Clinicopathologic changes associated with sarcoptic mange in rabbits include anemia and leukopenia and hepatic or renal amyloidosis in severe cases. Chronic cases can lead to anorexia, lethargy, emaciation, and death (Eshar 2010).

With *Notoedres*, lesions are more frequently confined to the head (Varga and Paterson 2020), but sometimes they can appear on the ventral abdomen, legs, and genital area, especially in younger animals (Mullen and O'Connor 2019; Sharma et al. 2021). The clinical signs of notoedric mange vary in severity and outcome, with alopecia, pruritis, yellow and crusty dermatitis, hyperkeratosis, secondary infections, emaciation, and even systemic debilitation and death in more severe chronic cases (Mullen and O'Connor 2019; Grunert et al. 2022).

3.5 Diagnosis

Diagnosis of sarcoptic or notoedric mange is easily done by collecting skin scrapings, from the different affected parts of the body, and examination under the microscope for mites. Microscopy should reveal crusts containing different developmental stages, i.e., eggs, larvae, nymphs, and adults (Darzi et al. 2007; Panigrahi et al. 2016; Choe et al. 2020). Scraping should be done at the periphery of the lesions just deeply enough to reach capillary blood (Casais et al. 2015; Foley et al. 2016).

If material is mounted in oil, the motility of the mites can be used to aid identification (Choe et al. 2020). If samples are not viewed immediately, the mites can curl up, making identification difficult. Potassium hydroxide is a better mounting material, which also clears and dissolves extraneous keratin and debris (Varga and Paterson 2020; Sharma et al. 2021).

Detection of one *Sarcoptes* or *Notoedres* mite is enough to classify an animal as positive (Casais et al. 2015). Microscopic diagnosis is based on the morphology of these rounded tortoise-like mites, with four comparatively short pairs of legs, focusing, among other features, on the position of the anus and the presence of the rings and scales on the dorsal surface (Galdhar et al. 2015; Choe et al. 2020).

S. scabiei has a round body with transverse ridges and triangular scales on dorsal surface and terminal anus, whereas *Notoedres* has concentric rings (thumbprint-like) on dorsal surface, absence of dorsal scales, and distinct dorsal anus (Darzi et al. 2007). If the animal is presented as a carcass, a piece of skin may be digested, and the mites floated on a high-density medium to concentrate mites (Foley et al. 2016).

An ELISA, based on the *S. scabiei* recombinant antigen, was able to detect specific antibodies from 3 weeks post-infestation in domestic rabbits experimentally infested, with very high sensitivity and specificity (Casais et al. 2015). Implementing molecular tools allows an early diagnosis and, therefore, a more effective use of treatments avoiding contagion of other animals, farms, and humans (Casais et al. 2015; Arlian and Morgan 2017). DNA extraction can be done either from the scraped material in saline or from a digested piece of skin (from a dead animal) (Foley et al. 2016).

Skin biopsy is not frequently used to diagnose sarcoptic or notoedric mange, but histopathologic examination, whenever performed, reveals hyperkeratosis with a mixed inflammatory cell infiltrate (Varga and Paterson 2020).

3.6 Treatment and Prevention

Currently, off-label treatments have been demonstrated to be efficacious for sarcoptic mange in rabbits with good prognosis and include ivermectin at 200–400 μg/kg SC repeated at 10–14-day intervals for three treatments (Hoppmann and Wilson Barron 2007), at weekly intervals for 4 weeks (Panigrahi et al. 2016; Divisha et al. 2020), or a single dose long acting (3.15%) formulation at 700 μg/kg (Sharun et al. 2019); selamectin at 6–18 mg/kg single dose of a topical spot-on solution on the neck (Kurtdede et al. 2007) or twice, 30 days apart, at a dose of

8–14 mg/kg (Farmaki et al. 2009); imidacloprid (10 mg/kg) plus moxidectin (1 mg/kg) solution local on the neck three times repeated after 30 days; doramectin (400 μg/kg, IM) as a single dose (Darzi et al. 2007) or 1% injectable solution at 0.2 mg kg at 10-day intervals for two treatments (Voyvoda et al. 2005); and fluralaner (25 mg/kg), as a single oral dose administration (d'Ovidio and Santoro 2021).

For notoedric mange, avermectins have been used with successful results, including ivermectin at 200–400 μg/kg SC at weekly intervals for 4 weeks (Panigrahi et al. 2016), two doses at fortnight interval (Galdhar et al. 2015; Sharma et al. 2021), or as a single dose (Borkataki et al. 2018) and doramectin (400 μg/kg, IM) as a single dose (Darzi et al. 2007). All in-contact animals should be treated (Varga and Paterson 2020). Appropriate drug withdrawal times must be considered in rabbits for food consumption (Harkness et al. 2010).

The development of drug resistance (Currie et al. 2004), presence of tissue residues, and potential adverse effects on the health and productivity of rabbits (El-Nahas and El-Ashmawy 2008; El-Maddawy and Abd El Naby 2017) are becoming a concern in mange control (Abd El-Ghany 2022). Moreover, several studies indicate that ivermectin residues present in the environment can become highly toxic when combined with other water pollutants such as pesticides, affecting aquatic organisms such as crustaceans and fish (Sanderson et al. 2007; Paola et al. 2022).

These worrying issues promoted the search efforts to discover innovative novel drugs that provide new acaricidal compounds for effective control of sarcoptic mange (Abd El-Ghany 2022). This includes different botanic extracts, including *Eupatorium adenophorum* (Nong et al. 2012, 2013b), *Acacia nilotica*, and *Psidium guajava* (Khan et al. 2022), and also propolis (Metwally et al. 2018).

Supportive treatment using vitamins with other therapeutic medicaments could improve the clinical recovery of rabbits against *S. scabiei* (Kumar et al. 2018).

Many studies showed that infestation with *S. scabiei* could induce some degree of protective immunity in rabbits (Casais et al. 2014). Although vaccination studies showed that it does not efficiently control sarcoptic mange in rabbits, the immunogenic properties of some compounds are promising for developing an effective vaccine in the future (Zhang et al. 2012; Liu et al. 2014).

The elimination of ectoparasites from the environment should be achieved by treatment and containment of all cohabiting animals (affected and unaffected rabbits), removing all kind of fomites such as feed bags, carts, trash cans, and clothing and notebooks, thorough mechanical scrubbing and, in severe cases, fumigation of the premises with hydrogen peroxide (Harkness et al. 2010). Additionally, a blowlamp can be used on the cages and surroundings (Darzi et al. 2007), or silicate dusts, alone or compounded with insecticides, can be applied on the bedding, aiming to abrade and desiccate ectoparasites, thus providing a mechanical alternative to potentially toxic chemicals (Harkness et al. 2010).

Some pyrethroids, with biocide and repellent effect, can be effective for environmental control of *S. scabiei* (Fang et al. 2015).

The application of restricted hygienic measures and thorough disinfection of rabbit cages and surroundings is crucial to prevent mite infestation (Abd El-Ghany

2022). Wild rodents must be excluded from the colony or housing room (Harkness et al. 2010).

In newly arriving rabbits to the farm, unless known to be free of ear mites, it should be considered to quarantine and treat them with acaricides, before being placed in an established clean colony (Harkness et al. 2010). Humans also must take all precautions during handling rabbits to avoid zoonotic transmission (Abd El-Ghany 2022).

3.7 Final Remarks

S. scabiei var. *cuniculi* and *N. cati* var. *cuniculi* are highly contagious burrowing mites that represent an important cause of skin disease in rabbits, resulting in considerable weight loss, productivity, and fur quality. These sarcoptiforme mites have characteristic globose bodies with short legs and inhabit the epidermis forming tunnels within the stratum corneum, completing their life cycle from egg to adult in 14–21 days.

Sarcoptic and notoedric mites are generally transmitted by direct contact with infested animals, but indirect contact is also possible, e.g., with the contaminated environment.

Disease is more common at higher host density, hence its higher occurrence among farm rabbits, leading to important prevalence worldwide. Human exposure must be considered, with zoonotic potential.

Clinical presentation is frequently characterized by intense pruritus, alopecia, erythema, scales, and crusts, with possible secondary bacterial dermatitis. Lesions are most commonly located on the face, neck, and ears of rabbits but can also spread to the limbs, trunk, ventral abdomen, and genital area.

Diagnosis of sarcoptic or notoedric mange is done by collecting skin scrapings, from the lesions, and examination under the microscope for mites (eggs, larvae, nymphs, and adults). Detection of *S. scabiei* antibodies and DNA extraction are high sensitivity and specificity tests and can be an additional diagnostic tool.

Currently, off-label treatments have been demonstrated to be efficacious for burrowing mites in rabbits, including ivermectin, selamectin, moxidectin, doramectin, and fluralaner. Supportive treatment using vitamins could improve clinical recovery. New studies with acaricidal compounds are being conducted for effective biocontrol of sarcoptic mange, along with promising vaccines. The application of restricted hygienic measures and thorough disinfection of rabbit cages and surroundings are crucial to treat and prevent mite infestation.

4 Fur Mites: *Cheyletiella parasitovorax* and *Leporacarus gibbus*

4.1 Introduction

Cheyletiella parasitovorax and *Leporacarus gibbus* (formerly known as *Listrophorus gibbus*) are nonburrowing fur mites that affect both wild and domestic rabbits worldwide (Harcourt-Brown 2002; Frank et al. 2013). The mites can just be seen with the naked eye in the fur or on microscopic examination of skin brushings. Mites can be found on any part of the body, but usually along the dorsum, and may not be associated with clinical signs and disease (Harkness et al. 2010; Varga and Paterson 2020).

Large number of *C. parasitovorax* can cause a very scaly (mites moving clear to white structures leading to the term "walking dandruff"), dry, pruritic dermatitis, with zoonotic potential (Hoppmann and Wilson Barron 2007); Varga and Paterson 2020; Abd El-Ghany 2022).

L. gibbus infestation is usually asymptomatic even in large numbers, although it may be associated with mild scurf and pruritus (Harcourt-Brown 2002; Varga and Paterson 2020), and can cause lesions in people (d'Ovidio and Santoro 2014). Grooming difficulties and hypersensitivity reaction could account for differences in pathogenicity of mites between individuals (Harcourt-Brown 2002).

4.2 The Pathogens

Several species of mites of the genus *Cheyletiella* are of veterinary and medical importance as ectoparasites of dogs, cats, or rabbits that may transfer to humans (Meredith 2006; Taylor et al. 2016). *C. parasitovorax* belongs to the family Cheyletiellidae (order Prostigmata), is a fur mite that normally parasites rabbits, and lives on the keratin layer of the epidermis. The body of the mite, up to 0.4 mm slightly elongated, has a "waist" and a bladelike chelicerae that are used for piercing their host and short, strong, opposable palps with curved palpal claws (Fig. 5). The legs are short, and tarsal claws are lacking and have comblike pulvilli at the ends of the legs (Harcourt-Brown 2002; Taylor et al. 2016).

All developmental stages of C. *parasitovorax* occur on the host and are completed in about 2 weeks. Ova are attached to hair shafts and there are two nymphal stages. The mites live in the hair and fur, only visiting the skin to feed on lymph and other tissue fluids. Adults are highly mobile and able to move about rapidly, and female mites can survive for at least 10 days off the host (Meredith 2006; Varga and Paterson 2020). Large numbers of *C. parasitovorax* mites can be found along the back and particularly above the tail base and on the neck of affected rabbits (Harcourt-Brown 2002; Kim et al. 2008a, b), but sometimes extending to the hindquarters and abdomen (Varga and Paterson 2020).

L. gibbus is a common fur mite from rabbits, belongs to the family Listrophoridae (order Astigmata), and browses on surface keratin and sebaceous secretions. The life

Fig. 5 *Cheyletiella* spp. (magnification 40×). Laboratory of Parasitology, UTAD

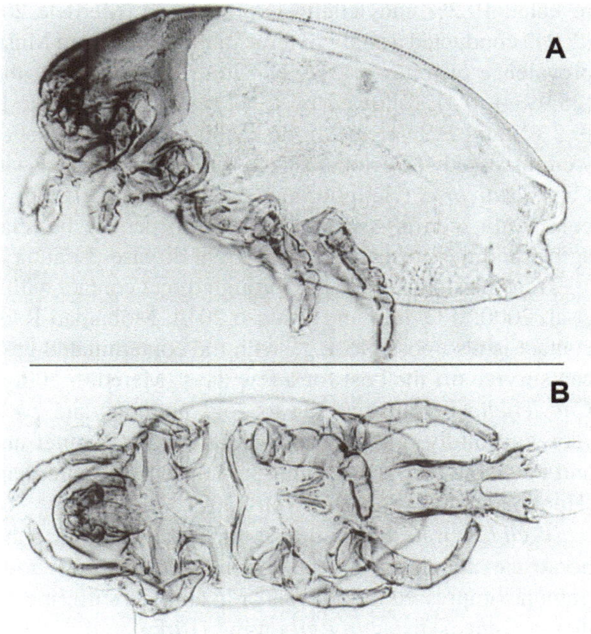

Fig. 6 *Leporacarus gibbus* (**a**) adult female, × 179 (length, 558 μm), and (**b**) adult male, × 174 (length, 605 μm). Adapted from Kirwan et al. (1998) with permission from John Wiley & Sons

cycle is completed on the host, and mites and eggs are found attached to hair shafts (Taylor et al. 2016; Varga and Paterson 2020). The brown mites are laterally compressed, with short legs and a distinct dorsal shield, and have mouthparts and legs modified with membranous flaps for grasping hairs (Fig. 6). There is sexual dimorphism; female mites are large and oval with a thumbprint pattern over the body, and males are slightly smaller and have two long adanal processes with distinct anal suckers (Harcourt-Brown 2002; Taylor et al. 2016). It is found predominantly on the abdomen and back of rabbits (Harkness et al. 2010).

Besides domestic and wild rabbits, *L. gibbus* can also parasite cats (Dumitrache et al. 2021) and has zoonotic potential (d'Ovidio and Santoro 2014).

4.3 Epidemiology

Being contagious skin parasites, mite infestation can be a problem reported in pets, petting farms, and farmed practices, and some species are potential zoonotic from rabbits to human (Merdele 2010; d'Ovidio and Santoro 2014; Hajipour et al. 2020; Mohamad-Radzi et al. 2021). In a study in pet rabbits in South Korea, *C. parasitovorax* was found in 80 of 140 rabbits (57.1%) and *L. gibbus* in 6 of 140 rabbits (4.3%). Clinical signs were observed in 76 of 80 infested rabbits, and prevalence was particularly higher in rabbits younger than 1 year (Kim et al. 2008a, b). In a rabbit breeding farm in Romania, 73 rabbits with skin lesions revealed 19.2% cheyletiellosis prevalence (Merdele 2010). Mohamad-Radzi et al. (2021) conducted a study in three rabbit farms from Malaysia and reported an overall prevalence of mites of 51.9%, with *S. scabiei* being the most frequent mite found (25.9% ± 0.4), followed by *C. parasitovorax* (single [9.3%] or mixed infestation [9.3%]) and *P. cuniculi*. In 50 rabbits from rabbit sellers and breeders in Iran that were assessed for ectoparasites, 18% were infested with *S. scabiei* and 38% with *C. parasitivorax* (Hajipour and Zavarshani 2020). For some authors, outbreaks are commonly seen in solitary rabbits and seem to be related to immunosuppression, aging, and the presence of concurrent disease (Lennox and Kelleher 2009).

Transmission is probably through direct contact with infested animals (Tsianakas et al. 2000; d'Ovidio and Santoro 2014; Mohamad-Radzi et al. 2021). But indirect contact is also possible, e.g., with the contaminated environment, since adult mites can survive off the host for a few days (Meredith 2006; Varga and Paterson 2020). *Cheyletiella* is transmissible to dogs and cats and zoonotic to people in whom it causes a mildly pruritic papular dermatitis (Wagner and Stallmeister 2000; Varga and Paterson 2020). In a multi-pet household, more than one pet is usually affected (Hoppmann and Wilson Barron 2007).

L. gibbus mite infestation is rarely reported in either laboratory or pet rabbits because of subclinical infestation in healthy rabbits; however, disease can be seen in immunosuppressed individuals or in animals suffering from an underlying condition that prevents satisfactory grooming (Birke et al. 2009; Zygner and Gójska-Zygner 2018; Deak and Turcu 2020).

4.4 Disease

The presence of fur mites may not be associated with disease, and they can be considered to be commensal ectoparasites, although hypersensitivity reactions have been reported (Harcourt-Brown 2002; Pritt et al. 2012),

Underlying grooming difficulties due to dental disease, long fine coat texture, obesity, spondylitis, or musculoskeletal disease prevent the rabbit from removing parasites and debris from the coat, thus allowing the parasites to proliferate and cause disease (Harcourt-Brown 2002; Deak and Turcu 2020). Young or immunosuppressed animals are also at risk (Kim et al. 2008a, b; Varga and Paterson 2020).

Cheyletiellosis can be a common parasitic skin disorder of rabbits, and many harbor the mite with no clinical signs. However, large numbers of *C. parasitovorax* mites can cause intense pruritus, alopecia, and dermatitis associated with erythema, particularly above the tail base and on the neck and in periauricular regions of affected rabbits (Hoppmann and Wilson Barron 2007; Merdele 2010; Varga and Paterson 2020).

The excessive white, large flaky scaling gives the fur a powdery appearance, especially over the dorsal midline of the back (Hoppmann and Wilson Barron 2007; Harkness et al. 2010).

L. gibbus infestation is usually asymptomatic in healthy animals, although in heavy infestations, aging, or immunosuppression, it may be associated with moderate scaling, erythema, pruritus, and alopecia (d'Ovidio and Santoro 2014; Lennox and Kelleher 2009).

Fur mites have zoonotic potential and usually, in rare cases, cause a mildly pruritic papular dermatitis in people that are in contact with infested animals (Wagner and Stallmeister 2000; Tsianakas et al. 2000; d'Ovidio and Santoro 2014).

4.5 Diagnosis

Diagnosis of fur mite infestation is made by parasitological examination of the rabbits' skin and fur, and they also are visible to the naked eye and tend to move (Harcourt-Brown 2002). *C. parasitovorax* and *L. gibbus* mites can be identified by microscopic examination of superficial skin scrapings, acetate tape preparations, hair plucking or brushing, or examination of dandruff collected by flea comb. Microscopy should reveal crusts containing different developmental stages, i.e., eggs, larvae, nymphs, and adults (Frank et al. 2013; Varga and Paterson 2020).

If material is mounted in oil, the motility of the mites can be used to aid identification (Choe et al. 2020). If samples are not viewed immediately, the mites can curl up, making identification difficult. Potassium hydroxide is a better mounting material, which also clears and dissolves extraneous keratin and debris (Varga and Paterson 2020). *Cheyletiella* mites are saddle-shaped with inward curving claws, hooklike mouth parts, and short, hairlike setae at the distal end of their limbs (Varga and Paterson 2020). *Leporacarus* mites are large, oval, color brown, and laterally compressed, with short legs, a distinct dorsal shield, and a thumbprint pattern over the body (Taylor et al. 2016).

Identification to genus is usually sufficient for diagnosis, but it is important to be aware of the potential for cross-transmission of the various species of *Cheyletiella* from other in-contact hosts (Taylor et al. 2016).

4.6 Treatment and Prevention

Treatment of cheyletiellosis and leparocariosis involves killing the mites, removing the scales and skin debris, and identifying and treating underlying grooming difficulties or health issues, such as obesity and dental or musculoskeletal disease (Harcourt-Brown 2002).

Currently, off-label treatments have been demonstrated to be effective for fur mites in rabbits, with good prognosis, which include ivermectin at 300–400 μg/kg SC every 14 days for two or three treatments (Tsianakas et al. 2000; Hoppmann and Wilson Barron 2007; Zygner and Gójska-Zygner 2018); selamectin at 11–16.6 mg/kg single dose of a topical spot-on solution on the neck (Kim et al. 2008a, b; Birke et al. 2009) or at 6–18 mg/kg twice, 28 days apart (Hoppmann and Wilson Barron 2007); imidacloprid (14.8–22.2 mg/kg) plus permethrin (74.1–111.1 mg/kg) single dose of a topical spot-on solution on the dorsal cervical area (Birke et al. 2009), although less rapidly than selamectin to eliminate the infestation; imidacloprid (10 mg/kg) plus moxidectin (1 mg/kg) single application of a spot-on formulation (d'Ovidio and Santoro 2014); and doramectin (200 μg/kg) as a single oral dose administration (dos Santos et al. 2017).

All in-contact animals should be treated (Varga and Paterson 2020). Appropriate drug withdrawal times must be considered in rabbits for food consumption (Harkness et al. 2010).

Contaminated environment should be addressed, since adult mites can survive off the host for a few days (Meredith 2006; Varga and Paterson 2020). The rabbit's litter box and bedding can be washed, when possible, with boiled water, and its litter recommended to be changed every day (Zygner and Gójska-Zygner 2018) or alternatively disinfected with permethrin at weekly intervals (Tsianakas et al. 2000; Harkness et al. 2010; d'Ovidio and Santoro 2014).

4.7 Final Remarks and Conclusions

The cosmopolitan parasite of rabbits *P. cuniculi* (syn. *P. ovis* var. *cuniculi*) causes infestation primarily in the ears and is of importance with respect to general animal health and hygiene and also economic concerns.

Nonburrowing mites of the genus *Psoroptes* feed superficially on lipid emulsion of skin lymph, cells, and exudates and complete their 21-day life cycle on the host. These oval mites have eight legs that projected beyond the body margin, pointed mouthparts, and three jointed pretarsi with funnel-shaped pulvilli.

Psoroptes mites are highly contagious and spread by direct contact with infested rabbits and fomites (especially bedding material). Psoroptic otocariosis in rabbits, also called psoroptic ear canker, is frequently associated with high temperatures and increased relative humidity, but also poor hygiene, feeding methods, and way of handling.

In clinical disease, a grayish-white to brown exudate builds up in the external auditory canal, consisting of desquamated epithelial cells, serum, inflammatory cells, mite feces, and adult mites. As disease progresses, the exudate becomes thicker and gets more extensive, often filling the ear canal and extending onto the pinnae, being very painful when touched.

Microscopic examination of samples of exudate or the yellow-gray scales from affected ears can identify the psoroptic adult mites, mite feces, or mite eggs.

Psoroptic mange in rabbits can be treated with a wide range of parasiticides, such as ivermectin, moxidectin, eprinomectin, doramectin, permethrin, and afoxolaner. Biological agents/plants with acaricidal properties, i.e., essential oils, have been proposed for effective clinical control of psoroptic mange. Ear crusts should not be initially removed, and analgesics can be administered as needed for pain and irritation. All in-contact animals should be treated and the environment cleaned thoroughly to remove all organic material. Quarantine and prophylactic treatment of animals whose history is unknown are advisable.

Nonburrowing fur mites *C. parasitovorax* and *L. gibbus* (formerly *Listorphorus gibbus*) affect both wild and domestic rabbits worldwide, are found on any part of the body usually along the dorsum, and may not be associated with clinical signs and disease.

Cheyletiella parasitovorax is slightly elongated and has a "waist," a bladelike chelicerae, and short and strong palps with curved palpal claws. *Leporacarus gibbus* are large and oval, laterally compressed, with short legs and a distinct dorsal shield.

Transmission is probably through direct contact with infested animals, but indirect contact is also possible, e.g., with the contaminated environment, since adult mites can survive off the host for a few days.

Cheyletiella mite infestation can be a problem reported in pets, petting farms, and farmed practices, and some species are potential zoonotic from rabbits to human causing a mildly pruritic papular dermatitis. *Leporacarus gibbus* mite infestation is rarely reported in either laboratory or pet rabbits because of subclinical infestation in healthy rabbits; however, disease can be seen in immunosuppressed individuals or in animals suffering from an underlying condition that prevents satisfactory grooming.

Clinical presentation is frequently characterized by flaky scaling, intense pruritus, alopecia, and dermatitis associated with erythema.

Diagnosis of fur mite infestation is made by parasitological examination of the rabbits' skin and fur, and they also are visible to the naked eye and tend to move. A large number of white *C. parasitovorax* moving mites are better known by the term "walking dandruff."

Treatment of cheyletiellosis and leparocariosis involves killing the mites with effective and safe drugs (i.e., ivermectin, selamectin, moxidectin, and doramectin), removing the scales and skin debris, and identifying and treating underlying grooming difficulties or health issues, such as obesity and dental or musculoskeletal disease. The application of hygienic measures and disinfection of rabbit bedding and surroundings are crucial to treat and prevent mite infestation.

5 Pediculosis, Pulicosis, and Ixodidosis: An Overview

The impact of pediculosis, pulicosis, and ixodidosis in rabbitries is low or residual. When lice, fleas, or ticks are present, they are usually limited to cause mild clinical signs as pruritus, unless intense infestation occurs. The major problem is their role in vector-borne diseases in rabbitries.

These parasitic diseases are usually prevented through biosecurity measures implementation in rabbitries. The treatment is mainly based on macrocyclic lactones, despite the possibility of drug resistance (Box 1).

Box 1 Ectoparasiticides for Treatment in Rabbits (Reviewed by Eshar 2019)

Macrocyclic lactones

- Ivermectin: 0.2–0.4 mg/kg SC every 10–14 days for three treatments (adult mites, lice, and ticks)
- Doramectin: 0.2–0.3 mg/kg SC once (*Psoroptes* spp.); 0.2 mg/kg PO once (*L. gibbus*; *Psoroptes* spp.)
- Eprinomectin: 0.2–0.3 mg/kg SC once or 0.5 mg/kg every 2–3 weeks according to treatment response (*Psoroptes* spp.)
- Moxidectin: 0.2–0.3 mg/kg SC every 10–14 days for three treatments (adult mites); 0.2 mg/kg PO every 10 days for two treatments (*Psoroptes* spp.)
- Selamectin: 12 mg/kg topically once (*Cheyletiella* spp.); 20 mg/kg topically once a week (fleas); 8–14 mg/kg topically every 30 days for two treatments (*Sarcoptes* spp.); 6–18 mg/kg topically once (*Psoroptes* spp.)

Imidacloprid (and associations)

- Imidacloprid: 10–16 mg/kg topically once or weekly as needed (adult fleas)
- Imidacloprid + moxidectin: topical 10 mg/kg + 1 mg/kg, respectively, every 4 weeks for three treatments (*Psoroptes* spp.)
- Imidacloprid + permethrin: topical 11–16 mg/kg once (*L. gibbus*)

Lufenuron: 30 mg/kg PO every 30 days (flea larva)
Fluralaner: 20 mg/kg PO once (*Psoroptes* spp.)

5.1 Pediculosis

The rabbit pediculosis is caused by the sucking louse *Haemodipsus ventricosus*. Its occurrence is rare, and when found, they are commonly located in the scapular zone. Nonetheless, the incidence of *H. ventricosus* (Eshar 2019) and other species of this genus, e.g., *H. lyriocephalus*, *H. setoncan* (Dik and Uslu 2007), and *H. leporis* (Dik and Uslu 2012), can be relevant in wild rabbits and other lagomorphs. The life cycle varies between 2 and 5 weeks (Mayer and Donnelly 2013).

Clinical signs of lice infestations include severe itching and erythema, fine hair, and zones of alopecia (Palmeiro and Roberts 2013). Anemia can also occur in very young rabbits or severe and advanced infestations. *H. ventricosus* can be directly detected on fur, once it is 1.5–2.5 mm in length (Harcourt-Brown 2002; Frank et al. 2013). The diagnosis can be confirmed by microscopic identification from trichograms and scrapings (Palmeiro and Roberts 2013).

The useful treatment is ivermectin (4 mg/kg, SC), at intervals of 7 days for 3–5 weeks. Imidacloprid can also be used (Paterson 2006).

Lice are vectors for tularemia (*Francisella tularensis*) and myxomatosis (Mehlhorn 2008; Mayer and Donnelly 2013; Varga 2014).

5.2 Pulicosis

The most commonly found flea in wild (González et al. 2021) and also domestic (Pinter 1999) rabbits is *Spilopsyllus cuniculi*. This is a European rabbit flea, but it is also frequent in North Africa (Ammam et al. 2022). The cat (*Ctenocephalides felis*) and dog (*C. canis*) fleas are also often found in pet rabbits or on familiar small-scale rabbitries. Fleas share their housing between hosts and environment, where they deposit eggs (Varga 2014). *S. cuniculi* reproduction (egg maturation) is related to rabbit pregnancy (host steroids association) (Rothschild and Ford 1964) and together with kits (in nest), they are more likely to be intensively infested (Mayer and Donnelly 2013).

Observable signs often include black spots on the skin, which when moistened and placed on cotton turn red. They correspond to flea feces, which are dried blood found where fleas suck the host's blood. It is uncommon to see live fleas unless the infestation is at an advanced stage. It is also possible to observe alopecia in the form of patches, which appear itchy. Pruritis is the most common clinical manifestation due to the flea bite.

Pulicosis is rarely fatal unless the infestation is of kits or young rabbits. The main caveat with fleas is that they are a vector for myxomatosis (El Hamzaoui et al. 2020) having the potential to transfer the infection from one rabbit to another or other parasites such as *Trypanosoma nabiasi* (Hamilton et al. 2005).

Sprays or shampoos containing pyrethrin or pyrethroids are usually ineffective and are not enough to definitively end the infestation, since the eggs laid by adult fleas hatch after the rabbit is treated. Avermectins, such as selamectin and ivermectin, can be used to treat and control fleas, at least partially. Fipronil is effective to kill flea but is contraindicated in rabbits due to neurotoxicity (Elhawary et al. 2018; d'Ovidio and Cortellini 2022). Environmental sanitation is the best measure to control and prevent flea infestation.

5.3 Ixodidosis

Ticks are not common in domestic rabbits, but the most frequent is *Haemaphysalis leporispalustris* (Labruna et al. 2000). Ticks may act as vectors of myxomatosis, tularemia, and rickettsioses (*Rickettsia rickettsia*) (Freitas et al. 2009; Palmeiro and Roberts 2013; Roth et al. 2017). In intense infestations, it can cause anemia, especially in young rabbits.

6 Conclusions

Overall, ectoparasitosis are a relevant group of diseases in rabbits that can cause an insidious impact on rabbit health and even significant losses in rabbitries. Ectoparasites can serve as vectors for other etiological agents leading to vector-borne diseases as well as a potential zoonotic threat. Even different therapeutic protocols can be suitable to treat and control these diseases, and the implementation of biosecurity measures in farms is the best option to prevent them, helping to reduce drug resistance.

Acknowledgments The authors thank Springer for the use of the AJE Digital Editing. This work was supported by the projects UIDP/CVT/00772/2020 (CECAV) and LA/P/0059/2020 (AL4AnimalS), funded by the Portuguese Foundation for Science and Technology (FCT).

References

Abd El-Ghany WA (2022) Mange in rabbits: an ectoparasitic disease with a zoonotic potential. Vet Med Int 2022:5506272. https://doi.org/10.1155/2022/5506272

Acar A, Kurtdede A, Ural K, Cingi CC, Karakurum MÇ, Yağçi BB, Sarı B (2007) An ectopic case of *Psoroptes cuniculi* infestation in a pet rabbit. Turkish J Vet Anim Sci 31:423–425

Amer S, Abd El Wahab T, El Naby Metwaly A, Feng Y, Xiao L (2015) Morphologic and genotypic characterization of *Psoroptes* mites from water buffaloes in Egypt. PLoS One 10:e0141554. https://doi.org/10.1371/journal.pone.0141554

Ammam I, Rahal M, Rahal K, Bitam I (2022) Epidemiological study and identification of some flea species infesting wild rabbits (*Oryctolagus cuniculus*) and cape hares (*Lepus capensis*) in northern Algeria. Iraqi J Vet Sci 36:159–165. https://doi.org/10.33899/ijvs.2021.129596.1665

Amorim MG, Azevedo SS, Riet-Correa F (2015) Seasonal prevalence and mean intensity of *Psoroptes ovis* infestation in goats in the Brazilian semiarid region. Braz J Vet Parasitol 24: 59–65. https://doi.org/10.1590/S1984-29612014094

Arlian LG, Morgan MS (2017) A review of *Sarcoptes scabiei*: past, present and future. Parasit Vectors 10:297. https://doi.org/10.1186/s13071-017-2234-1

Arther RG (2009) Mites and lice: biology and control. Vet Clin N Am Small Anim Pract 39:1159–1171. https://doi.org/10.1016/j.cvsm.2009.06.009

Arul Prakash M, Soundararajan C, Nagarajan K, Tensingh Gnanaraj P, Ramesh Saravanakumar V (2017) Sarcoptic mange infestation in rabbits in an organized farm at Tamil Nadu. J Parasit Dis 41:429–432. https://doi.org/10.1007/s12639-016-0821-2

Bates PG (1999) Inter- and intra-specific variation within the genus *Psoroptes* (Acari: Psoroptidae). Vet Parasitol 83:201–217. https://doi.org/10.1016/s0304-4017(99)00058-8

Birke LL, Molina PE, Baker DG, Leonard ST, Marrero LJ, Johnson M, Simkin J (2009) Comparison of selamectin and imidacloprid plus permethrin in eliminating *Leporacarus gibbus* infestation in laboratory rabbits (*Oryctolagus cuniculus*). J Am Assoc Lab Anim Sci 48:757–762

Borges DA, Oliveira PC, Dias JC, Assis RCP, Lima IP, Paula GB, Scott FB (2022) Efficacy of a dinotefuran, permethrin and pyriproxyfen combination product in the treatment of rabbits (*Oryctolagus cuniculus*) naturally infested by *Psoroptes ovis*. Vet Parasitol 303:109681. https://doi.org/10.1016/j.vetpar.2022.109681

Borkataki S, Islam S, Goswami P (2018) Efficacy of ivermectin against *Notoedres cati* var *cuniculi* in New Zealand white rabbits. Vet Pract 19:29–30

Bowman DD (2009) Georgis' parasitology for veterinarians. Saunders, St. Louis, p 451. ISBN 9781416057567

Bowman DD, Fogelson ML, Carbone LG (1992) Effect of ivermectin on the control of ear mites (*Psoroptes cuniculi*) in naturally infested rabbits. Am J Vet Res 53:105–109

Calderón-Arguedas O, Troyo A, Avendaño A, Aymerich R, Berrocal B, Coto-Morales T (2011) Infestación múltiple por ácaros ectoparásitos en conejos de crianza. Revista Ibero-Latinoamericana de Parasitologia 70:114–118

Cardells J, Lizana V, Martí-Marco A, Lavín S, Velarde R, Rossi L, Moroni B (2021) First description of sarcoptic mange in an Iberian hare (*Lepus granatensis*). Curr Res Parasitol Vector Borne Dis 1:100021. https://doi.org/10.1016/j.crpvbd.2021.100021

Casais R, Dalton KP, Millán J, Balseiro A, Oleaga A, Solano P, Goyache F, Prieto JM, Parra F (2014) Primary and secondary experimental infestation of rabbits (*Oryctolagus cuniculus*) with *Sarcoptes scabiei* from a wild rabbit: factors determining resistance to reinfestation. Vet Parasitol 203:173–183. https://doi.org/10.1016/j.vetpar.2014.03.001

Casais R, Millán J, Rosell JM, Dalton KP, Prieto JM (2015) Evaluation of an ELISA using recombinant Ssλ20ΔB3 antigen for the serological diagnosis of *Sarcoptes scabiei* infestation in domestic and wild rabbits. Vet Parasitol 214:315–321. https://doi.org/10.1016/j.vetpar.2015.07.011

Choe S, Kim S, Na KJ, Nath TC, Ndosi BA, Kang Y, Bia MM, Lee D, Park H, Eamudomkarn C, Jeon HK, Eom KS (2020) First infestation case of Sarcoptic mange from a pet rabbit *Oryctolagus Cuniculus* in Republic of Korea. Korean J Parasitol 58:315–319. https://doi.org/10.3347/kjp.2020.58.3.315

Currie BJ, Harumal P, McKinnon M, Walton SF (2004) First documentation of *in vivo* and *in vitro* ivermectin resistance in *Sarcoptes scabiei*. Clin Infect Dis 39:e8–e12. https://doi.org/10.1086/421776

Curtis SK, Housley R, Brooks DL (1990) Use of ivermectin for treatment of ear mite infestation in rabbits. J Am Vet Med Assoc 196:1139–1140

Darzi MM, Mir MS, Shahardar RA, Pandit BA (2007) Clinicopathological, histochemical and therapeutic studies on concurrent sarcoptic and notoedric acariosis in rabbits (*Oryctolagus cuniculus*). Vet Arh 77:167–175

Deak G, Turcu C (2020) A case of infestation with *Leporacarus gibbus* in a pet rabbit. Sci Parasitol 21:138–141

Dik B, Uslu U (2007) *Haemodipsus* species occurring on hares (*Lepus eurapeus*, L.): two new species in Turkish lice fauna. Turkiye Parazitolojii Dergisi 31:119–122

Dik B, Uslu U (2012) *Haemodipsus leporis* Blagoveshtchensky, 1966 (Phthiraptera: Anoplura: Polyplacidae) on a Hare (*Lepus europaeus*, L.): new record for Turkish Phthiraptera fauna. Turkiye Parazitolojii Dergisi 36:260–263. https://doi.org/10.5152/tpd.2012.62

Divisha R, Soundararajan C, Arul Prakash M (2020) Therapeutic management of concurrent sarcoptic and Psoroptic mange infestation in rabbits. J Entomol Zool Stud 8:1041–1043

Dos Santos RR, Coelho CN, Nunes TAP, Batista LC, Correia TR, Scott FB, Laguna AGV, Fernandes JI (2017) Efficacy of oral doramectin as treatment for *Psoroptes ovis* and *Leporacarus gibbus* in naturally infested rabbits. Pesqui Vet Bras 37:47–51. https://doi.org/10.1590/S0100-736X2017000100008

d'Ovidio D, Cortellini S (2022) Successful management of fipronil toxicosis in two pet rabbits. Open Vet J 12:508–510. https://doi.org/10.5455/OVJ.2022.v12.i4.13

d'Ovidio D, Santoro D (2014) *Leporacarus gibbus* infestation in client-owned rabbits and their owner. Vet Dermatol 25:46–e17. https://doi.org/10.1111/vde.12089

d'Ovidio D, Santoro D (2021) Efficacy of Fluralaner in the treatment of Sarcoptic mange (*Sarcoptes scabiei*) in 12 pet rabbits. Top Companion Anim Med 43:100528. https://doi.org/10.1016/j.tcam.2021.100528

Dumitrache MO, Györke A, D'Amico G, Mircean V (2021) First case report of dermatitis associated with *Leporacarus gibbus* in cat. BMC Vet Res 17:4. https://doi.org/10.1186/s12917-020-02681-0

Ebani VV, Mancianti F (2021) Entomopathogenic fungi and bacteria in a veterinary perspective. Biology 10:479. https://doi.org/10.3390/biology10060479

El Hamzaoui B, Zurita A, Cutillas C, Parola P (2020) Fleas and flea-borne diseases of North Africa. Acta Trop 211:105627. https://doi.org/10.1016/j.actatropica.2020.105627

Elhawary NM, Sorour SSGH, El-Abasy MA, Bazh EK, Sultan K (2017) A trial of doramectin injection and ivermectin spot-on for treatment of rabbits artificially infested with the ear mite "*Psoroptes cuniculi*". Pol J Vet Sci 20:521–525. https://doi.org/10.1515/pjvs-2017-0063

Elhawary NM, Sorour SS, Bazh EK, El Abasy MA, El-Madawy MA, Aziz A, Sultan K, Omar M (2018) Toxicity of fipronil in rabbits as a therapeutic drug for *Psoroptes cuniculi*: a preliminary observation. J Exp Appl Anim Sci 2:260–226

El-Maddawy ZK, Abd El Naby WSH (2017) Effects of ivermectin and its combination with alpha lipoic acid on expression of IGFBP-3 and HSPA1 genes and male rat fertility. Andrologia 50. https://doi.org/10.1111/and.12891

El-Nahas AF, El-Ashmawy IM (2008) Effect of ivermectin on male fertility and its interaction with P-glycoprotein inhibitor (verapamil) in rats. Environ Toxicol Pharmacol 26:206–211. https://doi.org/10.1016/j.etap.2008.03.011

Elshahawy I, El-Goniemy A, Ali E (2016) Epidemiological survey on mange mite of rabbits in the southern region of Egypt. Sains Malaysiana 45:745–751

Eshar D (2010) Prevalence of sarcoptic mange in pet rabbits (*Oryctolagus cuniculus*) in Israel. Israel J Vet Med 65:140–141

Eshar D (2019) Ectoparasites in rabbits. Clinician's Brief 17:12–18

Fang F, Bernigaud C, Candy K, Melloul E, Izri A, Durand R, Botterel F, Chosidow O, Huang W, Guillot J (2015) Efficacy assessment of biocides or repellents for the control of *Sarcoptes scabiei* in the environment. Parasit Vectors 8:416. https://doi.org/10.1186/s13071-015-1027-7

Farmaki R, Koutinas AF, Papazahariadou MG, Kasabalis D, Day MJ (2009) Effectiveness of a selamectin spot-on formulation in rabbits with sarcoptic mange. Vet Rec 164:431–432. https://doi.org/10.1136/vr.164.14.431

Fehr M, Koestlinger S (2013) Ectoparasites in small exotic mammals. The veterinary clinics of North America. Exotic Anim Pract 16:611–657. https://doi.org/10.1016/j.cvex.2013.05.011

Fichi G, Flamini G, Giovanelli F, Otranto D, Perrucci S (2007a) Efficacy of an essential oil of *Eugenia caryophyllata* against *Psoroptes cuniculi*. Exp Parasitol 115:168–172. https://doi.org/10.1016/j.exppara.2006.07.005

Fichi G, Flamini G, Zaralli LJ, Perrucci S (2007b) Efficacy of an essential oil of *Cinnamomum zeylanicum* against *Psoroptes cuniculi*. Phytomedicine 14:227–231. https://doi.org/10.1016/j.phymed.2006.01.004

Fisher M, Beck W, Hutchinson MJ (2007) Efficacy and safety of Selamectin (Stronghold®/Revolution™) used off-label in exotic pets. Int J Appl Res Vet Med 3:87–96

Foley J, Serieys LE, Stephenson N, Riley S, Foley C, Jennings M, Wengert G, Vickers W, Boydston E, Lyren L, Moriarty J, Clifford DL (2016) A synthetic review of Notoedres species mites and mange. Parasitology 143:1847–1861. https://doi.org/10.1017/S0031182016001505

Frank R, Kuhn T, Mehlhorn H, Rueckert S, Pham D, Klimpel S (2013) Parasites of wild rabbits (*Oryctolagus cuniculus*) from an urban area in Germany, in relation to worldwide results. Parasitol Res 112:4255–4266. https://doi.org/10.1007/s00436-013-3617-7

Freitas LH, Faccini JL, Labruna MB (2009) Experimental infection of the rabbit tick, *Haemaphysalis leporispalustris*, with the bacterium *Rickettsia rickettsii*, and comparative biology of infected and uninfected tick lineages. Exp Appl Acarol 47:321–345. https://doi.org/10.1007/s10493-008-9220-4

Galdhar CN, Khangal PS, Pawar ML, Rasal TD, Digraskar SU (2015) Clinico-biochemical and therapeutic studies on notoedric mange in pet rabbits. J Parasit Dis 39:113–116. https://doi.org/10.1007/s12639-013-0277-6

González MG, González J, Valcárcel F, Sánchez M, Tercero JM, Barrios L, Olmeda AS (2021) Seasonal dynamics of fleas (Siphonaptera: Pulicidae, Ceratophyllidae and Leptopsyllidae) on *Oryctolagus cuniculus* in a Meso-Mediterranean area of Central Spain. J Med Entomol 58: 1740–1749. https://doi.org/10.1093/jme/tjab033

Grunert RKA, Kunkel MR, Radisic R, Nemeth NM, Burrell CE, Shender LA, Guzman-Vargas V, Cunningham MW, Yabsley MJ (2022) Notoedric mange (*Notoedres centrifera*) in two species of free-ranging rabbits from Florida, USA. Vet Parasitol Reg Stud Rep 33:100748. https://doi.org/10.1016/j.vprsr.2022.100748

Guo X, Shang X, Li B, Zhou XZ, Wen H, Zhang J (2017) Acaricidal activities of the essential oil from Rhododendron nivale Hook. f. and its main compound, δ-cadinene against *Psoroptes cuniculi*. Vet Parasitol 236:51–54. https://doi.org/10.1016/j.vetpar.2017.01.028

Hajipour N, Zavarshani M (2020) Ectoparasites and Endoparasites of New Zealand White rabbits from north west of Iran. Iran J Parasitol 15:266–271

Halley BA, VandenHeuvel WJ, Wislocki PG (1993) Environmental effects of the usage of avermectins in livestock. Vet Parasitol 48:109–125. https://doi.org/10.1016/0304-4017(93)90149-h

Hamilton PB, Stevens JR, Holz P, Boag B, Cooke B, Gibson WC (2005) The inadvertent introduction into Australia of *Trypanosoma nabiasi*, the trypanosome of the European rabbit (*Oryctolagus cuniculus*), and its potential for biocontrol. Mol Ecol 14:3167–3175. https://doi.org/10.1111/j.1365-294X.2005.02602.x

Hansen O, Gall Y, Pfister K, Beck W (2005) Efficacy of a formulation containing imidacloprid and moxidectin against naturally acquired ear mite infestations (*Psoroptes cuniculi*) in rabbits. J Appl Res Vet Med 3:281–286

Harcourt-Brown F (2002) Infectious diseases of domestic rabbits. Textbook of rabbit medicine. Butterworth-Heinemann, Oxford, UK, pp 361–385. ISBN 978-0-7506-4002-2. https://doi.org/10.1016/B978-075064002-2.50019-9

Harkness JE, Turner PV, VandeWoude S, Wheler CL (2010) Harkness and Wagner's biology and medicine of rabbits and Rodents, 5th edn. Wiley-Blackwell, p 470. ISBN 978-1-118-70890-3

Hoppmann E, Wilson Barron H (2007) Ferret and rabbit dermatology. J Exotic Pet Med 16:225–237. https://doi.org/10.1053/j.jepm.2007.10.002

Ilić T, Stepanović P, Nenadović K, Dimitrijević S (2018) Improving agricultural production of domestic rabbits in Serbia by follow-up study of their parasitic infections. Iran J Vet Res 19: 290–297

Kanbur M, Atalay O, Ica A, Eraslan G, Cam Y (2008) The curative and antioxidative efficiency of doramectin and doramectin+vitamin AD3E treatment on *Psoroptes cuniculi* infestation in rabbits. Res Vet Sci 85:291–293. https://doi.org/10.1016/j.rvsc.2007.10.005

Khan A, Sohaib M, Ullah R, Hussain I, Niaz S, Malak N, de la Fuente J, Khan A, Aguilar-Marcelino L, Alanazi AD, Ben Said M (2022) Structure-based in silico design and in vitro acaricidal activity assessment of *Acacia nilotica* and *Psidium guajava* extracts against *Sarcoptes scabiei* var. *cuniculi*. Parasitol Res 121:2901–2915. https://doi.org/10.1007/s00436-022-07615-3

Kim SH, Jun HK, Song KH, Gram D, Kim DH (2008a) Prevalence of fur mites in pet rabbits in South Korea. Vet Dermatol 19:189–190. https://doi.org/10.1111/j.1365-3164.2008.00673.x

Kim SH, Lee JY, Jun HK, Song KH, Park BK, Kim DH (2008b) Efficacy of selamectin in the treatment of cheyletiellosis in pet rabbits. Vet Dermatol 19:26–27. https://doi.org/10.1111/j.1365-3164.2007.00629.x

Kirwan AP, Middleton B, McGarry JW (1998) Diagnosis and prevalence of *Leporacarus gibbus* in the fur of domestic rabbits in the UK. Vet Rec 142:20–21. https://doi.org/10.1136/vr.142.1.20

Kumar A, Kumar R, Kumari N (2018) A successful treatment report on rabbits infected with sarcoptic mange. Pharma Innov J 7:01–03

Kumar VV, Effendi MH, Lastuti ND, Triakoso N, Fiorenza MP (2022) Prevalence and severity of scabies (Sarcoptes scabiei) on rabbits in Kuala Lumpur City. J Parasite Sci 6:23–28. https://doi.org/10.20473/jops.v6i1.34286

Kurtdede A, Karaer Z, Acar A, Guzel M, Cingi CC, Ural K, Ica A (2007) Use of selamectin for the treatment of psoroptic and sarcoptic mite infestation in rabbits. Vet Dermatol 18:18–22. https://doi.org/10.1111/j.1365-3164.2007.00563.x

Kyung-Yeon E, Oh-Deog K (2010) Psoroptic otoacariasis associated with *Psoroptes cuniculi* in domestic rabbits in Korea. Pak Vet J 30:251–252

Labruna MB, Leite RC, Faccini JL, Ferreira F (2000) Life cycle of the tick *Haemaphysalis leporispalustris* (Acari: Ixodidae) under laboratory conditions. Exp Appl Acarol 24:683–694. https://doi.org/10.1023/a:1010768511790

Lennox AM, Kelleher S (2009) Bacterial and parasitic diseases of rabbits. The veterinary clinics of North America. Exotic Anim Pract 12:519–530. https://doi.org/10.1016/j.cvex.2009.06.004

Liu X, Walton S, Mounsey K (2014) Vaccine against scabies: necessity and possibility. Parasitology 141:725–732. https://doi.org/10.1017/S0031182013002047

Ljunggren EL (2005) Molecular analysis of *Sarcoptes scabiei*. Department of Biomedical Sciences and Veterinary Public Health. Thesis, Swedish University of Agricultural Sciences

Lu M, Cai Y, Yang S, Wan Q, Pan B (2018) A single subcutaneous administration of a sustained-release ivermectin suspension eliminates *Psoroptes cuniculi* infection in a rabbit farm. Drug Dev Ind Pharm 44:2000–2004. https://doi.org/10.1080/03639045.2018.1506474

Makepeace C, Nixon EJ, Burgess STG, Stubbings L, Wall R (2022) Sheep scab: comparison of spatial and temporal patterns determined by clinical diagnosis or ELISA. Parasit Vectors 15: 419. https://doi.org/10.1186/s13071-022-05564-5

Mayer J, Donnelly TM (2013) Ectoparasites. W.B. Saunders, pp 366–367. ISBN 9781416039693. https://doi.org/10.1016/B978-1-4160-3969-3.00162-1

McKellar QA, Midgley DM, Galbraith EA, Scott EW, Bradley A (1992) Clinical and pharmacological properties of ivermectin in rabbits and Guinea pigs. Vet Rec 130:71–73. https://doi.org/10.1136/vr.130.4.71

McTier TL, Hair JA, Walstrom DJ, Thompson L (2003) Efficacy and safety of topical administration of selamectin for treatment of ear mite infestation in rabbits. J Am Vet Med Assoc 223:322–324. https://doi.org/10.2460/javma.2003.223.322

Mehlhorn H (2008) Lice. Encyclopedia of parasitology. Springer, Berlin. https://doi.org/10.1007/978-3-540-48996-2_1750

Mehlhorn H (2016) Mites. In: Mehlhorn H (ed) Encyclopedia of parasitology. Springer, Berlin, Heidelberg. https://doi.org/10.1007/978-3-642-27769-6_1973-2

Melo RM, Fernandes JI, Vieira VP, de Ribeiro FA, Botelho MC, Verocai GG, Scott FB (2008) Efficacy of the pyrethroid permethrin on the control of *Psoroptes ovis* (Hering, 1838) (Acari: Psoroptidae) in naturally infested rabbits (*Oryctolagus cuniculus*). Braz J Vet Parasitol 17(Suppl 1):55–58

Merdele N (2010) Parasitical identification of *Cheyletiella* in a rabbit breeding farm. Lucrări Ştiinłifice Medicină Veterinară XLIII:57–60

Meredith A (2006) Skin diseases and treatment of rabbits. In: Paterson S (ed) Skin diseases of exotic pets. Blackwell, Oxford, UK, pp 288–311. ISBN 978-1-405-17330-8

Metwally DM, Al-Olayan EM, Alshalhoop RA, Eisa SA (2018) Biomarkers as predictive tools to test the in vivo anti-sarcoptic mange activity of propolis in naturally infested rabbits. Biosci Rep 38:BSR20180874. https://doi.org/10.1042/BSR20180874

Millán J, Casáis R, Delibes-Mateos M, Calvete C, Rouco C, Castro F, Colomar V, Casas-Díaz E, Ramírez E, Moreno S, Prieto JM, Villafuerte R (2012) Widespread exposure to *Sarcoptes scabiei* in wild European rabbits (*Oryctolagus cuniculus*) in Spain. Vet Parasitol 183:323–329. https://doi.org/10.1016/j.vetpar.2011.07.046

Miller WH, Griffin CE, Campbell KL, Muller GH (2013) Muller and Kirk's small animal dermatology, 7th edn. St Louis, Elsevier Health Sciences. ISBN 9781437711844

Mohamad-Radzi NN, Che-Amat A, Aziz NAA, Babjee SMA, Mazlan M, Hamid NFS, Lekko YM (2021) Preliminary detection of mites and coccidia with their zoonotic potential in meat-farmed rabbits in three districts in Selangor, Malaysia. J Parasit Dis 45:169–175. https://doi.org/10.1007/s12639-020-01291-9

Moroni B, Rossi L, Bernigaud C, Guillot J (2022) Zoonotic episodes of scabies: A global overview. Pathogens 11:213. https://doi.org/10.3390/pathogens11020213

Morton D, Verga M, Blasco A, Cavani C, Gavazza A, Maertens L, Mirabito L, Rosell J, Stauffacher M, Szendro Z (2005) The impact of the current housing and husbandry systems on the health and welfare of farmed domestic rabbits. EFSA J 267:1–31

Mullen GR, O'Connor BM (2019) Mites (Acari). In: Mullen GR, Durden LA (eds) Medical and veterinary entomology, 3rd edn. Academic Press, pp 533–602. ISBN 9780128140437. https://doi.org/10.1016/B978-0-12-814043-7.00026-1

Navarro-Gonzalez N, Serrano E, Casas-Díaz E, Velarde R, Marco I, Rossi L, Lavín S (2010) Game restocking and the introduction of sarcoptic mange in wild rabbit in North-Eastern Spain. Anim Conserv 13:586–591. https://doi.org/10.1111/j.1469-1795.2010.00390.x

Nong X, Fang CL, Wang JH, Gu XB, Yang DY, Liu TF, Fu Y, Zhang RH, Zheng WP, Peng XR, Wang SX, Yang GY (2012) Acaricidal activity of extract from *Eupatorium adenophorum* against the *Psoroptes cuniculi* and *Sarcoptes scabiei in vitro*. Vet Parasitol 187:345–349. https://doi.org/10.1016/j.vetpar.2011.12.015

Nong X, Ren YJ, Wang JH, Fang CL, Xie Y, Yang DY, Liu TF, Chen L, Zhou X, Gu XB, Zheng WP, Peng XR, Wang SX, Lai SJ, Yang GY (2013a) Clinical efficacy of botanical extracts from *Eupatorium adenophorum* against the scab mite, *Psoroptes cuniculi*. Vet Parasitol 192:247–252. https://doi.org/10.1016/j.vetpar.2012.10.005

Nong X, Ren YJ, Wang JH, Xie Y, Fang CL, Yang DY, Liu TF, Zhang RH, Chen L, Gu XB, Peng XR, Wang SX, Lai SJ, Yang GY (2013b) Clinical efficacy of botanical extracts from *Eupatorium adenophorum* against the *Sarcoptes scabiei* (Sarcoptidae: *Sarcoptes*) in rabbits. Vet Parasitol 195:157–164. https://doi.org/10.1016/j.vetpar.2013.02.020

Okumu OP (2014) Clinicopathologic survey and predisposing factors of diseases of domestic rabbits in selected areas in Kenya. MSc Thesis, Faculty of Veterinary Medicine, University of Nairobi, 190 pp

Osman SA, Hanafy A, Amer SE (2006) Clinical and therapeutic studies on mange in horses. Vet Parasitol 141:191–195. https://doi.org/10.1016/j.vetpar.2006.04.039

Palmeiro BS, Roberts H (2013) Clinical approach to dermatologic disease in exotic animals. The veterinary clinics of North America. Exotic Anim Pract 16:523–577. https://doi.org/10.1016/j.cvex.2013.05.003

Pan B, Wang M, Xu F, Wang Y, Dong Y, Pan Z (2006) Efficacy of an injectable formulation of eprinomectin against *Psoroptes cuniculi*, the ear mange mite in rabbits. Vet Parasitol 137:386–390. https://doi.org/10.1016/j.vetpar.2006.01.013

Pandey VS (1989) Effect of ivermectin on the ear mange mite, *Psoroptes cuniculi*, of rabbits. Br Vet J 145:54–56. https://doi.org/10.1016/0007-1935(89)90009-2

Panigahi PN, Gupta AR (2013) Therapeutic management of concurrent Sarcoptic and Psoroptic acariosis in rabbits. Intas Polivet 14:319–321

Panigrahi PN, Mohanty BN, Gupta AR, Patra RC, Dey S (2016) Concurrent infestation of Notoedres, Sarcoptic and Psoroptic acariosis in rabbit and its management. J Parasit Dis 40:1091–1093. https://doi.org/10.1007/s12639-014-0631-3

Paola DD, Iaria C, Marino F, Gugliandolo E, Piras C, Crupi R, Cuzzocrea S, Spanò N, Britti D, Peritore AF (2022) Environmental impact of pharmaceutical pollutants: synergistic toxicity of Ivermectin and Cypermethrin. Toxics 10:388. https://doi.org/10.3390/toxics10070388

Paterson S (2006) Chapter 23: Skin diseases and treatment of rabbits. In: Paterson S (ed) Skin diseases of exotic pets. Blackwell Science, Oxford, UK, pp 288–311. ISBN 9780632059690. https://doi.org/10.1002/9780470752432.ch23

Pegler KR, Evans L, Stevens JR, Wall R (2005) Morphological and molecular comparison of host-derived populations of parasitic *Psoroptes* mites. Med Vet Entomol 19:392–403. https://doi.org/10.1111/j.1365-2915.2005.00586.x

Pinter L (1999) *Leporacarus gibbus* and *Spilopsyllus cuniculi* infestation in a pet rabbit. J Small Anim Pract 40:220–221. https://doi.org/10.1111/j.1748-5827.1999.tb03064.x

Pritt S, Cohen K, Sedlacek H (2012) Parasitic diseases. In: Suckow M, Stevens K, Wilson R (eds) The laboratory rabbit, guinea pig, hamster and other rodents. Elsevier, London, pp 415–446. ISBN 978-0-12-380920-9. https://doi.org/10.1016/B978-0-12-380920-9.00015-8

Radi ZA (2004) Outbreak of sarcoptic mange and malasseziasis in rabbits (*Oryctolagus cuniculus*). Comp Med 54:434–437

Rafferty DE, Gray JS (1987) The feeding behaviour of *Psoroptes* spp. mites on rabbits and sheep. J Parasitol 73:901–906

Romero Núñez C, Flores Ortega A, Sheinberg Waisburd G, Martin Cordero A, Yarto Jaramillo E, Heredia Cárdenas R, Bautista Gómez LG (2020) Evaluation of the effect of afoxalaner with milbemycin 1 oxime in the treatment of rabbits naturally infected with *Psoroptes cuniculi*. PLoS One 15:e0230753. https://doi.org/10.1371/journal.pone.0230753

Roth T, Lane RS, Foley J (2017) A molecular survey for *Francisella tularensis* and *Rickettsia* spp. in *Haemaphysalis leporispalustris* (Acari: Ixodidae) in northern California. J Med Entomol 54: 492–495. https://doi.org/10.1093/jme/tjw202

Rothschild M, Ford B (1964) Maturation and egg-laying of the rabbit flea (*Spilopsyllus cuniculi* Dale) induced by the external application of hydrocortisone. Nature 203:210–211. https://doi.org/10.1038/203210a0

Saari S, Näreaho A, Nikander S (2019) Chapter 9 – Arachnida. In: Saari S, Näreaho A, Nikander S (eds) Canine parasites and parasitic diseases. Academic Press, New York, pp 187–228. ISBN 9780128141120. https://doi.org/10.1016/B978-0-12-814112-0.00009-X

Sanders A, Froggatt P, Wall R, Smith KE (2000) Life-cycle stage morphology of *Psoroptes* mange mites. Med Vet Entomol 14:131–141. https://doi.org/10.1046/j.1365-2915.2000.00223.x

Sanderson H, Laird B, Pope L, Brain R, Wilson C, Johnson D, Bryning G, Peregrine AS, Boxall A, Solomon K (2007) Assessment of the environmental fate and effects of ivermectin in aquatic mesocosms. Aquat Toxicol 85:229–240. https://doi.org/10.1016/j.aquatox.2007.08.011

Santos RR, Coelho CN, Nunes TA, Batista LC, Correia TR, Scott FB, Laguna AG, Fernandes JI (2017) Efficacy of oral doramectin as treatment for *Psoroptes ovis* and *Leporacarus gibbus* in naturally infested rabbits. Pesqui Vet Bras 37:47–51

Schoeb TR, Cartner SC, Baker RA, Gerrity LG (2007) Parasites of rabbits. In: Baker DG (ed) Flynn's parasites of laboratory animals, 2nd edn. Blackwell, Ames, IA, pp 451–500. ISBN 978-0-813-81202-1

Scott DW, Miller WH, Griffin CE (2001) Dermatoses of pet rodents, rabbits and ferrets. In: Scott DW, Miller WH, Griffin CE (eds) Muller and Kirk's small animal dermatology. W.B. Saunders, Philadelphia, PA, pp 1415–1458. ISBN 978-0-7216-7618-0. https://doi.org/10.1016/B978-0-7216-7618-0.50025-0

Seddiek SA, Khater HF, El-Shorbagy MM, Ali AM (2013) The acaricidal efficacy of aqueous neem extract and ivermectin against Sarcoptes scabiei var. cuniculi in experimentally infested rabbits. Parasitol Res 112:2319–2330. https://doi.org/10.1007/s00436-013-3395-2

Shang XF, Miao XL, Wang DS, Li JX, Wang XZ, Yan ZT, Wang CM, Wang Y, He XR, Pan H (2013) Acaricidal activity of extracts from Adonis coerulea Maxim. against Psoroptes cuniculi in vitro and in vivo. Vet Parasitol 195:136–141. https://doi.org/10.1016/j.vetpar.2012.12.057

Shang X, Wang Y, Zhou X, Guo X, Dong S, Wang D, Zhang J, Pan H, Zhang Y, Miao X (2016) Acaricidal activity of oregano oil and its major component, carvacrol, thymol and p-cymene against *Psoroptes cuniculi in vitro* and *in vivo*. Vet Parasitol 226:93–96. https://doi.org/10.1016/j.vetpar.2016.07.001

Shang X, Guo X, Yang F, Li B, Pan H, Miao X, Zhang J (2017) The toxicity and the acaricidal mechanism against *Psoroptes cuniculi* of the methanol extract of *Adonis coerulea* Maxim. Vet Parasitol 240:17–23. https://doi.org/10.1016/j.vetpar.2017.04.019

Sharma M, Kumar T, Gupta S, Jain V (2021) Notoedric mange successfully treated with ivermectin in rabbits. Int J Trop Insect Sci 41:1875–1881. https://doi.org/10.1007/s42690-020-00322-5

Sharun K, Anjana S, Sidhique SA, Panikkassery S (2019) Treatment of Sarcoptic mange infestation in rabbits with long acting injectable ivermectin. J Parasit Dis 43:733–736. https://doi.org/10.1007/s12639-019-01137-z

Sheinberg G, Romero C, Heredia R, Capulin M, Yarto E, Carpio J (2017) Use of oral fluralaner for the treatment of *Psoroptes cuniculi* in 15 naturally infested rabbits. Vet Dermatol 28:393–e91. https://doi.org/10.1111/vde.12429

Singh SK, Dimri U, Sharma MC, Swarup D, Kumar M, Tiwary R (2012) *Psoroptes cuniculi* induced oxidative imbalance in rabbits and its alleviation by using vitamins A, D3, E, and H as adjunctive remedial. Trop Anim Health Prod 44:43–48. https://doi.org/10.1007/s11250-011-9884-3

Sivajothi S, Reddy B, Rayulu VC (2014) Effect of Ivermectin against Psoroptic mange in rabbits. Int J Sci World 2:10–12. https://doi.org/10.14419/ijsw.v2i1.1893

Soundararajan C, Iyue M (2005) Incidence of mange infestation in rabbits. J Vet Parasitol 19:161–162

Swarnakar G, Sharma D, Sanger B, Roat K (2014) Infestation of ear mites *Psoroptes cuniculi* on farm rabbits and its anthropozoonozis in Gudli village of Udaipur District, India. Int J Curr Microbiol App Sci 3:651–656

Taylor MA, Coop RL, Wall RL (2016) Veterinary parasitology, 4th edn. Wiley-Blackwell, p 1032. ISBN 978-0-470-67162-7. https://doi.org/10.1002/9781119073680

Tsianakas P, Polack B, Pinquier L, Levy Klotz B, Prost-Squarcioni C (2000) La cheyletellose: une etiologie inhabituelle d'eruption vésiculo- bulleuse [*Cheyletiella* dermatitis: an uncommon cause of vesiculobullous eruption]. Ann Dermatol Venereol 127:826–829

Ulutas B, Voyvoda H, Bayramli G, Karagenc T (2005) Efficacy of topical administration of eprinomectin for treatment of ear mite infestation in six rabbits. Vet Dermatol 16:334–337. https://doi.org/10.1111/j.1365-3164.2005.00464.x

Varga M (2014) Chapter 7 - Skin diseases. In: Varga M (ed) Textbook of rabbit medicine, 2nd edn. Butterworth-Heinemann, London, pp 271–302. ISBN 9780702049798. https://doi.org/10.1016/B978-0-7020-4979-8.00007-8

Varga M, Paterson S (2020) Dermatologic diseases of rabbits. In: Quesenberry KE, Orcutt CJ, Mans C, Carpenter JW (eds) Ferrets, rabbits, and rodents, 4th edn. W.B. Saunders, pp 220–232. ISBN 9780323484350. https://doi.org/10.1016/B978-0-323-48435-0.00017-4

Voyvoda H, Ulutas B, Eren H, Karagenc T, Bayramli G (2005) Use of doramectin for treatment of sarcoptic mange in five angora rabbits. Vet Dermatol 16:285–288. https://doi.org/10.1111/j.1365-3164.2005.00446.x

Wagner R, Stallmeister N (2000) *Cheyletiella* dermatitis in humans, dogs and cats. Br J Dermatol 143:1110–1112. https://doi.org/10.1046/j.1365-2133.2000.03869.x

Wagner R, Wendlberger U (2000) Field efficacy of moxidectin in dogs and rabbits naturally infested with *Sarcoptes* spp., *Demodex* spp. and *Psoroptes* spp. mites. Vet Parasitol 93:149–158. https://doi.org/10.1016/s0304-4017(00)00357-5

Wei W, Ren Y, Shen N, Song H, Xu J, Hua R, Zhang H, Angel C, Gu X, Kuang L, Xie Y, Peng X, Xie X, Yang G (2019) Comparative analysis of host resistance to *Sarcoptes scabiei* var. *cuniculi* in two different rabbit breeds. Parasit Vectors 12:530. https://doi.org/10.1186/s13071-019-3764-5

Wen H, Pan B, Wang F, Yang Z, Wang Z, Liu S, Wang M (2010) The effect of self-licking behavior on pharmacokinetics of eprinomectin and clinical efficacy against *Psoroptes cuniculi* in topically administered rabbits. Parasitol Res 106:607–613. https://doi.org/10.1007/s00436-009-1704-6

White SD, Bordeau P, Meredith A (2002) Dermatologic problems of rabbits. Semin Avian Exotic Pet Med 11:141–150

Zahler M, Hendrikx WM, Essig A, Rinder H, Gothe R (2000) Species of the genus *Psoroptes* (Acari: Psoroptidae): a taxonomic consideration. Exp Appl Acarol 24:213–225. https://doi.org/10.1023/a:1006443722815

Zhang R, Jise Q, Zheng W, Ren Y, Nong X, Wu X, Gu X, Wang S, Peng X, Lai S, Yang G (2012) Characterization and evaluation of a *Sarcoptes scabiei* allergen as a candidate vaccine. Parasit Vectors 5:176. https://doi.org/10.1186/1756-3305-5-176

Zygner W, Gójska-Zygner O (2018) The first case of *Leporacarus gibbus* infestation in a rabbit from Poland. Acta Parasitol 63:210–213. https://doi.org/10.1515/ap-2018-0024

Mycotoxin Effects and Risk Management in Rabbit Production

Fernando Lima

Abstract

Mycotoxins are produced by fungi and can have a relevant negative impact on animal health. Most of the finished feed and feed raw materials are contaminated by at least one type of mycotoxin, even when they are present at very low concentrations. In the present times, mycotoxin co-contamination occurrence assumes an increasing relevance in the diets. Acute (e.g., outbreaks) or subacute poisoning can occasionally occur, but chronic exposure is more frequent leading to mycotoxin accumulation in the animal's body tissues. Clinical signs are usually unspecific and expressed as gastrointestinal disturbances, modulated by immunosuppressive condition, or related to organ damage. Due to the ubiquitous presence of mycotoxins in the environment, risk control programs are implemented in feed industry to decrease and control the impact of their contamination in final feed. This chapter highlights the main effects of mycotoxins in rabbits and provides insights for contamination risk management in feeds.

Keywords

Aflatoxins · Control risk · Contaminated feed · Poisoning · Fungi · *Fusarium*

1 Introduction

Today, mycotoxins are well known for their negative effects on animal health and immune system (Mavrommatis et al. 2021). Years back they were considered harmful, even lethal under high contamination level, which happens sporadically!

The original version of the chapter has been revised. A correction to this chapter can be found at https://doi.org/10.1007/978-3-031-44542-2_35

F. Lima (✉)
Schweizersholz, Switzerland

J. Simões, J. M. Monteiro (eds.), *Veterinary Care of Farm Rabbits*,
https://doi.org/10.1007/978-3-031-44542-2_32

With extended research on their hidden effects, we know, currently, that the reality is slightly different. Low to medium contamination can already disturb not just gut health (Liew and Mohd-Redzwan 2018) and immunity but affect animal's resilience to disease and impair proper performance (Antonissen et al. 2014; Kępińska-Pacelik and Biel 2021).

On intensive farming, this effect can turn to be more notorious (Xu et al. 2022) due to higher feed intake, more contact with gut system, and negative immune stimulation influence. Mycotoxins are unpredictable, not ubiquitous, but almost omnipresent on several raw material commodities we use in our feed mill operations (Gruber-Dorninger et al. 2019).

Mycotoxins are "invisible" secondary metabolites of molds which may persist in feed even when the molds that produced them are no longer present. When the clinical symptoms of mycotoxin poisoning can be observed, significant damage has already occurred. There are more often subclinical and indirect effects, rather than direct effects, due to its unpredictable frequent presence (Fink-Gremmels and Van der Merwe 2019; Awuchi et al. 2021).

Several mycotoxins are immunosuppressive, such as aflatoxins, trichothecenes, and ochratoxin. They affect immune response by reducing IgG and IgM (Oswald and Comera 1998). Their main effects on cellular immunity are decreased neutrophils (Brown et al. 2021). As example, the capacity of trichothecenes to inhibit protein synthesis apparently contributes to their potential to modulate immune function. Severe damage into actively dividing cells (e.g., bone marrow, lymph nodes, spleen, thymus, and intestinal mucosa) can occur after acute contamination (Pestka et al. 1990; Aupanun et al. 2017). Upon activation by chemokines released from pathogens or injured tissues, neutrophils transmigrate from the peripheral blood and then migrate toward sites of infection along gradients of chemokines. This process is named chemotaxis. Trichothecenes, for example, affect cell immunity by decreasing chemotactic migration of neutrophils and decreased phagocytosis by alveolar macrophages, inhibiting mitogen-induced blastogenesis, lymphocyte cytotoxicity, and platelet function (Pestka 2008; Janik et al. 2021; Li et al. 2022).

Overall, this understanding brings us a new question and awareness on how to counteract mycotoxin negative effects and control them to make commercial farms safer, more efficient, and sustainable. In this chapter, we pretend to describe what are the main effects of mycotoxins in rabbit meat and how to control their contamination in feeds.

2 Incidence of Mycotoxins, Climate Change, and Clinical Expression

From 2008 to 2017, 74,821 samples of finished feed and feed raw materials such as maize, wheat, barley, and soybean collected from 100 countries during a 10-year period were analyzed for aflatoxin B1 (AFB1), zearalenone (ZEN), fumonisins, ochratoxin A (OTA), deoxynivalenol (DON), and T-2 toxin (T-2) (Gruber-Dorninger et al. (2019). Of the samples tested for ≥ 3 mycotoxins by these researchers, 88% were contaminated with at least one mycotoxin. *Fusarium* mycotoxins DON, fumonisins, and ZEN were most prevalent and were detected in 64%, 60%, and 45% of all samples, respectively (Table 1). AFB1, T-2, and OTA were detected in 23%, 19%, and 15% of the samples, respectively (Table 1). Fumonisins and DON showed the highest median concentrations, namely, 723 μg/kg and 388 μg/kg, respectively (Table 1) (Gruber-Dorninger et al. 2019).

Climate change is one occurrence that, currently, poses a huge threat to agriculture around the world (Perrone et al. 2020; Zingales et al. 2022). High temperatures but also abnormal rainfall variations and extreme weather are the main factors involved in climate change. Under these conditions, toxigenic fungi have excellent opportunities for growth. Along with them come mycotoxins, which rank among the most dangerous contaminants in animal feed. If left uncontrolled, mycotoxins can have serious consequences for reproduction and other aspects of livestock performance.

There are field mycotoxins (*Fusarium* like fumonisin, trichothecene A-DON, nivalenol and trichothecenes B-T2 toxin and HT-2 toxin, and ZEN) and stockage mycotoxins like aflatoxins and ochratoxin. Both are produced by different fungi species, although it seems *Fusarium* is the most prevalent. Controlling fungi is possible in today's agriculture and feed processing but not the substrates produced by them which are mycotoxins. Most of them are heat and chemical resistant for common temperatures or substances used in industry.

Mycotoxin poisoning may give rise to acute or chronic incidences depending on the concentration of mycotoxin in feed, as well as the amount of feed intake. Also, the damage to organs is cumulative over a period of time. The symptoms of mycotoxicosis are sometimes like those of well-known diseases, i.e., unspecific. As an example, a high incidence of gastrointestinal upsets and diseases associated with depressed immune function (e.g., *Pasteurella*) may indicate that a mycotoxin problem exists (Richard 1991). There are some clinical signs which may appear in rabbits, such as severe pain in the abdomen, while radiograph series may reveal gut shutdown but no physical blockage and sometimes severe bloating as an effect of DON (Fioramonti et al. 1993). Anyhow, most prevalent issues come as subclinical and eventually triggering other sensible opportune pathogens by overloading some organs and defensive mechanisms.

Several of the experimental studies in rabbits prove their acute or subacute effects and also serve as animal model and were published mainly from the 1980s to 1990s decades of the last century. Citrinin (single oral dose of 125–150 mg per kg) induced fluid diarrhea (Hanika et al. 1983). Normally, mycotoxins also primarily induce feeding behavior and gut alterations (Abdelhamid 1990). Several symptoms and lesions than can occur due to mycotoxicosis in rabbits are reported in Box 1.

Table 1 Global mycotoxin occurrence in different commodities. Adapted from Gruber-Dorninger et al. (2019)

Mycotoxin	n[a]	Positive samples[b]		Median of positives (μg/kg)	1st quartile of positives (μg/kg)	3rd quartile of positives (μg/kg)	Maximum (μg/kg)
		n[a]	%				
All samples							
Aflatoxin B_1	51,475	11,941	23	4	2	17	10,918
Fumonisins[c]	46,477	27,890	60	723	240	1858	290,517
Zearalenone	61,413	27,559	45	55	25	147	105,000
Deoxynivalenol	59,107	37,940	64	388	200	885	84,860
Ochratoxin A	32,271	4858	15	3	2	7	2000
T-2 toxin	27,850	5289	19	22	8	40	3051
Finished feed							
Aflatoxin B_1	16,563	4251	26	6	2	23	10,918
Fumonisins[c]	16,285	11,825	73	555	198	1297	290,517
Zearalenone	19,171	10,676	56	41	20	102	9432
Deoxynivalenol	18,649	13,004	70	294	134	600	32,893
Ochratoxin A	11,990	2801	23	3	2	6	1582
T-2 toxin	9884	2246	23	10	4	22	1300
Maize							
Aflatoxin B_1	15,889	3835	24	4	1	22	6105
Fumonisins[c]	12,965	10,397	80	1300	520	2940	218,883
Zearalenone	15,860	7002	44	77	33	217	16,495
Deoxynivalenol	12,660	8486	67	520	260	1240	51,374
Ochratoxin A	6388	334	5	3	2	14	889
T-2 toxin	6087	727	12	25	11	53	978
Maize DDGS (maize dried distillers grains with soluble)							
Aflatoxin B_1	320	62	19	11	4	20	340
Fumonisins[c]	329	256	78	814	398	1870	26,828

Zearalenone	368	275	75	102	60	237	2896
Deoxynivalenol	381	316	83	1490	574	2579	84,860
Ochratoxin A	280	62	22	4	2	11	53
T-2 toxin	52	3	6	40	35	43	46
Maize silage							
Aflatoxin B_1	3104	188	6	2	1	4	342
Fumonisins[c]	3010	1114	37	138	45	416	7090
Zearalenone	3735	1508	40	84	34	201	6239
Deoxynivalenol	4206	2588	62	474	219	1092	34,861
Ochratoxin A	2830	161	6	3	2	6	69
T-2 toxin	1800	58	3	20	10	51	685
Soybean grains							
Aflatoxin B_1	916	186	20	1	1	2	74
Fumonisins[c]	794	135	17	68	29	223	7023
Zearalenone	1024	364	36	43	26	71	4336
Deoxynivalenol	975	284	29	416	160	640	5500
Ochratoxin A	718	86	12	3	2	7	46
T-2 toxin	557	102	18	29	23	37	317
Soybean meal							
Aflatoxin B_1	1692	490	29	2	1	4	109
Fumonisins[c]	1475	336	23	104	31	290	7210
Zearalenone	1767	1072	61	47	33	83	3720
Deoxynivalenol	802	247	31	119	25	424	5600
Ochratoxin A	606	82	14	4	2	10	141
T-2 toxin	975	324	33	33	25	44	754
Wheat							
Aflatoxin B_1	2210	221	10	1	1	3	161

(continued)

Table 1 (continued)

Mycotoxin	n[a]	Positive samples[b]		Median of positives (μg/kg)	1st quartile of positives (μg/kg)	3rd quartile of positives (μg/kg)	Maximum (μg/kg)
		n[a]	%				
Fumonisins[c]	2219	304	14	117	31	246	28,278
Zearalenone	4925	1624	33	34	20	75	23,278
Deoxynivalenol	5949	3866	65	369	218	865	49,307
Ochratoxin A	1973	172	9	3	2	5	364
T-2 toxin	1993	439	22	25	13	35	1300
Barley							
Aflatoxin B_1	727	64	9	1	1	2	120
Fumonisins[c]	776	65	8	53	17	366	10,485
Zearalenone	3129	637	20	25	20	58	8952
Deoxynivalenol	4046	2468	61	359	234	750	35,000
Ochratoxin A	730	46	6	3	2	9	150
T-2 toxin	1225	272	22	26	9	51	404
Rice							
Aflatoxin B_1	205	63	31	5	2	14	113
Fumonisins[c]	244	49	20	142	63	382	6895
Zearalenone	220	74	34	60	34	107	1530
Deoxynivalenol	226	60	27	266	87	436	3859
Ochratoxin A	230	32	14	3	2	5	20
T-2 toxin	54	5	9	9	8	26	30

[a]Sample number
[b]Positive samples are defined as > limit of detection, excluding aflatoxins below 0.5 ng/g and other mycotoxins below 1 ng/g
[c]Sum of fumonisins B_1, B_2, and B_3

Box 1 Symptoms and Lesions of Mycotoxicosis in Animals (Adapted from Mézes and Balogh 2009)

- Fever
- Gastrointestinal alterations
- Stomach ulcers
- Mouth sores
- Rectal prolapses
- Lung alterations
- Immune-suppression
- Central nervous system alterations
- Hemorrhages or bruising
- Internal bleeding
- Kidney or liver damage
- Heart muscle damage
- Tachycardia
- Hypertrophy of the adrenal cortex
- Reproductive alterations
- Tegument (skin) and eye problems
- Blood, bone marrow, and spleen abnormalities
- Vascular fragility

3 Mycotoxin Combined Effects

If we were able to ensure that the animals were exposed to only one mycotoxin at a time, following the regulatory guidelines would allow us to protect our animals in most of the cases. However, several worldwide surveys show that mycotoxin cooccurrence in feed is found very often and seems to be the rule rather than the exception in practical diets (Tolosa et al. 2021). Greco et al. (2012) observed that all (100%) of 42 representative feed samples for finished rabbits or *Chinchilla* from 7 companies in Argentina presented cooccurrence of mycotoxins. In this study, aflatoxins, fumonisins, and ZEN were recovered in 100% of the samples and OTA, T2 toxin, and DON in 95%, 98%, and 98% of the samples, respectively.

Regulations throughout the world do not consider the combined effects of mycotoxins. Not even is it easy to predict the combined clinical effect level, such as evidenced by Szabó et al. (2014, 2016). The cooccurrence of mycotoxins in crops and other raw feed enters in the food chain to contaminate food animals and successively people (Fig. 1).

Exposure to a mixture of mycotoxins may result in a combination of adverse effects or trigger the same effects, even on much less contamination than each one separately. This implies the surge of additive or synergistic effect, and in today's production, this represents a real nightmare for differential diagnosis and control

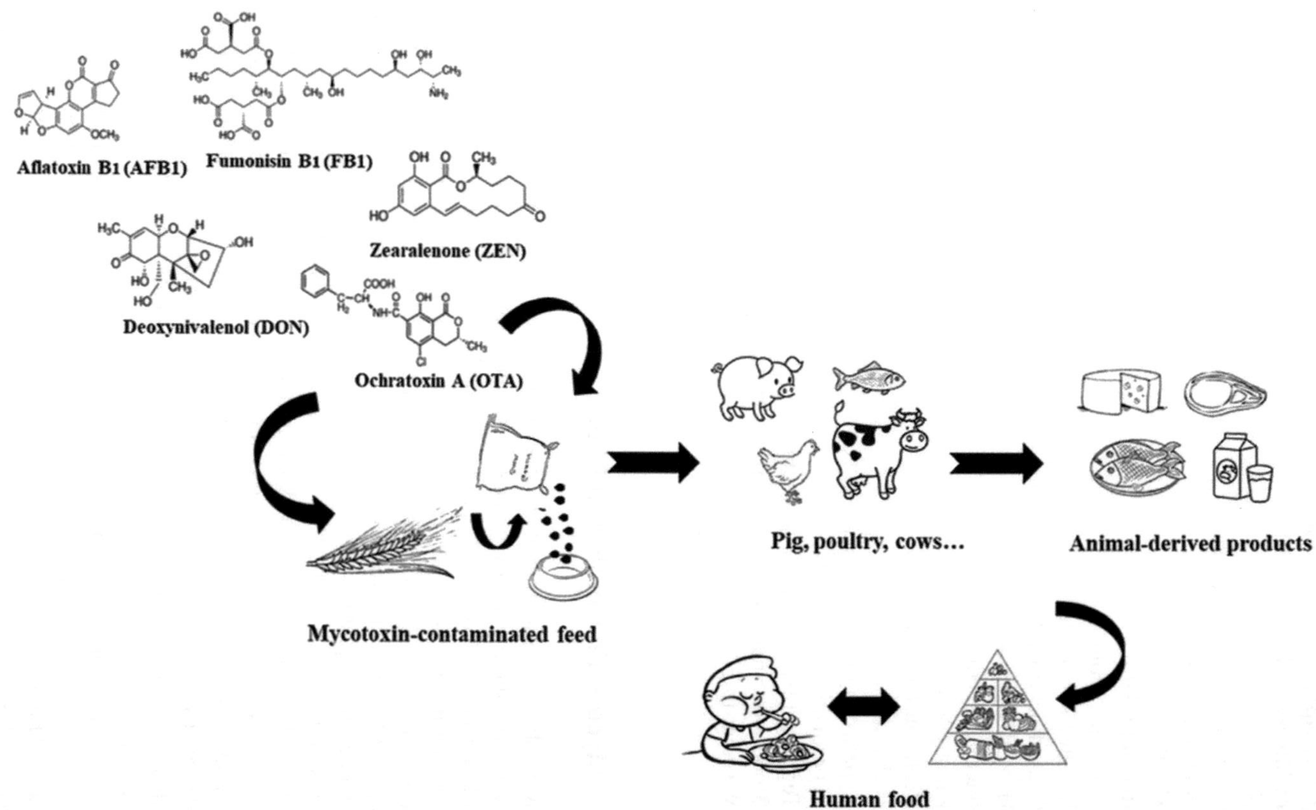

Fig. 1 Schematic representation of co-contamination of mycotoxins into the food chain. Adapted from Tolosa et al. (2021) with permission from Elsevier

(Smith et al. 2016). Synergic adverse effect can be defined as the toxicological effect of two or more mycotoxin present in the same sample that will be higher than the sum of toxicological effect of the individual mycotoxins. However, antagonism effect can be also observed (Szabó et al. 2014).

An example of potential synergism is between DON, ZEA, and T-2 on rabbits' reproductive performance. According to Tvrdá et al. (2016), and using in vitro *cultures*, DON ($\geq$10 μmol/L) can immediately increase the sperm motility, but a rapid decline occurs in the following stages; ZEN is highly toxic and causes a quick sperm motion decrease and complete stoppage of sperm kinetics at parameters 50 and 100 μmol/L. Similar to ZEN, T-2 also is highly detrimental for sperm motion but acts at very low doses ($\geq$1 μmol/L). In another study, the same research group (Tvrdá et al. 2022) observed that ZEN provokes alterations of the sperm plasma membrane, DON causes DNA integrity damage, and T-2 induces sperm necrotic changes and mitochondrial dysfunction.

Weaned rabbits, challenged with T-2 (10 mg/kg) and FB1 (2 mg/kg) contaminated feed during 28 days, showed more frequent liver and kidney pathological changes when exposed to both mycotoxins than those just with each one alone (Hafner et al. 2016). Nevertheless, additive or synergic effects of mycotoxins causing damage on hepatic lipid profile are not always evident in experimental studies, such as reported by Szabó et al. (2016).

4 Mycotoxin Risk Control

It's not possible to completely control mycotoxin production on the field or harvest but possible to add antifungal products to control fungi proliferation and consequently to decrease their contamination in raw feed (Habschied et al. 2021). Mycotoxins are toxins produced by fungi, and once animals ingest it with the feed, they get all those known negative effects already mentioned. There are more than 100,000 fungi species (Freire and da Rocha 2016) producing more than 500 different mycotoxins (Stein and Bulboacă 2017); even only a few part (e.g., *Aspergillus*, *Fusarium*, and *Penicillium* genera) is the main responsible for their production (Sweeney and Dobson 1998).

Part of mycotoxins are masked mycotoxins and not easy to diagnose with common analytical methods like ELISA (enzyme-linked immunosorbent assay) or HPLC (high performance liquid chromatography), but only with Spectrum 380 (LC-MS/MS; liquid chromatography–mass spectrometry) which can detect more than 700 mycotoxins and metabolites (Steiner et al. 2020; Sulyok et al. 2020). That is because plant's defense mechanisms add sugar (glycosylation), or other substances, to the mycotoxin and create a masked mycotoxin, e.g., DON-3-glucoside (Zhang et al. 2020). When the animal ingests contaminated feed containing masked mycotoxins, the cleavage of sugar by gut microbiota occurs with release of the mycotoxin (Gratz 2017). Most of the mycotoxins are chemically stable and thermostable and resist to processing in usual conditions.

The best and safest solution to control or mitigate mycotoxin contamination is to check periodically animal feed, and some commodities (Karlovsky et al. 2016), and use a proper protection in the feed, if not on full cycle, at least on those more sensitive phase feeds, like for does and kits.

Controlling mycotoxins is not just a matter of adding a normal binder available in the market. Mycotoxin binder products based only on adsorption are not enough; some mycotoxins such as trichothecenes, fumonisin, and ZEN need to be deactivated through enzymatic proven activity (Loi et al. 2017). Despite endotoxins, mineral adsorbents selectively bind only adsorbable mycotoxins such as aflatoxins and ergot alkaloids as well (Murugesan et al. 2015; Liu et al. 2022). This is related with different polarity and geometry (Kabak et al. 2006). Still all these deactivation processes should be taken in the gut where mycotoxin adsorption or enzymatic biotransformation into nontoxic metabolites should take place.

We would recommend using continuously mycotoxin prevention, even at lowest prevention, adjusting anytime if needed, to avoid accumulation in the organs and delayed detection that could compromise the farm profitability.

5 Conclusions

Mycotoxins are produced by several fungi and are ubiquitously disseminated in different commodities according to the environmental conditions. They are strong bioactive compounds to cause damage in animals' tissues and organs even at low contamination. Acute, subacute, and chronic poisoning can occur dependently of the mycotoxin load in the animal by ingestion and tissue accumulation. Additionally, feed co-contamination is relatively frequent mainly in intensive production systems supplied with complete feed.

The diagnosis of mycotoxicosis and detection of mycotoxins in feed and its raw materials pose challenges, and risk-free status cannot be achieved under field conditions. Regular mycotoxin screening of animal feed and commodities is crucial, and proper protection in the feed is considered to control risks and prevent mycotoxicosis in rabbit farms.

References

Abdelhamid AM (1990) Effect of feeding rabbits on naturally moulded and mycotoxin-contaminated diet. Archiv fur Tierernahrung 40:55–63. https://doi.org/10.1080/17450399009428382

Antonissen G, Martel A, Pasmans F, Ducatelle R, Verbrugghe E, Vandenbroucke V, Li S, Haesebrouck F, Van Immerseel F, Croubels S (2014) The impact of *fusarium* mycotoxins on human and animal host susceptibility to infectious diseases. Toxins 6:430–452. https://doi.org/10.3390/toxins6020430

Aupanun S, Poapolathep S, Giorgi M, Imsilp K, Poapolathep A (2017) An overview of the toxicology and toxicokinetics of fusarenon-X, a type B trichothecene mycotoxin. J Vet Med Sci 79:6–13. https://doi.org/10.1292/jvms.16-0008

Awuchi CG, Ondari EN, Ogbonna CU, Upadhyay AK, Baran K, Okpala COR, Korzeniowska M, Guiné RPF (2021) Mycotoxins affecting animals, foods, humans, and plants: types, occurrence, toxicities, action mechanisms, prevention, and detoxification strategies - a revisit. Foods 10: 1279. https://doi.org/10.3390/foods10061279

Brown R, Priest E, Naglik JR, Richardson JP (2021) Fungal toxins and host immune responses. Front Microbiol 12:643639. https://doi.org/10.3389/fmicb.2021.643639

Fink-Gremmels J, Van der Merwe D (2019) Mycotoxins in the food chain: contamination of foods of animal origin. In: Smulders FJM, Rietjens IMCM, Rose MD (eds) Chemical hazards in foods of animal origin. Food safety assurance and veterinary public health. Wageningen Academic Publishers, Wageningen, The Netherlands, pp 241–261

Fioramonti J, Dupuy C, Dupuy J, Bueno L (1993) The mycotoxin, deoxynivalenol, delays gastric emptying through serotonin-3 receptors in rodents. J Pharmacol Exp Ther 266:1255–1260

Freire FDCO, da Rocha MEB (2016) Impact of mycotoxins on human health. Fungal metabolites. In: Mérillon JM, Ramawat K (eds) Reference series in phytochemistry. Springer, Cham, pp 239–261

Gratz SW (2017) Do plant-bound masked mycotoxins contribute to toxicity? Toxins 9:85. https://doi.org/10.3390/toxins9030085

Greco MV, Pardo AG, Ludemann V, Martino PE, Pose GN (2012) Mycoflora and natural incidence of selected mycotoxins in rabbit and chinchilla feeds. Sci World J 2012:956056. https://doi.org/10.1100/2012/956056

Gruber-Dorninger C, Jenkins T, Schatzmayr G (2019) Global mycotoxin occurrence in feed: a ten-year survey. Toxins 11:375. https://doi.org/10.3390/toxins11070375

Habschied K, Krstanović V, Zdunić Z, Babić J, Mastanjević K, Šarić GK (2021) Mycotoxins biocontrol methods for healthier crops and stored products. J Fungi 7:348. https://doi.org/10.3390/jof7050348

Hafner D, Szabó A, D'Costa L, Szabó-Fodor J, Tornyos G, Blochné Bodnár Z, Ölbeiné Horvatovich K, Baloghné Zándoki E, Bóta B, Kovács M (2016) Individual and combined effects of feed artificially contaminated with Fumonisin B1 and T-2 toxin in weaned rabbits. World Mycotoxin J 9:613–622. https://doi.org/10.3920/WMJ2016.2067

Hanika C, Carlton WW, Tuite J (1983) Citrinin mycotoxicosis in the rabbit. Food Chem Toxicol 21: 487–493. https://doi.org/10.1016/0278-6915(83)90107-2

Janik E, Niemcewicz M, Podogrocki M, Ceremuga M, Stela M, Bijak M (2021) T-2 toxin—the most toxic Trichothecene mycotoxin: metabolism, toxicity, and decontamination strategies. Molecules 26:6868. https://doi.org/10.3390/molecules26226868

Kabak B, Dobson AD, Var I (2006) Strategies to prevent mycotoxin contamination of food and animal feed: a review. Crit Rev Food Sci Nutr 46:593–619. https://doi.org/10.1080/10408390500436185

Karlovsky P, Suman M, Berthiller F, De Meester J, Eisenbrand G, Perrin I, Oswald IP, Speijers G, Chiodini A, Recker T, Dussort P (2016) Impact of food processing and detoxification treatments on mycotoxin contamination. Mycotoxin Res 32:179–205. https://doi.org/10.1007/s12550-016-0257-7

Kępińska-Pacelik J, Biel W (2021) Alimentary risk of mycotoxins for humans and animals. Toxins 13:822. https://doi.org/10.3390/toxins13110822

Li J, Wang Y, Deng Y, Wang X, Wu W, Nepovimova E, Wu Q, Kuca K (2022) Toxic mechanisms of the trichothecenes T-2 toxin and deoxynivalenol on protein synthesis. Food Chem Toxicol 164:113044. https://doi.org/10.1016/j.fct.2022.113044

Liew WP, Mohd-Redzwan S (2018) Mycotoxin: its impact on gut health and microbiota. Front Cell Infect Microbiol 8:60. https://doi.org/10.3389/fcimb.2018.00060

Liu M, Zhao L, Gong G, Zhang L, Shi L, Dai J, Han Y, Wu Y, Khalil MM, Sun L (2022) Invited review: remediation strategies for mycotoxin control in feed. J Anim Sci Biotechnol 13:19. https://doi.org/10.1186/s40104-021-00661-4

Loi M, Fanelli F, Liuzzi VC, Logrieco AF, Mulè G (2017) Mycotoxin biotransformation by native and commercial enzymes: present and future perspectives. Toxins 9:111. https://doi.org/10.3390/toxins9040111

Mavrommatis A, Giamouri E, Tavrizelou S, Zacharioudaki M, Danezis G, Simitzis PE, Zoidis E, Tsiplakou E, Pappas AC, Georgiou CA, Feggeros K (2021) Impact of mycotoxins on animals' oxidative status. Antioxidants 10:214. https://doi.org/10.3390/antiox10020214

Mézes M, Balogh K (2009) Mycotoxins in rabbit feed: a review. World Rabbit Sci 17:53–62

Murugesan GR, Ledoux DR, Naehrer K, Berthiller F, Applegate TJ, Grenier B, Phillips TD, Schatzmayr G (2015) Prevalence and effects of mycotoxins on poultry health and performance, and recent development in mycotoxin counteracting strategies. Poult Sci 94:1298–1315. https://doi.org/10.3382/ps/pev075

Oswald IP, Comera C (1998) Immunotoxicity of mycotoxins. Revue De Medecine Veterinaire 149: 585–590

Perrone G, Ferrara M, Medina A, Pascale M, Magan N (2020) Toxigenic fungi and mycotoxins in a climate change scenario: ecology, genomics, distribution, prediction and prevention of the risk. Microorganisms 8:1496. https://doi.org/10.3390/microorganisms8101496

Pestka JJ (2008) Mechanisms of Deoxynivalenol-induced gene expression and apoptosis. Food Addit Contam Part A Chem Anal Control Expo Risk Assess 25:1128–1140. https://doi.org/10.1080/02652030802056626

Pestka JJ, Moorman MA, Warner RL, Witt MF, Forsell JH, Tai JH (1990) Immunoglobulin a nephropathy as a manifestation of vomitoxin (deoxynivalenol) immunotoxicity. In: Pohland AE, Dowell VR, Richard JL (eds) Microbial toxins in foods and feeds. Springer, Boston, MA, pp 427–440. https://doi.org/10.1007/978-1-4613-0663-4_40

Richard JL (1991) Mycotoxins as immunomodulators in animal systems. In: Bray GA, Ryan DH (eds) Mycotoxins, cancer and health, Pennington Center nutrition series, pp 196–220

Smith MC, Madec S, Coton E, Hymery N (2016) Natural co-occurrence of mycotoxins in foods and feeds and their in vitro combined toxicological effects. Toxins 8:94. https://doi.org/10.3390/toxins8040094

Stein RA, Bulboacă AE (2017) Mycotoxins. In: Dodd CER, Aldsworth T, Stein RA, Cliver DO, Riemann HP (eds) Foodborne diseases, 3rd edn. Books Press, pp 407–446. ISBN 9780123850072. https://doi.org/10.1016/B978-0-12-385007-2.00021-8

Steiner D, Sulyok M, Malachová A, Mueller A, Krska R (2020) Realizing the simultaneous liquid chromatography-tandem mass spectrometry based quantification of >1200 biotoxins, pesticides and veterinary drugs in complex feed. J Chromatogr A 1629:461502. https://doi.org/10.1016/j.chroma.2020.461502

Sulyok M, Stadler D, Steiner D, Krska R (2020) Validation of an LC-MS/MS-based dilute-and-shoot approach for the quantification of > 500 mycotoxins and other secondary metabolites in food crops: challenges and solutions. Anal Bioanal Chem 412:2607–2620. https://doi.org/10.1007/s00216-020-02489-9

Sweeney MJ, Dobson AD (1998) Mycotoxin production by *Aspergillus*, *Fusarium* and *Penicillium* species. Int J Food Microbiol 43:141–158. https://doi.org/10.1016/s0168-1605(98)00112-3

Szabó A, Szabó-Fodor J, Fébel H, Romvári R, Kovács M (2014) Individual and combined haematotoxic effects of Fumonisin B(1) and T-2 mycotoxins in rabbits. Food Chem Toxicol 72:257–264. https://doi.org/10.1016/j.fct.2014.07.025

Szabó A, Szabó-Fodor J, Fébel H, Mézes M, Bajzik G, Kovács M (2016) Oral administration of Fumonisin B1 and T-2 individually and in combination affects hepatic total and mitochondrial membrane lipid profile of rabbits. Physiol Int 103:321–333. https://doi.org/10.1556/2060.103.2016.3.5

Tolosa J, Rodríguez-Carrasco Y, Ruiz MJ, Vila-Donat P (2021) Multi-mycotoxin occurrence in feed, metabolism and carry-over to animal-derived food products: a review. Food Chem Toxicol 158:112661. https://doi.org/10.1016/j.fct.2021.112661

Tvrdá E, Ďuračka M, Halenár M, Lukáč N, Kolesárová A (2016) In vitro effects of selected trichothecenes on the rabbit spermatozoa motility behavior – a comparative study. Contemp Agric 65:21–26. https://doi.org/10.1515/contagri-2016-0013

Tvrdá E, Greifová H, Ďuračka M, Ondruška Ľ, Halenár M, Kolesárová A (2022) Comparative analysis of the detrimental *in vitro* effects of three fusariotoxins on the selected structural and functional characteristics of rabbit spermatozoa. Drug Chem Toxicol 45:2519–2527. https://doi.org/10.1080/01480545.2021.1962690

Xu R, Kiarie EG, Yiannikouris A, Sun L, Karrow NA (2022) Nutritional impact of mycotoxins in food animal production and strategies for mitigation. J Anim Sci Biotechnol 13:69. https://doi.org/10.1186/s40104-022-00714-2

Zhang Z, Nie D, Fan K, Yang J, Guo W, Meng J, Zhao Z, Han Z (2020) A systematic review of plant-conjugated masked mycotoxins: occurrence, toxicology, and metabolism. Crit Rev Food Sci Nutr 60:1523–1537. https://doi.org/10.1080/10408398.2019.1578944

Zingales V, Taroncher M, Martino PA, Ruiz MJ, Caloni F (2022) Climate change and effects on molds and mycotoxins. Toxins 14:445. https://doi.org/10.3390/toxins14070445

Insights into Clinical Pathology of Rabbits

Madalena Ferreira, Felisbina Queiroga, and Ana C. Silvestre-Ferreira

Abstract

With an estimation of 14 million pet rabbits worldwide, proper clinical care as well as an accurate understanding of the animal health status is necessary. Therefore, hematology, biochemistry, and urinalysis are essential tools for an overall health assessment and diagnosis of any underlying condition. However, in rabbits, the interpretation of hematology and serum and urinary biochemistry reference ranges can be quite challenging due to the lack of conformity from distinct reference sources and how wide they can be. In this chapter, the interpretation of the reference ranges and the clinical conditions associated with values outside of them are summarized. In a brief manner, hematological results are helpful to understand red blood cell or white blood cell disparities, linked to dehydration, anemias, inflammatory or infection responses, and even stress. Biochemistry informs about inadequate nutrition, renal and hepatic impairment, electrolyte inequality, and stress response. Lastly, urinalysis enlightens about kidney function as well as inadequate diet and electrolyte imbalance. The reference ranges here described are an amalgamation of different sources.

M. Ferreira (✉)
Departamento de Ciências Veterinárias, Universidade de Trás-os-Montes e Alto Douro (UTAD), Vila Real, Portugal
e-mail: aferreir@utad.pt

F. Queiroga · A. C. Silvestre-Ferreira
Departamento de Ciências Veterinárias, Universidade de Trás-os-Montes e Alto Douro (UTAD), Vila Real, Portugal

Centro de Ciência Animal e Veterinária (CECAV), Universidade de Trás-os-Montes e Alto Douro, Vila Real, Portugal

Laboratório Associado para a Ciência Animal e Veterinária – AL4AnimalS, Lisbon, Portugal

J. Simões, J. M. Monteiro (eds.), *Veterinary Care of Farm Rabbits*,
https://doi.org/10.1007/978-3-031-44542-2_33

Keywords

Hematology · Biochemistry · Urinalysis · Sample collection

1 Introduction

Over the last decades, rabbits are increasing its popularity as pets. Detection of disease can be difficult as rabbits exhibit confounding clinical signs or even hide the signs of disease. Laboratory tests can add additional information.

Hematology or serum biochemistry can provide valuable information on animal's health or welfare status, in order to detect disease in its early stage. The aim of this chapter is to give an insight on blood or urine sample collection and management methods, interpretation of the biochemical parameters and hemogram, as well as identification of blood cells and urinalysis in rabbits.

2 Sample Collection

2.1 Handling

Rabbits can be tranquil animals when accustomed to handling; however, sudden movements or loud noises can startle them and trigger anxiety or stress induced behaviors. Rabbits may struggle during handling or transportation by biting and scratching the handler or trying to flee by twisting and bucking their hindfeet which can result in spine injury leading to paresis or paralysis (Moore et al. 2015). Usually, some type of restraint is necessary during blood sampling, such as restraint device or bag, towel wrapping, or even anesthesia. Different restrain techniques have shown to interfere with varying biochemical and hematological parameters, so it's important to know the restraint method when interpreting the blood or biochemical work (Washington and Van Hoosier 2012).

2.2 Blood Collection Sites

There are several sites where blood can be collected from; however, as rabbits can be shifty animals, the collection area should be chosen accordingly. The blood volume of a healthy rabbit is approximately 55–65 mL/kg (Melillo 2007). In a single sampling, approximately 1% of the animal's body weight or 10% of the circulating blood volume can be drawn without compromising the rabbit's health (Washington and van Hoosier 2012). If various samplings are performed, only 1% of the total blood volume should be drawn every 24 h (Parasuraman et al. 2010).

The jugular vein is a site where a considerable amount of blood can be collected with ease in most rabbits; its large size and relatively easy access make it a satisfactory spot; However, it can be a stressful ordeal for the animal and may

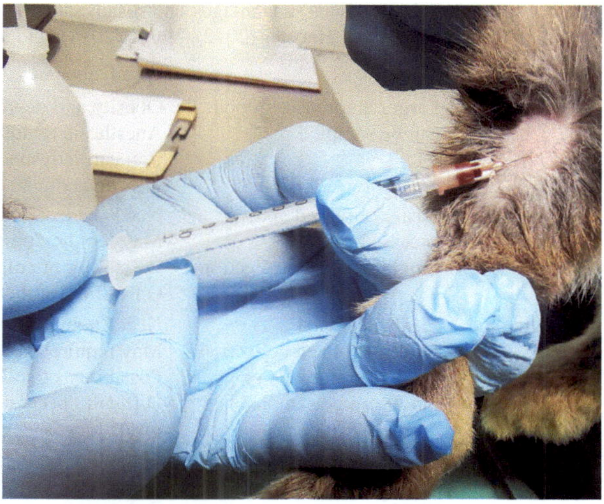

Fig. 1 Blood collection from the lateral saphenous vein (Courtesy of Dr. Roberto Sargo)

require chemical or physical immobilization. The visualization of the jugular vein can sometimes be obstructed by the dewlap of female rabbits and in cases of obesity (Melillo 2007). Good alternatives are the lateral saphenous vein (Fig. 1), as it has simple access and is somewhat less stressful, and the cephalic vein as it presents itself straight and easily accessible; both can be used for single or repeat collection and may not require anesthesia; however, the sample size is frequently small, and the cephalic vein presents the risk of collapsing due to its small size (Melillo 2007).

Other sites include the marginal ear vein and central ear artery which can be used for single or repeat blood collection, and although they are easily accessed in larger breeds, it can be difficult to manage in smaller rabbits who present smaller ears and can also exhibit insufficient blood flow in the area. From the marginal ear vein, only a small amount of blood can be sampled, 1–5 mL; however, from the central ear artery, a larger sample can be collected, up to 30 mL depending on the breed, but a topical anesthetic is recommended as it can be painful for the animal (Washington and Van Hoosier 2012).

When collecting from these sites, it's important to keep in mind that there's a chance of damaging the blood vessels and compromising blood flow to the area, namely, the pinna, resulting in head shaking and scratching that may further worsen the condition (Melillo 2007; Moore et al. 2015). Cardiac puncture involves profound anesthesia and is a terminal technique of blood sampling and therefore not suitable for pets. A large amount of blood can be collected for analysis, between 60 and 200 mL depending on the animal's size (Washington and Van Hoosier 2012). In Table 1, collection sites' advantages and disadvantages are summarized.

Table 1 Blood collection sites' advantages and disadvantages in rabbits. Adapted from Melillo (2007), Washington and van Hoosier (2012), Varga (2014), and Moore et al. (2015)

Site	Advantages	Disadvantages
Jugular vein	Larger amounts of blood can be sampled	Obesity and dewlaps my obstruct the site Anesthesia required Can cause dyspneic episodes in rabbits with URI infections due to neck extension
Lateral saphenous vein	Easily observed and accessed Single or repeat collection	Needs vein stabilization May require anesthesia Possibility of collapsing of the vein Only small amounts of blood can be sampled
Cephalic vein	Easily accessed Single or repeat collection Moderate amounts of blood can be sampled	Needs vein stabilization May require anesthesia Possibility of collapsing of the vein Difficult observation and access in smaller and dwarf breeds
Marginal ear vein	Easily observed and accessed Single or repeat collection Doesn't require anesthesia	Difficult to access in small or dwarf breeds Possibility of collapsing of the vein Only small amounts of blood can be sampled May cause thrombosis, hematomas, and bruising
Central ear artery	Easily observed and accessed Single or repeat collection Doesn't require general anesthesia Larger amounts of blood can be sampled	Requires topical anesthesia May cause hematomas and bruising May compromise blood flow to the pina if the blood vessel is damaged
Cardiac puncture	Very large amounts of blood can be sampled	Death of the animal Requires deep anesthesia

2.3 Blood Sample Processing

Rabbit's blood clots very fast, so it has to be sampled quickly but without inducing hemolysis. The aspiration of a few drops of an anticoagulant, such as heparin, into the collection needle and syringe can delay the coagulation process, without interfering significantly with biochemical parameters (Melillo 2007). The blood should then be placed in heparin coated tubes for hematology and tubes without anticoagulant for biochemistry followed by routine centrifugation and serum separation and analyzed preferably 1 h after collection; refrigeration at 4 °C may be done if the samples cannot be quickly processed (Moore et al. 2015).

Heparinized plasma can also be used for biochemistry. Blood smears should be done immediately after blood collection to avoid cell's morphology abnormalities caused by the anticoagulant. After routine staining (Romanowsky stains), a quick assessment of the animal's health status can be done through microscopical observation (Melillo 2007).

2.4 Urine Sample Collection

The technique for urine collection may vary according to sample contamination relevance. If lack of contamination is not a priority, urine can be collected by spontaneous micturition from a pan underneath the animal's housing (Washington and Van Hoosier 2012). Manual expression of the bladder can also be performed but should be done with the utmost care since the wall of rabbits' bladder is very thin and risks rupture if too much pressure is applied, especially if there's some type of urethral blockage (Melillo 2007).

For less contaminated samples, the best methods are catheterization or cystocentesis (Washington and Van Hoosier 2012). Cystocentesis should not be done repeatedly as the microtrauma caused by the piercing of the syringe in the bladder can create inflammation and induce the formation of calculi; furthermore, initial palpation is necessary to distinguish the bladder from other structures and avoid organ puncture or trauma (Washington and Van Hoosier 2012). To avoid the risk of trauma to adjacent structures, cystocentesis can be ultrasound-guided.

3 Reference Ranges for Hematology and Biochemistry

Hematology and biochemistry reference ranges in rabbits can vary significantly in routine clinical analysis; this is mainly due to the fact that initially reference ranges were stipulated using very particular group of individuals of the same breed, sex, and age and genetically homogenous (Varga 2014). In clinical practice, this is not the reality where rabbits of varying backgrounds are observed every day. Breed, sex, and age are all significant factors when assessing animals' health and can greatly influence some biochemical and hematological parameters (Melillo 2007; Özkan et al. 2012).

Different animals, sometimes with subclinical diseases and different laboratory methods, have resulted in a variable range of results (Melillo 2007; Özkan et al. 2012; Varga 2014). However, reference ranges should be set by each laboratory according to target population. In this work, the stated reference ranges result from variable sources that seemed to be the most generally accepted in rabbit's clinical practice.

4 Clinical Hematology

Hematology is the study of blood and its components, the organs responsible for their formation, and pathologies related to it. The cellular fraction of blood consists of red blood cells, white blood cells, and platelets, making up for 45% of the total volume (Washington and Van Hoosier 2012). Hematology is an indispensable tool for diagnosis, alongside the clinical assessment that provides a clear picture of the medical situation, allowing for a better oriented treatment and prognosis as there is a better understanding of individual's condition (Washington and Van Hoosier 2012).

4.1 Red Blood Cells

The biological process in which red blood cells (RBCs) are produced is designated erythropoiesis; after cells finish maturing, they are released into circulation. In mammals, generally speaking, these cells are round and concave, concentrate the majority of the hemoglobin in the edges, and are nonnucleated granting them plasticity to change shape to a certain degree. The hemoglobin molecules at their core cells are crucial for oxygenation of the blood and tissues (Palis 2014).

Rabbit's red blood cell diameter can vary from 5.0 to 7.8 μm; anisocytosis is a common finding in these animals and has no clinical significance; the red cell distribution width (RDW) is also higher in rabbits than in other species (Varga 2014). The low presence of reticulocytes (1–4% of the circulating RBC) and the observation of polychromasia (1–2% of the erythrocytes) in blood smears are associated with a short life span (57 days) and high erythrocyte turnover rate (Fig. 2a). The observation of a small number of nucleated RBCs (Fig. 2b) and the sporadic Howell-Jolly bodies is also a common finding and not linked with regeneration (Melillo 2007; Varga 2014). Male rabbits tend to have higher erythrocyte

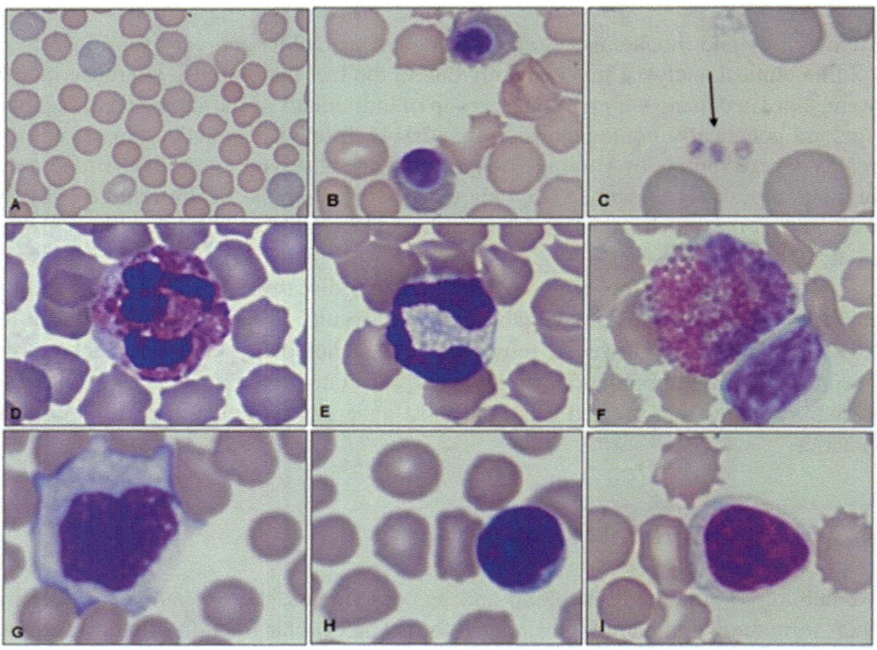

Fig. 2 Rabbit's blood smear, Diff-Quik stain, original magnification 1000×. (**a**) Erythrocytes with the presence of anisocytosis (variation in size) and polychromasia (variation in color). (**b**) Two nucleated red blood cells. (**c**) Platelets (arrow). (**d**) Heterophil. (**e**) Degranulated heterophil. (**f**) The two leukocytes represent an eosinophil (upper cell) and a lymphocyte (lower cell). (**g**) Monocyte. (**h**) Small lymphocyte. (**i**) Large lymphocyte

levels relative to female ones, and young rabbits of less than 12 weeks express lower erythrocyte levels than adults (Moore et al. 2015). Cold stress can lead to erythrocytosis while sample hemolysis causes erythropenia (Washington and van Hoosier 2012).

4.2 Hematocrit

Hematocrit (HCT) or packed cell volume (PCV) is the measurement of the volume of red blood cells in the total blood volume. It's often altered in animals suffering from anemia or dehydration. Age is also an interfering factor; some studies have shown that newborn rabbits have high PCV values that drop around 20 days old, only to increase to normal adult levels around 90 days old, while other studies state that newborn rabbits have similar hematocrits to adult ones. Differences between strain, breed, and gender seem to have no clinical value (Washington and Van Hoosier 2012).

Hematocrit below 30% is suggestive of anemia, especially coupled with low RBC and hemoglobin levels. The observation of reticulocytes is a telltale sign of a regenerative anemia that can be triggered by blood loss due to external or internal causes, for instance, intravascular hemolysis, hematuria due to urolithiasis, lead toxicity, uterine adenocarcinomas hemorrhage or endometrial aneurysms, trauma and parasitic infestations, and autoimmune hemolytic anemia (Melillo 2007; Washington and Van Hoosier 2012; Varga 2014). Nonregenerative anemia is usually linked to an underlying chronic disease of which common examples are rabbits' dental disease, endometritis and pyometra pododermatitis, otitis media, mastitis, renal disease, osteomyelitis, and pneumonia. Hematocrit levels above 45% are often a sign of dehydration or thermal stress (Melillo 2007; Washington and Van Hoosier 2012).

4.3 Hemoglobin

Hemoglobin (HGB) is a molecule at the core of red blood cells that is crucial for oxygen and carbon dioxide transport and therefore oxygenation of blood. In rabbits, its concentration is high up until around 20 days old, when it suffers a reduction, and then slowly increases and stabilizes on adult levels at 90 days old. Hemoglobin levels are higher in the morning and decrease toward the end of the day (Washington and Van Hoosier 2012).

4.4 Mean Corpuscular Hemoglobin, Mean Corpuscular Volume, and Mean Corpuscular Hemoglobin Concentration

Mean corpuscular hemoglobin (MCH) varies between infant and adult rabbits, showing higher levels in newborns; on the other hand, mean corpuscular hemoglobin concentration (MCHC) reveals no difference. Mean corpuscular volume (MCV) decreases from infant to adult age and is higher in pregnant rabbits (Washington and Van Hoosier 2012).

4.5 Reticulocytes

Reticulocytes are immature red blood cells that still present residual RNA and micro-organelles in their cytoplasm. They have larger diameters than RBC and basophilic staining. These cells are often associated with regenerative anemias; however, in rabbits a small percentage (1–4%) of reticulocytes is common and has no clinical significance (Washington and Van Hoosier 2012). Repeat blood samplings may increase reticulocyte count, and in infant rabbits (1–2 months), reticulocyte count is higher and steadies around 1 year of age (Moore et al. 2015).

4.6 Platelets

Platelets, also known as thrombocytes, are nonnucleated fragments of cells vital to maintain hemostasis representing key intervenors in the coagulation process (Gale 2011). Wright's stain revels oblong to oval shaped cells with a pale periphery and a purplish center, with sizes varying between 1 and 3 μm in diameter and appearing individually or in groups (Fig. 2c). Platelet number changes with age, illness, and stress. Thrombocytopenia can be observed in animals suffering from acute infections, while thrombocytosis has been described in rabbits exposed to cold stress (Washington and Van Hoosier 2012; Moore et al. 2015).

4.7 Total White Blood Cells

Total white blood cell (WBC) in rabbits varies with age, peaking at 3 months old with a rise in lymphocytes and later around 1 year old with an increase in neutrophils. There is also a diurnal variation with total WBC count being the lowest in late afternoon to evening (Washington and Van Hoosier 2012; Varga 2014). Unlike in other species, rabbits don't display a marked increase in total white blood cell count in response to inflammation or acute infections, although leukocytosis had been described in animals with lymphosarcoma (Melillo 2007; Varga 2014). Experiments have been conducted in which rabbits were experimentally infected with different pathogens (Toth and Krueger 1988; Toth and Krueger 1989), and although they displayed symptoms as fever, increased plasma cortisol levels, neutrophilia, and lymphopenia, there was no significant variation in total WBC count (Melillo 2007; Varga 2014).

4.8 Differential White Blood Cell Count and Stress

Diseased rabbits can show an alteration of the neutrophil-lymphocyte ratio, and this is perceptible through the differential leucocyte count. The ratio is usually 1:1, however, changes have been observed in stressed and ill animals (Melillo 2007; Varga 2014).

Stress can be a determining factor in neutrophil-lymphocyte ratio redistribution as rabbits are particularly susceptible to its effects. Adrenaline and cortisol discharges alter lymphocyte levels, the first causing lymphocytosis (acute stress) and the second lymphopenia (chronic stress). Long traveling distances or a loud environment in the waiting room can induce extended periods of stress in rabbits which is noticeable by high cortisol levels accompanied by relative neutrophilia and lymphopenia; this adjustment in the ratio can last for 24–48 h (Melillo 2007; Varga 2014). Disease in itself is a stress inducer in these animals, but calm handling and examination or sample collection does not seem to create significant changes in the ratio.

4.9 White Blood Cell Populations

White blood cells originated through leucopoiesis. Leucocytes are a nucleated, diverse group of cells that can be divided in two major groups, granulocytes and agranulocytes.

Granulocytes encompass neutrophils, eosinophils, and basophils. They originate from the same unipotent cell, the myeloblast, and possess segmented nuclei and small granules in their cytoplasm with specific chemical natures (Bain 2017). Together (alongside monocytes and agranulocyte), they represent the innate immunity, organism's first line of defense (Sheshachalam et al. 2014).

Agranulocytes entail monocytes and lymphocytes; these cells have no defined cytoplasmic granules, and their nuclei, which are not lobed, tend to occupy a larger portion of the cytoplasm. Monocytes originate from the same unipotent cell as granulocytes. However, they are morphologically distinct because they are stimulated by different colony-stimulating factors (CSFs) (Hong 2017).

4.10 Neutrophils

Neutrophils are some of the most numerous leucocytes in circulation. The release of chemical signals in the lesion or infected area and activation of the complement trigger a response from neutrophils known as chemotaxis, in which these cells move toward the damaged area (Bain 2017). Neutrophils have high mobility; with the capacity of diapedesis, they quickly act upon affected tissues, both as phagocyte and by releasing pro-inflammatory mediators contained in their cytoplasmic granules along with many other enzymes and proteins (Sheshachalam et al. 2014).

In rabbits, neutrophils are round cells 10–15 μm in diameter and may be referred to as heterophils due to the acidophilic staining properties of their granules, displaying a pink to red color (Melillo 2007). Their nucleus is lobed and polymorphous in shape and stains light purple-blue, while the cytoplasm stains pink (Fig. 2d) (Washington and Van Hoosier 2012; Moore et al. 2015). Minor heterophil degranulation may result from a quick Romanowsky stain such as Diff-Quick; this stain induced degranulation is thought to be a result of short fixation time (5–10 s). The degranulated cells will resemble heterophils; however, vacuoles will be present where the granules formerly were (Fig. 2e).

Unlike in other species where a rise in neutrophils levels leads to an increase in total leucocyte counts, this elevation is not apparent in rabbits (Varga 2014). A rise in neutrophils due to inflammation is often accompanied by a decrease in lymphocytes and in acute infections by a decrease in thrombocytes. Infections in rabbits tend to demonstrate an increase in neutrophil to lymphocyte ratio, with relative neutrophilia in conjunction with lymphopenia, but the presence of band neutrophils is rare and a left shift not observed (Melillo 2007; Washington and Van Hoosier 2012); however, there are reports of animals with sepsis presenting mature neutrophilia, normal neutrophilic count, or even neutropenia (Melillo 2007). Rabbits exhibit lower neutrophil levels in the morning and reach peak levels at the end of the day, and in stressed animals, it's common to find elevated cortisol levels alongside mature neutrophilia (Washington and Van Hoosier 2012).

4.11 Eosinophils

Eosinophils exist in small quantities in circulating blood and tissues; however, their number can increase exponentially in cases of allergic reactions (Spencer et al. 2014). Similarly, to neutrophils, eosinophils respond to chemotactic signals, act as phagocytes, and have a series of pro-inflammatory proteins contained in their granules that can swiftly be released into the system (Spencer et al. 2014). Their main role is the response and elimination of histamine like substance (Varga 2014).

Rabbits' eosinophils are slightly larger than heterophils, are 10–16 μm in diameter, and display many, large eosinophilic granules (Fig. 2f). The nucleus is bilobed or in the shape of a horseshoe and stains purple-blue, while the cytoplasm, covered by the large, round granules, displays an intense orange-pink color and a foamy texture (Melillo 2007; Washington and Van Hoosier 2012). In rabbits, low to no eosinophilic count is normal and has no association with any medical condition (Moore et al. 2015). Eosinophil levels are low in the morning and reach their highest levels toward the end of the day (Washington and Van Hoosier 2012). Chronic eosinophilia can be observed in animals with skin, lung, gastrointestinal (GI) tract, and uterus illnesses, while moderate to high eosinophilia is often associated with parasitism, although it's rare in this species, or trauma wound healing. High cortisol levels due to chronic stress can induce eosinopenia (Melillo 2007; Varga 2014).

4.12 Basophils

Although in other species basophil counts are usually low to zero, in rabbits, these cells are commonly found in circulation of healthy animals. Their exponential increased count is frequently associated with allergic/inflammatory reactions and hypersensitivity in other animals, but in rabbits, it's still not well understood (Melillo 2007). Basophils have many cytoplasmic granules, which are rich in histamine, heparin, serotonin, and cytokine, into circulation and tissues (Stone et al. 2010).

In rabbits, basophil diameter is 8–12 μm and displays many deeply basophilic granules in a light gray cytoplasm. The nucleus is less segmented than in eosinophils or neutrophils and its purple stain might not be perceptible due to the amount of cytoplasmic granules (Melillo 2007; Washington and Van Hoosier 2012). Like all the other granulocytes, basophil levels are low in the morning and reach their peak toward the end of the day (Washington and Van Hoosier 2012). Animals with chronic skin problems often show high basophilic and eosinophilic levels (Melillo 2007; Washington and Van Hoosier 2012).

4.13 Monocytes

Monocytes are primarily phagocytic cells. They are given different names according to the tissue they migrate to, and when disseminated in them, these white blood cells differentiate into macrophages and dendritic cells, phagocyting the foreign agents responsible for attack. Not only are these cells important members of the innate immune system but also of the adaptive one, presenting antigens to T lymphocytes (Espinoza and Emmady 2022).

Rabbits' monocyte diameter is 15 to 18 μm, making them the largest leucocyte. They display an abundant cytoplasm, with few granules and vacuoles, that stains blue-gray and a polymorphic, light purple nucleus that can be round, lobulated, horseshoe, or bean shaped (Fig. 2g) (Melillo 2007; Washington and Van Hoosier 2012). Monocytes tend to be low in the morning and reach their peak toward the end of the day (Washington and Van Hoosier 2012). In rabbits, monocytosis is associated with chronic inflammation (abscess, mastitis, or tympanic bulla empyema); however, the lack of monocytosis is not an indication that there isn't a chronic inflammation; rabbits with chronic osteomyelitis due to dental disease may have normal range monocyte values (Melillo 2007; Varga 2014).

4.14 Lymphocytes

Lymphocytes are the most abundant leucocyte in rabbits and are the representation of the adaptive immune system. They are spilt into two groups, B lymphocytes, the humoral immunity, and T lymphocytes, the cellular immunity. B lymphocytes play two essential roles, production of antibodies and development of memory cells (Janeway et al. 2001). T lymphocytes can be classified as cytotoxic (Tc), regulatory

Table 2 Hematological reference ranges for rabbits. Adapted from Washington and van Hoosier (2012), Varga (2014), and Moore et al. (2015)

Parameters	Units	Reference range
RBC	$\times 10^6/\mu l$	4.9–7.8
Hemoglobin	g/dL	10–17.4
HCT (PCV)	%	31–50
MCV	μm^3	57.5–75
MCH	pg/cell	17.1–23.9
MCHC	g/dL	28.2–37
Reticulocytes	%	1.7–6.3
WBC	$\times 10^3$ cells/μl	5.2–12.5
Neutrophils	%	20–75
Lymphocytes	%	30–85
Eosinophils	%	0–5
Basophils	%	0–10
Monocytes	%	0–10
Platelets	$\times 10^6/\mu l$	200–1000

(Treg), helper (Th), and lastly natural killer (NK) cells. T lymphocytes, as a whole, are mainly responsible for identification and binding to antigens and posterior signaling for its destruction (Janeway et al. 2001).

Small lymphocytes predominate in rabbits and are 7–10 μm in diameter; their acentric nucleus stains purple and is round/oval occupying most of the cytoplasm, which sometimes can only be seen in the shape of a light blue half-moon (Fig. 2h). Large lymphocytes, although less frequent, present with diameters of 10–15 μm; acentric round/oval purple nucleus, sometimes with indentions and encircled by an unstained halo (Fig. 2i); and a large blue cytoplasm where basophilic granules can be present (Melillo 2007; Washington and Van Hoosier 2012; Moore et al. 2015). Sometimes reactive lymphocytes can be present; they are lager cells with a more intense blue cytoplasm that have antigenic-stimulated origin.

In rabbits, lymphocytes are dispersed through the body but concentrate particularly in the blood, spleen, bone marrow, lymph nodes, and lymphatic tissues of the gastrointestinal tract. The circulating lymphocytes are not a direct measurement of lymphopoiesis but more of the number of cells entering or being eliminated from the body (Melillo 2007; Varga 2014). Lymphopenia has been observed in rabbits with dental disease as well as vestibular disease, gut impact, or abscess. Animals with viral infections, lymphoma, or lead poisoning may present lymphocytosis (Melillo 2007; Varga 2014). Contrary to every other leucocyte, lymphocyte levels tend to be higher in the morning and lower by the end of the day. Acute stress (adrenaline discharges) may induce an increase in lymphocyte levels, while chronic stress (cortisol discharges) may do the opposite (Washington and van Hoosier 2012).

In Table 2 are represented the reference ranges for the described hematological parameters for rabbits.

5 Clinical Biochemistry

Blood biochemical changes are often a sign of clinical pathologies; the establishment of reference ranges for each species helps in making the diagnosis, understanding the medical condition, and with the selection of the appropriate treatment.

The International Committee for Harmonization of Clinical Pathology Testing (IHCPT) is a committee, founded by ten scientific organizations, responsible for the publication of a paper determining the establishment of animal clinical pathology testing in toxicity and safety studies. In this publication are described the most relevant clinical biochemical parameters in the assessment of an animal's health status (Weingand et al. 1996). Those parameters are identified and explored through this chapter.

5.1 Alanine Aminotransferase

Alanine aminotransferase (ALT) is a transaminase enzyme that, in the alanine cycle, catalyzes the transport of an amino group from alanine to alpha-ketoglutarate producing pyruvate and glutamate or vice versa (Washington and Van Hoosier 2012).

Unlike in mammals such as dogs and cats in which this enzyme is liver-specific and has a prolonged half-life (45–60 h in dogs), in rabbits, this enzyme is not a good indicator of hepatic damage (Varga 2014). Rabbits present lower ALT hepatic activity, and the enzyme is found in other organs, such as the cardiac muscle, with a half-life of 5 h. Its short half-life and the fact that it's not organ specific make it an ineffective diagnostic marker for liver malfunction, but it can however be an indicator of cardiac muscle damage (Washington and Van Hoosier 2012; Özkan et al. 2012; Varga 2014). ALT is not affected by restrain and its increase is a sign of real tissue damage (Washington and Van Hoosier 2012). A minor increase of ALT in healthy rabbits seems to be associated with resins and solvents used in wood-based litter and a larger increase with the ingestion of aflatoxins, exposure to low levels of lead, and anesthesia by halothane (Washington and Van Hoosier 2012; Varga 2014). High levels of ALT in association with alkaline phosphatase (ALP), bilirubin, and γ-glutamyl transferase (GGT) are seen in cases of hepatic coccidiosis. Hepatic lipidosis, torsion of a hepatic lobe, or even late stages of hepatocyte neoplasms can all contribute to raised ALT levels (Washington and Van Hoosier 2012; Varga 2014).

5.2 Aspartate Aminotransferase

Aspartate aminotransferase (AST) is a transaminase that, in the aspartate cycle, catalyzes the transference of an amino group from aspartate to alpha-ketoglutarate forming oxaloacetate and glutamate and vice versa (Washington and van Hoosier 2012).

AST has a wide distribution across many tissues such as in the cardiac tissue, kidneys, and pancreas and in rabbits presents peak activity in the liver and skeletal muscle with a half-life of 5 h. Increased levels of ALT can be found alongside hepatic complications. Difficulty during the blood sampling, hemolysis, physical effort, restraint, and a stressful environment can lead to increased ALT results as well as halothane or isoflurane anesthesia (Washington and Van Hoosier 2012; Varga 2014).

5.3 γ-Glutamyl Transferase

γ-Glutamyl transferase (GGT) catalyzes the transference of γ-glutamyl groups to amino acids. In rabbits, GGT has low expression in the liver and is mostly found in the kidneys. With the majority of the enzyme being excreted in the urine, high levels of GGT can be detected in the excretions of animals with kidney damage (Melillo 2007; Bodnariu 2022). In the liver, GGT is mostly present in the epithelial cells of the bile ducts, and an increase of its levels is often associated with lesions or occlusion of the area and consequently hepatobiliary complications, not hepatocellular injury (Melillo 2007; Varga 2014).

5.4 Alkaline Phosphatase

Alkaline phosphatase (ALKP) is an enzyme which main purpose is to hydrolyze phosphate groups from molecules (Washington and Van Hoosier 2012).

ALKP has a broad distribution in organs and tissues such as in the liver, bones, intestinal tract, bile ducts, placenta, and kidney's tubules. Rabbits produce three ALKP isoenzymes, one in the intestine and the most abundant and two in the kidneys and liver (Varga 2014). ALKP is not affected by animal restraint so its increase is indicative of real injury (Washington and Van Hoosier 2012). Reference ranges can vary significantly, particularly with breed, in healthy animals; however, increased levels are often associated with bone fractures and osteoblastic activity. Biliary blockage due to neoplasia, hepatic coccidiosis, abscesses, or lipidoses is also responsible for ALKP increment (Melillo 2007; Varga 2014). A decrease of ALKP has been reported in pregnant rabbits (Özkan et al. 2012; Washington and Van Hoosier 2012).

5.5 Creatinine Kinase

Creatinine kinase (CK) is responsible for the phosphorylation of creatinine (Washington and Van Hoosier 2012).

CK has three main isoenzymes, CK-MM present in the muscle, CK-BB present in the brain, and CK-MB present in the heart. In rabbits, restrain may contribute to CK increase as well as any type of stressful commotion. Depending on the method of sampling, different isoenzymes may rise; for instance, CK-MB increases in cardiac puncture and CK-MM and CK-BB in draws from peripheral vessels. High levels of CK are often associated with necrosis of the muscle and illness (Washington and Van Hoosier 2012).

5.6 Lactate Dehydrogenase

Lactate dehydrogenase (LDH) is an enzyme that, during glycolysis, converts pyruvate and NADH in lactate and NAD+ and vice-versa, by transference of a hydride group (Washington and Van Hoosier 2012; Farhana and Lappin 2022).

Five isoenzymes have been reported for LDH. LDH-1 is found in cardiac tissue, LDH-2 is involved in macrophage and monocyte phagocytic system, LDH-3 has high activity in the lungs, LDH-4 exerts its function primarily in the kidneys, and LDH-5 is highly represented in the liver and skeletal muscle. Due to being produced in two organs, LDH-5 is usually not a very effective diagnostic marker (Washington and Van Hoosier 2012; Farhana and Lappin 2022). The increase of LDH is representative of tissue destruction and cell death, such as in cases of hepatitis or hypoxia. Troponin levels alongside LDH can be helpful tools in myocardial compromise diagnosis. In rabbits, restrain or any other stressful factor can influence LDH level upward as well as hemolysis during or after the sampling (Washington and Van Hoosier 2012).

5.7 Bile Pigments/Bilirubin

Bile pigments are produced as hemoglobin breakdown occurs in the liver, and its excretion occurs mainly in the urine. Rabbits produce large amounts of bile relative to their weight, the main pigment being biliverdin (Washington and Van Hoosier 2012). Only about 30% of biliverdin is converted into bilirubin due to low biliverdin reductase activity in these animals. Bilirubin suffers hepatic conjugation and is then secreted in the bile in monoconjugated form (Washington and Van Hoosier 2012). Unlike bilirubin, which can be measured in the blood, biliverdin has no known measuring assay. In rabbits, increased levels of bilirubin are often associated with bile obstruction which can be caused by hepatic coccidiosis (common in young rabbits), hepatic lipidosis, torsion of a hepatic lobe, or biliary neoplasia (more common in adult rabbits) (Melillo 2007; Washington and Van Hoosier 2012). Other complications such as hemolytic diseases or ingestion of aflatoxins can also induce hyperbilirubinemia. High bilirubin levels in viral hemorrhagic diseases and hepatic fibrosis are often accompanied by raised hepatic enzymes (Melillo 2007).

5.8 Bile Acids

Bile acids are a product of cholesterol metabolism in the liver and aid lipidic digestion and absorption when secreted in the intestine. They are then reabsorbed into circulation for hepatic retrieval and subsequent secretion (Washington and Van Hoosier 2012). Bile acids are not a routine biochemical analysis performed in rabbits. The evaluation is done comparing preprandial and postprandial bile acids in circulation, acting as a diagnostic marker for hepatic function. In rabbits, however, it's very difficult to induce a fast (Melillo 2007; Washington and Van Hoosier 2012). Persistent high levels of bile acids, around 100 mmol/L, have however been described in cases of liver disease, such as hepatic coccidiosis (Melillo 2007).

5.9 Cholesterol and Triglycerides

Cholesterol is an important constituent of the membrane of the cells. It's synthesized and metabolized hepatically or acquired nutritionally. Cholesterol is the precursor molecule from which bile acids, some lipid soluble vitamins, and steroid hormones are created. It's released in the bile in the intestinal tract after a meal and reabsorbed and recycled in the liver (Washington and Van Hoosier 2012). A fasting period is necessary to correctly determine cholesterol which makes it difficult in rabbits due to the ingestion of caecotrophs. Triglycerides consist of a glycerol molecule and three fatty acids and function as an energy source. Similar to cholesterol, triglyceride concentration rises postprandially, so measurements in rabbits are difficult due to the impossibility of fasting (Melillo 2007; Washington and Van Hoosier 2012; Varga 2014).

Female rabbits naturally exhibit higher cholesterol levels than males, and different ranges for cholesterol and triglyceride can be observed between animals. Animal age and the time of the day in which the sample is collected can also have influence, since a rise in cholesterol is observed in the afternoon (Washington and Van Hoosier 2012; Varga 2014). However, atypically high levels of cholesterol and triglycerides are often found in obese rabbits or with fat rich nutrition and in cases of hepatic illness. Hypercholesterolemia in anorexic rabbits suggests late-stage hepatic lipidosis and in healthy weight rabbits can be an indicator of diabetes mellitus, pancreatitis, nephrotic syndrome, or chronic renal failure. Hypocholesterolemia can be observed during pregnancy or subnutrition (Melillo 2007; Varga 2014).

5.10 Amylase

Amylase is an important digestive enzyme that breaks down carbohydrates into monosaccharides. In rabbits, this enzyme is produced in large quantities in the pancreas and residually in the salivary glands. It's also present in the caecotrophs rabbits consume (Varga 2014). Comparatively to other species, rabbits exhibit low serum concentrations of amylase, and an increase in it is associated with pancreatic duct obstruction, pancreatitis, treatment with corticosteroids, abdominal trauma, and peritonitis (Washington and Van Hoosier 2012). This enzyme is excreted by the kidneys, so renal failure can also be a cause for hyperamylasemia. Hemolyzed samples can lead to lower amylase concentrations (Washington and Van Hoosier 2012).

5.11 Urea and Creatinine

Urea is a nitrogenous by-product of protein catabolism, and it's excreted in the urine. Blood urea nitrogen (BUN) is often a marker of kidney function in animals. However, BUN levels are affected by several factors such as circadian rhythm (topping in late afternoon to evening), protein poor or rich diet, liver function, hydration, intestinal absorption, physical activity, and cecal flora urea usage (Bodnariu 2022). Breed and gender can also show influence and so reference ranges can vary (Melillo 2007; Washington and Van Hoosier 2012). On this account, urea may sometimes be an ineffective diagnostic tool in rabbits, small alterations of concentration are hard to interpret, and slight elevations are frequent in these animals (Varga 2014). Such elevation may be due to dehydration, urolithiasis, renal insufficiency, or ureteral ligation. Decreased levels can be caused by liver insufficiency, use of anabolic steroids, or muscle loss (Washington and Van Hoosier 2012).

Creatinine is nitrogenous by-product of creatine degradation in the muscles. It's excreted in the urine and functions as a marker of glomerular filtration. Circulatory creatinine levels are unaffected by diet, hydration, or activity, so elevated creatinine levels are mostly linked to impaired renal function, acting as better diagnostic tool than urea. Kidney's damage can be associated with neoplasia, glomerular nephritis, or pyelonephritis, among others. Severe muscle injury can also cause elevation of serum creatinine (Melillo 2007).

In rabbits, common prerenal azotemia triggers are dehydration, stress, and cardiac deficiency, all of which cause weak renal perfusion and diminish nitrogenous product excretion. GI hemorrhage where high quantities of protein are being digested also leads to prerenal azotemia (Melillo 2007; Varga 2014). Azotemia associated with damaged renal function is often accompanied with hyperkalemia or hypokalemia, hypercalcemia and hyperphosphatemia, nonregenerative anemia, and isosthenuric urine (Melillo 2007). Postrenal azotemia is often a result of urethra blockage or bladder sludge, preventing urine flow (Melillo 2007). *Encephalitozoon cuniculi* is a parasite that commonly affects rabbit's renal parenchyma leading to the development of granulomas and fibrosis (Washington and Van Hoosier 2012).

5.12 Glucose

Glucose is an indicator of carbohydrate metabolism and dependent on many factors such as nutrition, hormone secretion, fasting periods, restraint and stress, and anesthesia. Glucose usage by erythrocytes is also an important factor to account for when collecting the blood sample, since erythrocytes are highly dependent on blood glucose for energy (Washington and Van Hoosier 2012; Bodnariu 2022).

Rabbits are capable of maintaining constant glucose levels, since there is no fasting period in this species and volatile fatty acids resulting from bacterial fermentation in the gastrointestinal tract are constantly being absorbed and used as an energy source. Consumption of caecotrophs acts as another glucose source. These animals are capable of withstanding up to 4 days of starvation without glycemic changes (Melillo 2007; Varga 2014).

Naturally occurring diabetes mellitus in rabbits is rare, but laboratory induced elevated levels of glucose (>500 mg/dL) are found alongside other common diabetes symptoms such as glycosuria, polydipsia, polyuria, and polyphagia (Melillo 2007). Anesthesia with urethane, halothane, or isoflurane increases glucose levels (Washington and Van Hoosier 2012). Environmental stress such as hyperthermia, handling, and restraint are a common hyperglycemia trigger in rabbits as well as acute intestinal blockages (350 mg/dL). Terminal stage GI tract stasis causes a rise in glucose levels and has a poor prognostic, implying hepatic lipidosis. Hyperglycemia is also observed in hemorrhagic or traumatic shocks; however, common glucose raisers such as hyperadrenocorticism or acute pancreatitis have not been reported in pet rabbits (Melillo 2007; Varga 2014).

Hypoglycemia has been observed in anorexic animals and cases of nutrition deficiencies or malabsorption and carbohydrate metabolism defects. Septicemia, terminal mucoid enteropathy, liver failure, or other chronic diseases can be responsible for low glycemic levels (Melillo 2007; Washington and Van Hoosier 2012; Bodnariu 2022).

5.13 Total Proteins

Total proteins (TP) measure the sum of globulin and albumin levels. A series of factors can influence TP concentration, such as breed, age, nutrition, pregnancy, and environment, among others (Melillo 2007; Varga 2014).

Hyperproteinemia is often observed in dehydrated rabbits. Chronic diseases, shock, metabolic disorders, or difficulty during sampling causing hemolysis can also be a cause. Analysis of the globulin and albumin fractions as well as the hematocrit can help with the diagnosis. Malnutrition, malabsorption, and hepatic or renal impairment leads to low TP levels as well as hemorrhagic or exudative wounds such as burns or flystrike (Melillo 2007; Varga 2014).

5.14 Albumin

Albumin has exclusive hepatic synthesis, and it is serum's most predominant protein, responsible for maintaining osmotic pressure in the blood vessels. Albumin reference ranges can vary with breed, gender, and pregnancy (Washington and Van Hoosier 2012).

Hypoalbuminemia is often associated with liver disease. In rabbits, parasitism is a common reason for liver impairment, Eimeria stiedae causes hepatic coccidiosis and Cysticercus pisiformis migration through the liver leads to the development of scars and necrosis. Other hypoalbuminemia triggers can be glomerulonephropathy, protein-losing enteropathy, malabsorption, and cardiogenic ascites (Varga 2014). In pet rabbits, reduced ingestion of caecotrophs (due to skeletal problems, dental diseases, obesity, among others), dental diseases, and incorrect or poor diet are all usual causes of low albumin levels (Melillo 2007).

Hyperalbuminemia is not a common finding in rabbits and is not associated with any specific pathology. However, high levels of albumin alongside increased PCV can be a sign of dehydration (Varga 2014).

5.15 Globulins

Globulins are a range of proteins, subdivided in alpha, beta, and gamma globulins, that are separated into five fractions, $\alpha 1$, $\alpha 2$, $\beta 1$, $\beta 2$, and gamma globulins, through electrophoresis (Washington and Van Hoosier 2012; Varga 2014). Alpha globulins increase in the presence of acute infections. In rabbits, bacterial infections, growing abscesses, and fever can be triggers for elevated levels of alpha globulins. Beta globulins are binding proteins in the plasma; this fraction is comprised of acute-phase proteins among which is fibrinogen that in rabbit is linked with inflammation. Gamma globulins are for the most part antibodies, and raised levels of these immunoglobulins are often observed in chronic inflammations or infections (Melillo 2007; Varga 2014).

Serum or plasma electrophoresis is an important tool in the diagnosis of monoclonal gammopathy or polyclonal alterations, which can help better understand the nature of infection or inflammation (Varga 2014).

5.16 Electrolytes

Rabbit's complex GI tract processes and their difficulty correcting acid-base alterations in the kidneys make them very susceptible to electrolyte imbalances. In these animals, electrolyte changes are often associated with unusual losses, for instance, malabsorption and irregular secretions associated with GI illness. Electrolyte concentration can be influenced by blood sampling. Collection with plastic catheters as shown decreased sodium levels, as well as general anesthesia with pentobarbitone. On the other hand, potassium concentrations were higher in venous than arterial blood and in serum rather than in plasma (Varga 2014).

5.17 Sodium

In rabbits, sodium (Na^+) is not a very good diagnostic marker and reference ranges can vary between breeds (Washington and Van Hoosier 2012). Variations in sodium levels tend to indicate the osmolality of extracellular fluid and not overall body sodium concentration (Varga 2014).

Dehydration or fluid loss can lead to hypernatremia as well as diet with high sodium intake levels. Hyponatremia is often found in animals with acute or chronic renal failure where sodium-potassium exchange is compromised. Lipemia or hyperproteinemia can artifactually decrease sodium concentration levels (Melillo 2007; Varga 2014).

5.18 Potassium

Potassium (K^+) has great homeostatic importance, being one of the main ions involved in maintaining membrane potential. Unusually low or high concentrations of potassium are often lethal as cells exhibit compromised electrical activity (Melillo 2007). Around 95% of total body potassium is concentrated intracellularly, and this concentration gradient is maintained by a complex mechanism, affected by blood's pH and regulated by hormone, for instance, aldosterone, insulin, and catecholamines. Aldosterone is responsible for renal potassium excretion and insulin for the cell's potassium intake (Varga 2014). Catecholamines are directly responsible for resting membrane and action potentials.

In rabbits, hyperkalemia is often associated with acute renal failure or obstruction of urine flow. Severe tissue damage from trauma can also release elevated amounts of potassium into circulation. Metabolic acidosis produces an ion redistribution in cell's membranes, which increases potassium levels extracellularly. Intravascular hemolysis, difficulties during sampling, or late separation of the blood can also, artifactually, produce hyperkalemia (Melillo 2007; Varga 2014).

Hypokalemia in rabbits can cause muscle weakness and atrophy. It can be triggered by diet nutritional deficiencies, loss of fluids and nutrients through the GI tract, renal impairment, or the use of potassium free fluids for long periods of time (Varga 2014). Alkalosis, although rare in rabbits, induces an ion redistribution in the cell's membrane, increasing cellular potassium uptake that can lead to low circulating potassium levels. Hyperlipidemia or hyperproteinemia can have as a secondary effect artifactual hypokalemia. A stress induced rise in catecholamines can also result in low potassium levels (Melillo 2007; Varga 2014).

5.19 Chloride

Chloride (Cl^-) is an important extracellular ion that, alongside sodium, is responsible for the upkeep of osmolality and acid-base exchanges. Age and sex seem to not have influence in chloride levels; there are however some concentration variations between strains. Like many other ions, chloride is reabsorbed renally (Washington and Van Hoosier 2012).

Variations in chloride levels are commonly paralleled by changes in sodium levels. Hyperchloremia is often observed in dehydrated rabbits, with nutritionally high salt intake, or suffering from metabolic acidosis or respiratory alkalosis. Low levels of chloride are observed in animals with fluids loss, such as diarrhea, low dietary salt intake, or metabolic alkalosis (Evans 2009).

5.20 Bicarbonate

Bicarbonate (HCO_3^-) is essential for acid-base exchange and acts as a buffer in the blood. It has pancreatic secretion. Its concentration is affected by renal excretion and carbon dioxide formation in the blood due to respiratory metabolism (Evans 2009; Washington and Van Hoosier 2012).

Hyperbicarbonatemia can be a result of respiratory acidosis and metabolic alkalosis (Washington and Van Hoosier 2012).

5.21 Magnesium

Magnesium (Mg^{2+}) is a cation responsible for enzymatic function, neuromuscular activity, and bone metabolism. Usually, half of it is in its ionized form, while the remaining half is complexed with other molecules such as albumin (around 40%), phosphate, citrate, and bicarbonate (Evans 2009; Washington and Van Hoosier 2012). In rabbits, magnesium reference range is not altered by age, gender, or strain and is excreted, for the most part, in the urine.

Hypermagnesemia is observed in rabbits with severe dehydration, hyperthyroidism, trauma, or adrenocortical insufficiency (Washington and Van Hoosier 2012).

5.22 Phosphate

Inorganic phosphate (PO_4^-) is responsible for the functioning of enzymes, carbohydrate, and muscle metabolism as well as bone and teeth development. It's obtained through diet and filtered and reabsorbed renally. Phosphate concentrates mainly intracellularly, so hemolyzed samples present higher concentrations of this ion; alterations in acid-base metabolism may also incite shifts in extra- and intracellular phosphate levels (Melillo 2007; Varga 2014). In rabbits, this ion levels can significantly differ with strain.

Hyperphosphatemia is often found in rabbits with prerenal, renal, and postrenal complications, usually alongside azotemia. High phosphate levels are seen in animals with chronic kidney failure or soft tissue trauma. In animals with urolithiasis, nephrolithiasis, dental disease, or other signs of nutritional secondary hyperparathyroidism, it's important to evaluate both phosphate and calcium levels to assess mineral balance (Melillo 2007; Varga 2014).

Hypophosphatemia clinical relevance is yet to be understood; it seems to be linked with nutritional deficiency, decreased intestinal absorption, or metabolic disorders (Melillo 2007; Varga 2014).

5.23 Calcium

A major part of calcium (Ca^{2+}) reserves is stored in the bones conjugated with phosphate, while in circulation this cation can be found bound with serum proteins, in particular albumin, with anions or ionized (Varga 2014). Calcium is involved in the constitution of bones and teeth as well as responsible for muscle metabolism, enzyme activation, blood coagulation, and osmoregulation. Total serum calcium encompasses ionized and bound calcium. However, measurements of ionized calcium are typically more expensive and complicated and not done as often. Due to calcium's high capacity of binding to albumin, total serum calcium levels are proportional to albumin concentration (Melillo 2007; Varga 2014). Reference ranges can vary significantly within the species; young or pregnant rabbits tend to exhibit lower calcium levels due to its high requirement (Washington and Van Hoosier 2012).

In rabbits, calcium metabolism is a lot more demanding than in other species; all dietary calcium is passively absorbed (independent of vitamin D), so total blood calcium levels are therefore higher than for other species and are strongly influenced by diet and GI tract absorption, serum protein concentration, and renal excretion (Melillo 2007; Varga 2014). Vitamin D plays an important role in calcium distribution as well as absorption when nutritional calcium levels are low (Melillo 2007). Hypercalcemia frequently occurs in animals with impaired kidney function given that calcium urinary excretion rate in rabbits is 45% to 60% (<2% in other species); this also predisposes rabbits to urinary tract stones and sludgy urine (Melillo 2007; Varga 2014). Although rare, hypocalcemia has been described in rabbits and is often due to calcium nutritional deficiencies; however, other causes can be hypoalbuminemia, diarrhea, chronic renal failure, or hyperparathyroidism (Washington and Van Hoosier 2012).

In Table 3 are represented the reference ranges for the described biochemical parameters for rabbits.

6 Urine Examination

Rabbits are herbivores and therefore excrete alkaline (pH between 8 and 9), dense and turbid urine due to mineral deposits, especially calcium carbonate that is seen as a sediment in the sampling container. Rabbit's incapacity to concentrate urine as well as other species leads to large, daily excretion volumes, with an average of 130 mL/kg. Clear urine is associated with low calcium excretion which could be

Table 3 Biochemical reference ranges for rabbits. Adapted from Suckow et al. (2002), Melillo (2007), Evans (2009), Washington and van Hoosier (2012), and Varga (2014)

Parameters	Units	Reference range
AST	IU/L	14–113
ALT	IU/L	14–80
ALP	IU/L	10–140
CPK	IU/L	150–1000
GGT	IU/L	0–7.0
LDH	IU/L	30–140
Total bilirubin	mg/dL	0–0.8
Bile acids	mMol/L	0–40
Amylase	IU/L	200–500
Triglycerides	mg/dL	15–160
Cholesterol	mg/dL	10–80
Urea	mg/dL	10–30
Creatinine	mg/dL	0.5–2.5
HCO_3^-	mEq/L	16.2–38.0
Ca^{2+}	mg/dL	5.6–17
Cl^-	mEq/L	92–120
Mg^{2+}	mg/dL	2.0–5.4
PO_4^-	mg/dL	2.3–6.7
Na^+	mEq/dL	131–155
K^+	mEq/dL	3.5–6.9
Glucose	mg/dL	75–155
Total proteins	g/dL	5.4–8.3
Albumin	g/dL	2.4–4.6
Globulin	g/dL	1.4–2.2

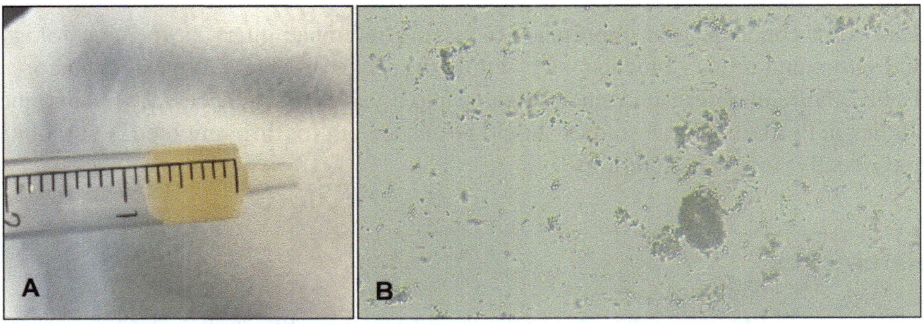

Fig. 3 Rabbit's urine. (**a**) Macroscopic appearance of rabbit's urine collected by cystocentesis. (**b**) Rabbit's urinary sediment with presence of numerous amorphous phosphate crystals, 400×

indicative of renal failure; however, urine from anorexic, pregnant, lactating, or young rabbits tends to be clear. Oxalate and ammonium magnesium phosphate crystals (Fig. 3b) are also common findings in rabbits' urine (Melillo 2007; Varga 2014). The elevated presence of mineral deposits makes specific gravity difficult to determine, but it's accepted to be between 1.003 and 1.036 with the average being

1.015. Elevated specific gravity, above 1.030, is linked to prerenal azotemia, while decreased specific gravity, below 1.013, is found in azotemia due to renal failure. Acidic urine is a pathological condition and is found in animals with hepatic lipidosis, acidosis due to anorexia, pregnancy toxemia, or fever (Melillo 2007; Varga 2014).

Urine color can vary from pale yellow to red, simulating hematuria which is seen in urinary tract infection or inflammation as well as urolithiasis and cystitis; however, urine's color is influenced by dietary pigments, and the distinction can be done with urinalysis dipstick tests or microscopic observation (Bodnariu 2022). Pseudohematuria can be observed in animals with uterine illnesses; usually it can be differentiated from hematuria as the blood only appears at the end of urination. Residual amounts of proteins, especially in young rabbits, and glucose are common findings in healthy rabbits' urine; ketone bodies on the other hand when identified in anorexic rabbits are a poor prognostic of progression of hepatic lipidosis and can also be a sign of toxemia in pregnant rabbits (Melillo 2007). Hyperglycemia may be stress-related; however, it can also be indicative of hepatic lipidosis or, although rare, diabetes mellitus. Significant proteinuria is often an early sign of renal disease (Melillo 2007). Urinary sediment should be observed microscopically for crystals, erythrocytes, or inflammatory cells as well as parasitic or bacterial infections. Gram staining may reveal gram positive *E. cuniculi* spores. Trace amounts of red and white blood cells are often observed and have no clinical relevance (Melillo 2007; Varga 2014; Bodnariu 2022). Rabbit's urine macroscopic aspect and sediment are represented in Fig. 3.

7 Conclusions

This work concludes the importance of having reference intervals for hematology and serum and urinary biochemistry, within the scope of rabbits' clinical evaluation. The literature on reference values is scarce and diverse, so it is necessary to continue studies in this area. These studies should be carried out for different breeds of rabbits and, ideally, for each laboratory.

References

Bain BJ (2017) A beginner's guide to blood cells, 3rd edn. Wiley Blackwell, London

Bodnariu A (2022) Clinical pathology of rabbits – interpretation of biochemistry, haematology and urinalysis results - veterinary practice. https://www.veterinary-practice.com/article/clinical-pathology-of-rabbits. Accessed 25 December 2022

Espinoza VE, Emmady PD (2022) Histology, monocytes [Updated 2022 Apr 28]. In: StatPearls [Internet]. StatPearls Publishing, Treasure Island, FL. https://www.ncbi.nlm.nih.gov/books/NBK557618/

Evans GO (2009) Animal clinical chemistry: a practical handbook for toxicologists and biomedical researchers, 2nd edn. CRC Press, Boca Raton. https://doi.org/10.1201/9781420080124

Farhana A, Lappin SL (2022) Biochemistry, lactate dehydrogenase. [Updated 2022 May 8]. In: StatPearls [Internet]. StatPearls Publishing, Treasure Island, FL. https://www.ncbi.nlm.nih.gov/books/NBK557536
Gale AJ (2011) Continuing education course #2: current understanding of hemostasis. Toxicol Pathol 39(1):273–280. https://doi.org/10.1177/0192623310389474
Hong CW (2017) Current understanding in neutrophil differentiation and heterogeneity. Immune Netw 17(5):298–306. https://doi.org/10.4110/in.2017.17.5.298
Janeway CA, Travers P, Walport M, Shlomchik MJ (2001) The components of the immune system, 5th edn. Garland Science, New York
Melillo A (2007) Rabbit clinical pathology. J Exotic Pet Med 16(3):135–145. https://doi.org/10.1053/j.jepm.2007.06.002
Moore DM, Zimmerman K, Smith SA (2015) Hematological assessment in pet rabbits: blood sample collection and blood cell identification. Vet Clin N Am Exotic Anim Pract 18(1):9–19. https://doi.org/10.1016/j.cvex.2014.09.003
Özkan C, Kaya A, Akgül Y (2012) Normal values of haematological and some biochemical parameters in serum and urine of New Zealand White rabbits. World Rabbit Sci 20:253–259. https://doi.org/10.4995/WRS.2012.1229
Palis J (2014) Primitive and definitive erythropoiesis in mammals. Front Physiol 5:3. https://doi.org/10.3389/fphys.2014.00003
Parasuraman S, Raveendran R, Kesavan R (2010) Blood sample collection in small laboratory animals. J Pharmacol Pharmacotherap 1(2):87–93. https://doi.org/10.4103/0976-500X.72350
Sheshachalam A, Srivastava N, Mitchell T, Lacy P, Eitzen G (2014) Granule protein processing and regulated secretion in neutrophils. Front Immunol 5:448. https://doi.org/10.3389/fimmu.2014.00448
Spencer LA, Bonjour K, Melo RC, Weller PF (2014) Eosinophil secretion of granule-derived cytokines. Front Immunol 5:496. https://doi.org/10.3389/fimmu.2014.00496
Stone KD, Prussin C, Metcalfe DD (2010) IgE, mast cells, basophils, and eosinophils. J Allergy Clin Immunol 125(2 Suppl 2):S73–S80. https://doi.org/10.1016/j.jaci.2009.11.017
Suckow MA, Brammer DW, Rush HG, Chrisp CE (2002) Biology and diseases of rabbits. Lab Anim Med 329–364. https://doi.org/10.1016/B978-012263951-7/50012-0
Toth LA, Krueger JM (1988) Alteration of sleep in rabbits by *Staphylococcus aureus* infection. Infect Immun 56(7):1785–1791. https://doi.org/10.1128/iai.56.7.1785-1791.1988
Toth LA, Krueger JM (1989) Hematologic effects of exposure to three infective agents in rabbits. J Am Vet Med Assoc 195(7):981–986
Varga M (2014) Chapter 2 - Clinical pathology. In: Varga M (ed) Textbook of rabbit medicine, 2nd edn. Edinburgh, Butterworth-Heinemann Elsevier, pp 111–136
Washington IM, Van Hoosier G (2012) Clinical biochemistry and hematology. Lab Rabbit Guinea Pig Hamster Other Rodents 2012:57–116. https://doi.org/10.1016/B978-0-12-380920-9.00003-1
Weingand K, Brown G, Hall R, Davies D, Gossett K, Neptun D, Waner T, Matsuzawa T, Salemink P, Froelke W, Provost JP, Dal Negro G, Batchelor J, Nomura M, Groetsch H, Boink A, Kimball J, Woodman D, York M, Fabianson-Johnson E, Lupart M, Melloni E (1996) Harmonization of animal clinical pathology testing in toxicity and safety studies. Toxicol Sci 29(2):198–201. https://doi.org/10.1093/TOXSCI/29.2.198

Necropsy and Sample Collection in Rabbits

Justina Prada and Isabel Pires

Abstract

Necropsy is a scientific investigation that requires skill, experience, and extensive knowledge. Therefore, it is of great responsibility for those who perform it and from which a diagnosis must be obtained. In the generic sense of the term, it is an examination performed to determine the cause of death or, if it is known, its exact nature, the extent of the lesions, and other alterations that may be present in the corpse. It consists of the clinical history and corpse examination: external habitus, internal habitus, and complementary examinations. Therefore, it should be carried out systematically and thoroughly to extract as much information as possible from it. This chapter describes the rabbit necropsy technique, emphasizing its anatomical particularities, and presents some of the most common lesions in the postmortem exam.

Keywords

Necropsy findings · Macroscopic lesions · Tissue sampling · Gross anatomy · Diagnosis

J. Prada (✉) · I. Pires
Department of Veterinary Sciences, University of Trás-os-Montes and Alto Douro, Vila Real, Portugal

Animal and Veterinary Research Centre (CECAV), University of Trás-os Montes and Alto Douro, Vila Real, Portugal

Associate Laboratory for Animal and Veterinary Science (AL4AnimalS), Lisbon, Portugal
e-mail: jprada@utad.pt

J. Simões, J. M. Monteiro (eds.), *Veterinary Care of Farm Rabbits*,
https://doi.org/10.1007/978-3-031-44542-2_34

1 Introduction

Necropsy is the examination performed after the animal's death to determine the cause and circumstances of its death. In the production animals, it acquires greater importance as it allows, besides diagnosing the disease, the elaboration of therapeutic schemes to prevent morbidity and mortality on the farm (Ortiz 1989; Aluja and Casas 2002; Maxie and Miller 2016).

The necropsy should be performed after the animal's death as soon as possible. Postmortem alterations may make identifying lesions in the different organs and systems difficult and compromise the diagnosis. If it is impossible to perform the necropsy immediately, the animal should be refrigerated and not frozen. Putrefaction also makes it impossible to perform complementary exams such as histological and microbiological exams (Fiette et al. 2017). Thus, in diseases associated with high morbidity and mortality, the euthanasia of a sick animal should be considered, which will allow a timely necropsy and the identification of the nosological entity in question (Petisca and Montano 1962; Maxie and Miller 2016).

Whether performed on the farm or in an appropriate room, all procedures related to the manipulation of cadavers should respect the basic safety rules for the operator, the environment, and the other animals, namely, when performed on the farm. The necropsy should be performed at a time and place where there are minimum distraction and movement of people, to minimize the possibility of contamination (Nietfeld 2010; Küker et al. 2018). Thus, the necropsy should be performed in an isolated room, with access to running water and sewage, on a stainless steel worktable. The room should have walls and floors that are easy to wash and disinfect (Petisca and Montano 1962; King et al. 2013).

The practice of veterinary medicine, including necropsy, requires recognition of risk. Although "zero risk" is not possible in procedures with biological systems, knowledge of the risks inherent in this examination can reduce them to the minimum acceptable. These include operator cutting, exposure to formaldehyde vapors, and exposure to zoonotic agents (King et al. 2013; McDonough and Southard 2017).

This chapter aims to describe the technique of necropsy in rabbits, with emphasis on their anatomical particularities, and to give a summary description of the most common lesions found in the postmortem examination.

2 Material and Equipment

The material needed for the necropsy includes latex or rubber gloves, steel necropsy knives, scalpel # 4 with a blade, duck-billed and rat-tooth forceps, dissection scissors, enterotomy scissors, a ruler, and scales. In addition, two pairs of disposable gloves and a mask should be worn whenever the zoonotic disease is suspected (Oliveira et al. 2004; King et al. 2013). The operator should wear a disposable suit that covers the entire body, including the arms.

Material for complementary exams should also be available. For example, for histological exams, 10% buffered formaldehyde and watertight, double-lid, wide-mouthed bottles of different sizes are required, as well as adequate material for labeling the packages (King et al. 2013).

The material for microbiological examination should be collected with sterile material, placed in sterile Petri dishes, and then transported to a microbiology laboratory as soon as possible (Ortiz 1989; Nietfeld 2010).

For parasitological examination, in addition to flasks for collecting fecal material, 70% alcohol is required if parasites are observed and collected (Ortiz 1989).

3 Necropsy Technique

The necropsy should be methodical and systematic in an established order to observe all organs. Thus, the necropsy should begin before any incision is made on the animal, seeking to obtain data about its death. These should include the date of death and a brief, objective, and concise clinical history that includes previous symptoms, morbidity, and mortality associated with the disease, the total number of animals on the farm, characteristics of the farm, vaccination and deworming schemes, and other information potentially relevant to the diagnosis. The necropsy exam requisition must also include clinical suspicion and complementary exams. In the case of euthanasia, the method used should be recorded (Ortiz 1989; Aluja and Casas 2002; King et al. 2013). Box 1 summarizes the order of events that a necropsy should follow.

Box 1 Order of Necropsy

- Identification of the animal
- External examination
- Incision in the forelimbs and hind limbs
- Midline incision of the corpse, with skin folding
- Incision to open the abdominal cavity
- Incision to open the thoracic cavity
- Evisceration and examination of each of the organs of the thoracic cavity and neck
- Evisceration and examination of the organs of the abdominal cavity
- Opening the cranial cavity and examining the brain and eyes
- Opening and examination of the spinal canal, bones, joints, and bone marrow
- Collection of samples for complementary exams

The external and internal examination of the animal is performed after identification. The necropsy continues with complementary exams and ends with the elaboration of a report with the description of the macroscopic lesions observed, concluding with a presumptive macroscopic diagnosis (Petisca and Montano 1962). See findings of external (Plate 1) and internal (Plate 2 and Plate 3) examinations at necropsy.

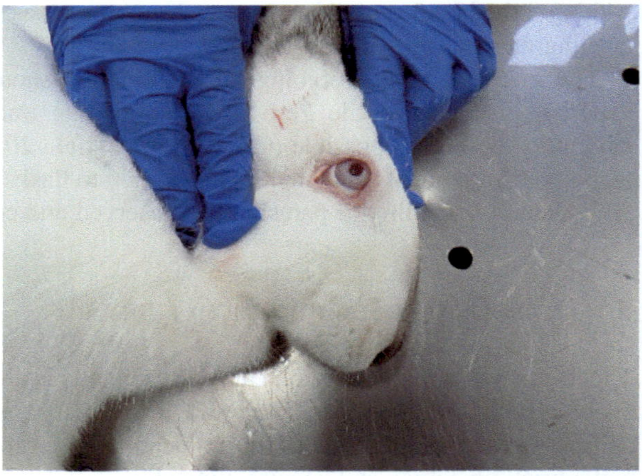

Plate 1 Necropsy findings at external examination. (**a**) Epistaxis – viral hemorrhagic disease. (**b**) Purulent rhinitis. (**c**) Purulent conjunctivitis – myxomatosis. (**d**) Perianal region skidded by soft, greenish stools – enteritis. (**e**) Auricular myxomatosis. (f) Dental malocclusion

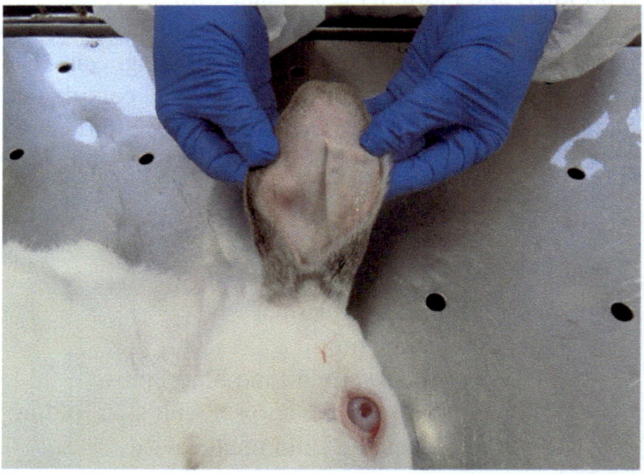

Plate 2 Necropsy findings at internal examination (I). (**a**) Subcutaneous abscess; *Pasteurella* sp. (isolate). (**b**) Necrotic peritonitis. (**c**) Mucoid enteritis. (**d**) Presence of parasitic vesicles in the abdominal cavity – cysticercosis. (**e**) Gastric ulcers. (**f**) Red colon coloring: nonspecific colitis

3.1 External Examination

Body condition should be assessed by taking into account breed, sex, and age characteristics and should consider the weight of the animal.

Assessment of external habitus includes visual examination and palpation of the animal. Changes in body conformation may be associated with abdominal distension due to dilatation of the gastrointestinal tract related to food or gas accumulation, effusions, or advanced gestation (Percy and Barthold 2007). External surfaces and structures, namely, skin and hair, nails, eyes, pinna, oral cavity, external genitalia, and perineum, should be assessed visually (Figs. 1, 2, 3, and 4). In the case of the skeleton and joints, it should also include palpation, noting any changes in conformation or mobility.

The size, consistency, and adherence to adjacent tissues should be recorded in swellings. Early detection of abscesses will allow manipulation that does not

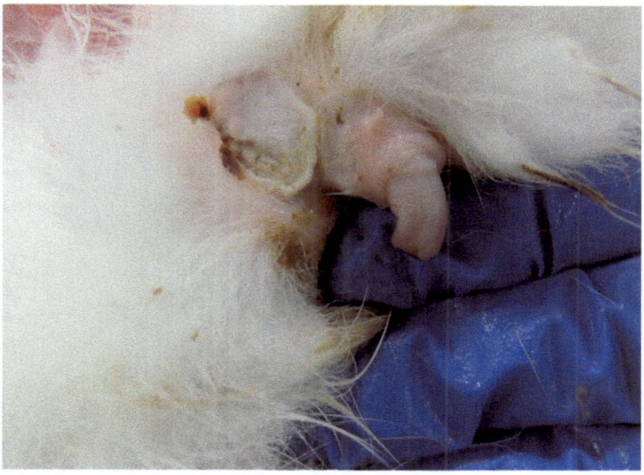

Plate 3 Necropsy findings at internal examination (II). (**a**) Single pulmonary abscess. (**b**) Fibrinous pleuritis. (**c**) Multifocal pulmonary abscesses: suppurative pneumonia. (**d**) Multiple microabscesses in the lung – embolic pneumonia. (**e**) Diffuse red staining: hemorrhage. (**f**) Embolic-purulent nephritis: kidney abscesses

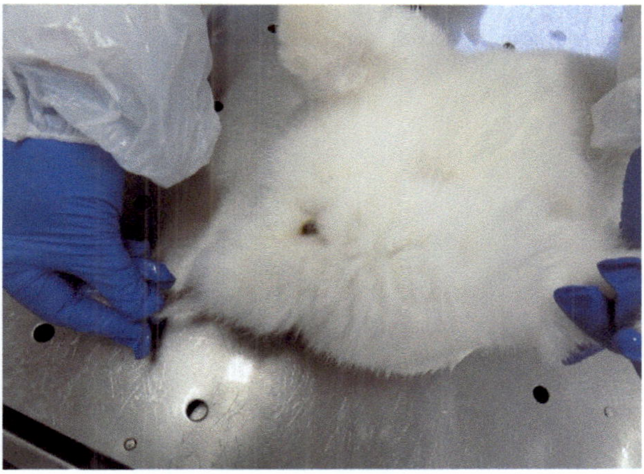

Fig. 1 Observation of the ocular mucosa

compromise the aseptic collection of material for subsequent isolation of the etiologic agent (Küker et al. 2018). Cadaveric changes such as stiffness or chromatic spots should be recorded. In addition, tattoos, scars, wounds, bone malformations, discharge, etc. should also be described (King et al. 2013).

On the skin, in addition to assessing its elasticity, one should search for alopecia, erosions or ulcers, plaques, changes in color or thickness, and ectoparasites. The detection of fur soiled with feces in the perianal region may suggest the presence of diarrhea and possible enteritis (Küker et al. 2018). In the external natural cavities (nasal cavities, lips, oral mucosa, ocular mucosa, external ear, anus, vulva, and foreskin), the content and different changes in symmetry, color, moisture, erosions, exudates, discharge, nodules, and the presence of foreign bodies should be examined. Given the importance of nasal respiration, in these animals, any changes in the nose region should be appreciated (Keeble and Meredith 2006; Percy and Barthold 2007).

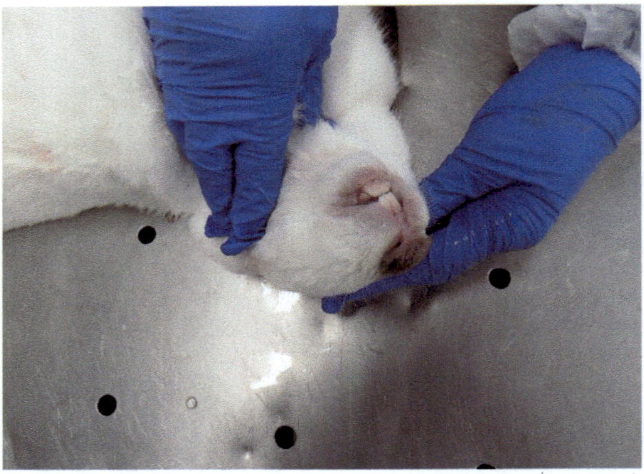

Fig. 2 Observation of the pinna

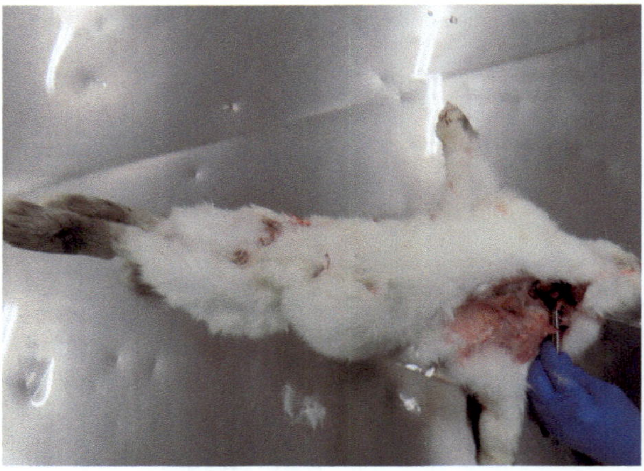

Fig. 3 Observation of the penis and penile mucosa

The oral cavity examination should also include an evaluation of the teeth (Fig. 5). Since rabbits have continuously growing incisors (elodonts), a poor dental occlusion or low fiber diet can lead to overgrowth of the teeth, associated with chewing problems, soft tissue damage, and decreased feed intake. On the other hand, congenital tooth changes, although relatively frequent, do not cause problems unless they are associated with malocclusion (Keeble and Meredith 2006; Percy and Barthold 2007).

The mucosa and the papillary body on the tongue should be sectioned transversely to appreciate the muscle.

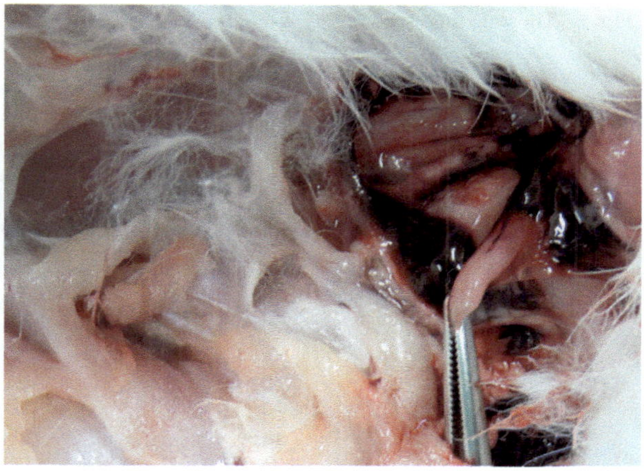

Fig. 4 Observation of the anal region

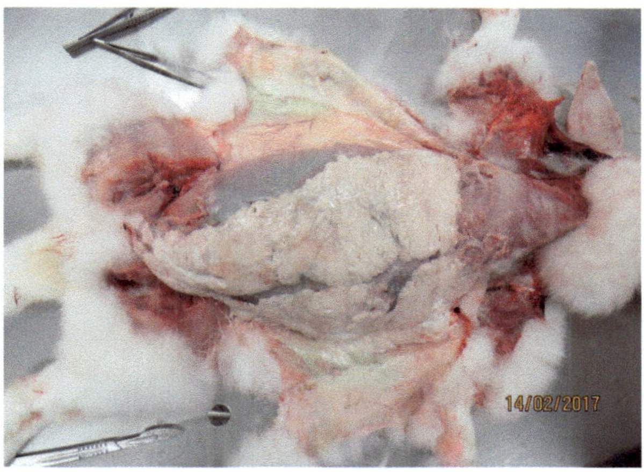

Fig. 5 Observation of the mouth and teeth

3.2 Internal Examination

The animal should be placed in dorsal decubitus, cutting at the right and left axillary regions. After the initial incision, it should be sectioned from the inside out so that contact with the hair does not affect the knife's cutting edge. The pectoral muscles and brachial plexus are sectioned to fold the forelimbs, examining the different anatomical structures, including the lymph nodes (Figs. 6 and 7) (Petisca and Montano 1962).

The hind limbs are then disarticulated at the hip joint level, and the bone surfaces and the amount, color, and transparency of the synovial fluid are evaluated (Alves et al. 2016). In smaller animals, the body weight is insufficient to stabilize the body during the necropsy. In these cases, the animal should be placed on a cork board or other material, and the extremities attached to the board with needles or pins (McDonough and Southard 2017).

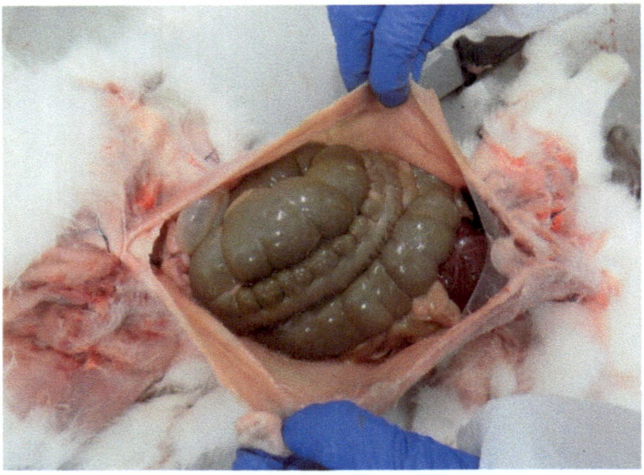

Fig. 6 Thoracic limb section

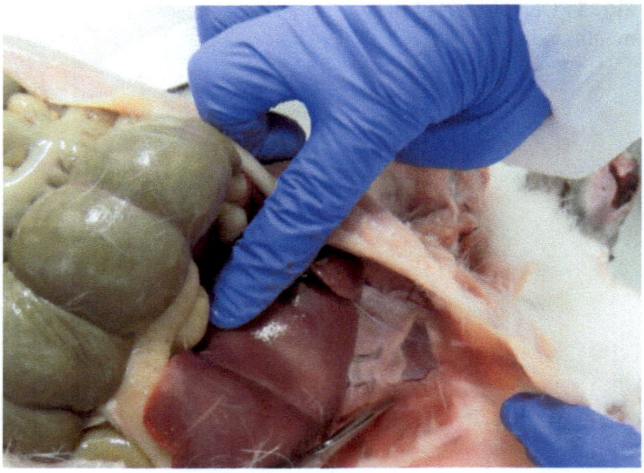

Fig. 7 Examination of the prescapular lymph node

Next, a midline incision is made, starting at the mandibular symphysis; continuing down the neck, thorax, and abdomen; and ending at the pubis, folding the skin lateral to each side (McDonough and Southard 2017) (Fig. 8). At this stage, the subcutaneous tissue, mammary gland, salivary glands, lymph nodes, penis, and foreskin should be observed. The thyroid gland is isolated next to the trachea (Petisca and Montano 1962). The lymph nodes are evaluated for size and cut surface (by making a longitudinal cut), where changes in color and consistency are investigated (Valli et al. 2016). In the subcutaneous tissue, signs of death, such as greenish-black stains or cadaveric emphysema, are easily observed in this area (King et al. 2013).

The mammary gland should be evaluated for volume, shape, consistency, and cut surface, and the content should also be valued when cut. The mammary gland occupies much of the lactating female's subcutaneous abdominal and ventral thoracic area (Percy and Barthold 2007).

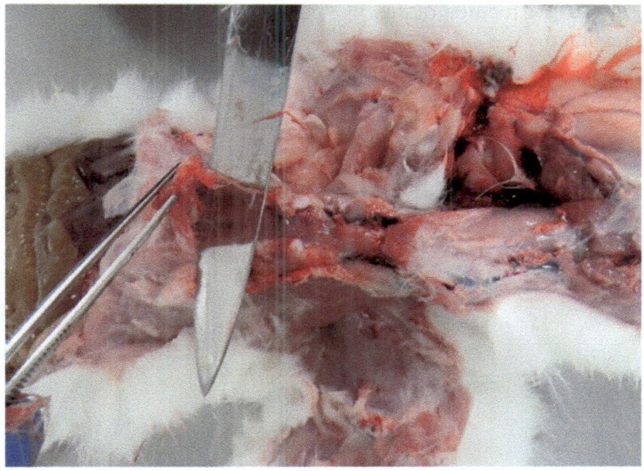

Fig. 8 Skin folding and observation of the mammary gland

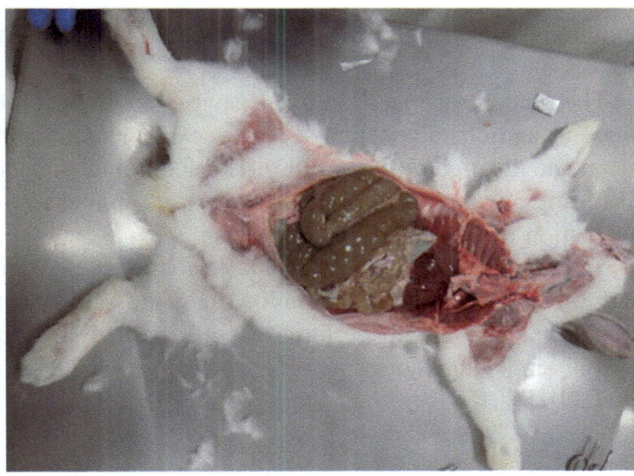

Fig. 9 Opening of the abdominal cavity

In males, the incision should surround the penis; at this point, the testes are examined. They are removed after cutting the scrotum and surrounding tunica. They should then be sectioned along the central axis, and the thickness of the layers, volume, consistency, and coloration should be inspected. The testes descend at about 12 weeks of age, but the inguinal canals remain open, so they may be retracted into the abdomen. The testicles may undergo seasonal involution, which may involve abdominal retraction, with testicular size reduced to half its normal size (King et al. 2013).

3.3 Opening of the Abdominal Cavity

A small incision should be made near the sternum to open the abdominal cavity. This should be continued through the midline to the perineal region (Fig. 9). The

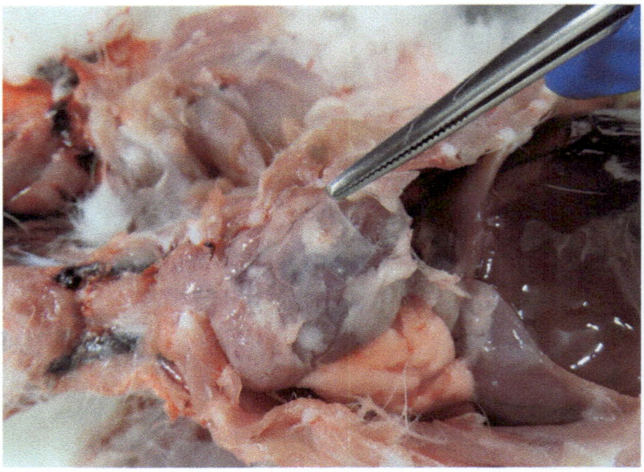

Fig. 10 Perforation of the diaphragm

abdominal wall should then be folded laterally along the costal margin (Nietfeld 2010).

The content of the abdominal cavity should be observed, noting the presence, amount, and characteristics of fluids, exudates, and adhesions between the peritoneum and the organs (Ortiz 1989). The abdominal organs should be evaluated in situ, recording anatomical changes. Fat deposits should also be observed for nutritional evaluation of the animal (Ortiz 1989).

The diaphragm should be evaluated for concavity. This is usually concave on the abdominal side, and its alteration may indicate pneumothorax, hydrothorax, or pyothorax. Diaphragm perforation, in a normal situation, should be accompanied by a hissing sound corresponding to the air entering the chest cavity (Fig. 10) (Uzal et al. 2016).

3.4 Opening of the Thoracic Cavity

The thoracic cavity should be opened posteriorly to the abdominal cavity by sectioning the costochondral junctions from the first to the last rib and folding the sternum (Fig. 11). A cut is then made between the muscles of the first and second ribs, the diaphragm is debrided at its insertions, and the thorax is forced apart by traction on each hemithorax (King et al. 2013).

The inspection of the chest cavity will comprise the study of its contents, the topography of the organs, and the evaluation of the parietal and visceral pleura (King et al. 2013) (Fig. 12). The content may be liquid or gaseous. A liquid content may be due to the presence of exudate or transudate (hydrothorax), blood (hemothorax), or lymph (chylothorax). The exudate can be serous, serofibrinous, fibrinous, or purulent. On the other hand, a gaseous content (pneumothorax) is challenging to evaluate (Nietfeld 2010).

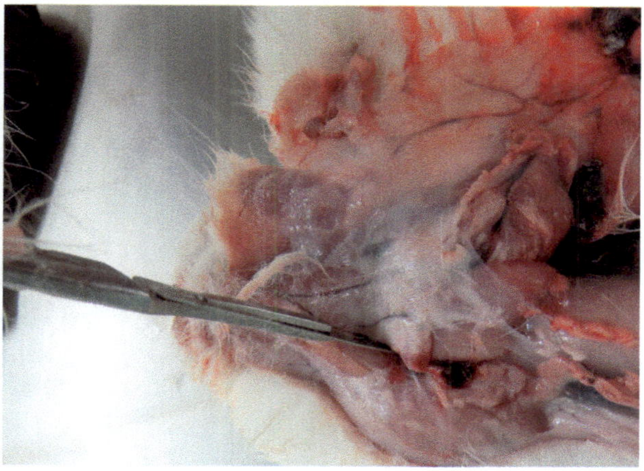

Fig. 11 Section of the sternum

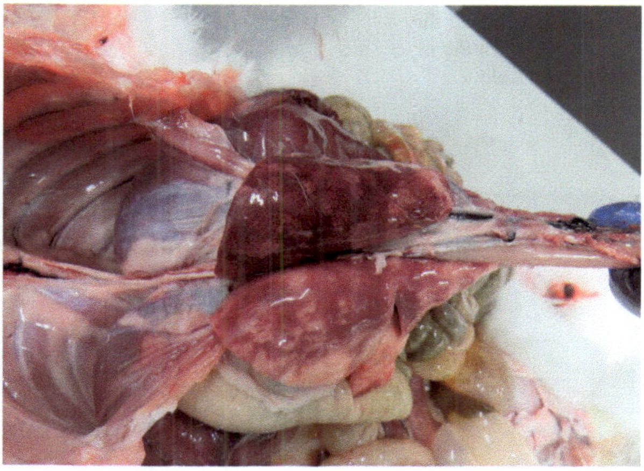

Fig. 12 Observation of organs in situ

Observation of the anatomical position of the organs should be thorough, recording variations in position and other abnormalities of the lungs, heart, esophagus, pleura, and mediastinum. The pleura should be evaluated for color changes (presence of hemorrhage or substance deposition), opacity, thickening, and adhesions in both visceral and parietal pleura (Nietfeld 2010; McDonough and Southard 2017). At this stage, the pericardium is opened in situ, and its content is examined by making an incision in the leaflet at the level of the cardiac apex (Fig. 13) (Nietfeld 2010; Robinson and Robinson 2016). The parietal and visceral surfaces of the pericardium are observed. It is normal to have a small amount of colorless fluid inside the pericardial sac. The content of the pericardium can be fluid (hydropericardium or hemopericardium), purulent, or fibrinous (Robinson and Robinson 2016).

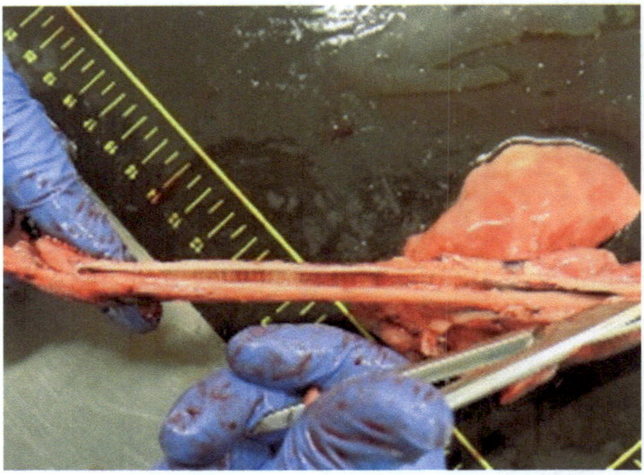

Fig. 13 Pericardial sac opening

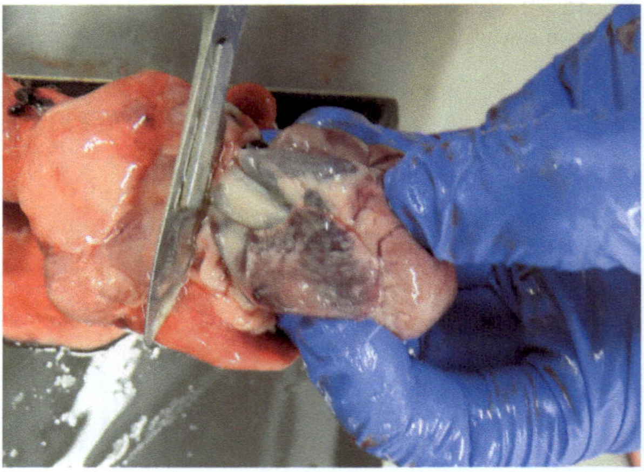

Fig. 14 Slicing of the musculature, on the inner side of the mandible

3.5 Evisceration and Evaluation of the Organs in the Thoracic Cavity

The evisceration of the organs of the thoracic cavity is done en bloc, starting from the tongue, sectioning the soft tissues from the medial side of the body of the mandible on both sides, continuing in a caudal direction (Fig. 14) (Oliveira et al. 2004).

The fascia and pharyngeal musculature are incised, manually downward and backward, tractioning the tongue. The soft palate and hyoid bone cartilages are sectioned, debriding the adhesions with the rest of the structures and removing everything, by posterior traction, until the entrance to the thorax, where the phrenic-pericardial ligaments are sectioned. The esophagus and the great vessels (aorta and vena cava) are cut near the diaphragmatic hiatus, allowing the pharynx, larynx, trachea, esophagus, lungs, and heart to be removed together (Fig. 15). The

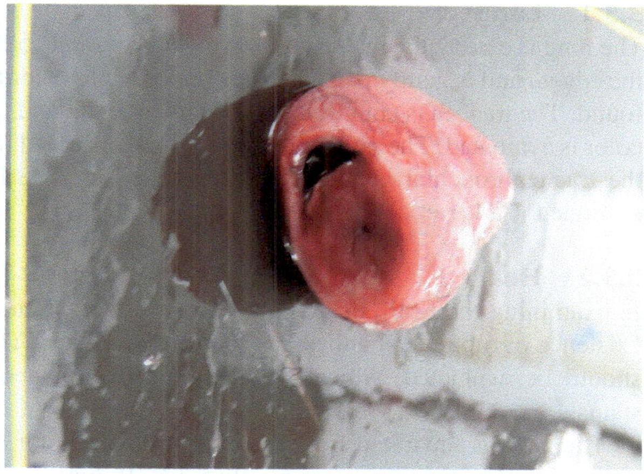

Fig. 15 Thoracic organ evisceration

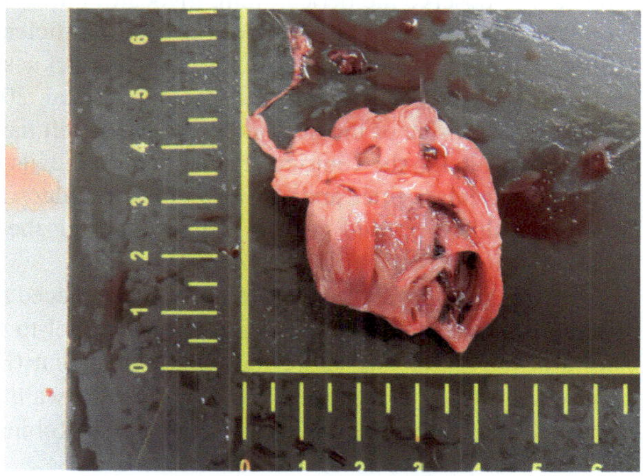

Fig. 16 Opening of the tracheal rings

jugular veins are examined, and the vertebral bodies are palpated to check for asymmetries (King et al. 2013).

After the longitudinal section along its entire length, the pharynx, larynx, and esophagus should be evaluated for content, mucosa, muscle, and conformation. Tonsils are observed for variations in size and coloration (Aluja and Casas 2002).

External inspection of the trachea and large bronchi should consider malformations, diameter, and external appearance. The cricoid cartilage is then incised, and the face-caudal incision is continued through the tracheal rings, opening along the entire length (Fig. 16). A frothy content indicates edema, while the presence of blood suggests aspiration, hemorrhages, inflammation, or ulcers (Caswell and Williams 2016).

3.5.1 Lung

The lung is externally evaluated for weight, size, and collapse degree (more visible at the edges) and is gently palpated. It is important to record the location of the changes found. The trachea is cut through the large bronchi, and the degree of uniformity and color is evaluated. The pulmonary arteries are observed for emboli and parasites, and the tracheobronchial ganglia are also looked at. The lung should then be cross-sectioned (Küker et al. 2018).

3.5.2 Heart

In examining the heart, anatomical changes, weight and size, shape, color, and the existence of ruptures or adhesions should be reported. The fat in the coronary sulcus should be evaluated, which provides information about the animal's nutritional status.

The heart is separated from the lungs by sectioning near the great vessels, as far away from their origin as possible (Fig. 17) (Robinson and Robinson 2016). Then, it is opened to observe the pericardium, myocardium, and endocardium. A transverse cut is made at the apex at the necessary and sufficient distance to open the two ventricular cavities inferiorly (Fig. 18), and the thickness of each ventricular wall is measured (Petisca and Montano 1962; Oliveira et al. 2004).

Next, the left side of the heart is placed in the left hand, and the right ventricle is sectioned parallel to the ventricular septum until it exits into the vena cava, crossing the atrioventricular orifice and cutting all the muscular wall (Fig. 19). The pulmonary arteries should be opened at their origin, exposing the respective valves (Oliveira et al. 2004; Robinson and Robinson 2016).

To open the left side of the heart, the organ is placed so that it is to the right of the operator, and the left ventricle is sectioned parallel to the interventricular septum, exiting at the aorta. The entire muscular wall is cut (Fig. 20). After opening, the cavities are washed to remove blood clots, and the walls and valvular surfaces of the endocardium are examined (Oliveira et al. 2004; Robinson and Robinson 2016).

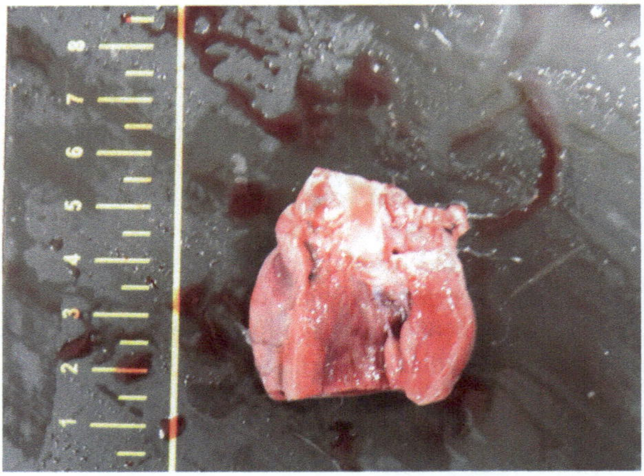

Fig. 17 Cross section of the heart base

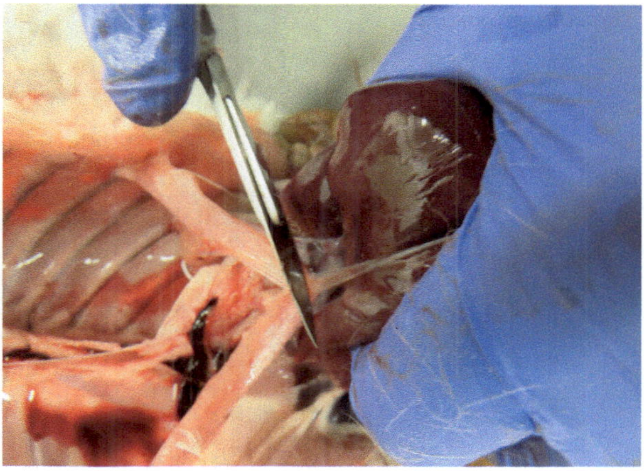

Fig. 18 Exposure of the two ventricles

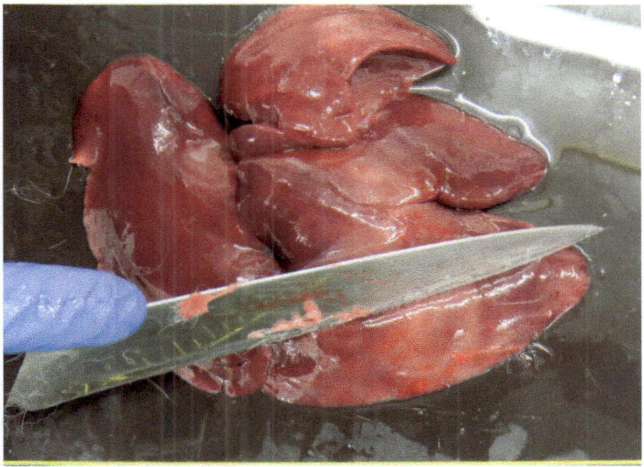

Fig. 19 Right ventricle

Rigor mortis begins earlier in the myocardium than the skeletal muscle and reaches maximum development in the left ventricle. As a consequence, the expulsion of blood must be complete. Rigor in the right ventricle is less efficient, and emptying is incomplete. It is normal, in these animals, to observe abundant blood clotting in the right ventricle without evidence of postmortem contraction. It should also be noted that the right atrioventricular valve in this species is bicuspid (Robinson and Robinson 2016).

3.6 Evisceration and Evaluation of Organs in the Abdominal Cavity

When eviscerating organs, the vascular pathways should be inspected, evaluating the size and presence of thrombi, the vasculonervous pathways, the ganglionic chain,

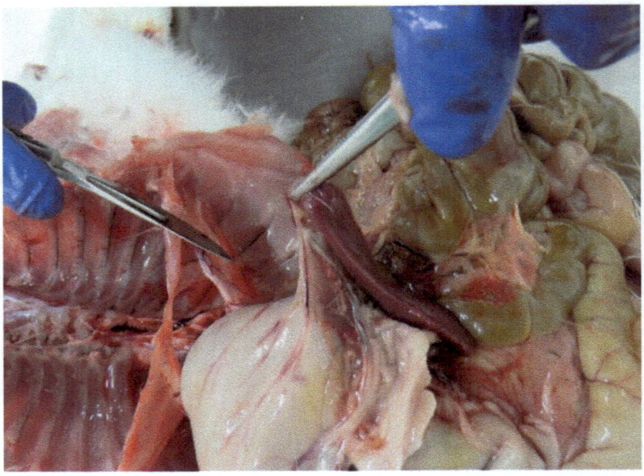

Fig. 20 Left ventricle

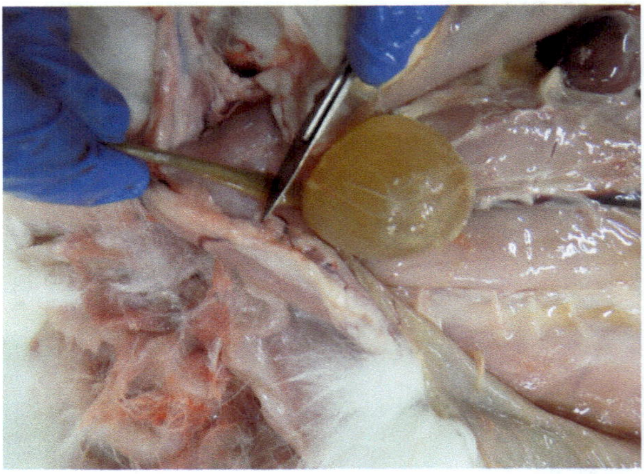

Fig. 21 Extraction of the liver

and the fat. The removal of organs should be done methodically, and their weight, size, shape, consistency, color, cut surface, presence of lesions and their distribution, and in the case of tubular organs, their diameter, wall (including the muscular layer and serosa), and content should be evaluated in general. After removing all the organs, the vertebral bodies should be palpated, and the abdominal aorta and iliac lymph nodes inspected (Oliveira et al. 2004; Alves et al. 2016).

3.6.1 Liver and Gallbladder

After sectioning the greater and lesser omentum, the liver is extracted, cutting the insertions of the ligaments that connect it to the abdominal viscera (Fig. 21).

Evaluation of the liver should include, in addition to the above parameters, the shape of the lobes, edges, cut surface, and hilum. Cross sections are made over the entire surface of the liver (Fig. 22), and the cut surface is examined for color, appearance, fluid, and the possible presence of cavities or parasitic tracts. The

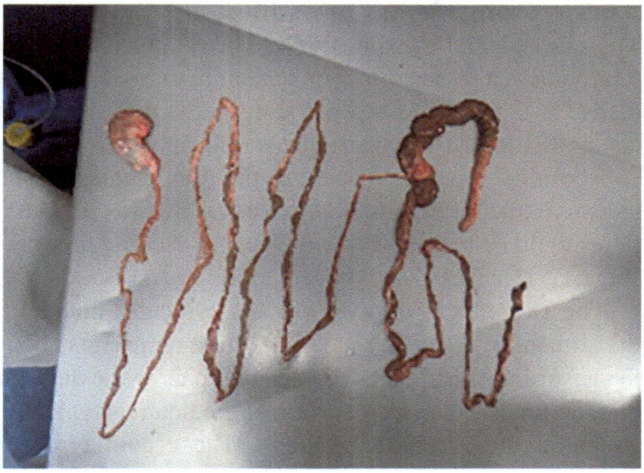

Fig. 22 Section of the liver

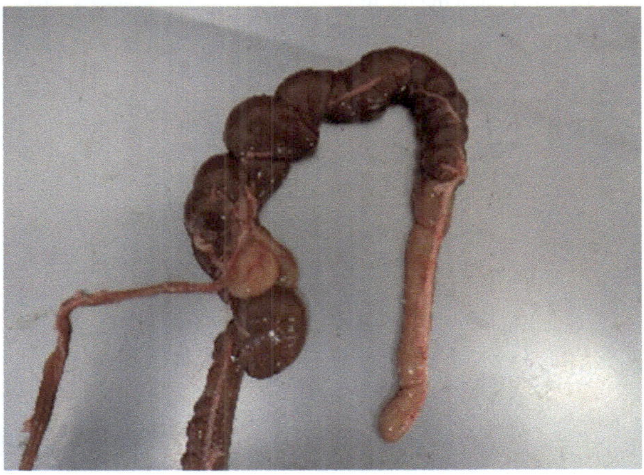

Fig. 23 Spleen extraction

large bile ducts are opened, and the gallbladder and great vessels are examined. In the hilum examination, the portal vein is evaluated, along with the hepatic bile duct and the hepatic lymph nodes. The anatomical conformation and volume of the gallbladder are a function of the contents. Internal examination is performed by incision from the bile duct to the blind cul-de-sac (Uzal et al. 2016; McDonough and Southard 2017).

3.6.2 Spleen

The spleen should be removed by sectioning, along the hilum, connective tissue, fat, and gastrosplenic ligament (Fig. 23). In addition to the other parameters, size can provide important information for diagnosing the nosological entity. Splenomegaly is associated with hyperemia, congestion, or inflammatory states. Palpation checks for consistency and several cross sections should be made to evaluate the macroscopic distribution of the white and red pulp (Valli et al. 2016).

3.6.3 Stomach and Intestines

The stomach and intestines should be removed by cutting at the level of the rectum, removing in a rostral direction the entire digestive tract up to the cardia (Fig. 24). The digestive tube should be open only in the final stage of the necropsy, to avoid fecal contamination of the instruments and other tissues. The pancreas is an ill-defined collection of lobules adjacent to the duodenum and should also be evaluated. The digestive tract in rabbits is large and relatively complex, so its opening and observation should be methodical and take into account the species' anatomical peculiarities (Fig. 25) (Uzal et al. 2016; McDonough and Southard 2017).

The stomach is thin-walled and typically houses 15% of the food present in the alimentary tract. The small intestine represents about 12% of the total volume of the gastrointestinal tract. The bile and pancreatic ducts have separate openings in the duodenum. The terminal portion of the ileum is typically thicker, forming a spherical structure, the *sacculus rotundus*. The cecum typically holds about 40% of the

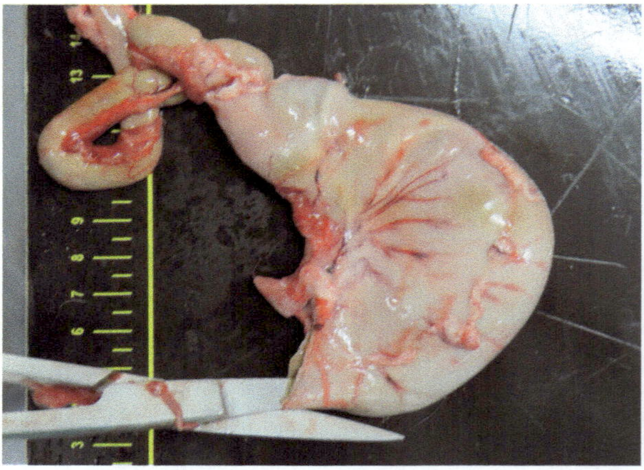

Fig. 24 Section of the rectum

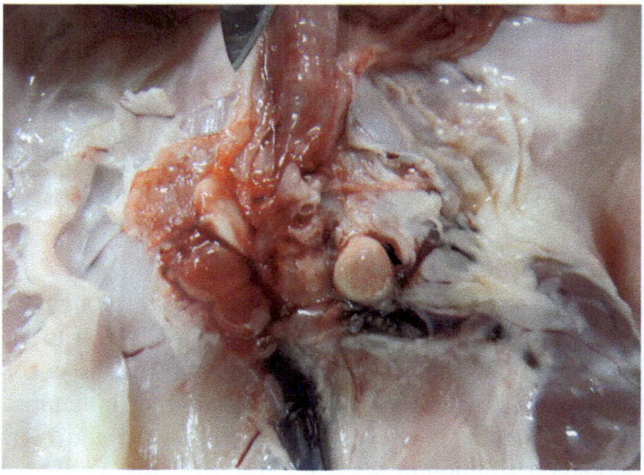

Fig. 25 Stomach and intestines

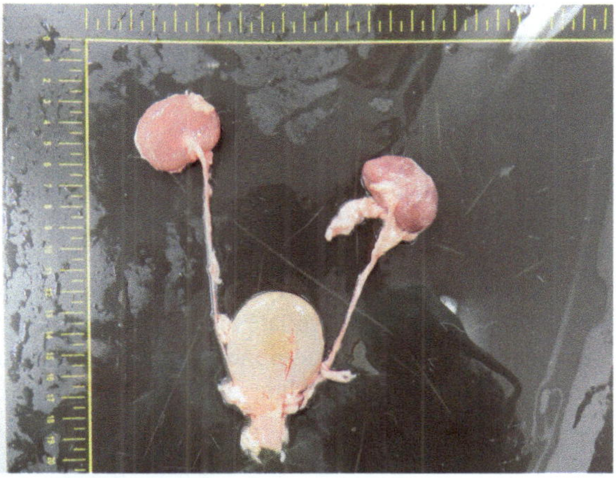

Fig. 26 Observation of the cecum

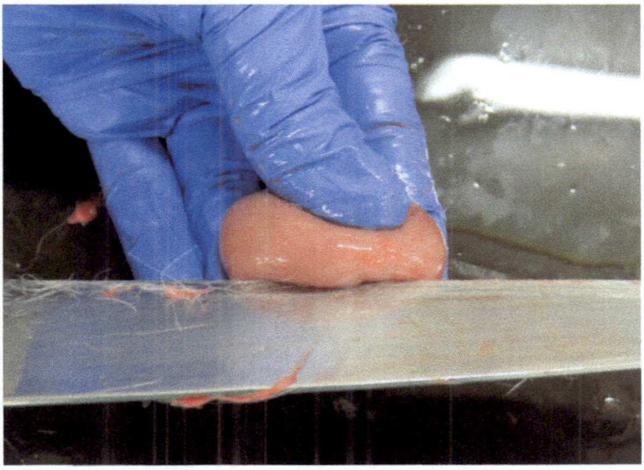

Fig. 27 Opening of the stomach through the greater curvature

gastrointestinal contents and contains an appendix at the distal end (Fig. 26). The colon should also be evaluated in all of its major segments, from the cranial to the distal portion (Keeble and Meredith 2006; Percy and Barthold 2007). In evaluating these organs, one should include their shape, *lumina*, and wall size (coloration, thickness, and presence of ulcerations).

After opening the stomach, which should be done through the larger curvature (Fig. 27), starting at the cardia, and continuing the cut through the mesenteric edge of the intestine, the characteristics of the contents should be recorded. The stool evaluation can be determined in detecting inflammation, management problems, and other diseases (Percy and Barthold 2007; Uzal et al. 2016).

Particular attention should be given to the gut-associated lymphoid tissue (GALT), located in the lamina propria and submucosa and representing about 50% of the total mass of lymphoid tissue in the rabbit. It includes Peyer's patches in the ileum, the lymphoid tissue of the *sacculus rotundus*, the cecal tonsil (on the wall of the proximal cecum), and the cecal appendix. Its reactivity denotes inflammatory situations at the digestive level (Percy and Barthold 2007; Uzal et al. 2016; Valli et al. 2016).

3.6.4 Urinary System and Adrenal Glands

The adrenal glands, located cranial to the kidneys, should be removed and then sectioned longitudinally (Fig. 28). The size, color, consistency, and cut surface should be evaluated, noting any changes (Cianciolo and Mohr 2016).

The kidneys, uni papillary, should be observed in situ for their size, which will correspond to the approximate size of three vertebrae. They will then be carefully separated from the subperitoneal strip along with the ureters and, if possible, removed en bloc, including the bladder, after sectioning the bladder neck

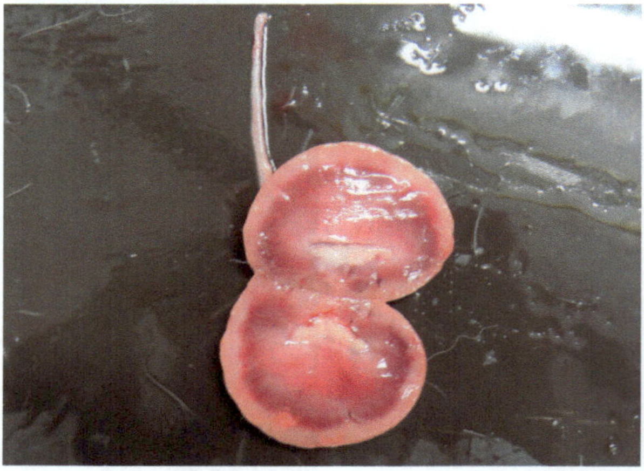

Fig. 28 Observation of the adrenal glands

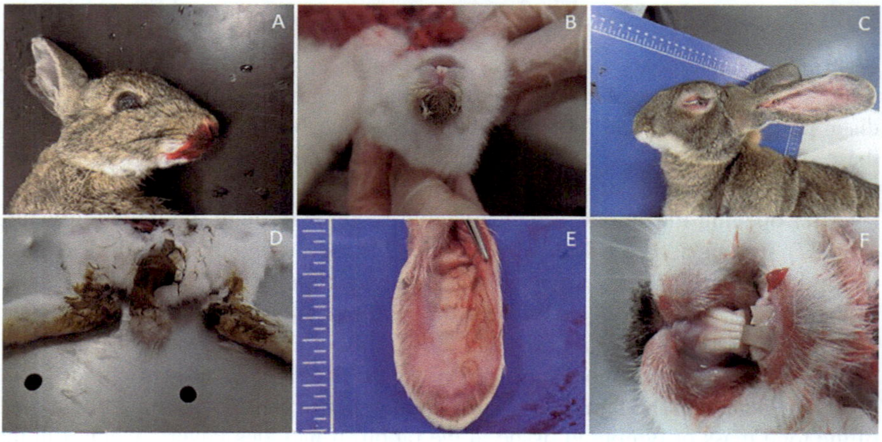

Fig. 29 Kidneys, ureters, and bladder

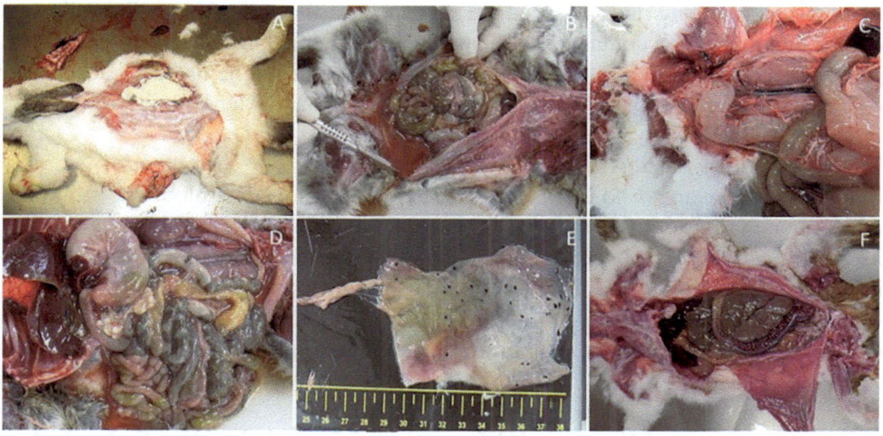

Fig. 30 Longitudinal section of the kidney

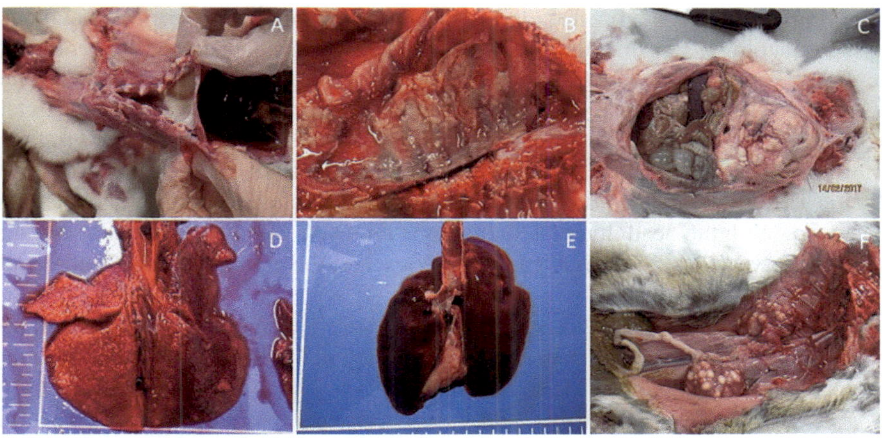

Fig. 31 Observation of the kidney after sectioning

(Fig. 29). The ureters should be palpated, along their entire pathway, before separating them from the kidney to look for possible obstructions or concretions. Their diameter, content, and mucosal surface are observed (Fiette and Slaoui 2011; Cianciolo and Mohr 2016).

The kidney should be observed for its shape, size, weight, consistency, cortical surface, and decapsulation. Then cut along the major axis to the renal pelvis (Fig. 30), and assess the cortex-medulla ratio, which should be 1:2 or 1:3, and the pelvis (Fig. 31). Decapsulation, done with the help of tweezers, allows the evaluation of the cortical surface (Nietfeld 2010).

In the bladder, the state of repletion, volume, and external appearance is observed. The organ is then sectioned from the neck to the bottom of the blind sac and the mucosa, and contents are observed. In this species, the urine is usually yellow to orange and turbid and contains sediment. Hyperpigmented urine is not necessarily a pathological situation and should be distinguished from hematuria caused, for example, by tumors, polyps, or inflammation (Keeble and Meredith 2006).

3.6.5 Genital Apparatus

In females, the ovaries and uterus are dissected and removed en bloc. The rabbit has a bicornuate uterus with two uterine horns and two cervixes. It should be opened by its greater curvature exposing the mucous membrane and its contents. After sectioning, the ovaries should be observed by their largest diameter. In males, the penis, testicles, and attached glands should be observed externally and, after sectioning, for uniformity of the cut surface, color, and content (McDonough and Southard 2017).

3.7 Opening of the Cranial Cavity and Examining the Encephalon and Eyeball

The cranial cavity is not opened, except in cases with previous or suspected nervous symptomatology. For this, the skin and the largest muscular masses are removed, and then a transversal cut is made in the skull over the frontal bone, caudal to the zygomatic process. This is followed by two sagittal cuts, medial to the occipital condyles. The last transverse cut is in the occipital bone, above the *foramen magnum*. The calotte is removed with a Virchow's skull breaker, exposing the brain. The *tentorium cerebelli*, if present, and the lining by the dura mater are removed. With the head inverted, the brain is removed with gentle traction, cutting the olfactory bulbs, the internal carotid arteries, and the cranial nerves (Cantile and Youssef 2016).

The circumvolutions' volume, color, consistency, weight, and appearance are evaluated. The brain is then immediately immersed in formaldehyde and is cut only after it is fixed (Petisca and Montano 1962).

The eyeballs are removed by gently cutting the tissue around the orbit. Then, a deep cut is made around the orbit, close to the bone, with moderate traction, to pull out the eyeball. The optic nerve is cut, and the whole thing is removed en bloc. The eye should be fixed immediately, and since decomposition is exceptionally rapid, it should be removed at the beginning of the necropsy (King et al. 2013).

3.8 Opening and Examining the Spinal Canal, Bones, Joints, and Bone Marrow

3.8.1 Spinal Canal

To open the spinal canal, the animal is placed in a sphinx position, completely skinned, removing all the muscle masses adhering to the rachis. In each vertebra to be opened, two longitudinal cuts are made parallel to each other, which must reach the vertebrae to remove the spinous apophysis and part of the arch. Once the spinal cord is exposed, a cut is made in the dura, the spinal nerves are cut at their point of entry into the conjugation holes, and the cord is carefully removed (Cantile and Youssef 2016).

3.8.2 Joints and Bone Marrow

During the necropsy, at least five joints should be opened: the right hip joint, the left and right knees, the right shoulder, and the atlantooccipital joint. The articular fluid is evaluated for color, viscosity, and volume, and the articular surfaces are evaluated for appearance, regularity, and presence of continuity solutions or articular rats. Bone marrow examination is done by cutting the femur or other long bone at the proximal portion, removing the marrow from the fractured bone and collecting it into a mesh cassette, or making appositions on a glass slide (Fiette et al. 2017).

4 Collection of Material for Histopathological Examination

All organs with macroscopic changes should be collected for histopathological examination. When harvesting, the topography of each organ should be respected, and its different anatomical components should be included. Care should also be taken to harvest the lesion and the transition zone with normal tissue. If no noteworthy macroscopic changes are observed, lung, heart, liver, intestine, and kidney fragments should be collected (Petisca and Montano 1962; King et al. 2013). The fragments should be fixed in 10% buffered formalin, in a volume ten times the volume of the samples (King et al. 2013).

5 The Necropsy Report

The necropsy report is a document issued at the time of the necropsy that should include all relevant observations. It should be written simply, clearly, directly, and concisely. It should have a description of the lesions and, when possible, their interpretation and be concluded with a presumptive macroscopic diagnosis (King et al. 2013; Pires 2016).

The report should follow a logical order, usually coinciding with the order of the necropsy. After the animal identification and clinical history, it should include a description of the body's condition. Data on the external habitus, internal habitus, and macroscopic diagnosis follow this. It should also mention eventual harvests for other complementary exams (Pires 2016).

The description of lesions should include their identification and description of their location, shape, distribution, size, color, consistency and texture, and other unique characteristics. Usually, the report is completed after the necropsy and should be written in the present tense. Except in forensic cases, it should not include unnecessary information, such as the technique's description or normal anatomical features (Pires 2016; Küker et al. 2018).

6 Conclusions

Necropsy is a scientific investigation of high diagnostic value. It should be done in a systematic and thorough way so that the maximum conclusions can be drawn from it. In rabbits, necropsy has some specificities, given the anatomical particularities of the species. The correct manipulation of organs and samples is also of crucial interest in order to obtain a conclusive diagnosis. Compliance with the safety rules of the operator is also of utmost importance, not only to avoid the spread of infectious diseases but also to avoid the contamination of the operator.

References

Aluja AS, Casas FC (2002) Técnicas de Necrópsia en Animales Domésticos, 2nd edn. Editorial el manual Moderno, Santa Fé de Bogotá. ISBN 9789684269385

Alves A, Pires I, Gama A, Prada J (2016) Técnica de necrópsia em mamíferos. In: Payan-Carreira R, Pires MA (eds) Descrição anatomopatológica em medicina veterinária. UTAD, Vila Real, pp 19–39

Cantile C, Youssef S (2016) The nervous system. In: Maxie MG (ed) Jubb, Kennedy and Palmer's pathology of domestic animals, vol 1, 6th edn. W.B. Saunders, St. Louis, MO, pp 250–406. ISBN 9780702053177. https://doi.org/10.1016/B978-0-7020-5317-7.00004-7

Caswell JL, Williams KJ (2016) Respiratory system. In: Maxie MG (ed) Jubb, Kennedy and Palmer's pathology of domestic animals, vol 2, 6th edn. W.B. Saunders, St. Louis, MO, pp 465–591.e4. ISBN 9780702053184. https://doi.org/10.1016/B978-0-7020-5318-4.00011-5

Cianciolo RE, Mohr FC (2016) Urinary system. In: Maxie MG (ed) Jubb, Kennedy and Palmer's pathology of domestic animals, vol 2, 6th edn. W.B. Saunders, St. Louis, MO, pp 376–464.e1. ISBN 9780702053184. https://doi.org/10.1016/B978-0-7020-5318-4.00010-3

Fiette L, Slaoui M (2011) Necropsy and sampling procedures in rodents. Methods Mol Biol 691: 39–67. https://doi.org/10.1007/978-1-60761-849-2_3

Fiette L, Slaoui M, Bauchet AL (2017) Procedures of necropsy and tissue sampling. Methods Mol Biol 1641:71–100. https://doi.org/10.1007/978-1-4939-7172-5_3

Keeble EK, Meredith AL (2006) Self-assessment colour review of rabbit medicine and surgery, 1st edn. CRC Press, London. ISBN 9780367853969. https://doi.org/10.1201/9780367853969

King JM, Dodd DC, Roth L, Newson ME (2013) The necropsy book: a guide for veterinary students, residents, clinicians, pathologists, and biological researchers, 7th edn. College of Veterinary Medicine, Cornell University, New York

Küker S, Faverjon C, Furrer L, Berezowski J, Posthaus H, Rinaldi F, Vial F (2018) The value of necropsy reports for animal health surveillance. BMC Vet Res 14:191. https://doi.org/10.1186/s12917-018-1505-1

Maxie MG, Miller MA (2016) Introduction to the diagnostic process. In: Maxie MG (ed) Jubb, Kennedy and Palmer's pathology of domestic animals, vol 1, 6th edn. W.B. Saunders, St. Louis, MO, pp 1–15.e1. ISBN 9780702053177. https://doi.org/10.1016/B978-0-7020-5317-7.00001-1

McDonough SP, Southard T (eds) (2017) Necropsy guide for dogs, cats, and small mammals. Wiley, Iowa. ISBN 9781119115656. https://doi.org/10.1002/9781119317005

Nietfeld JC (2010) Field necropsy techniques and proper specimen submission for investigation of emerging infectious diseases of food animals. Vet Clin North Am Food Anim Pract 26:1–13. https://doi.org/10.1016/j.cvfa.2009.10.005

Oliveira JP, Alves A, Pires I, Gama A, Pinto-da-Costa JE (2004) Apontamentos de Tanatologia Veterinária- Carnivoros Domésticos. UTAD, Vila Real, pp 1–59

Ortiz AG (1989) La necropsia em los mamíferos domésticos. Interamericana McGraw-Hill, Madrid, Spain. ISBN 84-7605-414-9

Percy DH, Barthold SW (2007) In: Percy DH, Barthold SW (eds) Rabbit, pathology of laboratory rodents and rabbits, 3rd edn. Blackwell Publishing, Oxford, UK, pp 253–304. ISBN 9780813821016. https://doi.org/10.1002/9780470344613
Petisca JLN, Montano AT (1962) A técnica de Necrópsia em Medicina Veterinária. Livraria Luso-espanhola, Lisbon
Pires I (2016) A linguagem no relatório de necrópsia. In: Payan-Carreira R, Pires MA (eds) Descrição anatomopatológica em medicina veterinária. UTAD, Vila Real, pp 77–93
Robinson WF, Robinson NA (2016) Cardiovascular system. In: Maxie MG (ed) Jubb, Kennedy and Palmer's pathology of domestic animals, vol 3, 6th edn. W.B. Saunders, St. Louis, MO, pp 1–101.e1. ISBN 9780702053191. https://doi.org/10.1016/B978-0-7020-5319-1.00012-8
Uzal FA, Plattner BL, Hostetter JM (2016) Alimentary system. In: Maxie MG (ed) Jubb, Kennedy and Palmer's pathology of domestic animals, vol 2, 6th edn. W.B. Saunders, St. Louis, MO, pp 1–257.e2. ISBN 9780702053184. https://doi.org/10.1016/B978-0-7020-5318-4.00007-3
Valli VEO, Kiuppel M, Bienzle D (2016) Hematopoietic system. In: Maxie MG (ed) Jubb, Kennedy and Palmer's pathology of domestic animals, vol 3, 6th edn. W.B. Saunders, St. Louis, MO, pp 102–268.e1. ISBN 9780702053191. https://doi.org/10.1016/B978-0-7020-5319-1.00013-X

Correction to: Mycotoxin Effects and Risk Management in Rabbit Production

Fernando Lima

Correction to:
Chapter 32 in: J. Simões, J. M. Monteiro (eds.),
Veterinary Care of Farm Rabbits,
https://doi.org/10.1007/978-3-031-44542-2_32

The original version of the chapter was inadvertently published with an incorrect affiliation and email address of the author "Fernando Lima". The affiliation has been updated and the email address was removed.

From:
F. Lima
DSM Nutritional Products Ltd, Kaiseraugst, Switzerland
e-mail: fernando.lima@dsm.com

To:
F. Lima
Schweizersholz, Switzerland

The updated version of this chapter can be found at
https://doi.org/10.1007/978-3-031-44542-2_32

J. Simões, J. M. Monteiro (eds.), *Veterinary Care of Farm Rabbits*,
https://doi.org/10.1007/978-3-031-44542-2_35

Appendix Meat Rabbits' Inspection: Abnormal Findings at Slaughterhouse – Courtesy of Dr. Anabela Cristo Trindade Ferreira (Sanitary Inspector)

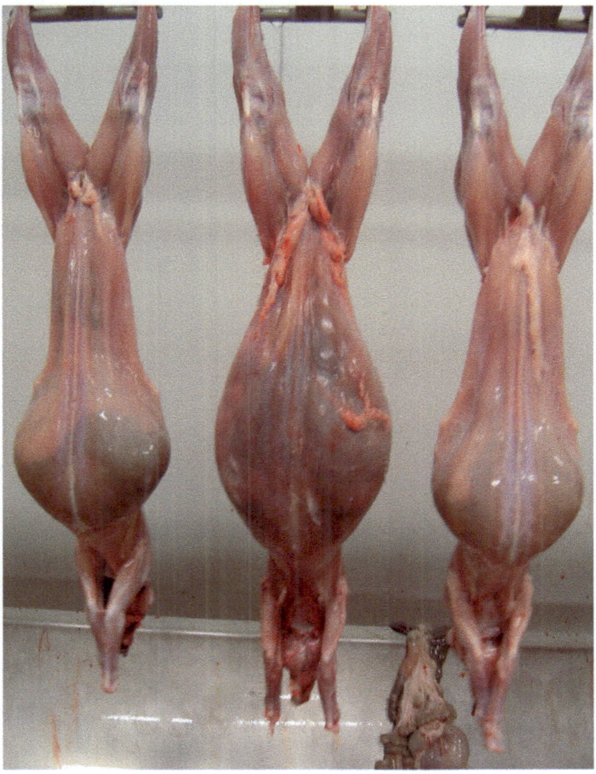

Plate 1 Fattening rabbit carcass showing febrile syndrome at *postmortem* examination (middle carcass looking congested and febrile)

J. Simões, J. M. Monteiro (eds.), *Veterinary Care of Farm Rabbits*,
https://doi.org/10.1007/978-3-031-44542-2

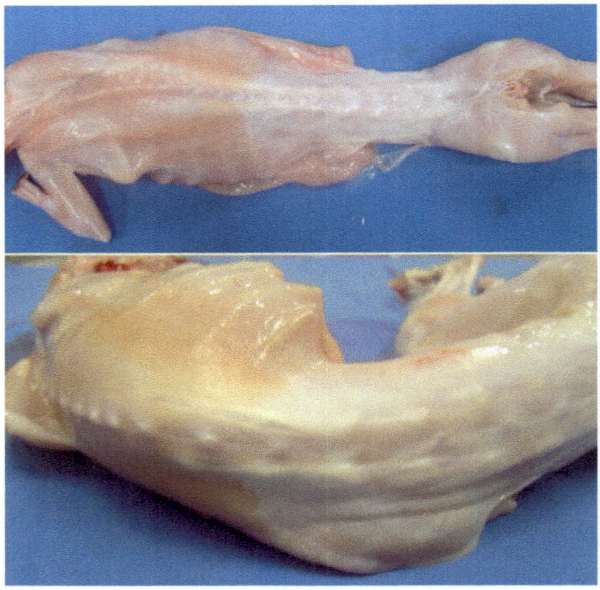

Plate 2 Caquexia in fattening rabbits

Plate 3 Dirty carcass (fecal contamination)

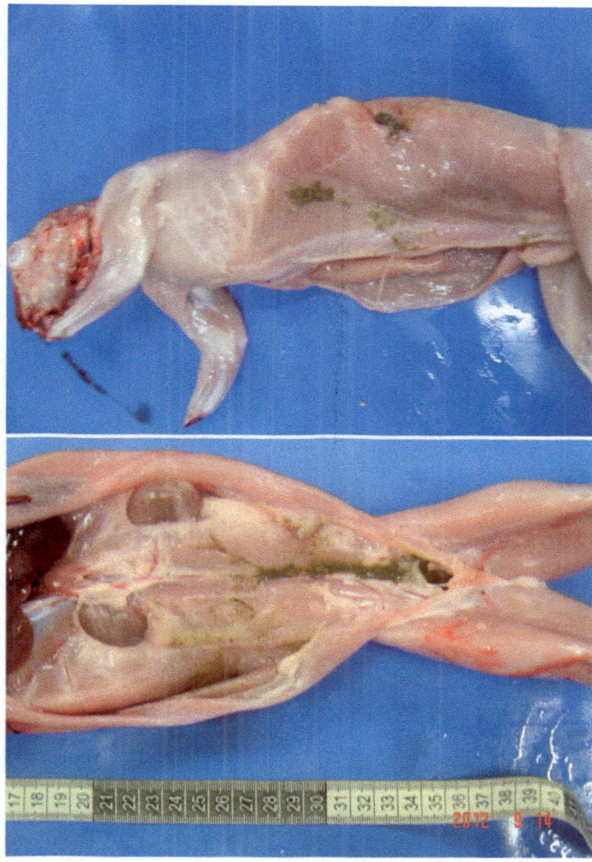

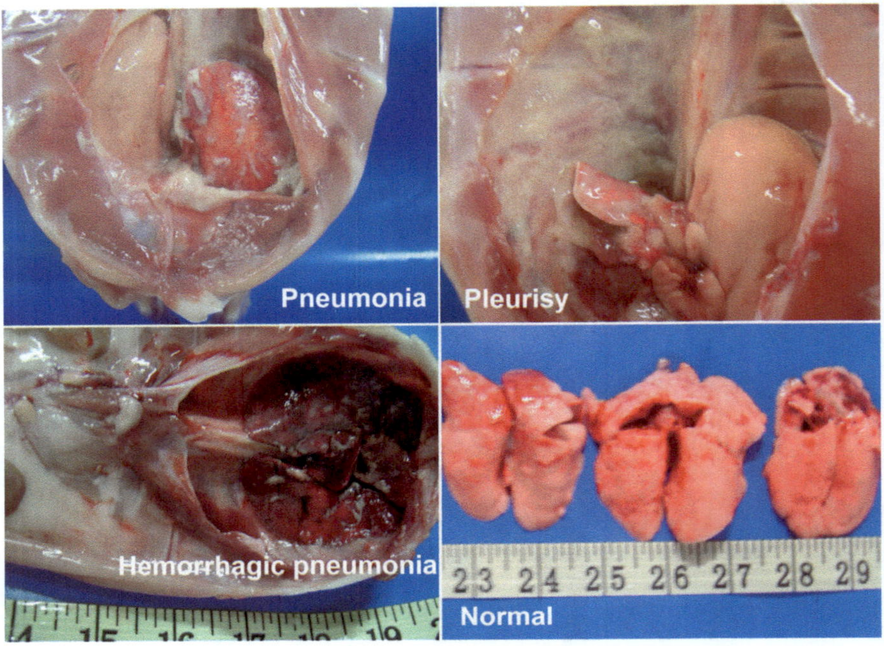

Plate 4 Pneumonic lesions in fattening rabbits and lung patterns without pathology

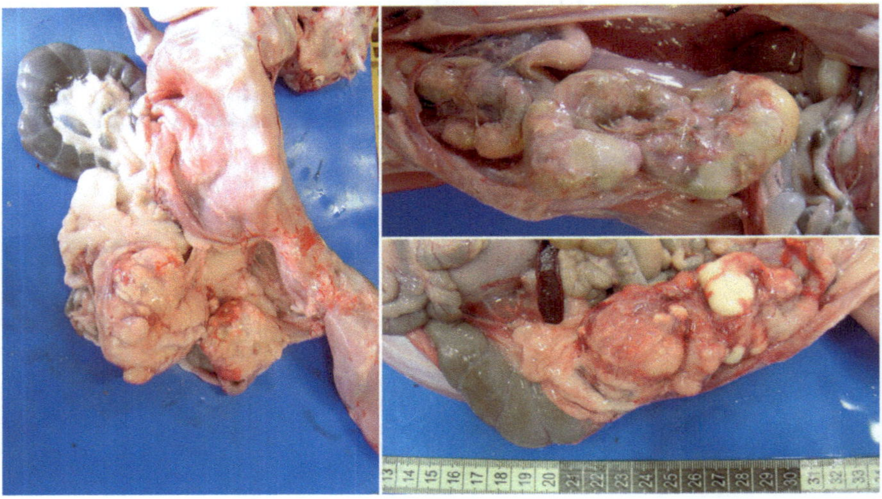

Plate 5 Pyometra in rabbit

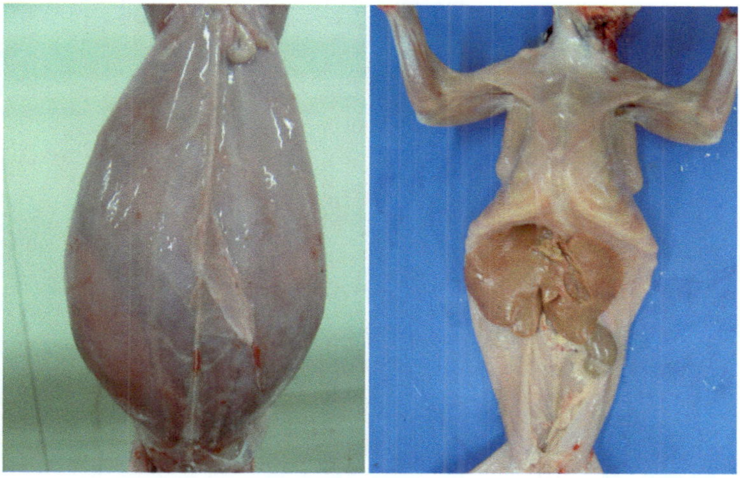

Plate 6 Ascites (left) and liver diffuse alterations (right)

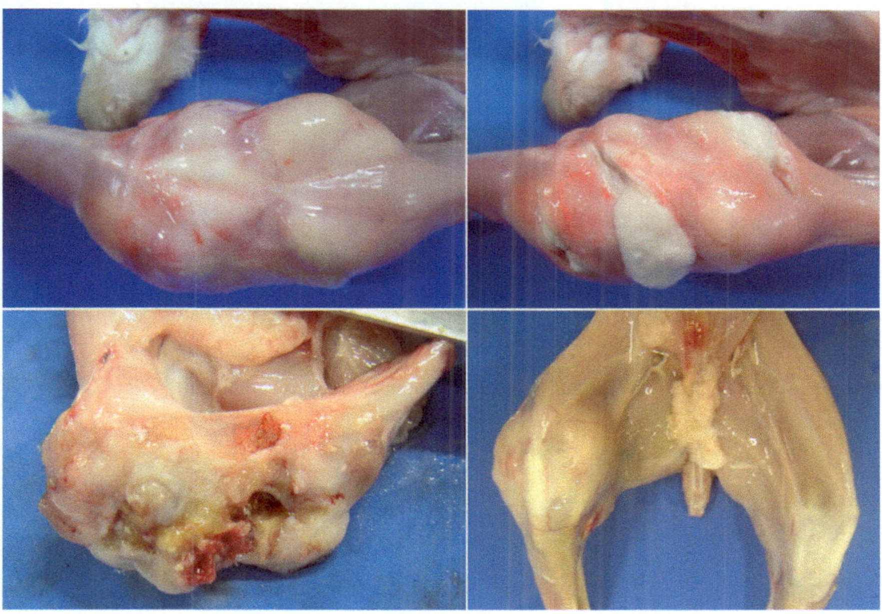

Plate 7 Purulent and fibrin-purulent arthritis

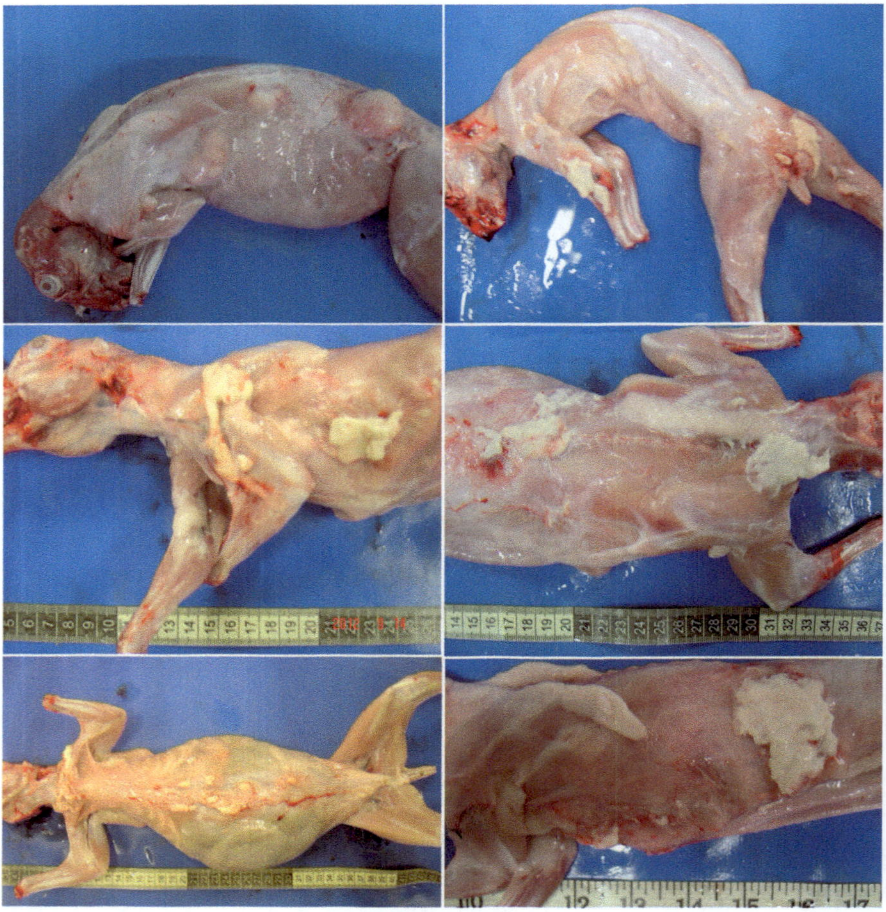

Plate 8 Multiple abscesses (lumbar, thoracic, abdominal, cervical, umbilical)

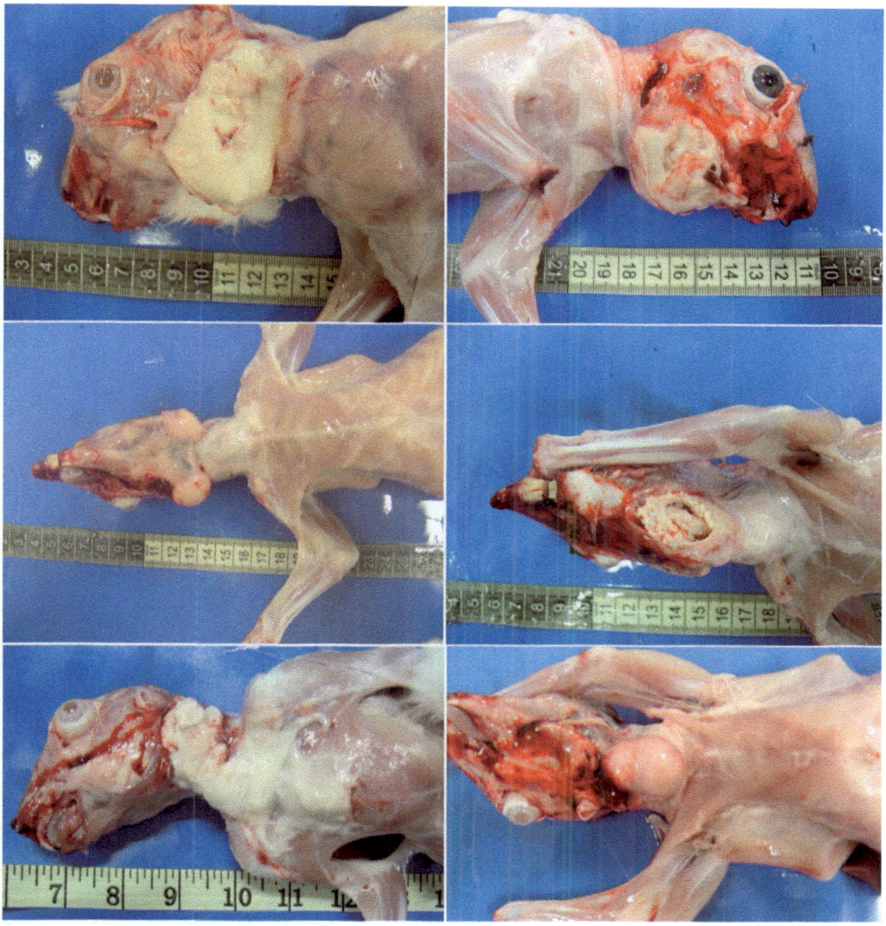

Plate 9 Cervical abscesses

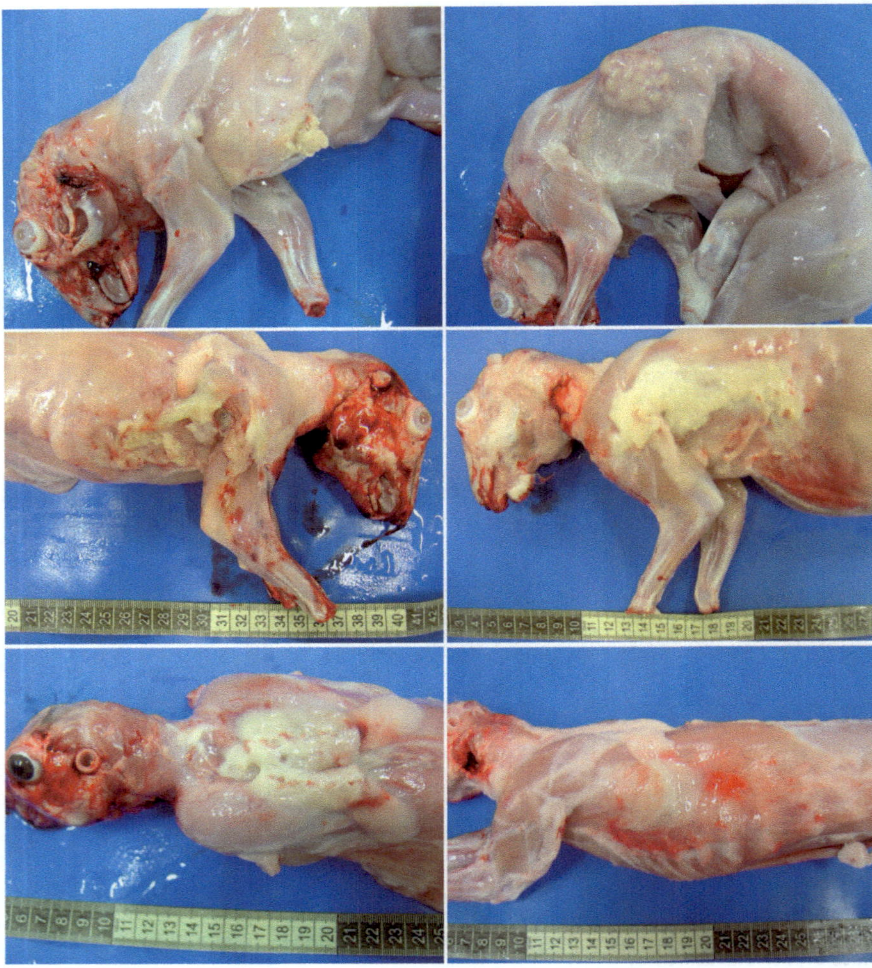

Plate 10 Thoracic abscesses

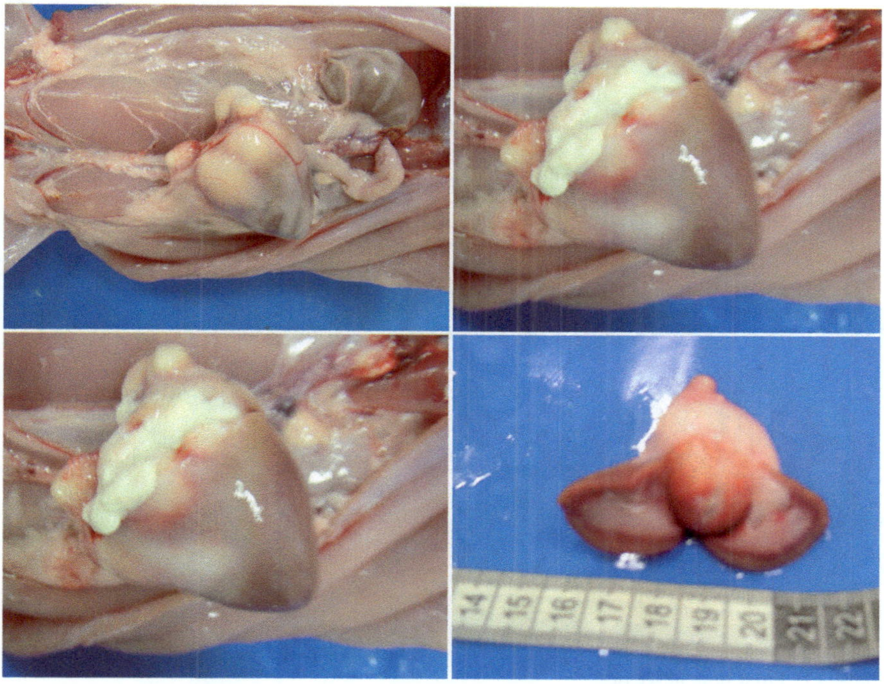

Plate 11 Cardiac abscesses

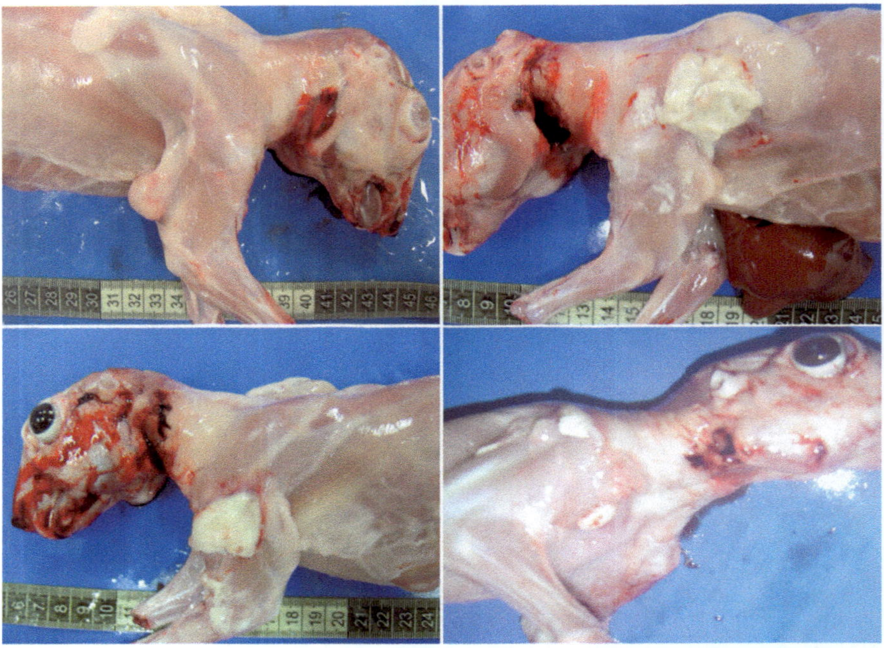

Plate 12 Front limb abscesses

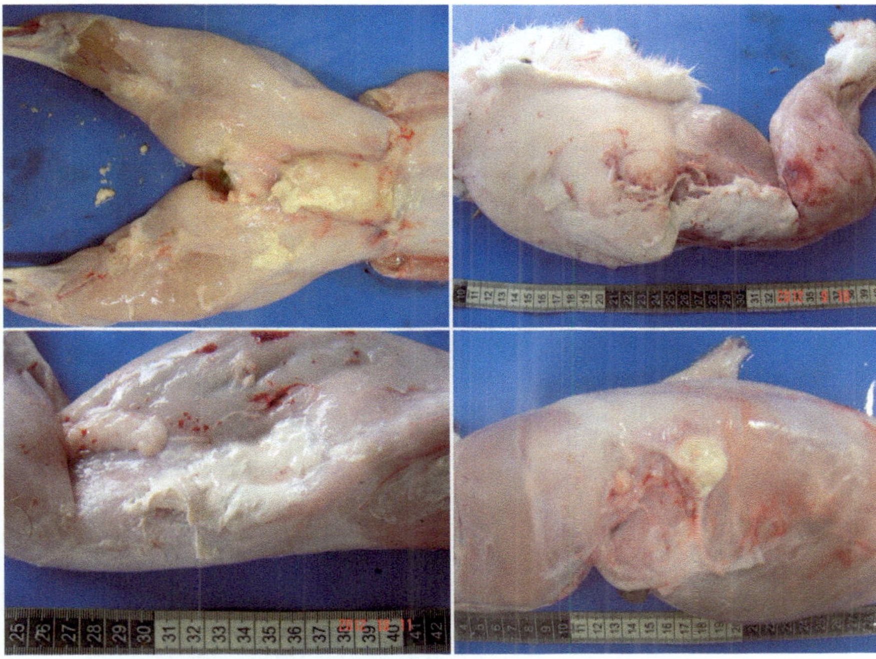

Plate 13 Lumbar abscesses

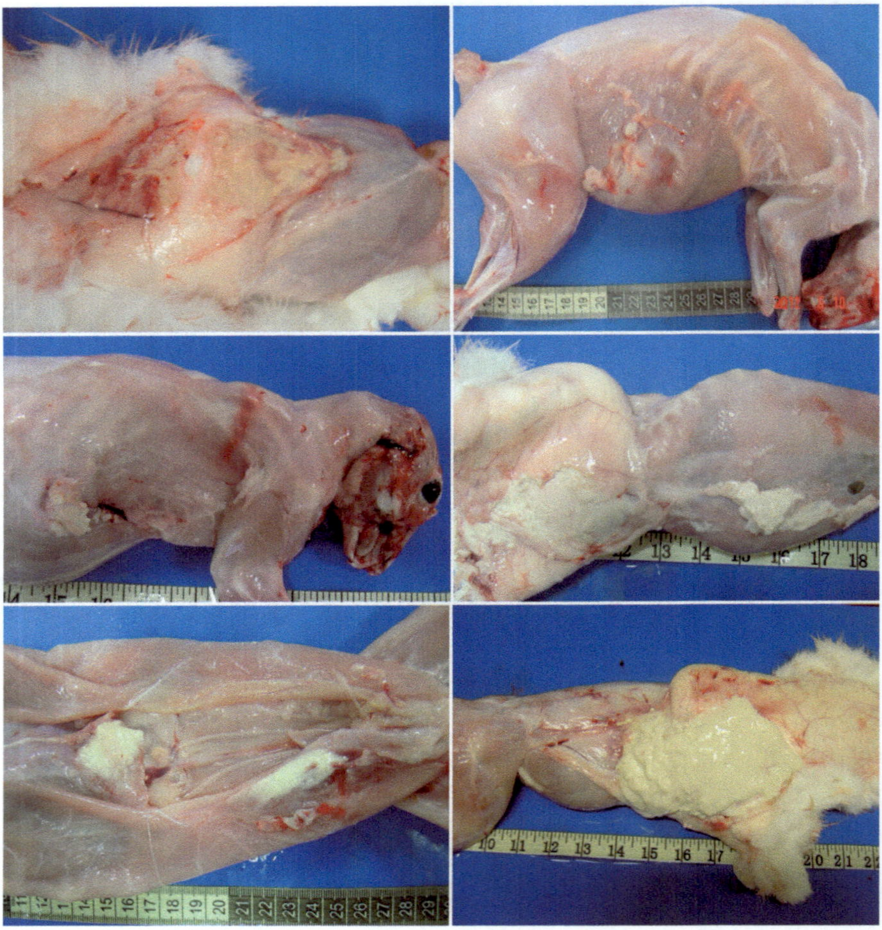

Plate 14 Abdominal abscesses

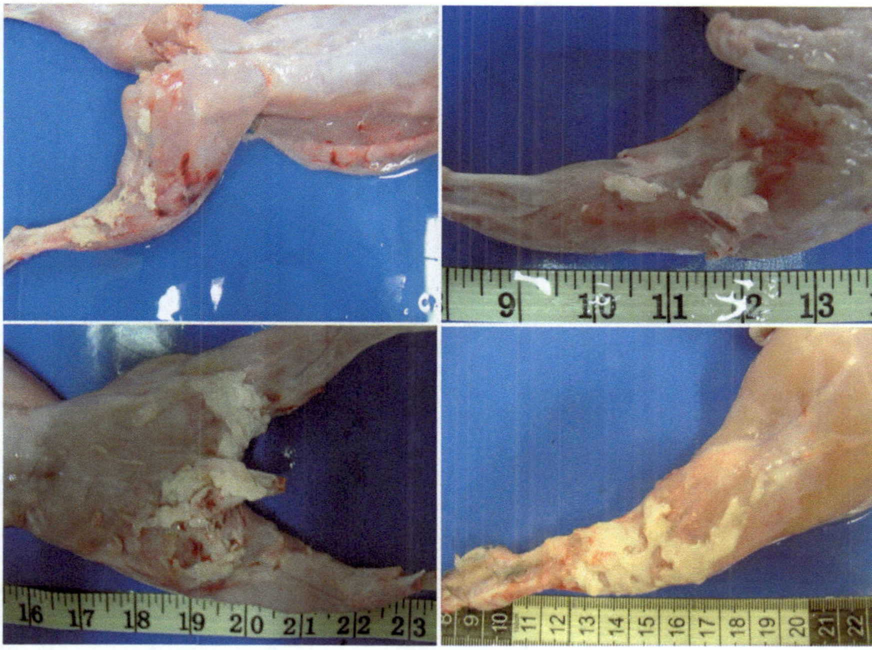

Plate 15 Hind limb abscesses

Plate 16 Umbilical abscesses

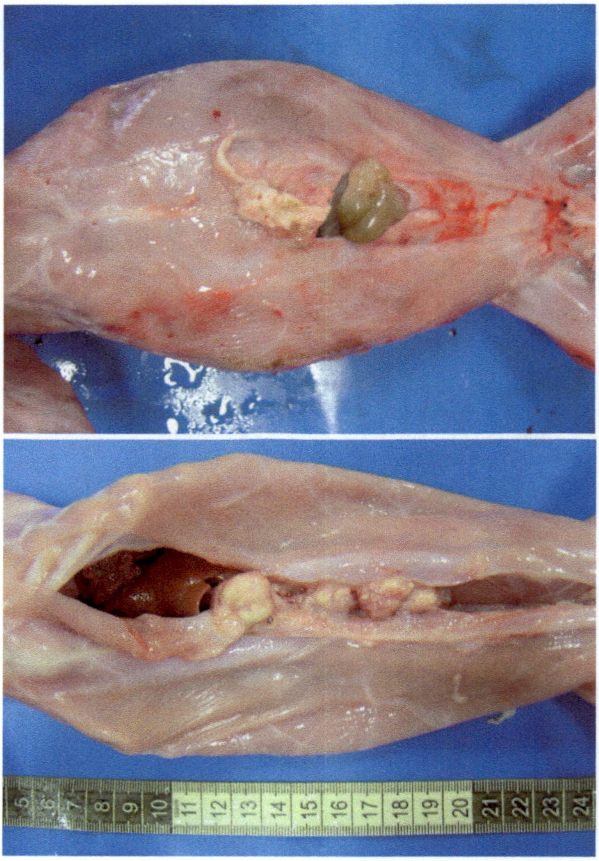

Plate 17 Genital abscesses (male)

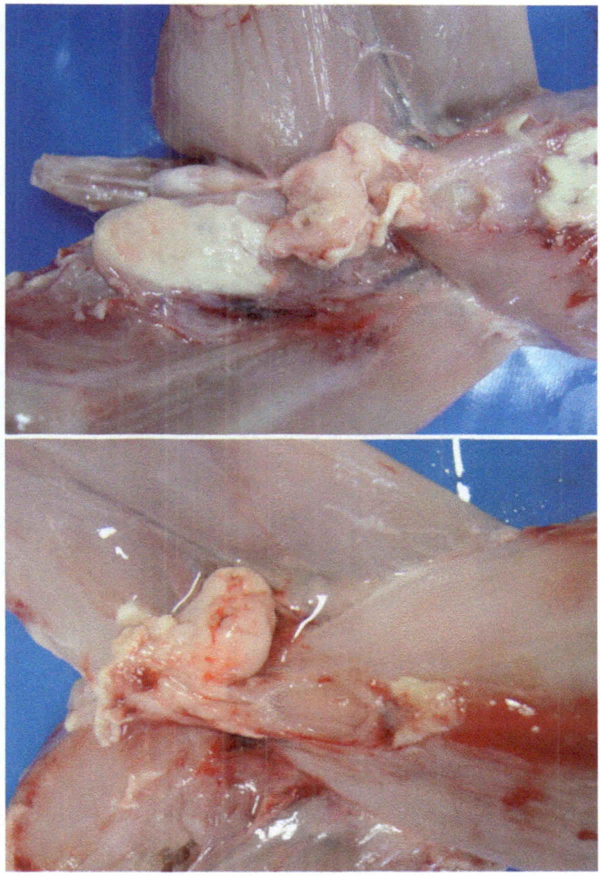

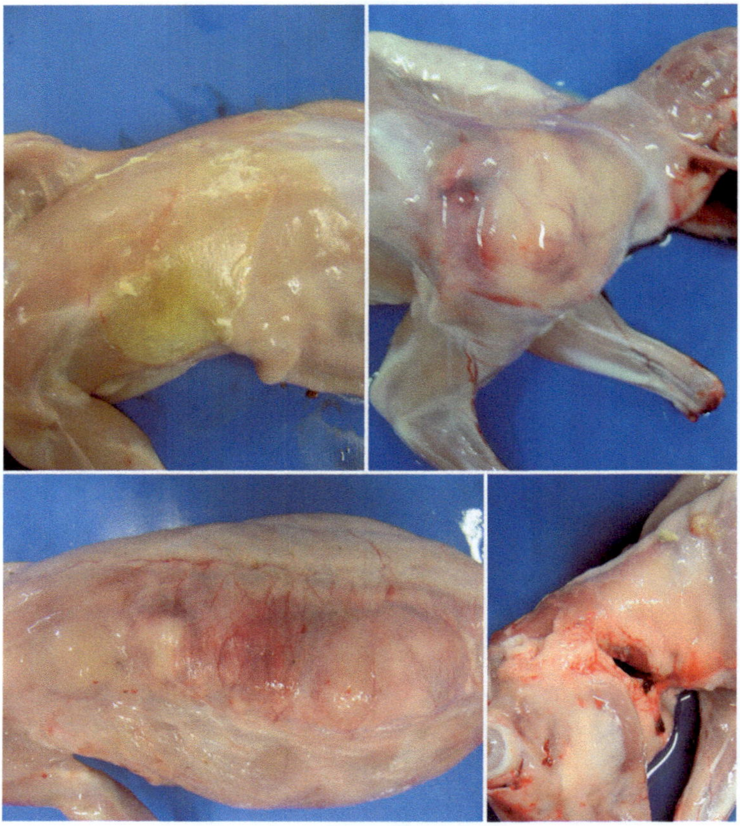

Plate 18 Vacinal (iatrogenic) abscesses

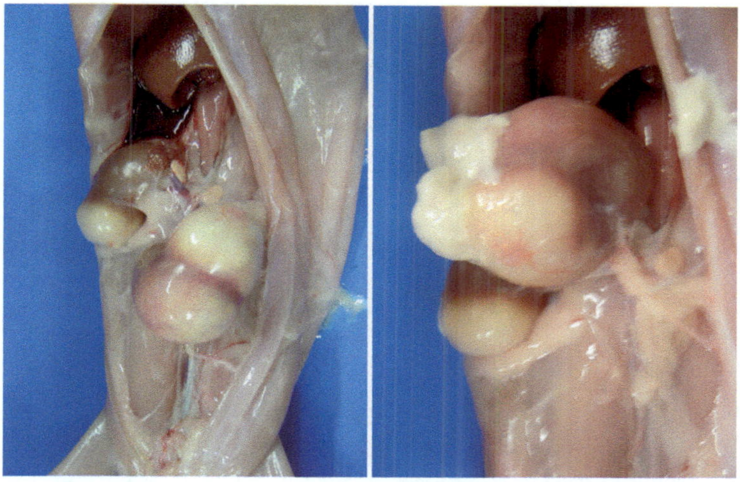

Plate 19 Renal abscesses

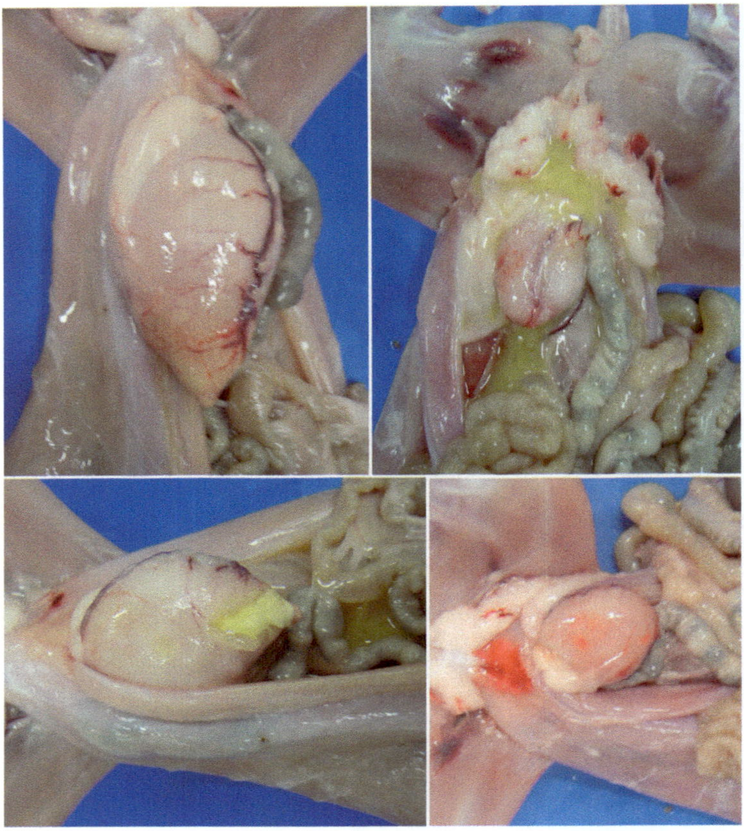

Plate 20 Cystitis

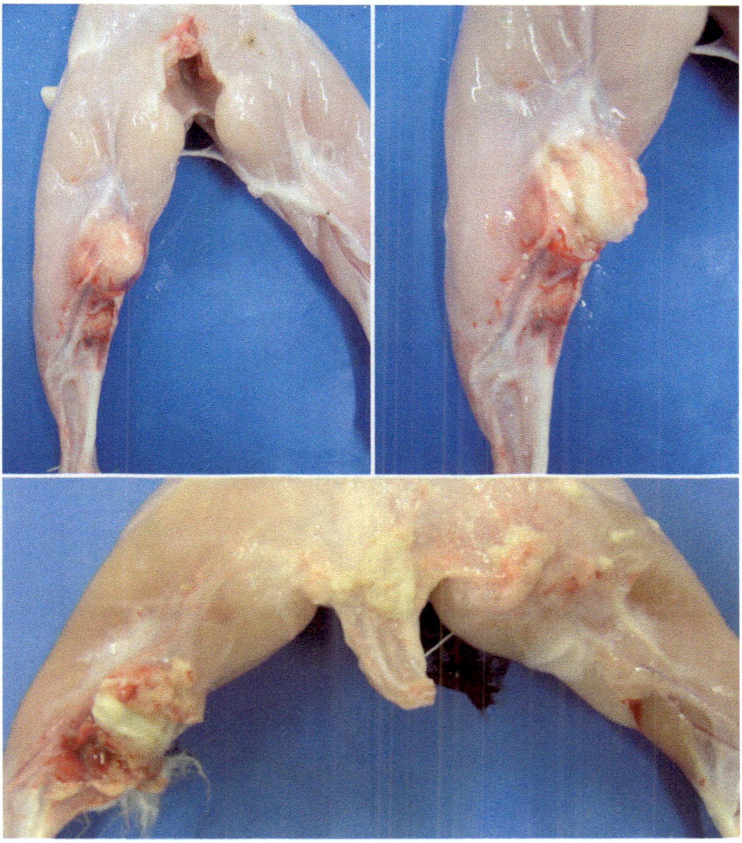

Plate 21 Purulent lymphadenitis

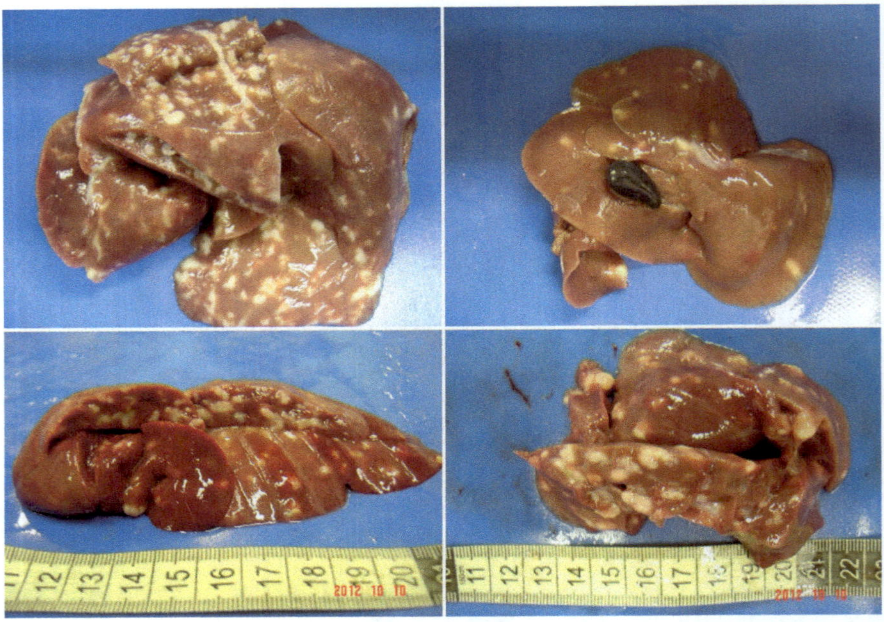

Plate 22 Generalized parasitic lesions of the liver

Index

J. Simões, J. M. Monteiro (eds.), *Veterinary Care of Farm Rabbits*,
https://doi.org/10.1007/978-3-031-44542-2

MIX
Papier aus verantwortungsvollen Quellen
Paper from responsible sources
FSC® C105338

If you have any concerns about our products,
you can contact us on
ProductSafety@springernature.com

In case Publisher is established outside the EU,
the EU authorized representative is:
Springer Nature Customer Service Center GmbH
Europaplatz 3, 69115 Heidelberg, Germany

Printed by Libri Plureos GmbH
in Hamburg, Germany